AEROSOLS

An Industrial
and Environmental Science

AEROSOLS

An Industrial
and Environmental Science

GEORGE M. HIDY

Environmental Research & Technology, Inc.
Westlake Village, California

1984

ACADEMIC PRESS, INC.

(Harcourt Brace Jovanovich, Publishers)

Orlando   San Diego   San Francisco   New York   London
Toronto   Montreal   Sydney   Tokyo   São Paulo

ACADEMIC PRESS, INC.
Orlando, Florida 32887

United Kingdom Edition published by
ACADEMIC PRESS, INC. (LONDON) LTD.
24/28 Oval Road, London NW1 7DX

Library of Congress Cataloging in Publication Data

Hidy, George M.
    Aerosols, industrial and environmental science.

    Includes bibliographical references and index.
    1. Aerosols. I. Title.
TP244.A3H53   1984   660.2'94515   84—2743
ISBN 0—12—347260—1

PRINTED IN THE UNITED STATES OF AMERICA

84 85 86 87   9 8 7 6 5 4 3 2 1

To Dana, Anne, Adrienne, and John
—To Prin, Ogden, Red, Leon, and Rosebud

CONTENTS

Chapter 4   GENERATION OF PARTICULATE CLOUDS

Chapter 5   MEASUREMENT OF AEROSOL PROPERTIES

Chapter 6   APPLICATIONS TO TECHNOLOGY

Chapter 7   ATMOSPHERIC AEROSOLS

Chapter 8   EFFECTS ON THE EARTH'S ATMOSPHERE

Chapter 9   HEALTH EFFECTS OF INHALED AEROSOLS

PREFACE

When Milton Kerker convinced me to undertake the writing of this book, I viewed the project with enthusiasm. With many years of experience in some elements of aerosol science and technology and a working knowledge of the field, I felt the venture was achievable with reasonable effort. To my surprise, I found the diversity of literature on aerosol science and technology to be far in excess of expectations. Consequently, I seriously underestimated the task of bringing together the wealth of information accumulated in this field, but the project has been completed. The results of several years of effort constitute the whole of this book. Although it is not perfect by any means, it does serve to introduce the interested reader to the aerosol field at a level comparable with H. L. Green and W. R. Lane's famous volume of the 1950s, *Particulate Clouds, Dusts, Smokes and Mists*. Comparison between the contents of their book and this manuscript will provide some perspective on the dramatic increase in knowledge about aerosols through the past three decades. Aerosol science and technology has kept pace with the general expansion of information in the physical sciences and engineering. Its traditional stimulation has come from applications to industrial challenges and to the environmental and health-related sciences.

Consistent with Green and Lane's approach, this book is intended to be at least partially encyclopedic in scope as a survey of knowledge. The book is intended to serve scientists and engineers who are concerned both with the underlying principles of aerodynamic and physical chemical behavior of suspended particles and with the nature of the application of these principles to a wide variety of uses. The applications range from consideration of pest control, combustion, and powder technology to environmental concerns for the potential hazards of suspended particles in ambient air.

Since aerosol science and technology remains basically experimental or observational in character, much of the book is devoted to description of measurement techniques and results in terms of a framework of classical mechanics and macroscopic chemistry.

The book could be used as a text for graduate students in specialized courses on aerosol or colloid chemistry, atmospheric processes, and chemical, mechanical, or environmental engineering. However, an instructor would want to select carefully the material to be presented in a restricted period of time. For a basic course in colloidal science, Chapters 1–5 would be most appropriate. For a course useful to atmospheric scientists, Chapters 1–3, 7, and 8 would be appropriate. An engineering curriculum should consider Chapters 1–5, 6, and 10. Training of environmental scientists or those interested in regulatory considerations would focus on Chapters 5, 7, 9, and 10.

Although I hoped that this work would be comprehensive in its survey of knowledge applicable to aerosol behavior, the book has a certain bias of viewpoint from my own experience and interests. The approach taken in organization, selection, and emphasis of material reflects the influence of many of my friends and colleagues. There is strong emphasis on elements of the fluid dynamic models of aerosol particles and rigid spheres, an area in which I worked for many years with my friends Sheldon Friedlander and James Brock. There is also considerable effort devoted to atmospheric phenomena involving suspended particles, an area of central interest to me for many years. With the current concerns for clean air and environmental conservation, this application follows naturally from atmospheric science. To those workers whose studies may be neglected inadvertently in my selection and treatment, my apologies before you read the book. Given the limitations of my own time and energy, I hope that readers will be sufficiently stimulated by one viewpoint to seek a broader penetration into the aerosol literature than that contained here. In this way, they may discover a diverse expression of viewpoints and judge for themselves a direction appropriate to their individual work that touches or advances this field of science and technology.

If nothing else, I believe the book will serve to introduce the reader to the wide variety of physics and chemistry that has been used to characterize and interpret aerosol behavior. With assimilation of the knowledge contained here, the reader should be prepared to contribute actively to the continued evolution of this component of modern science and technology.

ACKNOWLEDGMENTS

I am indebted to the many co-workers whose research made this book possible. Without the benefit of the knowledge, cooperation, and resources of a large number of people, a comprehensive book dealing with aerosol science and technology would not be feasible. I acknowledge especially the continued association, for more than twenty years, with Sheldon Friedlander and N. A. Fuchs. Through Sheldon's closely related research and Dr. Fuchs' pioneering leadership, I have retained the interest in this field to permit the effort needed to undertake this work.

It is appropriate to recognize especially the contributions of my friend Kenneth Whitby. His sudden death in 1983 saddened the aerosol community. However, his legacy of work remains with us and is well referenced in this book.

I want to express my gratitude to Doris (Sharp) Wilson. As my secretary and friend for many years, she assisted crucially in the many tasks involved in the preparation of the manuscript and the correspondence required to confirm information sources. I am also indebted to Marcia Henry and Harry Bowie who have assisted me with the literature research and graphics needed to prepare the manuscript.

I am grateful to Milton Kerker and Richard Countess who read the draft of the manuscript and provided many suggestions for its improvement before publication.

It has been my privilege to be a part of the development of the aerosol field and to work personally with B. Appel, K. Bell, J. R. Brock, C. S. Burton, R. Cadle, S. Calvert, J. Calvert, R. J. Charlson, R. Countess, B. V. Derjaguin, J. Durham, A. Goetz, D. Grosjean, J. Hales, S. L. Heisler, R. C. Henry, B. Herman, P. Hobbs, D. Hochrainer, R. B. Husar, J. L. Katz, M. Kerker, C. and N. Knight, A. Lazrus, M. Lippmann, B. Y. H. Liu, J. P. Lodge, Jr., P. McMurry, T. Mercer, V. Mohnen, P. K. Mueller, L. Newman, B. Ottar, O. Preining, H. Pruppacher, H. Reiss, J. Rosinski, J. Seinfeld, G. Slinn, P. Squires, W. Stöber, G. Sverdrup, O. Vittori, A. Waggoner, J. G. Watson, J. Wesolowski, W. Wilson, G. Wolff, and G. Zebel.

NOMENCLATURE

The following nomenclature is used in the text most frequently.

Symbol	Meaning	Symbol	Meaning
A	Stokes–Cunningham correction [Eq. (2.6)]	b_{sg}	Light-scattering coefficient for gases (m^{-1})
A_F	Attractive force coefficient [Eq. (4.23)]	b_{sp}	Light-scattering coefficient for particles (m^{-1})
A_p	Acoustic amplitude factor [Eq. (3.58)]	b_i	Regression coefficient [Eq. (7.5)]
A_{ij}	Loading factor for principal components [Eq. (7.3)]	$\mathcal{B}$	Condensation–coagulation scale parameter [Eq. (3.65)], or flame theory transport scale [Eq. (6.4)]
a	Characteristic external radius or diameter of body (cm)		
$\mathcal{A}$	Inverse product of the particle mass and the mobility, $(m_p B)^{-1}$ (sec^{-1}).	C	Contrast
		$C_{\mathcal{D}}$	Drag coefficient
		C_{at}	Bradley–Hamaker attractive force constant [Eq. (4.22)]
B	Particle mobility (sec/gm) or luminance $(lm/m^2\ sr)$	C_M	Modulation contrast
B_F	Attractive force retardation coefficient [Eq. (4.23)]	C_p	Heat capacity at constant pressure (cal/g mol °K)
b or b_{ij}	Coagulation coefficient (cm^3/sec)	C_v	Heat capacity at constant volume (cal/g mol °K)
b_{ext}	Light extinction coefficient (m^{-1})	c	Speed of sound (m/sec)
		c_i	Molar concentration, species i liter (mole/liter)
b_{ag}	Light absorption coefficient for gases (m^{-1})	c_m	Momentum slip coefficient
b_{ap}	Light absorption coefficient for particles (m^{-1})	c_p	Specific heat at constant pressure (cal/g °K)

Symbol	Meaning	Symbol	Meaning
c_v	Specific heat at constant volume (cal/g °K)	F', $\mathbf{F}'$	Normalized force, F/m_p (cm/sec^2)
c_t	Thermal slip coefficient	F_{at}	Attractive force between particles (dyn)
c_{tm}	Isothermal slip coefficient		
CN	Condensation nuclei concentration (no./cm^3)	FR	Fractional reduction for linear rollback
CCN	Cloud condensation nuclei (no./cm^3)	$\mathcal{f}$	Frequency (sec^{-1})
$\mathcal{C}$	Coagulation or reaction efficiency (probability)	$G(d_p)$	Total light extinction per unit volume (m$^{-1}/\mu$m^3 cm^{-3})
$\mathcal{c}$	Speed of light (m/sec)	G	Velocity ratio, q_G/q_0 or E_0QB/U_∞, or group combustion number
D_{AB}	Binary gas diffusion coefficient for A diffusing into B (cm^2/sec)		
D_i, D_p	Particle (Stokes–Einstein) diffusivity (cm^2/sec)	Gr	Filtration scale factor, $\tau_p g/q_0$ [Eq. (5.39)]
D_j	Generalized dispersion factor [Eq. (10.1)]	$\hat{G}_i$	Gibbs free energy (kcal/mol; kcal/molecule or kcal/embryo)
D_t	Turbulent diffusion coefficient (cm^2/sec)	$\Delta\hat{G}^*$	Gibbs free energy to form critical-size embryo (kcal/mol or kcal/molecule)
d	Characteristic diameter (cm)		
d_A	Diameter of molecule A (Å)	g	Gravitational acceleration constant (cm/sec^2)
d_a	Aerodynamic equivalent diameter (μm)	$g(v,n_i,\mathbf{r},t)$	Composition–size probability density function
d_e	Stokes equivalent diameter (μm) for nonspherical particle	$\mathcal{G}$	Shearing rate in fluid
d_p, d_i	Particle diameter (μm)	H	Velocity ratio, $2Q'QB/U_\infty a$; impactor jet nozzle to plate distance; source stack height; height of control device
d_{ac}	Probable droplet diameter (μm) from acoustic generator [Eq. (4.14)]		
$\mathcal{D}$	Drag force (dyn)	ΔH	Enthalpy change (heat of reaction)(kcal/mol)
$\mathcal{D}^*$	Dimensionless drag force on filter ($\mathcal{D}/q_o\mu_g$)	h_∞	Specific enthalpy (kcal/g)
E, E_o	Electrical field (V)	$\mathcal{H}$	Henry's-law constant (dyn/cm^2)
ΔE	Change in chromaticity	h	Heat transfer coefficient (cal/cm^2 sec)
E_j	Emission rate (kg/sec)		
EF	Enrichment factor	$\hbar$	Planck's constant (erg sec)
EOF	Empirical orthogonal function	I	Interception parameter R/a or light intensity [Eq. (5.9)](erg/cm^2 sec)
ERV	Expiratory reserve volume		
e	Unit electrical charge (esu; coulomb)	$\mathbf{I}$	Radiation flux (erg/cm^2 sec)
e_k	Embryo unit of kth size	I_c	Combustion intensity
e_{ij}	Relative entrainment [Eq. (3.57)]	I, I_i	Particle current (no. cm^3 sec)
$\mathcal{E}$	Total (overall) collection efficiency (%)	i	Electrical current (A)
		IH	Filter inhomogeneity factor
$\mathcal{E}_d$	Energy dissipated per unit mass (erg/g) [Eq. (4.5)]	IFN	Ice-forming nuclei
F, $\mathbf{F}$	External force (dyn)	$\mathcal{J}$	Particle interaction parameter

Symbol	Meaning	Symbol	Meaning
J	Light source function [Eq. (5.36)](erg/cm^3 sec)	m_c	Interfacial tension ratio, $\sigma_{cg} - \sigma_{cl}/\sigma_{lg}$
j_ν	Heat or mass flux (e.g., g/cm^2 sec)	$\dot{m}_F$	Mass burning rate (g/sec)
JND	Just noticeable difference in contrast C	MD	Modulation depth
$\mathbb{K}$	Shape factor (Table 2.2)	$\mathfrak{M}_\nu$	Moments of size distribution
K	Burning rate constant [Eq. (6.2)](cm^2/sec)	$\mathfrak{M}_{ij}$	Symmetry factor
K_a	Empirical factor in Duetsch equation [Eq. (10.35)]	m	Index of refraction
K_{abs}	Light absorption efficiency	N, N_∞	Total particle concentration (no./cm^3)
K_c	Burning rate coefficient [Eq. (6.20)](cm/sec)	N_o	Total number of particles per unit volume initially present, or present at ground level (no./cm^3)
K_{ext}	Light extinction efficiency		
K_H	Hydrodynamic factor for cylindrical fibers [Eq. (5.42)]	N_R	Total number concentration of rain or cloud drops (no./cm^3)
$\overset{*}{K_p}$	Plate column scale factor	N_t	Number of cyclone turns to remove particles of size R
K_{scat}	Light-scattering efficiency		
K_v	Constant in prevailing visibility–particle mass relation [Eq. (8.5a)]	n_ν	Number density of species, ν (no./cm^3)
K_{Vc}	Venturi throat parameter	$n(v,\mathbf{r},t)$,	Size distribution function based on particle volume (no./cm^3 μm^3)
k	Boltzmann constant (erg/molecule °K)		
k_p, k_i, k_g	Thermal conductivity (cal/sec cm °K)	n_e	Charge accumulation
$\mathscr{k}_p$	Mass transfer coefficient or deposition velocity (cm/sec)	$n_R(R,\mathbf{r},t)$	Size distribution function based on particle radius (no./cm^3 μm)
$\mathscr{k}_1, \mathscr{k}_2$	Adsorption coefficients for chemisorption and desorption [Eq. (6.11)]	NO_x	Nitrogen oxides, nitric oxide (NO); nitrogen dioxide (NO$_2$)
$\mathscr{k}^+$	Dimensionless mass transfer coefficient ($\mathscr{k}_p/u^*$)	NMHC	Nonmethane (often photochemically reactive) hydrocarbon vapors
L	Length scale; throat length for impactor nozzles (cm)	n_i	Moles of species i
$\mathscr{L}_c$	Average column packing diameter (cm)	P	Total pressure (dyn/cm^2) or scale parameter, Re^2/Stk,
$\mathscr{L}_h$	Sieve hole diameter (cm)	ΔP	Pressure drop (dynes/cm^2)
L_p	Prevailing visibility (km)	P_e	Penetration of control device $(1 - \mathscr{E})$
L_v	Visual range (km)		
l	Stopping distance (cm)	P_{jk}	Principal component [Eq. (7.4)]
M	Particle mass concentration (μg/m^3)	$p_v(n_i, P, T)$	Partial pressure of species (v)(dyn/cm^2)
M_A	Molecular weight of species A (g/mol)	$p_0(n_i, T)$	Partial pressure in equilibrium with a flat liquid surface (dyn/cm^2)
$\hat{M}$	Mass of dust per unit filter surface		
m_i, m_p, m_A	Particle mass or gas molecule mass (g/molecule)	$p_s(T)$	Vapor pressure in equilibrium with a droplet (dyn/cm^2)

Symbol	Meaning	Symbol	Meaning
$\wp$	Precipitation rate (cm/hr), or sheering scale parameter	S_E	Surface area of electrostatic precipitator (cm²) [Eq. (10.36)]
P	Phase function [Eq. (5.32)]	S_f	Effective filter fiber mat surface (cm²) [Eq. (10.33)]
Q, Q'	Electrical charge (esu; coulomb)	S_j	Source contribution [Eq. (10.1)]
Q_l, Q_g	Liquid or gas volume flow rate (liter/min)	s_i	Surface area of particle i
Q_s	Saturation charge (esu; coulomb)	SO_x	Sulfur oxides, as SO_2 and SO_4^{2-}
$q, \mathbf{q}$	Aerosol component velocity (cm/sec)	SC	Solubility coefficient [Eq. (7.8)]
q_E	Electrical migration velocity (cm/sec)	SU	Suction coefficient [Eq. (5.2)]
q_g	Gas velocity (cm/sec)	SBE	Scenic beauty estimate
q_G	Gravitational sedimentation velocity (cm/sec)	$\tilde{S}$	Supersaturation ratio, p_s/p_0
q_0	Face velocity for filters (cm/sec)	T	Temperature (°C, °K)
		T_s	Surface temperature (°K)
q_m	Mainstream velocity (cm/sec)	t	Time (sec)
		t_e	Saturation time, $\pi e B_i N_0$ [Eq. (5.6)]
q_s	Sampler velocity (cm/sec)	t_g	Penetration time (sec) [Eq. (10.29)]
$\bar{q}_i, \bar{q}_p, \bar{q}_g$	Average thermal velocity (cm/sec)		
R, R_p	Particle radius (μm)	t_H	Characteristic time of spray droplet motion (sec)
R_a	Aerodynamic particle radius (μm)	t_s	Characteristic time of spray droplet disintegration (sec)
R_c	Flame radius (cm)		
R_{CL}	Combustible-droplet cloud radius (cm)	TP	Thoracic particle concentration (μg/m³)
R_e	Stokes equivalent radius (μm)	TSP	Total suspended particulate concentration (μg/m³)
R_s	Sphericity	$\mathfrak{I}$	Coagulation time, $3\pi\mu_g/8kTN_0$ [Eq. (3.61)]
$\bar{R}$	Mean number; surface or volume radius (μm)		
R^*	Radius of critical sized embryo (μm)	U_∞	Free-stream fluid velocity (cm/sec)
R_{ij}	Sum of radii, $R_i + R_j$ (μm)	$\bar{u}$	Mean wind or gas speed (cm/sec)
R_{min}	Minimun radius for cyclone removal [Eq. (5.51)](cm)	u^*	Friction velocity $(F/\rho_g)^{1/2}$
$r, \mathbf{r}$	Radial coordinate	V	Volume fraction of particles $(\frac{4}{3}\pi\mathfrak{M}_3)(\mu\text{m}^3/\text{m}^3)$
r_0	Pore radius (μm)	VM	Volatile matter in coal
$\mathcal{R}$	Universal gas constant (1.987 cal/g mol °K; 8.314 × 10⁷ g cm²/sec² g mol °K)	VMD	Volume median diameter (μm)
		VAQI	Visual air quality index
$\varkappa$	Reaction rate	v_p, v_i	Particle volume (μm³)
S	Total surface area per unit volume $(4\pi\mathfrak{M}_2)(\mu\text{m}^2/\text{cm}^3)$, or distance from the impactor nozzle to plate (cm)	$\bar{v}$	Average particle volume $(V/N)(\mu\text{m}^3)$
		v_m	Molecular volume of condensed species (cm³)
		$\bar{v}_\nu$	Molar volume of solution

Symbol	Meaning	Symbol	Meaning
W	Width of impactor jet nozzle, or optical shape factor (cm)	Δ_a	Rainfall parameter [Eq. (7.13)]
W_{res}	Limiting resolution of microscope	Δ_{ij}	Fuch's concentration depletion factor [Eq. (3.51)]
W_o, W_1	Diffusion or chemical resistance in particle combustion	δ	Fluid boundary layer thickness (cm)
X	Color tristimulus coordinate [Eq. (8.9)]	ε_o	Permittivity of gas
X_{ik}	Value of i variable [Eq. (7.3)]	$\varepsilon_p, \varepsilon_s, \varepsilon_f$	Dielectric constant of particles or collectors
x	Mole fraction or Cartesian coordinate (1), or dimensionless wave number, 2	ε_v	Rainout efficiency
		ε_t	Turbulent energy dissipation rate (cm²/sec³)
$\bar{x}$	Normalized color tristimulus [Eq. (8.10)]	ε_v	Void fraction
Y	Color tristimulus coordinate [Eq. (8.9)]	ζ_a	Refill factor [Eq. (3.56)]
y	Cartesian coordinate (2)	ζ_o	Filter pressure loss coefficient [Eq. (10.30)]
y_ν	Mass fraction of species ν (g/gm)	η, η_v	Collection efficiency (%); self-preserving spectrum size scale, vN/V
$\bar{y}$	Normalized color tristimulus [Eq. (8.10)]		
Z	Color tristimulus coordinate [Eq. (8.9)]	κ	Debye reciprocal length (cm⁻¹)
Z_c	Collision factor [Eq. (3.38)]	Λ	Washout rate coefficient or scavenging coefficient
Z_{ik}	Generalized variable for statistical analysis [Eq. (7.3)]	λ	Wavelength of light (cm)
z	Cartesian coordinate (3)	$\lambda_p, \lambda_i, \lambda_g$	Mean free path for particles or gas (cm)
z_i	Number of charges on a particle	λ_ν	Latent heat of vaporization
z_o	Roughness length (cm)	μ_g	Dynamic viscosity
$z_\pm$	Number of ions per molecule	μ_ν^*	Dimensionless moment of size distribution, $\int_0^\infty \eta^\nu \Psi(\eta) d\eta$
$\bar{z}$	Normalized tristimulus component [Eq. (8.10)]	$\hat{\mu}$	Chemical potential (cal/molecule or cal/embryo)
$\alpha_c, \alpha_m, \alpha_t$	Accommodation coefficients for condensation (evaporation), momentum, and thermal energy	ν	Kinematic viscosity μ/ρ
		ξ	Sound intensity; washout ratio; packing density $(1 - \varepsilon_v)$ of filters or packed beds
$\bar{\alpha}(a)$	Polarizability tensor (scalar)		
β	Vapor flux parameter [Eq. (3.14)]	ρ_ν	Mass density (gm/cm³) or concentration (gm/cm³)
β_k'	Vapor flux parameter [Eq. (3.15)](mol/cm² sec)	σ	Surface free energy or surface tension (dyn/cm)
γ	Expansion ratio c_p/c_v; exponent in power-law form of size distribution; stoichiometric ratio in flames	σ_{cg}, σ_{cl}	Surface energy between substrate (c), gas (cg), or liquid (cl)(dyn/cm)
		σ_g	Standard deviation (geometrical)
		σ_{AB}	Coefficient as a function of m_ν and d_ν [Eq. (2.18)]

Symbol	Meaning	Subscripts	Meaning
σ_y, σ_z	Dispersion coefficients in Gaussian plume model, $2D_t x/\bar{u}$	$A,B,\ldots,\nu$ g l s ∞	Molecular component Gas Liquid Surface property Free-stream condition
τ	Particle relaxation time $(\tau^{-1}$; residence time in atmosphere (sec;hr)		

Symbol	Meaning	Named dimensionless numbers	Meaning
τ, τ_a	Optical depth (thickness) for aerosols (m)		
τ^+	Dimensionless relaxation time $\tau u*2/\nu$		
$\hat{\tau}$	Thermal force factor [Eq. (2.11)]	Br	Brown number $(\bar{q}_i/\bar{q}_g)$
Φ	Rate of exchange or transfer, $jS(\text{g/sec})$	Kn La Le	Knudsen number (λ_g/R_p) Langmuir number $(\mu_l \bar{q}_p/\sigma)$ Lewis number $(k/\rho_g c_p D_{AB})$
ϕ	Contact angle	Ma	Mach number $(|\,q_i - q_g\,|/\bar{q}_g)$
χ	Condensation–coagulation similarity ratio [Eq. (3.65)]	Nu Pe	Nusselt number $(2\hat{k}R/k_g)$ Peclet number $(U_\infty a/D_\nu)$
χ_ν	Rainwater concentration of species ν (mg/liter)	Pr Re	Prandtl number $(c_p \mu_g/k_g)$ Reynolds number $(qR_p/\nu_g;$ $U_\infty a/\nu_g)$
Ψ	Self-preserving size distribution function nV/N^2	Sc Sh	Schmidt number (ν_g/D_ν) Sherwood number $(2\hat{k}R/D_{AB})$
Ω_{ij}	Electrical correction factor for coagulation [Eq. (3.47)]	Stk	Stokes number $(2U_\infty \rho_p R^2/9\mu_g a)$
ω	Angular rotation speed (sec^{-1})	We	Weber number $(\rho_g q_p^2 R_p/\sigma)$
$\bar{\omega}_o$	Single light-scattering albedo		

CHAPTER 1

INTRODUCTION

Contained in one corner of man's storehouse of knowledge is an expanding collection of information about the behavior of tiny particles dispersed in gases. Generically such suspensions have been called *aerosols, aerocolloids,* or *aerodisperse* systems. They include clouds of suspended matter ranging from dust and smoke to mists, smogs, or sprays. The science and technology of aerosols has matured rapidly in the twentieth century as part of the increasing interest in their chemistry and physics. But the history of this part of colloidal chemistry dates back to much earlier times. One might guess that man has been concerned about airborne particles since the time he first choked on smoke from his campfires; indeed, particulate matter has played a major role in the development of the knowledge of air pollution. The optical experiments of Tyndall in 1869, followed by Rayleigh's (1871) theory of light scattering, may have signaled the beginnings of modern aerosol science. This work was followed by Aitken's (1884) studies of particle mechanics, and Wilson's (1897) classical work on nucleation. At the turn of the century came Einstein's (1905) theories of Brownian motion, which bridged the link between a microscopic approach to particle behavior akin to that of large molecules and the evolution of a continuum theory of fluids. Since then, the science has progressed rapidly as described in reviews such as Whytlaw-Gray and Patterson (1932), Fuchs (1964), and Hidy and Brock (1970).

Aerial dispersions vary widely in physical and chemical properties, depending on the nature of the suspended particles, their concentration in the gas, their size and shape, and the spatial homogeneity of dispersion. Both

1

liquid and solid material can be suspended in a gas through a variety of mechanisms. Aerosols produced under laboratory conditions, or by specific generating devices, may have very uniform properties that can be investigated relatively easily by known methods. However, natural aerosols are mixtures of materials from many sources which are highly heterogeneous in composition and physical properties. These natural dispersoids are difficult to characterize and have required a long effort for their study.

1.1 CLASSIFICATION AND DEFINITIONS

1.1.1 Dusts. These are suspensions of solid particles brought about by mechanical disintegration of material which is suspended by mixing in a gas. Examples include (a) clouds of particles from breakup of solids in crushing, grinding, or explosive disintegration; and (b) disaggregation of powders by air blasts. Dust clouds are often dramatic, in the form of storms rising from the earth's surface and traveling hundreds of miles. A dramatic example of production of a natural dust occurred recently during the volcanic eruption of Mt. St. Helens in Washington State (Fig. 1.1). Generally, dusts are quite heterogeneous in composition and have poor colloidal stability with respect to gravitational settling because they are made up of large particles. Yet the lower range of their particle size distribution typically may extend to submicroscopic sizes.

1.1.2 Smokes. In contrast to dusts, smokes cover a wide variety of aerial dispersions dominated by residual material from burning, or from condensation of supersaturated vapors. Such clouds generally consist of smaller particles than dusts, and are composed of material of low volatility, in relatively high concentrations. Because of the small size of the particles, smokes are more stable to gravitational settling than dusts and may remain suspended for extended periods of time. Examples include particulate plumes from combustion processes, from chemical reactions between reacting gases such as ammonia and hydrogen chloride or ozone and olefinic hydrocarbon vapors, oxidation in a metallic arc, and the photochemical decomposition of such materials as iron carbonyl. A principal criterion defining smokes is one of particle size; the distribution in diameter is constrained to from ten micrometers (μm) to less than a tenth of a micrometer. Until a few years ago, smoke from chimneys was a common sight, but many modern cities have all but eliminated this identifier of pollution. Yet the pall of smoke not identified with individual sources remains (Fig. 1.2).

When smoke formation accompanies traces of noxious vapors, it may be called a *fume,* for example, metallic oxide developing with sulfur in a melting or smelting process. The term fume also has more general use to describe a particle cloud resulting from mixing and chemical reactions of vapors diffusing from the surface of a pool of liquid.

Fig. 1.1. Volcanic dust cloud from Mt. St. Helens eruption in May 1980. The cloud covered much of the northwestern United States, and dispersed millions of tons of finely divided ash up to several centimeters thick hundreds of kilometers eastward. [USGS Photography, #8053-136, by R. Krimmel.]

1.1.3 Mists. Suspensions of liquid droplets by atomization or vapor condensation are called *mists*. These aerial suspensions often consist of particles larger than one micrometer in diameter, and relatively low concentrations are involved. With evaporation of the droplets or particle formation by condensation of a vapor, higher concentrations of very small particles in the submicrometer size range may be observed. Historically, mists have been referred to as large particle suspensions with particle size being the principal property distinguishing them from smokes. If the mist has sufficiently high

Fig. 1.2. Smoke and haze over an industrial complex in Los Angeles.

particle concentration to obscure visibility, it may be called *fog*. *Hazes* in the atmosphere usually contain relatively high concentrations of small particles with absorbed liquid water. The term *smog* (smoke mixed with fog) refers to a particulate cloud normally observed over large urban areas, where pollutants mix with haze and chemical reactions contribute to the particulate mix.

1.1.4 Aerosols. The term *aerosol* has been associated with F. G. Donnan in connection with his work during World War I. This name was used to describe clouds of microscopic and submicroscopic particles in air, as for example, chemical smokes for military applications. An aerosol was regarded as an analogy to a liquid colloidal suspension, sometimes called a *hydrosol*. These aerosols possess a degree of stability with respect to settling in a gravitational field. This in effect was the first criterion applied for defining an aerosol.

Gravitational settling in itself is not adequate for defining an aerosol. Additional criteria have emerged and have been applied. For example, the thermal or Brownian motion of particles is thought to be an important characteristic of aerosol particles. Brownian motion becomes a factor for particles less than half a micrometer in diameter. Brownian motion essentially provides a link between the idealized behavior of molecules and small particles. The mechanical theory of large molecules applies well to very small particles in the submicroscopic size range (Hidy and Brock, 1970). This idealized model is central to the evolution of a large segment of particle science and technology. Indeed it forms the basis for the sampling of particles, the description of their depositional behavior in the lungs, and their control by air cleaning devices.

1.1.5 Particle Mechanics—The Gas-Kinetic Model. Idealization of particle behavior in a gas medium involves a straightforward application of fluid dynamics. Mechanical constraints on aerosol particle dynamics can be defined by using certain basic parameters. Model particles are treated as smooth, inert, rigid spheres in near-thermodynamic equilibrium with their surroundings. The particle concentration is very much less than the gas molecule concentration. In such an idealized model, the "state" parameters of interest are listed in Table 1.1. The subscript i refers to the ith class, and g refers to the gas. Typical ranges of the parameter for particles found in atmospheres are included in the table. The idealization requires that the size (radius) ratio $R_g/R_i < 1$ and the mass ratio $m_g/m_i \ll 1$. Application of Boltzmann's dynamic equations for aerosol behavior requires further that the length ratios $R_g/\lambda_g \ll 1$ and $R_g/L_g \ll 1$, where λ_g is the mean free path of the gas and L_g is a typical aerodynamic length, for example, a spherical collector diameter or pipe diameter. The theory extends to incorporate electrical effects, as well as a coagulation or sticking capacity and gas–condensed phase interaction.

TABLE 1.1

Basic Parameters for Describing the Dynamics of Aerosol Particles[a]

	Particles	Gas	Atmospheric aerosols	Air (sea level)
Number density (cm^{-3})	n_i	n_g	10^2–10^5	10^{19}
Mean temperature (°K)	T_i	T_g	$T_i \approx T_g$	~240–310
Mean velocity (cm/sec)	q_i	q_g	10^{-2}–10^3	0–10^3
Mean free path (cm)	λ_i	λ_g	>10^2	$\lambda_g \approx 6 \times 10^{-6}$ (N_2)
Particle radius (cm)	R_i	R_g	10^{-6}–10^{-3}	1.9×10^{-8} (N_2)
Particle mass (g)	m_i	m_g^b	10^{-18}–10^{-9}	4.6×10^{-23} (N_2)
Particle charge (units of elementary charge)	$\pm Q_i$	Weakly ionized	~0–100 units	Weakly ionized single charge
Aerodynamic length	L_i	L_g	$L_i \approx R_i$	$L_g \to \infty$
Coagulation or reaction probability	$\mathscr{C}$	0	$0 \le \mathscr{C} \le 1$	0
Interaction parameter[c]	ϑ_i	ϑ_g		

[a] From Hidy and Brock (1970); reprinted with permission from Pergamon Press.

[b] In a mixture of gases, m_g is the average gas molecular mass.

[c] Interaction parameters are accommodation coefficients for particles of type i and gas g on a boundary s of the aerosol system. These may also include such coefficients for interaction between the gas and the particles themselves.

Virtually all of the mechanical theory emerges from a simplification, called the *single-particle regime*. In this situation, particles are assumed to interact only "instantaneously" in collision; otherwise they can be assumed to behave as a body moving in a medium of infinite extent. The dimensionless parameters describing particle behavior in the single-particle regime are listed in Table 1.2. The four basic parameters are the *Knudsen number*, the *Mach number*, the *Schmidt number*, and the *Brown number*. The *Knudsen number* Kn_i specifies the extent to which the system departs from that given by continuum dynamics. As the $Kn_i \to 0$, the dynamics of particles is described by a continuum theory, where the gas acts as a viscous fluid "sticking" to the particle surface. In the extreme $Kn_i \to \infty$, the dynamics of a particle is characterized by a rarified medium in which gas molecules collide with the particle surface. At normal temperature and pressure, essentially all of the dynamics of submicroscopic particles takes place in a noncontinuum regime. The Mach number Ma indicates the effective particle speed relative to the gas, compared with the gas intensity of its thermal agitation (thermal speed). In most applications, this ratio is very small. The Schmidt and Brown numbers Sc and Br indicate the degree of importance of thermal agitation of the particles. In the Schmidt number, Brownian diffusion of the particle is measured in relation to motion in the fluid medium. The Brown number characterizes the ratio of the particle thermal speed to that of the surrounding gas molecules.

TABLE 1.2

Dimensionless Ratios of Dynamic Significance for Aerosols in the Single-Particle Regime[a]

Knudsen number	$Kn \equiv \lambda_g / R_i$
Mach number[b]	$Ma \equiv \lvert q_i - q_g \rvert \bar{q}_g$
Schmidt number	$Sc \equiv R_i^2 n_g \lambda_g$
Brown number[b]	$Br \equiv \bar{q}_i / \bar{q}_g$

[a] From Hidy and Brock (1970); reprinted with permission from Pergamon Press.

[b] Here $\bar{q}_g$ and $\bar{q}_i$ are the mean thermal speeds of gas and particle, respectively.

In most treatises, aerosols are not defined in terms of the basic ratios in Table 1.2. The theory has emerged principally as an application of continuum fluid dynamics. Thus, the mechanical definition has used the Reynolds number ($Re \equiv q_i R_i / \nu_g$) as a characteristic constraint, where ν_g is the kinematic viscosity of the gas. It is important to note that

$$Re \approx 4Ma/Kn$$

for particles. In the classical definition of an aerosol, the gravitational sedimentation velocity q_G is small. At normal pressures and temperatures, this requires that $q_G \lesssim 0.1$ cm/sec. Practically, this limits particles to about 1 μm radius if they are to be classed as aerosol particles by a criterion of sedimentation stability. In practice, aerosol particles are included for purposes of definition to larger sizes in excess of 10 μm radius. This in turn requires that *the Reynolds number be small, of order unity or less,* implying that the Mach number must approach zero by its definition.

The simplification of the single-particle regime extends to assemblies or collections of particles by adding the following restrictions:

$$n_i/n_g \ll 1,$$

$$\lambda_g n_i \ll 1,$$

$$R_i n_j^{1/3} \ll 1,$$

$$n_i^{-1/3} L_{vi}^{-1} \ll 1,$$

$$Q_i Q_j \kappa / kT \ll 1,$$

where Q is the electrical charge, κ the Debye reciprocal length, k Boltzmann's constant, and T the absolute temperature. These constraints are fulfilled for most practical applications, except for extreme conditions such as fluidized powder beds.

1.1.6 Physical and Chemical Properties. The second group of criteria that describe aerosols uniquely are certain physical and chemical properties which may depart from those of a bulk state. These are independent of the mechanical constraints cited above, and represent a potentially unique resource for technological development. One simple measure of expected differences is the surface to volume ratio. Aerocolloids generally have surface to volume ratios far in excess of 10^3 cm^{-1} in contrast to bulk systems, in which this ratio approaches 10^2 cm^{-1} or less.

The description of the dynamics of aerosol particles requires knowledge of various physicochemical properties of the particles themselves. Although these properties are specified empirically in existing mechanical models, their nature must still be known. For example, the description of the dynamics may entail describing a priori for the particle its mass density, surface energy, freezing point (if liquid), heat of vaporization or sublimation, solubility, heat of adsorption for gases, vapor pressure, viscosity or elastic properties, thermal conductivity, diffusivity of components, magnetic and electric properties, chemical reactivity, radioactivity, and momentum and energy properties.

Suppose the particle composition can be determined or it is assumed that the substances making up the particle are the same as in macroscopic amounts. Then the properties of the bulk state[†] can be translated into the equivalent properties of the particle. Bulk properties are readily determined by standard analytical methods. These techniques generally will not involve direct measurements on aerosol particles. This additional assumption of bulk state behavior, however, may not be justified generally for small aerosol particles. The physical properties of small aerosol particles of a given substance can be quite different from the corresponding properties of the same substance in the bulk state.

Deviations in behavior of small aerosol particles from that of the bulk state are widely recognized for many physical properties such as vapor pressure, freezing point, and crystal structure. Yet there has been a lack of a systematic classification of these deviations. Also, although deviations are expected to occur for small particles, there have been few experimental measurements of such deviations, owing partly to the difficulties of such measurements.

We may classify a deviation from the bulk state for properties of small aerosols as either *extrinsic* or *intrinsic*. Examples of properties falling into these categories are listed in Table 1.3. *Extrinsic deviations* are associated with characteristics of aerosols which are not inherent, but are caused by external agents such as the mode of formation of the particle or the absence of phase transition nuclei in the particles. Thus, extrinsic deviations are associated more with a lack of control and lack of knowledge in the

---

[†] We may define the bulk state here in terms of the state of a substance in a particle with $R \gtrsim 10$ μm. For smaller radii the curvature of the surface may place strains on material, or create dislocations which introduce the potential for distinct condensed states.

TABLE 1.3

Deviations of Physical Properties of Small Particles from Bulk State
Either Observed Experimentally or Predicted in Theory[a]

Extrinsic deviations

   Freezing temperature
   Condensation temperature
   Physical adsorption
   Specific surface area and density

Intrinsic deviations

   Vapor pressure
   Heat of evaporation
   Freezing temperature
   Surface tension, or surface energy
   Solubility
   Equilibrium between a floating particle and a surface film
   Azeotropic composition
   Chemical equilibrium in a heterogeneous reaction involving the particle
   Optical properties, including Raman and fluorescent scattering
   Crystal structure
   Work function

---

[a] From Hidy and Brock (1970); reprinted with permission from Pergamon Press.

particle generation process than with any fundamental cause. *Intrinsic deviations* may occur in several ways. One type of intrinsic deviation is associated with the radius of curvature of small particles. For example, it is well known that a liquid droplet with a given radius has a higher vapor pressure than another droplet of the same composition but larger radius. For sufficiently small particles, another type of intrinsic deviation occurs. Here the intermolecular energy of interaction of molecules making up a very small particle is altered by the fact that a given molecule does not interact with an extremely large number of other molecules, but instead can interact only with a limited number within the particle. Further, if the aerosol particle is still smaller, of almost molecular size, molecular fluctuations will be so large that it is probably no longer meaningful to speak of the physical properties of a particle of such small size in the usual macroscopic sense.

One type of extrinsic deviation is found in the lowering of the freezing point or the raising of the boiling point for small liquid droplets from that for the bulk state. Such effects are usually attributed to the absence of phase transition nuclei. The absence of such nuclei stems from the fact that the bulk material from which the aerosol particles are formed probably contains only minute traces of foreign material (nuclei) per unit volume, so that there is only a very small probability that any small aerosol particle will contain even one nucleus. This circumstance results in the situation that nearly all aerosol particles formed by vapor condensation and subsequent cooling well below the melting point of the parent material are likely to be in a metastable liquid state. As an example, sodium chloride has been observed to exist as a relatively stable subcooled liquid at room temperature, hundreds of degrees below the melting point of crystalline sodium chloride.[†]

Another example of an extrinsic deviation recently has been reported by Jäger and Kieschke (1968). In examining absorption spectra from freshly formed aerosols composed of $Fe(CO)_5$, 30–200 Å in size, they found spectra corresponding to excited states of carbon monoxide, and bands possibly associated with a molecular oxygen transition. The oxygen atom excitation had energy levels of 7–9 eV, suggesting that the excitation was not due to chemical reactions or incident photons only. It is possible that the spectral absorption also was related to gas molecules adsorbed on the iron surface, or to large surface energy of the small particles. When the small particles coagulate, or surface crystallites are relocated, large amounts of energy may be released and transferred to gas molecules adsorbed on the particle surface. In this way, certain high excitation levels may be populated in a manner differing from that predicted by thermal equilibrium.

Basic extrinsic properties of natural and man-made aerosols are beginning to be examined in more detail. For example, Corn and Reitz (1968) have reported some interesting measurements of specific surfaces and densities of samples of urban aerosols taken in the Pittsburgh area of Pennsylvania. They found that the specific surface would increase by about a factor of 2 after degassing these samples at 200°C. Thus natural aerosols may easily contain a variety of adsorbed molecular species which can contribute potentially to anomalies in the expected behavior of the particles based on their bulk composition.

Other types of extrinsic deviation are found in the special properties imparted small particles by the manner of preparation of the particles. For example, production of small particles by grinding in a mill has been observed to produce alteration in the heat of adsorption of gases by the particles. In general, the method of particle production may introduce defects or microscopic impurities which differ from what is found in the parent material. Often extrinsic and intrinsic deviations for the same physical property, as in the example of supercooling of sodium chloride cited above, may both occur in a particle. This fact makes the study of intrinsic deviations very difficult.

Intrinsic deviations are perhaps most widely known through the effect of the radius of curvature of small particles on many physical properties such as vapor pressure, freezing point, surface tension, heat of evaporation, etc. Intrinsic deviations not directly associated with the radius of curvature have been observed by x-ray crystallographic studies of very small crystallites (R_i < 0.01 μm). In these studies, the lattice spacings observed in the small

---

[†] Of course, small particles also exhibit supercooling caused by the curvature of the particle.

crystallites differed significantly from the lattice spacings observed for the bulk state of the parent material. The effect of such alterations on various physical properties has not been studied. In general, one expects that for particles of radius less than the order of 0.01 μm, intrinsic deviations of this sort must occur; however, it is obviously very difficult to observe such deviations experimentally. The result is that there is virtually no information available other than that suggested above.

An illustration of an attempt to look for deviations in surface energy of small spheres about 0.05 μm in radius was reported by Blackman *et al.* (1968). They found for spherical particles of silver, lead, and bismuth evaporated under a microscope that the particle surface energy in these three cases was essentially that of the bulk material. This seems consistent with the calculations for water.

Another type of intrinsic property is derived from the theory of light scattering in particles. Kerker (1982) has reviewed the phenomenon of Raman and fluorescent scattering from molecules suspended in small dielectric particles. Scattered light is affected by the morphology and optical properties of the particle and the distribution of optically active molecules within it. The light scattered and its angular distribution are quite different from those found when the molecules are distributed within the same material in bulk.

In Table 1.4 are presented some examples of the magnitude of known intrinsic deviations. There is very little information on the intrinsic deviations, and the entire subject of intrinsic deviations perhaps represents one of the more poorly understood areas in the physics and chemistry of aerosols. Fortunately, the deviations investigated so far are small until the particle approaches molecular cluster dimensions (<0.01-μm radius).

In summary, the variety phenomena which have been studied in the realm

TABLE 1.4

Variation of Some Properties of Aerosol Particles with Size[a,b]

Property	Particle radius (μm)					
	>10	10	1	0.1	0.01	0.001
Vapor pressure of water at 5°C, p/p_0	1.0000	1.0001	1.0012	1.012	1.124	3.32
Heat of vaporization of water at 4°C, λ/λ_0	1.0000	0.999994	0.99994	0.9994	0.994	0.94
Surface tension of water at 4°C, σ/σ_0	1.0000	—	0.9996	0.996	0.968	0.755
Solubility of crystalline substances	1.0000	—	—	—	~1	~2

[a] Tabulation of thermodynamic variables as a function of particle radius given in micrometers. The subscript 0 refers to bulk conditions.

[b] From Hidy and Brock (1970); reprinted with permission from Pergamon Press.

of aerocolloids is restricted to the consideration of the behavior of small particles suspended in a gas such that a particulate cloud exists for an extended period and such that

(a) $Re_i \approx 4Ma/Kn_i \lesssim 1$,

(b)   the assumptions of the single-particle mechanical regime apply,

(c)   the surface to volume ratio of the particulate system exceeds 10^3 cm^{-1}, and

(d)   certain deviations in physicochemical properties of the particles from the bulk state may be expected.

In practice, these limitations are not rigorously adhered to, for suspended particles are rarely rigid spheres, but often are polycrystalline or agglomerates of several particles. In addition, mechanical considerations extend to conditions where particle radius is significantly larger than ten micrometers, and $Re_i \approx 100$ or more without undue loss of fidelity in characterization of particle behavior.

1.2 FUNDAMENTAL AND PRACTICAL APPLICATIONS

Examples of aerosol suspensions are encountered in a wide variety of natural and industrial situations. They may be found, for example, in astronomy, atmospheric processes, manufacturing, occupational hazard control, and in air pollution control technology. Aerosols have also been studied in medicine for their influence on disease.

Aerosol technology for practical purposes covers a broader size range than called for by their rigorous scientific definition. Normally, the range of interest is less than 100 μm in diameter. A summary of particle dispersoids, methods of measurement, gas cleaning equipment, and mechanical parameters is given in Fig. 1.3. Much of this book covers elements of topics noted in this figure.

A striking and important feature of aerosols is well illustrated in Fig. 1.3, which shows that this science concerns particle sizes ranging over *five orders of magnitude* from molecules to fog droplets. By analogy, this roughly corresponds to a domain from sand grains to tall buildings in a city. Thus the microscopic world of fine-particle suspensions should be as diverse and rich in nature as our everyday macroscopic environment.

In nature, aerosol particles are found over a range of at least four decades in size from tens of angstroms to more than ten micrometers. The fascinating world of interplanetary and interstellar dusts and micrometeors represents an esoteric application of the science. Over the past fifty years, investigators have explored the character of such particles in detail and have examined the interplay between aerosol particles and the formation of water clouds and precipitation in the earth's atmosphere. Since air pollution contributes large quantities of particulate matter to the atmosphere, much effort has

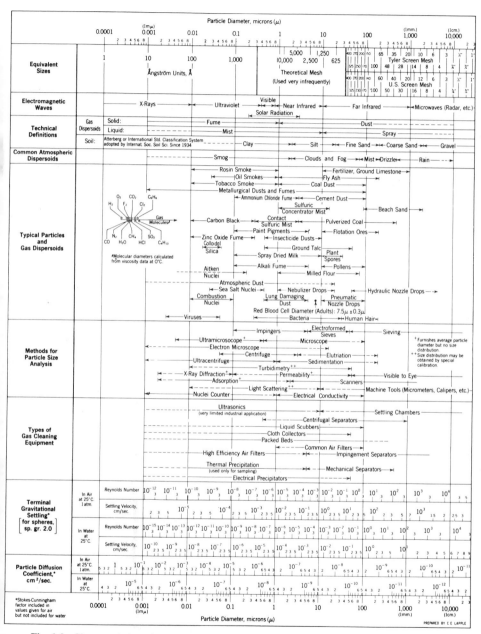

Fig. 1.3. Characteristics of particles and particle dispersoids. [From Lapple (1961); reprinted with permission of Stanford Research Institute.]

been devoted to describing the natural and anthropogenic origins of such material. The extent of man's responsibility for inadvertent weather modification remains an important long-term unresolved question.

Medical studies of the influence of aerosols on human health have led to major improvements in knowledge of the respiratory system. The adverse role of tobacco smoking and exposure to industrial aerosols in the work environment has extended to studies of pollution effects in the ambient atmosphere. On the benefits side, aerosol technology actually has been used productively in the treatment of certain respiratory diseases, and has application in the administration of medication. Since pollution is being considered for increased control measures, a major industry has emerged for gas cleaning equipment, with application both to factory conditions and to atmospheric emissions.

The behavior of airborne particle suspensions is of considerable importance in technology. Pulverization of minerals and other materials to powders, the spray drying of suspensions ranging from milk to detergents, the manufacture of particles for composite materials, and the preparation of paint pigments and dyes are only a few such applications. Combustion technology involving spray injection of fuels and particle formation in rocket engines is another expanding field. Finally, the use of household sprays, paints, and agricultural sprays represents a major technology in which aerosol science has played a role. Taken in toto, the benefits of aerosol technology have made many new industrial products possible; only time will reveal whether these will outweigh the environmental issues created by emerging industrialization.

1.3 SCOPE AND ORGANIZATION OF THIS BOOK

To give the reader a comprehensive view of modern aerosol science and technology, this book begins with a discussion of the foundations of the science. The dynamic theory of aerosols as rigid spheres is introduced and summarized in Chapters 2 and 3. This is followed by two chapters, 4 and 5, which deal with laboratory science. Chapter 4 treats the production of particle suspensions. Then Chapter 5 discusses the methods of particle sampling and measurement and physical or chemical characterization. The theory of particle motion and collision described in Chapter 2 provides the fundamental basis for design of particle collectors and particle deposition. The theory finds application in measurement technology described in Chapter 5. The theory of particle collections is discussed in Chapters 2 and 3, beginning with particle diffusion by Brownian motion, particle formation and growth, and finally coagulation processes. In Chapter 4, the production of particle clouds is described by means of molecular agglomeration (condensation) processes, breakup and disintegration, and chemical reactions. Methods for controlled production of aerosols are outlined with examples of generation techniques developed in recent years. Chapter 5 follows with a description of methods for collection and measurement of properties of particles. The theory of impaction and filtration collectors is included here. Techniques for characterization are also included, ranging from sizing by microscopy and electrical mobility to light scattering. Chapter 5 closes by considering the variety of techniques developed for chemical characterization of collected particles. Methods introduced range from bulk, wet chemical techniques to the ingeneous application of physical methods developed especially in the past twenty-five years.

Following Chapter 5, the second half of the book is devoted to detailed discussion of several major applications of aerosol science. In Chapter 6, applications to combustion; industrial production use of particulate suspensions, powders and composite materials; consumer and agricultural sprays; and large-volume controlled particle generation for medical application. This chapter ends by calling attention to the hazards of dust cloud explosions and deflagration. The remainder of the book concentrates on the nature and environmental effects of aerosols in the earth's atmosphere. Chapter 7 is devoted to a description of atmospheric aerosols, including their physical and chemical properties. The natural origins of particulate matter are surveyed. Then the airborne particles associated with man's activities are discussed. The concepts of dynamical processes shaping the size distribution of atmospheric aerosol particles is included to complete this material. In Chapter 8, the effects of aerosol particles on the earth's atmosphere are covered. These include a discussion of visibility and visual impairment, potential climate change from modification of the radiative energy balance, and water cloud formation and weather modification.

The theme of adverse effects of airborne particles enters in Chapter 9 with a discussion of the human respiratory system. The mechanisms and consequences of collection of particles in the lungs are described. The significance of respiratory disease in relation to particles covers the deposition process, lung clearance mechanisms, and the biology of chemical interactions during assimilation into the body. Given the clear need for control of particulate matter on the grounds of health hazards as well as potential for atmospheric effects, the book closes in Chapter 10 with a discussion of the regulatory framework for control of aerosols. Included in Chapter 10 is a description of modern methods for control of emissions of particulate matter. The importance of particle size to efficiency of particle removal is a central theme of this section. Finally, the possibility of controlling particle formation by atmospheric chemical processes is introduced.

The reader will find that assimilation of the many subjects covered in the book will greatly assist an appreciation of the diverse and complicated character of aerosols. Even though a remarkable record of achievement in this field exists, aerocolloidal science and its applications are only now reaching a truly mature stage. One can expect continuing steady progress in theory and experiment in the future which undoubtedly will result in an expanding technology over the next century.

REFERENCES

Aitken, J. (1884). *Trans. R. Soc. Edinburgh* **32,** 393.
Blackman, M., Lisgarten, H., and Skinner, L. (1968). *Nature (London)* **217,** 1245.
Corn, M., and Reitz, R. (1968). *Science* **159,** 1350.
Einstein, A. (1905). "Investigations on the Theory of Brownian Motion." Dover, New York.
Fuchs, N. A. (1964). "Mechanics of Aerosols." Pergamon, Oxford.
Hidy, G. M., and Brock, J. R. (1970). "The Dynamics of Aerocolloidal Systems." Pergamon, Oxford.
Jäger, H., and Kieschke, H. (1968). *Nature (London)* **217,** 934.
Kerker, M. (1982). *Aerosol Sci. Tech.* **1,** 275.
Lapple, C. E. (1961). *SRI J.* **5,** 94.
Rayleigh, L. (1871). *Philos. Mag.* **41,** 447.
Tyndall, J. (1869). *Philos. Mag.* **37,** 223.
Whytlaw-Gray, R., and Patterson, H. S. (1932). "Smoke." Arnold, London.
Wilson, C. T. R. (1897). *Philos. Trans. R. Soc. London, Ser. A* **189A,** 265.

CHAPTER 2

THE DYNAMICS OF SMALL PARTICLES

Historically much work in aerosol science has focused on the motion of particles in a fluid medium and the associated heat and mass transfer to those particles. Recent theory has recognized that significant differences exist in momentum, heat, and mass transfer depending on the continuous nature of the medium. As discussed in Chapter 1, this is characterized by the Knudsen number Kn. In this chapter, transport processes to small particles in a suspending gas are discussed in detail over a full range of Kn. The simplification adopted for the theory is the single-particle regime. This approach is applied to describing molecular transport processes, then is extended to account for particle–particle interactions later in Chapter 3.

In general, the exchange of momentum between a gas and a particle involves interaction of heat and mass transfer processes to the particle. Thus, the forces acting on a particle in a nonuniform[†] gas may be linked with temperature and concentration gradients as well as velocity gradients in the suspending medium. The assumption that the Knudsen number approaches zero greatly simplifies the calculation of particle motion in a nonuniform gas. Under such circumstances, momentum transfer, resulting in particle motion, will be influenced only by aerodynamic forces associated with surface friction and pressure gradients. In such circumstances, particle motion can be estimated to a good approximation by the classical theory of a viscous fluid

[†] *Nonuniformity* corresponds to a deviation from thermodynamic equilibrium in the suspending, multicomponent gas, involving gradients in velocity, temperature, and gas-component concentration.

medium where Kn = 0. Heat and mass transfer can be considered separately in terms of convective diffusion processes in a low Reynolds number regime. In cases in which noncontinuum effects must be considered (Kn > 0), the coupling between influences of gas nonuniformities becomes important, and so-called phoretic forces play a role in particle motion, but heat and mass transfer again can be treated somewhat independently. Phoretic forces are those that are associated with temperature, gas-component concentration gradients, or electromagnetic forces.

It is often the practice to treat particle motion as the basic dynamical scale for transport processes. Thus, let us begin with this convention and consider first the steady rectilinear motion of spherical particles in a gas.

2.1   TRANSPORT IN STEADY RECTILINEAR MOTION

2.1.1   Stokes' Law and Momentum Transfer.   When a spherical particle exists in isolation in a stagnant, suspending gas, its velocity can be predicted from viscous fluid theory for the transfer of momentum to the particle. Perhaps no other result has such wide application to aerosol mechanics as Stokes' (1851) theory for the motion of a solid particle in a stagnant medium. The model estimates that for Kn → 0, Re ≪ 1, the drag force $\mathcal{D}$ acting on the sphere is

$$\mathcal{D} = 6\pi\mu_g R U_\infty, \tag{2.1}$$

where U_∞ is the fluid velocity far from the particle, R is the particle radius (also R_p), and μ_g is the gas viscosity.

The particle mobility B is defined as $B \equiv U_\infty/\mathcal{D}$. The particle velocity $\mathbf{q}$, equivalent to U_∞ in this case, is given in terms of the product of the mobility and an external force $\mathbf{F}$. Under such conditions, the particle motion is called *quasistationary*. That is, the fluid–particle interactions are slow enough that the particle behaves as if it is in steady motion even if it is accelerated by external forces (see also Section 2.3). Mobility is an important basic particle parameter; its variability with particle size is shown in Table 2.1 along with other important parameters described later.

The analogy for transport processes is readily interpreted from Eq. (2.1) if we consider the generalization that "forces" or fluxes of a property are proportional to a diffusion coefficient, the surface area of the body, and a gradient in the property being transported. In the case of momentum, the transfer rate is related to the frictional and pressure forces on the body. The diffusion coefficient in this case is the kinematic viscosity of the gas ($\nu_g \equiv \mu_g/\rho_g$, where ρ_g is the gas density). The momentum gradient is $\rho_g U_\infty/R$.

If the particles fall through a viscous medium under the influence of gravity, the drag force balances the gravitational force, or

$$\tfrac{4}{3}(\rho_p - \rho_g)g\pi R^3 = 6\pi\mu_g R q_G,$$

TABLE 2.1

Some Characteristic Transport Properties[a] of Aerosol Particles of Unit Density in Air at 1 atm Pressure and 20°C (cgs Units)

Particle radius	Mobility B_p	Diffusivity D_p	Drag coeff. per unit mass $\mathfrak{a}$	Mean thermal speed $\bar{q}_p$	Mean free path $\lambda_p = \bar{q}_p \mathfrak{a}^{-1}$
1×10^{-3}	2.94×10^5	1.19×10^{-8}	8.13×10^2	4.96×10^{-3}	6.11×10^{-6}
5×10^{-4}	5.96×10^5	2.41×10^{-8}	3.24×10^3	1.41×10^{-2}	4.34×10^{-6}
1×10^{-4}	3.17×10^6	1.28×10^{-7}	7.59×10^4	0.157	2.07×10^{-6}
5×10^{-5}	6.71×10^6	2.71×10^{-7}	2.85×10^5	0.444	1.54×10^{-6}
1×10^{-5}	5.38×10^7	2.17×10^{-6}	4.44×10^6	4.97	1.12×10^{-6}
5×10^{-6}	1.64×10^8	6.63×10^{-6}	11.7×10^6	14.9	1.20×10^{-6}
1×10^{-6}	3.26×10^9	1.32×10^{-4}	7.35×10^7	157	2.14×10^{-6}
5×10^{-7}	1.26×10^{10}	5.09×10^{-4}	1.52×10^8	443	2.91×10^{-6}
1×10^{-7}	3.08×10^{11}	1.25×10^{-2}	7.79×10^8	4970	6.39×10^{-6}

[a] Definitions are given in the Nomenclature listing.

where g is the gravitational force per unit mass. The settling velocity is

$$q_G \approx \tfrac{2}{9} R^2 \rho_p g / \mu_g, \tag{2.2}$$

since $\rho_p \gg \rho_g$. Thus, the fall velocity is proportional to the cross-sectional area of the particle, and the ratio of its density to the gas viscosity.

If the particle Reynolds number approaches or exceeds unity, the theory must be modified, and takes a form based on Oseen's 1927 formula

$$\mathfrak{D} = 6\pi \mu_g R U_\infty (1 + \tfrac{3}{8} Re + 19/640 Re^2 + \cdots), \qquad Re = U_\infty R / \nu_g \lesssim 1. \tag{2.3}$$

This reflects the well-known fact that the drag force on a sphere deviates from Stokes' law as a function of the Reynolds number of the particle. In terms of the drag coefficient $C_{\mathfrak{D}}$, the drag force is written as

$$\mathfrak{D} = \tfrac{1}{2} \rho_g \pi R^2 C_{\mathfrak{D}} U_\infty^2. \tag{2.4}$$

The results of experimental measurements for spheres in a fluid indicate that the drag coefficient can be expressed as

$$C_{\mathfrak{D}} = (12/Re)(1 + 0.250 Re^{2/3}), \tag{2.5}$$

where the multiplier of the first term in parentheses is the drag coefficient for Stokes flow.

If the Knudsen number is not assumed to be zero, then the drag force on the particle must be corrected for slippage of gas at the particle surface. Experiments of Millikan and others showed that the Stokes drag force could be corrected in a straightforward way. Using the theory of motion in noncontinuous flow, the mobility takes the form

$$B = A/6\pi\mu_g R, \qquad \text{Re} > 0, \quad \text{Kn} \neq 0 \tag{2.6}$$

Here the numerator is called the *Stokes–Cunningham factor*. The coefficient A is

$$A = 1 + 1.257\text{Kn} + 0.400\text{Kn}\,\exp(-1.10\text{Kn}^{-1}), \tag{2.7}$$

based on experiments. Corrections for noncontinuum effects for Reynolds number of unity or larger have not been reported, but one would expect that they could be placed in a similar mathematical form for the mobility correction.

2.1.2 Particle Shape Factors.

In general, aerosol particles are not spherical, but are irregular in shape. Because of the tradition of interpreting particle motion in terms of the behavior of spheres, most studies employ these particles as a reference. Interpretation of observations then are often reported in terms of *dynamic shape factors*. These have been defined in two ways. The first defines an *aerodynamic* radius (or diameter), which is the size of a particle of arbitrary shape and density whose terminal velocity would be equivalent to that of a sphere of unit density. The second method relates an equivalent radius to the Stokes sphere through the terminal settling velocity:

$$q_G = \tfrac{2}{9}\rho_g R_e^2/\mu_g, \tag{2.8}$$

where $R_e = (3v_p/4\pi)^{1/3}$, v_p being the particle volume. Then a shape factor $\mathbb{K}$ can be defined in terms of the Stokes equivalent:

$$\mathbb{K} = R_e^2/R^2.$$

The factor $\mathbb{K}$ for several geometrical shapes are summarized in Table 2.2. Dynamic shape factors measured for agglomerates of spheres and irregular particles have been reported for example, by Stöber (1972) and Kotrappa (1972). For specific values, the reader should refer to these references.

2.1.3 Motion in External Force Field—Phoretic Forces.

In many circumstances, particles accumulate electrical charge, and are subject to electrical forces. If a particle has an electrical charge of Q and is placed in a uniform electrical field E_0, the electrical force balances the frictional force in steady motion. Thus, the "terminal velocity" in an electrical field based on Stokes' law is

$$q_E = QE_0 B = QE_0/6\pi\mu_g R, \qquad \text{Re} \to 0, \quad \text{Kn} = 0. \tag{2.9}$$

Particle motion induced by gravity can be controlled using an electrical field. Millikan's (1923) famous experiment made use of this to deduce the value of

TABLE 2.2

Shape Factor $\mathbb{K}$ for Particles of Different Shape Used in the Stokes Settling Velocity Equation[a]

Particle shape	$\mathbb{K}$	Remarks
Sphere	1	
Hollow sphere	$1.44(\Delta R/R)^{1/3}$	R sphere radius, ΔR wall thickness, $\Delta R/R \ll 1$
Hemispherical cap	$1.23(\Delta R/R)^{1/3}$	R cap radius, ΔR wall thickness, $\Delta R/R \ll 1$; settling direction: equatorial plane down
Polyhedra		
Cube–octahedron	0.964	Sphericity $R_s = 0.906$
Octahedron	0.939	$R_s = 0.846$
Cube	0.921	$R_s = 0.806$
Tetrahedron	0.853	$R_s = 0.670$
Ellipsoids		Ellipsoids of revolution, length $2c$, equatorial diameter $2R$
	$(\parallel c)$ $5X^{1/3}/(4 + X)$	Prolate, $X = c/R > 1$
	$(\perp c)$ $2.5X^{1/3}/(1.5 + X)$	Oblate, $X = c/R < 1$
Needle	$(\parallel c)$ $1.5X^{-2/3}(\ln 2X - 0.5)$	Prolate ellipsoid of revolution, large
	$(\perp c)$ $1.5X^{-2/3}(\ln 2X + 0.5)$	$X = c/R \gg 1$
Disk[b]	$(\parallel c)$ $0.375X^{1/3}f_1(X)$	Oblate ellipsoid of revolution, small
	$(\perp c)$ $0.375X^{1/3}f_2(X)$	$X = c/R \ll 1$
Cylinder	$(\parallel h)$ $1.72X^{-2/3}(\ln 2X - 0.72)$	Approximations for a straight circular cylinder, height h, base radius R,
	$(\perp h)$ $0.86X^{-2/3}(\ln 2X + 0.5)$	$2X = h/R$
Half ring		
(horizontal)	$0.75X^{-2/3}(\ln 2X + 0.56)$	Half-circular ring bent from a long
(vertical)	$X^{-2/3}(\ln 2X - 0.68)$	prolate ellipsoid, $X = c/R \gg 1$
Ring		
(horizontal)	$0.75X^{-2/3}(\ln 2X + 0.75)$	Circular ring bent from a long prolate
(vertical)	$X^{-2/3}(\ln 2X - 2.09)$	ellipsoid, $X = c/R \gg 1$

[a] Assimilated by Lerman (1979); reprinted with permission of Wiley.
[b] $f_1(X) = 2X/(1 - X^2) + [2(1 - 2X^2)/(1 - X^2)^{3/2}]\tan^{-1}[(1 - X^2)^{1/2}/X]$,
$f_2(X) = -X/(1 - X^2) - [(2X^2 - 3)/(1 - X^2)^{3/2}]\sin^{-1}(1 - X^2)^{1/2}$.

a unit of electrical charge e. The principle of this experiment was to levitate a small oil droplet in an electrical field by first balancing the gravitational force on the droplet with an electrical force. In other words, $q_G = q_E$, or

$$\tfrac{2}{9}R^2 g\rho_p/\mu_g = QE_0/6\pi\mu_g R.$$

Then the electrical field required to hold the particle stationary against gravity is

$$E_0 = \tfrac{4}{3}\pi R^3 g\rho_p/Q. \tag{2.10}$$

By observing different sized particles moving in a vertical electrical field, one can determine the value of the particle mobility for a wide range of Knudsen numbers.

The parameter Kn can be increased both by changing particle size and by gas pressure since the mean free path of the gas is proportional to pressure. The settling velocity can be determined by watching droplet fall rates without the electric field. Then the fall rate can be observed under the simultaneous action of gravitational and electrical forces acting in different directions. Such experiments will yield the velocity q_E. By charging particles with different amounts of charge, one can vary Q, which is ze, where z is the number of unit charges e. Then the particle mobility can be calculated as a function of Kn (e.g., Fuchs, 1964).

Particles can experience external influences induced by forces other than electrical or gravitational fields. Differences in gas temperature or vapor concentration can induce particle motion. Electromagnetic radiation can also produce movement. Such phoretic processes were observed experimentally by the late nineteenth century. For example, in his experiments on particles, Tyndall (1870) described the cleaning of dust from air surrounding hot surfaces. This clearance mechanism is associated with the thermal gradient established in the gas. Particles will move in the gradient under the influence of differential molecular bombardment on their surfaces, giving rise to the thermophoretic force. This mechanism has been used in practice to design thermal precipitators for particles. Although this phenomenon was observed and identified with thermal gradients, no quantification of the phenomenon was made until the 1920s. Epstein (1929) and Einstein (1924) discussed theories for the phenomenon; Watson (1936) and Paranjpe (1936) made measurements of the thickness of the dust-free space in relation to other parameters. Later, the theory was refined for the free-molecule regime by Waldmann (1959) and by Bakanov and Derjaguin (1960). The theory for the continuous–noncontinuous transition regime was extended by experiments of Brock (1967) and others (e.g., Schmitt, 1959). The relation for the *thermal* or *thermophoretic force* $\mathbf{F}_t$ on a spherical particle is

$$\mathbf{F}_t = -\frac{32}{15} R^2 \frac{k_g}{\bar{q}_g} \nabla T \exp\left(\frac{-\hat{\tau}R}{\lambda_g}\right), \qquad 0.25 \lesssim \lambda_g/R \leq \infty, \quad \mathrm{Ma} \ll 1. \quad (2.11)$$

where $\bar{q}_g$ is the mean thermal speed of the gas, and

$$\hat{\tau} = 0.9 + 0.12\alpha_m + 0.21\alpha_m(1 - \alpha_t k_g/2k_p)$$

for monatomic gases, based on experiments. Here α_m and α_t are the accommodation coefficients for momentum and heating, and k_g and k_p are the gas and particle thermal conductivities. In the limit $\mathrm{Kn} \to \infty$, the thermal force is directly proportional to the particle cross-sectional area and the temperature gradient in the gas. The magnitude of the force depends on an experimental factor which varies with Kn for accommodation of thermal energy and momentum of molecules colliding with the particle surface.

The accommodation coefficients vary from zero to unity and are a measure of transfer of molecular energy and momentum to the particle. If the accommodation coefficient is zero, the energy or momentum of collision is not absorbed. If they are unity the property is completely absorbed at the surface. Unfortunately, they are generally not known for particles. A few measurements reviewed by Brock (1967) suggest that $\alpha_m \simeq 0.8$–0.9 for oils or sodium chloride. For monatomic gases at room temperature $\alpha_t \simeq \alpha_m$. Some examples of thermal and momentum accommodation coefficients are listed in Tables 2.3 and 2.4. They can vary over a wide range depending on the condensed material and the gas.

The thermal force is caused by heat transfer from the gas to the particle during molecular motion. If the particle receives heat from sources other than molecular motion, say from absorption of electromagnetic radiation, an additional force can develop as a result of a temperature gradient established in the particle. The force generated by radiation is called the *photophoretic force* $\mathbf{F}_p$, in the full molecule regime:

$$\mathbf{F}_p \approx \frac{-\pi R^3 P \mathbf{I}}{6\left(\dfrac{1}{2\rho_g \bar{q}_g \mathcal{R} + k_p T}\right)}, \qquad \text{Kn} \to \infty, \qquad (2.12)$$

where $\mathbf{I}$ is the radiation flux, P is the gas pressure, and $\mathcal{R}$ is the gas constant. The theory for photophoresis has not been worked out for finite Kn. Other than a laboratory curiosity, it has yet to find an important application in nature, or in particle technology.

The relationship given in Eq. (2.11) represents the theory corrected for the thermal force as derived for the regime of flow on a rarified gas as Kn $\to \infty$. This theory is reasonably well developed and considers molecular collisions

TABLE 2.3

Some Typical Data for Momentum Accommodation Coefficients[a]

System	α_m
Air on old shellac or machined brass	1.00
Air on oil	0.895
CO_2 on oil	0.92
Air on glass	0.89
He on oil	0.87
Air on Ag_2O	0.98
He on Ag_2O	1.0

[a] From Paul (1962); by courtesy of American Institute of Aeronautics and Astronautics.

TABLE 2.4

Some Values of Thermal Accommodation Coefficients [a]

System	α_t
Air on machined bronze	0.91–0.94
Air on polished cast iron	0.87–0.93
Air on etched aluminum	0.89–0.97
Air on machined aluminum	0.95–0.97
O_2 on bright Pt	0.808
Kr on Pt	0.699
N_2 on Wo	0.35
He on Na	0.090
Ne on K	0.199
He on Wo	0.017

[a] From Wachmann (1962); by courtesy of the American Institute of Aeronautics and Astronautics.

at the particle surface as distributed by molecular collisions with each other near the particle surface. When these molecular collisions become a dominant factor in the surface transport processes, the fluid is viewed more as a continuum, and the theory must be developed from the alternate extreme where Kn → 0 (Hidy and Brock, 1970). Correction for the continuum fluid–particle interactions yields a separate formula for the thermal force (Brock, 1967):

$$\mathbf{F}_t = \frac{-12\pi\mu_g R^2 c_{tm} \mathrm{Kn}[k_g/k_p + c_t\mathrm{Kn})(1 + 1.33 a c_m\mathrm{Kn}) - 1.33 a c_m\mathrm{Kn}] \, \nabla T_g}{(1 + 3c_m\mathrm{Kn})(1 + 2k_g/k_p + 2c_t\mathrm{Kn})}.$$

(2.13)

This relation is derived from an approximation solution to the Boltzmann equation for the molecular flux expressions for particles in the slip flow regime, where $0 \le \mathrm{Kn} \le 0.2$. The coefficients c_{tm}, c_t, and c_m are respectively the thermal creep coefficient, the temperature jump coefficient, and the isothermal slip coefficient. The constant a is a second-order fitting parameter taken from experiments to be approximately unity.

For monatomic molecules,

$$c_{tm} \approx 3\mu_g/4\rho_g T_g\lambda_g \tag{2.14a}$$

and

$$c_t = \hat{P}_t(2 - \alpha_t)/\alpha_t, \tag{2.14b}$$

where the coefficient $1.87 \le \hat{P}_t \le 2.48$,

$$c_m = \hat{P}_m(2 - \alpha_m)/\alpha_m, \qquad 1.00 \le \hat{P}_m \le 1.274. \tag{2.14c}$$

From the work of many investigators $c_m \approx 1.2$.

Equation (2.13) is similar to Epstein's (1929) early theory, but appears to be capable of fitting observations over a wide range of thermal conductivity. The slip coefficients are not known for particles, so Eq. (2.13) is difficult to use for a priori estimates of the thermal force. We note also that Eq. (2.13) indicates that the thermal force goes to zero with Kn. Thus, in the continuum theory there can be no thermophoretic effect.

In analogy to their motion in a thermal gradient, particles may experience a force associated with nonuniformities of composition in the suspending gas mixture. This force is called the *diffusion* or *diffusiophoretic force*. For the case of equimolar counterdiffusion in a dilute binary mixture of component A in B, $n_A \gg n_B$, $\alpha_{mA} = \alpha_{mB} = 1$, and the force acting on a spherical particle is

$$\mathbf{F}_d = \frac{8}{3} R^2 n_g (2\pi k T)^{1/2} \left(1 + \frac{\pi}{8}\right) D_{AB} \nabla x_A \left\{ (m_B^{1/2} - m_A^{1/2}) \right.$$

$$\left. + \left[\frac{d_A^2 m_A^{1/2}}{d_{AB}^2} - \left(\frac{2 m_A m_B}{m_A + m_B} \right)^{1/2} \right] \frac{0.311 R}{\lambda_g} \right\}, \qquad 0.5 \lesssim \frac{\lambda_g}{R} \lesssim \infty. \qquad (2.15)$$

Here x_A is the mole fraction of species A and ∇x_A is the gradient in mole fraction of species A. The diffusion coefficient for the binary mixture is D_{AB}; m_B and m_A are the molecular mass of species A and B; d_A and d_B are the molecular diameters of A and B, $d_{AB} = (d_A + d_B)/2$; and the mass free path of the gas mixture is

$$\lambda_g = (\sqrt{2} \pi n_g d_{AB}^2)^{-1}.$$

The parameter n_g is the molecular number concentration of gas molecules. Equation (2.15) is quite complicated, but the theory quantifies the analogy between the thermal and diffusion force. Basically the latter is proportional to the concentration gradient of the diffusing species, the cross-sectional area of the particle, the product of the molecular mass differences, and a Kn factor. The last factor in brackets represents a Kn correction factor to the free-molecule theory where Kn $\to \infty$. For diffusion of gas component A through stagnant component B, with $n_A \ll n_B$ and $\alpha_{mA} = \alpha_{mB} = 1.0$, the diffusion force has the form

$$\mathbf{F}_d = -\frac{8}{3} R^2 n_g (2\pi m_A k T)^{1/2} \left(1 + \frac{\pi}{8}\right) \frac{D_{AB}}{x_B} \nabla x_A$$

$$\times \left[1 - 0.071 \left(\frac{2 m_B}{m_A + m_B} \right)^{1/2} \frac{R}{\lambda_g} \right]. \qquad (2.16)$$

This equation is similar to Eq. (2.13), with the first right-hand term the free-molecule formula and the square-bracketed term on the right the Kn correction factor. In the slip flow regime, for the case of equimolar counterdiffusion and neglecting the effect of velocity slip,

$$\mathbf{F}_d = -6\pi \mu_g R \sigma_{AB} D_{AB} (\nabla x_A)_\infty, \qquad 0 \leq \text{Kn} \leq 0.3. \qquad (2.17)$$

The gradient $(\nabla x_A)_\infty$ is taken at a large distance from the particle surface. And for diffusion of A through stagnant B,

$$\mathbf{F}_d = -6\pi\mu_g R(1 + \sigma_{AB}x_B)(D_{AB}/x_B)(\nabla x_A)_\infty, \qquad 0 \leq Kn \leq 0.25. \quad (2.18)$$

The coefficient σ_{AB} is an empirical coefficient introduced by Waldmann and Schmitt (1967) of the form

$$\sigma_{AB} = 0.95(m_A - m_B)/(m_A + m_B) - 1.05(d_A - d_B)/(d_A + d_B).$$

This formula is based on experiments on particle motion of several different gases mixed with nitrogen. Values of σ_{AB} range from -0.10 to $+0.90$ for the binary mixtures investigated by Waldmann and Schmitt.

2.1.4 Heat and Mass Transfer.

Particles suspended in a nonuniform gas may be subject to absorption or loss of heat or material by diffusional transport. If the particle is suspended without motion in a stagnant gas, transport to or from the body can be estimated from heat conduction or diffusion theory. For steady state conditions, continuum theory indicates that the mass concentration or temperature field around a spherical particle is

$$\frac{\rho_A - \rho_{A\infty}}{\rho_{As} - \rho_A} = \frac{T - T_\infty}{T_s - T_\infty} = \frac{R}{r}, \qquad (2.19)$$

where the radial direction from the center of the sphere is r and the subscripts s and ∞ indicate the particle surface value and the value far from the sphere. Using this relation, one finds that the net rate of transfer of heat to the particle surface in a continuum is

$$\Phi = 4\pi r^2 D_T \left(\frac{\partial T}{\partial r}\right)_{r=R} = 4\pi R D_T(T_\infty - T_s), \qquad (2.20)$$

where D_T is the thermal diffusivity $k_g/\rho_g c_p$ and c_p is the specific heat of the particle. For mass transfer of species A through B to the sphere,

$$\Phi_A = 4\pi R D_{AB}(\rho_{A\infty} - \rho_{As}), \qquad (2.21)$$

where D_{AB} is the binary molecular diffusivity for the two gases. This relation is basically that derived by Maxwell (1890). His equation applied to the steady state evaporation from or condensation of vapor component A in gas B on a sphere. For the case of condensation, Eq. (2.21) may be rewritten in the form

$$\Phi = (4\pi R D_{AB} M_A/\mathcal{R}T)(p_{A\infty} - p_{As}), \qquad (2.22)$$

where M_A is the molecular weight of species A and p_A is its vapor pressure. For evaporation the sign of the density or vapor pressure difference is reversed.

The applicability of Maxwell's equation is limited in describing particle growth or depletion by mass transfer. Strictly speaking, mass transfer to a small droplet cannot be a steady process because the radius changes, causing a change in the transfer rate. However, when the difference between vapor concentration far from the droplet and at the droplet surface is small, the transport rate given by Maxwell's equation holds at any instant. That is, the diffusional transport process proceeds as a quasistationary process.

When the particle is moving relative to the suspending fluid, transport of heat or matter is enhanced by convective diffusional processes. These are measured in terms of transfer coefficients, defined as

Heat transfer:

$$\ell_h = \frac{1}{4\pi R^2(T_s - -T_\infty)} \int_S \left[-k_g\left(\frac{dT}{dr}\right)_{r=R}\right] dS;$$

Mass transfer: (2.23)

$$\ell = \frac{1}{4\pi R^2(\rho_{As} - \rho_{A\infty})} \int_S \left[-D_{AB}\left(\frac{d\rho_A}{dr}\right)_{r=R}\right] dS.$$

These transfer coefficients are normally expressed in dimensionless form. For heat transfer, the Nusselt number $Nu \equiv 2\ell_h R/k_g$; for mass transfer, the Sherwood number $Sh \equiv 2\ell R/D_{AB}$. Where there is no motion, $Sh = Nu = 2$. If there is particle motion,

$$Nu = f_1(Re, Pr, U_\infty t/R), \qquad Sh = f_2(Re, Sc, U_\infty t/R), \qquad (2.24)$$

where the Prandtl number $Pr \equiv c_p\mu_g/k_p$, and the Schmidt number $Sc \equiv v_g/D_{AB}$, where v_g is the kinematic viscosity. The Reynolds number and the Prandtl and Schmidt numbers are often combined to give the Peclet number, $Pe \equiv RePr$, or $ReSc$. The Peclet number, $U_\infty R/D_{AB}$, for heat or transfer is analogous to the Reynolds number for momentum transfer.

For Stokes flow, the Nusselt or Sherwood number can be written in terms of an expansion in the Peclet number (Acrivos and Taylor, 1962):

$$Nu = 2 + Pe + Pe^2 \ln Pe + 0.0608Pe^2 + \cdots, \qquad Pe < 1. \quad (2.25)$$

For larger Pe the empirical Frössling equation is often used; it has the form

$$Nu \approx 2(1 + k_1 Pr^{1/3} Re^{1/2}). \qquad (2.26)$$

This relation reduces to the form of Eq. (2.20) as $Pe \to 0$, but does not correctly fit the theoretical expression of Acrivos and Taylor. Most experiments for heat and mass transfer to spheres at $Re \geqslant 100$ give $k_1 \simeq 0.39$. However, Kinzer and Gunn's (1951) experiments for $Pe < 1$ suggest $k_1 \simeq 0.65$.

When the theory of transport is developed taking into account noncontinuum effects, the theory must be corrected as in the case of momentum transfer. The rates of heat and mass transfer to spheres are given, for negligible Mach number (Ma) and $0.25 \lesssim \lambda_B/R \leq \infty$, by

$$\Phi_A = [2\pi R^2 \alpha_c n_A (2kT/\pi m_A)^{1/2}](1 + \alpha_c C_1 R/\lambda_B)^{-1} \tag{2.27a}$$

for radial molecular mass transfer of species A into B from a sphere. The first term in the product on the right is the rate of transfer by free-molecule flow (Kn → ∞). The term on the far right of the product is the Kn correction for the transition regime. The coefficient α_c is the condensation or evaporation coefficient, which is analogous to the thermal or momentum accommodation coefficient. Some example values are listed in Table 2.5. The coefficient C_1 has the form given by Brock (1967); $C_1 = 0.807 W_B^{1/2}(d_{AB}/d_B)^2 I$, where the reduced mass $W_B = m_B/(m_A + m_B)$, and I is a numerical factor which varies with W_B; for example,

W_B:	1.000	0.800	0.600	0.400	0.200
I:	0.295	0.245	0.201	0.162	0.130

TABLE 2.5

Typical Data for the Evaporation Coefficient[a]

Material	α_c	Temperature range (°C)
Beryllium	≈1	898–1279
Copper	≈1	913–1193
Iron	≈1	1044–1600
Nickel	≈1	1034–1329
Ammonium chloride	$(3.9–29.0) \times 10^{-4}$	118–221
Iodine	0.055–1	−21.4 to 70
Mercurous chloride	0.1	97, 102
Sodium chloride	0.11–0.23	601–657
Potassium chloride	0.63–1.0	—
Water (liquid)	0.04–>0.25	15–100
Benzene	0.85–0.95	6
Camphor (synthetic)	0.139–0.190	−14.5–5.5
Carbon tetrachloride	≈1	—
Diamyl sebacate	0.50	25, 35
Di-*n*-butyl phthalate	1	150–350
Ethyl alcohol	0.0241–0.0288	12.4–15.9
Glycerol	Unity, 0.052	18–70
Naphthalene	0.036–0.135	40–70
Tetradeconal	0.68	20

[a] From Paul (1962); courtesy of the American Institute of Aeronautics and Astronautics.

The analogous extension of the continuum theory for $0 \leq Kn \gtrsim 0.25$ is

$$\Phi_A = [4\pi R D_{AB}(\rho_s - \rho_\infty)](1 + c_{dA}\lambda_g/R)^{-1}. \qquad (2.27b)$$

The bracketed term in the product on the right is the continuum rate of mass transfer; the slip correction is the last term on the right. Here the coefficient $c_{dA} = [(2 - \alpha_{cA})/\alpha_{cA}](2D_{AB}/\bar{q}_A\lambda_g)$. In general, there are no rigorous theoretical estimates for c_{dA}, but values between 1.2 and 1.5 are consistent with experiment.

The heat transfer expressions are

$$\Phi_H = [4\pi R^2\alpha_t(2kT^-/\pi m_g)^{1/2}n^-k(T^+ - T^-)](1 + \alpha_t C_2 R/\lambda_g)^{-1} \qquad (2.28a)$$

in the free-molecule to transition regime, where $0.25 \leq Kn \leq \infty$, $Ma < 1$.

The theory for the radial heat flux from a sphere in the free-molecule to transition regime is written in terms of a flux of molecules n^- leaving the surface, and a temperature difference between impinging and reflected molecules $T^+ - T^-$. The coefficient $C_2 \equiv \frac{7}{144}\pi(C_v + \frac{1}{2}k)/2k$ for polyatomic gases, where C_v is the heat capacity at constant volume. The coefficient $C \approx 0.24$ for air.

For the continuum model, corrected for slippage of gas about a sphere, the rate of heat transfer is given by

$$\Phi_H = 4\pi k_g R(T_s - T_\infty)(1 + c_t Kn)^{-1}, \qquad (2.28b)$$

in analogy to Eq. (2.27b). The formula for c_t is given in Eq. (2.14b).

2.1.5 Condensation and Evaporation. Perhaps the most widely used application of the heat and mass transfer relations concerns the condensation and evaporation of particles. Using Maxwell's equation, the growth in a quasistationary state of a pure spherical droplet in a supersaturated vapor is estimated as

$$\frac{dm_p}{dt} = \Phi = \frac{4\pi\rho_p D_{AB} R v_m}{k}\left(\frac{p_\infty}{T_\infty} - \frac{p_s}{T_s}\right), \qquad Kn \to 0, \quad Re \to 0. \quad (2.29)$$

where v_m is the molecular volume of the condensing species, p is the vapor pressure of A, and k is the Boltzmann constant.

For depletion by evaporation in a quasistationary state, Eq. (2.29) applies, but the sign is reversed, owing to the nonsaturation of the vapor. For droplets of low volatility, the average droplet temperature or gas temperature can be used, but high-volatility liquids may generate temperature differences associated with heating or cooling during mass transfer.

In general, the growth or depletion of droplets by phase change should be estimated taking into account both heat and mass transfer. By combining the heat conduction equation and Maxwell's equation, an estimation can be made of the temperature difference between the droplet and the medium.

For a condensing droplet, the coupling of heat and mass transfer for quasi-steady state conditions leads to the relationship

$$T_s - T_\infty = (\lambda_v D_{AB} v_m / k_g k \bar{T})(p_s - p_\infty), \tag{2.30}$$

where λ_v is the latent heat of vaporization of the liquid and $\bar{T}$ is the mean value of the temperatures T_s and T_∞. This relation was derived by Maxwell in his 1890 work; it indicates that evaporation or growth by condensation by pure vapor under quasisteady conditions is accompanied by temperature changes which are independent of particle size.

When the droplets are very small, the diffusion rate requires a correction for the droplet curvature. For small values of the vapor pressure difference, a correction can be made using the Kelvin–Gibbs formula. For a pure liquid droplet in its own vapor, the partial pressure difference is written

$$p_\infty - p_s = p_\infty - p_0 \exp \frac{2\sigma v_m}{RkT} = p_\infty - p_0 \exp\left(\frac{R^*}{R} \ln \mathcal{S}\right), \tag{2.31}$$

where p_0 is the vapor pressure over a flat surface of the liquid, $\mathcal{S}$ is the supersaturation ratio p_s/p_0, σ is the surface tension of the droplet, and R^* is the "critical" droplet radius. The critical size is the nucleus size which will just be stable at thermodynamic equilibrium with its own vapor and will not evaporate. If the droplet is a solution, then the vapor pressure–particle size relationship becomes more complex, and should be accounted for. If the droplet contains an insoluble fraction, wetting also may be a factor in the growth rate. These factors are discussed further in Chapter 3.

If a droplet contains traces of surface-active material such as organic acids or alcohols, the evaporation or condensation can be reduced. The effect of a surface layer of material on a volatile droplet is to reduce the surface vapor concentration below saturation, reducing the vapor pressure driving force. The change in mass transfer rate can be related to the surface film by adding a transfer resistance, which may be defined in terms of the depletion in vapor concentration just outside the droplet because of the film (Davies, 1978). The rate of change in mass of a droplet can be written

$$\frac{dm_p}{dt} = \frac{4\pi R^2 D_{AB} \rho_p v_m}{k} \left(\frac{p_\infty}{T_\infty} - \frac{p_s}{T_s}\right)(R + D_{AB} \ell_e^{-1})^{-1}, \tag{2.32}$$

where ℓ_e^{-1} is the evaporation resistance (sec/cm). If $D_{AB} \ell_e^{-1}$ is much smaller than the droplet radius, this relation reduces to Maxwell's equation. If $D_{AB} \ell_e^{-1}$ is larger than R, then this expression reduces to the gas kinetic relation, if $\ell_e = 4/\alpha_c \bar{q}_A$, where $\bar{q}_A$ is the thermal velocity of vapor molecules. Thus the film resistance can be linked on a molecular scale to the accommodation coefficient. Davies (1978) has calculated that a resistance from a surface film can increase the lifetime of a micrometer-size volatile droplet by order of magnitude, as might be expected.

The evaporation rate of droplets under conditions where noncontinuum effects are important can be determined by semiempirical curve fitting for quasi-stationary state conditions where the latent heat changes are small. Basically the mass exchange equations for the free-molecule and continuum regimes constrain the model [Eqs. (2.27a) and (2.27b)]. Several investigators have proposed such formulas. They can be written in the form

$$\Phi/\Phi_c = 1/[1 + f(Kn)], \tag{2.33}$$

where Φ_c is given by Maxwell's equation; Fuchs and Sutugin (1971) give

$$f(Kn) = [(1.333 + 0.71 Kn^{-1})/(1 + Kn^{-1})]Kn. \tag{2.34}$$

Another version of this expression based on an approximate solution of the Boltzmann equation in tabular form has been reported recently by Loyalka (1973).

Very little data have been reported for which formulation could be verified. However, Davis *et al.* (1978) have published data for the evaporation rates of dioctyl phthalate and dibutyl sebacate (DBS) over a wide range of Kn. Their data for DBS are shown in Fig. 2.1 with the theory from different investigators. The form in the figure corresponds to the ratio of the rate of transfer normalized to the kinetic theory value Φ_k rather than Φ/Φ_c.[†]

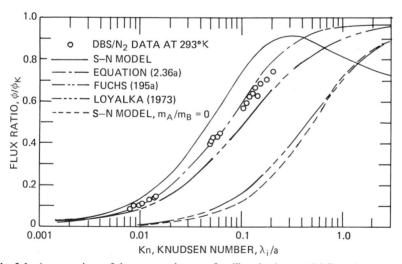

Fig. 2.1. A comparison of the evaporation rate for dibutyl sebacate (DBS)(A) in N_2(B) with the Fuchs–Sutugin (1971) model [Eq. (2.36a)]. Shown for comparison are the results from other theoretical forms including the Sitarski–Nowakowski (1979) (S–N) model, Fuchs (1959), and Loyalka (1973). Also included is the curve for the S–N model with $m_A/m_B = 0$. [From Davis *et al.* (1980), with permission of the author and Gordon & Breach.]

[†] Using the Fuchs and Sutugin value of $\frac{4}{3}$ for the ratio $D_{AB}/\lambda_A \bar{q}_A$,

$$\Phi/\Phi_k = 1.33 Kn \left[1 + \frac{1.33 Kn + 0.71}{1 + Kn^{-1}} \right]^{-1}. \tag{2.33a}$$

2.2 ACCELERATED MOTION OF PARTICLES—DEPOSITION PROCESSES

The behavior of particles in steady rectilinear flow represents the simplest configuration of interest in aerosol mechanics. Of great practical importance is the accelerated motion of particles, since acceleration-induced fluid in media is at the root of deposition or collection processes. Taking into account the fact that the particles in accelerated motion do not follow the gas speed, the drag force is

$$\mathbf{D} = (6\pi\mu_g R/A)(\mathbf{q}_g - \mathbf{q}_p).$$

Then the force balance on a particle in motion is

$$\underset{(1)}{\frac{4}{3}\pi R^3 \rho_p \frac{d\mathbf{q}_p}{dt}} = \underset{(1)}{\frac{6\pi\mu_g R}{A}(\mathbf{q}_g - \mathbf{q}_p)} + \underset{(2)}{\frac{4}{3}\pi R^3 \rho_g \left(\frac{d\mathbf{q}_g}{dt}\right)} + \underset{(3)}{\frac{2}{3}\pi R^3 \rho_g \left(\frac{d\mathbf{q}_g}{dt} - \frac{d\mathbf{q}_p}{dt}\right)}$$

$$\underset{(4)}{+ \, 6R^2(\pi\rho_g\mu_g)^{1/2} \int_{t_0}^{t} dt' \left[\left(\frac{d\mathbf{q}_g}{dt}\right) - \left(\frac{d\mathbf{q}_p}{dt}\right)\right] \Big/ (t - t')^{1/2}}$$

$$\underset{(5)\quad(6)}{+ \, \mathbf{F}_e + \mathbf{F}_{NU}.} \tag{2.35}$$

In Eq. (2.35), t_0 denotes the starting time for particle acceleration, and $\mathbf{q}_g$ represents the gas velocity near the particle, but far enough away not to be disturbed by the relative motion of the body. The terms in Eq. (2.35) have the following meanings: term (1) on the right is the viscous resistance given by Stokes' law. The second term (2) is due to the pressure gradient in the gas surrounding the particle, caused by acceleration of the gas by the particle. The third term (3) denotes the force required to accelerate the apparent mass of the particle relative to the ambient gas. The fourth term (4), *the Basset term,* accounts for the effect of the deviation from steady state in gas flow pattern. The last two terms are the forces associated with external force fields (5), $\mathbf{F}_e$, and from nonuniformities in the suspending medium (6), $\mathbf{F}_{NU}$.

The equation of motion can be summarized in the form

$$\frac{d\mathbf{q}_p}{dt} + \alpha(\mathbf{q}_p - \mathbf{q}_g) = \mathbf{F}(t) + \mathbf{F}_e' + \mathbf{F}_{NU}', \tag{2.36}$$

where $\alpha = (m_p B)^{-1}$ and $\mathbf{F}' = \mathbf{F}/m_p$. The term $\mathbf{F}(t)$ is a contribution to particle acceleration from an "external" perturbation function acting on the particle. This has the form of the *Langevin equation,* well known in the theory of Brownian motion.

2.2.1 Relaxation Time to Reach Steady Motion.

An important characteristic scale in particle mechanics is the time to achieve steady motion. To find the parameter, we can apply Eq. (2.36) for simple conditions of the deceleration of a particle by friction in a stationary gas. In the absence of external forces, the velocity of a particle traveling in the x direction is calculated by

$$\frac{dq_p}{dt} + \alpha q_p = 0,$$

or

$$q_p = q_0 \exp(-\alpha t) \tag{2.37}$$

if the initial velocity is q_0.

The distance traveled by the particle is

$$x = \int_0^t q_p \, dt' = \alpha^{-1} q_0 [1 - \exp(-\alpha t)]. \tag{2.38}$$

The significance of α^{-1} is clear;   it is a constant which is the *relaxation time* scale τ for stopping a particle in a stagnant fluid. Values of the reciprocal relaxation time are listed for different particle sizes with the mobility B in Table 2.1. Similarly, one can show that α^{-1} represents the time it takes for a particle falling in a gravitational field to achieve its terminal speed. Note that the terminal speed $q_G = g\alpha^{-1}$. At $t/\alpha \rightarrow \infty$, the distance over which the particle penetrates, or the *stopping distance* 1 is $q_0 \alpha^{-1}$.

2.2.2 Curvilinear Particle Motion.   When particles change their direction of movement, as for example around blunt bodies or bends in tubing, inertial forces tend to modify their paths relative to the suspending gas. This is shown schematically in Fig. 2.2. As shown, particles may depart from the path of gas molecules (streamlines) to allow deposition of the body. This is the principle underlying inertial collectors.

The motion of the particle undergoing curvilinear flow can be calculated from the vector force of Eq. (2.35) if the particles are large enough that their thermal (Brownian) motion can be disregarded. The trajectory of a particle moving in a gas can be estimated by integrating the equation of motion for the particle over a time period given by increments of the ratio of the radial distance traveled divided by the particle velocity, i.e., $|r/q_p|$. Interpretation of the equation of motion, of course, requires knowledge of the flow field of the suspending gas; one can assume that the particle velocity equals the fluid velocity at some distance r far from the collecting body.

Perhaps the simplest case to illustrate the difference between particle flow and fluid flow is the deposition from a particle cloud traveling horizontally through a straight channel of length L into a gravity field. The gas flow is two dimensional, laminar, and horizontally uniform, with mean velocity $\bar{q}_x$. The flow is sufficiently slow that vertical convective currents are small compared with the gravitational settling velocity. For the purpose of this and subsequent calculations, let us assume that the particles are infinitely small

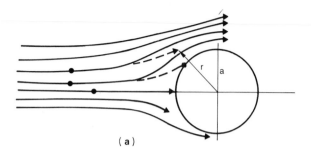

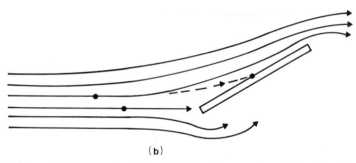

Fig. 2.2. Schematic diagram of particle and fluid motion around a cylinder (a) and an inclined flat plate (b). Streamlines are shown as solid lines, while the dotted lines are aerosol particle paths.

spheres, or the particle radius approaches zero, but thermal motion is disregarded. For fully developed flow, the horizontal component of particle velocity is

$$q_x = (6z/d - 3z^2/d^2)\bar{q}_x,$$

where z is the vertical coordinate and d is the channel width. Disregarding particle inertia in the horizontal component, the particle trajectories are determined by

$$\frac{dx}{dt} = q_x, \qquad \frac{dz}{dt} = q_z - q_G,$$

where q_G is the settling velocity. This pair of equations can be solved by introducing a fluid dynamic stream function, and the fraction η of the total number of particles removed on the bottom of the channel is written

$$\eta = q_G L/d\bar{q}_x. \tag{2.39}$$

The parameter η is normally called the *precipitation or collection efficiency*.
   For flow in a round tube, an analogous result is obtained:

$$\eta_0 = 3q_G L/8R_T\bar{q}_x,$$

where R_T is the tube radius. In the case of viscous gas flow in horizontal tubes, Thomas (1958) obtained

$$\eta = (2/\pi)[2\eta_0(1 - \eta_0)^{1/2} + \arcsin \eta_0^{1/3} - \eta_0^{1/3}(1 - \eta_0^{2/3})^{1/2}]. \quad (2.40)$$

Such calculations may be extended to include external electrical forces. Precipitation of aerosol particles in electrically charged vertical or horizontal channels has been used to study particle mobility.

The effect of including a finite particle radius effectively increases the potential collection efficiency over the estimate given for the case in which the radius is negligible.

2.2.3   Inertial Deposition on Blunt Bodies.

A blunt body in an aerosol stream can act as a particle collector. This is easily demonstrated by suspending a wire or a sphere in dusty air. After exposure a coating of particles will be found on the side of the body facing the stream. Like the channel flow example, the deposition of particles can be calculated using the equation for particle trajectories. Particle paths which touch the body will give rise to particle collection. The collection efficiency can be defined in terms of a distance S from the axis of the obstacle to the outermost trajectory of particles that strike the surface compared with the radius of the collector. In other words, the efficiency is

$$\eta = S/a,$$

where a is the cylinder radius. The trajectories of the particles can be calculated by a step-by-step integration of vector equations of motion in the form

$$m_p \frac{d^2\mathbf{r}}{dt^2} = B^{-1}\left(\mathbf{q}_g - \frac{d\mathbf{r}}{dt}\right). \quad (2.41)$$

This relation often is written in the dimensional form

$$\text{Stk} \frac{d^2\mathbf{r}'}{dt'^2} + \frac{d\mathbf{r}'}{dt'} = \mathbf{q}_g'(R'), \quad (2.42)$$

where $\mathbf{r}' = \mathbf{r}/R$, $\mathbf{q}_g' = a/U_\infty + t' = U_\infty t/R$. The *Stokes number* is

$$\text{Stk} \equiv \frac{U_\infty}{Ca} = \frac{\ell}{a} = \frac{2U_\infty \rho_p R^2}{9\mu_g a}. \quad (2.43)$$

It is equivalent to the ratio of the stopping distance ℓ and the radius of the obstacle. For "point" particles in Stokes flow, the Stokes number is the only criterion other than geometry that determines *similitude* for the shape of the particle trajectories. To ensure hydrodynamic similarity, in general, the collector Reynolds number also must be preserved, as well as the ratio $I \equiv R/a$, called the interception parameter. Then the collection efficiency can be written in the form

$$\eta = f(\text{Stk, Re, I}). \quad (2.44)$$

Here the interception parameter effectively accounts for a small additional increase in the collection efficiency for a given Stk associated with the finite size of the particles.

The inertial collection of particles is referred to as *impaction*. The influence of finite particle size relative to the collector is called *interception*.

Where Stk $\to 0$, or as the particle mass becomes small, the particles tend to follow the motion of the gas. In steady flow, the particles then never reach the obstacle, or $\eta = 0$. Thus, there should be a certain minimum value of the Stokes number below which particle entrainment in the gas dominates inertia. The existence of such a regime apparently was first established by Langmuir (1948). In the limiting case of high Reynolds number for flow around obstacles, where inviscid or potential flow is a useful approximation, one can estimate analytically the values of the critical Stokes number Stk_{cr}. Such calculations have been reported by Levin (1953); his results are shown in Table 2.6 for $I \to 0$.

To illustrate the form of the relationship between the collection efficiency and the Stokes number, consider an aerosol streaming around a sphere. Calculations can be made for two extreme cases of gas flow around the sphere, first inviscid motion, where Re $\to \infty$; second for viscous flow, where Re is finite. The extreme for the second regime is the case of viscous flow, where Re $\lesssim 1$. In the inviscid flow case, the streamlines around the body are symmetrical around it; in viscous flow where Re $\ll 1$ the streamlines also are symmetrical, but are pushed outward relative to the inviscid case as a consequence of viscous effects. In reality, high Reynolds number flows "separate" on the lee side of the body and form a turbulent wake. However, the

TABLE 2.6

Limiting Values of the Stokes Number for Collection of Particles on Bodies[a]

Obstacle in flow	Stk_{cr}
1. Infinitely long ribbon of width $2a$	
(a) Without separation	$\frac{1}{4}$
(b) With separation	$4/(\pi + 4)$
2. Elliptical cylinder, major to minor axis ratio X	$\frac{1}{4}(1 + X)$
3. Circular cylinder	$\frac{1}{8}$
4. Sphere	$\frac{1}{12}$
5. Circular disk	$\pi/16$
6. Flat jet striking plane at right angles[b] with $H/W \to \infty$	$2/\pi$

  [a] Estimates based on potential flow near a stagnation point as derived by Levin (1953).

  [b] H/W is the ratio of the distance from the jet to the plate and W is the jet width or diameter.

inviscid flow pattern remains a useful approximation in the regime near the forward stagnation point of the collector, where most of the particles impact.

Calculations for the theoretical efficiency curves for particles in Stokes flow have been made, for example, by Dorsch *et al.* (1955), and are shown for the extremes in Fig. 2.3. Also included are curves for intermediate collector Reynolds number. The curves in Fig. 2.3 indicate that the $Stk_{cr} \approx 0.08$ estimated by Levin is in agreement with the aerodynamic calculations, and for inviscid flow, the rough approximation that $\eta \approx 0.5$ for Stk of about unity is apparent. The efficiency curve for the inviscid approximation is considerably higher than the calculation for viscous flow for Stk < 100. Experiments reported by Ranz and Wong (1952) and later by Walton and Woolcock (1960), for $70 \lesssim Re \lesssim 5700$, generally agree with the inviscid approximation.

Curves of collection efficiency versus Stk for a circular cylinder have similar shape to those for a sphere. The inviscid and viscous limits for a cylinder are shown in Fig. 2.4. The inviscid limit corresponds to the dimensionless parameter P being very large; the viscous flow limit is $P \to 0$. The curves are based on calculations of Brun *et al.* (1955). Again the Stk_{cr} derived by Levin is in agreement with the detailed calculations. Limited observations of Ranz and Wong for sulfuric acid mist collection in the range $55 < Re < 5520$ agree satisfactorily with the inviscid approximation.

2.2.4 Electrical Forces and Particle Deposition.

There are circumstances where particle deposition is determined by electrical forces induced between the collector and the particles flowing past it. The deposition rate on the obstacle can be estimated from electrical theory. For conditions of a nondivergent or solenoidal electrical field, the particle concentration far from

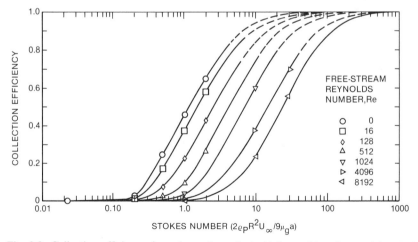

Fig. 2.3. Collection efficiency for spheres in an inviscid flow with point particles. [After Dorsch *et al.* (1955).]

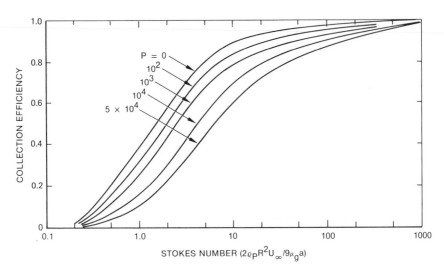

Fig. 2.4. Collection efficiency for cylinders in an inviscid flow with point particles. The rate of deposition, particles per unit time per unit length of cylinder, is $2NU_\infty a$, where N is the particle concentration in the mainstream. The parameter P is the dimensionless group $4Re_p^2/Stk = 18\rho_g^2 U_\infty a/\mu_g \rho_p$.

the obstacle will remain constant. For steady state conditions, the particle concentration near the body's surface will remain in proportion to that found far away. Thus the total flow of particles toward the surface of the collector will equal the integral of the flux of particles over any surface of distance and from the center of the collector.

In the absence of Brownian motion, the total number of particles hitting the surface can be expressed as

$$\Phi = N \int_S (q_r \text{ and } BF_r)dS, \tag{2.45}$$

where q_r is the component of particle flow velocity normal to the surface of the body, F_r is the component of external (electrical) force directed normal to the surface, and dS is an element of surface area. In the case of an electrical field, $F_r E_{0r} Q$, where E_{0r} is the radial component of the field strength, and Q is the charge on the particle.

If the particles are small the component of particle flow velocity q_r will be negligible, and Eq. (2.45) becomes

$$\Phi = NBQ \int_S E_{0r} \, dS = 4\pi BQQ',$$

where Q' is the total charge on the collector surface. The efficiency of collection can be written in terms of the net particle flux Φ/S divided by the total number of particles flowing past the body, or

$$\eta = \Phi/NU_\infty S = -4\pi QQ'B/U_\infty S_0, \tag{2.46}$$

where S_0 is the cross-sectional area of the obstacle normal to the direction of the free-stream velocity U_∞.

Deposition from aerosol clouds on spheres by electrical forces was investigated by Kraemer and Johnstone (1955). They calculated particle trajectories induced by electrical forces, and estimated collection efficiencies by interception, neglecting impaction. Idealized situations were estimated for (a) the Coulombic force between dielectric spheres and charged particles, (b) the force between unipolar charged particles and the charge induced by them on a grounded sphere, and (c) the induced force between a charged and uncharged particle. The interactive forces can be written in a form normalized to the Stokes resistance:

(a)   Coulomb interaction:

$$\mathcal{F}_E = QQ'/6\pi a^2 R\mu_g U_\infty;$$ (2.47a)

(b)   Induction to a grounded sphere:

$$\mathcal{F}_G = Q_E^2 NA_0^2/3\mu_g RU_\infty a,$$ (2.47b)

where A_0 is the radius of the aerosol cloud; and

(c)   Induction to a charged sphere:

$$\mathcal{F}_T = (\varepsilon_s - 1)R^2Q^2/(\varepsilon_s + 2)3\pi a^5\mu_g U_\infty,$$ (2.47c)

where ε_s is the dielectric constant of the sphere.

A typical theoretical curve of Kraemer and Johnstone is shown in Fig. 2.5. It is compared with experimental measurements for dioctylpthalate particles of mean radius 0.27–0.59 μm. The particles were either charged with units of the same sign (unipolar) or different sign (bipolar) at average charge values of 0.15–1.37 units. They were collected on spheres of 0.33–0.55 cm radius from a gas flow of 1.5–6.9 cm/sec. The series of experiments was run with the three different dimensionless forces predominant as shown in Fig. 2.5. The experiments indicated that interception tended to increase the collection efficiency with $\mathcal{F}$ in the low ranges of $\mathcal{F}$, compared with that expected from the theory. A significant departure from the theory was found for $\mathcal{F}$ values less than 10^{-2}, but above this the simple model appears to work satisfactorily.

When electrical forces are present, but do not necessarily dominate particle collection, the estimation of the collection efficiency depends on the aerodynamics involved, as well as the electrical parameters that govern particle trajectories around conducting or nonconducting bodies. These have been studied in detail by Natanson (1957), Zebel (1968), and Hochrainer *et al.* (1969).

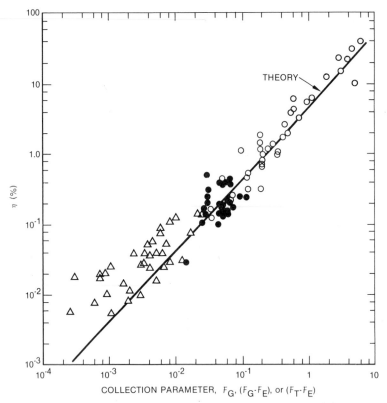

Fig. 2.5. Collection efficiencies for particles deposited on spheres by electrostatic forces. ○, collector charged, aerosol charged by corona, parameter $(\mathscr{F}_G - \mathscr{F}_E)$. △, collector charged, aerosol with naturally occurring charge, parameter $\mathscr{F}_T - \mathscr{F}_E$. ●, collector grounded, aerosol charged by corona, parameter $\mathscr{F}_G$. [Reprinted with permission from Kraemer and Johnstone (1955). Copyright 1955 American Chemical Society.]

At large Re, two dimensionless parameters have been used to characterize regimes of particle motion around cylinders, for example. These are

$$G = \frac{E_\infty QB}{U_\infty} = \frac{\text{Particle velocity in electrical field}}{\text{Undisturbed gas velocity}}, \qquad (2.48)$$

where the field strength E_∞ is oriented parallel to the flow direction U_∞, and

$$H = \frac{2Q'QB}{U_\infty a} = \frac{\text{Velocity of a charged particle near a charged cylinder}}{\text{Undisturbed gas velocity}}.$$

$$(2.49)$$

Examples of differences in particle motion around a charged cylinder are shown for different rates of G and H in Fig. 2.6. These cases illustrate the great variations in idealized inviscid flow around the collector with electrical forces uninvolved. Even in conditions where the fluid flow does not contain

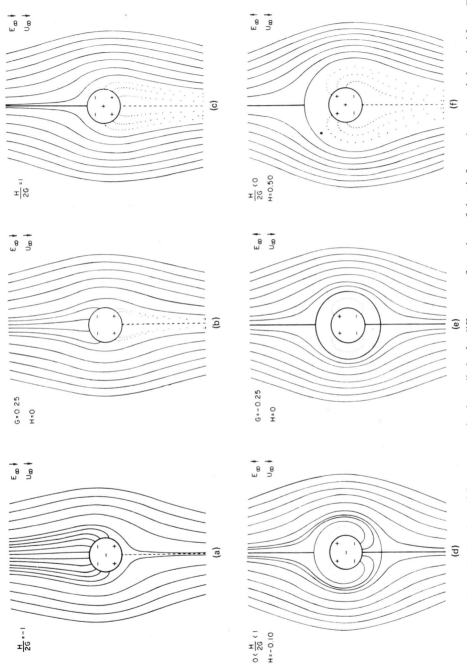

Fig. 2.6. Potential flow of dielectric particles past a conducting cylinder for different configurations of electric forces. An external electrical force E_∞ is imposed parallel to the aerosol flow direction U_∞. The intensity of the electrical forces is characterized by the parameters G and H. [From Hochrainer *et al.* (1969).]

a wake, a particle trajectory "wake" emerges under certain conditions, for example, in (b), (c), and (f) in Fig. 2.6. Cases of collection on the backside of the collector without a fluid wake also become possible (d), or a "particle free space" around the collector may be induced (e).

The experiments of Hochrainer et al. (1969) agree qualitatively with several of the theoretical conditions in Fig. 2.6. Deviations between experiment and theory were observed at the forward stagnation point of the cylinder. The deviations were probably associated with difficulties in flow stabilization and measurement methods near the forward stagnation point.

In view of these factors, investigators should beware of indiscriminant application of the dynamic theory involving only aerodynamics, if electrical fields are involved.

2.3 DIFFUSION AND BROWNIAN MOTION

2.3.1 Diffusion in a Stagnant Medium.
So far we have concentrated on the behavior of particles in translational motion. If the particles are sufficiently small, they will begin to experience an agitation from random molecular bombardment in the gas which will create a thermal motion analogous to that of the surrounding gas molecules. The agitation and migration of small colloidal particles has been known since the work of Robert Brown in the early nineteenth century. This thermal motion is likened to the diffusion of gas molecules in a nonuniform gas. The applicability of Fick's equations for the diffusion of particles in a fluid has been accepted widely after the work of Einstein (1905) and others in the early 1900s. The theory of Brownian diffusion has been reviewed in detail by Hidy and Brock (1970) and will not be repeated here.

For an aerosol, the basic equation for convective diffusion of the particles is

$$\frac{\partial N}{\partial t} + q_p \cdot \nabla N = \nabla \cdot D_p \nabla N \tag{2.50}$$

where N is the number concentration of particles. The parameter D_p is the *Stokes–Einstein diffusivity*,

$$D_p = BkT = kT\tau^{-1}/m_p. \tag{2.51}$$

Given the definition of a mean free path for particles as $\lambda_p = \bar{q}_p \tau^{-1}$, the average thermal speed of a particle is

$$\bar{q}_p = (8kT/\pi m_p)^{1/2},$$

and

$$D_p = \tfrac{1}{8}\pi \bar{q}_p \lambda_p.$$

Some characteristic values of these aerosol transport properties of particles in air are listed in Table 2.1 for reference.

The simplest case for which particle diffusion has been analyzed is the situation where there is no gas motion, or the second term on the left of Eq. (2.50) is zero. The mathematical solutions for Eq. (2.50) in such cases with appropriate boundary conditions are readily available in Carslaw and Jaeger (1959) or Crank (1959). A form of these solutions which is often useful concerns calculations of the flux of particles to the surface in different geometries. For example, the diffusional flux of particles to a vertical wall located at position x_0 is given by

$$j = D_p \left(\frac{\partial N}{\partial x} \right)_{x = x_0},$$

with $N(x, t) = N_\infty \, \mathrm{erf}[(x - x_0)/4D_p t]$ from an appropriate solution of Eq. (2.50). N_∞ is the particle concentration far from the wall. Thus

$$j = N_\infty (D_p/\pi t)^{1/2}. \tag{2.52}$$

By a similar argument, the number of particles striking the entire inner surface per unit time of a sphere of radius a is (Pich, 1976)

$$\Phi_p = 8a\pi D_p N_\infty \sum_{n=1}^{\infty} \exp\left(\frac{-n^2 \pi^2 D_p t}{a^2} \right); \tag{2.53}$$

for an infinitely long cylinder of radius a,

$$\Phi_p = 4\pi D_p N_\infty \sum_{n=1}^{\infty} \exp\left(\frac{-D_p \alpha_n^2 t}{a^2} \right), \tag{2.54}$$

where α_n are the positive roots of the Bessel function $J_0(\alpha_n, a) = 0$.

In the case of particle diffusion from a stagnant medium to the *outer* surface of a sphere of radius a, the total number of particles absorbed per unit time is

$$\Phi_p = 4\pi D_p \left(r^2 \frac{\partial N}{\partial r} \right)_{r=a} = 4\pi D_p a N_\infty \left(1 + \frac{a}{(D_p t)^{1/2}} \right). \tag{2.55}$$

Over a time interval Δt, the number of particles collected on the sphere per unit area is

$$\int_0^t \Phi_p \, dt = 4\pi a D_p N_\infty \left(\Delta t + \frac{2a(\Delta t)^{1/2}}{(\pi D_p)^{1/2}} \right).$$

Thus when $a^2/D_p \gg \Delta t$, the number of particles deposited per unit area is the same as for a flat wall. Conditions where $a^2/D_p \ll \Delta t$ give a steady state solution for the particle concentration $N = N_\infty(1 - a/r)$, yielding the steady state flux,

$$\Phi_p = 4\pi D_p a N_\infty. \tag{2.56}$$

In the case of the infinitely long circular cylinder of radius a, the flux to the cylinder is

$$\Phi_p = \frac{8N_\infty D_p}{\pi} \int_0^\infty \exp(-D_p\gamma^2 t) \frac{dy}{\gamma} [J_0^2(a\gamma) Y_0^2(a\gamma)]. \tag{2.57}$$

For short times,

$$\Phi_p \simeq 2\pi D_p N_\infty[(\pi t')^{-1/2} + \tfrac{1}{2} - \tfrac{1}{4}(t'/\pi)^{1/2} + (\tfrac{1}{8})t' \cdots]. \tag{2.58a}$$

For long times,

$$\Phi_p \simeq 2\pi D_p N_\infty \left(\frac{1}{\ln(4t')} - 1.154 - \frac{C_3}{(4 \ln(4t') - 2)^2} \cdots\right),$$

$$t' = \frac{\ddot{D}_p t^2}{a^2}, \ C_3 = 0.577. \tag{2.58b}$$

2.3.2 Diffusion in Laminar Flow. When there is a mean flow of the aerosol, the solutions to the convective diffusion equation account for that flow. In cases where the gas motion is uniform, the solutions for particle diffusion with $q = 0$ in Eq. (2.50), "conduction" solutions, can be extended using a Galilean transformation of coordinates, for example, $x' = (x - qt)$. In most circumstances of interest, however, there is a gradient in gas velocity involved, or there is a shearing flow.

In such circumstances the total mass flux is given by the equation

$$\mathbf{j} = \mathbf{q}(\mathbf{x}, t)N - D_p \nabla N, \tag{2.59}$$

where the first term on the right is the particle flux associated with convective motion of the fluid, and the second term on the right denotes the diffusion flux of particles by Brownian motion. In any arbitrary fluid volume the number of particles passing through the surface S of that volume is

$$\Phi_p = \int_S \mathbf{j} \, dS.$$

The theory of convective diffusion in flowing fluids is well developed for the steady state case, where $\partial N/\partial t = 0$. The aerodynamic boundary layer model (e.g., Schlichting, 1960) is normally applied to aerosol collection problems. This theory deals with estimation of the diffusional flux to various collecting bodies, by considering approximate solutions to Eq. (2.50), with the steady state conditions, and assuming that the particles do not lag behind the fluid. That is, $\mathbf{q}_p = \mathbf{q}$. In a dimensionless form, the diffusion equation can be written as

$$\mathbf{q}_1 \cdot \nabla_1 N_1 = (1/\text{Pe}) \nabla_1^2 N_1, \tag{2.60}$$

where the parameters are normalized to a main stream flow far from a surface, or $N_1 = N/N_\infty$, $\mathbf{q}_1 = \mathbf{q}/U_\infty$, $\nabla_1 = L \nabla$, and the *Peclet number* Pe =

LU_∞/D_p. The length scale L is characteristic of the geometry of the collector, say a sphere radius, or a fluid dynamically defined thickness such as the thickness of a boundary layer along a flat plate. The surface condition is given by

$$N = 0, \qquad z/L = R/L \equiv I,$$

where z is a measure normal to the collector surface. For a sphere or cylinder the length scale is the radius.

The dimensionless concentration can be expressed as

$$N_1 = f(\mathbf{r}/L, \text{Re}, \text{Pe}, I). \tag{2.61}$$

Thus two convective diffusion regimes are similar if the Reynolds, Peclet, and interception numbers are the same. Since Pe $\gg$ 1 for conditions of convective particle diffusion, this leads to simplification in aerosol problems (e.g., Hidy and Brock, 1970).

The local rate of particle transfer by diffusion to the surface of a body is given by

$$j = \frac{D_p N_\infty}{L} \left(\frac{\partial N_1}{\partial r_1}\right)_{r_1=I}.$$

In analogy to fluid heat and mass transfer (p. 27), the local mass transfer coefficient, k is defined as J/N_∞.[†] In dimensionless form

$$kL/D_p = f(\text{Re}, \text{Pe}, I) \tag{2.62}$$

for the steady state case. The dimensionless number is the Sherwood number Sh for particle diffusion.

Within this general framework, consider a flow of aerosol along a flat plate as shown in Fig. 2.7. As the gas moves over the surface, the friction at the surface decelerates neighboring layers of fluid, creating a velocity gradient.

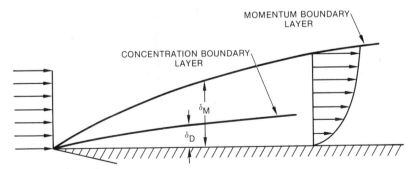

Fig. 2.7. Schematic diagram of the development of momentum and concentration boundary layers over a flat plate at high values of the Peclet number.

[†] This term is sometimes called the deposition velocity. (See also p. 51.)

At the same time, particles diffuse from the cloud to the surface and collect there, creating a gradient in particle concentration. Conventionally, a length scale L for this system is defined in terms of a momentum thickness δ_M which is defined as the distance away from the surface at which the gas velocity is 99% of the free-stream value U_∞. Similarly, a diffusion boundary thickness is the distance δ_D from the surface at which the particle concentration has reached 99% of its free-stream value N_∞. In laminar flow, where no turbulence exists, the relation between the two length scales is

$$\delta_D \approx (D_p/\nu_g)^{1/3}\delta_M ,$$

where ν_g is the gas kinematic viscosity. Since the particle Schmidt number, $\nu_g/D_p = 10^3$ or larger for aerosols, the thickness of the diffusion layer has to be about one-tenth that of the momentum layer. That is, the fluid velocities are about 10% of the free-stream flow when the particle concentration is at its free-stream value over a collector.

In the case of the flat plate, the diffusional flux to the plate has the form

$$j_p = 0.34 D_p N_\infty \left(\frac{U_\infty}{\nu_g x}\right)^{1/2}\left(\frac{\nu_g}{D_p}\right)^{1/3} , \qquad (2.63)$$

or the local particle flux is proportional to the Schmidt number to the one-third power.

An important application of the theory is the diffusional deposition in tubes or pipes. Ideally, the flow of an aerosol through a tube begins with fluid velocity uniform across the tube cross section. As the fluid friction at the wall takes effect, the fluid velocity distribution reaches a steady parabolic shape, given by Poiseuille's equation, for a pipe of radius a,

$$q(r) = \bar{q}(1 - r^2/a^2),$$

where $\bar{q}$ is the average flow velocity, equal to $a^2\Delta p/4\mu_g L$, with $\Delta p/L$ the pressure drop per unit pipe length L. Near the pipe entrance, where the velocity is nearly uniform, the particle flux to the walls can be estimated by the boundary layer approximations for a flat plate. Further downstream, where the velocity profile becomes parabolic, the calculation of diffusional wall deposition requires a more complicated solution (Friedlander, 1977). The average mass transfer coefficient k_{av} over a length x is found to be

$$2 k_{av} a/D_p = \frac{a}{2x_1} \ln(N_0/N_{av}), \qquad (2.64)$$

where $x_1 = (x/a)Pe$ and N_0 is the initial particle concentration. This relationship is shown in Fig. 2.8 with the normalized local mass transfer coefficient $2 k a/D_p$ and the normalized concentration N_{av}/N_0.

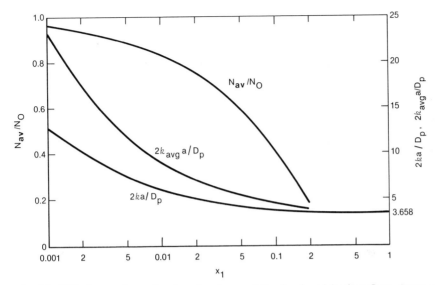

Fig. 2.8. Diffusion to the walls of a pipe from a fully developed laminar flow. Average concentration and local and average mass transfer coefficients are shown as a function of distance from the tube entrance. [From Friedlander (1977); reprinted with permission of the author and John Wiley & Sons.]

These results can be used to characterize the deposition of particles of different size in tubes. Particles entering and leaving a tube can be measured from the curve in Fig. 2.8. From measured values of the reduction in concentration, x_1 can be estimated. Then D_p (or R_p) can be calculated from x, a, and $\bar{q}$ (see also Section 5.7.4).

The theory of convective particle diffusion also can be applied to the collection of particles on blunt obstacles such as spheres and cylinders. In the case where the interception parameter $I \to 0$, and for a blunt-body radius a, the theory (e.g., Hidy and Brock, 1970) projects a diffusion rate of

Sphere:

$$\Phi = 4\pi D_p N_\infty a(1 + 0.64 \mathrm{Pe}^{1/3}), \qquad (2.65)$$

$$\mathrm{Sh} \equiv 2\mathscr{k}a/D_p \approx 1.26 \mathrm{Pe}^{1/3};$$

Cylinder:

$$\Phi = 2\pi N_\infty a C_4 \mathrm{Pe}^{1/3}/[2(2.002 = \ln 2\mathrm{Re})]^{1/3} \qquad (2.66)$$

(where C_4 is a constant given as 1.305), and

$$\mathrm{Sh} \equiv 2\mathscr{k}a/D_p \approx 1.26 \mathrm{Pe}^{1/3}/[2(2.002 - \ln 2\mathrm{Re})]^{1/3}. \qquad (2.67)$$

The collection efficiency has a direct relation with the Sherwood number. In the case of cylinders, for example,

$$\text{Sh} = \frac{2\hbar a}{D_p} = \frac{j}{2\pi RN} = \frac{2\eta \text{Pe}}{\pi} \tag{2.68}$$

or

$$\eta = \pi \hbar / U_\infty = 3.68[2(2.002 - \ln 2\text{Re})]^{-1/3}\text{Pe}^{2/3}. \tag{2.69}$$

When I is finite, similarity theory requires that

$$\eta I \text{ Pe} = \pi \hbar R/D_p = f(I\text{Pe}^{1/3}[2(2.002 - \ln 2\text{Re})]^{1/3}). \tag{2.70}$$

As a last consideration, the mechanism for diffusional deposition should be merged with a model of impaction to provide a complete picture of collection efficiency. When the mathematical equations for particle motion are combined with those for diffusion, a solution based on similitude arguments cannot be obtained. However, calculations have been made for viscous flow around a sphere of different radii using boundary layer approximations, and

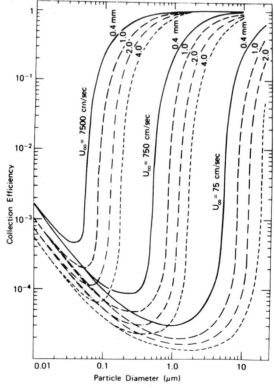

Fig. 2.9. Aerodynamic collection efficiency as a function of particle diameter for flow around a sphere, and collector diameter. Curves are shown for each free-stream velocity and for collector diameters from 0.4 to 4.0 mm [From Hidy and Heisler (1978); reprinted with permission of John Wiley & Sons.]

inviscid flow around the collector in the diffusion range. Interception has been included in the calculations. The range of collector diameter considered is typical of water spray scrubber design. The results of Shaw and Friedlander are shown in Fig. 2.9. On the right side of the curve, the impaction range shows an increase in collection efficiency with free-stream velocity as expected from earlier arguments. The collection efficiency goes through a minimum with decreasing particle size which differs somewhat with free-stream velocity; then the collection efficiency increases again for small particles as diffusion becomes important. The minimum in collection efficiency can be found between 0.1 and 1 μm diameter according to these calculations, depending on the free-stream velocity. The minimum in efficiency also shifts to smaller size with decreasing collector diameter. The collector diameter is shown near the top of the curves for each case in Fig. 2.9.

Finally, let us consider the influence of electrical forces on particle collection. For illustration the case of the spherical collector is included. Here the velocity ratios G and H are used to characterize the migration associated with electrical forces (p. 40).[†] For $I \rightarrow 0$, Zebel (1968) has calculated the deposition rate assuming that the deposition is dominated by accumulation near the forward stagnation point of the collector. Shown in Table 2.7 are his results for a sphere falling in an electrical field oriented parallel to the direction of motion of the sphere. Here a wide range of efficiency is predicted which depends on the strength of electrical forces.

2.4 DEPOSITION IN A TURBULENT FLUID MEDIUM

The deposition of particles from a turbulent medium is of considerable practical interest as a natural extension of turbulent transport theory, and from the need for design of collection devices. Turbulent deposition also plays an important role in the removal of aerosol particles from the earth's atmosphere.

When particles are suspended in a turbulent fluid, they are subjected to a random, fluctuating motion which is analogous to thermal agitation in a stagnant medium. Thus, we can derive an equation for convective transport in a turbulent gas equivalent to that of Eq. (2.50). In this application, however, the equation applied to an average behavior, such that the "diffusivity" is identified with an eddy motion rather than with Brownian diffusion (see, for example, Hidy and Brock, 1970). The theory for particle dispersion in turbulent media such as the earth's atmosphere is deduced from solutions to the turbulent diffusion equation. The deposition of particles from a turbulent medium also can be estimated from an analogous model involving boundary layer approximations (Schlichting, 1960).

As in the case of laminar flow, the deposition of particles suspended in a turbulent fluid on an adjacent surface can occur by inertial effects, gravitational settling, Brownian diffusion, or by the influence of external forces. For electrically neutral particles more than 1 μm in diameter, deposition is

---

[†] G remains the same for spheres, but H is redefined as $H = QQ'B/U_\infty a^2$.

TABLE 2.7

Collection Efficiencies and Sherwood Numbers for the Deposition of Charged Particles on a Falling Sphere in an Electrical Field. [a,b]

G	H/G	η	Sh	Remarks
A. No Brownian diffusion				
>0	≥ 3	0		Positively charged point particles; repulsive forces on falling sphere
>0	$-3 < H/G < 3$	$\dfrac{(3G - H)^2}{3G(G + 1)}$		Positive charge on point particles; repulsive and attractive forces on falling sphere
>0	$H/G \leq -3$	$\dfrac{-4H}{G + 1}$		Positive charge on point particles; attractive forces on falling sphere
<0	>0	$-4H/G + 1$		Negatively charged point particles with attractive and repulsive forces on the falling sphere, or attractive forces only on the sphere
<0	≤ 0	0		Negative charge on point particles with negative charge on the sphere; dust-free space will appear around falling sphere
B. With diffusion				
0	0	$2.52\text{Pe}^{-2/3}$		Uncharged particles
$>0 \ll 1$	$\ll \dfrac{1.5\text{Pe}^{-2/3}}{G}$	$2.52\text{Pe}^{-2/3}$ $+ \tfrac{1}{3}(3G - H)$		Positive charged particles; weak electrical forces compared with aerodynamic forces
$\gg \tfrac{1}{2}\text{Pe}^{-2/3}$	0	$\dfrac{6G}{G + 1}$		Strongly positive charged particles
>0	$\gg \dfrac{1.5\text{Pe}^{-2/5}}{G}$	$\dfrac{3G}{G + 1}$	$\tfrac{3}{2}\text{Pe}_i G$	$H = 0$, strongly positive-charged particles
>0	>3	0	0	$H = 0$, strongly positive-charged particles
>0	$-3 < H/G < 3$	$\dfrac{3G - H}{3G(G + 1)}$	$\dfrac{(3G - H)\text{Pe}}{6G}$	$H = 0$, strongly positive-charged particles
>0	≤ -3	$\dfrac{-4H}{G + 1}$	$-2H\text{Pe}$	$H = 0$, strongly positive-charged particles; attractive electrical force over entire surface of falling sphere
<0	$\gg -[1 + (0.875)\text{Pe}^{-2/3}]$	$\dfrac{-4H}{G + 1}$	$-2H\text{Pe}$	$H < 0$, strong negative charging on particles

[a] The field direction is oriented parallel to the direction of motion of the large sphere.
[b] Based on theory of Zebel (1968).

effected by inertial forces. The particle deposition rate from a turbulent fluid is enhanced over that of a laminar flow as a consequence of eddy diffusion to the laminar sublayer next to the surface. The development of a theory for deposition which accounts for the turbulence has been controversial since the initial attempt of Friedlander and Johnstone (1957).

The theory is written to provide an estimate of the mass transfer coefficient or *deposition velocity* in dimensionless form, $k^+ = k/u^*$, which is reported as a function of dimensionless particle relaxation time τ^+,

$$\tau^+ = \tau u^{*2}/\nu_g, \tag{2.71}$$

where u^* is the friction velocity equal to $(F/\rho_g)^{1/2}$; F is the frictional stress at the surface. The existing theories for turbulent deposition, including that of Friedlander and Johnstone (1957), are based on a so-called "diffusion–free flight" model. In this model, particles are assumed to be transported by turbulent diffusion from the turbulent core of the fluid through the boundary layer to a diffusion sublayer whose thickness is approximately one stopping distance from the wall. At this point the particle follows a free path to the wall.

The differences in various theoretical models center on the estimate of the particle velocity at the beginning of free flight to the wall. Friedlander and Johnstone assumed a value of $0.9u^*$, while others assumed different values. Davies (1966) took the initial free flight velocity as the root-mean-square (rms) fluctuating velocity of the fluid at the point where free flight begins. Beal (1970) assumed a free flight velocity of half the axial velocity of the fluid, but Sehmel (1971) assumed a form proportional to $\frac{1}{2}\tau u^{*3}$. Liu and Ilori (1973) assumed that the particle eddy diffusivity near the wall is proportional to the sum of eddy diffusivity of the fluid and the product of the rms fluctuating velocity near the wall and the relaxation time.

The experiments of Montgomery and Corn (1970) suggested that the early Friedlander–Johnstone "diffusion–free-flight" hypothesis and other models, except that of Davies, were in reasonably good agreement with experiments for a range of behavior characterized by a dimensionless relaxation time less than 30.

Liu and Agarwal (1974) reported some new experiments for deposition on the walls of a vertical pipe. Their results for the inertial deposition range are shown in comparison with four of the theoretical models in Fig. 2.10. The mass transfer coefficient or deposition velocity is shown as a function of the dimensionless relaxation time. Their results confirm Montgomery and Corn's conclusions. The theory of Liu and Ilori shows equally good agreement with experimental data near $\tau^+ = 2$. The deposition velocities estimated by Davies' model are considerably lower over the entire range shown. A slight curvature in Beal's result is observed below $\tau^+ > 0.3$, departing from the Friedlander–Johnstone model. Yet in Davies' model, Brownian diffusion has an influence at considerably higher values of τ^+, evidently because of the low inertial deposition estimated by his theory.

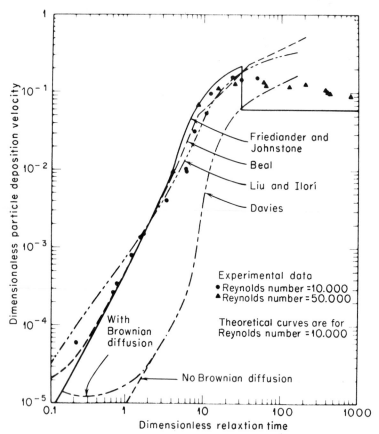

Fig. 2.10. Comparison between theories and experiments for the dimensionless deposition velocity and relaxation time of particles. [From Liu and Agarwal (1974); reprinted with permission of Pergamon Press.]

In the range of $\tau^+ \gtrsim 20$, the measurements in Fig. 2.10 show a weak decline in dimensionless deposition velocity with increase in τ^+ (or particle size). This has not been accounted for in the theories until recently. Reeks and Skyrme (1976) have explained this behavior by considering a stochastic model for particle motion in the air near the collecting surface. They found that the observed decrease is due to the fact that the increased fractional penetration of the fluid boundary layer with increasing particle mass becomes insufficient to compensate for the reduced rate of turbulent particle transport to that region.

The Reeks and Skyrme model gives a simple form for ℓ^+ in its range of application ($\tau^+ \gtrsim 5$). The expression is given in terms of the rms velocity and the fractional penetration of particles into the fluid boundary layer. The inertial dependence of the particle velocity is given as a function of the particle response to turbulent velocity fluctuation in the neighboring fluid by relating velocity spectral densities of the particle and the fluid in the equation of particle motion. The dimensionless deposition velocity is

$$k^+ = A'q' \ \mathrm{erfc}[\mathfrak{a}^+\theta/(2q')^{1/2}], \tag{2.72}$$

where $q' = \langle q_p \rangle / \langle q_g \rangle$, $\theta = \delta u^*/v_g q_g$, $\mathfrak{a}^+ = 1/\tau^+$: $\langle q_p \rangle$ is the rms particle velocity and $\langle q_g \rangle$ is the rms fluid velocity. The two arbitrary constants A' and θ are given as 0.56 and 6.25 based on the Liu and Agarwal data.

Deviations from the linear theory given by Eq. (2.72) are expected to be significant with large τ^+, where the linear form of the equation of particle motion will be perturbed by particle Reynolds number influences. For small values of τ^+ lower than $\tau^+ \approx 3$ deposition from inertial effects decreases rapidly, Brownian diffusion becomes important, and the Reeks and Skyrme model again breaks down.

Although refinements in the theory of turbulent deposition appear warranted to include other processes over a broad range in τ^+, there is little doubt that the picture of inertial deposition for uncharged particles gives a reasonable model of the process.

These comparisons show that while our present understanding of the turbulent deposition process is imperfect, and there is need for a unified theory covering a broader range of τ^+ values, our notion of the basic turbulent deposition process in the absence of electrical forces is essentially correct. In particular, there is little doubt that the observed phenomenon is essentially an inertial effect resulting from the finite mass of the particles and the action of fluid turbulence.

There are many situations in which turbulent deposition on rough surfaces is of interest. For equivalent friction velocity, surface roughness should increase the deposition rate compared with a smooth surface. Deposition velocities for grass surfaces were measured several years ago by Chamberlain (1966, 1967). A compilation of available data and theoretical evidence has been reported by Sehmel and Hodgson (1974). An example of their results is shown in Fig. 2.11. Curves for different particle size, density, and surface roughness are shown for comparison. Surface roughness is measured according to the roughness length[†] z_0. The curves take into account Chamberlain's results. Shown in solid lines in Fig. 2.11 is the deposition velocity dependent on gravitational action alone. The effect of surface roughness is to dramatically change the deposition velocity for particles in the 0.1–1.0 μm diameter range. With surface roughness larger than 3–10 cm, the results suggest that the deposition velocity approaches a nearly constant value near 1 cm/sec with little variation with respect to particle size. This

---

[†] The roughness length or height for a surface is proportional to the height of roughness elements such as sand grain diameter or grass blade length. The proportionality is determined experimentally by scaling a velocity profile (Schlichting, 1960).

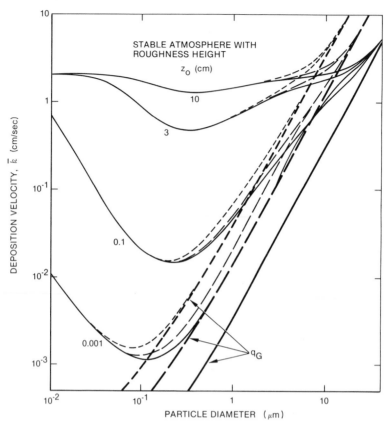

Fig. 2.11. Curves of deposition velocity from Sehmel and Hodgson (1974) for dry deposition of particles on various solid surfaces. The data encompassed roughness heights only up to about 0.1 cm, and therefore the extrapolation to greater roughness heights is tentative. The deposition velocity plotted is the flux divided by the extrapolated concentration at 1 m; q_G is the gravitational settling speed for particles of indicated density ρ_p; the case with $u^* = 30$ cm sec^{-1} is shown (average air speed $\simeq$ 10 m sec^{-1}). ρ_p (g/cm^3): ——, 1.0; — —, 4.0; ---, 11.5.

result is subject to considerable controversy since no data are available to verify such a conclusion.

Recently, Browne (1974) has extended Davies' theory to deposition on rough surfaces without gravitational settling. Browne's estimates are strongly surface roughness and Reynolds number dependent. However, for the case of large roughness heights, which are qualitatively equivalent to Chamberlain's and Sehmel's experiments, a broad range of nearly constant values of deposition velocity approximately 0.1 cm/sec are found for particles between 0.08 and 2 μm diameter for Re = 10,000. Browne's calculations suggest that one should expect a limit more like 1 cm/sec for the deposition velocity over rough surfaces with very large Re, as indicated in Fig. 2.11.

The available experimental and theoretical results for turbulent deposition on rough surfaces are limited. More work in this area is warranted since this aspect of aerosol dynamics has widespread practical implications, especially for deposition from the earth's atmosphere.

REFERENCES

Acrivos, A., and Taylor, T. (1962). *Phys. Fluids* **5**, 387.
Bakanov, S., and Derjaguin, B. V. (1960). *Discuss. Faraday Soc.* **30**, 130.
Beal, S. K. (1970). *Nucl. Sci. Eng.* **40**, 1.
Brock, J. R. (1962). *J. Colloid. Sci.* **15**, 546.
Brock, J. R. (1967). *J. Colloid Interface Sci.* **23**, 448.
Browne, L. W. B. (1974). *Atmos. Environ.* **8**, 801.
Brun, R., Lewis, W., Perkins, P., and Serafini, J. (1955). *Natl. Advis. Comm. Aeronaut., Rep.* No. 1215, pp. 139–183.
Carslaw, H., and Jaeger, S. (1959). "Conduction of Heat in Solids," 2nd ed. Oxford, London.
Chamberlain, A. C. (1966, 1967). *Proc. R. Soc. London, Ser. A* **290A**, 236; also **296A**, 45.
Crank, J. (1959). "Mathematics of Diffusion." Oxford, London.
Davies, C. N. (1966). "Aerosol Science" (C. N. Davies, ed.), p. 408. Academic Press, New York.
Davies, C. N. (1978). In "Fundamentals of Aerosol Science," p. 135. Wiley (Interscience), New York.
Davis, E. J., Ravindran, P., and Ray, A. (1980). *Chem. Eng. Commun.* **5**, 251.
Davis, E. J., Ray, A. K., and Chang, R. (1978). *AIChE Symp. Ser.* **74**, 190–203.
Dorsch, R., Saper, P., and Kadow, C. (1955). *Natl. Advis. Comm. Aeronaut., Rep.* No. 3587, pp. 1–29.
Einstein, A. (1905). "Investigations on the Theory of Brownian Motion." Dover, New York (republished 1956).
Einstein, A. (1924). *Z. Phys.* **27**, 1.
Epstein, P. (1929). *Z. Phys.* **54**, 537.
Friedlander, S. K. (1977). "Smoke, Dust and Haze." Wiley (Interscience), New York.
Friedlander, S. K., and Johnstone, H. (1957). *Ind. Eng. Chem.* **49**, 1151.
Fuchs, N. A. (1959). "Evaporation and Droplet Growth in Gaseous Media." Pergamon, Oxford.
Fuchs, N. A. (1964). "Mechanics of Aerosols." Pergamon, Oxford.
Fuchs, N. A., and Sutugin, A. G. (1971). *Int. Rev. Aerosol Phys. Chem. No. 2, 1971* pp. 1–60.
Hidy, G. M., and Brock, J. R. (1970). "The Dynamics of Aerocolloidal Systems," Chaps. 3–6. Pergamon, Oxford.
Hidy, G. M., and Heisler, S. L. (1978). *Recent Dev. Aerosol Sci., Symp. Aerosol Sci. Technol., 1976* p. 142.
Hochrainer, D., Hidy, G. M., and Zebel, G. (1969). *J. Colloid Interface Sci.* **30**, 553.
Kinzer, G., and Gunn, R. (1951). *J. Meteorol.* **81**, 71.
Kotrappa, P. (1972). *Assess. Airborne Part., Proc. Rochester Int. Conf. Environ. Toxic., 3rd, 1970* pp. 331–360.
Kraemer, H., and Johnstone, H. (1955). *Ind. Eng. Chem.* **47**, 2726.
Langmuir, I. (1948). *J. Meteorol.* **5**, 175.
Lerman, A. (1979). "Geochemical Processes—Water and Sediment Environments," p. 262. Wiley (Interscience), New York.
Levin, L. (1953). *Dokl. Akad. Nauk SSSR* **91**, 1329.
Liu, B. Y. H., and Agarwal, J. K. (1974). *J. Aerosol Sci.* **5**, 145.
Liu, B. Y. H., and Ilori, T. A. (1973). *ASME Symp. Flow Stud. Air Water Pollut., Atlanta.*
Loyalka, S. K. (1973). *J. Chem. Phys.* **58**, 354.

Maxwell, J. (1890). *In* "The Scientific Papers of Clerk Maxwell" (W. D. Niven, ed.). Cambridge Univ. Press, London and New York.

Millikan, R. (1923). *Phys. Rev.* **21,** 217.

Montgomery, T. L., and Corn, M. (1970). *J. Aerosol Sci.* **1,** 185.

Natanson, G. (1957). *Dokl. Akad. Nauk SSSR* **112,** 696.

Oseen, C. (1927). "Neure Methoden und Ergebnisse in der Hydrodynamik." Akademische Verlag, Leipsig, Germany.

Paranjpe, MiK. (1936). *Proc. Indian Acad. Sci., Sect. A 4A,* 423.

Paul, B. (1962). *ARS J.* **32,** 1321.

Pich, J. (1976). *Atmos. Environ.* **10,** 131.

Ranz, W., and Wong, J. (1952). *Ind. Eng. Chem.* **44,** 1371.

Reeks, M. H., and Skyrme, G. (1976). *J. Aerosol Sci.* **7,** 485.

Schlichting, H. (1960). "Boundary Layer Theory," 4th ed. McGraw-Hill, New York.

Schmitt, K. (1959). *Z. Naturforsch. A* **14A,** 870.

Sehmel, G. A. (1971). *J. Colloid Interface Sci.* **37,** 891.

Sehmel, G. A. (1973). *J. Aerosol Sci.* **4,** 145.

Sehmel, G., and Hodgson, W. (1974). "Atmospheric-Surface Exchange of Particle and Gaseous Pollutants" (R. Englemann and G. Sehmel, eds.). U.S. Atomic Energy Symp. Ser. No. 38. U.S. Govt. Printing Office, Washington, D.C.

Sitarski, M., and Nowakowski, B. (1979). *J. Colloid Interface Sci.* **72,** 113.

Stöber, W. (1972). *Assess. Airborne Part., Proc. Rochester Int. Conf. Environ. Toxic., 3rd, 1970* pp. 249–289.

Stokes, G. G. (1851). *Trans. Cambridge Philos. Soc.* **9,** 8–106.

Thomas, J. (1958). *J. Air Pollut. Control. Assoc.* **8,** 32.

Tyndall, J. (1870). *Proc. R. Inst. G. B.* **6,** 3.

Wachmann, H. (1962). *ARS J.* **32,** 2.

Waldmann, L. (1959). *Z. Naturforsch. A 14A,* 589.

Waldmann, L., and Schmitt, K. (1967). *In* "Aerosol Science" (C. N. Davies, ed.), p. 137. Academic Press, New York.

Walton, W., and Woolcock, A. (1960). "Aerodynamic Capture of Particles." Pergamon, Oxford.

Watson, H. H. (1936). *Trans. Faraday Soc.* **32,** 1037.

Zebel, G. (1968). *J. Colloid Interface Sci.* **27,** 294.

CHAPTER 3

PARTICLE CLOUDS—THE SIZE
DISTRIBUTION FUNCTION

So far, we have been concerned with the mechanics of single particles in a moving gas, or clouds of noninteracting particles of single size. Now let us turn to the theory of clouds of particles of different size. Virtually all of the theory for aerosol behavior focuses on conditions where the cloud is dilute enough to be considered in the single-particle regime. In general, particle clouds may undergo (a) differential separation by the action of external forces such as the gravitational force; (b) particle loss to surrounding surfaces by settling or diffusional deposition; (c) coagulation by particle collisions; or (d) particle growth or reduction by condensation, evaporation, or by chemical reactions. These processes can be considered separately or together in increasing complexity depending on the application.

3.1 PARTICLE SIZE DISTRIBUTIONS

The theory for particle clouds proceeds from consideration of the dynamics of the particle size distribution function or its integral moments. This distribution can take two forms. The first is a discrete function in which particles are allowed only in sizes that are multiples of a singular species. As an example, consider the coagulation of a cloud of particles initially of a unit size. Then, after a time, all subsequent particles will be ith aggregates of the single particle, where $i = 1, 2, 3, \ldots$ represents the number of unit particles per aggregate. Physically, the discrete size distribution is appealing since it describes well the nature of the particulate cloud. The second function, the *continuous distribution*, is usually a more useful concept in practice. This function is defined in terms of the differential dN, equal to the number of particles per unit volume of gas at a given point $\mathbf{r}$ in space at time t in the particle volume range $v + dv$. The distribution function then is

$$dN = n(v, \mathbf{r}, t)\, dv. \tag{3.1}$$

Although this form can account for the distribution of particles of arbitrary shape, the theory has been developed only for spheres. In this case, one can also define the distribution function in terms of the particle radius (or diameter):

$$dN = n_R(R, \mathbf{r}, t)\, dR, \tag{3.2}$$

where dN is the concentration of particles in size range R and $R + dR$; n_R is the distribution function in terms of radius. The radius may be geometrical, or it may be used as the aerodynamic equivalent.

The two distribution functions are not equal, but can be related (Friedlander, 1977):

$$n_R = 2\pi R^2 n. \tag{3.3}$$

The moments of the size distribution function are useful parameters. These have the form

$$\mathfrak{M}(\mathbf{r}, t) = \int_0^\infty n_R R^\nu\, dR.$$

The *zeroth moment* ($\nu = 0$),

$$\mathfrak{M}_0 = \int_0^\infty n_R\, dR = N, \tag{3.4}$$

represents the total number concentration of particles at a given point and time. The first moment normalized by the zeroth moment gives a number-average particle radius:

$$\mathfrak{M}_1/\mathfrak{M}_0 = \overline{R} = \int_0^\infty n_R R\, dR \left/ \int_0^\infty n_R\, dR \right. \tag{3.5}$$

The second moment gives the surface area of the aerosol cloud:

$$S = 4\pi\mathfrak{M}_2 = 4\pi \int_0^\infty n_R R^2\, dR. \tag{3.6}$$

The surface mean radius is $\mathfrak{M}_1/4\pi\mathfrak{M}_2$. The third moment is proportional to the total volume concentration of particles, or

$$V = \tfrac{4}{3}\pi\mathfrak{M}_3 = \tfrac{4}{3}\pi \int_0^\infty n_R R^3\, dR, \tag{3.7}$$

where V is the volume fraction of particles in cubic centimeters of material from each cubic centimeter of gas. If the particle density is uniform, the average particle volume is

$$\bar{v} = V/N = 4\pi \mathfrak{M}_3/3\mathfrak{M}_0 ;$$

the volume mean radius is $\bar{R} = 3\mathfrak{M}_1/4\pi M_3$.

In general, particle distributions are broad in size–concentration range, and so they often are displayed on a logarithmic scale. For example, data are frequently reported as log n_R versus log R. Another display is $dV/d(\log R)$ versus log R. The area under the distribution curve plotted in this way is proportional to the mass concentration of (constant-density) particles over a given size range, independent of size. For $v = \frac{4}{3}\pi R^3$,

$$\frac{dV}{d(\log R)} = \frac{2.3\pi^2 R^6 n(v,\ t)}{3}. \tag{3.8}$$

The shape of the size distribution function for aerosol particles is often broad enough that distinct parts of the function make dominant contributions to various moments. This concept is useful for certain kinds of practical approximations. In the case of atmospheric aerosols, the number distribution is heavily influenced by the 0.005–0.1 μm radius range, but the surface area and volume fraction, respectively, are dominated by the 0.1–1.0 μm range and larger. As will be discussed in Chapter 7, advantage can be taken of the separation by moments to analyze certain features of the distribution function.

The cumulative number distribution curve is another useful means of displaying particle data. This function is defined at a point **r** as

$$N(R,\ t) = \int_0^R n_R(R,\ t)\ dR. \tag{3.9}$$

It corresponds to the number of particles less than or equal to the radius R. Since $n_R = dN(R)/dR$, the distribution function can be calculated in principle by differentiating the cumulative function. The characterization of particles in terms of a size distribution has occupied an important part of the literature on statistics. Some useful aspects of size distributions are included for the reader in the Appendix.

3.1.1 The Size–Composition Probability Density Function. The concept of the size distribution function has been generalized to take into account collections of particles with differing chemical composition (e.g., Friedlander, 1970). Let dN now be the number of particles per unit volume of gas containing molar quantities of each chemical species in the range between n_i and $n_i + dn_i$, with $i = 1, 2, \ldots, k$, where k is the total number of chemical species. Assume that the chemical composition is distributed continuously in each size range. The full size–composition probability density function is

$$dN = Ng(v,\ n_2,\ \ldots,\ n_k,\ \mathbf{r},\ t)dv\ dn_2 \ldots dn_k. \tag{3.10}$$

Here n_i, the number of moles of a given species, has been eliminated from g by the relation

$$v = \sum_i n_i \bar{v}_i$$

where $\bar{v}_i$ is the partial molar volume of species i. Since the integral of dN over all v and n_i is N,

$$\int_v \cdots \int \int_{n_k} g(v, n_2, \ldots n_k, \mathbf{r}, t)\, dv\, dn_2 \cdots dn_k = 1.$$

Furthermore, the size distribution function can be retrieved by integration over all chemical species:

$$n(v, \mathbf{r}, t) = N \int_{n_2} \cdots \int_{n_k} g(v, n_2, \ldots, n_k, \mathbf{r}, t) dn_2 \cdots dn_k.$$

Friedlander's generalized distribution functions offer a useful means of organizing the theory of aerosol characterization for chemically different species; to date, however, the available data have not been sufficiently comprehensive to warrant such formalism in practice.

3.2 DYNAMICS OF PARTICLE DISTRIBUTIONS

Collections of particles in a gas will change with time as a result of the action of several processes. These include diffusion, gravitational settling, and migration associated with external forces such as electrical or phoretic forces. A second group includes particle formation, growth or depletion by condensation, evaporation or chemical reactions, and coagulation. The former grouping involves differential motion, but the latter shifts particles from one size range to another. The changes taking place generally are discussed in terms of the size distribution function or its moments.

A general dynamic equation (GDE) can be derived describing the evolution of the aerosol particles (Friedlander, 1977). This equation applies to a volume of gas containing the suspended particles. The first group mentioned above represent basically *external* processes which involve movement of particles across walls of the gas volume. The second group involves processes taking place within the boundaries of the gas volume.

The change in a discrete distribution function of particles of i species in time has the following form, which is interpreted term by term:

$$\frac{\partial n_i}{\partial t} = -\nabla \cdot n_i \mathbf{q} + \nabla \cdot D_p \nabla n_i$$

$$\underset{\substack{\text{Change} \\ \text{in time}}}{} = \underset{\substack{\text{Convective} \\ \text{transport}}}{\left[\quad\right]} + \underset{\substack{\text{Brownian} \\ \text{diffusion}}}{\left[\quad\right]}$$

$$+ \left(\frac{\partial n_i}{\partial t}\right)_{\text{growth}} + \left(\frac{\partial n_i}{\partial t}\right)_{\text{coag}} - \nabla \cdot \mathbf{q}_F n_i \qquad (3.11a)$$

$$+ \underset{\substack{\text{Growth or reduction} \\ \text{by gas–particle interaction}}}{\left[\quad\right]} + \underset{\text{Coagulation}}{\left[\quad\right]} + \underset{\substack{\text{Migration by} \\ \text{external forces}}}{\left[\quad\right]}$$

Equation (3.11a) is basically the general dynamic equation (GDE) for an aerosol cloud. The term $[\partial n_i/\partial t]_{\text{growth}}$ is defined as follows:

$$\left[\frac{\partial n_i}{\partial t}\right]_{\text{growth}} = \begin{bmatrix} \text{input from } i-1 \\ \text{by condensation} \end{bmatrix} + \begin{bmatrix} \text{input from } i+1 \\ \text{by evaporation} \end{bmatrix}$$

$$- \begin{bmatrix} \text{output from } i \\ \text{by evaporation} \end{bmatrix} - \begin{bmatrix} \text{output from } i \\ \text{by condensation} \end{bmatrix} = I_i - I_{i+1} \quad (3.12)$$

where I_i is the *particle current* and n_i is the concentration of particles of ith class. In the case of vapor condensation, for example, $I = 1$ for condensing molecule A, and the particle current is

$$I_i = n_{i-1}s_{i-1}\beta - n_i s_i \beta_i', \quad (3.13)$$

or the excess rate at which particles pass from size $i - 1$ to i by condensation over the rate of passage from i to $i - 1$ by evaporation. Here s_i is the surface area of the ith particle, β equals the flux to the surface of condensing molecules, and β_i' is the evaporation flux. From the kinetic theory of gases, the flux of vapor to and from particles is written

$$\beta = p_A/(2\pi m_A kT)^{1/2} \quad (3.14)$$

and

$$\beta_i' = [p_0/(2\pi m_A kT)^{1/2}] \exp(2\sigma v_{\text{m}}/R_k T) \quad (3.15)$$

where p_0 is the vapor pressure of A at equilibrium with a flat surface of temperature T, σ is the surface tension, and v_{m} is the molecular volume of the condensed species.

The rate of change by coagulation is given by

$$\left(\frac{\partial n_i}{\partial t}\right)_{\text{coag}} = \frac{1}{2}\sum_{k+j=i} b(v_k, v_j)n_k n_j - n_i \sum_{j=2}^{\infty} b(v_j, v_i)n_j \quad (j, k, i \geq 2). \quad (3.16)$$

Collisions between single molecules are excluded. Here, the collision parameter $b(v_j, v_k)$ depends on the kinematics of collision for different mechanisms. The first term on the right-hand side of Eq. (3.16) represents a gain in i particles by collision between j and k particles. The second term on the right is the loss of i particles by collision with any other particle.

The expressions for the collision parameter for mechanisms inducing relative motions between nearby particles include Brownian motion, differential or shearing gas motion, differential motion by external forces. These processes are discussed in more detail below in Section 3.3.

The GDE for a continuous size distribution comes from rewriting Eq. (3.11a) in terms of a continuous distribution function, or

$$\frac{\partial n}{\partial t} + \nabla \cdot n\mathbf{q} = \nabla \cdot D_p \nabla n + \frac{\partial I}{\partial v} + \frac{1}{2} \int_0^v b(\tilde{v}, v - \tilde{v}) n(\tilde{v}) n(v - \tilde{v}) \, d\tilde{v}$$

$$- n(v) \int_0^\infty b(v, \tilde{v}) n(\tilde{v}) \, d\tilde{v} - \nabla \cdot \mathbf{q}_F n, \qquad (3.11b)$$

where $\mathbf{q}_F$ is the particle velocity associated with an external force field. For nondivergent or incompressible flow,

$$\nabla \cdot n\mathbf{q} = \mathbf{q} \cdot \nabla n. \qquad (3.17)$$

No general mathematical solutions for the GDE exist, but a number of approximations can be made depending on application to certain dynamic regimes (e.g., Hidy and Brock, 1970; Drake, 1972; Peterson *et al.*, 1978).

3.2.1 Dynamic Equations for Moments. The GDE can be useful to give relationships for certain moments of the distribution function. Consider the zeroth moment, or the relation for the change in total number of particles per unit volume N larger than some volume v:

$$\frac{\partial N}{\partial t} + \mathbf{q} \cdot \nabla N + I_d = \nabla^2 \int_v D_p n \, dv$$

$$+ \frac{1}{2} \int_v^\infty \int_0^v bn(\tilde{v}) n(v - \tilde{v}) \, d\tilde{v} \, dv$$

$$+ \int_v^\infty \int_0^\infty bn(v) n(\tilde{v}) \, d\tilde{v} \, dv - \nabla \cdot \int_v^\infty \mathbf{q}_F n \, dv. \quad (3.18)$$

Here I_d is the particle current flowing into the lower end of the size spectrum, or

$$I_d = \int_v \frac{\partial I}{\partial v} \, dv.$$

In a large closed chamber, where there is minimal loss by diffusion and external forces including gravity are small, Eq. (3.18) becomes

$$\frac{\partial N}{\partial t} \approx I_d + \left(\frac{\partial N}{\partial t}\right)_{coag}. \qquad (3.19)$$

In other words, measurements of $N(t)$ in such chambers can give useful information about the competing effects of particle growth and change by coagulation.

In the absence of growth or nucleation, the total number concentration in such a vessel decreases continuously.

The GDE for the first moment of the volume distribution, the volume fraction V, gives effectively the changes with time in mass concentration for spherical particles of uniform density ρ_p. This expression has the form (Friedlander, 1977)

$$\frac{dm_p}{dt} = \frac{\partial \rho_p V}{\partial t} + \rho_p \mathbf{q} \cdot \nabla V = \rho_p \int_{v_d}^{\infty} I \, dv + \rho_p (Iv)_{v_d} \tag{3.20}$$

$$\underbrace{\phantom{\rho_p \int_{v_d}^{\infty} I \, dv}}_{\substack{\text{Growth of} \\ \text{particles larger} \\ \text{than a minimum} \\ \text{of site } v_d}} \underbrace{\phantom{\rho_p (Iv)_{v_d}}}_{\substack{\text{Formation of} \\ \text{new particles} \\ \text{by nucleation}}}$$

$$+ \nabla^2 \int_0^{\infty} \rho_p D_p vn \, dv - \nabla \cdot \int_0^{\infty} \rho_p \mathbf{q}_F v \, dv.$$

$$\underbrace{\phantom{\nabla^2 \int_0^{\infty} \rho_p D_p vn}}_{\text{[Diffusion]}} \qquad \underbrace{\phantom{\nabla \cdot \int_0^{\infty} \rho_p \mathbf{q}_F v}}_{\substack{\text{Migration by external} \\ \text{force field}}} \tag{3.20}$$

In the case of the large chamber experiment where no wall losses take place by diffusion settling and there is no particle formation, the volume fraction or mass concentration remains constant with coagulation since $\partial V/\partial t = 0$.

3.3   NUCLEATION AND GROWTH BY CONDENSATION

The production and growth of particles in the presence of condensable vapors represents a major process in aerosol dynamics. A considerable body of literature has accumulated on the subject, beginning with the thermodynamics of phase transition, and continuing with the kinetic theory of molecular cluster behavior.

The process of phase change to form clouds of particles may be induced such that supersaturation is achieved by (a) an adiabatic expansion of a gas, (b) the mixing of a warm, moist gas with a cool gas stream, or (c) chemical reactions producing condensable species. In the absence of particles in a condensable vapor, particles will be formed by *homogeneous nucleation* on molecular clusters in a supersaturated vapor. When vapor supersaturation takes place in a multicomponent system, mixed particles may form from *heteromolecular nucleation*. When particles exist in a supersaturated vapor mixture, they act as *nuclei* for condensation. Perhaps best known of these processes is the formation of clouds in the earth's atmosphere by water condensation on dust or water-soluble aerosol particles.

The theory of nucleation begins with consideration of vapor–liquid equilibrium. The vapor pressure p_0 over a flat liquid surface at equilibrium is given by the Clapeyon equation,

$$\frac{d(\ln p_0)}{dT} = \frac{\lambda_v}{\mathcal{R}T^2}, \tag{3.21}$$

where λ_v is the molar heat of vaporization. At *saturation*, the vapor pressure p_s is its equilibrium value at the temperature of the liquid beneath the vapor.

Condensation may take place when the vapor is supersaturated, or when the supersaturation ratio

$$\mathcal{S} \equiv p_s/p_0 > 1.$$

The vapor pressure p_s over a pure liquid droplet at equilibrium depends on its radius of curvature. The *Kelvin equation* gives this relationship as

$$\ln p_s/p_0 = 2\sigma v_m/RkT. \tag{3.22}$$

Thus, a supersaturation ratio greater than unity is expected for small droplets at equilibrium with a condensable vapor. The logarithm of this ratio is proportional to the product of the surface tension and the molecular volume of the liquid, and inversely proportional to the particle radius.

In the case of droplets composed of mixtures of constituents, the equilibrium vapor pressure relations are more complicated, but can be derived from thermodynamic relationships. As an example, consider a case of the vapor pressure of a solvent, water, over a droplet containing a nonvolatile solute such as sodium chloride or ammonium sulfate. The relationship analysis to Kelvin's equation for such a dilute, binary solution is

$$\ln \frac{p_{AS}}{p_{A0}} = \frac{2\sigma \bar{v}_A}{RkT} - \frac{3n_B \bar{v}_A}{4\pi R^3}, \tag{3.23}$$

where p_{A0} is the vapor pressure of the pure solvent at temperature T, $\bar{v}_A$ and $\bar{v}_B$ are the molar volumes of solvent and solute, respectively, and n_B is the moles of solute. This equation is sometimes called Kohler's equation. This relationship is shown in Fig. 3.1 for the case of the binary solution and for a droplet of solvent alone according to Kelvin's relation, Eq. (3.22). Thus in this case the vapor pressure reducing effect of the solute dominates at small particle size, whereas the radius of curvature influence is strongest at larger size. The maximum supersaturation sustainable corresponds to a particle of about 0.1 μm radius in this situation. For dilute aqueous solutions, the

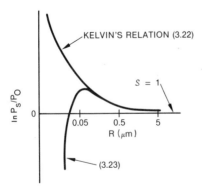

Fig. 3.1. Equilibrium vapor pressure curves for droplets composed of a pure liquid according to the Kelvin reaction, and of a solvent as the same liquid with a fixed mass of nonvolatile solute.

coefficient multiplying $1/R$ is approximately $3.3 \times 10^{-5}/T$, and the term multiplying $1/R^3$ is approximately $4.3z_{\pm} \, m_B/M_B$ where m_B is the mass of salt in solution, $z_{\pm}$ are the number of ions per molecule in solution, and M_B is the molecular weight of salt.

3.3.1   Homogeneous Nucleation.

Thermodynamics indicates that once supersaturation in a vapor is achieved, an unstable state exists in which phase change may take place. The Kelvin relationship suggests that in the absence of nuclei, very large supersaturation ratios may be experienced before particles will form in the vapor. However, thermodynamics does not indicate their rate of formation, or their rate of growth in the supersaturated state. The theory of nucleation basically extends thermodynamic analysis to address these questions.

In the initial stages, the process of formation of particles in a supersaturated vapor is seen as a random accumulation of molecular clusters as the vapor becomes increasingly supersaturated. The presence of such tiny clusters has been verified experimentally. For example, Miller and Kusch detected and quantified cluster formation of sodium chloride in a molecular beam experiment in 1956. The size distributions of argon clusters were reported by Milne and Greene (1967). Castleman (1974) has extended such beam studies to investigate ion–molecule cluster dynamics.

In analogy to the theory of chemical reactions of polymers, the formation of clusters begins with considering the equilibrium between agglomerates of different size in a pure vapor. The reactions of interest are formally written as

$$e_{i-1} + e_1 \rightleftarrows e_i, \tag{3.24}$$

where e_i is a cluster or embryo of i molecules, and e_1 is a single molecule. The formation of the molecular clusters has an "activation" energy barrier associated with it. The estimation of this barrier, the free energy of embryo formation, is examined. According to Frenkel (1955), the work required to form a spherical embryo of i molecules, or the free energy of formation of a spherical embryo in a system containing n_g molecules of vapor and n_1 molecules in the liquid phase is

$$\Delta \hat{G}_i = 4\pi R_i^2 \sigma + i(\hat{\mu}_1 - \hat{\mu}_g) \tag{3.25}$$

where σ is the surface tension or surface free energy per unit area of the embryo, $\hat{\mu}_1$ the chemical potential of a molecule in the condensed phase, and $\hat{\mu}_g$ the chemical potential of the vapor molecules.

For an isothermal process at an arbitrary vapor pressure p,

$$d\hat{\mu}_1 = v_1 \, dp, \qquad d\hat{\mu}_g = v_g \, dp$$

or

$$d(\hat{\mu}_1 - \hat{\mu}_g) = (v_1 - v_g)dp, \tag{3.26}$$

where v_l is the volume of a vapor molecule in the liquid phase and v_g is the volume of a vapor molecule in the gas phase. If the two phases are in equilibrium, the chemical potential is the same for molecules in each phase, or $\hat{\mu}_l = \hat{\mu}_g$. Thus in our example, the liquid will exist in equilibrium with a saturated vapor with pressure over a plane liquid surface. The vapor under consideration is at an arbitrary pressure, so the change in chemical potential $(\hat{\mu}_l - \hat{\mu}_g)$ can be obtained by integration of Eq. (3.26).

In the case of the vapor being a perfect gas, we have

$$\hat{\mu}_l = \hat{\mu}_g = -kT \ln(p_s/p_0), \tag{3.27}$$

noting that $v_g (=kT/p_s) \gg v_l$. From Eq. (3.27), the chemical potential difference $\hat{\mu}_l - \hat{\mu}_g$ is positive when the vapor is unsaturated, but negative for a supersaturated vapor. When the vapor is unsaturated, the free-energy change increases rapidly with increasing embryo size. However, when the vapor is supersaturated, the free energy of the embryo formation displays a maximum value, denoted as $\Delta\hat{G}^*$, then decreases rapidly with embryo size. The radius where $\Delta\hat{G}$ is a maximum is designated the *critical radius* where the embryos exist in equilibrium with the vapor at a given supersaturation. This radius, R^*, is basically given for a perfect gas vapor by Kelvin's relation, Eq. (3.22).

The distribution of embryos at equilibrium for the system of reactions (3.24) can be estimated by calculating the equilibrium constant for each of the reactions in this series. At equilibrium,

$$\beta s_{i-1} n_{i-1} = \beta_i' s_i n_i.$$

Then using (3.14) and (3.15), one finds that

$$\frac{n_{i-1}}{n_i} = \frac{1}{S} \exp\left(\frac{2\sigma v_m(\tfrac{4}{3}\pi v_m)^{1/3}}{i^{1/3} kT}\right). \tag{3.28}$$

Multiplying equations of this form together, for $i = 2$ and larger, and combining the distribution of embryos gives

$$n_i = \frac{p_0}{kT} S^i \exp\left(\frac{-3\sigma v_m(\tfrac{4}{3}\pi v_m)^{1/3}}{kT} i^{2/3}\right). \tag{3.29}$$

This is basically a discrete size spectrum for an aerosol consisting of extremely small particles in equilibrium with a supersaturated vapor. The form of the distribution for cases where the vapor is unsaturated ($S < 1$) and supersaturated ($S > 1$) can be calculated, and is shown in Fig. 3.2. For unsaturated conditions, the largest number of particles is the single molecules, with decreasing classes of ith-size particles. When the vapor is supersaturated, the distribution still carries a maximum of single molecules, but contains a minimum number at a "critical" cluster size i^*th radius equivalent to the maximum in free energy $\Delta\hat{G}^*$, or

$$R^* = 2\sigma v_m/kT \ln S.$$

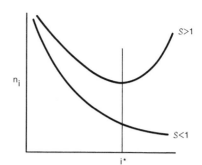

Fig. 3.2. Discrete size distribution at equilibrium for clusters formed by homogeneous nucleation. For $\mathcal{S} > 1$, an infinite mass of material must be present in the cluster phase at equilibrium.

This is significant because it is the size that all embryos must pass in order to become stable in the condensed phase. Smaller particles will evaporate; those larger will continue to grow.

The number of embryos of critical size is

$$n_i^* = n_1 \exp[-16\pi\sigma^3 v_m^2/3(kT)^3(\ln \mathcal{S})^2]. \tag{3.30}$$

The exponential term essentially represents the ratio of the free energy of formation of the critical-sized droplet and kT, $\Delta\hat{G}^*/kT$. Using the equilibrium distribution for n_i^e, one can estimate the particle current using (3.27), or

$$I_i = n_{i-1}^e s_{i-1} \beta(n_{i-1}/n_{i-1}^e - n_i/n_i^e). \tag{3.31}$$

For $i > 10$, the current can be written as a continuous function of i, or

$$I(i) = -\beta s \frac{\partial n}{\partial i} - \frac{\beta s n}{kT} \frac{\partial \Delta\hat{G}}{\partial i},$$

where

$$\Delta\hat{G}_i/kT = (36\pi)^{1/3}\sigma v_m^{2/3} i^{2/3}/kT - i \ln \mathcal{S}. \tag{3.32}$$

An approximate form for the embryo current was derived by Becker and Döring (1935), and later Frenkel (1955), who made a series of so-called threshold approximations and assumed quasi-steady state where

$$\frac{\partial I}{\partial k} = -\frac{\partial n}{\partial t} \approx 0,$$

or

$$I(i) = 0.$$

The result for I at steady state becomes (Becker and Döring, 1935)

$$I \approx C \exp\left(\frac{\Delta\hat{G}^*}{kT}\right)$$

$$\approx \frac{\alpha_c 2p_{A0}(n_A v_m)^{2/3}}{(2\pi m_A kT)^{1/2}} \left(\frac{\sigma v_m^{2/3}}{kT}\right)^{1/2} \exp\left(-\frac{16\pi\sigma^3 v_m}{3(kT)^3(\ln \mathcal{S})^2}\right). \tag{3.33}$$

This formula indicates that the embryo current is a strong function of vapor supersaturation, as shown for water in Fig. 3.3. Using this feature of the relationship, investigators tested the theory by seeking data on the critical supersaturation ratio, where $I = 1$ cm^{-3}-sec^{-1}. Results of some early studies with an expansion cloud chamber[†] are listed in Table 3.1 with theoretical estimates of the critical value of the supersaturation. The agreement between the theory and this test is surprisingly good considering the approximate nature of the theory and the coarse nature of the early experiments.

Experiments to test nucleation theory have been refined in recent years. They have employed both the expansion chamber method (Kassner and Schmitt, 1967) and a diffusion chamber device (Katz et al., 1975). The expansion chamber results largely bear out the conclusions of the nucleation theory for inorganic vapors in helium or nitrogen, and for some inorganics including water. The results display remarkably good agreement for organic, nonpolar compounds. The agreement is much less satisfactory for certain highly polar fluids such as water (e.g., Heist and Reiss, 1973).

Concern for the weaknesses in nucleation theory has led investigators to examine its limitations from (a) matching a macroscopic approach (e.g., Hill, 1960), (b) steady state assumptions (e.g., Courtney, 1961), and (c) preexponential factors (e.g., Lothe and Pound, 1966; Katz, 1970). Despite the criticisms, the B-D model remains in widespread use, because it contains in

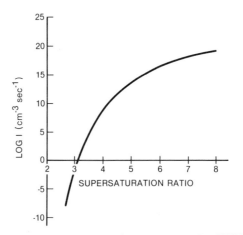

Fig. 3.3. Droplet current I for supersaturated water vapor at $T = 300°$K calculated from Eq. (3.33). The critical saturation ratio, corresponding to $I = 1$ cm^{-3} sec^{-1}, is about 3.

[†] These experimental methods are discussed further as a part of aerosol generation methods later in Section 4.1.

TABLE 3.1

Comparison of Measured and Calculated Values of Critical Supersaturation Ratio[a]

Vapor	Number of molecules in stable nucleus	Nucleus radius R_i (Å)	S(crit) calculated	S(crit) measured
Water, 275.2°K	80.0	8.9	4.2	4.2 ± 0.1
Water, 261.0°K	72.0	8.0	5.0	5.0
Methanol, 270.0°K	32.0	7.9	1.8	3.0
Ethanol, 273.0°K	128.0	14.2	2.3	2.3
n-propanol, 270.0°K	115.0	15.0	3.2	3.0
Isopropyl alcohol, 265.0°K	119.0	15.2	2.9	2.8
n-butyl alcohol, 270.9°K	72.0	13.6	4.5	4.6
Nitromethane, 252.0°K	66.0	11.0	6.2	6.0
Ethyl acetate, 242.0°K	40.0	11.4	10.4	8.6–12.3

[a] Data of Volmer and Flood (1934) taken in an expansion cloud chamber. (Reprinted courtesy of Pergamon.)

a relatively simple way the main features believed to be important in the process.

3.3.2 Heteromolecular Nucleation. The theory for homogeneous nucleation accounts for new particle formation in a pure vapor. There are many situations where nucleation takes place in a multicomponent system, where one or more components are supersaturated. This group is called *heteromolecular* nucleation. The calculation for particle formation has been worked out for a binary system, extending the classical model described above. An early analysis was made by Flood (1934) and was later extended by Reiss (1950) and by Mirabel and Katz (1974).

The theory for heteromolecular nucleation in a binary vapor mixture was worked out in a generalized way by Reiss (1950). The calculation takes the same basic form as the theory for the homogeneous nucleation from a pure vapor; that is, the droplet current is given by Eq. (3.33). However, the free energy of the critical embryo is that required to form a droplet having a size and composition such that it is in equilibrium with the surrounding gas phase.

The free energy of formation of an embryo of arbitrary size and composition is

$$\Delta \hat{G} = n_A(\hat{\mu}_{Al} - \hat{\mu}_{Ag}) + n_B(\hat{\mu}_{Bl} - \hat{\mu}_{Bg}) + 4\pi R^2 \sigma, \qquad (3.34)$$

where n_A and n_B are the number of moles of components A and B in the embryo. The chemical potentials $\hat{\mu}_{Al}$ and $\hat{\mu}_{Bl}$ are those for a macroscopic amount of a liquid phase of the same composition as an embryo, and $\hat{\mu}_{Ag}$ and

$\hat{\mu}_{Bg}$ are the chemical potentials of A and B in the vapor. The maximum of $\Delta\hat{G}$ or $\Delta\hat{G}^*$ for a fixed number of molecules of A and B (for fixed vapor pressures of constituents) can be found by solving the pair of equations

$$\left(\frac{\partial\Delta\hat{G}}{\partial n_A}\right)_{n_B} = 0, \qquad \left(\frac{\partial\Delta\hat{G}}{\partial n_B}\right)_{n_A} = 0.$$

For ideal mixtures, Reiss (1950) showed that these two equations lead to the existence of a saddle point in the three-dimensional system of $\Delta\hat{G}$ versus n_A and n_B. The saddle point is the barrier that the embryo must overcome to grow and stabilize during the nucleation process. Solving for the partial derivatives, we obtain

$$\left(\frac{\partial\Delta\hat{G}}{\partial n_A}\right)_{n_B} = \hat{\mu}_{Al} - \hat{\mu}_{Ag} + \frac{2\sigma\bar{v}_A}{R^*}$$

$$- \frac{3x^*}{R^*} v_{So} \left.\frac{d\sigma}{dx}\right|_{x=x^*} = 0, \qquad (3.35a)$$

$$\left(\frac{\partial\Delta\hat{G}}{\partial n_B}\right)_{n_A} = \hat{\mu}_{Bl} - \hat{\mu}_{Bq} + \frac{2\sigma\bar{v}_A}{R^*}$$

$$- \frac{3v_{So}(1 - x^*)}{R^*} \left.\frac{d\sigma}{dx}\right|_{x=x^*} = 0. \qquad (3.35b)$$

Here, $x = n_B/n_A + n_B$; $\bar{v}_A$ and $\bar{v}_B$ are partial molar volumes and v_{So} is the molar volume of solution, $= (1 - x)\bar{v}_A + x\bar{v}_B$. Equations (3.35) permit the calculation of values of x^* and R^*, the composition and the critical embryo radius at the saddle point. The free-energy change $\Delta\hat{G}^*$ is obtained from the mole fraction and using the relation

$$\frac{4}{3}\pi R^{*3} = n_A^*\bar{v}_A + n_B^*\bar{v}_B = (n_A^* + n_B^*)\, v_{So}$$

for n_A^* and n_B^*. These are substituted with R^* into Eq. (3.35).

The Reiss form for the frequency factor, analogously to Eq. (3.33), is

$$C = \frac{\beta_A\beta_B}{\beta_A \sin^2 \phi + \beta_B \cos^2 \phi}\, (n_A + n_B)\, s \left(-\frac{P}{Q}\right)^{1/2}, \qquad (3.36)$$

where $\beta_\alpha = p_\alpha/(2\pi m_\alpha kT)^{1/2}$ for the gaseous species, n_A and n_B are the number densities for A and B, s is the surface area of the embryo, and ϕ is the angle between the axis n_A and the direction of passage through the saddle point; P and Q are second derivatives of $\Delta\hat{G}^*$ with respect to rotated axes for n_A and n_B given by

$$P = \frac{\partial^2\Delta\hat{G}}{\partial n_A^2}\cos^2 \phi + 2\frac{\partial^2\Delta\hat{G}}{\partial n_A\partial n_B}\cos \phi \sin \phi + \frac{\partial^2\Delta\hat{G}}{\partial n_B}\sin^2 \phi,$$

$$Q = \left(\frac{\partial^2 \Delta \hat{G}}{\partial n_A^2}\right) \sin^2 \phi + \frac{\partial^2 \Delta \hat{G}}{\partial n_A \partial n_B} \sin \phi \cos \phi + \frac{\partial^2 \Delta \hat{G}}{\partial n_B^2} \cos^2 \phi.$$

The angle of rotation is given by

$$\phi = \tfrac{1}{2} \tan^{-1}\left[2 \frac{\partial^2 \Delta \hat{G}}{\partial n_A \partial n_B} \bigg/ \left(\frac{\partial^2 \Delta \hat{G}}{\partial n_A^2} - \frac{\partial^2 \Delta \hat{G}}{\partial n_B^2}\right)\right].$$

Calculations have been made for the sulfuric acid–water vapor system by Doyle (1961) and later for this system and the nitric acid–water vapor system (Mirabel and Katz, 1974). The activities a_α of the vapors were approximated by ideal-gas values or $a_\alpha = p_\alpha/p_\alpha^0$, or the ratio of the partial pressure to the equilibrium vapor pressure of the pure component.

Results of the Mirabel and Katz (1974) calculation are shown in Fig. 3.4 for the sulfuric acid–water system and the nitric acid–water system. The figure shows the relation between the acid activity in the vapor and relative humidity for nucleation of one embryo per cubic centimeter–second. The calculation shows that the activity for nucleation decreases with humidity, and that sulfuric acid will nucleate at activities two to three orders of magnitude less than those for nitric acid. Note also that in heteromolecular nucleation the concept of supersaturation does not apply in these cases. With two condensable vapors present, nuclei will be produced at activities well below those required for pure vapors.

3.3.3   Heterogeneous Nucleation.

The theory of new particle formation has been extended to include nucleation processes on foreign nuclei present in a condensable vapor system. These nuclei may consist of ions, as well as soluble or insoluble particles. The early experiments of Wilson (1900), for example, demonstrated that the presence of ions could substantially reduce

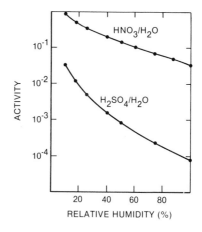

Fig. 3.4. Activity of the acid in the vapor phase needed to achieve a nucleation rate of 1 nucleus/cm³ sec as a function of relative humidity for $H_2SO_4 + H_2O$ and $HNO_3 + H_2O$ at 25°C. [From Mirabel and Katz (1974); with permission from the American Physical Society.]

the supersaturation required to produce droplets. It is well known that nucleation of water vapor in the atmosphere takes place on dust or salt particles at conditions very near saturation of the vapor.

There is only an equilibrium-based theory for nucleation on ions. The classical equilibrium-based theory of homogeneous nucleation is easily extended by accounting for the modification in free energy of formation of critical-sized embryos. For ion nucleation, Volmer and colleagues (1939) proposed the formula for $\Delta\hat{G}^*$ as

$$\Delta\hat{G}^* = -\frac{4}{3}\pi R^{*3}kT \ln S + 4\pi R^{*2}\sigma + \frac{Q_1^2}{2}\left(1 - \frac{1}{\varepsilon}\right)\left(\frac{1}{R^*} - \frac{1}{R_1}\right),$$

where Q_1 is the charge on the ions and ε is the dielectric constant of the embryo.

The last term on the right accounts for the electrostatic energy of the ions, assuming that the dielectric constant of the medium is unity. The radius of critical size is determined as before by requiring that $\partial\hat{G}/\partial R = 0$, or

$$\ln S(\text{crit}) = \frac{2\sigma v_m}{kTR^*} - \frac{Q_1 v_m}{8\pi R^{*4}kT}\left(1 - \frac{1}{\varepsilon}\right). \tag{3.37}$$

Comparison of this relation with the Kelvin–Gibbs equation (3.22), indicates that the influence of ions tends to decrease the supersaturation ratio needed to maintain a droplet of critical size.

For the arguments used to develop the equation for homogeneous nucleation, the droplet current when ions are present takes the form

$$I = [\alpha_c p_0/(2\pi m kT)^{1/2}](4\pi R^*)^2 Z_c N_c \exp(-\Delta\hat{G}^*/kT), \tag{3.38}$$

where N_c is the ion concentration and the factor

$$Z_c = \left(\frac{4\pi\sigma R^{*2} - (1 - 1/\varepsilon)Q_1^2/R^*}{9\pi i^{*2}kT}\right)^{1/2}$$

has a value of order 10^{-2}.

Equation (3.38) for the droplet current is of essentially the same form as the homogeneous nucleation expression, Eq. (3.33). The current is a product of the net rate of vapor diffusion to a nucleus of size R^* and the number distribution of ions larger than R^*. The coefficient Z_c is sometimes called a nonequilibrium correction for the theory. The correction arises when the calculation takes into account the flow of nucleation as the difference between net growth of embryos to size R^* less the decrease in size smaller than R^* per unit time. In the case of water vapor nucleation at 265°K, for example, the critical supersaturation measured by Wilson was found to be 3.2 for negative ions present in an expansion chamber, in contrast with 4.1 for the case without ions. Although qualitative observations have been made of the influence of ions on nucleation, the quantitative agreement between the simple theory and experiment is considered poor. The reasons for the disagreement are not known, but investigators have examined the effects of the dielectric constant of small particles, as well as the differences associated with different signs of charge on droplets, influencing surface electrical effects.

When foreign nuclei such as dust are present, the initiation of condensed phase again can be estimated using the free-energy argument (e.g., Fletcher, 1966). In the case of condensation on an insoluble particle, the embryo can be visualized schematically as shown in Fig. 3.5. Here a spherical liquid cap forms on a substrate, with a contact angle ϕ between the liquid and the solid. For equilibrium at the surface, the line common to the three phases—substrate c, liquid l, and vapor g—is restricted by

$$m_c \equiv \cos \phi = (\sigma_{cg} - \sigma_{cl})/\sigma_{lg}.$$

The chemical potential of embryo formation as a spherical cap then is

$$\Delta \hat{\mu} = \tfrac{1}{3}\pi r^3 (2 + m_c)(1 - m_c)^3 (\hat{\mu}_l - \hat{\mu}_g) + 2\pi r^2 (1 - m_c)$$
$$+ (\sigma_{cl} - \sigma_{cg}) 2\pi r^2 (1 - m_c) \quad (3.39)$$

where r is the radius of curvature of the embryo. Applying the criterion that the free energy of formation of a critical-sized cap should be maximum, the critical radius of curvature from the above relation has a form analogous to the Kelvin relation, and

$$\Delta \hat{G}_c^* = [16\pi\sigma_{lg}^3 v_m^2/(3kT \ln S)^2] f(m_c), \quad \text{where } f(m_c) = \tfrac{1}{4}(2 + m_c)(1 - m_c)^2.$$
$$(3.40)$$

The term $f(m_c)$ represents a correction for the free energy of formation given in the case of homogeneous nucleation. Since $0 \le f(m_c) \le 1$, any foreign material acting as a nucleus will reduce the free energy required to form an embryo of critical size.

The calculation of droplet current in this case can be made if the distribution of critical-sized embryos is related to the critical cap radius r^* and the rate of impact of vapor molecules per unit area of surface, or

$$I \approx [\alpha_c p_0/(2\pi mkT)^{1/2}]4\pi r^{*2} n_{ads} \exp(-\Delta \hat{G}_c^*/kT), \quad (3.41)$$

when n_{ads} is the number of vapor molecules adsorbed per unit surface area of nuclei substrate.

Fig. 3.5. Schematic diagram of spherical embryo on solid substrate.

The theory for nucleation on a planar, insoluble substrate has been tested for water by several investigators, whose work has been reviewed by Pruppacher and Klett (1978). An example of Mahata and Alofs' (1975) recent comparison is shown in Fig. 3.6. Here the experimental results for the critical supersaturation for the onset of water nucleation on a planar, water-soluble, partially wettable substrate are compared with the theoretical estimate. The comparison is given as a function of contact angle. The theory and experiment agree reasonably well up to a critical supersaturation $S - 1$ of 10%, then the departure is found which cannot be accounted for using corrections for surface tension. The reason for the deviation at large contact angle is not known at present.

The heterogeneous nucleation model can be extended to spherical nuclei. Pruppacher and Klett (1978) discuss further the theory and its limitations for the interested reader.

When the nucleus is soluble, condensed-phase formation is complicated by the dissolution process. Consider for example water vapor condensation.

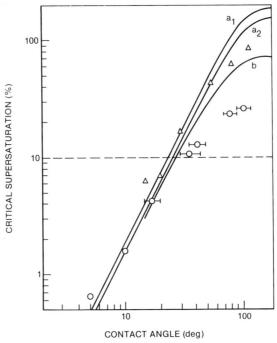

Fig. 3.6. Critical supersaturation for onset of water nucleation on a planar, water-insoluble, partially wettable substrate, as a function of contact angle of water on substrate. Comparison of experiment with theory, with size correction and surface tension. Experiment: ⊢O⊣ and O, Mahata and Alofs (1975); △, Koutsky *et al.* (1965). Theory: a_1: $I = 10^8$ cm^{-2}-sec^{-1}; a_2: $I = 1$ cm^{-2}-sec^{-1}; b: $I = 1$ cm^{-2}-sec^{-1}. [From Mahata and Alofs (1975), with changes by Pruppacher and Klett (1978); reprinted with permission of Reidel Publishing Co.]

TABLE 3.2

Relative Humidity and Concentration for Saturated Solutions at 20°C[a]

Salt	Relative humidity (%)	Solubility (g/100 g H_2O)
$(NH_4)_2SO_4$	81	75.4
NaCl	75.7	36
NH_4NO_3	62 (25°C)	192
$CaCl_2 \cdot 6H_2O$	32	74.5

[a] Based on data from Dean (1973) and Stokes and Robinson (1949); from Friedlander (1977).

A small, dry, hygroscopic salt particle exposed to increasing relative humidity will remain solid up to a characteristic relative humidity less than 100% at which it absorbs water and dissolves to form a saturated solution. The relative humidities for dissolution of large salt crystals are listed for some salts in Table 3.2. The values of humidity will vary with crystal size by the Kelvin effect. Some equilibrium growth curves of Winkler (1967) for salts are shown in Fig. 3.7. The abscissa is the ratio of mass of water collected to mass of dry salt versus the relative humidity, plotted on the ordinate. The dissolution of

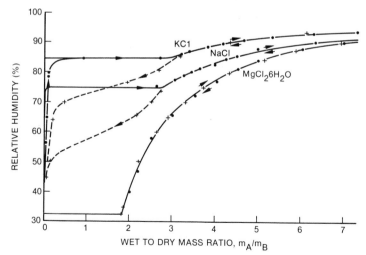

Fig. 3.7. Equilibrium growth curves of pure salt deposits (25°C). The arrows indicate the direction of equilibrium conditions for increasing relative humidity versus decreasing relative humidity. ——, calculated curves from deliquescence relation; $\cdots$, measured points (increasing humidity); + + +, measured points (decreasing humidity). [From Winkler (1967); by courtesy of the author.]

sodium chloride is shown to take place at 75% RH, where the mass of the crystal increases by a factor of 2 in forming a droplet. With additional humidity increase, the droplet continues to grow to maintain equilibrium with the surrounding vapor. If the surroundings dry out, we see that the droplet decreases in size along a curve which shows a different change than during growth. This "hysteresis" suggests that hygroscopic particles retain water during decrease in size for a medium decreasing in relative humidity. In any case, the nucleation process on a soluble particle then involves the assumption of maintaining thermodynamic equilibrium with the vapor throughout the transition process from a solid to a droplet.

The nucleation of new phase on soluble particles follows directly from the calculation of the supersaturation versus particle size relation [Eq. (3.23)]. From Fig. 3.1, or sets of similar curves, it is seen that as the supersaturation ratio increases with very small particles, the droplet of solution at equilibrium with the vapor increases with increasing radius, but is metastable until a maximum supersaturation is reached. After the maximum in supersaturation ratio is exceeded, the embryo size will increase even though the supersaturation ratio decreases to approach unity, providing an unlimited supply of condensable vapor exists. Fletcher (1966) has derived the critical radius and supersaturation ratio from Eq. (3.23), giving

$$R^* = \left(\frac{9}{8} \frac{z_\pm n_B \mathcal{R} T}{\pi \sigma}\right)^{1/2} \tag{3.42}$$

and

$$\left(\frac{p_{AS}}{p_{A0}}\right)_{crit} \approx 1 + \frac{8}{9}\left[\frac{2\pi M_A^2}{z_\pm n_B \rho_l^2}\left(\frac{\sigma}{\mathcal{R} T}\right)^3\right]^{1/2}, \tag{3.43}$$

where p_{AS} is the partial pressure of solvent A over the solution and p_{A0} is the partial pressure of A at equilibrium with the solution; $z_\pm$ is the number of ions in solution, n_B is the number of moles of solute dissolved (B), and ρ_l is the mass density of the solvent.

3.3.4 Growth Laws. The droplet current calculated by nucleation models represents a limit of initial new phase production. The initiation of new phase takes place rapidly once a critical supersaturation is achieved in a vapor. The phase change occurs in seconds or less, normally limited only by vapor diffusion to the surface. In many circumstances, we are concerned with the evolution of the particle size distribution well after the formation of new particles or the addition of new condensate to nuclei. When the growth or evaporation of particles is limited by vapor diffusion or molecular transport, the growth law is expressed in terms of the vapor flux equation, given for example by Eq. (2.33). However, other growth processes may occur. These include those associated with chemical reactions to form condensed species taking place either at the particle surface, or within the particle volume (Schwartz and Freiberg, 1981). The growth by surface reactions, or a vapor-diffusion-limited process, will be proportional to the particle surface area or radius squared for spheres. Volume reactions will be controlled by particle volume or proportional radius cubed for spheres. The idealized growth laws for gas to spherical particle conversion are summarized in Table 3.3.

In the previous discussion, the droplet current I consisted of two terms, one representing diffusion to the particle surface and the other migration in volume or v-space. The vapor diffusion process, proportional to $\partial n/\partial i$, or $\partial n/\partial v$, leads to the spreading in volume of a group of particles initially the same size. This term is large for homogeneous nucleation where molecular clusters are formed. However, for growth of larger particles this term is small and the particle current can be approximated by

$$I(v,t) \approx n\, \partial v/\partial t, \tag{3.44}$$

where $\partial v/\partial t$ is given in Table 3.3. According to this expression, all particles of the same initial size grow to the same final size in a fixed time interval if such processes dominate the size spectrum.

TABLE 3.3

Growth Laws for Gas to Particle Conversion Component A is the Condensing Species in Volume Change per Unit Time[a]

Mechanism	Growth law	Reference equation
Molecular bombardment $(R \ll \lambda_g)$	$\dfrac{4\alpha_c \pi R^2 v_m (p_\infty - p_s)}{(2\pi m_A kT)^{1/2}}$	(2.27)
Molecular diffusion $(R \gg \lambda_g)$	$\dfrac{4\pi D_{AB} R^2 v_m (p_\infty - p_s)}{kT}$	(2.29)
Surface chemical reactions	$\dfrac{4\alpha_c \pi R^2 v_m p_\infty}{(2\pi m_A kT)^{1/2}}$	(2.27a)
Condensed-phase reaction[b]	$\dfrac{4\pi R^3}{3\rho_p}\left(\sum_\alpha \nu_\alpha M_\alpha\right) r$	

[a] From Friedlander (1977), with permission of Wiley.

[b] ν_α are the stoichiometric coefficients for the reaction of α species with molecular weight M_i; r is the reaction rate, $(1/\nu_\alpha v)dn_\alpha/dt$, where v is the particle volume.

3.4 COAGULATION PROCESSES

Once particles are present in a volume of gas, they will collide and agglomerate by different processes. As described above, this coagulation process leads to substantial changes in particle size distribution with time. Coagulation can be induced by any mechanism which involves a relative velocity between particles. Such processes include Brownian motion, shearing flow of fluid, turbulent motion, and differential particle motion associated with external force fields. The theory of particle collisions is quite complicated even if each of these mechanisms is isolated and treated separately. The theoretical models are summarized below, but are considered in detail elsewhere (Hidy and Brock, 1970; Pruppacher and Klett, 1978).

3.4.1 Brownian Motion. As a beginning, let us consider the idealized situation of uncharged spherical particles colliding by Brownian motion in a stagnant or uniform flowing gas. This case was first considered in 1916 by Smoluchowski for the continuum limit. The geometry of collision is drawn schematically in Fig. 3.8. Applying a model for particle diffusion to the surface of another in a fixed reference frame, the collision rate in quasi-steady state is given by

$$\text{Rate} = 4\pi D_p \left(r^2 \frac{\partial n}{\partial r} \right)_{r=R_i+R_j} = 4\pi D_p (R_i + R_j) N_\infty . \tag{3.45}$$

From this, the collision parameter is defined in terms of the collision frequency (No./cm^3-sec) from Eq. (2.55),

$$b_{ij} n_i n_j ,$$

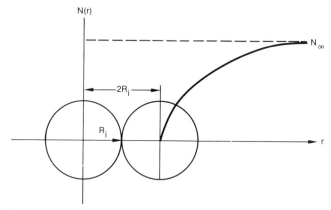

Fig. 3.8. Geometry of collision model as diffusion to a central sphere.

where n_i is the number of ith particle size species per unit volume. Then

$$b_{ij} = 4\pi(D_i + D_j)(R_i + R_j) = \frac{2kT}{3\mu_g}\left(\frac{1}{v_i^{1/3}} + \frac{1}{v_j^{1/3}}\right)(v_i^{1/3} + v_j^{1/3}) \quad (3.46)$$

under conditions where $Kn \to 0$. This relation can be extended to include the effects of electrostatic charges. For unpolarized spheres, using the Debye–Hückel model, the collision rate can be written as

$$b_{ij}n_in_j\Omega_{ij},$$

where the effect of charging is

$$\Omega_{ij} \approx \frac{z_iz_je^2 \exp(-\kappa R_{ij})}{R_{ij}kT\{\exp[(z_iz_je/R_{ij}kT \exp(-\kappa R_{ij})] - 1\}}, \quad (3.47)$$

where e is a unit of charge, z_i is the number of charges on particle i, $R_{ij} = R_i + R_j$, and κ is the Debye reciprocal length

$$\kappa \equiv \left(\frac{4\pi e^2}{kT}\sum_{i=1} n_iz_i^2\right)^{1/2}. \quad (3.48)$$

The influence of electrical charging on coagulation during Brownian motion is readily illustrated by the data presented in Table 3.4. Here calculations for the relative collision rate between particles of different radii and charge are given. For particles of like charge, collisions are suppressed strongly with 10 units of like charges, but enhanced with 10 units of opposite charge.

The strong influence of charging on aerosol coagulation has been borne out by workers who have had considerable difficulty in comparing the theory with experiment because particles are nearly always charged to some extent (e.g., Mercer, 1978).

Experimental studies of coagulation in charged clouds have yielded ambiguous results. Early experiments on bipolar charged aerosols indicated

TABLE 3.4

Electrostatic Effects on Coagulation[a,b]

R_i (μm)	R_j (μm)	z_i	z_j	Ω_{ij}	z_i	z_j	Ω_{ij}
1.0	1.0	+1	+1	0.990	+1	−1	1.01
1.0	0.1	+1	+1	0.983	+1	−1	1.03
0.1	0.1	+1	+1	0.869	+1	−1	10.2
1.0	1.0	10	10	0.173	+10	−10	1.15
1.0	0.1	10	10	0.028	+10	−10	1.28
0.1	0.1	10	10	0.93×10^{-11}	+10	−10	28.7

[a] $T = 20°C$, $N = 10^4$ cm^{-3}.
[b] From Hidy and Brock (1970), with permission from Pergamon.

that, for particles in the micrometer and submicrometer range, charging up to 4–6 units per particle had no detectable effect on coagulation. However, a study of Gillespie (1953) has indicated that bipolar charging of $z_i \approx 6\text{--}14$ units on particles of ~ 1 μm radius can increase coagulation rates by an order of magnitude. Fuchs (1964) has remarked that these results may not be reliable because of complications resulting from electrostatic deposition of particles on the walls of Gillespie's chamber (Gillespie, of course, did attempt to account for such losses in analyzing his data). Careful experiments recently reported by Weinstock (1967) of more monodisperse dioctylphthalate aerosols ($R_i \approx 0.7$ μm, $z_i = 1\text{--}3$) have indicated no observable effect of charging. This is certainly in agreement with early results, and indicates that a simple Coulombic charge interaction model is an adequate theory for the effect of weak charging.

Measurement of the coagulation of unipolar charged aerosols is complicated by the effect of electrostatic dispersion. As a result of the concentration of like charge, there is a tendency for the entire aerosol cloud to expand because of the mutual repulsion induced locally between particles. Because of this effect, virtually no reliable experimental studies concerning unipolar charging and aerosol coagulation are available. However, it still may be possible to test for decreased coagulation of unipolar charged aerosols by a method similar to Gillespie's or Weinstock's, taking into account losses to the walls of the apparatus.

Strictly speaking, Smoluchowski's continuum model will rarely apply to aerosols since Kn $\neq$ 0. Coagulation of very small particles in the noncontinuum limit can readily be calculated from kinetic theory. For binary collision, of charged particles, the collision parameter is

$$b_{ij} = \mathfrak{M}_{ij}[8\pi kT(M_i + M_j)/m_i m_j]^{1/2}\Omega_{ij}R_{ij}^2, \qquad \text{Kn} \to \infty,$$

where

$$\Omega_{ij} = \exp[-z_i z_j e^2 \exp(-\kappa R_{ij})/kTR_{ij}]$$

and $\mathfrak{M}_{ij}$ is a symmetry factor, equal to $\frac{1}{2}(i = j)$ or 1 $(i \neq j)$. This expression has also been generalized to include phoretic forces (Brock and Hidy, 1965).

In the case of $0.1 \leq$ Kn ≤ 10, the transition regime, the collision theory has not been worked out in detail. An extension of the free-molecule theory for Kn > 10 has yielded a very complicated expression (Hidy and Brock, 1970). For uncharged particles, Fuchs (1964) has suggested a simple form which involves an extension of the continuum calculation corrected for diffusion in a rarified medium. This is basically the same diffusion theory he adopted for evaporation (Fuchs, 1959).

Wagner and Kerker (1977) have discussed Fuchs' (1959) model, presenting the coagulation factor b from Eq. (3.16) in a dimensionless form,

$$\beta(R_i, R_j) = (3\mu_g/8kT)b_{ij}. \tag{3.50}$$

Fuchs' form, which empirically interpolates the value of β for small and large Kn, is

$$\beta(R_i, R_j) = \beta_s(R_i, R_j) \left(\frac{R_{ij}}{R_{ij} + \Delta_{ij}} + \frac{4D_{ij}}{\mathcal{C}\bar{q}_{ij}R_{ij}} \right) \qquad (3.51)$$

where $\beta_s(R_i, R_j)$ is the form for correction at small Kn,

$$\beta_s = \frac{1}{4}(R_i + R_j) \left(\frac{A(\mathrm{Kn}_i)}{R_i} + \frac{A(\mathrm{Kn}_j)}{R_j} \right);$$

Δ_{ij} is Fuchs' concentration depletion factor, which depends on Kn, $\Delta_{ij} = \Delta_i + \Delta_j$, and

$$\Delta_i = (1/6R_i\lambda_i)[(2R_i + \lambda_i)^3 - (4R_i^2 + \lambda_i^2)^{3/2}] - 2R_i;$$

$\mathcal{C}$ is the sticking or coalescence efficiency, and $\bar{q}_{ij} = (\bar{q}_i^2 + \bar{q}_j^2)^{1/2}$.

A form of the expression for β is plotted as a function of Kn in Fig. 3.9, with Kn varying with pressure. Shown in the figure are the results of careful experiments with monodisperse, 0.2 μm diam di(2-ethylhexyl) sebacate particles. For the range of Kn between approximately one and ten, the results agree to within 1 to 11% of Fuchs' model for a coalescence efficiency of unity. Other experiments have shown a greater departure from the theory, as discussed by Mercer (1978). Mercer concluded from review of a number of experiments that the increased collision rate in the Knudsen transition regime generally is confirmed, but they do not suggest a basis for distinguishing between the Fuchs semiempirical model and the extended

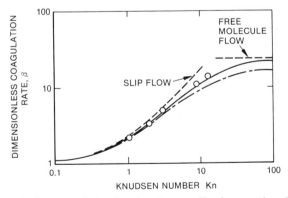

Fig. 3.9. Dimensionless coagulation function β versus Knudsen number. Fuchs' interpolation formula is shown for collision efficiencies of unity (upper curve) and 0.75 (lower curve). The Smoluchowski law with Cunningham slip flow correction and the free-molecule limit are shown by dashed curves. The experimental results are indicated by circles. ——, coalescence efficiency = 1.0; —·—, coalescence efficiency = 0.75. [From Wagner and Kerker (1977); reprinted with permission of the authors and the Optical Society of America.]

kinetic theory model. Large experimental differences appear between most coagulation experiments, and within the same experiments. They are identified qualitatively with various difficulties, including wall losses by diffusion and gravitational setting, and inaccuracies in particle size determination along with electrical effects. The work of Wagner and Kerker (1977) is one of the few exceptions; similar careful work will be needed to evaluate coalescence models.

3.4.2 Shearing Flow and Turbulence.

When a differential velocity is induced in a fluid, particles following the fluid will begin to approach one another. Perhaps the simplest example of this situation is an aerosol exposed to a laminar shearing, with a fluid velocity gradient dq/dx. Smoluchowski (1916) calculated the collision factor in such circumstances to be

$$b_{ij} = \frac{32}{3} (R_i + R_j)^3 \frac{dq}{dx}. \tag{3.52}$$

In a turbulent gas, particles will move relative to one another as a result of (a) differential velocities associated with spatial inhomogeneities from the eddying fluid motion, and (b) differential velocities associated with particle inertia relative to the air flow. Since the particle inertia depends on their size, the first mechanism may involve particles of equal or unequal size. The second mechanism involves only particles of unequal size. Saffman and Turner (1956) developed a theory for these processes. Their results in a continuum medium for particle motion with air motion are

$$b_{ij} = 1.7 (R_i + R_j)^3 (\varepsilon_t/\nu_g)^{1/2}, \tag{3.53}$$

where ε_t is the turbulent energy dissipation rate in centimeters squared per second cubed.

In cases where there is particle motion relative to the air,

$$b_{ij} \approx 5.7(R_i + R_j)^2 (\mathfrak{a}_i^{-1} - \mathfrak{a}_j^{-1})\varepsilon_t^{3/4}\nu_g^{-1/4}. \tag{3.54}$$

Saffman and Turner considered the limitations of these mechanisms and found that the two effects became of the same order if $R_i - R_j \approx 3$ μm and $\varepsilon_t = 5$ cm^2/sec^3, or when $R_i - R_j \approx 1$ μm and $\varepsilon_t \approx 1000$ cm^2/sec^{-3}. Except for small particles, where $R < 1$ μm, collisions in clouds of mainly uniform particles are controlled by motion relative to the air. With increasing polydispersity, particle motion with the air becomes most important.

3.4.3 Differential Motion by External Forces.

Collisions between particles can also be induced by differences in motion associated with the action of external forces. The theory developed is a continuum approximation which models the action of gravity. For particles falling according to Stokes settling velocity, the collision parameter is given approximately as (Fuchs, 1964; Friedlander, 1957).

$$b_{ij} \approx \frac{2\pi\rho_i g}{9\mu_g} R_j^4(1 + I)^2 \left\{\left[1 - \frac{3}{2(1 + I)} + \frac{1}{2(1 + I)^3}\right] - I^2(1 + I)^2\right\}, \quad (3.55)$$

where

$$I \equiv R_i/R_j < 1.$$

In the literature for collision models, the results have been presented relative to the simplest case where the smaller particles are swept out of a volume corresponding to a cylinder of diameter equal to that of the collecting body. Thus, interception alone dictates a collision efficiency $\eta = (1 + I)^2$.[†] Correction for the flow of small particles around the collecting body is then the term in large parentheses on the far right in Eq. (3.55).

Verification of the inertial collision theory for falling spheres involves very difficult experiments which require sufficient accumulation of material on the collecting body to be detectable, accounting for complicating factors of Brownian motion and phoretic forces and provision for an approach to fall velocity differences at terminal fall speeds. Given these requirements, few experiments have been attempted. Pruppacher and Klett (1978) have reviewed them. Among the more notable results are those of (a) Kerker and Hampl (1974), who observed the collection of AgI in water droplets for Re $\gtrsim$ 200; and (b) Wang and Pruppacher (1977), who investigated the capture of indium-acetylacetonate by water droplets for Re $\gtrsim$ 300. The Kerker and Hampl study indicated a decrease in collection efficiency with droplet radii 0.094 cm or larger as compared with the theory, taking into account flow around the upstream part of the collecting surface. Wang and Pruppacher found that the collection efficiency increased, following the theory satisfactorily, for droplets up to 300 μm in radius. For larger droplets, the collision efficiency decreased with increasing radius. Wang and Pruppacher (1977) have hypothesized that the observed behavior for large drops found in their work and that of Kerker and Hampl results from the formation of a turbulent wake behind the falling droplets. The eddying motion in the wake evidently decreases the capture rate over that obtained with a model considering only smooth flow around the sphere.

Given the uncertain experimental verification of the inertial capture theory, care should be taken in application of the collision factor estimates using Eq. (3.55) or similar equations in the literature.

Using the theoretical expressions, the relative effects of different collision mechanisms can be estimated as a function of particle size. An illustration is

[†] More precisely, the collision efficiency is defined as the ratio of the actual cross section swept by the collecting sphere (πy_c^2) and the geometrical cross section $\pi(R_i + R_j)^2$; y_c is the offset normal to the line of centers where particles just touch each other in counterflow.

shown in Figure 3.10, for conditions typical of a weakly turbulent medium ($\varepsilon_t = 5$ cm^2/sec^3) and a moderately turbulent medium ($\varepsilon_t = 1000$ cm^2/sec^3). The calculations indicate that collisions by Brownian motion dominate processes below about 1 μm radius, while gravitational sedimentation becomes important at particle radii larger than 1 μm. Turbulent shear becomes a factor for large particle behavior under conditions characterized by dissipation rates in excess of tens of centimeters squared per second cubed. These conditions are not fulfilled except in highly turbulent media, such as in pipe flow, or within a few meters of the ground in the earth's atmosphere. Here

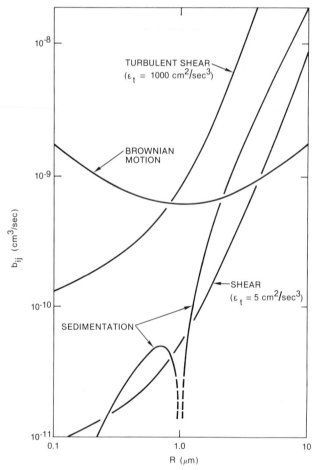

Fig. 3.10. Comparison between collision mechanisms for a particle of 1 μm radius as a function of the radius of the second interacting particle from Eqs. (3.46), (3.53), and (3.55). Spheres are suspended in turbulent media with dissipation rates ε_t of 5 and 1000 cm^2/sec^3. [From Hidy (1973); reprinted with permission of Plenum Press.]

the frictional interactions between air flow and the rough stirred tank reactors or turbulent pipe flow can generate turbulent energy dissipation rates in excess of 10^3 cm^2/sec^3, making turbulent coagulation important over a wide range of particle size.

3.4.4   Aerodynamic and Electrical Forces.   In some circumstances, the interaction between particles in collision may involve aerodynamic forces and electrical forces induced by near-neighbor interactions. These circumstances have been studied mainly in the context of particles larger than 1 μm radius, in the context of exploring droplet dynamics in water clouds in the earth's atmosphere (Pruppacher and Klett, 1978).

Aerodynamic forces arise when particles approach each other in a viscous fluid. They tend to act on both particles such that the collision efficiency is reduced from that expected by interception. Calculations have been made to account for these effects.

Hocking (1959) first attempted to find the influence in the Stokes regime of the falling particles. His analysis sought the critical trajectories for spheres approaching one another initially at their terminal settling velocities. Particles were separated horizontally by distances up to $x/R_j \approx 50$. Critical trajectories were those that just allowed the spheres to touch during their travel past each other. Hocking's calculations were carried out using approximations for the terms accounting for aerodynamic interaction, but excluding electrical forces. Hocking's calculations gave the surprising result that no collision could take place between two different sized spheres less than about 19 μm radius. In addition, the limit that $\eta \to 0$ when $I \to 0$ is not consistent with this theory. In view of these theoretical issues, others repeated the calculations for spheres in Stokes' or Oseen's approach to approximate solution for Re $\to 0$. The results of the recent calculations are illustrated in Fig. 3.11. The models for slip-corrected Stokes interaction corresponding to those of Jonas (1972) are compared with Hocking and Jonas' (1970) continuum Stokes flow theory, and the Klett and Davis (1973) results for a modified Oseen flow approximation. The theories are similar for the case of a 30-μm particle flowing past smaller spheres; the largest differences appear for the case of a 10-μm sphere. The results indicate that for $I \to$ 1, the collision efficiency is independent of the size of the spheres, as shown by the curves approaching the same ordinate on the right in Fig. 3.11. This implies that in this limit the sphere accelerations become infinitesimally small since there are no fluid inertial effects. Within the zero Reynolds number regime, model geometric and dynamic similarity exists as $I \to 1$, and the absolute size of equal spheres does not enter in this extreme. In the absence of other information, the slip-corrected Stokes theory or the Oseen approximation is probably most applicable. Note that Hocking's early result that particles fail to collide for radii less than 19 μm is no longer supported by the newer calculations.

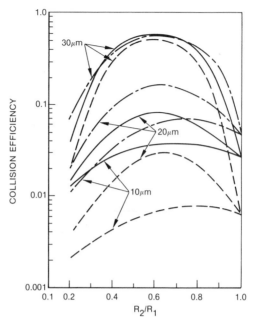

Fig. 3.11. Calculated collision efficiencies for spheres interacting in Stokes flow or in Oseen flow (Re → 0). The larger spheres of different radii are shown by the labeled curves. ——, Jonas (1972); ---, Hocking and Jones (1970); —·—, Klett and Davis (1973), $Re_1 = Re_2 = 0$. [From Pruppacher and Klett (1978); reprinted permission of the authors and Reidel Publishing Co.]

Study of the effect of wake capture and other fluid inertial effects finds the collision efficiency increases throughout the range of I until it approaches unity over the full range of particles if $R_j \gtrsim 50$ μm. There is a tendency for the theoretical collision efficiency to exceed unity for $I \to 1$ with inertial effects in the fluid.

De Almeida (1976) has extended the calculations for collision efficiency to account for turbulence in the fluid. He found that the efficiency of small collector drops increased sharply for dissipation rates between 0 and 1 cm^2/sec^3. However, the effect was not noticeable for spheres ≥ 30 μm radius.

Like the case of inertial capture at high Re, the verification in the viscous flow regime with low Re requires very carefully designed experiments. Some measurements for the collision and coalescence of water droplets in the micrometer size range have been reported for both extreme cases of interception parameter. For small I ratios the theory of Klett and Davis (1973) is in good agreement with theory, but the theory of Schlamp et al. (1976) is somewhat larger than experimental values. For large ratios of I, results of some investigations, Telford et al. (1955) and Abbott (1974), agree well with theory. Others, including the measurements of Woods and Mason (1965) and

Beard and Pruppacher (1968), do not support the theory. The reasons for these differences remain unresolved.

As might be expected, electrical forces acting on interacting spheres can have a substantial influence on theoretical collision efficiencies. Experiments of Telford *et al.* (1955), for example, have indicated that bipolar charging with opposite sign of $\sim$50-μm-radius droplets to 10^{-4} esu above natural ionization levels can increase the coagulation rate substantially, while unipolar charging effectively stopped coagulation. Theoretical arguments of Pauthenier and Cochet (1953), neglecting aerodynamic interaction and droplet inertia, have indicated that coagulation of falling droplets of $\sim$10 μm radius or less may be affected substantially by charging or by electrical induction forces with charges greater than 4×10^{-4} esu on one colliding droplet. More recently, Sartor (1961) has estimated the electrical fields necessary to induce collision of $\sim$20-μm-radius droplets considering an arbitrary initial distance between particles. His calculation made use of modifications in Stokes' equations to include the effect of electrical forces.

Perhaps the most extensive calculations of electrical forces and their influence on collisions are found in the work of Krasnogorskaya (1967) and Schlamp *et al.* (1976). An example of the calculations of the latter authors is shown in Fig. 3.12. These curves are for 19.5-μm spheres colliding with smaller particles in air in a gravity field. The curves shown indicate the increase expected from Stokes' model combined with Davis' (1964) expressions for the electrostatic forces on the spheres. The cases examined are for externally imposed electrical fields oriented parallel to the gravity vector, and charges on particles of the same or opposite sign. The calculations suggest that collision efficiencies for charged small droplets in strong electrical fields such as those found in thunderstorms can be up to one hundred times those without electrical forces. By the same token, weak electrical fields probably have a minor influence on the collision efficiency.

3.4.5 Coagulation in Acoustic Sound Fields.   Sound vibrations can induce coagulation between particles. The acoustic agglomeration mechanisms have been discussed extensively by Mednikov (1965), who postulated that the most important mechanism for particle agglomeration is the so-called *orthokinetic interaction*. This is a simple process in which small particles are collected by large ones because of the relative oscillating motion caused by the imposed acoustic field. This is basically the same process as, for example, the inertial capture of small droplets by large drops if the motion of the oscillation particles is treated as a quasi-steady state motion. For a given sound wave, small particles have little inertia. Thus small particles tend to oscillate together with the gas medium whereas large ones tend to remain stationary in the fluid. The relative motions would ideally lead to the complete capture of all small particles with radius R_j or smaller, located inside the agglomeration volume, by the large particles in one cycle of oscillation.

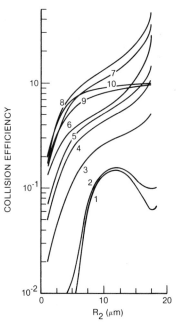

Fig. 3.12. Effect of electric fields and charges on the collision efficiency of a 19.5-μm water drop in air at 800 mbar and 10°C. Curves 1, 3, 4, 5, 7, 8: $\hat{Q}_1 = 0$, $\hat{Q}_2 = 0$, $E = 0$; 500, 1000, 1236, 2504, 3000 V/cm, respectively. Curves 2,6: $E = 0$, $\hat{Q}_1 = +0.2$, $+2.0$ esu/cm^2, $\hat{Q}_2 = -0.2$, -2.0 esu/cm^2, respectively. Curves 9, 10: $\hat{Q}_1 = +2.0$ esu/cm^2, $\hat{Q}_2 = -2.0$ esu/cm^2, $E = 1236$, 2504 V/cm, respectively, where $\hat{Q} = Q/a^2$, $R_1 = 19.5$ μm. [From Schlamp et al. (1976); by courtesy of the American Meteorological Society.]

When it was proposed, many researchers objected to Mednikov's concept of orthokinetic interaction. Among them, St. Clair (1949) especially questioned the theory's validity because he could see no way for additional particles to refill the agglomeration volume once it is swept clean in one cycle. Mednikov and others rebutted this objection by proposing two refill mechanisms—parakinetic and attractional (hydrodynamic) interactions—as the processes primarily responsible for rapid refill of the dust-free agglomeration volume after every oscillating cycle.

In addition, several secondary refill mechanisms were also proposed. These include various types of acoustic drift, acoustic turbulence, and so on. These are the processes that push particles a large distance from the agglomeration volume to the immediate neighborhood of the volume. From there the primary refill mechanisms take over and send the particles into the volume after each oscillation.

The sonic agglomeration constant can be written as

$$b_{ij} = \zeta_a \pi (R_i + R_j)^2 \eta e_{ij} A_p n_i , \tag{3.56}$$

where subscripts i and j correspond to large and small particles, respectively; ζ_a is the refill factor; n_i is the concentration of large particles; and e_{ij} is the relative entrainment given by

$$e_{ij} = f(\tau_i - \tau_j)(1 + f^2\tau_i^2)^{-1/2}(1 + f^2\tau_j^2)^{1/2}, \qquad (3.57)$$

where f is the frequency of the sound waves. A_p in Eq. (3.56) is the amplitude of the particle oscillating velocity,

$$A_p = (2\xi/c\rho_g)^{1/2}, \qquad (3.58)$$

where ξ is the sound intensity and c is the speed of sound. The efficiency η is that for collision of spheres, for example, Fig. 2.3.

Two important conclusions can be drawn from the orthokinetic model. First, based on Eqs. (3.56) and (3.58) one can write

$$b \sim \xi^{1/2}.$$

Thus the acoustic agglomeration constant is directly proportional to the square root of the sound intensity.

In addition, if the orthokinetic interaction is the only mechanism responsible for particle agglomeration, it is easy to determine an optimum frequency, which is defined as the frequency yielding the maximum coagulation. Mathematically, this can be done by taking $\partial e_{ij}/\partial f = 0$, which gives an optimum frequency

$$f_{opt} = \frac{9}{4\pi} \frac{\mu_g}{\rho_p R_i R_j}. \qquad (3.59)$$

Alternatively, Shaw (1978) has suggested that, for unit-density particles in air, the optimum frequency is related to a critical radius R_{crit} such that particles smaller than R_{crit} oscillate, while particles larger than R_{crit} remain stationary. The relation is

$$R_{crit}^2 f_{opt} \simeq 6\text{–}10 \quad \mu m^2 \ kHz.$$

The orthokinetic model for acoustic coagulation has been extended to incorporate aerodynamic forces. Shaw (1978) has summarized the results for estimating the collision taking into account acoustic and gravitational agglomeration. For the case where the ratio of the relaxation time (τ) to the velocity difference is small, the efficiency is

$$\eta = (1 + I)^{-2}\{1 + I - (I^6 + C_1^6)^{-1/16} - [(I - 1)^{1.5} + C_2^{1.5}]^{-2/3}\}$$

$$\times \left[\frac{1.587}{R_{ef}} + \frac{32.73}{R_{ef}^2} + \frac{344}{R_{ef}^2(R_{ef}/20)^{1.56}}\right.$$

$$\left.\exp\left(-\frac{R_{ef} - 10}{15}\right) \sin \pi \left(\frac{R_{ef} - 10}{63}\right)\right]^2, \qquad (3.60)$$

where R_{ef} is the effective radius of an acoustic equivalent particle whose settling velocity is the same as the average velocity of the large particle (j)

$(2\Omega/c\rho_g)^{1/4} R_j$; C_1 and C_2 are constants such that $\eta = 0$ when $I = 0$ and $\eta = 64$ when $I = 1$; they are determined by iteration for each R_i.

The estimate of coagulation rate from the hydrodynamic model has been tested by comparison with experiments. An example reported by Shaw (1978) is shown in Fig. 3.13. Data are shown with the theoretical curves for the acoustic agglomeration constant for three ranges of applied acoustic frequency. For acoustic intensities of 145 dB, the results show the most dramatic increases in coagulation for particles above a few micrometers in radius. However, enhancement of coagulation is also experienced in the submicrometer range. The simple theory is satisfactory for making first esti-

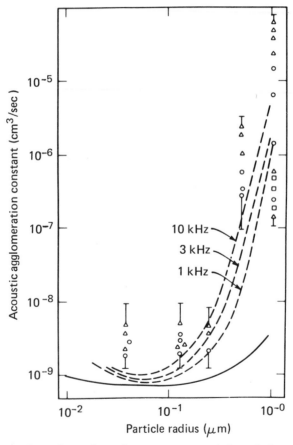

Fig. 3.13. Comparison of experimental measurements and theoretical predictions of the acoustic agglomeration constant. The dashed line is the theoretical prediction of the acoustic agglomeration constant; the solid line is the coagulation constant without acoustic field; experimental data are taken at three frequencies (0–1, △–3, □–10 kHz) and 145 dB intensity. [From Shaw (1978); reprinted courtesy of the author and John Wiley and Sons.]

mates of acoustic agglomeration for design purposes. A further discussion of the theory and application is included in Shaw's (1978) review.

3.5 SOLUTIONS FOR THE GENERAL DYNAMIC EQUATION (GDE)

In general, the combination of processes leading to new particle formation, particle growth, and coagulation, with losses by settling or diffusion, provides a complicated GDE describing the evolution of the aerosol size distribution. Historically, it was not possible to obtain mathematical solutions to the GDE. Therefore, workers concentrated on simplified forms to derive an understanding of the behavior of aerosol clouds. In Chapter 2, certain simple groups of solutions for monodisperse aerosols diffusing in a flowing medium were discussed. In the remaining sections of this chapter, we shall summarize additional theories for describing particle clouds. First, the simplification of a coagulating cloud without a source or a loss of particles is discussed. Then the combination of growth and coagulation is treated. Finally, this case is extended to include a source and loss term as the most general GDE form investigated to date.

3.5.1 Coagulation and Size Distributions.

As in the case of homogeneous nucleation discussed above, analytical solutions for the GDE can be obtained in the case when coagulation completely dominates the particle dynamics. Perhaps best known of these is Smoluchowski's (1916) case investigating the collision of spheres in a continuum. Solving the discrete form of the GDE, Eq. (3.16), for an initially monodisperse aerosol of concentration N_0, one obtains, considering only coagulation,

$$\frac{n_k}{N_0} = \frac{(t/\Im)^{k-1}}{(2 + t/\Im)^{k+1}}, \tag{3.61}$$

where $\Im = 3\pi\mu_g/8kTN_0$ and N_0 is the initial number concentration of particles.

The corresponding result for the change in the total number of particles in the cloud is

$$\frac{N}{N_0} = \frac{1}{1 + t/\Im}.$$

This can be rewritten as

$$\frac{1}{N(t)} - \frac{1}{N_0} = \Im N_0 t. \tag{3.62}$$

In much of the early literature on the coagulation of sols, this form was used to test data and to deduce the coagulation constant for comparison with the theory of Brownian motion. For example, a plot of $1/N$ versus t would yield a slope proportional to the coagulation constant (see also Whytlaw-Gray and Patterson, 1932).

With more complicated forms of the collision parameter, the size distribution will differ, but the qualitative behavior should be similar to that predicted by the initially monodisperse model. That is, the initial concentration will drop roughly exponentially, but polymer concentrations will increase to a maximum and then decrease. Agglomerate maxima will occur sequentially in time according to multiples of the monomer spheres. The behavior of the solutions to the GDE for different single-collision parameters and combinations is discussed at length by Drake (1972).

3.5.2 Self-Preservation.

An interesting class of asymptotic solutions to the GDE for coagulation-dominated cases are those described as "self-preserving." This form is based on inspectional analysis, and was deduced from analyzing the GDE in dimensionless variables. The self-preserving "similarity" transformation was developed theoretically by Friedlander and Wang (1966), and in the computer experiments of Hidy (1965). The similarity transformation for the particle size distribution assumes that the fraction of particles in a given size range is a function only of particle volume normalized by the average particle volume V/N. Thus, the distribution function is written as

$$n(v, t) = (N^2/V)\Psi(\eta_v), \tag{3.63}$$

where $\eta_v = vN/V$. Both N and V are functions of time, but under the limiting condition of coagulation only, the average volume V is nearly constant, and the size distribution for different times has the same shape, differing only by a scale factor. These distributions are termed self-preserving, and are in the form of Eq. (3.63). The self-preserving form for the size distribution of an aerosol coagulating by Brownian motion in a continuum is shown in Fig. 3.14. Two different solutions to the coagulation equation are shown for the self-preserving spectrum. A simple approximation for the upper end of the spectrum is indicated, and two moments of the dimensionless spectrum are included. The approximation for the upper end should hold best where $\Psi(\eta_v)$ is a maximum and at larger η_v. The approximation of $\exp(-\eta_v)$ is the form where the coagulation factor b_{ij} is constant.

In the case of Kn = 0, the collision factor using the Stokes–Cunningham correction does not yield a self-preserving spectrum. The asymptotic distribution function for Kn $\to \infty$ has also been found; the forms for Kn extremes are similar in the two cases, the differences being related to the collision parameter.

The (dimensionless) time to reach self-preserving conditions, at least for the continuum case, was found in the numerical experiments of Hidy (1965):

$$2\pi D_i R_i N_0 t \approx 3.$$

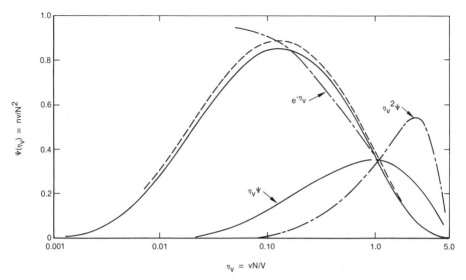

Fig. 3.14. Self-preserving size distribution: ——, numerical solution; - - -, Hidy.

As an example, the case where the initial distribution is monodisperse gives a time of 2×10^7 sec in air for $N_0 = 10^4$ cm^{-3}, and $R_i = 0.1$ μm at $T = 20°$C. This is a very long time for relatively dilute aerosols but could be much shorter for more concentrated particle suspensions.

Experimental evidence for self-preservation in a size distribution was found in the behavior of cigarette smoke during coagulation (Friedlander and Hidy, 1969). The distribution for the free-molecule regime fit experimental results for particles coagulating on a large vessel (Husar and Whitby, 1973). After 600 sec, the size spectrum appeared to reach approximate self-preservation, beginning with $N_0 \approx 3 \times 10^6$ cm^{-3} and (number) mean radius of about 0.08 μm.

Wang and Friedlander (1967) also found that self-preservation could be achieved for a linear combination of coagulation by Brownian collisions and collision by shearing motion, provided that the ratio

$$\mathcal{G} = 3\mu_g \mathcal{G} V / 2\pi k T N$$

remains constant, where $\mathcal{G}$ is the shearing rate in the fluid.

In the case where the size distribution develops in a steady state, such that $\partial n / \partial t = 0$, self-preserving distributions of a form equivalent to

$$n(v, t) = \text{const } Vn^{-3/2} \tag{3.64}$$

should emerge for any arbitrary collision parameter (Liu and Whitby, 1968). Achievement of a steady state, of course, is not possible unless both gain and loss processes are in balance. Since coagulation provides for gain and for loss only in a given size ranges with the overall volume fraction constant, a steady state condition for $n(v, t)$ cannot occur with coagulation alone.

3.5.3 Growth and Coagulation. The next level of complexity in treating aerosol cloud dynamics corresponds to the case of combined growth by vapor condensation and coagulation. In an example of this case with Brownian coagulation, where Kn → 0, Pich $et\ al.$ (1970) found that similarity in shape of size distributions could be preserved if the parameter

$$\chi = (3\mu_g/4kT)(2/V^{2/3}N^{1/3})\mathcal{B}(\mathcal{S} - 1) \tag{3.65}$$

remained constant, where $\mathcal{B} = 3^{1/3}(4\pi)^{2/3}D_{AB}p_0v_m/kT$. This parameter is a measure of the relative rates of condensation and coagulation. For small values of χ, coagulation is rapid compared with the condensation process. An interesting special case emerges from this analysis, corresponding to changes holding the supersaturation ratio constant in time. The theory indicates that the total particle surface area per unit volume should remain constant in such circumstances, owing to a balance between coagulation and condensation. At the same time, the number concentration of particles decreases while the volume fraction increases. Such a situation was observed in experiments measuring the growth and coagulation of small particles generated by exposing filtered air to solar radiation (Husar and Whitby, 1973). The observations of number concentration, surface area, and volume fraction with time are shown in Fig. 3.15. In the latter stages of evolution of the aerosol, the particle surface area per unit volume S approaches a constant while the total number concentration N decreases and the volume fraction V increases. These workers suggested that the production of particles in this case was the result of photochemically generated traces of condensable gas species in the irradiated, filtered air. Husar and Whitby (1973) also found that the size distribution of the particle mixture evolving in this experiment took a self-preserving form corresponding to free-molecule conditions of large Kn.

3.5.4 Generalization to Include Particle Sources and Losses. Recently, Peterson $et\ al.$ (1978) have reported progress in developing analytical solutions for the particle size distribution governed by diffusional growth and coagulation, with a source and loss contribution. These authors have developed a variety of mathematical solutions to the GDE which follow from earlier work reviewed by Hidy and Brock (1970) and Drake (1972). The solutions discussed cover idealized cases of constant coagulation coefficient, or those proportional to the sum of particle volumes. Condensation is idealized as constant growth or proportional to volume. Removal is in a form proportional to member concentration, and the source term proportional to a parameter $v^* \exp(-v/v^*)$, where v^* is a scale for the range of volumes over which nucleation occurs. None of these parameters is fully realistic physi-

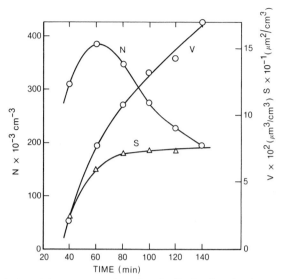

Fig. 3.15. Evolution of the moments of the size distribution function for aerosols generated in an irradiated chamber from reactive gases. The peak in the number concentration N probably results when formation by homogeneous nucleation is balanced by coagulation. Total aerosol volume V increases with time as gas-to-particle conversion takes place. Total surface area S increases at first and then approaches an approximately constant value due, probably, to a balance between growth and coagulation. [Reprinted with permission from Husar and Whitby (1973). Copyright 1973 American Chemical Society.]

cally, but they offer a tractable mathematical means to obtain analytical solutions for the GDE.

The work of Peterson *et al.* (1978) has been complemented by Gelbard and Seinfeld (1978). These workers have reported an efficient method for numerical integration of the GDE for the case of source-reinforced coagulating and condensing aerosols. The method of collocation using two finite-difference schemes is used. The approach may offer opportunities to obtain new insight into the complicated dynamic interactions shaping aerosol size distributions. However, this and other methods have not been pursued to a point where they can be applied in a general, tractable way similar to the results derived from the self-preservation hypothesis.

REFERENCES

Abbott, C. E. (1974). *JGR, J. Geophys. Res.* **79,** 3098.
Beard, K. V., and Pruppacher, H. R. (1968). *JGR, J. Geophys. Res.* **73,** 6407.
Becker, R., and Döring, W. (1935). *Ann. Phys.* **24,** 719.
Brock, J. R., and Hidy, G. M. (1965). *J. Appl. Phys.* **36,** 1857.
Castleman, A. W. (1974). *Space Sci. Rev.* **15,** 147.
Courtney, W. (1961). *J. Chem. Phys.* **35,** 2249.
Davis, M. H. (1964). *Q. J. Mech. Appl. Math.* **17,** 499.

De Almeida, F. C. (1976). *J. Atmos. Sci.* **33,** 571.

Dean, J. A. (ed.) (1973). "Lange's Handbook of Chemistry," p. 458. McGraw-Hill, New York.

Doyle, G. (1961). *J. Chem. Phys.* **35,** 795.

Drake, R. L. (1972). *Int. Rev. Aerosol Phys. Chem.* **2.**

Fletcher, N. H. (1966). "The Physics of Rain Clouds." Cambridge, London.

Flood, H. (1934). *Z. Phys. Chem., Abt. A* **170A,** 286.

Frenkel, J. (1955). "Kinetic Theory of Liquids." Dover, New York.

Friedlander, S. K. (1957). *AIChE J.* **3,** 43.

Friedlander, S. K. (1970). *J. Aerosol Sci.* **1,** 295.

Friedlander, S. K. (1977). "Smoke, Dust and Haze." Wiley, New York.

Friedlander, S. K., and Hidy, G. M. (1969). *Proc. 7th Int. Conf. Condens. Ice Nuclei, Academia, Prague* pp. 21–25.

Friedlander, S. K., and Wang, C. S. (1966). *J. Colloid Interface Sci.* **22,** 126.

Fuchs, N. A. (1959). "Evaporation and Droplet Growth in Gaseous Media." Pergamon, Oxford.

Fuchs, N. A. (1964). "Mechanics of Aerosols." Pergamon, Oxford.

Gelbard, F., and Seinfeld, J. H. (1978). *J. Comput. Phys.* **28,** 357.

Gillespie, T. (1953). *Proc. R. Soc. London, Ser. A* **216A,** 569.

Heist, R. H., and Reiss, H. (1973). *J. Chem. Phys.* **59,** 665.

Hidy, G. M. (1965). *J. Colloid Sci.* **20,** 123.

Hidy, G. M. (1972). *Assess. Airborne Part., Proc. Rochester Int. Conf. Environ. Toxic., 3rd, 1970* p. 81.

Hidy, G. M. (1973). *In* "Chemistry of the Lower Atmosphere" (S. I. Rasool, ed.), Chap. 3. Plenum, New York.

Hidy, G. M., and Brock, J. R. (1970)."The Dynamics of Aerocolloidal Systems." Pergamon, Oxford.

Hill, T. S. (1960). "Introduction to Statistical Thermodynamics." Addison-Wesley, Reading, Massachusetts.

Hocking, L. M. (1959). *Q. J. R. Meteorol. Soc.* **85,** 44.

Hocking, L. M., and Jonas, P. R. (1970). *Q. J. R. Meteorol. Soc.* **96,** 722.

Husar, R. B., and Whitby, K. T. (1973). *Environ. Sci. Technol.* **7,** 241.

Jonas, P. R. (1972). *Q. J. R. Meteorol. Soc.* **98,** 681.

Joutsky, J., Walton, A., and Baer, B. (1965). *Surface Sci.* **3,** 165.

Kassner, J., Jr., and Schmitt, H. R. (1967). *J. Chem. Phys.* **44,** 4166.

Katz, J. (1970). *J. Chem. Phys.* **52,** 4733.

Katz, J., Scoppa, C. J., Kumar, N. G., and Mirabel, P. (1975). *J. Chem. Phys.* **62,** 448.

Kerker, M., and Hampl, V. (1974). *J. Atmos. Sci.* **31,** 1368.

Klett, J. D., and Davis, M. H. (1973). *J. Atmos. Sci.* **30,** 107.

Krasnogorskaya, N. (1967). *Izv. Akad. Nauk SSSR, Atmos. Ocean. Phys.* **1,** 200.

Liu, B. M., and Whitby, K. T. (1968). *J. Colloid Interface Sci.* **26,** 161.

Lothe, J., and Pound, G. (1966). *J. Chem. Phys.* **45,** 630.

Mahata, P. C., and Alofs, D. J. (1975). *J. Atmos. Sci.* **32,** 116.

Mednikov, E. P. (1965). "Acoustic Coagulation and Precipitation of Aerosols." Consultants Bureau, New York (transl. from Russian by C. V. Larrick).

Mercer, T. T. (1978). *In* "Fundamental Aerosol Science" (D. T. Shaw, ed.), p. 85. Wiley, New York.

Miller, R. C., and Kusch, P. (1956). *J. Chem. Phys.* **11,** 860.

Milne, T. A., and Greene, F. T. (1967). *J. Chem. Phys.* **47,** 4095.

Mirabel, P., and Katz, J. L. (1974). *J. Chem. Phys.* **60,** 1138.

Pauthenier, M., and Cochet, R. (1953). *RGE, Rev. Gen. Electr.* **62,** 255.

Peterson, T. W., Gelbard, F., and Seinfeld, J. H. (1978). *J. Colloid. Interface Sci.* **63,** 426.

Pich, J., Friedlander, S. K., and Lai, F. S. (1970). *Aerosol Sci.* **1,** 115.

Pruppacher, H., and Klett, J. D. (1978). "Microphysics of Clouds and Precipitation." Reidel Publ. Co., London.

Reiss, H. (1950). *J. Chem. Phys.* **18**, 840.

Saffman, P., and Turner, J. S. (1956). *J. Fluid Mech.* **1**, 16.

Sartor, J. D. (1961). *J. Geophys. Res.* **66**, 3.

Schlamp, R. J., Grover, S. N., Pruppacher, H. R., and Hamielec, A. E. (1976). *J. Atmos. Sci.* **33**, 1747.

Schwartz, S. E., and Freiberg, J. (1981). *Atmos. Environ.* **15**, 1229.

Shaw, D. (1978). *In* "Recent Developments in Aerosol Science" (D. T. Shaw, ed.), p. 279. Wiley, New York.

Smoluchowski, M. (1916). *Z. Phys.* **17**, 557.

St. Clair, H. W. (1949). *Ind. Eng. Chem.* **41**, 2434.

Stokes, R. H., and Robinson, R. A. (1949). *Ind. Eng. Chem.* **41**, 2013.

Telford, J., Thorndike, N. S., and Bowen, E. G. (1955). *Q. J. R. Meteorol. Soc.* **81**, 241.

Volmer, M. (1939). "Kinetik der Phasenbildung." Steinkopft, Dresden.

Volmer, M., and Flood, H. (1934). *Z. Phys. Chem.* **170A**, 2713.

Wagner, P., and Kerker, M. (1977). *J. Chem. Phys.* **66**, 638.

Wang, C. S., and Friedlander, S. K. (1967). *J. Colloid Interface Sci.* **24**, 170.

Wang, P. K., and Pruppacher, H. R. (1977). *J. Atmos. Sci.* **34**, 1664.

Weinstock, W. (1967). *J. Colloid. Interface Sci.* **36**, 1857.

Whytlaw-Gray, R., and Patterson, H. (1932). "Smoke." Arnold, London.

Wilson, C. T. R. (1900). *Philos. Trans. R. Soc. London, Ser. A* **193A**, 289.

Winkler, P. (1967). Diplom. Thesis, Meteorol. Inst., Univ. of Mainz, Mainz, Germany.

Woods, J., and Mason, B. J. (1965). *Q. J. R. Meteorol. Soc.* **91**, 35.

C H A P T E R 4

GENERATION OF PARTICULATE CLOUDS

A large amount of effort has gone into the investigation and development of aerosol generation devices. Over the years, a wide variety of methods for production of aerosols has emerged; these methods depend on technological requirements for the aerosol. For basic studies, these include (a) control of the particle size distribution, (b) stability of operational performance for key periods of time, and (c) control of volumetric output. The generation devices also have been investigated extensively in themselves to verify the physico-chemical processes of particle formation.

The generation of aerosol requires the production of a colloidal suspension in one of four ways: (a) by condensing out small particles from a super-saturated vapor—the supersaturation may come from either physical or chemical transformation, (b) by direct chemical reaction in a medium such as a flame, (c) by disrupting or breaking up bulk material, or (d) by dispersing fine powders into a gas. In each of these broad groupings, a wide variety of ingenious devices have been designed, some of which employ hybrids of two or more of these groups.

The means for production of particles during condensation is represented well by the generator introduced by LaMer (Sinclair and LaMer, 1949). This device was specifically built to produce a laboratory aerosol with controlled physical properties, using a low-volatility liquid, such as dioctyl phthalate. The device generated particles from a vapor supersaturated by mixing a warm, moist vapor with a cooler gas. Since then, a number of refinements to the Sinclair–LaMer generator have emerged. These are reviewed by Kerker (1975). Many other generators using the condensation process have appeared; some of these achieve vapor supersaturation by adiabatic expansion in the vapor, others by the mixing process. Aerosols also have been formed from condensation of a supersaturated vapor produced by chemical reaction. Some examples include reactions in combustion processes, photochemical processes, and through discharges between volatile electrodes. An example of a hybrid of condensation and breakup or vaporization is the exploding-wire technique described, for example, by Kasiosis and Fish (1962).

The second case, involving molecular aggregation, takes place by direct chemical reactions akin to polymerization. The best-known example of this is the process of carbon particles in a premixed acetylene–oxygen flame (e.g., Homann, 1967). Evidently particle formation in this case does not involve condensation from a supersaturated vapor, but proceeds directly through the pyrolysis of the acetylene, forming in the process unstable polyacetylenes as intermediates in the flame.

In the third case, where disintegration of coarse bulk material into colloids is involved, three main types of devices have been used. The first of these is the air blast or aerodynamic atomizer in which compressed gas ejects at high speed into a liquid stream emerging from a nozzle. This type of breakup is found, for example, in paint spray guns, Venturi atomizers, and other practical sprayers. A second class of atomizer depends on centrifugal action wherein a liquid is fed into the center of a spinning disk, cone, or top and is centrifuged to the outer edge. Provided that the rate of flow of liquid into the spinning device is well controlled, sprays produced in this way are rather uniform in size in contrast to the results of other methods of atomization. A third type of atomizer is the hydraulic design in which a liquid is pumped through a nozzle and, upon its exit from the orifice, breaks up into droplets. Disintegration here depends largely on the physical properties of the liquid and the ejection dynamics at the nozzle orifice rather than on the intense mixing between the liquid and the surrounding gas. Perhaps best known of this class of device is the swirl chamber atomizer which has been used in agricultural equipment, oil-fired furnaces, internal combustion engines, and gas turbines. In addition to these three main classifications, special methods are available including the electrostatic atomizer and the acoustic atomizer. The former makes use of a liquid breakup by the action of electrostatic forces, while the latter applies high-intensity sonic or ultrasonic vibrations to disrupt a liquid.

The fourth group, involving the generation of dust clouds by dispersal of fine powders, is a straightforward method in principle. All such generators depend on blowing apart a bed of finely divided material by aerodynamic forces, or by a combination of air flow and acoustic or electrostatic vibrations. The size of particulate suspensions produced in this way is limited by the minimum size of material ground up by mechanical or other means, and the nature of cohesive and adhesive forces acting between particles in agglomerates. Generally dust clouds produced by powder dispersal will not be less than a few micrometers in radius.

Aerosols composed of solid particles, or nonvolatile liquids with sizes much less than those attainable by atomization of pure liquids or by dispersal of powders, may be produced by atomizing salt solutions. Breakup of suspensions of a volatile carrier liquid in which solid particles, an immiscible liquid, or a nonvolatile solute are suspended will yield very small particles after evaporation of the volatile liquid.

Several examples, particularly of devices for generating aerosols by disintegration or by dispersal, are reviewed through the 1950s by Green and Lane (1964). Recently, Kerker (1975) and Liu (1976) have surveyed many aspects of aerosol generation in current practice.

This chapter begins by extending the discussion of the macroscopic aspects of nucleation to the condensation device. The various means of generating dispersed condensed phase from gases are discussed. Later, the disintegration processes for liquids and powders are described, and several types of generators are discussed which illustrate methods presently used for laboratory experiments and controlled studies of airborne particle behavior.

4.1 PARTICLE FORMATION FROM SUPERSATURATED VAPORS

There are basically two thermodynamic processes by which supersaturation can be achieved in a vapor. The first involves adiabatic expansion of the gas, cooling to temperatures and pressures beyond saturation conditions. The second is mixing of a warm, moist gas with a cool, dry gas.

4.1.1 Adiabatic Expansion. In this process expansion may be carried out in an expansion chamber or in a steady process involving flow in the diverging section of a nozzle, generating supersonic flow. As a good approximation, the temperature and pressure in such systems can be related by the expression for a perfect gas:

$$P_2/P_1 = (T_2/T_1)^{\gamma/\gamma-1}, \tag{4.1}$$

where the subscripts refer to conditions before (1) and after (2) expansion. The ratio of the specific heats at constant pressure and volume is γ. Since $\gamma > 1$ for gases, a reversible adiabatic expansion leads to a decrease in both temperature and pressure of an initially unsaturated gas to the point where considerable supersaturation can be reached. A path of adiabatic expansion is shown in the pressure–temperature diagram, Fig. 4.1. Supersaturation is reached in the final state. Condensation will take place when $S > 1$; the level of supersaturation realized in the device will depend on the presence of nuclei and ions, and the rate of expansion.

Cloud chamber experiments of the type carried out by Wilson at the end of

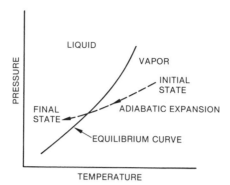

Fig. 4.1. Reversible, adiabatic expansion from an initially unsaturated state carries the vapor across the saturation curve into a region where the stable state is a liquid.

the nineteenth century (summarized in his 1927 Nobel lecture) demonstrate the nature of the condensation process at various saturation ratios with and without foreign nuclei. The air in a chamber is first saturated with water vapor. Upon rapid expansion of the chamber contents, both pressure and temperature fall, carrying the system into a supersaturated state (Fig. 4.1). At first, condensation takes place on small particles initially present in the air. By repeatedly expanding the chamber contents and allowing the drops to settle, the vapor–air mixture can be cleared of these particles.

With a clean system, no aerosol forms unless the expansion exceeds limits corresponding to a saturation ratio of about four. At this critical value, a shower of drops forms and falls. The number of drops in the shower remains about the same no matter how often the expansion process is repeated, indicating that these condensation nuclei are regenerated.

Further experiments show a second critical expansion ratio corresponding to a saturation ratio of about eight. At higher saturation ratios, dense clouds of fine drops form, the number increasing with the supersaturation. The number of drops produced between the two critical values of the saturation ratio is small compared with the number produced above the second limit.

Wilson interpreted these results such that the nuclei that act between the critical saturation limits are air ions normally present in a concentration of about 1000 cm^{-3}. We now know that these result largely from cosmic rays and the decay of radioactive gases emitted by the soil. Wilson supported this interpretation by inducing condensation at saturation ratios between the saturation limits by exposing the chamber to x rays, which produced large numbers of air ions. Wilson proposed that the vapor molecules themselves serve as condensation nuclei when the second limit is exceeded, leading to the formation of very high concentrations of very small particles.

The original experiments were carried out with water vapor. Similar results were found with other condensable vapors, but the value of the critical saturation ratio changed with the nature of the vapor.

Wilson used the droplet tracks generated in the cloud chamber at the lower condensation limit to determine the energy of atomic and subatomic species. Other workers subsequently became interested in the phenomenon occurring at the upper supersaturation limit when the molecules themselves were believed to be serving as condensation nuclei. This is the case corresponding to homogeneous nucleation in a pure vapor, or heteromolecular nucleation in a mixed vapor. Although the expansion chamber has not been used as a practical aerosol generator, it has served an important function in cosmic ray research.

Expansion chambers continue to be used to examine details of the nucleation and cloud forming processes. Perhaps the most elaborate device presently in existence is the one at the University of Missouri, Rolla. This facility is described by Kassner and co-workers (Kassner *et al.*, 1968). There are at least two large chambers in this laboratory which are computer controlled for the expansion process, and are arranged for extensive monitoring of this process. The devices have been used to explore aspects of water and ice nucleation applicable to the initial stages of cloud growth in the atmosphere.

Particle production by adiabatic expansion of a vapor also can be achieved in a steady-flow device employing a diverging nozzle flow. The use of a flow system tends to circumvent some ambiguities of static chamber experiments. In particular, the investigations of homogeneous nucleation have been criticized as nonquantitative because of uncertainties in the transient heat transfer process in the chamber, and the inability to detect embryos until they reach macroscopic size. Kassner and Schmitt (1967) made progress in developing the principles underlying expansion chamber operation as applied to tests of the Becker–Döring theory. The alternative for other workers was to consider steady-flow methods, where transients are avoided.

Essentially two steady-flow condensation techniques using expansion have been studied. These are (a) supersaturation by adiabatic expansion in a converging–diverging nozzle, and (b) supersaturation in a low-density nozzle beam (a molecular beam). Both of these methods offer certain advantages over the chamber experiments. Since the nucleation process is spread out spatially instead of in time, detailed measurements can be made of the "time" history of condensation assuming a steady state condition. This information combined with lengthy computations based on the theoretical behavior of fluids in such devices provides greater insight into the rate processes taking place than in the cloud chamber experiment. Furthermore, supersaturation may be achieved sufficiently rapidly that homogeneous nucleation can occur in both methods despite the presence of significant numbers of foreign nuclei.

The onset of nucleation in a supersonic nozzle is generally detected by a sharp change in light scattering as the gas expands in the diverging section. The condition of nucleation may also be detected by measuring the static pressure along the longitudinal axis of the nozzle. Typically, the static pressure of the gas in the supersonic flow tends to decrease continuously with expansion of the gas. At the location where droplets are observed to first form, a "jump" in static pressure can be detected if the vapor concentration in the gas mixture is high enough. The location where the pressure increase is observed is the same as the zone where droplets are detected optically. Some typical experimental results of Binnie and Woods (1938) for the distribution of the ratio of static pressure to stagnation pressure P_0 in a converging–diverging nozzle are shown in Fig. 4.2. The continuous decrease in pressure for flow of dry air is indicated by the dotted line. The onset of condensation of nitrogen in the air is marked well by the pressure "bump" 10 cm from the inlet shown by the open dots. Because the zone of the onset of nucleation is a sharp front accompanied by a pressure increase, this region has been called a condensation shock.

It is possible to calculate theoretically the static pressure distribution for the supersonic flow of dry gas in the converging–diverging nozzle. Using a heat balance, combined with the rate equation for formation of nuclei, one can estimate the location of the zone of first appreciable nucleation theoretically. Oswatitsch (1942) applied the theory of compressible flow through a nozzle to estimate the change in thermodynamic properties of the gas mix-

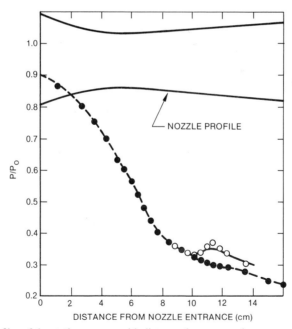

Fig. 4.2. Profiles of the static pressure with distance downstream in a supersonic nozzle flow based on data of Binnie and Woods (1938) compared with calculations of Oswatitsch (1942). [From Dunning (1961).]

ture. He then made calculations of the nucleation rate based on the Becker–Döring (1935) theory and obtained the solid curve shown in Fig. 4.2. The experiments of Binnie and Woods (1938) shown in the figure are believed to be quite accurate, and Oswatitsch considered that the deviations near 11 cm shown between his theoretical calculations exceeded the experimental errors. However, he speculated that perhaps the values of the surface tension used in the calculations were at fault. Despite this possibility, the agreement between the theory and experiment for the converging–diverging nozzle seemed encouraging because of the intricate calculations required for integrating the rate theory with the fluid dynamics.

Further efforts to make use of adiabatic expansion in nozzles have been reported by Wegener and Pouring (1964). They studied the formation of ice crystals by condensation in a nozzle flow. The measurements of static pressure in the nozzle combined with observations of the onset of cloud formation failed to agree quantitatively with the known rate theories based on the steady state model. However, the results could be correlated qualitatively with the theory. Like Oswatitsch, these workers believed that the lack of quantitative agreement may be due to the lack of knowledge about the material properties of small ice crystals at temperatures of 200–220°K.

Supersonic nozzle flows containing particle suspensions have been used recently to investigate certain dynamic features of gas–particle systems other than phase change. (Dahneke, 1978; Schwartz and Andres, 1976). These nozzle flows have certain properties which are analogous to those of molecular beams. Features such as particle size segregation, electrical charging, and gas–particle velocity differentials have been examined, as well as the collision and sticking of particles on flat surfaces.

4.1.2 Supersaturation by Mixing. A second method for producing supersaturation relies on the simultaneous processes of heat and mass transfers. Supersaturation can be achieved either by molecular diffusion mechanisms in a stagnant medium, or by convective diffusion in a flowing medium. The former is the basic principle underlying the diffusion cloud chamber (Reiss and Katz, 1967), which is an analogy to the static expansion chamber. Diffusion processes also have been studied in turbulent jets and for fogs developing when cool air flows over warm, moist surfaces (Amelin, 1948).

Condensation can result when a hot gas carrying a condensable vapor is mixed with a cool gas. This process can occur in stack gases as they mix with ambient air or with exhaled air that is saturated at body temperature when it comes from the lungs. As mixing with ambient air takes place, the temperature drops, favoring condensation, but dilution tends to discourage condensation. Whether saturation conditions are reached depends on the relative rates of cooling and dilution during the mixing process. In the absence of condensation, the concentration distribution in the fluid is determined by the equation of convective diffusion for a binary gas mixture:

$$\rho_g \frac{\partial y}{\partial t} + \rho_g \mathbf{q} \cdot \nabla y = \nabla \cdot \rho_g D_{AB} \nabla y, \qquad (4.2)$$

where ρ_g is the mass density of the fluid, y is the mass fraction of the diffusing species, and D_{AB} (D) is the vapor diffusion coefficient. The temperature distribution is determined by the energy equation:

$$\rho_g c_p \frac{\partial T}{\partial t} + \rho_g c_p \mathbf{q} \cdot \nabla T = \nabla \cdot k_g \nabla T, \qquad (4.3)$$

where c_p is the specific heat at constant pressure and k_g is the thermal conductivity.

A system can be visualized with a hot, moist gas mixing with gas at a lower temperature. Examples include hot air jetting into a cool gas. Another would consist of warm, moist air passing over a cold ground or water surface. The boundary conditions for the jet geometry, for example, can be written
At the jet orifice:

$$y = y_0, \qquad T = T_0;$$

In the ambient air:

$$y = y_\infty, \qquad T = T_\infty.$$

When c_p is constant, the equations for the concentration and temperature fields and the boundary conditions are satisfied by the relation

$$(y - y_\infty)/(y_0 - y_\infty) = (T - T_\infty)/(T_0 - T_\infty), \qquad (4.4)$$

provided $k_g/c_p = \rho_g D$. The dimensionless group $k_g/\rho_g c_p D$, known as the Lewis number, is usually of order unity for gas mixtures.

The $k_g/\rho_g c_p D = 1$, the relation between concentration and temperature, Eq. (4.4), is independent of the laminar or turbulent nature of the flow. It applies both to the instantaneous and time-average concentration and temperature fields, but only in regions in which condensation has not yet occurred.

According to Eq. (4.4), the path of the condensing system on a mass fraction versus temperature diagram is a straight line determined by the conditions at the orifice and in the ambient atmosphere. For $y \ll 1$, the partial pressure is approximately proportional to the mass fraction. The path is shown in Fig. 4.3 with the vapor pressure curve. From this relationship, it is possible to place limits on the concentrations and temperatures that must exist at the jet orifice for condensation to occur. The mixing line must be at least tangent to the vapor pressure curve. For $k_g/\rho_g c_p D = 1$, the relationship between y and T will in general depend on the flow field.

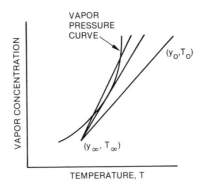

Fig. 4.3. Air–vapor mixtures under three different source conditions (y_0, T_0) mixing with air of the same ambient conditions (y_∞, T_∞). No condensation occurs for the source on the right, while condensation can occur, depending on availability of nuclei and mixing rates, for the one on the left. The middle line shows a limiting situation where the operating line for mixing just touches the equilibrium vapor pressure curve.

   An experimental study of the behavior of a condensing jet has been carried out by Hidy and Friedlander (1964). These investigators passed a collimated light beam through the center line of a condensing jet to reveal the internal structure of the fog. The cross section of the fog-filled region was roughly that of a wedge with its tip located vertically downstream from the edge of the nozzle. The region where condensation first appeared fell within the turbulent mixing zone between the emerging jet and the ambient air. No fog was observed in the mixing zone near the edge of the nozzle, and it was concluded that this region was supersaturated. The extent of the fog-free region could be increased by decreasing the concentration or velocity, or by increasing the temperature. Like the condensation shock observed in supersonic nozzle flow, a similar nucleation front is observed at the zone of high apparent supersaturation in a turbulent free jet. If one knows the temperature and vapor distribution in the jet one can, of course, calculate the location of the zone of "critical" supersaturation based on the theory of homogeneous nucleation. Amelin (1948) first attempted this, and found that the critical supersaturation calculated for a free jet based on the Becker–Döring theory agreed qualitatively, at least, with the region of first appreciable nucleation determined optically in a free jet. Higuchi and O'Konski (1960) made use of such a jet to test the Becker–Döring theory of nucleation in materials of rather low volatility. In an analogous manner to the method of Oswatitsch, Higuchi and O'Konski applied the theory of jet mixing to estimate the average distributions of vapor pressure and temperature in the jet. Using an integral technique for estimating the overall rate of nucleation by the theory, the experimental measurements of the rate of droplet formation could be compared by changes in light scattering. Since the surface free energy of small particles is of key interest in the theory, Higuchi and

O'Konski chose to compare the bulk values of the surface tension with those estimated from the apparent values of $\Delta\hat{G}^*$ from the experiments. Their results are shown in Table 4.1. Accepting the validity of the jet technique, the agreement between theory and experiment is qualitatively satisfactory except for the case of triethylene glycol. The difference in this case was interpreted as the result of possible effects of hydrogen bonding on the behavior of a very small cluster containing about 30 molecules.

The study of nucleation in a turbulent jet by Hidy and Friedlander (1964) suggested that the method used by Higuchi and O'Konski may not be as quantitative as had been hoped. The influence of intense fluctuations in vapor properties induced in a turbulent medium are difficult to interpret by mean values alone. At any instant spatial and temporal variations in temperature and vapor pressure may cause supersaturation ratios to deviate substantially from those predicted by a mixing theory like the one used by Higuchi and O'Konski.

The mixing principle has been used in a well-controlled series of experiments using the diffusion cloud chamber. This device is a particularly attractive experimental system for the study of nucleation kinetics; it is compact and produces a well-defined, steady supersaturation field. The chamber is cylindrical in shape, perhaps 30 cm in diameter and 4 cm high. A heated pool of liquid at the bottom of the chamber evaporates into a stationary carrier gas, usually hydrogen or helium. The vapor diffuses to the top of the chamber, which cools, condenses, and drains back into the pool at the bottom. Since the vapor is denser than the carrier gas, the density is greatest at the bottom of the chamber, and the system is stable with respect to convection. Both diffusion and heat transfer are one dimensional, with transport occurring from the bottom to the top of the chamber. At some position in the chamber, the temperature and vapor concentrations reach levels corresponding to supersaturation. The variation in the properties of the system are calculated by a computer solution of the one-dimensional equations for

TABLE 4.1

Comparison of Results of Higuchi and O'Konski (1960) with the Becker–Döring (B–D) Theory for the Surface Tension of Embryos of Critical Size

Aerosol	Temperature (°K)	Super-saturation	Surface tension literature (dyn cm⁻¹)	Surface tension estimated from B–D theory	Number of molecules in cluster of critical size
Dibutyl phtalate	332	70	29.4	28.8	21
Triethylene glycol	324	25	42.8	36.8	29
n-octadecane	319	63	26.1	24	25
Sulfur	325	420	67.6	54	14

heat conduction and mass diffusion [the one-dimensional forms of Eqs. (4.2) and (4.3)]. The supersaturation ratio is calculated from the computed local partial pressure and vapor pressure.

The goal of an experiment is to set up a "critical" chamber state; that is, a state which just produces nucleation at some height in the chamber where the vapor is critically supersaturated and droplets are visible. This occurs when the temperature difference across the chamber has been increased to the point where a rain of drops forms at an approximately constant height. Drop formation in this way must be distinguished from condensation on ions generated by cosmic rays passing through the chamber. An electrical field is applied to sweep out such ions, which appear as a trail of drops.

For each critical chamber state, the distribution of the saturation ratio and temperature can be calculated as shown in Fig. 4.4. The set of curves for the critical chamber states based on measurements with a condensable vapor are deduced. The experimental saturation ratio passes through a maximum with respect to temperature in the chamber. Condensation occurs not at the peak supersaturation but at a value on the high temperature side because the critical supersaturation decreases with increasing temperature. Hence the family of experimental curves should be tangent to the theoretical curve. The results obtained are quantitatively consistent with elements of the homogeneous nucleation theory for nonpolar, organic condensates studied so far (e.g., Katz *et al.*, 1975). However, agreement between theory and experiment is less satisfactory for water (Heist and Reiss, 1973).

4.1.3   Vapor Condensation and Monodisperse Aerosols.

As a practical measure, hot jet mixing offers a simple and widely used method for generating aerosol clouds of thermally stable, low-volatility materials. To produce a particle suspension of nearly uniform size (monodisperse clouds), the condensation process has to be controlled carefully. In current practice, monodisperse aerosols are almost always used to calibrate the particle characterization instruments. They are important for performance studies of gas-cleaning devices and in investigations of fundamental aerosol behavior.

The generation of monodisperse aerosols with particles of similar size and

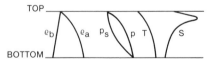

Fig. 4.4. Variation with height of the properties of a mixture in the diffusion cloud chamber. Shown are the mass densities of the carrier gas, ρ_b, and the vapor, ρ_a; the equilibrium vapor pressure p_s; the partial pressure of the vapor p; the temperature T; and the saturation ratio $\mathcal{S}$. The highest temperature, vapor pressure, and gas density are at the chamber bottom, above the heated pool. The distributions with respect to chamber height are calculated by integrating expressions for the steady state fluxes of heat and mass through the chamber.

physicochemical characteristics is desirable in many experimental applications since the effects of particle size can be evaluated. Many of the methods for producing monodisperse aerosols have been reviewed, for example, by Raabe (1970), Mercer (1973), and Kerker (1975). These methods include the growth of uniform aerosol particles or droplets by controlled condensation; or the electrostatic formation of uniform droplets (which may dry to solid particles) by controlled dispersion of liquids, by dispersion of liquid jets with periodic vibration, or with a spinning disk or top. Another popular method for producing monodisperse aerosols is the nebulization of a suspension of monodisperse particles, but this involves some inherent difficulties. It is not a simple matter to produce suitable monodisperse aerosols for specific applications. The variety of types of such aerosols is still somewhat limited and the specialized equipment usually requires careful design and operation.

The definition of monodispersity suggested by Fuchs and Sutugin (1966) provides a guide for evaluation of monodispersity of particle size distributions. If the coefficient of variation (ratio of the standard deviation to the mean) of the distribution of sizes is less than 0.2 (20%), the aerosol may be satisfactorily described as having "practical monodispersity." For a lognormal distribution, this is basically equivalent to a geometric standard deviation $\sigma_g \lesssim 1.2$. In general, this criterion is not very stringent, and investigators may prefer to achieve greater size uniformity. However, even with very effective devices for producing particles or droplets of uniform size, it is sometimes necessary to tolerate a small fraction of odd size particles or doublets (two primary particles which have coalesced).

It has long been possible to generate aerosol particles of fairly uniform size using biologically produced particles such as spores, pollen, bacteria, viruses, and bacteriophages. Interest in the characteristics of screening smokes and in the design of gas mask filters during World War I led to the development of condensation aerosol generators; these were easier to set up and operate in nonbiological laboratories. By seeding a condensable vapor with nuclei and then allowing condensation to take place under carefully controlled conditions, aerosols of nearly uniform size were produced. The method is of great practical importance in producing test aerosols from a variety of liquids and from solid materials, such as salts, as well.

Monodisperse aerosols were first prepared by condensation by Sinclair and LaMer (1949) by carefully regulated vapor condensation on suitable nuclei. Basically, the method involves the introduction of dry gas laden with nuclei into a moist, warm, nuclei-free residual vapor. This mixture is passed into a reheater where any residual condensed material is vaporized and well mixed as a nuclei-containing vapor. The mixture passes up a long uniformly cooled chimney where the vapor cools uniformly and the vapor condenses on the nuclei, forming an approximately monodisperse, finely divided aerosol containing submicrometer particles.

A modern version of the Sinclair–LaMer generator has been described by

Huang *et al.* (1970). Their design is shown in Figure 4.5a; this generator has improved reproducibility and stability for particle production compared with earlier versions. The device, consisting of a 500-ml three-necked flask (10) containing about 10 ml of aerosol material, was immersed in an oil bath (6) maintained at constant temperature by an immersion heater (7) controlled by a thermoregulator (8). Temperature control to within ±0.3°C was achieved by inserting into the oil bath a second immersion heater which, with proper setting of a Variac resistor, maintained the bath a few degrees below the desired temperature; final temperature control was then maintained by the first heater. Typical boiler temperature ranges were 55–70°C for octanoic acid up to as high as 115–143°C for linolenic acid particle generation. Condensation nuclei are generated by heating the nucleating material, usually AgCl or NaCl, to an appropriate temperature. This is contained in a glazed porcelain combustion boat which was placed in a combustion tube which, in turn, is inserted into an electric furnace. The furnace temperature, which ranged from 400 to 700°C, is controlled by a Variac resistor. A regulator was employed in order to furnish a steady voltage. With this, temperature control of ±4°C can be achieved.

The carrier gas, helium, issuing from a tank (1), passes through a flowmeter (3) and a fritted glass filter (4) and then through the furnace (5), whence it carried the nuclei into the aerosol boiler (10). The flow rate ranges typically from 0.5 to 2.5 l/min. As the mixture of nuclei and vapor passes up the cooling chimney (11), the vapor condenses on the nuclei, forming aerosol particles. The aerosol then flows through a glass tube, a portion of which is maintained at an elevated temperature with an electric heating tape (13), and through another chimney before it moves into a viewing chamber or a light-scattering cell.

Huang *et al.* (1970) note that the incorporation of a reheating unit (13), and thereafter a cooling chimney, greatly increases the monodispersity of the aerosols. Particle size increases with increasing boiler temperature, with decreasing furnace temperature, and with decreasing flow rate, as would be anticipated if the size were determined by the ratio of condensable vapor to number of nuclei. Generators of this type can produce particles in the 0.3–1.0 μm radius range with $\sigma_g \lesssim 1.15$.

A second type of condensation generator has been designed by Rapaport and Weinstock (1955). This device utilizes a nebulizer for formation of a primary polydisperse aerosol which is subsequently evaporated on passing through a heated tube. It is thought that the residual nonvolatile particles serve as nuclei so that a homogeneous aerosol is produced upon recondensation of the vapor. This procedure eliminates the need for prolonged heating of the aerosol material with the attendant possibility of decomposition and also produces greater quantities of aerosol.

Liu *et al.* (1966) introduced a number of refinements to the Rapaport–Weinstock generator. They assumed that the condensable liquid normally

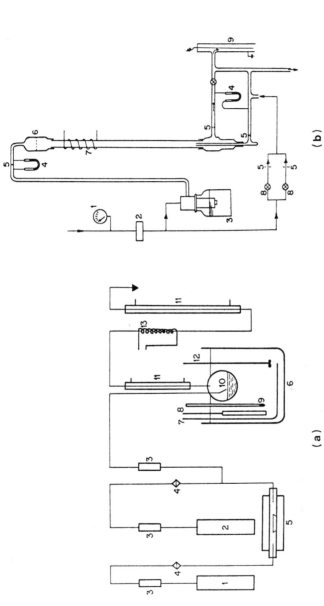

(a)

(b)

Fig. 4.5. (a) Modified Sinclair–LaMer condensation aerosol generator: (1) carrier gas for nuclei, (2) carrier gas for diluting nuclei, (3) flow meter, (4) fritted glass filter, (5) tube furnace with combustion boat containing nucleating material, (6) oil bath, (7) heater, (8) thermal regulator, (9) thermometer, (10) boiler, (11) cooling chimney, (12) stirrer, (13) reheater. [After Huang *et al.* (1970); courtesy of M. Kerker.] (b) Liu *et al.* (1966) condensation aerosol generator: (1) pressure gauge, (2) absolute filter, (3) collision spray generator, (4) flow meter, (5) orifice, (6) mesh screen, (7) heating tape, (8) valve, (9) electrostatic precipitator. (After M. Kerker; with permission of Elsevier.)

contains sufficient nonvolatile impurities to form the residual particles that serve as condensation nuclei, and so they did not add anything to this material. The particle size is controlled by using a solution of low-boiling-point liquid dioctyl phthalate in a volatile solvent, e.g., ethyl alcohol. This type of generator is shown schematically in Fig. 4.5.

The Liu generator uses a type of atomizer called a nebulizer (3) to generate small droplets of the condensable fluids. Filtered air is bubbled through the liquid to be nebulized. The droplets of spray move into an evaporation section, where they are completely vaporized (5)–(7) (there is usually a tiny residual of particles that serve as nuclei in the vapor so as not to require their addition). There the vapor is cooled uniformly in the final stage to condense vapor on the nuclei, providing a steady supply of approximately monodisperse aerosol of submicrometer size. The most uniform part of the aerosol is that flowing in the center of the exit tube, where the radial temperature profile is flat. It is this portion of the stream that is sampled to provide an almost monodisperse aerosol. By diluting the aerosol material in the nebulizer with ethyl alcohol, Liu *et al.* (1966) were able to generate aerosols with diameters ranging from about 0.036 to 1.1 μm. The larger particles are more uniform than the smaller, the values of σ_g increasing from 1.22 at 0.6 μm to 1.50 at 0.036 μm.

4.1.4   Role of Gas-Phase Chemical Reactions.   Supersaturation of a condensable species can also result from chemical reactions in a gas to form a product of low vapor pressure. There are several reactions which can produce particles. Perhaps best known is the reaction

$$NH_3 + HCl \rightarrow NH_4Cl,$$

which produces a finely divided smoke. Other gas-phase reactions include the production of sulfuric acid from sulfur dioxide oxidation in air, and production of organic particles from olefinic hydrocarbons in air. Finely divided particles of metal oxides are also generated by a variety of chemical reactions in plasma jets, in flame processes, and in electric arcs. The production of sulfuric acid particles and organic particles in air has important application to haze in the earth's atmosphere, and consequently will be discussed mainly in Section 7.3. Likewise the generation of metal oxide smokes has important industrial application; these methods are covered further with fine-particle technology in Section 6.7. Before we pass on to particle production disruption and breakup, a few examples of laboratory studies for particle generation by chemical reactions will serve to introduce the potential for this approach.

In the photochemical condensation method, a substance suitably illuminated yields an aerosol product of very low vapor pressure. Observations on photochemically produced aerosols were made many years ago by Tyndall (1869). He directed a beam of light through a tube containing a mixture of air, butyl nitrate vapor, and hydrochloric acid gas, and observed that the light induced a chemical reaction resulting in the formation of an aerosol; as the particles slowly increased in size they gave rise to a striking phenomenon of light scattering. The well-known *Tyndall beam* originated from this work.

An elegant method of generating smokes of ferric oxide by photochemical decomposition of dilute iron carbonyl vapor in air has been developed by Jander and Winkel (1933). On irradiation with ultraviolet light, iron carbonyl vapor combines with atmospheric oxygen according to the equation

$$4Fe(CO)_5 + 13O_2 = 2Fe_2O_3 + 20CO_2.$$

Using the simple procedure described by Jander and Winkel, ferric oxide smokes of concentration 10–100 mg/m^3 are easily generated in a flask.

Dimethyl mercury is another compound which, on irradiation with ultraviolet light, is dispersed as an aerosol of very fine particles. This substance has a vapor pressure at room temperature high enough to provide a sufficient concentration of vapor, while its stability in the presence of oxygen, moisture, and ordinary light makes it easy to use. Because of these properties any residual undispersed dimethyl mercury remaining after the irradiation in no way affects the aerosol. Varying the initial concentration of mercury dimethyl and time of irradiation, Harms and Jander (1936) have shown that aerosols of predictable and stable number concentration and particle size can be reproduced consistently.

There may be conditions in which polymeric particles are formed without going through condensation from a supersaturated vapor. One case may be nuclei formation from irradiated organic mixtures such as those used by Wen *et al.* (1978).

The oxidation of sulfur dioxide can occur by a variety of reactions, which range from its photolysis in air to interactions with photochemically derived intermediates including free-radical species such as the hydroxyl radical (OH), and the alkoxy radical (RO). Once an oxidized sulfur dioxide intermediate is formed, it will react rapidly with water vapor present to produce sulfuric acid vapor (Calvert *et al.*, 1978; see also Section 7.3). Sulfuric acid has a low vapor pressure, and in the presence of water vapor can undergo heteromolecular nucleation at very low vapor activities, as we saw in Chapter 3. Sulfuric acid aerosol production via chemical reactor can be achieved readily by SO$_2$ oxidation mechanisms. Several experiments of this type have been reported, including those reviewed by Middleton and Kiang (1978). Examples of the methods using photochemical oxidation reactions are summarized in Table 4.2. Each produced varying quantities of sulfuric acid particles from reactions of SO$_2$ in the parts per million concentration range.

Given the sulfuric acid–water vapor system, Boulard *et al.* (1978) have reported laboratory experiments testing the estimates of the theory for heteromolecular nucleation to estimate particle production rates. An example of their results is shown in Fig. 4.6. As expected from the theory, the

TABLE 4.2

Experimental Conditions for SO_2 Oxidation by Photochemical Processes[a]

Investigators	RH (%)	Wavelength (Å)	SO_2 (ppm)	Other gases (1 atm)	Temperature (°C)	Experimental system	Irradiation time (sec)
Quon et al. (1971)	13–77	3000–4000	0.2–0.65	Filtered air	23	Diffusion system	0–240
Cox (1973)	1–80	2900–4000	5–500	$N_2 : O_2$ (4:1)	35	Flow system residence time ≈ 150 sec	Constant
Friend et al. (1973)	<20	2200–5000	0.1–1.0	Filtered air	−55, ~25	Flow system residence time ≈ 2100 sec	Constant
Boulard et al. (1975)	0–90	2900–4000	0.2–0.3	Filtered air	~25	Diffusion system	Constant

[a] From Middleton and Kiang, (1978); courtesy of Pergamon Press.

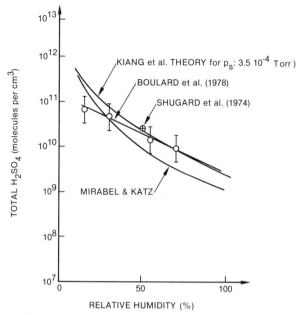

Fig. 4.6. Comparison of the experimental data of Boulard *et al.* (1978) with theories of Kiang *et al.* (1973) and Mirabel and Katz (1974). The Kiang *et al.* model uses a saturation vapor pressure for sulfuric acid of 3.5×10^{-4} Torr; the theory depends critically on the vapor pressure data used for the calculation. (Reprinted with permission of John Wiley & Sons.)

experiments indicate that the number density of molecules of H_2SO_4 required for nucleation (rate of 1 cm^{-3}/sec^1) decreases with increasing relative humidity. For sulfuric acid vapor pressure estimates found in the literature (Gmitro and Vermeulen, 1964), the theory is consistent with the reported experiments.

The presence of hydrocarbon vapor and nitrogen oxides with SO_2 in irradiated air mixtures can show distinctly different patterns of particle production. An example from Kocmond *et al.* (1977) is shown in Fig. 4.7. These experiments were conducted in a large irradiation chamber containing air and traces of hydrocarbons, nitric oxide, and sulfur dioxide at parts per million concentrations. The production of particulate matter during irradiation of these mixtures with ultraviolet and visible light over a period of several hours is measured by a mean surface diameter of particles. The larger the mean diameter the more condensate produced. The interaction of hydrocarbons (HCs) and NO produces NO_2, O_3, and other intermediates in photochemical chain reactions. The experiments show that of the cyclic olefins, cyclohexene is the strongest particle producer, followed by *m*-xylene and hexene in the system. Interestingly, the introduction of SO_2 actually decreases the production of mean surface diameter over the

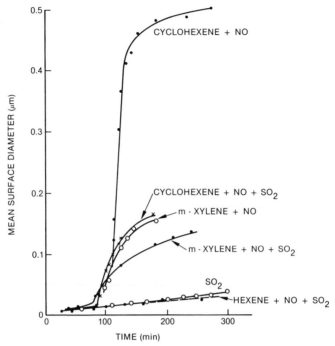

Fig. 4.7. Mean surface diameter versus time for several HC + NO, HC + SO$_2$, and SO$_2$ experiments in air. [From Kocmond *et al.* (1977), with permission of John Wiley & Sons and the authors.]

HC–NO systems. The reason for this may be related to nucleation of finely divided sulfuric acid particles reducing the mean diameter.

The smog chamber results of Kocmond *et al.* contrast with those of Lipeles *et al.*(1977).These experiments were conducted in an irradiated tubular flow reactor with residence times of less than a minute compared with several hours in the smog chamber. Some results for dry HC–NO$_2$ mixtures in filtered air with added ammonia and sulfur dioxide are listed in Table 4.3. The results readily illustrate that the irradiated olefin nitrogen dioxide mixtures are prolific particle producers even at low reactant concentrations and (constant) reaction times of a minute or less. Dodecene and hexadiene produce particle concentrations similar to that of hexene, but the latter requires much higher initial concentrations. The carbon-ringed terpene compound α-pinene is also a prolific particle producer. Although the addition of SO$_2$ to the 1-dodecene system does not markedly change the particle production rate, the addition of ammonia has a suppressing effect in the case of α-pinene. These results illustrate the complexity of the chemical process taking place in such systems; the chemical processes are poorly understood at present.

TABLE 4.3

Filter Collections of Aerosol Samples at a Constant Distance Down a Tubular Reactor[a,b]

Hydrocarbon	Initial hydrocarbon concentration (ppm)	Nuclei concentration (per cm³)	Weight gain (μg)	Mass concentration (μg/m³)
1-hexene	50	4×10^4	30	10.4
	50	4×10^4	54	18.8
1-hexene + NH$_3$	50	5×10^4	45	16.6
	50	5×10^4	65	22.6
1-dodecene	15	10^5	63	21.9
	15	10^5	74	25.7
	15	10^5	91	31.6
	15	1.8×10^5	140	48.6
	15	3×10^5	41	14.2
1-dodecene + SO$_2$	12	10^6	91	31.6
1-dodecene + SO$_2$ + NH$_3$	12	10^7	2	—
	12	10^7	0	—
1,5-hexadiene	70	3×10^4	38	13.2
1,3-hexadiene	15	2×10^5	55	19.1
	15	2×10^5	53	18.4
α-pinene	20	7×10^5	216	75.0

[a] 16-h collections at 3 l/min; for all cases [NO$_2$] = 0.8 ppm; where applicable [SO$_2$] = 5.7 ppb; [NH$_3$] = 2.0 ppm.

[b] From Lipeles *et al.* (1977); courtesy of John Wiley & Sons.

From the data in Fig. 4.7, we can readily see that aerosol particle mixtures produced in the irradiated HC–NO$_x$–SO$_2$ systems are very finely divided material less than 0.5 μm in diameter, as an upper limit. Aerosol particle formation can be expected to be a by-product of these kinds of gaseous chemical reaction systems. As will be seen later in Section 7.3, they are an important factor in atmospheric aerosols.

Aerosol production by condensation can be achieved effectively in combustion media and plasma discharges. When smokes are generated by combustion, the basic mechanism operating is again usually that of vapor condensation, though because of the high temperatures involved and the rapidity of the reactions taking place, it is impossible to analyze the process in any detail. In the reaction zone, where the molecules which condense to form the smoke first appear, they collide with one another to form nuclei, which then grow by accretion of additional molecules to form primary particles. Two types of primary particles can be distinguished: (a) crystalline primary particles, which are formed when the ratio of activation energy to absolute temperature is low enough to permit rearrangement of the molecules into a crystalline form; and (b) amorphous primary particles, which are formed in cases where this ratio is too high to permit rearrangement into a crystalline form. When magnesium ribbon is burned in air, the primary particles are nearly perfectly cubic crystals of magnesium oxide which are too small to be resolved by optical microscope but are readily seen with the aid of an electron microscope. The primary particles of carbon smoke, on the other hand, are revealed by the electron microscope as amorphous particles which are approximately spherical in shape. In either case, after the particles leave the reaction zone, coagulation sets in and aggregates are formed. When carbon smoke is produced by incomplete combustion the small particles tend to form long chains or filaments.

The burning of fuels results in the creation of complex aerosols some of which are of considerable theoretical interest and practical importance. In the relatively simple case of the combustion of gases, a particulate cloud may be formed in the absence of an adequate excess of air, resulting in a smoky flame. The tendency of organic substances to smoke on burning freely in air is well known. Parker and Wolfhard (1950) have suggested that carbon formation in flames may occur as a result of high-molecular-weight hydrocarbons formed by pyrolysis. The molecular weight and concentration of these high-molecular-weight species increase until the saturation vapor pressure is exceeded, whereupon condensation occurs and fine droplets are formed. The droplets contain nuclei of graphite, and graphite crystallites grow until each droplet is converted into carbon and hydrogen is largely eliminated. Later work on acetylene–oxygen flames by Homann (1967) suggests that the carbon particles are formed directly as polymers of acetylenes rather than requiring supersaturation (see also Section 6.5).

Combustion in a carrier gas is occasionally employed as a method of aerosol generation in the laboratory; highly dispersed ferric oxide aerosols, for example, are produced by burning, in air, a stream of carbon monoxide laden with iron pentacarbonyl vapor (Schweckendiek, 1950).

When an electric arc is struck between electrodes of a suitable metal, in a current of air, metal vapor evolved at a very high temperature is cooled in the air stream and condenses to form a smoke. Oxidizable metals—for example, cadmium, lead, copper, manganese, chromium, magnesium, and aluminum—readily give oxides, while platinum, silver, and gold yield metal smokes. The smokes from copper and iron consist of mixed oxides. In this method of generating smokes, nucleation of the vapor is facilitated by the copious supply of ions existing in the neighborhood of the arc, and it is probable that most of the original embryos are formed by condensation on the ions rather than by self-nucleation. Such nuclei will, of course, be electrically charged, and since the initial concentration is high and is reduced comparatively slowly by the diluting air, we should expect that aggregates of a wide range of complexity, electrical charge, and size would be present in arc smokes. Microscopic examination shows this to be the case, the complexes consisting of long chains, frequently of hundreds of fine particles, most of which are too small to be resolved by the optical microscope, so that electron microscopy must be used to observe their structure and form. Electron diffraction shows the composition and nature of the surface regions of these fine particles more sensitively than x-ray diffraction.

An unusual example of the extremely small particles of rather uniform size that can be produced in a plasma ablation reaction has been reported by Boffa *et al.* (1976). Basically, the plasma jet is used as an external heat source to ablate particulate material from a porous matrix. The temperature of the ablated surface is controlled by a gas flow transpiring through the matrix which quenches and dilutes the particles formed. Particles can be generated from porous material of sufficiently fine structure and uniform porosity. Examples include graphite, tungsten, and ceramic materials. Variables ranging from 0.02 to 20 μm in diameter can be produced, with logarithmic or geometric standard deviation of $1.13 < \sigma_g < 1.32$. Sample size distribution for a carbon particle aerosol from Boffa's experiment is shown in Fig. 4.8. The data were obtained for the number, surface, and volume distribution with an electrical mobility analyzer, described later in Section 5.2. The result is a very finely divided aerosol which is quite monodisperse.

4.2   DISINTEGRATION OF LIQUIDS

To subdivide bulk liquids into particles of colloidal dimensions, sufficient energy has to be added to the bulk material to overcome the intermolecular forces holding the material together. When a liquid is atomized, the energy expended in atomization is used for (a) forming new surface, (b) overcoming viscous forces in changing the liquid shape, and (c) balancing losses resulting from inefficient input of energy into the liquid. The energy required to form a new surface upon dividing the bulk phase into droplets of radius R_1 is $3\sigma/R_1\rho_1$ per unit mass. For water droplets of about 1 μm radius this is approximately 0.12 J/g, or about 0.05 cal/g. In addition to this amount there has to be included a small amount of energy used in creating any surface formed during atomization which exceeds that of the final surface. Since the time of droplet formation generally has to be rather short, the rate of deformation of liquid has to be large, and the amount of energy dissipated by the action of viscosity during deformation will be larger.

If a highly idealized model is assumed for droplet formation by atomization, it is possible to estimate the minimum energy required to reshape the liquid. Suppose we adopt Monk's (1952) model, which assumes that the liquid streams out into a thread or film which collapses under surface tension to yield droplets of average radius half the thread thickness. Assuming that the thread forms after the liquid enters a conical region of transition in which its velocity increases at a constant rate, the minimum energy dissipated per unit mass $\mathscr{E}_d$ of liquids is

$$\mathscr{E}_d \approx 8\mu_1 d_1^2 Q_1 / 3\pi d_2^2 L \rho_1, \tag{4.5}$$

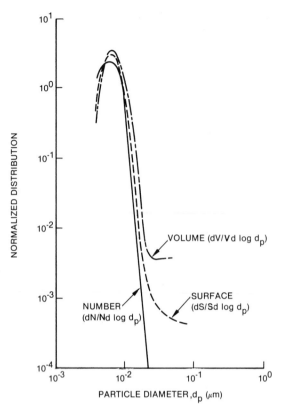

Fig. 4.8. Example of size distributions for *ultrafinely divided, approximately monodisperse carbon* produced from ablation of a porous graphite, with argon flow. [Data from Boffa *et al.* (1976).]

where d_1 and d_2 are the diameters of the entry and exit of the transitional cone, L is the cone length, Q_1 is the volume flow rate of liquid, and μ_1 is the liquid viscosity.

For $L = 1$ cm, $d_1 = 1$ cm, $Q_1 = 1$ cm^3/sec, and $\mu_1 = 1$ cP, calculated values of $\rho_1 \varepsilon_d$ have been tabulated in Table 4.4 for the case of water. Monk has suggested that values for the energy needed to overcome viscous forces in the liquid below the dashed lines in Table 4.4 may constitute conditions where viscous energy losses will not limit the atomization process. Green and Lane (1964) note, however, that Monk's calculations are rather crude and should be regarded only as a qualitative illustration of the importance of viscosity to the generation of fine droplets from bulk liquids.

It is difficult to estimate the losses occurring in applying energy to breaking up a liquid, except in specific configurations. However, it seems clear that energy transfer from a gas to a liquid during atomization by aerody-

TABLE 4.4

Minimum Energy $\mathcal{E}_d$ (cal) Dissipated in Atomizing 1 cm of Water at a Rate of 1 cm³/sec[a]

Thread diameter (d_2) (μm)	Number of threads assumed		
	1	10	100
1	2×10^6	2×10^5	2×10^4
5	3.2×10^4	320	32
10	200	20	2.0
20	13	1.3	0.13

[a] From Monk (1952).

namic forces is inefficient. However, this inefficiency is offset, at least in part, by the large amounts of energy applied by a compressible fluid, such as a gas, as compared with that retained in a liquid under pressure.

4.2.1 Mechanics of Droplet Formation. Droplets may be produced by a number of different processes, many of which are quite complicated and not well understood. Two of the simplest processes for droplet production are (a) the formation of liquid pendants from the tip of a capillary tube, with subsequent detachment; and (b) the development of wavelike instabilities on the surface of a filament or sheet of liquid, with subsequent breakup and droplet evolution. The former mechanism is reviewed by Lane and Green (1956), and the latter represents a classical problem of vibrational instability in jets investigated years ago by Lord Rayleigh (1879). The mechanisms of instability of thin liquid sheets have also been treated more recently by Hagerty and Shea (1955) and by Dombrowski and Johns (1963).

The size of droplets falling at a slow rate from circular tips is directly proportional to the external radius of the tip. Furthermore, cylindrical filaments will display beadlike swelling and contractions on their surfaces, resulting in an instability of wave number

$$2\pi/9.016a,$$

where a is the mean radius of the filament. Then unstable disturbances will increase in amplitude until the filament breaks up into droplets having principally a diameter of about $9a$. Because it is quite difficult to obtain capillary tubes of radius less than tens of micrometers, both of these simple processes have limited direct applicability to the production of swarms of very small droplets in the micrometer and submicrometer size range. However, we shall find below that many experimental observations suggest that Rayleigh's instability mechanisms also play an important role in more complex atomization processes involving aerodynamic forces.

Like the case of breakup of capillary jets, the disintegration of thin sheets of liquid takes place as a result of the development of unstable waves on the sheet. To understand better the breakup process involved in several nozzles, Hagerty and Shea (1955) studied classes of inviscid instabilities on a planar liquid sheet of uniform thickness. Two classes of wave modes were identified; (a) the sinuous type, where waves on both sides of the sheet are in phase; and (b) the dilational class, in which waves on the two sides are 180° out of phase. Considering only a balance between surface tension and pressure forces, Hagerty and Shea developed a theory for the unstable waves patterned after Rayleigh's results. Assuming that these waves break up to form droplets the size of the wavelength of the most unstable wave modes, a theoretical mean droplet size could be estimated in terms of physical properties of the fluids. The theory has been used to correlate data for droplet formation during the breakup of liquid sheets. Hagerty and Shea found that the lowest stable modes were sinuous waves whose minimum stable frequency was

$$f = \rho_g \langle q_l \rangle^3 / 2\pi\sigma \tag{4.6}$$

where $\langle q_l \rangle$ is the mean flow of liquid in the sheet. Generally droplets produced by film breakup have a minimum mean radius of 40–60 μm.

More recently Dombrowski and Johns (1963) have reported an analysis of the aerodynamic instability and disintegration of viscous sheets with particular application to the performance of fan-spray nozzles. Their conceptual picture of the disintegration of a liquid sheet is shown in Fig. 4.9. Here a sinuous wave is shown initially growing in amplitude, then breaking up first into long filaments with axes roughly parallel to the wave crests. Later the filaments develop axial instabilities of dilational waves, possibly related to instabilities in Rayleigh's capillary jet. These ripples grow and finally break up into droplets. Thus Dombrowski and Johns' model involves two classes of wave formation taking place in sequence to form first ligaments then droplets. A liquid sheet issuing from a nozzle undergoes wavelike transitional instability before disintegrating into droplets. Taking into account the viscuous and inertial forces, as well as pressure and surface tension, these workers found relationships for the most probable ligament diameter and most probable drop diameter. Based on their results, the droplet radius formed from the modes of maximum instability of the film and the ligaments (lig) is given by

$$[R_i] = (3\pi/\sqrt{2})^{1/3} R_{lig}(1 + 3\mu_l/\rho_l\sigma 2R_{lig})^{1/6}, \tag{4.7}$$

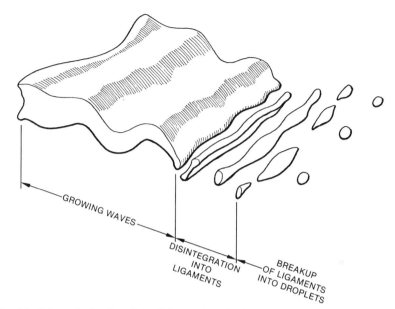

Fig. 4.9. Schematic drawing of the disintegration of a flat liquid sheet traveling through a still gas. [From Hidy and Brock (1970), with permission of Pergamon Press.]

where the ligament radius is

$$R_{\text{lig}} = 0.9614[(hx)^2\sigma^2/\rho_g\rho_l\langle q_l\rangle^4]^{1/6}[1 + 2.60(hx)\rho_g^4\langle q_l\rangle^7/72\rho_p^2\sigma^5]^{1/5}. \quad (4.8)$$

Here h is the sheet thickness and x is the distance downstream along the sheet.

Dombrowski and Johns' equations agree reasonably well for observed values of the surface–volume mean particle radius found for wax sprays by Hasson and Mizrabi (1961). In these data, the mean radius was about 25 μm.

4.2.2   Influence of Electric Effects.   The breakup of droplets issuing from a capillary tip may be modified somewhat by electrically charging the issuing liquid filament and applying an electrostatic field. Droplets falling from a finely drawn-out tube become smaller when a high potential is applied to the liquid. If this electrical potential reaches a few thousand volts, the droplets become exceedingly small and are expelled from the tip of the tube at a high velocity, and so they appear to form a continuous thin stream. The droplets get smaller and fall faster from the tip as the electrical potential is raised.

Under certain conditions of electrical potential and hydrostatic pressure, the liquid flowing from the capillary tip forms an inverted cone. From the apex of this cone streams a fine thread of liquid which breaks up into a brushlike aerosol cloud. If the electrical potential is increased further, this thread may disappear, leaving only a plume of fine mist. The potential at which this occurs depends on the radius, shape, and wall thickness of the capillary tube in the field gradient, and on properties of the liquid. Fine mists of ~20 μm diameter and smaller can be produced with liquids of low conductivity such as distilled water, alcohol, dibutyl phthalate, etc. Experiments have indicated that conducting liquids and nonpolar organic liquids of low dielectric constant, such as carbon tetrachloride or benzene, cannot be dispersed by this method even at very high electrical potentials.

Because of the interest in electrical atomization for producing fuel oil sprays entering combustion chambers and for ion propulsion of rockets in space, considerable work has been reported on this problem, particularly for atomization into a vacuum. Lapple (1969) has reviewed this work in detail.

Most investigators believe that electrostatic atomization is achieved as a result of a "Rayleigh-like" instability of the liquid thread issuing from the tube. However, Schultz and Bronson (1959) reported some evidence that atomization in this case involves rupturing of molecular bonds during high rates of discharge in their high-vacuum experiments. The instability of electrified liquid surfaces, analogous to Rayleigh's hydrodynamic instability, was investigated some time ago by Zeleny (1916).

Suppose we consider a spherical surface of radius R_p with charge Q. Such a sphere will experience an outward pressure of $Q^2/8\pi R_p^4$ which, by opposing surface tension, will result in detachment of a droplet from a tip sooner than an uncharged sphere. If the applied potential is sufficiently high, a spray will be formed whose droplets will be governed at equilibrium by the relation

$$2\sigma/R_p - P - E_0/8\pi R_p^2 = 0$$

where P is the (excess) pressure in the droplet and E_0 is the applied potential. Equilibrium becomes unstable when a small decrease in radius gives an outward pressure exceeding that of surface tension, $2\sigma/R_p$. Zeleny found that the criterion for instability of a charged droplet depends on

$$E_0^2 = f(S,S')R_p\sigma \tag{4.9}$$

where the function f depends on the shape of the droplet and the rate of axial changes, given, respectively, for example, by its surface S and the rate of change in S with time, S'. Thus when the electrical pressure at the liquid surface reaches a critical value, depending on the surface tension of the liquid and the radius of curvature of the droplet, the sphere will become unstable. Then slight disturbances will be amplified, and eventually breakup will take place. This condition should be reached first at the lower extreme of the drop where the intensity of electrical field is greatest, and the liquid here is stretched into a thread which breaks apart into droplets, forming the brushlike cloud. Zeleny's analysis has been extended by Drozin (1955) to incorporate the dielectric constant of the liquid.

Lane and Green (1956) remark that the dispersal of droplets into the conical shape appears to be the result of changes in the direction of the electrical force acting on the thread and the mutual repulsion of the charged particles.

Experimental data suggest that the greater the conductivity, the easier the atomization. Poor atomization of liquids of very high conductivity is difficult to understand. However, Lapple (1969) has noted that it is likely that at least two modes of atomization may exist, the first being Zeleny's instability mechanism and the second being a more intensive disruption for which the mechanism has not been elucidated. The latter may involve corona formation or breaking of molecular bonds as proposed by Schultz and Bronson (1959). While the general guidelines have been established for electrostatic atomization, the details of the physical mechanisms are incomplete, and at present there appear to be no methods for predicting, in detail, atomization performance either from the level of charging or in terms of the size distribution of droplets formed. However, if a mechanism like Rayleigh's instability controls the formation of droplets from a jet in an electrical field, the most probable drop size may be estimated from classical stability analysis.

Peskin and Raco (1964) have investigated theoretical relationships for droplet instability in an electric field. They have developed semiempirical relationships for the most probable droplet diameter formed in a capillary jet with different applied electrical field strengths. Their theory has been verified at least in part by experiments with an amyl alcohol jet in a longitudinal electrical field greater than 1800 V/cm.

Electrostatic dispersion provides another method for generating uniform droplets for certain solutions. Vonnegut and Neubauer (1952) and others (e.g., Raabe, 1976) have studied this approach using a filament of electrically charged liquid released from a small capillary. Droplets of uniform size and charge can be produced and must be discharged soon after generation and are difficult to control because of electrostatic behavior.

4.2.3   Sprays by Centrifugation.   When a liquid is allowed to flow out over a spinning surface, it will flow to the edge of the surface, where it will be flung outward and break up to form droplets. Atomization of liquids by this means was investigated rather thoroughly more than twenty years ago. Some of the first experiments were carried out on flat spinning disks by Walton and Prewett (1949), and on spinning cups by Hinze and Milborn (1950) and Ryley (1959). Later other geometries were tested, including vaned and grooved disks (Pattison and Aldridge, 1957).

The studies of Hinze and Milborn (1950) indicated that sprays could be formed from liquid flowing away from the edge of a spinning disk via three mechanisms: (a) by direct drop formation at the edge of the disk, (b) by the formation and breakup of liquid ligaments stretched away from the edge of the disk, and (c) by the formation and disintegration of a liquid sheet spreading radially from the edge of the disk. The latter two situations could be prompted by increasing the liquid flow at the axis of the disk; by increasing the angular speed but decreasing the disk radius; or by increasing the liquid density, its viscosity, or the surface tension. Experiments generally showed that spray homogeneity could be maintained best by minimizing the liquid flow to aid in direct droplet formation.

Mechanisms of droplet formation by centrifugation have been elucidated further by Fraser *et al.* (1963). These studies involved the exploration of flow characteristics of liquid sheets emanating from spinning cups. These authors have derived expressions for conditions of sheet formation and for the variation of sheet thickness from near the cup to the zone of disintegration. These results were verified by experimental measurements.

As a result of their experiments, Fraser *et al.* (1963) found that two principal mechanisms of disintegration in spinning liquid sheets could be distinguished. When sheets are formed at low peripheral speeds and low liquid flow rates, a relatively undisturbed sheet is observed which extends from the cup lip until an equilibrium radius is achieved where the contracting action of surface tension at the free edge equals the centrifugal force. As liquid collects at this outer circular periphery a thick rim develops which is unstable and breaks down into long filaments or threads. These elongated masses of fluid tend to break up into droplets as they spin away from the liquid sheet. At low liquid flow rates, and low spinning rates, the droplets disengage themselves from protrusions very close to the peripheral layer, forming a nearly monodisperse spray of primary droplets essentially at the cup rim. At somewhat higher spinning rates, liquid filaments develop at the rim layer. The elongated threads are then flung outward away from the cup, eventually breaking up into droplets.

The second mechanism occurs at higher peripheral speeds and high liquid flow rates. Under such conditions the shearing of the stagnant gas in the surroundings tends to induce growing waves on the radially propagating sheet of liquid. These disturbances extend from the cup lip and are normal to the tangentially flowing liquid. Sheet disintegration proceeds in the form of waves shed in half wavelengths, and these in turn break up into droplets. The wave disturbances may be sinuous or dilational, but the former appear to be most unstable, as observed by Hagerty and Shea (1955).

Experiments on the behavior of fluids breaking up during centrifugation have been analyzed for simple relations to predict the most probable size of droplets generated. If a liquid is fed at the center of a spinning disk, a spray of nearly uniform droplet size can be produced by centrifugation off the edge of the disk. The size of droplets depends on the speed of rotation. For a given disk geometry, Walton and Prewett (1949) found that

$$\omega R_p (d\rho_l/\sigma)^{1/2} = \text{const} \qquad (4.10)$$

where ω is the angular speed of the disk, R_p is the principal droplet radius, and d is the disk diameter. For droplet production in the aerosol range, formed by high speed rotation, experiments have shown that the constant in Eq. (4.10) is about 2.2, but for larger droplets ~1 mm in radius, the constant is ~1.6.

Walton and Prewett found from experiments that in examining the main droplet size produced, the edge profile of the disk was of minor importance over the size range studied. The principal drop size was independent of liquid viscosity from 0.01 to 15 P. If the liquid feed rate was too high, the spray became heterogeneous, and a range of particles consisting of the principal size and a variety of satellite droplets merged to form a continuous spectrum of droplets.

Lane and Green (1956) have described the formation of droplets by a spinning disk based on observations by Straus (1950), for example. When a nearly monodisperse spray is produced, a liquid torus forms at the outer edge. By the action of surface tension this bulge of fluid gathers into uniformly distributed beadlike protuberances which stream from the edge. As each protuberance extends outward, a droplet forms which is connected by a filament to the disk. The filament necks down until it is detached from the droplet. The filament then shrinks up into an irregularly shaped droplet which breaks apart into two satellites. Straus has indicated that the process is quite similar to the formation of a droplet on a stationary tip except that, in this case, beads form on the filament which separate and gather into two or more satellite particles.

The droplet–tip analogy may be applied assuming that droplet formation is slow and that the disk thickness exceeds the equivalent tip diameter necessary for R_p to remain constant for a given value of surface tension. In this limit one finds for droplets falling from a capillary tip

$$R_p(2g\rho_l/\sigma) = 2.0. \tag{4.11}$$

If g is replaced by $\frac{1}{2}\omega^2 d$, then (4.11) applies to the spinning disk, as given in Eq. (4.10).

Earlier, Walton and Prewett appreciated the implications of the droplet–tip mechanism, but they preferred to interpret droplet formation as a consequence of Rayleigh instability on the surface of the filaments formed during the spinning motion. Application of Rayleigh's theory for the principal modes of instability yields an expression equivalent to Eq. (4.11) except that the constant on the right-hand side is 2.3 instead of 2.0.

The experiments of Straus (1950) qualitatively confirmed the prediction of his theoretical model. And his results agreed substantially with Walton and Prewett, except for the fact that he found the edge geometry had a significant influence on the drop size.

Further measurements of the constant in Eq. (4.10) have been reported by Dunskii and Nikitin (1965) for two different oils over a wide range of operating parameters. Although no details on the edge geometry of their disk were given, Dunskii and Nikitin found that the atomization constant varied from

1.87 to 2.14 with an arithmetic mean of 2.0 over a range of rotation rates from 4,000 to 16,000 rpm. These data apply to oil flow rates from 0.03 to 0.10 cm³/sec; for transformer oil, $\mu_l = 1.95$ P and $\sigma = 32.2$ g/sec², and for diesel fuel, $\mu_l = 0.231$ P and $\sigma = 30.6$ g/sec². These results largely confirm the earlier work of Walton and Prewett and of Straus. Furthermore, these data support a droplet capillary tip mechanism for controlling the generation of the main droplets by spinning disks.

Under certain operating conditions secondary droplets can be produced in abundance from a spinning disk atomizer. Dunskii and Nikitin (1965) have made some interesting suggestions about the mechanism generating these satellite droplets. To follow their arguments, let us apply dimensional reasoning to the performance of the spinning disk atomizer. Suppose it is assumed that droplet radius R_p is governed by the particle velocity q_p (related to the rotational speed ω), the disk parameter d, the surface tension σ, the gas and liquid densities ρ_g and ρ_l, and the viscosities μ_l and μ_g. With eight variables and three fundamental units, five independent dimensionless parameters should describe the behavior of these atomizers. Dunskii and Nikitin used the following five parameters[†]:

$$\text{Weber number:} \qquad \text{We} = \rho_g q_p^2 R_p / \sigma,$$

$$\text{Langmuir number:} \qquad \text{La} = \mu_l q_p / \sigma,$$

$$\text{Density ratio:} \qquad \qquad = \rho_l / \rho_g,$$

$$\text{Viscosity ratio:} \qquad \qquad = \mu_l / \mu_g,$$

$$\text{Time ratio:} \qquad \qquad = t_H / t_S.$$

Here t_H is the characteristic time of motion of a droplet with high relative velocity, and t_S the characteristic time for disintegration of a droplet.

The experiments of Dunskii and Nikitin indicate that the Weber number is not the controlling parameter for formation of secondary droplets. Instead they determined that the Langmuir number—the ratio of the viscous force to surface tension—is important in this process. Physically one interprets the mechanism of disintegration of droplets as follows: As liquid runs off in filaments from the edge of a spinning disk, larger droplets form with velocities much larger than the surrounding air. The droplets decelerate rapidly by the action of friction, and the dynamic pressure acting on the droplet falls even more quickly. Any deformation of a droplet begun at high pressure will be exhibited as the dynamic pressure drops. If the duration of action of high dynamic pressure on the droplet is less than the time required for disintegration, breakup will not take place. As the deforming forces acting on the droplets decrease, the droplets will tend to take a more or less spherical

---

[†] Dunskii and Nikitin also included the Reynolds number for the liquid, but this should not be independent of the other groups listed above.

shape by the restoring force of surface tension. Thus the breakup of primary drops will depend on the ratio of characteristic times τ_H/τ_S.

Using the calculation for the ballistic trajectory for a large droplet, with the viscous drag given an empirical relation for $Re_p > 1$, Dunskii and Nikitin estimate

$$t_H = \frac{2R_p^2\rho_p}{9\mu_g} \ln \frac{q_0}{q_x} \left(\frac{1 + 0.771(R_pq/\mu_g)^{2/3}}{1 + 0.771(R_pq_0/\mu_g)^{2/3}}\right)^{2/3}$$

where q_x is the horizontal component of the relative velocity q_p and q_0 is the value of q_x at the disk edge. For $q_0/q_x = 2$,

$$\frac{t_S}{t_H} \approx \frac{3.54\rho_l}{\rho_g} \bigg/ La\left[1 + \left(\frac{La}{We\rho_l\mu_l/\rho_g\mu_g}\right)\right] \tag{4.12}$$

Applying Levich's (1962) estimate of t_S ($\approx 2\mu_l R_p/\sigma$), Dunskii and Nikitin have evaluated several sets of data for secondary droplet formation in spinning disk atomizers. Their calculations of the dimensionless parameters based on the results of several investigators are listed in Table 4.5. They found that in experiments where subdivision and formation of secondary droplets was observed, the time ratio τ_H/τ_S was very large, hundreds of times greater than in cases where no subdivision was found. Clearly, then, the Weber number alone does not control secondary droplet production, but from Eq. (4.12) such droplet formation depends on both La and We, as well as the density and viscosity ratios.

The spinning disk generation method can be adopted to produce droplets of uniform size according to the model of Walton and Prewett. Indeed, devices have been built using this principle which look like that in Fig. 4.10. The primary droplets thrown off at the perimeter of a spinning disk were of uniform size. Liquid is fed to the center of the disk and flows to the edge by centrifugal forces, where it accumulates until the centrifugal force, which increases with increasing liquid at the edge, overcomes the surface tension and disperses the liquid. This dispersion also produces some secondary fragments (satellite droplets) which are easily separated dynamically from the larger primary droplets. This is usually done by a separate flow of air near the disk, into which the satellites move and beyond which the larger primary droplets are thrown. A spinning disk generator, shown schematically in Fig. 4.10, can attain disk speeds up to 100,000 rpm. The drop radius produced by a spinning disk is given theoretically by Eq. (4.10). Many investigators have developed and successfully used spinning disk monodisperse aerosol generators for a variety of experimental applications including aerosol studies and inhalation experiments (see also Kerker, 1975; Raabe, 1976). Although the spinning disk and spinning top have probably been the most generally successful methods for producing monodisperse aerosols, the production of particles smaller than 0.5 μm geometric diameter is not practical

TABLE 4.5

Critical Values of Characteristic Parameters for Spinning Disk Atomization[a]

Experiments	We	La	ρ_l/ρ_g	μ_l/μ_g	τ_H/τ_S	Secondary breakup
Volynskii (1948)	5.35–7.00	0.185–0.206	670–11,300	0.0076–0.0640	16,400–470,000	Yes
Lane (1951)	5.10	0.151–0.480	834	0.0640	8,000–19,000	Yes
Bukhman (1954)	1.70–5.20	0.115–0.155	834	0.0640	22,000	Yes
Merrington and Richardson (1947)	7.50	0.138	834	0.0640	Very large[b]	Yes
Merrington and Richardson (1947)	7.50	—	1,050	69	Very large[b]	Yes
Basanaev et al. (1949)	0.05	0.010	10	0.85	Very large[b]	Yes
Walton and Prewett (1949)	5.20	104	872	1.58	11.5	No
Dunskii and Nikitin (1965)	13.0	82.0	743	1.39	16.4	No

[a] From Dunskii and Nikitin (1965).

[b] For tests of Merrington and Richardson (1947) and Basaneav et al. (1949); the quantity τ_H/τ_S cannot be determined from Eq. (4.12), which is only applicable for droplets moving with a large initial velocity relative to the medium.

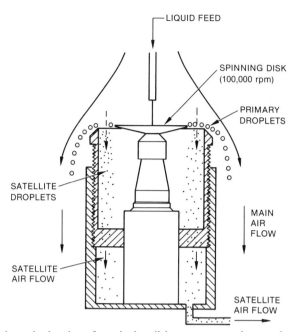

Fig. 4.10. Schematic drawing of a spinning disk generator used to produce monodisperse aerosols of both soluble and insoluble forms from solutions of suspensions. Air flow into the *satellite collector* is adjusted so that the inertia of the primary particles allows them to enter the main air flow.

because droplets produced with these devices are generally in the 20–30 μm diameter range and even purified water has trace impurities which may yield resultant particles as big as half a micrometer. Monodisperse aerosol particles produced with spinning disk and spinning top devices are usually near or larger than 3.5 μm in aerodynamic equivalent diameter. As with other methods, the resulting aerosol particles are highly charged from frictional effects and must be passed through a suitable electrical neutralizer soon after being produced.

   Another method for generating sprays using centrifugal forces involves swirl nozzles. In practical devices a hollow chamber is used where a liquid is introduced, tangentially under pressure at the base of the cone, and is spun inside the chamber before issuing through a hole at the apex of the cone. If the liquid pressure is high enough and if the liquid is not too viscous, a vortex with a hollow core of gas is established. The liquid streams out of the orifice as an unbroken tulip-shaped film at low pressures, while at higher pressures the liquid disintegrates into droplets close to the orifice. In the tulip regime, droplets eventually break away from the ragged downstream edge of the film. Lane and Green (1956) have noted that breakup of the liquid issuing from such nozzles is not the result of turbulent shearing flow in the liquid jet. Evidently the formation of droplets here is linked closely with the thinning of the effluent sheet under the influence of surface tension, viscous forces, and interaction with surrounding gas.

The production and dispersal of droplets depends on the thickness of the sheet of liquid, which in turn is controlled by the flow in the conical chamber. It has been found that theories for the fluid dynamics in a swirl chamber cannot neglect viscosity. However, an approximate analysis based on boundary layer theory gives results which correspond at least qualitatively to observations. Taylor (1950) has proposed such a model. He estimated the thickness of the layer in the conical nozzle by a boundary layer approximation for the fluid flow (e.g., Schlichting, 1960). With this analysis the velocity of the liquid toward the vertex (vertex angle 2α) was estimated, neglecting the axial component of velocity outside the viscous layer. From his calculations, Taylor found that the thickness δ' of the boundary layer was

$$\delta' \approx b(C_1 b/\sin \alpha)[(\nu_1 \sin \alpha)/\omega]^{1/2}, \tag{4.13}$$

where C_1 is a dimensionless constant ranging from 1.14 to 2.66, ν_1 is the liquid kinematic viscosity, and b is the radius of the spray cone, equal to about ten times the orifice radius a. For this model, Green (1953) considered the constant $C_1 \approx 20a$ in practice; ω is the rotational speed.

By neglecting the contribution of the cone flow, this relation may be used to obtain an indication of the average drop size produced by the nozzle, R_p, assuming that $R_p \approx \delta'$, the thickness of the film leaving the orifice. Making use of Eq. (4.13) in a different form, Lane and Green (1956) have indicated that this relation agrees fairly well with experimental values of (Sauter) mean droplet radius[†] from 25 to 125 μm as determined by Watson (1948).

4.2.4   Atomization by Acoustic Disturbance.

Experiments have indicated that a mist of liquid droplets can be generated by an intense beam of high frequency sound waves. The beam is focused on the liquid surface with a concave reflector or some other type of ultrasonic radiator. When the intensity of the ultrasonic beam is large enough in the focal region, a liquid spout called an ultrasonic fountain rises into the gases overhead, and a dense fog evolves from the base of the spout. Such fogs contain particles which are variable in size down to a radius of a few micrometers. McCubbin (1953), for example, has measured the droplet spectrum in water fogs generated by acoustic waves. He found that the droplet radii were grouped mainly around 2–3 μm.

Fog formation by acoustic vibration has been attributed to cavitation, the formation and collapse of cavities induced in the liquid by the intense waves. However, Green and Lane (1964) have remarked that acoustic fog formation may proceed by a different mechanism for thin layers of liquid spreading

---

[†] Radius determined by dividing the droplet volume by the surface area.

over a violently oscillating surface. If the surface of such a layer resting over a vibrating transducer is examined closely, one observes that the surface is covered by tiny ripples. Because of variations in the liquid film and the boundary reflections, the ripple pattern generally may be quite complex. If the amplitude of these capillary ripples becomes sufficiently large, their crests may "break" and shed very tiny droplets. The size of droplets should then be related to the ripple wavelength or the vibration frequency. Experiments by Bisa et al. (1954) provide some evidence for such a mechanism of atomization. Working at frequencies of 1.2–5.4 mHz, these workers showed that uniform patterns of crossed capillary waves developed on a liquid surface when atomization occurred, and that the median radius of the droplets generated was directly proportional to the capillary wavelength. The latter could be estimated from Kelvin's capillary equation using the existing frequency and properties of the atomized liquid as calculated by Lang (1962).

Peskin and Raco (1963) have pointed out that Kelvin's theory assumes no coupling between the surface velocities in the liquid and the forcing function driving the surface. As a result, properties of the atomization system, like the liquid film thickness and the forcing amplitude, do not enter into the expression for the capillary wave number. Peskin and Raco have taken into account such coupling in an analysis using linearized stability equations.

Performance curves for ultrasonic atomizers have been calculated where the most probable drop size is given in terms of transducer signal amplitude and frequency, liquid film thickness, surface tension, and fluid density. A bounding relation at the low frequency extreme of disruption denotes a limiting value for the radius of particles produced. For large film thickness, the Peskin and Raco (1963) model reduces to an estimate of the most probable droplet diameter d_{ac}:

$$d_{ac} \approx (4\pi^3 \sigma / \rho_1 f_0^2)^{1/3}, \qquad (4.14)$$

where f_0 is the acoustic excitation frequency.

Equation (4.14) corresponds to a correlation that has been reported by several investigators based on experimental results (Hidy and Brock, 1970). The expression is essentially the result obtained by applying Kelvin's capillary wave theory, and it represents a limiting regime in Peskin and Raco's theory neglecting film thickness and acoustic wave amplitude. Unlike the Kelvin theory, however, the results of these investigators predict production of a nearly constant drop size at higher frequencies for a fixed value of the ratio of the film thickness to transducer amplitude.

Ultrasonic atomization offers advantages over other more conventional methods in that an aerosol of high concentration and reasonably narrow size range is produced which can be controlled easily by varying the frequency of the acoustic waves. In pneumatic atomization discussed below, particle size may be decreased only by diminishing concentration because the air flow has to be increased. In ultrasonic atomization the aerosol concentration is limited only by the ultrasonic power input or the air flow past the vibrating liquid surface. With unrestricted power the limiting concentration of liquid that may be suspended depends ideally only on the fallout rate. Thus, the diameter of a droplet d_{ac} should be inversely proportional to the cube root of the applied frequency, and directly proportional to the cube root of the liquid surface tension. Although the rate of disruption of a liquid stream depends upon both the viscosity and surface tension of the liquid, the characteristics of the droplets formed depend only upon the diameter of the stream and the frequency of the vibrations as a first approximation. This is true because the minimum length of such a stream that can be separated into a separate droplet is equal to the circumference of the stream; a shorter length cannot be separated since the surface tension along the circumference of the stream will tend to elongate a shorter segment and prevent separation. For most liquids, it is satisfactory to assume that the stream diameter is equivalent to the orifice diameter. If the liquid is emitted with a flow rate Q_1, then a spherical droplet formed of each separated cylindrical segment will have a diameter given by

$$d_{ac} = (6Q_1/\pi f_0)^{1/3}. \tag{4.15}$$

The frequency required to produce the minimum droplet size is given by a maximum frequency f_{max} for disruption of the stream (Raabe, 1976):

$$f_{max} = 4Q_1/\pi^2 d_{ac}^3$$

such that the optimum frequency of disruption is

$$f_{opt} = 4Q_1/\pi^2 d_{ac}^2 \sqrt{2}; \tag{4.16}$$

the minimum diameter droplet is given by $d_{min} \approx 1.77 d_{ac}$ and the optimum diameter is given by $d_{opt} \approx 1.88 d_{ac}$. Hence the minimum or optimum diameters of droplets that can be generated in practice depend only on the stream diameter (orifice diameter). Measurement of volumetric flow rate from the orifice Q_1 can be made for any combination of orifice diameter and reservoir pressure, so that droplet sizes can be calculated.

Fulwyler and Raabe (1970) have described the application of an acoustic droplet generator to the production of monodisperse aerosols (Raabe, 1976). In the Fulwyler device an audio oscillator and amplifier provide a high frequency power signal which is converted to mechanical vibrations by an ultrasonic transducer linked by a coupling rod to a small liquid reservoir. The reservoir is pressurized (≈ 30 psig) to emit the liquid through a small orifice (≈ 10 μm) as a fine stream. This stream is uniformly disrupted by the ultrasonic vibrations into droplets that vary less than 1% in volume. To produce an aerosol of these uniform droplets, the droplet stream needs to be directed through the exit of the device to minimize coalescence. This can be accomplished by concentric mixing or cross flow of air streams.

An interesting modification of the acoustic principle for generating aero-

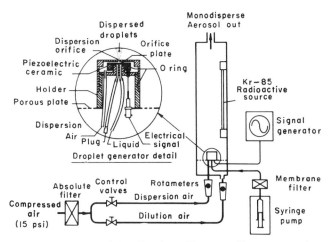

Fig. 4.11. Schematic diagram of the vibrating-orifice monodisperse aerosol generator. [Reprinted with permission from Berglund and Liu (1973). Copyright 1973 American Chemical Society.]

sols is a generator design reported by Berglund and Liu (1973). A schematic diagram of this instrument is shown in Fig. 4.11. The key to the generator is a vibrating orifice, which is driven at acoustic frequencies by a piezoelectric plate. The device then produces a spray by a rapidly varying orifice along with direct disruption using sound waves from the oscillating crystal. The output of the generator is controlled over a range by air and liquid flow rates as well as by the crystal oscillating frequency and orifice diameter. The particles produced are found to be of uniform size in the micrometer range. Since the aerosol produced contains highly charged particles, a radioactive ^{85}Kr source is attached to the chimney as a charge neutralizer. This latter technique is widely used in practice for particle charge neutralization in laboratory studies.

4.2.5 Mechanisms of Aerodynamic Atomization. Aerodynamic or air blast atomization involves the shattering of a coarse liquid stream by interaction with a high speed gas stream. The droplet breakup process is well illustrated in a classic series of photos in Fig. 4.12. This disintegration process is analogous to but more complex than the ones described above. Consequently, the detailed mechanisms for droplet generation from bulk liquid, in this case, are important in the overall disintegration mechanism. Initially, the droplet is distorted by the air flow from the right in Fig. 4.12 to form a spherical cap (Lane, 1951). Distortion continues as the droplet blows out into a hollow shell attached to a nearly circular rim. The shell is unstable; as it oscillates the layer bursts, producing a spray of small droplets. Finally the rim becomes a thin ringed ligament which becomes unstable to disturbances and breaks apart into a series of large droplets. Aerodynamic forces act

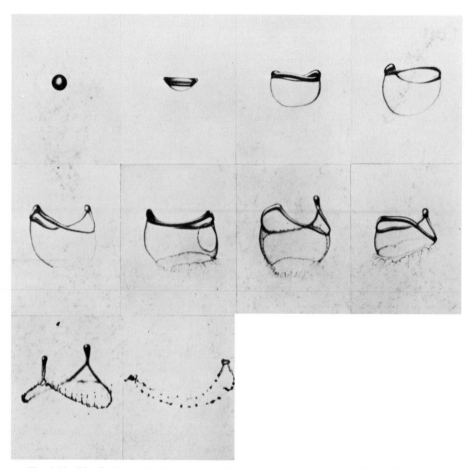

Fig. 4.12. Distribution and disintegration of a water drop in steady gas flow. The 1.3-mm-radius drop is falling through an upward-directed air flow. The sequence of photos begins on the upper right, down through the right-hand column, then to the upper center photo, and downward to the left column. At a critical air velocity of 22.5 cm/sec, the drop flattens out and blows into a hollow shell attached to a nearly circular rim. The shell then bursts, producing a spray of droplets. The rim breaks up at a later stage to form larger droplets. (Photo courtesy of H. Green and W. Lane; British Crown copyright; reproduced by permission of the Controller of Her Brittanic Majesty's Stationery office.)

continuously on fluid protuberances to deform them until they are drawn out into fine ligaments from the original droplet film or its rim. As in the cases described previously, the ligaments collapse into droplets under the influence of surface tension. As the gas velocity increases, the radii of the ligaments formed decrease, their lifetimes become shortened, and smaller droplets result. This decrease in size reaches a limit when the ligaments tear apart almost as soon as they form.

The mechanism adopted to explain the formation of droplets during the disintegration of the filamentous rings in an air flow originates from Rayleigh's theory of instabilities in a fluid thread. Castleman (1931) used Rayleigh's theory (1878, 1879) to predict the rate of collapse of a ligament which is subjected to an oscillating disturbance. The rate of shattering is greatest when the oscillating frequency is maximum:

$$f = (\sigma/\rho_l a^3)\mathfrak{F}(L/a). \tag{4.17}$$

Here a is again the radius of the unstable filament and $\mathfrak{F}$ is a function of the ratio the length of the filament L to its radius. The maximum value of $\mathfrak{F}$ is 0.343 at $L/a = 4.5$. Using Eq. (4.17), it is possible to estimate the time scale for droplet formation during atomization. If we assume that the filament volume equals the droplet volume, for 100-μm-radius droplets, $f \approx 10^2$ sec^{-1} for water at 20°C, or the collapse time of a filament of 10 μm radius is $\sim 10^{-2}$ sec.

The Rayleigh–Castleman mechanism provides an explanation for 10-μm-radius atomization to form large droplets by air blast. However, droplets of much smaller sizes than those produced by ligament collapse are found in sprays. Droplets much less than 100 μm in radius are likely to be produced by the stretching and rupture of thin liquid films, which may originate near the liquid nozzle or from the distortion of very large droplets. Such a mechanism is seen in the breakup of isolated droplets of millimeter radius in an air stream as indicated in Fig. 4.12.

The formation of finely divided droplets from disintegrating films is analogous to droplet formation from bubbles bursting on a liquid surface. This process has been investigated experimentally by several investigators including Tomaides and Whitby (1976). They have shown that two processes are involved during bubble formation and collapse of a liquid film. The bubble film breaking process produces a broad range of particles less than approximately 10 μm in diameter as indicated in the lower size range of Fig. 4.13. The disturbances on the film with bubble collapse produce a capillary wave disturbance which exudes a jet at its center. If its ejection amplitude from the liquid surface is large, the jet can break up, producing large droplets in the 100 μm size range; this range is indicated by a secondary peak in the particle size distribution shown in Figure 4.13. Thus the droplet disintegration process from air blast is envisaged as a bubble film bursting process followed by ring ligament disruption, leading to two distinct particle size ranges.

The experiments of Lane (1951) showed that the critical velocity $q_g - q_p$ required to break droplets apart could be estimated by balancing the aerodynamic pressure at the stagnation point of the droplet with the pressure resulting from surface tension:

$$\tfrac{1}{2}K_D\rho_g(q_g - q_p)^2 = 2\sigma/R_p. \tag{4.18}$$

Here q_g is the critical velocity of the gas stream, q_p is the speed of the liquid

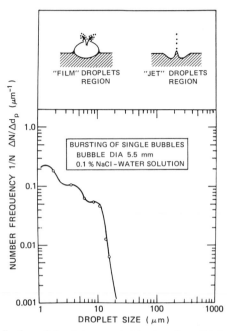

Fig. 4.13. Size distribution of droplet generated by bursting of 1.4-mm-diam bubbles on a surface of 0.1% sodium chloride solution. [From Tomaides and Whitby (1976), with permission of Academic Press.]

droplet at the moment of bursting, and K_D is a constant determined experimentally based on the drag coefficient of a distorted droplet. For the water droplets,

$$(q_g - q_p)^2 R_p = 6.12 \times 10^3.$$

A relation similar to this equation has been suggested for the critical velocity, for example, by Hinze (1948), who indicated that the Weber number We $= \rho_g (q_g - q_p)^2 R_p / \sigma$ should reach a critical value for disintegrating droplets. Hinze estimated the critical Weber number to be about 6.5 for liquids of low viscosity. This compares favorably with the value of 5.3 deduced from Eq. (4.18) and $We_{crit} = 5.3$–7.0 from data of Volynskii (1948). Lane and Green (1956) also have pointed out that the critical Weber number depends somewhat on the variation of relative velocity with time. Higher values of $We_{crit} \approx 10$ have been observed for freely falling droplets. High viscosity will also tend to increase the critical We.

In designing air blast atomizers, it was found many years ago that the polydispersivity of a mist from such devices could be reduced by adding a baffle in the aerosol particle stream. Installation of a baffle will remove larger droplets from the spray formed by ordinary air jets. Green and Lane (1964) have noted that it is sometimes claimed that the fineness of aerosol produced by baffled atomizers results from secondary shatter by impingement of the primary spray on the walls of the baffle. However, it appears now that the baffle only serves as a barrier to collect coarser material, while the finer material travels out through the device. It is noted, however, that the fineness of spray from baffled atomizers may be controlled to a degree by varying the distance between the jet and the baffle surface.

An interesting combination of air blast atomization and centrifugal atomization has been reported by Fraser *et al.* (1963). Using their spinning cup, combined with an air jet directed normally onto the free liquid sheet, Fraser *et al.* found that breakup could be induced in a controlled way. They found that the sheet of spinning liquid originating from the cup does not disintegrate immediately on impact with the air stream, but is deflected away from the gas flow. Waves are initiated at the point of impact and the sheet breaks down into droplets through unstable ligaments. The resulting droplet distribution depends primarily on the thickness of the liquid sheet. Thus the thinner the sheet the finer the atomization for this device.

The method of disruption of liquids by air blast atomization also can be used to produce aerosol particles with size ranges considerably smaller than tens of micrometers in diameter. By atomizing solutions of soluble salts or mixtures of miscible liquids it is possible to generate clouds of submicrometer size particles after evaporation of larger droplets produced by the atomization process.

When subjected to transient pulses of high speed gas streams, drops appear to break up in a manner distinctly different from the steady-flow case described above. Instead of the drop blowing out into a thin hollow bag suspended from a thicker rim, it is deformed in the opposite direction as a surface convex to a gas flow. According to Lane and Green (1956), the edges of the convex saucer are drawn out into a thin sheet that will disintegrate into a spray of droplets if the gas speed is fast enough. Taylor (1949) has made some calculations for the shape which a drop might develop when accelerated in a gas stream before disintegration. These calculations indicate that the drop should be flattened into a planoconvex lenticular body of radius about twice the radius of the original sphere. This prediction qualitatively represents the observations taken in the laboratory.

Measured gas velocities for disintegration by transient blasts are lower than in steady flows. The difference in critical velocity increases as the droplets get smaller. Taylor (1949) has suggested that this may be expected if the drop behaves as a vibrating system. If, for example, we make an analogy to a loaded spring, the force F_T applied suddenly is equal to a force F_s applied steadily to produce the same extension. For a maximum extension, which would be analogous to the maximum distortion, a droplet could allow without breaking up, $F_T = \frac{1}{2}F_s$. Since the force tending to distort the droplet is proportional to the square of the relative velocity,

$$q_T/q_s = (F_T/F_s)^{1/2} \approx 0.71. \qquad (4.19)$$

In Table 4.6 several cases for breakup of water drops are listed. The ratio 0.71 is approached for droplets of radius 1 mm or less, but is exceeded for larger droplets having oscillation periods much larger than the duration of air blast.

Taylor's (1949) theoretical calculations have demonstrated that the thickness of the liquid sheet shed by a gas flow of given velocity from a drop of a given density and viscosity may be estimated by applying boundary layer theory. At comparatively small gas speeds the predicted thickness of the sheet is in satisfactory agreement with estimates from photographs, but at high gas speeds the calculated value is larger than that deduced from the droplet size observed in the mists produced. If the surface sheet of the drop is removed by tangential stresses, while at the same time the droplet is accelerated by the pressure distribution in the gas, the life history of the droplet may be deduced in principle by incorporating the rate of removal of liquid and the drag coefficient for the lens-shaped droplet.

Disruption of droplets by aerodynamic forces may be linked to the formation of unstable waves on their surface by Kelvin's capillary wave mechanism. If the droplet radii produced by the shearing flow of air over the droplet are of the order of the wavelengths of the most unstable waves, a theory analogous to Peskin and Raco's (1963) would apply. Some years ago Taylor (1949) examined this question and found that the most probable droplet size formed by such an instability mechanism would be

$$R_p \sim \pi\sigma/\rho_g q_g^2,$$

TABLE 4.6

Critical Velocities for Shatter of Water Drops in Transient Blasts and Steady Streams of Air[a]

Diameter of drop (mm)	Transient (q_T) (m/sec)	Steady (q_S) (m/sec)	q_T/q_S	Period of oscillation of drop (sec)
4.0	12.0	12.5	0.96	0.023
3.0	12.1	14.1	0.83	0.015
2.0	12.6	17.5	0.72	0.008
1.0	16.0	24.7	0.65	0.003
0.5	24.0	35.0	0.69	0.001

[a] From Lane and Green (1956).

where the proportionality factor is a function of ρ_l, ρ_g, and q_g^2 (Lane and Green, 1956). Taylor's estimate gives much smaller droplet radii than encountered in practice. The discrepancy, however, may be linked to the fact that q_g is much smaller near the droplet surface than the free-stream value normally taken for estimating R_p on theoretical grounds.

Simultaneously with mechanisms of breakup, some droplets will collide and recoalesce. The evidence of such agglomeration in atomizers, based on calculations of Dunskii (1956), for example, suggests that the collision process may be an important consideration in the performance of these devices, but further studies would be needed to clarify this point.

Because of the complex mechanism for droplet formation by aerodynamic forces, one can expect a wide range of droplet sizes to be present in mist formed by this mechanism. This is borne out by many experimental studies (Green and Lane, 1964). From the spinning cup–air blast atomizer work of Fraser et al. (1963), prefilming devices can produce smaller drop sizes than jet atomizers, particularly when the film thickness is controlled.

4.2.6 Particle Production by Nebulization.   A nebulizer is an atomizer used to produce aerosols of fine particles by atomization of liquids to produce droplets, the majority of which are less than 10 μm in diameter, as opposed to those from a spray atomizer, which may range up to 100 μm or larger. Both compressed-air and ultrasonic nebulizers produce aerosols from liquids. Often, the particle residue remaining after the droplets evaporate form the desired aerosol of fine particles. Either solutions or suspensions may be aerosolized with nebulizers.

Fine aerosols of both soluble and insoluble materials may be produced by nebulization of solutions or suspensions. For example, spherical insoluble particles of plastics may be made from solutions with suitable organic solvents, or soluble forms may be made from aqueous solutions of electrolytes such as sodium chloride. Changes may be made in the chemical state of the particles formed after solvent evaporation by heating and in the size distribution by selective collection of a portion of the particles.

Both crystalline and amorphous forms of aerosol particles may be produced from aqueous solutions. Often the type of particle produced will depend upon the conditions of drying as well as the chemical nature of the materials. If drying is too rapid, low-density particles, which are essentially hollow shells, may be formed by the encrustation of the surfaces of the drying droplets. More often, drying is seriously hindered by the hygroscopicity of the solute, the presence of immiscible liquids or evaporation inhibitors, the insufficiency of energy available to the droplet, the saturation of the air, and other factors. Adequate drying usually requires that the primary droplet aerosol be mixed with filtered, dry air. Warming of an aerosol to speed drying, by passing it through a heated tube, may be necessary in some experiments. Reduction of electrostatic charge to Boltzmann equilibrium is usually essential to prevent selective loss of charged particles.

When droplets evaporate, the residue particles become the aerosol. Since nebulizers produce droplets of many sizes, the aerosols formed after evaporation are polydisperse with geometric standard deviation about equal to the value for the droplet distribution produced by the nebulizer. The size of a given particle depends upon the droplet diameter and the cube root of the solution concentration.

Nebulization of an aqueous colloidal suspension forms insoluble particles of the aggregates of the colloidal micelles. This method has the advantage of requiring no organic solvents. If the micelles are small and in high concentration, the resultant particles will be nearly spherical in shape and their size distribution may be predicted by assuming that the suspension behaves as a solution. For example, a colloidal suspension of ferric hydroxide may be aerosolized to produce relatively insoluble and spherical particles of ferric oxide for inhalation experiments. Water physically trapped or chemically bonded in such particles may be removed by heating. If the colloidal micelles are comparable in size to the droplets produced by the nebulizer, or if the concentration is small, physical factors and the statistics of the random pickup of micelles by droplets will determine the size distribution of the resultant aerosol. Unfortunately, aerosols produced from colloids sometimes have inherent porosity because of the interstices between micellar components. Aerosols of bacteria and viruses can also be produced by nebulization of suitable suspensions.

The nebulization of suspensions of uniform particles which have been grown chemically or separated from polydisperse particles has been a simple and useful means of generating monodisperse aerosols. Suspensions of polystyrene latex spheres of fairly uniform size grown by emulsion polymerization as developed by Bradford *et al.* (1956) have been commonly used for this purpose. Other types of monodisperse hydrosols have also been made. In addition, the polystyrene latex spheres can be labeled with radionuclides.

It is not possible at reasonable dilutions to guarantee that only individual particles will be aerosolized and droplets containing more than one of the suspended particles will not become undesirable aggregates upon evaporation. Raabe (1968) has reported the dilution (DIL) for polystyrene latex suspensions required to generate an aerosol with singlet ratio, (SR) defined as the ratio of droplets with but one monodisperse particle to all droplets that contain one or more monodisperse particles:

$$\mathrm{DIL} = F(\mathrm{VMD})^3 \exp[4.5(\ln \sigma_g)^2]\{1 - 0.5 \exp[(\ln \sigma_g)^2]\}/(1 - \mathrm{SR})d_1^3$$

$$(4.20)$$

for SR > 0.9 and $\sigma_g < 2.1$; F is the fraction by volume of the particles in the original stock suspension and d_1 the diameter of the monodisperse spheres, with the volume median diameter (VMD) and geometric standard deviation σ_g of the droplet distribution from the nebulizer.

Equation (4.20) can be used to calculate the maximum number concentration N_{max} of monodisperse particles of any type in a liquid suspension to be nebulized so that the singlet ratio is not less than R:

$$N_{max} = 6(1 - SR) \exp[-4.5(\ln \sigma_g)^2]/\pi(VMD)^3\{1 - 0.5 \exp[(\ln \sigma_g)^2]\}$$
$$\text{(no. of droplets/cm}^3\text{).   (4.21)}$$

Here the VMD is the nebulizer droplet median diameter in centimeters.

Raabe (1976) also indicated that more than 90% and probably more than 99% of the droplets produced at these low suspension concentrations contain no spheres when the singlet ratio SR is taken at 0.95. These empty droplets dry to form ultrafine particles consisting of a residue of the impurities in the liquid; these secondary aerosols can be undesirable since they are considerably more numerous than the monodisperse particles themselves.

Monodisperse suspensions of polystyrene and poly(vinyl toluene) latex spheres (Dow Diagnostics, Indianapolis, Indiana) have been available in sizes from 0.088 to 3.5 μm and have been widely used for preparing monodisperse aerosols. The monodispersity of these particles is excellent, with σ_g < 1.05. The density of polystyrene latex has been measured to average 1.05 g/cm^3, and the density of poly(vinyl toluene) latex at 1.027 g/cm^3. These particles are grown by emulsion polymerization and are stabilized in aqueous suspensions with an anionic surfactant having an active sulfonate radical. The particles carry a negative charge in aqueous suspensions, which contributes to their stability. The compositions of some typical latex suspensions are given in Table 4.7. Solids normally represent 10% of the aqueous suspension as supplied by the manufacturer.

If the 0.264 μm particles (batch LS-057-A in Table 4.7) are generated with a Lovelace nebulizer (VMD = 5.8 μm, σ_g = 1.8) with no dilution, the 0.177% emulsifier and inorganics in the suspension would increase the diameter of a single particle which happens to be aerosolized in a 5.8 μm droplet from 0.264 to 0.714 μm. Dilution to 30,000 to 1 to generate an aerosol with 95% singles makes a 0.264 μm particle aerosolized in a 5.8 μm droplet yield a latex particle essentially unchanged in diameter. For the larger particles, it is convenient to separate the particles from the liquid by centrifugation and resuspend them in purified and filtered water before use.

Monodisperse aerosols of polystyrene and poly(vinyl toluene) particles are highly charged when generated by nebulization. A reduction of charge to Boltzmann equilibrium with a suitable aerosol neutralizer is essential in studies using these aerosols to minimize selective particle loss to walls, or coagulation. The presence of a background aerosol produced from the numerous empty droplets containing dissolved material must be given appropriate consideration as well.

Important factors in describing the operation of nebulizers include (a) the output rate and concentration of usable aerosol, (b) the volumetric rate of air, (c) the evaporation losses of liquid which are independent of usable

TABLE 4.7

Typical Operating Characteristics of Selected Nebulizers with Reference, Gauge Pressure
ΔP, Output Concentration[a] V before Evaporation, Volumetric Flow Rate Q, and Droplet
Distribution VMD and σ_g[b]

Nebulizer	Reference	ΔP [psi (gauge)]	V (μl/l)	Q (l/min)	VMD (μm)	σ_g
Dautrebande D-30	Mercer *et al.*	10	1.6	17.9	1.7	1.7
	(1968a)	20	2.3	25.4	1.4	1.7
		30	2.4	32.7	1.3	1.7
Lauterbach	Mercer *et al.*	10	3.9	(1.7)	3.8	2.0
	(1968a)	20	5.7	(2.4)	2.4	2.0
		30	5.9	(3.2)	2.4	2.0
Collison	May (1973)	20	7.7	7.1	(2.0)	(2.0)
		25	6.7	8.2	2.0	2.0
		30	5.9	9.4		
		40	5.0	11.4		
DeVilbiss #40	Mercer *et al.*	10	16	10.8	4.2	1.8
	(1968a)	15	15.5	13.5	3.5	1.8
		20	14	15.8	3.2	1.8
		30	12	20.5	2.8	1.8
Lovelace	Raabe (1972)	15	27	1.3		
[Baffle screw set		20	40	1.5	5.8	1.8
for optimum operation		30	31	1.6	4.7	1.9
at 20 psi (gauge)]		40	21	2.0	3.1	2.2
		50	27	2.3	2.6	2.3
Retec X-70/N	Raabe (1972)	20	56	5.0	5.1	2.0
		20	53	5.4	5.7	1.8
		30	54	7.4	3.6	2.0
		40	53	8.6	3.7	2.1
		50	49	10.1	3.2	2.2
DeVilbiss Ultrasonic	Mercer *et al.*	—	150	41	6.9	1.6
(setting #4)	(1968b)					

[a] Measure of aerosolization efficiency.

[b] Corrected, as required, to operation at 76 cmHg and 21°C; values in parentheses are estimates of Raabe (1976); courtesy of Academic Press.

aerosol, (d) the droplet size distribution, (e) the volume of liquid required for
proper operation, and (f) the maximum unattended operating time. Cognizance of these factors in relation to a particular application determines the
choice of a nebulizer. They have been reviewed in detail by Kerker (1975)
and Raabe (1976).

The output rate of usable aerosol is usually described as volume of liquid
at initial formation associated with droplets that leave the generator and may
be expressed in milliliters or microliters per minute. The concentration of
droplets from most nebulizers is usually between 10^6 and 10^7 cm^{-3} of air;
however, since the volumetric rate of air may be several liters per minute,
the droplet output rate may be well in excess of 10^9 min^{-1}.

Evaporation losses increase the concentration of the solute in solution or particles suspended in the aerosolized liquid. Evaporation occurs primarily from the surface of the liquid and from droplets that evaporate slightly and then hit the wall of the nebulizer to be returned to the reservoir. This change in concentration causes an increase in the sizes of the particles that are formed when the droplets dry.

Nebulizers produce droplets of many sizes and resultant aerosol particles after evaporation are concomitantly polydisperse. The droplet distributions described for nebulizers are the initial distributions at the instant of formation; droplet evaporation begins immediately even at saturation humidity since the vapor pressure on a curved surface is elevated. The rate of evaporation depends upon many factors including surface tension, energy availability, degree of saturation of the air, solute concentration, solute hygroscopicity, the presence of immiscible liquids or evaporation inhibitors, and the size of the droplets.

The specific design of nebulizers depends on the application and the ingenuity of the investigator. The designs all employ one form or another of air blast atomization, with assistance from electrical forces or acoustic disturbances. Kerker (1975) and Raabe (1976) have discussed their design and operating characteristics. They have summarized the performance of some of the devices commonly employed. Raabe's summary is reproduced in Table 4.8. Basically all of the examples cited provide a volume mean diameter of generated particles in the micrometer range with a σ_g of less than two for a lognormal fit of size distribution.

4.3 AEROSOLS BY DISPERSAL OF POWDERS

4.3.1 Formation of Powders by Attrition or Impulsive Forces.
Dust clouds may be produced by the gaseous suspension of particulate material originating either from the disruption of bulk solids or from the dispersion of

TABLE 4.8

Relative Concentrations of Solids in Latex Suspensions, which Represent approximately 10% of the Aqueous Suspension[a]

Batch reference number	Particle diameter (μm)	Composition of solids (%)		
		Polymer	Emulsifier	Inorganics
LS-040-A	0.088	92.59	5.56	1.85
LS-052-A	0.126	97.09	2.43	0.48
LS-057-A	0.264	98.23	1.13	0.64
LS-63-A	0.557	98.78	0.26	0.95
LS-449-E	0.796	99.04	0.25	0.70
LS-464-E	1.305	99.15	0.29	0.56

[a] From Raabe (1976); courtesy of Academic Press.

finely divided powders. Examples of the first group are the dust clouds that evolve when rocks are disintegrated by processes of grinding or blasting.

Mechanical destruction of a solid substance may be carried out by applying shearing or tensile stresses to the material. This may be done slowly in milling processes, or by crushing material by means of rollers or ball grinders. More rapid disruption may be effected by impact crushing, or by blasting with explosives. In both cases the applied forces disintegrate the solid by splitting or cracking along zones of weakness in the material, like cleavage planes in polycrystalline substances.

Apparently, there is a limit to the extent to which solids may be fragmented by *comminution processes*. Green and Lane (1964) have pointed out that if a material is ground indefinitely, no change in the particle size distribution will be achieved after a certain time no matter how much longer the grinding is prolonged. This means that even though particles of all sizes continue to be broken up, some particles, probably the smallest ones, reagglomerate with each other or with larger particles as quickly as they are formed so that a kind of dynamic equilibrium evolves. This is illustrated in Figure 4.14, where an approach to "equilibrium" surface area of two powders is shown with grinding duration (Gregg, 1968). In the case of calcite, the

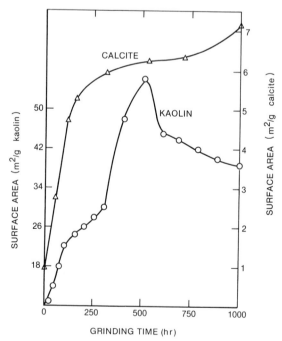

Fig. 4.14. Effect of prolonged grinding on the surface area of the powders. [From Gregg (1968).]

surface area levels out, but shows a sign of increasing again, while Kaolin clay peaks and actually decreases with additional time.

The studies of Bowden and Tabor (1950) suggest a possible mechanism in which smaller particles unite with larger ones by welding resulting from frictional heating. The continual rubbing together of particle surfaces probably produces sufficiently high temperatures to soften small particles and to allow welding together. Since particles increase in size as well as decrease in size during grinding, the particle size distribution arising from continued disruption results from a dynamic equilibrium rather than a static one.

At first thought one might expect that the shock of a high explosive charge in a solid material would be especially effective in disintegration. However, it has been found that in fact the size distribution resulting from explosive shattering of minerals is about the same as less violent processes of disruption. Green and Lane (1964) note that large numbers of tiny particles are present in dust raised from mining blasts, but these are spherical in shape and are believed to be released from the explosive itself rather than the broken stone. Explosion involves breakup at very high velocities, but the mechanism of disintegration remains the same; i.e., fragmentation with release of finely divided material along fresh surfaces. Only the rate of disintegration is vastly different in the explosive case. The fragmentation by explosion is more complete, of course, but the sizes of finer particles are quite similar to industrial dusts.

Particulate clouds also may be formed by explosive breakdown caused by gases generated within a solid material. The copious quantities of dust emitted from volcanoes are believed to originate in this way. The pressures in volcanic magma saturated with steam and hot gas at high pressure are released suddenly to the atmosphere, and the molten or partially solidified material explodes violently.

4.3.2 Cohesive and Adhesive Forces. Particles in dusts or powders tend to stick together remarkably well. It is difficult to break aggregates of particles up to produce clouds of nonagglomerated material. The ability of particles to stick together indiscriminately is the result of relatively weak attractive forces between molecules as well as attractive electrostatic forces. These forces have been termed cohesive and adhesive, depending on the heterogeneity of material at the boundary between particles. The distinction between cohesion and adhesion in the literature on fine powders is somewhat fuzzy, but we shall adopt the following conventions, which are consistent with classical definitions in physics. *Cohesion* is the tendency of parts of a body of *like composition* to hold together. This implies that cohesive forces arise between like molecules in a solid or between small particles of the same composition. *Adhesion,* on the other hand, refers to attraction across the boundary or interface between two *dissimilar* materials. Thus adhesive forces are likely to be the most common attractive forces in all but artificially generated aerosols. The mechanisms for adhesion of particles to surfaces has been reviewed in detail recently by Corn (1966) so only a brief summary of this topic is included below.

The combined effect of the London–van der Waals interaction between two particles containing several million molecules may be appreciable despite the fact that these forces are very weak between individual molecules and fall off with the seventh to eighth power of the distance between the molecules. An estimate of the order of magnitude of the attractive forces between two microscopic spheres has been carried out by integrating the London–van der Waals attraction over all pairs of interacting molecules. The calculations of Bradley (1932) and Hamaker (1937) indicate that the net attractive force in a vacuum between two spheres of radii R_1 and R_2 is

$$F_{at} = \pi^2 n_A^2 C_{at} R_1 R_2 / 6 r_m^2 (R_1 + R_2) \qquad \text{dyn,} \qquad (4.22)$$

where n_A is the number concentration of atoms, C_{at} is a constant, and r_m is the minimum distance between particles. Similar estimates have been made for nonsymmetrical particles. From Eq. (4.22) one readily calculates that the net force between two 0.25 μm spheres, for example, separated by ~5 Å, is ~10^{-4} dyn.

Calculations have also been made for the influence of nonsphericity of particles on the magnitude of attractive force. For example, Vold (1954) has found estimates of F_{at} for flat plates, rectangular rods, and cylinders. Particle orientation is important in these cases because these particles tend to seek an orientation of smallest potential energy, which is normally that in which the largest plane areas are opposed.

It was pointed out some time ago (e.g., Casimir and Polder, 1948) that the London theory has certain severe limitations in that its predictions will err at very short interatomic distances where the atomic wave functions overlap, and at large distances where electromagnetic retardation develops. For the attraction force per unit area between two flat plates close together, the retardation effect is

Nonretarded:

$$F_{at} = A_F / 6 \pi r^3;$$

Retarded: $\qquad\qquad\qquad\qquad\qquad\qquad\qquad\qquad\qquad\qquad\qquad\qquad (4.23)$

$$F_{at} = -B_F / r^4$$

where r is the distance between the plates and A_F is a constant ~10^{-12} erg. Casimir and Polder (1948) have estimated the retardation coefficient to be

$$B_F = \frac{23\, \hbar c}{8\pi} \left(\sum_{1,2} n_{AS}^\alpha \right)^2,$$

where $\hbar$ is Planck's constant, c is the velocity of light, n_{AS} are numbers of atoms per unit area in plates 1 and 2, and α is the polarizability of atoms in plates 1 and 2. Black *et al.* (1960) have indicated that B_F can be related to the dielectric constant ε of the plate material, or

$$B_F = 0.162 \frac{\hbar c}{\pi^3} \left(\frac{\varepsilon - 1}{\varepsilon - 2} \right)^2.$$

This relation yields a value of $B_F \approx 10^{-19}$ which is several orders of magnitude smaller than the parameter A_F. A remarkable macroscopic theory of Lifschitz (1954) also gives a form for B_F:

$$B_F = \frac{\pi \hbar c}{480} \left(\frac{\varepsilon - 1}{\varepsilon + 1} \right)^2 \phi(\varepsilon)$$

for large distances r. Here $\phi(\varepsilon)$ is given as a tabulated function of the dielectric constant ε. Lifschitz's result is nearly equivalent to Casimir and Polder's above.

Several investigators (Black *et al.*, 1960; Kitchener and Prosser, 1957) have looked at the retarded attraction between two flat quartz plates, and a micron size sphere. They found satisfactory agreement between the experimental results and the theory of Casimir and Polder and of Lifschitz.

Corn (1966) has summarized the results of various workers for the term $\pi^2 n_A^2 C_{at}$ in Eq. (4.22). These results are listed in Table 4.9. The large variation and high values obtained by Overbeek and Spaarnay (1954) have been attributed to electrostatic enhancement of intermolecular forces, and figures have been obtained agreeing with the lower values in Table 4.9. Thus on the basis of these results, the term $\pi^2 n_A^2 C_{at}$ of order 10^{-13} erg for clean, pure surfaces in a dry gas free of electrostatic charge.

In a later study not included in Corn's review, Tabor and Winterton (1968) have reported direct measurements of van der Waals forces between two mica cylinders. Their results indicate that the normal van der Waals interaction predominates only at separations of less than 100 Å. The retarded forces tend to dominate at separations greater than 200 Å. Tabor and Winterton's measurements for mica–mica interaction indicate that $A_F \approx 10^{-12}$ erg, in agreement with Hamaker's constant, while $B_F \approx 10^{-19}$, as calculated for the retarded forces by Lifschitz.

The theory provides an order-of-magnitude estimate for the attractive forces between irregular particles in a bed. However, it is reasonable, based on Eq. (4.22), to expect that both the adhesive and cohesive forces are proportional to R_p, the particle radius, if r_m is constant. This is verified by an extensive review of Fuchs (1964). Other than this information, scarcely anything is known about the nature of particle contact and attractive forces in powders or dusts. It is often assumed that the equilibrium distance is a few angstroms, but the lack of any information about submicrometer structure and surface irregularity of particles prevents making quantitative estimates

TABLE 4.9

Summary of Investigations of the Value of $\pi^2 n_A^2 C_{at}$ in Eq. (4.22)[a]

Author	Gaseous media	Materials	Minimum separation (Å)	$\pi^2 n_A^2 C_{at} \times 10^{12}$ (ergs)
Bradley (1936)	Vacuum	Borate and	3	4.7
		quartz spheres	3	2.2
Overbeek and Sparnaay	Air	Glass plates 1	200	0.01–22
(1954)		Glass plates 2	200	0.015–15
		Glass plates 1	2500	1.1–30
		Glass plates 2	5000	60–230
		Quartz plates	3000	11–30
Derjaguin *et al.* (1956)	Air	Quartz sphere and plate	1000	0.05
Bailey and Courtney-Pratt (1955)	Air	Cleaved mica	5–25	0.1–10 (estimated)
Prosser and Kitchener (1956)	Air	Quartz sphere and plate	—	0.05
Howe *et al.* (1955)		Pyrex and glass plate	Close approach	0.1–10 (estimated)
Theoretical (Casimir and Polder, 1948)	Vacuum			1.0

[a] After Corn (1966).

of the effective distance of separation. Molecular attraction, however, will become increasingly important as particles become smaller. If r_m remains constant, for example, the attractive forces vary with R_p. But forces tending to separate loose agglomerated particles vary with R_p^3 (for vibration), or as R_p^2 (aerodynamic forces). Thus, small particles will be more difficult to pull apart than larger ones.

The calculations of Bradley (1936), Hamaker (1937), Casimir and Polder (1948), and Lifschitz (1954) elucidate the role of interatomic forces between bodies, but other attractive forces may also come into play. Generally, particles can carry substantial electrical charge. For particles charged with opposite signs, attractive forces will exist. Thus in some cases electrical effects may contribute to the mutual adhesion of particles forming aggregates. The larger the particle, the larger the charge it may carry. The importance of electrical charge is readily appreciated in the remarks of Black *et al.* (1960) dealing with experimental difficulties in obtaining measures of the attractive forces between plates. Furthermore, we may note Beisher's work (1939), which showed the obvious importance of electrical forces when the force between 0.5 μm particles carrying only one electron unit of charge will consist mainly of electrostatic interaction and not molecular interaction. Other examples of qualitative electrostatic influences on adhesion have been cited by Corn (1966).

In his treatment of electrification, Loeb (1958) has noted that there is a well-defined mechanism for charge accumulation on small particles called spray electrification. This mechanism probably stems from the orientation of electrical dipoles in liquid substances at a gas–liquid interface with negative polarity outward. Yet the question of electrification during solid–solid contact is poorly understood. The electrification phenomenon in solid particles is complicated by the influence of moisture, electrolytic effects, and the formation of diffusive double layers, or by films of impure materials. In addition, the sliding, rolling, or static nature of solid-surface contacts may affect the charging process. The nature and state of the surface along with mechanical and thermal stressing can alter charging. Because of the complexity of electrification between solids, little of a quantitative nature is known at present about electrical forces and adhesion.

By making a crude estimate on the basis of electrostatic theory, however, one can obtain an idea of the relative magnitude of the electrostatic forces and the intermolecular forces. Suppose a particle carrying a charge Q_i is separated a distance h from an adhering particle in a dielectric medium with dielectric constant ε. The electrical force will be

$$F_{el} = \varepsilon Q_i^2/h^2.$$

With an "equilibrium" charge of 15 esu for a particle of 0.5 μm radius and $h \approx 10$ Å, $F_{el} \approx 5 \times 10^{-3}$ dyn. With Eq. (4.22) and $\pi^2 n_A^2 C_{at} \approx 10^{-13}$, $F_{at} \approx 10^{-4}$ dyn for this case. Thus the electrostatic forces may easily exceed the intermolecular forces as an adhesive attraction. Such a conclusion may also be reached from Derjaguin and Smilga's (1967) calculations of the adhesive force between metal semiconductor crystals based on electronic theory. Their calculations suggest that the electric force of adhesion for such materials is 10^8–10^9 dyn/cm^{-2}. For a zone of contact of 100 Å, the net force of adhesion would be $\sim 10^{-4}$ dyn.

Derjaguin and Smilga's theory indicates the importance of surface conditions in controlling high adhesion. They have suggested ways of conditioning adhesion by chemical reactions on the surface of contacting bodies. Indeed, modification of the adhesive force has been accomplished by irradiation of contacting surfaces with a beam of electrons (Dukhoskoi et al., 1967).

From the discussion above, it is clear that future work on the fundamental nature of adhesion must incorporate studies of electrical interactions between solids, as well as intermolecular interactions.

If a surface is wet or dirty, an additional attractive force may develop as a result of capillary effects of surface films. It has been found, for example, that increase in humidity always increases the adhesion of particles to a surface. A film of water apparently forms by capillary condensation (Fig. 4.15) between the surface and the particles, which adds to the attractive interaction between the two bodies.

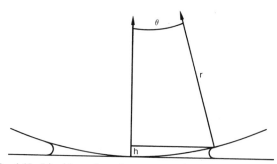

Fig. 4.15. Liquid meniscus between a sphere and a flat surface.

The magnitude of the capillary force or surface tension can be estimated considering the diagram in Fig. 4.15. Suppose a sphere lies on a flat surface and a liquid meniscus is formed around the zone of contact. When the radius of the film is small relative to the radius of curvature of the sphere ($r = R_p$), the distance h (Fig. 4.15) is

$$h = R_p(1 - \cos \theta) \approx 0.5R_p\theta^2,$$

where θ, the angle between the vertical and the radial line through the top of the meniscus, is $\sim\frac{1}{2}h$. Then the negative pressure with the liquid film is $2\sigma/h$ (σ is the surface tension of the liquid film). And, according to Laplace's formula, the adhesive force is estimated as

Sphere–plane:

$$F_{at} = (2\sigma/h)R_f^2 \approx 4\pi R_p\sigma, \qquad (4.24)$$

where R_f is the radius of the film at the point of contact. For two equal-size spheres in contact, h is twice as large as the value in Fig. 4.14 so that

Sphere–sphere:

$$F_{at} \approx 2\pi R_p\sigma. \qquad (4.25)$$

MacFarlane and Tabor (1948) verified this relation by direct experimental measurements of the attractive force between wetted spheres. Equations (4.24) and (4.25) represent the maximum force between two surfaces exerted by a film of moisture. It is not realized until the relative humidity of the environment reaches 100%. From these results we find that the adhesive force associated with wetted surfaces will be as high as the order of σR_p, which can exceed 10^{-3} dyn between two smooth spheres, for example. Thus the effect of surface films may be larger than either the attraction of intermolecular forces or electrical forces.

In addition to the influences of intermolecular forces, electrical forces, and surface tension, Corn (1966) has pointed out that other known mechanisms also alter the adhesive properties; for example, polarization of charge on particles may be modified by an external electrical field. The microscopic surface structure of bodies is important, as well as the nature and extent of surface contamination. Particle surface contact plays an important role in adhesion. This is demonstrated very well by data of Bowden and Tabor (1950) listed in Table 4.10. These workers found that a wide range of net adhesive force can exist when glass spheres are placed on a flat surface at 100% relative humidity. The larger the height of the surface irregularities, the smaller the adhesion because of reduction in the surface of contact.

4.3.3 Entrainment of Particles from Surfaces. In practice the production of natural and artificial dust clouds involves the disruption of a bed of particles by various means such as blowing a gas over the top of the bed or passing gas up through the bed. Before taking up these cases, it is useful to consider the mechanisms of entrainment of particles from a smooth flat surface.

To dislodge a single particle from a surface several possible processes may be invoked. In all cases, however, the principal requirement for removal is that the dispersing force acting on the particle must exceed the adhesive force of attraction. We have seen that strength of adhesion (as well as cohesion) depends on several factors including the nature of the particle material, its size and shape, surface roughness, and the relative humidity of the surrounding gas. In addition, the presence of electrostatic charge must be taken into account.

Particles adhering to a surface can be entrained into surrounding air by several means including (a) mechanical concussion or vibration, (b) explosive dislodgment, or (c) aerodynamic dispersement.

Dislodgment by mechanical means such as vibration requires that the normal component of the adhesive attraction be overcome. As we have seen above, adhesion depends on many factors. However, Corn and Stein (1967)

TABLE 4.10

Adhesion of Glass Spheres to Glass Plate in Atmosphere of 100% Relative Humidity[a]

Glass plate surface	Mean approximate height of surface irregularities (Å)	Adhesion as percent of value observed with highly polished surfaces
Highly polished	150	100
500 carborundum paper	1,000	79
320 carborundum paper	4,000	51
150 carborundum paper	100,000	0

[a] Based on data of Bowden and Tabor (1950); from Corn (1966).

have cited typical values of the attractive force for glass spheres on a glass slide from which may be estimated the typical mechanical force for entraining particles into air. For example, for an 18.5-μm-radius glass sphere sticking to a clean, smooth flat glass surface, the component of cohesive force normal to the slide is 0.60 dyn to remove 50% of the particles in air at 35% relative humidity, 20°C, and 1 atm pressure. For a particle density of 2 g/cm^3, the acceleration equivalent to this force is 1.25×10^7 cm sec^2. If the surface on which the particle is lodged moves at 1 m/sec, the deceleration to a standstill of the particle in removal from the surface would have to take place in $\sim 10^{-4}$ cm; at a surface speed of 10 m/sec, the deceleration distance is $\sim 10^{-2}$ cm. Corn and Stein note that under these circumstances, removal of particles is not easily carried out by deceleration. However, particles can easily be dispersed by transfer of momentum from a moving body. Dust may be swept away by a brush or a hand. Transfer of the 0.66 dyn and dispersal of the 18 μm particles may be accomplished as in the action of a shoe scraping across a floor.

Vibration of a substrate may also induce entrainment. For constant acceleration α,

$$\alpha = 4\pi f_0^2 A_0$$

where f_0 is the frequency of vibration and A_0 the vibration amplitude. To remove an 18 μm glass sphere from the glass surface the frequency is 565 Hz for a 1-cm-amplitude vibration. Such rapid oscillations are not likely to occur commonly.

Several years ago Billings and co-workers (1960) found experimentally that the explosive release of energy could dislodge dust particles from a slag wool filter. A blast wave of 127 mmHg overpressure with a wind speed of 42 m/sec and a dynamic pressure of 10 cm of water was enough to remove adhering particles from the filter fibers. The exact mechanism of dislodgment is still uncertain. Particle removal may be due in part to aerodynamic effects, as discussed below, and possibly to vibration of the substrate. However, the work of Gerrard (1963) dictated that energy transfer from the blast wave also may be associated with reflection of the shock wave from the substrate and subsequent raising of the particles with the reflected wave. This appears to be analogous to vibrating the substrate.

Perhaps most common of all mechanisms for dispersing powders is the blowing away of particles by a gas flow. Prediction of the onset of particle pickup from a powder bed is difficult because of uncertainties in specifying the velocity distribution near the particles as well as the cohesive and adhesive force. In the somewhat simpler case of aerodynamic removal of single particles from surfaces, more is known, based on the studies of Bagnold (1941) of sand grain migration induced by wind. Above the range in particle size where cohesive bonding is important, particles on a smooth surface go through successive stages of movement induced by the air flow overhead.

Under the action of the wind the particles roll over other grains. The continuing rollover, with a bouncing action, was referred to by Bagnold as *saltation.*[†]

Particles evidently leave horizontal surfaces by wind action in the following way. The grains first begin to slide or roll along until their speed is sufficient for small bumps on the surface, or on the particles themselves, to cause them to jump. Since the air flow near the surface generally will be turbulent, particles may jump out of the viscous sublayer very close to the surface. Then the particle may either be caught up by turbulent fluctuations associated with eddying motion or, if it is spinning, may be affected by the so-called Magnus force,[‡] and may be carried clear of the surface. Fine particles held mainly by molecular or electrostatic attraction probably break loose in a similar way provided that the attractive force between the particle and the surface is overcome by the aerodynamic lifting. Since the aerodynamic forces oriented tangentially to a smooth surface will be greater than those normal to the surface, one expects that the particle will begin its dislodgment by sliding or rolling. This will inevitably give rise to a reduction in the number of points of contact between the particle and the surface and, as a consequence, will cause a reduction in the net adhesive force. Thus the probability of a particle leaving the surface after beginning to roll is considerably enhanced.

The calculation of the force required to carry away particles from a smooth surface is uncertain. It is difficult to estimate quantitatively the adhesive and electrostatic forces, and the values of static friction are not well known. Though the static friction of large bodies is easy to measure, there are considerable difficulties in performing similar measurements for bodies smaller than about 0.5 mm in radius.

The net force of static friction, which is the tangential component of the adhesive force, is proportional, like the normal component, to the area of contact. The extent of this area depends largely on the plastic deformation of microprojections at the points of contact. Bowden and Tabor (1950) have shown that for larger bodies, the contact area is approximately proportional to the normal force on which the frictional theory is based. However, when the size of the body decreases, the normal force (its weight) decreases rapidly as well, and the influence of the adhesive force becomes more important. The actual contact area, and consequently the frictional force, is proportional to the sum of the normal forces. Since the adhesive force roughly is proportional to R_p, the coefficient of friction will increase with decrease in R_p.

Experiments of Löffler (1967), using two independent methods for mea-

---

[†] For additional discussion relating to blowing dust in the atmosphere, see Section 7.4.3.

[‡] The Magnus force is an apparent force acting on a spinning body. The spin may be induced on a smooth body by a velocity gradient.

suring the force of adhesion, indicated that F_{ad} depends on particle size, as well as the deposition velocity onto a surface and relative humidity. For 5–15 μm dry particles filtered onto monofile polyamide fibers, F_{ad} as determined by centrifugation or blowoff ranged from 10^{-4} to 5×10^{-2} dyn, with the two experimental methods in agreement.

Differences of roughness in zones of contact of large bodies having the same normal force will be averaged out, but for fine particles the normal component of the adhesive force and the contact area will depend largely on the microgeometry of the surfaces and should vary greatly from particle to particle, from one orientation to another, and from one point on the surface to another. Experiments on the removal of dusts from flat surfaces have shown considerable differences in results, which tend to confirm the inherent variability in adhesive forces. The above remarks also explain why, in a given set of measurements of dust removal, relations between flow rate and fraction of particles blown off change such that complete removal requires several times greater gas flow than that at which removal begins.

Because of the lack of knowledge of surface geometry, as well as of the adhesive and cohesive forces, estimation of the force required to blow off particles from powder beds is even more uncertain. Based on the consensus of many measurements, however, Fuchs (1964) gives a rough rule of thumb which has been given for the adhesive or cohesive force between particles in beds:

$$F_{ad} \approx 2R_p \quad \text{to} \quad 5R_p \quad \text{dyn.}$$

(R_p in cm). This magnitude is of the same order as the frictional component, but appears to be substantially greater than predicted by the Bradley–Hamaker theory. One would expect that this intermolecular force, given by Eq. (4.22), would be the minimum force required to lift a particle out of a powder bed.

4.3.4 Dispersal of Powders by Air Jets. When gas flows at relatively low speeds over layers of powder, particles become detached from the surface and are carried away by the turbulent motion. As particles lift away from the powder bed, the bed is compacted by the gas flow and begins to assume a rounded, streamlined shape until erosion no longer occurs. To renew the removal of particles from the powder layer, much greater gas speeds are required. After the onset of the higher gas flow, similar erosion and streamlining occur. However, at sufficiently high gas speeds not only are individual particles removed, but whole aggregates may be shed by the bed and immediately disintegrated by the action of the air stream. Under these conditions bulges in the surface of the bed are produced and the surface becomes more and more uneven, aiding, in turn, further particle removal. As Fuchs (1964) has pointed out, this constitutes the transition from surface to bulk atomization of the powder. The latter occurs, for example, in nozzles used to inject coal dust into furnaces, in the pneumatic separation of powders, and in the dispersal of insecticide dusts.

Bulk atomization of powder beds by a gas jet is closely akin to the fluidization of powders enclosed in a vertical pipe. Fluidization has been studied extensively in recent years because of the interest in chemical technology in using such systems for carrying catalysts for chemical reaction. One comprehensive survey of understanding of fluidization mechanisms has been given by Davidson and Harrison (1963).

Qualitatively the process of fluidization of a powder bed in a tube is described as follows: gas is passed through the bed via a porous bottom plate until critical velocity is reached, when the bed begins to expand in depth. Up to critical gas velocity the pressure drop increases steadily; after reaching critical flow the pressure drop becomes nearly constant. At low gas velocities, where particles remain stationary, the pressure drops follow the Blake–Kozeny relation

$$\Delta P = K_p q_g L \mu_g V^2 / R_p^2 (1 - V)^3, \tag{4.26}$$

where K_p is a factor depending on particle shape ($\sim$37.5), L is the depth of the bed, V is the volume fraction of particles, and q_g is the gas velocity above the bed. The critical gas flow q_{cr} is given by the situation when the pressure gradient of the medium equals the gradient in hydrostatic pressure of the powder, $gV(\rho_p - \rho_g)$, or

$$q_{cr} = \Delta P R_p^2 (1 - V)^3 / K_p L \mu_g V^2. \tag{4.27}$$

At this point the particle bed begins to expand and the flow resistance (given by ΔP) becomes essentially constant because the product LV remains constant.

In finer powders, where cohesion between particles plays an important role, the Blake–Kozeny formula is only qualitatively correct. The ratio $q_g/\Delta P$ increases more rapidly than Eq. (4.26) predicts. This is related to the observation that fluidized beds do not expand uniformly, but break down into separate cells with slitlike open passages between them. Much of the gas goes through the passages, so q_g is greater than predicted by Eq. (4.26).

After a certain amount of expansion corresponding to 5–20% of the original volume, particles begin to move and the resistance of the bed decreases a little, possibly as a result of reduced friction between the particles. Beyond this stage, fluidization takes place in which particles are at least partially suspended in the turbulent gas. In this regime the gas bubbles through the expanded bed, and the layer depth continues to expand but fluctuates wildly. The bubbling causes rather vigorous agitation of the powder, promoting a circulation rising at the center of the tube and settling along the walls. This state has been aptly likened to the boiling of a liquid. As the gas flow is further raised, particles whose settling rate is less than the flow velocity begin to leave the "fluid phase" and are drawn out into a "gaseous" or "aerosol" phase. The particle concentration in the aerosol phase keeps increasing until the boundary between phases fades away, and the powder is completely dispersed in the gas.

Once a fluidized bed begins to "leak" aerosol particles, increased gas velocity continues to yield higher particle concentrations. The increase is linear for experiments performed to date.

The successive stages of aerosol formation in a powder bed being fluidized by gas flow give a picture of the interaction between the fluid and the particles in dispersal. Similar interaction no doubt can be associated with any dispersal by high speed jet air blowing horizontally along a layer of powder. The main uncertainties in fluidization processes lie in the influences of particle shape and nature of the material just as in the previous discussion of aerodynamic entrainment at relatively low gas speeds. In the "boiling" state dispersibility again depends on the cohesive and adhesive properties of the powder as well as on the size and shape of the particles.

Apart from particle size, shape, and mechanical properties, plasticity and the density of packing of particles, or their number of contacts with a neighbor, are all important. Fuchs (1964) has noted, for example, that it is more difficult to disperse powders of soft plastic materials than hard ones. Monodisperse powders atomize more readily than polydisperse powders because in the latter more intergranular space is filled, so more points of contact exist between individual particles. It is especially difficult to disperse powders whose size range is large since their packing density is so much greater than that of isodisperse powders. Decrease in packing density or in the bulk density of a powder, on exposure to sonic vibrations, for example, apparently substantially facilitates dispersion.

Apart from studies of fluidization the investigation of mechanisms for powder dispersal is virtually untouched in the literature. Even empirical information on the relative dispersibility of different powders is scarce. One classical study of this kind by Anderson (1939) is available.

In Anderson's experiments 2 cm³ of powder were poured through a slit into a vertical tube of height 250 cm and diameter 4.5 cm. The particles separated partially during fall to the bottom of the tube. The percentage of powder not settled on the tube bottom in 6 sec was measured. Anderson estimated that individual particles would not fall through the tube to the bottom in this time, so his numbers represent a percentage of dispersed powder, or *dispersibility*. Fuchs (1964) has remarked that almost certainly some of the unsettled fraction also contained some aggregates so that Anderson's experiments were only of qualitative significance. Nevertheless, one can readily see from his data in Table 4.11 that typically there exists quite a wide range in dispersibility for various classes of particulate material of roughly the same size.

Of all the powders listed in Table 4.11, lycopodia spores apparently have the greatest fluidity. This may be attributed to the uniformity of size, light-

TABLE 4.11

Dispersibility of Some Powders[a]

	Particle radius limits (μm)	Dispersibility (%)
Lycopodium	12	100
Wood charcoal dust	0–25	85
Wood charcoal dust	0–7	23
Aluminum powder	0–15	66
Talc	0–20	57
Carbon black	0–15 (?)	47
Potato starch	0–35	27
Graphite dust	0–25	17
Pulverized slate	0–25	13
Cement	0–45	5.5
Prepared chalk	0–6	1.5
Polydisperse silica dust (coarse)	—	21
Polydisperse silica dust (fine)	—	8
	11.5	68
Isodisperse silica dust	8	83
	5.6	45
	7	50
Porcelain dust with fine fractions	2.7	52
	1.1	21
	0.45	12
Porcelain dust without fine fractions removed	—	5

[a] From Anderson (1939).

ness, and spherical shape of the particles as well as to the presence on their surfaces of tiny ridges, which will favor a loose structure in the powder.

REFERENCES

Amelin, A. (1948). *Kolloidn. Zh.* **10,** 168.

Anderson, A. (1939). *Kolloid. Z.* **86,** 70.

Bagnold, R. (1941). "Physics of Blown Sand and Desert Dunes." Metheun, London.

Bailey, A., and Courtney-Pratt, J. (1955). *Proc. R. Soc. London, Ser. A* **227A,** 500.

Basanaev, M., Teverovskii, E., and Tregubova, E. (1949). *Dokl. Akad. Nauk. SSSR* **66,** 821.

Becker, R., and Döring, W. (1935). *Ann. Phys.* **24,** 719.

Beisher, D. (1939). *Kolloid. Z.* **89,** 215.

Berglund, R. N., and Liu, B. Y. H. (1973). *Environ. Sci. Technol.* **7,** 147.

Billings, C., Silverman, L., Dennis, R., and Levenbaum, L. (1960). *J. Air Pollut. Control Assoc.* **10,** 318.

Binnie, A., and Woods, M. (1938). *Proc. Inst. Mech. Eng.* **138,** 29.

Bisa, K., Dirongh, K., and Esche, R. (1954). *Siemans Z.* **28,** 341.

Bitron, M. (1955). *Ind. Eng. Chem.* **47,** 23.

Black, W., de Jangh, J., Overbeek, J., and Sparnaay, M. (1960). *Trans. Faraday Soc.* **56,** 1597.

Boffa, C. V., Mazza, A., and Rosso, D. A. (1976). *In* "Fine Particles: Aerosol Generation, Measurement, Sampling, and Analysis" (B. Y. H. Liu, ed.), p. 11. Academic Press, New York.

Boulard, D., Madelaine, G., Vigla, D., and Bricard, J. (1975). *Water, Air, Soil Pollut.* **4**, 435.

Boulard, D., Madelaine, G., Vigla, D., and Bricard, J. (1978). *In* "Recent Developments in Aerosol Science" (D. T. Shaw, ed.), p. 61. Wiley (Interscience), New York.

Bowden, F., and Tabor, D. (1950). "Friction and Lubrication of Solids." Oxford, London.

Bradford, E. B., Vanderhoff, J. W., and Alfrey, T., Jr. (1956). *J. Colloid Sci.* **11**, 135–149.

Bradley, R. (1932). *Trans. Faraday Soc.* **32**, 1088.

Bradley, R. (1936). *Philos. Mag.* **13**, 853.

Bukhman, S. (1954). *Izv. Akad. Nauk. Kaz. SSR* No. 11.

Calvert, J. G., Su, F., Bottenheim, J., and Strausz, O. P. (1978). *Atmos. Environ.* **12**, 197.

Casimir, H., and Polder, D., (1948). *Phys. Rev.* **73**, 360.

Castleman, R. (1931). *J. Res. Natl. Bus. Stand. (Wash.)* **6**, 309.

Corn, M. (1966). *In* "Aerosol Science" (C. N. Davies, ed.), p. 359. Academic Press, New York.

Cox, R. A. (1973). *J. Aerosol Sci.* **4**, 473–483.

Dahneke, B. (1978). *In* "Recent Developments in Aerosol Science" (D. Shaw, ed.), p. 187. Wiley (Interscience), New York.

Davidson, J., and Harrison, D. (1963). "Fluidized Particles." Cambridge, London.

Derjaguin, B. V., and Smilga, V. (1967). *J. Appl. Phys.* **38**, 4609.

Derjaguin, B. V., Abrikosona, I., and Lifschitz, E. (1956). *Q. Rev. Chem. Soc.* **10**, 295.

Dombrowski, N., and Johns, W. (1963). *Chem. Eng. Sci.* **18**, 263.

Drozin, V. (1955). *J. Colloid Sci.* **10**, 158.

Dukhoskoi, E., Dragelskii, I., and Ailin, A. (1967). *Sov. Phys. Dokl. (Engl. Transl.)* **12**, 730.

Dunning, W. (1961). *Discuss. Faraday Soc.* **30**, 9.

Dunskii, V. (1956). *Zh. Tekh. Fiz.* **26**, 1262.

Dunskii, V., and Nikitin, N. (1965). *Inzh. Fiz. Zh.* **9**, 54.

Fraser, R., Dombrowski, N., and Rontley, J. (1963). *Chem. Eng. Sci.* **18**, 315, 323.

Friend, J. P., Leifer, R., and Trichon, T. (1973). *J. Atmos. Sci.* **30**, 465.

Fuchs, N. A. (1964). "The Mechanics of Aerosols." Pergamon, New York.

Fuchs, N. A., and Sutugin, A. G. (1966). *In* "Aerosol Science" (C. N. Davies, ed.), pp. 1–30. Academic Press, New York.

Fulwyler, M. S., and Raabe, O. G. (1970). The Ultrasonic Generation of Monodisperse Aerosols. Presented at the Am. Ind. Hygiene Assoc. Conf., Detroit, Michigan.

Gerrard, J. (1963). *Br. J. Appl. Phys.* **14**, 186.

Gmitro, J. I., and Vermeulen, T. (1964). *AIChE J.* **10**, 741.

Green, H. (1953). *In* "Flow Properties of Disperse Systems" (J. Hermans, ed.), p. 299. North-Holland Publ., Amsterdam.

Green, H., and Lane, W. (1964). "Particulate Clouds: Dusts, Smokes and Mists," 2nd ed. Van Nostrand-Reinhold, Princeton, New Jersey.

Gregg, S. J. (1968). *Chem. Ind. (London)* p. 611.

Guichard, J. C. (1976). *In* "Fine Particles: Aerosol Generation, Measurement, Sampling, and Analysis" (B. Y. H. Liu, ed.), p. 174. Academic Press, New York.

Hagerty, W., and Shea, J. (1955). *J. Appl. Mech.* **22**, 509.

Hamaker, H. (1937). *Physica* **4**, 1058.

Harms, J., and Jander, G. (1937). *Kolloid. Z.* **77**, 267.

Hasson, D., and Mizrahi, J. (1961). *Trans. Inst. Chem. Eng.* **38**, 415.

Heist, R., and Reiss, H. (1973). *J. Chem. Phys.* **49**, 665.

Hidy, G. M., and Brock, J. R. (1970). "The Dynamics of Aerocolloidal Systems." Pergamon, Oxford.

Hidy, G. M., and Friedlander, S. K. (1964). *AIChE. J.* **10**, 115.

Higuchi, W., and O'Konski, C. (1960). *J. Colloid Sci.* **15**, 14.

Hinze, J. (1948). *Appl. Sci. Res.* **1A**, 273.

Hinze, J., and Milborn, H. (1950). *J. Appl. Mech.* **17,** 145.

Homann, K. (1967). *Combust. Flame* **11,** 265.

Howe, P., Benton, D., and Puddington, I. (1955). *Can. J. Chem.* **33,** 1375.

Huang, C. M., Kerker, M., Matijevic, E., and Cooke, D. D. (1970). *J. Colloid Interface Sci.* **33,** 244.

Hubrecky, H. (1958). *J. Appl. Phys.* **29,** 572.

Jander, G., and Winkel, A. (1933). *Kolloid. Z.* **63,** 5.

Kasiosis, F., and Fish, B. (1962). *J. Colloid Sci.* **17,** 155.

Kassner, J., Jr., and Schmitt, R. (1967). *J. Chem. Phys.* **44,** 4166.

Kassner, J., Jr., Carstons, J., and Allen, L. (1968). *J. Atmos. Sci.* **25,** 919.

Katz, J. L., Scoppa, C. J., Kumar, N. G., and Mirabel, P. (1975). *J. Chem. Phys.* **62,** 448.

Kerker, M. (1975). *Adv. Colloid Interface Sci.* **5,** 105.

Kiang, C. S., Stauffer, D., Mohnen, V. A., Bricard, J., and Vigla, D. (1973). *Atmos. Environ.* **1,** 1279.

Kitchener, J., and Prosser, A. (1957). *Proc. R. Soc. London, Ser. A* **242A,** 403.

Kocmund, W., Kittelson, D. B., Yang, J. Y., Demer, K. L., and Whitby, K. T. (1977). *In* "Fate of Pollutants in the Air and Water Environments" (I. H. Suffet ed.), Part 2, p. 101. Wiley (Interscience), New York.

Lane, W. (1951). *Ind. Eng. Chem.* **43,** 1312.

Lane, W., and Green, H. (1956). *In* "Surveys in Mechanics" (G. Batchelor and R. Davies, eds.), p. 163. Cambridge Univ. Press, London and New York.

Lang, R. (1962). *J. Acoust. Soc. Am.* **34,** 6.

Lapple, C. E. (1969). *Adv. Chem. Eng.* **8,** 26.

Levich, V. (1962). "Physicochemical Hydrodynamics." Prentice-Hall, Englewood Cliffs, New Jersey.

Lifschitz, E. (1954). *Dokl. Akad. Nauk. SSSR* **97,** 643.

Lipeles, M., Landis, D. A., and Hidy, G. M. (1977). *In* "Fate of Pollutants in the Air and Water Environments" (I. H. Suffet, ed.), Part 2, p. 69. Wiley (Interscience), New York.

Liu, B. Y. H., ed. (1976). "Fine Particles: Aerosol Generation, Measurement, Sampling, and Analysis," p. 87. Academic Press, New York.

Liu, B. Y. H., Whitby, K. T., and Yu, H. (1966). *J. Rech. Atmos.* **2,** 397.

Loeb, L. (1958). "Static Electrification." Springer-Verlag, Berlin and New York.

Löffler, F. (1967). *Staub* **26,** 10.

McCubbin, T. (1953). *J. Acoust. Soc. Am.* **25,** 1013.

MacFarlane, J., and Tabor, D. (1948). *Proc. 7th Congr. Appl. Mech.* pp. 335.

May, K. R. (1973). *J. Aerosol. Sci.* **4,** 235.

Mercer, T. T. (1973). "Aerosol Technology in Hazard Evaluation." Academic Press, New York.

Mercer, T. T., Tillery, M. I., and Chow, H. Y. (1968a). *Am. Ind. Hyg. Assoc. J.* **29,** 66.

Mercer, T. T., Goddard, R. F., and Flores, R. L. (1968b). *Ann. Allergy* **26,** 18.

Merrington, A., and Richardson, E. (1947). *Proc. Phys. Soc., London* **59,** 1.

Middleton, P., and Kiang, C. S. (1978). *Atmos. Environ.* **12,** 179.

Mirabel, P., and Katz, J. L. (1974). *J. Chem. Phys.* **60,** 1138.

Monk, G. (1952). *J. Appl. Phys.* **23,** 288.

Oswatitsch, K. (1942). *Z. Angew. Math. Mech.* **22,** 1.

Overbeek, J. T. G., and Spaarnay, M. (1954). *Discuss. Faraday Soc.* **18,** 12.

Parker, W. G., and Wolfhard, H. (1950). *J. Chem. Soc.* Part 3, p. 2038.

Pattison, J., and Aldridge, J. (1957). *Engineer* **203,** 514.

Peskin, R., and Raco, R. (1963). *J. Acoust. Soc. Am.* **35,** 1378.

Peskin, R., and Raco, R. (1964). *AIAA J.* **2,** 781.

Prosser, A., and Kitchener, J. (1956). *Nature (London)* **178,** 1339.

Quon, J. E., Siegel, R. P., and Hulburt, H. M. (1971). *Proc. 2nd Int. Clean Air Congr., 1970* pp. 330–335.

Raabe, O. G. (1968). *Am. Ind. Hyg. Assoc. J.* **29,** 439.

Raabe, O. G. (1970). *In* "Inhalation Carcinogenesis," pp. 123–172. UAEC Div. of Tech. Inf., Oak Ridge, Tennessee.

Raabe, O. G. (1972). "Operating Characteristics of Two Compressed Air Nebulizers Used in Inhalation Experiments." Report LF-45. Lovelace Foundation, Albuquerque, New Mexico.

Raabe, O. G. (1976). *In* "Fine Particles: Aerosol Generation, Measurement, Sampling, and Analysis" (B. Y. H. Liu, ed.), p. 60. Academic Press, New York.

Rapaport, E., and Weinstock, S. E. (1955). *Experientia* **11,** 363.

Rayleigh, L. (1878). *Proc. London Math Soc.* **10,** 4.

Rayleigh, Lord (1879). *Proc. R. Soc. London* **24,** 71.

Reiss, H., and Katz, J. L. (1967). *J. Chem. Phys.* **46,** 2496.

Ryley, D. (1959). *Br. J. Appl. Phys.* **10,** 180.

Schlichting, H. (1960). "Boundary Layer Theory," 4th ed. McGraw-Hill, New York.

Schultz, R., and Bronson, L. (1959). *2nd Symp. Adv. Propulsion Concepts, Boston.*

Schwartz, M. H., and Andres, R. P. (1976). *J. Aerosol Sci.* **7,** 281.

Schweckendiek, O. E. (1950). *Z. Naturforsch A* **5A,** 397.

Shugard, W. J. (1974). *J. Chem. Phys.* **61,** 5298.

Sinclair, D., and LaMer, V. K. (1949). *Chem. Rev.* **44,** 245.

Straus, R. (1950). Ph.D. Dissertation, London Univ., England.

Tabor, D., and Winterton, R. (1968). *Nature (London)* **219,** 1120.

Taylor, G. I. (1949). U.K. Ministry of Supply Report, London.

Taylor, G. I. (1950). *Q. J. Mech. Appl. Math.* **3,** 129.

Tomaides, M., and Whitby, K. T. (1976). *In* "Fine Particles: Aerosol Generation, Measurement, Sampling, and Analysis" (B. Y. H. Liu, ed.), p. 235. Academic Press, New York.

Tyndall, J. (1869). *Philos. Mag.* **37,** 384.

Vold, M. (1954). *J. Colloid Sci.* **9,** 451.

Volynskii, M. (1948). *Dokl. Akad. Nauk. SSSR* **62,** 301.

Vonnegut, B., and Neubauer, R. (1952). *J. Colloid Sci.* **7,** 616.

Walton, W., and Prewett, W. (1949). *Proc. Phys. Soc., London, Sect. B* **62B,** 341.

Watson, E. (1948). *Proc. Inst. Mech. Eng.* **158,** 187.

Wegener, P., and Pouring, A. (1964). *Phys. Fluids* **1,** 352.

Wen, F. C., McLaughlin, T., and Katz, J. L. (1978). *Science* **200,** 796.

Wilson, C. T. R. (1927). *In* "Nobelstiftelsen; Nobel Lectures–Physics," Vol. 2 (1922–1941), p. 194. Elsevier, New York.

Zeleny, J. (1916). *Proc. R. Cambridge Philos. Soc.* **18,** 71.

CHAPTER 5

MEASUREMENT OF AEROSOL PROPERTIES

The measurement of the physical and chemical properties of particle suspensions has been a central theme of aerosol science since its beginnings. The variety of devices and methods adopted for such purposes represents a diverse collection of instrumentation designed for specific applications. The reason for this diversity is that no single technique or group of techniques provides a means for characterizing properties over the extremely wide range of particle size, shape, and chemical composition found in nature, or in the laboratory. The devices range from simple instruments for measurement of light transmission, or porous filters to collect material for determining the mass concentration, to very sophisticated sensors or collectors to characterize the particle size distribution and chemistry.

To adequately characterize the dynamic properties of a chemically reactive aerosol, a very large amount of information is required. However, aerosol properties generally are only determined in a limited way because of constraints in available microtechniques. The commonly observed and most important characterizing parameters are listed in Table 5.1. With air pollution monitoring and the driving force of progress in development of theory, heavy emphasis has been placed on the size distribution and its moments, as well as the chemical composition of particles and the suspending gas. The measurement of properties of the gas component of the aerosol medium is beyond the scope of this book, and is not considered further.

Measurement of particles or particle collections is done by one of two approaches: by (a) in situ, or quasi-in situ continuous observation, or (b) collection on a medium and subsequent laboratory investigation of the accu-

TABLE 5.1

Measurements Characterizing Aerosol Properties

Property	Units	Common measure
Particle size characterization as "spheres"		
Number–size distribution	$\#/cm^3$ μm	Continuous
Surface–size distribution	$\mu m^2/cm^3$ μm	Continuous
Volume–size distribution	$\mu m^3/cm^3$ μm	Continuous
Particle size characterization		
Mass–size concentration	$\mu g/cm^3$ μm	Impactor
Size/shape morphology	$\#/cm^3$ μm	Microscopy
Size–settling spectrum	$\mu g/cm^2$ sec μm	Centrifuge
Size–chemical composition distribution	$\mu g/cm^3$ μm	Collection–laboratory analysis
Integrated physical properties		
Total mass concentration	$\mu g/m^3$	Collection–gravimetry
Total surface area	cm^2/m^3	Collection–BET
Volume fraction	$\mu m^3/m^3$	Collection
Average density	g/cm^3	Mass concentration/volume fraction
Light–scattering coefficient	b_{sp} (m^{-1})	Nephelometry/light scattering
Light absorption coefficient	b_{ap} (m^{-1})	Light absorption
Mobility	sec/g	Fluid motion in external force field
Surface energy	erg/cm^2	Bulk surface tension
Vapor pressure	atm or mmHg	—
Hygroscopicity	g/m^3 RH	Mass change with relative humidity
Accommodation coefficient	Dimensionless	—
Chemical characterization		
Inorganic elements (S, Fe, Pb, Zn, Mn, Na, K, . . .)	$\mu g/m^3$	X-ray fluorescence analysis (XRFA); neutron activation analysis (NAA)
Common inorganic ions (SO_4^{2-}, NO_3^-, Cl^-, NH_4^+)	$\mu g/m^3$	Wet-ion chemistry; ion chromatography
Total, elemental and organic carbon	$\mu g/m^3$	Thermo gravimetric-CH_4 detection
Water content	$\mu g/m^3$	Microwave spectroscopy
Suspending gas properties		
Reactive components	ppm	—
Water vapor content	Absolute humidity; ppm	Dew-point hygrometry

mulated particulate material. No single method provides a self-consistent, complete physical picture of particle suspensions. For example, the first group basically assumes that the particles can be treated as inert spheres during the measurement process. The second assumes accumulation of material on a medium or substrate without modification of the particles. This is known to be a less than satisfactory assumption for particles reacting with the suspending gas, but no better techniques have been developed.

If one focuses on the particle size distribution function, a central framework for describing aerosols, one can conveniently classify the measurement instruments according to properties of the size distribution function. Organization of instrumentation gives perspective on the ideal requirements as contrasted with practical limits available with present technology. An idealized hierarchy suggested by Friedlander (1977) is illustrated in Fig. 5.1. At the top of the list is the ideal modern aerosol analyzer, which gives continuous resolution in particle size with chemical composition. The ideal instrument operated at full capacity would measure and read out directly the

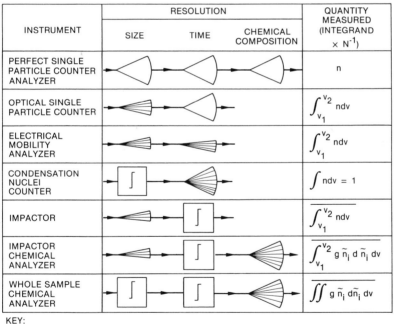

KEY:

◁ RESOLUTION AT SINGLE PARTICLE LEVEL

◀ DISCRETIZING PROCESS

∫ AVERAGING PROCESS

Fig. 5.1. Characteristics of aerosol measurement instruments. Note $\bar{n}_i$ is the molar concentration of species i. [From Friedlander (1977); by permission of the author and John Wiley & Sons]

particle distribution as a function of size and chemical properties. No such device exists even for the simplest kinds of particle suspensions. In practice, a variety of instruments are available which report size distribution functions or integral average properties of the distribution function. Some are listed in Fig. 5.1. These include the single-particle counter, which measures particles over discrete size ranges using differences in light-scattering properties with particle size. Next are devices such as the electrical mobility analyzer, which measures particles in size groups by counting particles in a given size range over a discrete time interval. Finally, there is a series of devices which integrate the distribution function and give information about certain moments of the distribution. These include idealized total particle counters such as a nuclei counter, a particle collector like an impactor which segregates the sample by size over certain discrete size and time intervals, and the total mass–chemical analyzer such as a filter submitted to the laboratory for gravimetry and chemical assay.

Once collected, particles may be sized by a variety of means, which are summarized in Table 5.2. Optical and electron microscopy are probably best known, but they involve tedious scanning of many samples to obtain sufficient counts to provide meaningful particle statistics. The other techniques are suitable only for solid particles and not for liquids.

Microscopy remains the principal standard method of particle sizing and shape and morphological classification. Though often tedious and time consuming in its application, it remains a standard by which (a) individual particles can be classified with confidence, and (b) most particle sizing methods are referenced. The preparation of samples for particle characterization by microscopy has been discussed by Silverman *et al.* (1971).

The magnification of a microscope is a function of the focal lengths of the lenses in the optical system. Any magnification is achievable in principle by the selection of focal lengths. However, a practical limit for an imaging

TABLE 5.2

Methods of Size Analysis for Particles Collected on a Substrate

Principle	Function measured	Approximate submicrometer size range
Optical imaging (microscopy)	#/surface area	>100 nm
Electron image formation	#/surface area	>5 nm
Conductivity change in electrolyte	Volume	>200 nm
Gas adsorption based on Harkins– Jura or Brunauer–Emmett–Teller method	Surface area	
X-ray diffraction line broadening	Crystallite size	<~500 nm

device exists because of diffraction effects and optical distortion. The image of a point produced by an ideal lens is not a point but a diffraction pattern appearing as a circular disk surrounded by alternating dark and light rings of diminishing intensity. Distortion enters in any practical system where basic constraints are present on manufacture.

In examining particles by microscope, the sizing depends on one's being able to distinguish an edge of a particle from another on the opposite side. The ability of the microscope to make such distinctions is measured in terms of its resolving power, the closest distance to which two objects can approach and still be recognized as being separate. This limiting resolution of the instrument is the radius of the central disk of the object's diffraction pattern, which is

$$W_{res} = 0.61\lambda/(m \sin \theta), \tag{5.1}$$

where λ is the wavelength of light, m is the refractive index of the medium, and θ is the half angle of the light rays coming from the object. To maximize resolution, the wavelength should be small, and the index of refraction and half angle large. The refractive index can be increased over that of air ($m = 1$) by using oil of $m \approx 1.5$. The highest numerical aperture $m \sin \theta$ attainable in practical systems is about 1.4. Thus, with $\lambda = 0.5$ μm, the best resolution achievable for optical microscopy is about 0.2–0.3 μm.

The resolving power of microscopes has been enhanced significantly by using very short wavelength radiation from high speed electron beams. Using electric and magnetic fields as "lenses," microscopes have been devised with limits of resolution of about 10 Å. The range of resolving power for microscopy in current practice is listed in Table 5.3. Since there is heating from intense exposure to light and the electron beam operates in an evacuated medium, volatile particles will evaporate, limiting the application of this method for solid aerosol particle collections. Sometimes particles will leave a characteristic residue which may be identified with the original particle, this can be used for classification, but not for accurate sizing.

With this background in mind, the remainder of this chapter is devoted to examination of particle measurement principles in detail. The discussion begins by considering the limitations of aerosol flow in sampling. Then the principles of optical and electrical charging are reviewed, with application to instrumentation. Before discussing the methods for chemical analysis, collection by inertial forces and filtration are described as a major application of certain mechanical principles noted in Chapters 2 and 3.

5.1 SAMPLING DESIGN

There are many pitfalls in measuring the properties of aerosols. One of the more critical is sampling of particulate matter without disturbing the aerial suspension. There are some optical devices which can make measurements

TABLE 5.3

Applications of Microscopy Magnification Given Limiting Human Eye Resolution of 0.2 mm

Type of instrument	Resolving power (μm)	Magnification ($\times$)	Type of particle that can be observed
Eye	200	1	Large dust particles, filters in sunlight
Magnifying glass	25–100	2–8	Dust particles, fine powders
Low-power compound optical microscope	10–25	8–20	Pollen, soil dust, fibers, sea salt
Medium-power compound optical microscope	1–10	20–200	Soil dust, fly ash
High-power compound optical microscope; electron microscope	0.25	800	Cigarette smoke, fly ash
Electron microscope; high-power compound optical microscope using ultraviolet light	0.10	2,000	Fine particle of large particles, combustion products, chemical reaction products
Electron microscope	0.001–0.05	4,000–200,000	Nuclei, metallic reaction products

of an aerosol in situ without disturbance. However, most devices require that a small sample be taken from the gas–particle suspension. Because of inertial forces acting on particles, it can be deduced readily that syphoning part of the fluid as a sample must be done carefully to avoid preferential withdrawal of different size particles. Particle deposition on the walls of the sampling tube, as well as possible reentrainment, must be minimized and accounted for. Care also must be taken to avoid condensation or chemical reactions in the sampling duct. Problems of this kind are especially severe in sampling high temperature, moist gases from a stack or from a chemical reactor. Condensation can be avoided with a probe heated to the temperature of the sampled gas if the pressure difference has been minimized. Where pressure differences in the sampler are large, control of pressure may be important. Chemical reactions on the wall of the sampling tube are often difficult to control, but can be minimized by using inert glass or Teflon-lined tubes.

The ideal condition for sampling is to draw the gas–particle suspension into the instrument at a speed nearly that of the external flow. In the ideal, sampling should be done *isokinetically,* or with the sampler velocity q_s equal to the mainstream velocity q_m. For a straight tube probe, different sampling conditions are illustrated in Fig. 5.2. Only in the isokinetic case will the

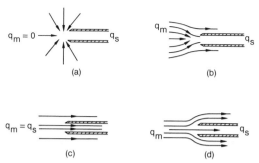

Fig. 5.2. Patterns of gas flow at the entrance to a sampling probe for different ratios of sampling to gas velocities: (a) stagnant gas, (b) $q_s > q_m$, (c) $q_s = q_m$, (d) $q_s < q_m$. [After May (1967); courtesy of Cambridge University Press.]

inertial deposition at the sampler tip be minimized and preferential size separation be small.

For a probe oriented in the direction of flow considered uniform in the mainstream and the probe, dimensional arguments indicate that the ratio of mainstream and probe concentrations, or suction coefficient SU, is a function of the ratio of mainstream to sampler velocity and Stokes number:

$$SU \equiv N_s/N_m = f(q_m/q_s, \text{Stk}), \tag{5.2}$$

where N_m and N_s are the particle number concentration in the mainstream and the sampler for particles larger than 1 μm is a gas at normal pressure and temperature. The normal design goal is the maintenance of this ratio near unity.

The dependence of the ratio N_s/N_m on the velocity ratio q_s/q_m has been determined experimentally, and is shown in Fig. 5.3 for particles of different diameter. Here we see that for a stagnant medium, the velocity ratio q_m/q_s goes to zero and the sample tends to be representative with $N_s \approx N_m$. The sample probe acts as a point sink for particles as in Fig. 5.2a. In this case, inertial effects can be disregarded. With increasing velocity ratio, the concentration ratio first decreases, then increases. For $q_m/q_s < 1$ (Fig. 5.2d) inertial effects push particles around the sampling orifice. Further increase in the velocity to the isokinetic condition again yields a representative sample. But higher mainstream velocities push particles preferentially into the tube (Fig. 5.2b). Because the effect involves inertial forces, it is more sensitive to larger particles. Experiment suggests that unit-density particles sampled in a tube of diameter approximately 1 cm and velocities of 500 cm/sec result in an essentially ideal representation for particles less than a few micrometers in diameter.

Recently Zebel (1978) has reviewed aspects of isokinetic sampling in terms of idealized models for calculating the suction coefficient. He notes that a

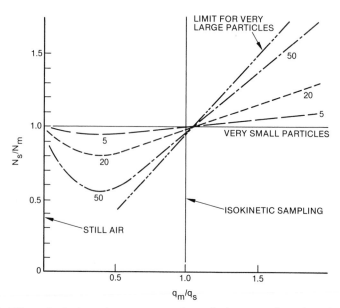

Fig. 5.3. Effect of velocity ratio on concentration ratio for a sampling tube oriented in the direction of the mainstream flow. The curves are approximate representations of the data of various experiments for unit-density particles of diameters (in μm) as indicated. The displacement of the point $N_s/N_m = 1$ from $q_m/q_s = 1$ results from the finite particle diameter. The curves apply to a nozzle of 1–2 cm diameter sampling at about 5 m/sec. [From May (1967); courtesy of Cambridge University Press.]

point-source flow analogy based on potential flow theory has application as an approximation. Calculations of Levin (1957) give for the suction coefficient

$$SU = 1 - 0.8Stk + 0.08Stk^2; \quad Stk < 0.5; \qquad (5.3)$$

here $Stk = \tau_p(q_s + q_p)/\ell_0$; the characteristic length $\ell_0 = [Q_g/4\pi(q_s + q_p)]^{1/2}$, where Q_g is the aspiration rate of the sampling tube. According to Zebel, numerical calculation from Eq. (5.3) give errors of 2.5% for $Stk = 0.5$. Davies (1976) has noted that Eq. (5.3) yields suction coefficients that are too high. Corrections for the introduction of gravitational settling would require an additional independent parameter.

Zebel (1978) also has investigated the application of approximate solutions for the suction of aerosol through a slit near the stagnation point of a blunt body or the case of a round orifice on a wall with a crosswind. An approximate formula is

$$SU = 1/(1 + 1.09Stk), \qquad (5.4)$$

where the Stokes number is calculated with respect to the gas velocity.

Davies (1976) has obtained a similar formula for a thin-walled tube in the case of isoaxial sampling, but the Stokes number is based on flow velocity in the tube.

Deposition on the walls of a straight sampling tube takes place by sedimentation or by turbulent deposition for large particles and by Brownian diffusion or phoretic motion for small particles if temperature, electrical force gradients, or (gas) concentration gradients are present. Such losses can be estimated by using appropriate calculations for tube flow as summarized in Chapter 2. Continued wall deposition may be followed by reentrainment of particles, which may be in the form of agglomerates; the reentrainment process is complimented by surface forces and is not well quantified for prediction.

5.2 ELECTRICAL CHARGING AND THE MOBILITY ANALYZER

The electrical mobility of aerosol particles in an electric field makes it possible to separate particles electrically from the suspending gas stream. If the electric mobility is a single-valued function of particle size, it is possible to use mobility classifiers as size analyzers. Although these principles have been known for some time, it seems that Rohman in 1923 was the first to attempt the measurement of an aerosol mean size by electrical means.

The rapid advances in electronics during the past few decades, combined with an increasing need for fine-particle measurement, have resulted in a number of electrical measurement techniques and instruments being brought into general use.

5.2.1 Particle Charging.   Electrical charging of particles can occur by three mechanisms: (a) field charging, (b) diffusion charging, and (c) contact charging. Charging can occur with units of one sign (unipolar) or with units of different sign (bipolar). When a dielectric or conducting particle is placed in an electric field, the lines of force tend to concentrate near the particle. Particles then become charged by collisions, with ions traveling along the lines of force to impinge on the particle surface. The particle accumulates charge until the local field developed by the charge on the particle causes distortion of the field lines. When this occurs, ions no longer intercept the particle and no further charging takes place. The magnitude of the limiting or saturation charge can be calculated from electrostatic theory assuming a spherical particle. As an example, the saturation charge accumulated over a period of time in an electric field is given by

$$Q_s = [1 + 2(\varepsilon_p - 1)/(\varepsilon_p + 2)]E_0R^2. \tag{5.5}$$

Thus the magnitude of the charge depends mainly on the particle radius and the magnitude of the electric field E_0 in free space, modified by the dielectric constant ε_p of the collected material.

The time required to reach saturation varies from ion density and the region in which charging takes place. For a field of constant ion density N_0, the particle charge varies as

$$Q(t) = Q_s(1 + 1/tt_e)^{-1}, \tag{5.6}$$

where $t_e = \pi e B_i N_0$; B_i is the ion mobility.

Diffusion of Ions. In the absence of an electric field, the thermal motion of ions will result in collisions with particles, causing charging. This process is called *diffusion charging;* it has been studied extensively partly because of controversy about the details of the mechanism for the free-molecule and continuum regimes, and partly for practical interest in the charging process of very small particles.

In the most common and useful approach, small ions generated by a radioactive source or by a corona discharge are allowed to charge aerosol particles to be detected or collected.

Production of small ions for aerosol charging is not difficult. However, making a uniform and constant concentration by exposing the aerosol to the ionized stream for a controlled length of time is much harder.

There are three basic kinds of diffusional charging which have been identified, each of which results in a different magnitude of charge collection and a different relationship between particle electric mobility and size.

The first kind of charging exists when the ion mixture is unipolar but there is no electric field. Under these conditions, ions diffuse to the particle until the repelling force from the net charge on the particle reduces the probability of acquiring further charge.

The second kind of charging exists when a particle is immersed in a bipolar mixture of small particles in the absence of an electric field. Under these conditions of bipolar diffusion charging, individual particles fluctuate in charge as the particle acquires positive or negative ions as a result of collisions with the particle.

A third kind of charging occurs when an electric field is superimposed on the unipolar ion cloud. The additional velocity of the ions along the field lines which pass through the particle can increase the equilibrium charge for given ion concentrations above those for unipolar diffusion charging, especially for particles larger than a few tenths of a micrometer in diameter. Basically, this mechanism is a mixture of field and diffusion charging.

The theory of diffusion charging can be developed readily if the charging process is not too rapid, so that the ion distribution can be approximated by a Boltzmann equilibrium distribution for molecules in a force field. In the discussion below, only unipolar charging is presented since it is most important to particle collection.

The electrical potential distribution around a spherical particle can be calculated from Poisson's equation, with the surface charge on the particle given by Coulomb's relation for a point charge at the center of the particle.

The change in charge of the particle then is given by the rate of collisions of ions with the particle surface. For a finite time, the accumulation of charge has been estimated for the free-molecule regime as

$$z_i = \frac{R_p kT}{e^2} \ln \left[1 + \left(\frac{2\pi}{m_i kT} \right)^{1/2} 2N_0 R_p e^2 t \right], \tag{5.7}$$

where m_i is the ion mass. Thus, the charge acquired depends on the product $N_0 t$ as well as the particle radius. For a very long time, z_i approaches infinity logarithmically. This, of course, cannot be true because of limitations in units of charge a particle can absorb. In practice, values of $N_0 t$ are $\lesssim 10^8$ ion sec/cm^3 based on experimental data.

On empirical grounds, Pui (1975) found that Eq. (5.7) could be written in the form

$$z_i = \frac{2R_p kT}{e^2} \ln \left[1 + \frac{N_0 t}{\dfrac{1}{\pi R_p^2 \bar{q}_i} + \dfrac{1}{4\pi R_p D_i}} \frac{e^2}{2R_p kT} \right], \tag{5.8}$$

where $\bar{q}_i$ is the mean thermal speed of the ions and D_i is the ion diffusivity. Whitby (1976) has reported that this equation fits available data on the unipolar charging rate by diffusion satisfactorily. This result is shown in Fig. 5.4 for different values of $N_0 t \leq 10^8$ ion sec/cm^3. One can see that up to approximately 0.1 μm diameter, the theory gives a maximum of a few units of charge per particle. The relation between charge and particle diameter also is approximately linear.

Contact Charging. The third mechanism, *contact electrification*, refers to electric charge transferred between two bodies as a result of their contact without sliding or rubbing. Strictly speaking, triboelectric charging involves rubbing. There is little evidence that aerosol particles are charged by the latter effect, though unquestionably contact electrification may take place.

Harper (1967) has discussed elements of contact electrification in some detail. The charge transfer that occurs when two metal spheres touch results from a flow of electrons when a difference in contact potential exists. The buildup of charge continues until the resulting electrical potential equals the difference in work functions of the metals. Additional charge can flow via quantum-mechanical tunneling even if a small potential gap is present between the two bodies. Further, as the two bodies separate, air can break down, giving rise to a back flow of charge.

The theory for contact charging of metals can be generalized to include semiconductors (Harper, 1967). In this case, charge flows to the interior of the bodies in contrast with restriction to the surface for metals. Since the

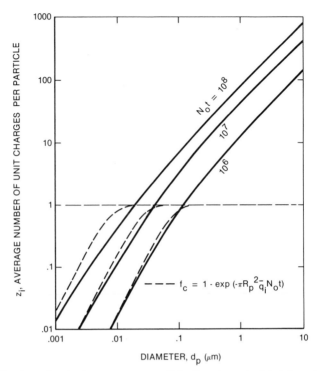

Fig. 5.4. Number of unit charges versus particle diameter calculated from Eq. (5.8), for unipolar diffusion charging at three different $N_0 t$ values. The fraction of particles charged, f_c, calculated from the inset equation is shown for comparison. [After Whitby (1976); courtesy of Academic Press.]

charge transfer depends on duration of contact, high resistivity semiconductors do not exchange appreciable charge in brief contact.

In the case of insulators having even higher resistivity, no flow of charge occurs to the interior of bodies. Surface states then are important as the source of charge for transfer. The surface states are complicated functions of the physicochemical condition of the surface, in which adsorbed ions in electric double layers may play a role. Charge evidently involves ions rather than electrons in this case.

The oxidized or contamination state of metallic particles will alter charge transfer. If the oxidized surface layers are thin, charge may be transferred by conduction or tunneling. The effects of moisture on the surface also may complicate the process.

Cheng and Soo (1970) have reported a theory for the charge exchange between two elastic spheres. The charge transferred in collision was found to be proportional to the electrical capacities of the spheres; the difference in work functions of the surfaces; and a factor accounting for the duration and area of contact, the electrical conductivities of spheres, and a charge trans-fer length. The latter is defined in terms of current density and potential at the point of contact. In this model, the total charge transferred from a collision between particles and a probe is proportional to the number of aerosol particles and inversely proportional to the particle radius squared. For resistivities less than 10 Ω cm, charge transfer is predicted to be inde-pendent of the velocity of particles relative to the probe. For resistivities >10 Ω cm, the total charge transfer is proportional to the particle radius R_p^{-n}, $1 < n < 2$ (John, 1976).

From these considerations, it is clear that both diffusion charging and contact charging have the potential for development of size measurement devices. Field charging is not particularly useful for such applications.

5.2.2   Electrical Charging and Particle Size Measurement.

Two methods have been adopted to develop particle measurement devices. These involve diffusion charging and contact charging. First, let us consider the diffusion charging approach. Apart from such operational problems as charging stabil-ity and uniformity of charging in different parts of the charger, there are three characteristics of ion charging which affect the usefulness of a diffu-sion charging method for aerosol sizing by electric methods.

First, the relationship between electric mobility and particle size must be established. This basically provides the means for calculating the migration velocity of particles under the influence of an electrical force, which in turn gives the basis for locating a particle collection in an instrument. The aver-age charge for different kinds of charging and resulting mobilities is shown in Fig. 5.5. Note that the mobility versus particle diameter curves are single valued for the bipolar diffusion and the unipolar diffusion but are not for the field charging. Also, note that the slope of the unipolar diffusion curve is essentially zero above 1 μm. Thus, sizing methods using unipolar diffusion charging have very poor resolution above about 0.6 μm at atmospheric pressure.

For the second characteristic, the fraction charged must be known. Parti-cles which do not acquire a charge during their passage through a charger cannot be influenced by subsequent electric fields and, therefore, cannot be measured by electrical migration.

For any electrical instrument measuring aerosol particle concentration (neutral as well as charged particles), the indicated concentration in a given size range must be corrected by dividing by the fraction charged f_c. Since the error in the corrected concentration will be proportional to Δf_c, where Δf_c is the uncertainty in f_c, errors in corrected concentrations become unaccept-ably large at values of $f_c \leq 0.03$ according to Whitby (1976). Thus, the fraction charged, in combination with aerosol losses, becomes the principal factor which limits the lower useful size detection of an instrument.

In the third characteristic, the discrete nature of electric charge must be

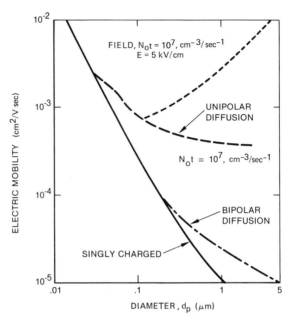

Fig. 5.5. Electric mobility of particles charged in four different ways. Note that for field charging, because of the minimum mobility at about 0.1 μm, two sizes may have the same mobility. [After Whitby (1976); reprinted with permission of Academic Press.]

accounted for. From Fig. 5.4, it can be seen that the average number of charges per particle ranges from less than unity to more than unity at about 0.04 μm in the case of unipolar diffusion charging. This transition is about 0.3 μm for bipolar diffusional charging. The discreteness of the charge accumulated at these sizes limits the resolution that can be achieved. Husar (1973) estimated this limit as corresponding to a geometric standard deviation σ_g in charge accumulation of 1.3 for unipolar diffusion charging of 0.04-μm-diameter spheres. Similar calculations by Knutson (1971) for bipolar diffusion charging yielded $\sigma_g = 1.3$ for 0.3-μm-diameter particles. In principle, calculation can correct for this effect on a measured size distribution, but the methods have not been evaluated yet. Whitby (1976) suggests that they show promise for resolution of monodisperse aerosol suspension with geometric standard deviations of 1.1 or less. If measurements are made using only the singly charged particles, then the resolution is as good as the resolution of the mobility analyzer itself. This requires, however, that the fraction of aerosol carrying unit charge be known accurately.

Aerosol concentration measurement using electrical effects requires a method for detecting the charged aerosol. This is usually done by measurement of an electrical current on a grounded collector with attachment and charge transfer of a particle. In addition, electrical sizing methods require a precipitator or classifier by which the particles of different electric mobilities are separated before detection. Various approaches have been discussed by Whitby (1976). Some of these are:

1. *Condenser Ion Counter.* The particles are collected on a metal condenser; the condenser of this type of counter may consist either of parallel plates or of concentric cylinders. The current collected from one of the electrodes is measured as a function of applied potential difference by means of a picoammeter. Although this instrument is deceptively simple, to obtain current change and hence the number of particles falling within a given mobility range requires that the current versus voltage curve be differentiated twice. This seriously limits its accuracy for size distribution measurement purposes. It has been used mostly for measuring small and intermediate ions of high mobility because having the picoammeter connected to one electrode of the precipitator also limits the maximum voltage that can be applied to about 700 V and hence limits the lowest mobility that can be measured.

2. *Denuder.* The Denuder developed by Rich *et al.* (1959) uses a condensation nuclei counter as the detector at the end of an electrical precipitator to measure the fraction of the aerosol charged. If the fraction charged is related in a unique manner to the particle size, as it is for a Boltzmann equilibrium bipolar charge distribution, then the instrument can be used as a particle size distribution analyzer. This technique is found to be limited to the size range from about 0.02 to 1 μm. If the aerosol to be measured has a large number of particles smaller than 0.02 μm, then the change in nuclei count as the voltage is varied is too small to provide useful estimates of aerosol size distribution.

3. *Ion Capture Aerosol Concentration Measurement.* If essentially neutral aerosol particles pass through a corona charger, the aerosol particles are charged by the flow of ions from the center to the outer electrode. The capture of the ions by the aerosol particles reduces the effective electrical mobility of the charges so that they are carried out of the device instead of reaching the outer electrode. This reduction of current is a measure of aerosol concentration. The exact relationship between the current and the particle concentration depends on the method of charging and the particle size distribution. This is basically the principle of the Whitby mobility analyzer (Whitby and Clark, 1966).

The mobility analyzer principle has been used in a device called the electrical aerosol analyzer (EAA) (Knutson and Whitby, 1975). The instrument is configured such that the aerosol is first sampled into a particle charger, where the particles are exposed to unipolar positive ions from a corona discharge and become electrically charged.

The diffusion charging curve in Fig. 5.5 shows the relationship between the electrical mobility and the size of the particles under different charging conditions. It is observed that there is a monotonic functional relationship between particle mobility and size. From this relationship, the size distribution of the aerosol can be calculated from the mobility distribution measured by the mobility analyzer.

An example of a mobility analyzer is shown in Fig. 5.6. The device has the form of a cylindrical condenser with clean air and aerosol flowing down the tube in a laminar gas stream. The charged aerosol particles are deflected through the clean gas core by the voltage applied on the center electrode. For a given voltage on the center rod, particles above a certain mobility (i.e., a certain size) are precipitated, while those with lower mobility and larger particle size escape precipitation and are sensed by the electrometer current sensor at the bottom. By sequentially changing the voltage on the center rod and measuring a corresponding electrometer current, the mobility and size distribution of the particle suspension can be determined. The standard operating condition of the commercially available instrument built by the Thermo-Systems, Inc., provides a total of eleven voltage steps dividing the size range of the instrument into ten equal geometrical intervals of four intervals per decade in particle size. The size interval boundaries are located at 0.0032, 0.0056, 0.01, 0.0178, 0.032, 0.056, 0.1, 0.178, 0.32, 0.56, and 1.0

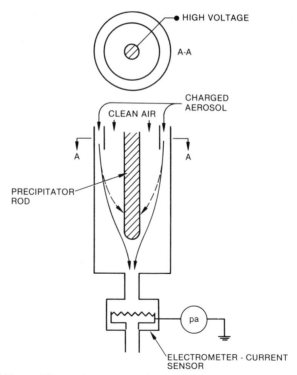

Fig. 5.6. Whitby mobility analyzer arranged to measure charge carried by particles not collected by the precipitator. [After Whitby (1976); courtesy of Academic Press.]

μm. The complete voltage sequence can be scanned in about two minutes, thus allowing a size distribution analysis to be made in the same time period.

To eliminate the end effects of the parallel-plate designs, axial symmetry has been adopted in design, and is shown in the configuration in Fig. 5.6. To obtain good resolution with a reasonably high ratio of aerosol flow to clean air flow, the charged aerosol particles are introduced next to the outer wall. Charged-particle collection for current measurement is accomplished with a separate carefully shielded and isolated filter connected to the picoammeter. Separation of the precipitator and current collection functions permits the use of precipitator voltages near breakdown, and therefore the collection of particles with electrical mobilities less than 10^{-4} cm/sec per V/cm. With good electrical shielding, the precipitator voltage may be varied without shorting of the electrometer input. This permits completely automatic operation as in commercial instruments (Liu *et al.*, 1974c).

The absolute calibration of the electrical mobility analyzer for both aerosol concentration sensitivity with particle size and for size distribution resolution has been a difficult task because of the lack of suitable particulate standards in the size range from 0.003 to 1 μm diameter. However, recent development by Liu and Pui (1974) of a particle generator using classification by electrical mobility has made it possible to make such calibrations with acceptable accuracy over most of the size range of the aerosol analyzer.

The sizing resolution of the EAA is essentially a function of three factors: the inherent mobility resolution of the instrument, the discrete nature of the electric charge, and the slope of the mobility versus particle diameter curve. The Liu and Pui studies indicate that the current design of the mobility analyzer functions close to theory with respect to its mobility resolution, and with respect to what can be accomplished considering the discrete nature of particle charging. From this work and those of others, Whitby (1976) suggests that the maximum size range over which such an instrument can be made to operate with useful resolution is about two orders of magnitude, centered at ~0.05 μm diameter.

Instrumentation based on contact electrification has been reported. In these devices the flow of aerosol is directed around and through an insulated probe. The transfer of charge from particle–probe collisions results in an electrical current which can be monitored with an electrometer. Schütz (1966) built an early version of this type employing a spherical probe. Prochzaka (1964) used a tubular electrode of a semiconductor (magnesium hydrosilicate) in configurations where the electrode was a cylindrical vortex chamber, and a Venturi nozzle. Schütz's devices were tested and an empirical relationship between current i (in A) and mass concentration M (in mg/m^3) was obtained in the form

$$i = aM^n,$$

where n ranges between 1.26 and 1.30, and a is inversely proportional to particle diameter. Prochzaka's data showed a linear relation between operating current and dust loadings up to 3 g/m³. John's (1976) work indicates that these devices are sensitive to the chemical composition and the resistivity of particles, so they do not appear to be useful for monitoring of complex particle mixtures.

5.3   LIGHT INTERACTIONS AND THE OPTICAL PARTICLE COUNTER

Small particles can scatter and absorb light. This phenomenon has been used to investigate aerosol behavior extensively since Tyndall's work in the nineteenth century. In recent years, instruments have been built to take advantage of light interactions to deduce particle size distributions. To appreciate how such devices work, we introduce first certain basic principles of light interaction with airborne material.

Basically, the scattering and absorption of light by individual particles depends on their size, shape, and index of refraction, as well as the wavelength of incident light. The total scattering and absorption from a beam of light by an aerosol cloud corresponds to the summation of the scattering from all particles of different size and refractive index. If the particle cloud is dilute enough, the effects of multiple light scattering can be disregarded, and the summation of single particles suffices to describe the interaction.

The light attenuation process may be analyzed by considering a single particle of arbitrary size and shape, irradiated by a plane electromagnetic wave as shown in Fig. 5.7. The effect of the presence of the particle is to diminish the amplitude of the plane energy wave. At a distance large compared with the particle diameter and the wavelength (λ), the scattered energy appears as a spherical wave, centered on the particle and possessing a phase different from that of the incident beam. The total energy lost by the plane wave, the extinction energy, is equal to the scattered energy in the spherical wave plus the energy absorption.

The light intensity I is the conventional measure of the energy of scattered light, and in cgs units, has units of ergs per centimeter squared–second. The intensity of scattered light is proportional to the intensity of the incident light beam I_0 and the radial distance expressed as

$$I = \frac{I_0 F(\theta, \varphi, \lambda)}{(2\pi r/\lambda)^2}. \tag{5.9}$$

In general, $F(\theta, \varphi, \lambda)$ depends on the wavelength of the incident beam and on the size, shape, and optical properties of the particle, but not on r, the radial distance from the particle. For spherical particles, there is no φ dependence. The relative values of F can be plotted in a polar diagram (scattering diagram) as a function of the angle θ for a plane in the direction of the incident beam.

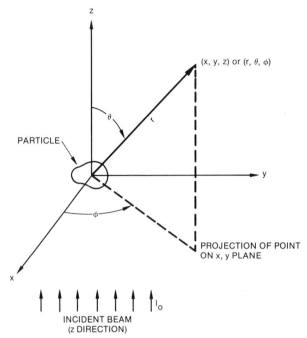

Fig. 5.7. Geometry for calculating the effect of a particle on attenuation of a light beam. The direction of scattering at any distance r is characterized by the scattering angle, θ measured relative to the direction of the incident beam and the azimuth angle φ.

The scattering function can be determined from theory for certain important special cases. The performance of optical single-particle counters depends on the variation of the scattering function with angular position.

The intensity function in itself is not sufficient to characterize the scattered light. Needed also are the polarization and phase of the scattered light, which are discussed in references such as van de Hulst (1957) and Kerker (1969). For air pollution applications including instrument design, the parameters of most interest are the intensity function and the scattering efficiency.

The scattering efficiency or cross section is a dimensionless quantity which is defined as the energy scattered by the particle divided by the energy intercepted by the particle, based on its geometric cross section:

$$K_{\text{scat}} = \frac{\int F(\theta, \varphi) \sin \theta \, d\theta \, d\varphi}{(2\pi/\lambda)^2 \pi R_{\text{p}}^2}. \tag{5.10}$$

Similarly, the absorption efficiency is defined as the fraction of the incident beam absorbed per unit cross-sectional area of particle. The total energy removed from the incident beam, the extinction, is the sum of the energy scattered and absorbed. The corresponding *extinction efficiency* is

$$K_{\text{ext}} = K_{\text{scat}} + K_{\text{abs}}. \tag{5.11}$$

5.3.1 Scattering and Absorption for Particles. Light is characterized as an electromagnetic wave with electric and magnetic field vectors. Consider a case of a plane wave, linearly polarized, incident on a small particle. The wavelength of light in the visible range is about 0.5 μm. For particles much smaller than this wavelength, say less than 0.05 μm, the local electric field produced by the wave is approximately uniform at any instant. This applied electric field induces a dipole in the particle. Since the electric field oscillates, the induced dipole oscillates and radiates in all directions. This type of scattering is called *Rayleigh scattering* after its discoverer.

The dipole moment induced in the particle is proportional to the instantaneous electric field vector **E**; the proportionality constant being the polarizability tensor $\bar{\bar{\alpha}}$. If it is assumed that the particles are isotropic, α can be treated as a scalar quantity. From the energy of the electromagnetic field produced by the oscillating dipole, the intensity of the scattered radiation is derived as

$$I = [(1 + \cos^2 \theta)k^4\alpha^2/2r^2]I_0, \tag{5.12}$$

where the wave number $k = 2\pi/\lambda$. This expression applied to particles of arbitrary shape. The scattering is symmetrical with respect to the direction of the incident beam in this case.

Since the intensity of the scattered light varies inversely with the fourth power of the wavelength, blue light (short wavelength) is scattered preferentially to red.

For an isotropic spherical particle, electromagnetic theory yields

$$\alpha = [3(m^2 - 1)/4\pi(m^2 + 2)]v, \tag{5.13}$$

where m is the refractive index of the particle and v is its volume. When scattering without absorption occurs, the efficiency factor is obtained by substituting Eq. (5.13) into (5.12) and integrating:

$$K_{scat} = \tfrac{8}{3}x^4[(m^2 - 1)/(m^2 + 2)]^2, \tag{5.14}$$

where $x = 2\pi R_p/\lambda$ is the dimensionless optical particle size parameter.

Both scattering and absorption are taken into account by writing the refractive index as a complex number, where the sum of the squares of the real and imaginary parts of m is the dielectric constant of the particle. The imaginary part of m is related to absorption; it vanishes for nonconducting particles. For metals in the optical frequency range, both the real and imaginary parts of m are of order unity.

The efficiency for Rayleigh scattering from absorbing particles is given by (Van de Hulst, 1957)

$$K_{scat} = \tfrac{8}{3}x^4 \, \mathcal{Re}[(m^2 - 1)/(m^2 + 2)]^2, \tag{5.15}$$

where $\Re e$ indicates that the real part of the expression is taken. The absorption efficiency can be shown to be given by

$$K_{abs} = -4x \; \Im m[(m^2 - 1)/(m^2 + 2)], \qquad (5.16)$$

where $\Im m$ indicates that the imaginary part is taken.

For particles much larger than the wavelength of the incident light ($x \gg 1$), the scattering efficiency approaches two. That is, a large particle removes from the beam twice the amount of light intercepted by its geometric cross-sectional area. For light interacting with a large particle, the incident beam can be perceived as a set of separate light rays. Of those rays passing within the geometric cross section of the sphere, some will be reflected at the particle surface and others refracted. The refracted rays may emerge again, potentially after several internal reflections and refractions. The reflected light and the refracted light eventually emerging are part of the total scattering by the particle. Any incident beam that does not emerge is lost by absorption within the particle. Hence all of the energy incident on the particle surface is removed from the beam by scattering or absorption, accounting for an efficiency factor of unity.

Another source of scattering from the incident beam also must be accounted for. The portion of the beam not intercepted by the sphere forms a plane wave front from which a region corresponding to the cross-sectional area of the sphere is missing. This is equivalent to the effect produced by a circular obstacle placed normal to the beam. The result is a diffraction pattern "within the shadow area" at large distances from the obstacle. The appearance of light within the shadowed area is the reason why diffraction is sometimes likened to the bending of light rays around an obstacle. The intensity distribution within the diffraction pattern depends on the shape of the perimeter and size of the particle relative to the wavelength of the light. It is independent of the composition, refractive index, or reflective nature of the surface. The total amount of energy that appears in the diffraction pattern is equal to the energy in the beam intercepted by the geometric cross section of the particle. Hence the total efficiency factor based on the cross-sectional area is equal to two.

Rayleigh scattering for $x \ll 1$ and the large-particle extinction law for $x \gg 1$ provide useful limiting relationships for the efficiency factor. Aerosol light scattering, however, is often limited by particles whose size is of the same order as the wavelength of light in the optical range from 0.1 to 1 μm in diameter. In this range, Rayleigh's theory is not applicable since different parts of the particle interact with different portions of an incident wave. Yet such particles are still too small for the large-particle scattering theory to be applicable. For such situations the theory of Mie is applied. Expressions for the scattering and extinction are obtained by solving Maxwell's equations for the regions inside and outside a homogeneous sphere with suitable boundary conditions. Mie found that the efficiency factors are functions of x

and m alone. The calculations must be carried out numerically and the results have been tabulated for certain values of the refractive index (e.g., Penndorf, 1957).

For water, the index of refraction is 1.33, but for organic liquids it is often approximately 1.5. The scattering efficiency for these two values of m are shown in Fig. 5.8 as a function of the dimensionless particle diameter x. For $x \to 0$, the theory of Rayleigh is applicable. Typically, the curves show a sequence of maxima and minima, the maxima corresponding to the reinforcement of transmitted and diffracted light, and the minima to interference. The curves show a sequence of maxima and minima of diminishing amplitude typical of nonabsorbing spheres with $2 > m > 1$. Taking the abscissa of the curve for $m = 1.5$ to be $2x(m - 1)$, extinction curves for the range $2 > m > 1$ are reduced to approximately the same curve.

For absorbing spheres, the curve for K_{ext} can be simpler in form, rising rapidly to reach a maximum at small values of x and then falling slowly to approach two at large values of x. Figure 5.9 shows the extinction efficiency for the case of carbon spheres. For such particles, nearly all of the scattering is due to diffraction, while almost all of the geometrically incident light is absorbed. For small values of x, the extinction is due primarily to absorption, but for large x, scattering and absorption are of almost equal importance. The refractive index for absorbing spheres usually varies with wavelength, and this results in variation of K_{ext} as well.

The angular dependence of the light scattering can be calculated from the Mie theory. The form of the angular dependence is used in the design of

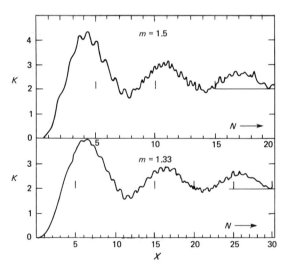

Fig. 5.8. Extinction efficiencies for nonabsorbing particles from the theory of Mie for $m = 1.5$ and 1.33. [After van de Hulst, from "Light Scattering by Small Particles," 1957 H. C. van de Hulst, 1981 Dover Publications.]

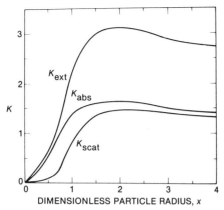

Fig. 5.9. Extinction efficiency for carbon particles with $m = 2.00(1 - 0.33i)$, temperature not specified. [From McDonald (1962).]

optical particle counters. By choosing a favorable geometry for the detector of the scattered light, the resolution of the instrument with respect to particle size can be optimized.

Using combinations of layers in particles with different dielectric constants, concentric spheres can give enhanced scattering over a dispersion of equivolume particles, or the opposite effect (Kerker, 1982). A striking case is derived from Mie theory wherein particles can become invisible if the polarizability is zero. Kerker (1982) has suggested an example of an organic sphere with dielectric constant of 2 covered by a layer of silver with dielectric constant of -2 (at 354 nm).

5.3.2 Single-Particle Optical Analyzers. Particle sizing by means of light scattering from single particles was known more than 30 years ago. In the meantime, the subject has been steadily developed. Since about 15 years ago, optical particle counters using white light illumination have been commercially available. Since the invention of the laser principle several attempts have been made to replace the white light illumination of scattering devices by coherent and monochromatic laser light illumination.

To examine the influence of the optical properties and the shape of a particle upon the response of a light-scattering device, it is helpful to introduce some definitions of equivalency in particle size. Particle scattering cross section K is related to the geometrical radius or diameter d_p, by some measured relation which depends on a calibration refraction index and the spherical shape, equivalent of the particles. For a nonspherical particle, a light-scattering diameter is defined which is related to equivalent spheres, but the particle shape is described by an optical shape factor. Thus, an optical counter measures a light extinction cross section and not particle size

directly. Unless the counter is calibrated for a given aerosol particle material, its optical size generally will not be the same as the geometrical size.

The light extinction system shown in Fig. 5.10a is used to exemplify the
operating principle of the single-particle optical counter (OPC). Light from
a lamp is condensed onto an aperture with the light source focused in the
plane of that aperture. The emanating light cone is condensed into the view
volume, where a sharp image of the defining aperture is formed. The aerosol
flow is ducted to the light beam in this plane of focus. A view volume may
then be defined as the region bounded by the cross-sectional area of the
aerosol stream and the length of the aperture image. The light emanating
from the view volume is collimated, condensed, and focused onto a photo-
sensitive detector, such as a photomultiplier or a photodiode.

In an OPC with an incandescent light source, a light extinction system
such as that shown in Fig. 5.10a is only practical for particles larger than
several micrometers in diameter. Since most particles in an ambient or industrial environment are smaller in size, this light extinction technique is

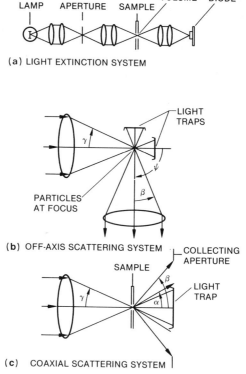

Fig. 5.10. Geometries of illumination and collection in conventional single-particle optical
counters.

rarely used for aerosols, but is frequently used for the characterization of particles in a liquid.

Most commercially available OPCs measure the light scattered out of the incident beam rather than the light intensity reduction of that beam. The optical system of the common scattering instruments is schematically represented in Figs. 5.10b and 5.10c. In general, the illumination corresponds to that shown in Fig. 5.10a; the details of light collection and detection, however, may differ in how the scattered light is focused onto the light detector, i.e., photodiode or photomultiplier. Each commercial system has its own arrangement of illumination and acceptance angles. Specifications of the following four angles are graphically represented in Figs. 5.10b and 5.10c. The light interaction in the view volume is fully described by γ, the half angle of the illuminating cone; α, the light trap half angle; β, the collecting aperture half angle; and ψ, the inclination between illuminating and collecting cone axes.

The optical system in an OPC is sometimes characterized by the scattering angle limits. Referring to Fig. 5.10, scattering of light from illuminating angle γ to collecting angle β gives an upper limit of $\gamma + \beta$; scattering from γ to light trap angle α gives the lower limit of $\alpha - \gamma$. For instance, an instrument with $\gamma = 15°$ and viewing angles $\alpha = 35°$ to $\beta = 90°$ has a scattering range of $20°$–$105°$.

When particles are illuminated by a monochromatic light source such as a laser, the Mie scattering curve of scattered light flux shows periodic amplitude oscillations (Fig. 5.9) with respect to particle size parameter $2\pi R_p/\lambda$. The counter response, therefore, may not be a single-valued function of particle size. Incoherent, incandescent light usually washes out the scattering oscillations, and is normally used in most commercial OPCs.

When a powerful light source is used, volatile particles may partially evaporate during the time they remain in view volume. This presents an inherent limitation which is often disregarded in interpreting results of measurements.

From recent work, the advantages and disadvantages of white light and laser light radiation have been worked out (Gebhart et al., 1976). The outstanding characteristics of laser light are the high stability of the output, the degree of spatial and temporal coherence, and the high flux density. The high stability renders unnecessary a reference light source. The spatial coherence or the low beam divergence leads to a high intensity at the focal point of the optical system and, therefore, the need of parabolic mirrors for light focusing and complicated lens systems are eliminated. In the case of focal plane illumination, the intensity of a typical 5 mW He–Ne laser radiation is four to five orders of magnitude greater than the intensity from the most intense incandescent light sources. Therefore, the sizing range of usual optical particle counters can be extended to smaller particles by utilizing lasers. Since a laser beam can be focused to a very tiny, diffraction-limited spot, the dimensions of the sensing volume can be kept very small. For this reason, aerosols of high concentrations (up to 10^6 cm^{-3}) can be analyzed with negligible coincidence losses with laser systems.

The temporal coherence or monochromacy is advantageous for instruments analyzing the angular dependence of the scattered light, but it is a drawback for particle counters, since the response curves of monochromatic counters exhibit oscillations in the size range above the wavelength of incident light.

With such considerations, one concludes that for size analysis of particles larger than the light wavelength white light illumination is advisable in order to avoid ambiguous response of the instrument. For size analysis of particles smaller than the wavelength either laser or white light illumination can be used. However, laser light illumination is preferable in order to achieve a higher sensitivity and resolving power of the instrument.

The performance of OPCs has been studied extensively over the past few years. Their limitations and operating characteristics are now well understood (e.g., Willeke and Liu, 1976). Several features are of particular interest, including (a) the *size resolution* or the counter's ability to distinguish between neighboring particle sizes; (b) the detection limit, or the smallest size to which the counter can respond; and (c) counting accuracy and errors, involving coincidence of particles entering the light beam, and statistical proportion of counting over a given sampling time.

The size resolution ideally depends on the monotonic nature of the response curve. A theoretical response curve for one optical counter configuration is shown in Fig. 5.11. Curves for different indexes of refraction are included, including an absorbing material (carbon). An important limitation of the OPC is illustrated in this figure. Since the counter response is strongly dependent on the index of refraction of particles of a given size, the size discrimination will be quite variable for an aerosol of heterogeneous composition.

OPCs are especially useful for continuous measurement of particles of uniform physical properties. However, as discussed earlier, uncertainties

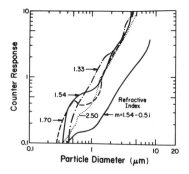

Fig. 5.11. Theoretical response of an optical counter configured after the Bausch & Lomb 40-1A instrument. The curve for a complex index refraction corresponds to carbon particles. [From Cooke and Kerker (1975); with permission of the Optical Society of America.]

develop in measurement of particle clouds which are heterogeneous in composition because the refractive index may vary from particle to particle. Thus, in making atmospheric aerosol measurements, workers have assumed an "average" refractive index characteristic of the mixture to estimate a calibration curve, or have reported data in terms of the equivalent particle diameter for a standard aerosol, such as suspended polystyrene latex spheres.

The practical resolution of an OPC involves electronic recording of the counter response to an individual particle; this is done through a multichannel analyzer (MCA). The capability of an instrument to distinguish two monodisperse aerosols of different mean size then depends on the channel cross-sensitivity.

The OPC response to monodisperse particles depends on the uniformity of the light intensity in the illuminated optical view volume, the uniformity of sensitivity of the photodetector surface, and on the baseline electronic and optical noise. The MCA therefore responds to monodisperse particles with pulses distributed over a range of channels.

The shape of the counter signal is approximately normally distributed, and may be characterized by the overall variance σ_0^2 (σ_0 is the standard deviation). The variance in signal from nonuniformity in illumination and detection, σ_s^2, and the variance in baseline electronic and optical noise, σ_b^2, are essentially independent of each other, so that

$$\sigma_0^2 = \sigma_s^2 + \sigma_b^2. \tag{5.17}$$

The relative standard deviation of the output pulse distribution for monodisperse aerosols is then

$$\frac{\sigma_0}{H_0} = \left[\left(\frac{\sigma_s}{H_0} \right)^2 + \left(\frac{\sigma_b}{H_0} \right)^2 \right]^{1/2},$$

where H_0 is the pulse height at the peak of the distribution for monodisperse aerosols of particle diameter. The electronic noise is approximately constant and is equal to the counter output when no particles are passing through the optical view volume. Since the counter voltage output increases with particle size, approximately by one to two decades per decade of particle size, either electronic noise or Rayleigh scattering from the gas molecules sets the lower limit of particle size an OPC can detect, which is generally 0.1 μm in diameter.

The variance in signal σ_s^2 is found from measurements with large particles (large H_0), for which the ratio σ_b/H_0 tends toward zero, and consequently $\sigma_0/H_0 = \sigma_s/H_0$. The ratio σ_s/H_0 approaches 0.08 for large test particles when measured, for example, with the Bausch & Lomb 40-1 counter (e.g., Liu *et al.*, 1974a). These measurements also showed that σ_s increased to a higher value than expected from Eq. (5.17) when the particle size was reduced. A probable explanation of this discrepancy is that the smaller particles experience a greater lack of uniformity in illumination and detection than the large particles do, and that σ_s is actually a variable, depending upon particle size, rather than a constant, as implied in Eq. (5.17). Also, small particles may scatter a finite number of photons per unit time, thus increasing the fluctuations in output signal.

From the above discussion, the resolution of an OPC depends on the ability of the counter to produce uniform pulses upon exposure to monodisperse aerosols. It also depends on the slope of the calibration curve of counter response versus particle size, a steep one being the most desirable. Optical counters of the same model and produced by the same manufacturer may not have the same resolution owing to normal manufacturing tolerances and differences in alignment of the optics.

When calibrating OPCs with nonspherical or absorbing particles, such as monodisperse coal dust, the MCA may show a pulse height distribution with a spread which is considerably wider than the intrinsic properties of the instrument would dictate. Liu *et al.* (1974b) attribute the difference to the irregular shape of the monodisperse particles, and to a lesser extent to refractive index variations of the particles. This spreading effect further decreases the ability of the instrument to resolve small particle size differences. In general, the smallest variance is obtained with spherical particles of uniform index of refraction and the worst with irregularly shaped particles, because the scattering intensity received in the collecting aperture from irregularly shaped particles depends upon the orientation of the particle in the view volume.

The detection limit of the OPC depends on instrument noise, the Rayleigh scattering by suspending gas molecules, and stray light resulting from imperfections in optics. Rayleigh scattering gives the theoretical limit of detection. Carefully designed research instruments using laser light sources begin to approach this ideal, which is of order 0.1 μm diameter. However, most commercial devices will not detect particles less than 0.4 μm in diameter.

Errors in counting occur with *coincidence,* or when there is more than one particle in the optical view volume. This basically determines the maximum particle concentration the device can measure accurately. When particle concentrations are too high, dilutors are sometimes used to improve OPC accuracy (see also Willeke and Liu, 1976).

At the opposite extreme, the accuracy of the counter is limited by the probability of correct counting over a sample time interval by the instrument when only a few particles are present in the gas. For a sample of commercial instruments, the practical limits to measurement cover a number concentration range of three or four decades. The number concentration of atmospheric aerosol particles is about three to four decades lower for particles of 10 μm diameter compared with those of 0.5 μm diameter. Thus for sampling over a large size and large number concentration range, it may be necessary to use more than one OPC optimized to a given size range.

5.4 HYBRID CONTINUOUS SIZE ANALYZERS

For accurate measurements of size distributions of atmospheric aerosols, where the size–number concentration range is very large, hybrid continuous analyzer systems have been built. This type of approach was attempted first by Whitby et al. (1972) combining an EAA and an OPC. In a later study, the EAA, and two OPCs were employed, optimizing measurements over the range from 0.005 μm diameter to greater than 10 μm diameter. The configuration of the system mounted in a transportable laboratory has been described by Semb et al. (1980).

The hybrid system was set up such that the EAA and the two OPCs were coupled to a sampling flow for several instruments. Since sampling losses in the inlet manifold could be large for particles of diameter >1 μm, the sensing optics for the OPC were mounted in the open air on the roof of the laboratory. In this way, sampling losses were minimized. The output from the three devices was transferred into a minicomputer. In the computer, size distributions were automatically calculated in engineering units over selected time intervals. These were recorded on magnetic tape for future use, or could be printed out automatically on a teletype terminal.

An example of particle size distributions measured by the hybrid system is shown in Fig. 5.12. The number–diameter distribution is the basic function calculated from the response of the instruments. The surface and volume–size distribution were calculated from the number–size distribution assuming spherical particles. These corresponding distributions are seen in the figure. The number–size distribution (not shown) has an estimated maximum of 0.5 μm diameter in this case, but fails to reveal the modes of the surface and volume–size distributions. The surface and volume distribution modes appear as in the calculated structure. These are found to be a common feature of atmospheric particle size distribution, as discussed in Chapter 7.

5.5 MOMENTS OF THE SIZE DISTRIBUTION AND RELATED MEASUREMENTS

So far, we have discussed direct measurement techniques which provide continuous information about particle size distributions. In Friedlander's (1977) concentrated organization of instrumentation (Fig. 5.1), a second grouping includes methods which measure certain integral moments of the particle size distribution, or related observations. Devices include those that detect total number concentration (number per cubic centimeter), mass concentration (micrograms per cubic meter), and volume light extinction or scattering (reciprocal meters). Other observations of potential interest such as total surface area per unit volume or particle volume fraction can be detected by suitable integration of the output from number–size distribution analyzers.

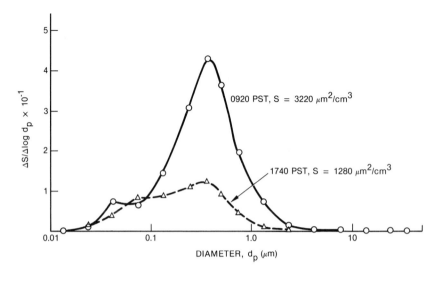

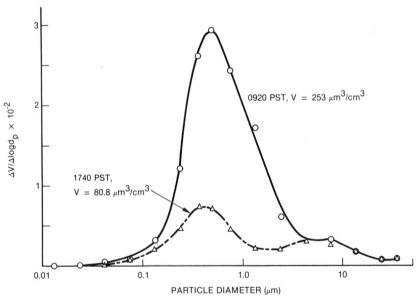

Fig. 5.12. Example of size distribution measurements in Los Angeles air from a three-analyzer system composed of a Thermosystems Inc. EAA, a Royco 220 OPC, and a Royco 245 OPC optimized for measurement of particles between 5 Å and 30 μm diameter. Data shown in the form of surface [(a) graph with $\Delta S/\Delta \log d_p$ on ordinate] and volume [(b) graph with $\Delta V/\Delta \log d_p$ on ordinate] size distributions, corresponding to the calculated second and third moments of the number–size distributions. 0920 PST: moderate photochemical smog based on midday oxidant level; $b_{sp} = 17 \times 10^{-4}$ m^{-1}. 1740 PST: light photochemical smog, $b_{sp} = 4.05 \times 10^{-4}$ m^{-1}. [From Hidy *et al.* (1973).]

In this section, the "moment" observations involving number concentration and light extinction are considered. Then we return to application of electromagnetic radiation principles to cover certain remote-sensing techniques of importance, as related to light extinction. The measurement of mass concentration is treated separately as a particle collection process in Section 5.6.

5.5.1 Number Concentration and Condensation Nuclei Counters.

A basic measurement of traditional importance in aerosol physics is the total number concentration of particles. The most reliable measurement of number concentration is believed to be direct counting of particles collected in a microscope substrate. This method is extremely tedious, and does not provide a range of time resolution. Intermittent, qualitative measurements were first devised in the 1900s using portable expansion cloud chambers. The basic idea was to collect an aerosol in a closed volume for a fixed time. Rapid expansion of the volume in the presence of a condensable vapor generates a supersaturation condition such that particles present act as nuclei for droplets formed in the volume and are detected using light scattering. The early instruments were used extensively to survey the geographical distribution of aerosol particle concentrations in the earth's atmosphere. These devices are believed to "activate" all particles in the atmosphere, and give at least a qualitative estimate of particles to tens of angstroms in diameter. For a condensable vapor of water, the particles measured by an expansion chamber realizing supersaturation ratios much greater than unity are identified as Aitken nuclei (AN) or condensation nuclei (CN).

Later, continuous expansion counters were developed. One was produced in the 1940s, the General Electric Company device invented by Rich *et al.* (1959). This instrument was used in the Second World War as a submarine detector, and later as an acid gas detector. The former application involved observation of direct exhaust particles emitted from the sea surface. In the latter, workers found that water-soluble acid gases act as effective nuclei in the expansion chamber instrument so that gas concentrations can be estimated semiquantitatively (Washimi, 1973).

The original continuous Rich counters were heavy, bulky devices. More recent instrumental designs have considerably reduced their size and weight, making them truly portable devices (e.g., the Rich 100 counter of the Environment One Company).

Several years ago, cloud physics researchers became concerned with the distinction between Aitken nuclei measurements and cloud droplet condensation nuclei (CCN) which are activated in water vapor supersaturations of a few tenths of a percent. Continuous CCN instruments were devised (Squires, 1972) using a vapor diffusion–supersaturation principle similar to the diffusion cloud chamber described on p. 107. These instruments have been used extensively for examining atmospheric CCN concentrations in many different locations.

Continuous nuclei counters employing the vapor diffusion principle are designed along the lines of the device shown in Fig. 5.13. In this particular instrument, the aerosol flows into a saturator using an alcohol as the condensable vapor maintaining a supersaturation level. As nuclei are activated and droplets grow, the particle-laden stream passes through an illuminated volume, where the number concentration is measured by a calibrated scattering photometer system.

The calibration of nuclei counters to yield intercomparable, quantitative measurements is difficult because of uncertainties in generation of aerosol particles over a broad size–nucleus activation range. Studies through the early 1970s relied on on-site instrument intercomparisons for calibration. However, Liu and Pui (1974) recently have reported a rational procedure using the EAA and available aerosol generators which can provide a reasonable calibration for the continuous CN counters.

5.5.2 Light Transmission and Nephelometry. The extinction of a light beam or incident radiation associated with a cloud of particles basically involves a measure of a moment of the particle number–size distribution roughly proportional to the surface concentration. The extinction of a light beam is a well-developed basis for semiquantitative measurement of particles suspended in gases. Devices for these purposes, including smoke photometers, have been available commercially for many years (e.g., Green and

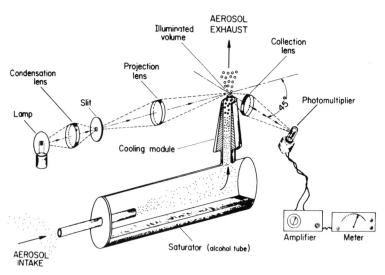

Fig. 5.13. Schematic diagram of the continuous-flux condensation nuclei counter. [From Bricard *et al.* (1976); reprinted with permission of Academic Press.]

Lane, 1964). They may take many forms, one of which is the *transmissometer* or *opacity* instrument.

Transmissometers in simplest form involve a light source and a detector located some distance away along the axis of the light beam. Among their applications, these devices are commonly used at airports to provide data in visual ranging conditions, and are used to measure particle loadings in a smoke stack. A modern instrument design with a double-pass light beam for measuring the opacity of a stack gas is shown in Fig. 5.14. When calibrated for the type of particles present in the aerosol in question, a semiquantitative measure of the particulate emissions can be made, with knowledge of the volumetric gas flow.

To understand the principles of light transmission through a cloud of particles, let us return to the theory of electromagnetic radiation transfer in a medium containing particles.

Consider an aerosol illuminated by a collimated light source of a given wavelength. The experimental arrangement is shown schematically in Fig. 5.10a, which is a simplification of the double-beam instrument in Fig. 5.14. At concentration levels of interest in aerosol measurement practice, particles are separated by distances large compared with their diameter and are distributed in space in a random fashion. Light is scattered in a given direction from an incident beam such that the total energy of the scattered wave per unit area or the intensity of the scattered wave in a given direction will be equal to the sum of the intensities of the individual particles in that direction. This type of behavior can be referred to as collective single-particle scattering. Its application represents a simplified calculation of the total scattering by particulate systems.

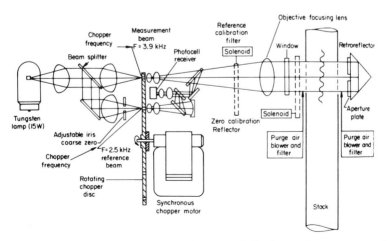

Fig. 5.14. Typical double-pass in situ transmissometer design for opacity measurement in flue gases. [After Nader (1976).]

If there are dN particles in the diameter range d_p to $d_p + d(d_p)$ per unit volume of air, this corresponds to a total particle cross-sectional area of $\frac{1}{4}\pi d_p^2\, dN\, dz$ over the light path length dz per unit area normal to the beam. The attenuation of light over this length is given by the relation

$$-dI = I \left(\int_0^\infty \frac{\pi d_p^2}{4} K_{ext}(x,\,m) n_d(d_p) d(d_p) \right) dz, \qquad (5.18)$$

where $dN = n_d(d_p) d(d_p)$. Hence the quantity

$$b_{ext} = -\frac{dI}{I\, dz} = \int_0^\infty \frac{\pi d_p^2}{4} K_{ext}(x,\,m) n_d(d_p) d(d_p) \qquad (5.19)$$

represents the fraction of the incident light scattered and absorbed by the particle cloud per unit length of path. It is called the *extinction coefficient* or sometimes the *attenuation coefficient* or *turbidity*, and it is a key measure of the optical behavior of particulate systems. In terms of the separate contributions for particle scattering and absorption,

$$b_{ext} = b_{sp} + b_{ap}, \qquad (5.20)$$

where each term is a function of wavelength of incident radiation.

The particle light-scattering coefficient has been measured by a variety of different instruments. One simple device that was invented many years ago is the integrating nephelometer (Buettell and Brewer, 1949). Charlson and colleagues (Ahlquist *et al.*, 1967) have used this instrument extensively in the study of atmospheric aerosols. Their version is shown schematically in Fig. 5.15. The unit derives its name from the fact that it integrates the light scattered from the sampling volume over nearly all angles along the axis of

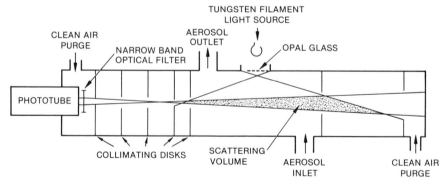

Fig. 5.15. Schematic diagram of the integrating nephelometer. Sometimes a partial shutter is installed under the opal glass. Without the shutter, the instrument integrates the particle scattering coefficient over $\sim 7°$ to $170°$ to measure b_{sp}. With the shutter in place, the instrument integrates over $\sim 90°$ to $170°$ to measure b_{sp}. [After Ahlquist and Charlson (1967); reprinted courtesy of the Air Pollution Control Assn.]

the detector. The device has a tungsten light source where radiation intensity follows a cosine law, and a light collection design to define the shape of the scattering volume. The values of b_{sp} obtained can be related to certain properties of the size distribution. With peripheral modifications, it also has been used to deduce the deliquescence properties of aerosol particles (Charlson *et al.*, 1974).

The contributions to b_{ext} from a given particle size range depend on the extinction cross section and on the particle size distribution function. The integral in Eq. (5.19) can be rearranged as follows:

$$b_{ext} = \int_{-\infty}^{\infty} \frac{db_{ext}}{d \log d_p} \, d \log d_p, \tag{5.21}$$

where

$$\frac{db_{ext}}{d \log d_p} = \frac{3}{2} \frac{K_{ext}}{d_p} \frac{dV}{d \log d_p}$$

is strongly weighted by particles between 0.1 and 1.0 μm in diameter.

If the aerosol is composed of a mixture of particles from different sources with different refractive indices, the value of the extinction coefficient must be summed over its contributions to the aerosol from each source. When the chemical properties of the aerosol are continuously distributed and the refractive index depends on composition, it would be proper to make use of the size–composition probability density function, but such calculations have not yet been made.

The reduction in the intensity of the light beam passing through an aerosol is obtained by integrating (5.19) between any two points L_1 and L_2 along the beam:

$$I_2 = I_1 \exp(-\tau), \tag{5.22}$$

where the *optical thickness* $\tau = \int_{L_1}^{L_2} b_{ext} \, dz$ is a dimensionless quantity; b_{ext} is placed within the integral to show that it can vary in space with the aerosol concentration. Equation (5.22) is a form of *Lambert's law;* its differential form is Eq. (5.19).

In the calculations, only single-particle scattering has been taken into account. In reality, each particle is also exposed to light scattered by other particles, but if the aerosol is sufficiently dilute and the path length sufficiently short, multiple scattering can be disregarded.

In most cases of practical interest, the incident light is distributed with respect to wavelength. The contribution to the integrated intensity I from the wavelength range λ to $\lambda + d\lambda$ is given by

$$dI = I_\lambda \, d\lambda, \tag{5.23}$$

where I_λ is the intensity distribution function. The loss in intensity over the visible range, taking into account only single scattering, is determined by integrating Eq. (5.23) over a wavelength range:

$$d \left(\int_{\lambda_1}^{\lambda_2} I_\lambda \, d\lambda \right) = - \left(\int_{\lambda_1}^{\lambda_2} b_{ext}(\lambda) I_\lambda \, d\lambda \right) dz. \tag{5.24}$$

Here λ_1 and λ_2 can refer to the lower and upper ranges of the visible spectrum and b_{ext} is now regarded as a function of the wavelength interval.

Substituting Eq. (5.24) and (5.21) into Eq. (5.19), we obtain the extinction over the visible wavelengths of light:

$$\bar{b}_{ext} = \int_{\lambda_1}^{\lambda_2} b_{ext}(\lambda) f(\lambda) \, d\lambda = \int_{-\infty}^{\infty} G(d_p) \frac{dV}{d \log d_p} \, d \log d_p, \tag{5.25}$$

where

$$G(d_p) = \frac{3}{2d_p} \int_{\lambda_1}^{\lambda_2} K_{ext}(x, m) f(\lambda) \, d\lambda. \tag{5.26}$$

The term $f(\lambda) \, d\lambda$ is the fraction of the incident radiation in the range λ to $\lambda + d\lambda$, and $f(\lambda)$ has been normalized with respect to the total intensity in the range between λ_1 and λ_2. The quantity $G(d_p)$ represents the extinction over all wavelengths between λ_1 and λ_2 per unit volume of aerosol in the size range between d_p and $d_p + d(d_p)$. It is independent of the particle size distribution function. For a refractive index $m = 1.5$, $G(d_p)$ has been evaluated for the standard distribution of solar radiation at sea level, using Mie scattering functions. The result is shown in Fig. 5.16 as a function of diameter for nonabsorbing particles.

Friedlander (1977) has noted interesting features exhibited by this curve: The oscillations of the Mie functions (Fig. 5.8) are no longer present because of the integration over the wavelength of incident light, but they would remain in the monochromatic light case. For $d_p \to 0$ in the Rayleigh scattering rings, $G(d_p) \sim d_p^3$. For large d_p, $G(d_p)$ vanishes since K_{scat} approaches a constant value (two) at all wavelengths; as a result, $G(d_p) \sim d_p^{-1}$ for $d_p \to \infty$. The most efficient size for light scattering on a mass basis corresponds to the peak in this function, which for $m = 1.5$ occurs in the size range between 0.5 and 0.6 μm. Particles of 0.1 μm diameter and smaller at the low extreme and particles of 3 μm diameter on the upper extreme contribute only one-tenth the scattering on an equal volume basis. The volume distribution $dV/d \log d_p$ of atmospheric aerosols often shows a peak in the 0.1–1 μm size range. This indicates the importance of this portion of the submicrometer range to total light scattering in the earth's atmosphere.

If particles are composed of absorbing material, curves analogous to those in Fig. 5.16 become more complicated. Examples of calculated particle light scattering per unit mass (taken proportional to unit volume) versus diameter for four different spheres are shown in Fig. 5.17. Absorbing carbon and iron particles are compared with silica and water, which are nonabsorbing mate-

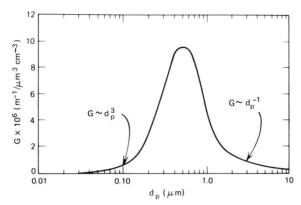

Fig. 5.16. Light scattering per unit volume of aerosol material as a function of particle size, integrated over all wavelengths for a refractive index $m = 1.5$. The incident radiation is assumed to have the standard distribution of solar radiation at sea level [R. E. Bolz and G. E. Tuve, eds., *Handbook of Tables for Applied Engineering Science*, Chemical Rubber Co., Cleveland (1970), p. 159]. The limits of integration on wavelength were 0.36–0.680 μm. The limits of visible light are approximately 0.350–0.700 μm. [From Friedlander (1977); reprinted with permission of John Wiley & Sons.]

rials, for the case of monochromatic light of $\lambda = 550$ nm. The total extinction per unit mass is included with scattering and absorbing contributions. The maximum in extinction with particle diameter differs significantly between the absorbing materials and the nonabsorbing particles. Note also that the maximum in extinction per unit mass of carbon is divided roughly into three absorbing parts, and two scattering, while iron is roughly one to one in absorbance to scattering.

As discussed in Chapter 3, the size distribution of aerosols can be represented by a power-law relationship in the light-scattering subrange, $0.1 < d_p < 3$ μm:

$$n_d(d_p) \sim d_p^{-\gamma}. \qquad (5.27a)$$

This expression can be written without loss of generality in terms of an average particle diameter $\bar{d}_p = (6V/\pi N_\infty)^{1/3}$, or

$$n_d = (A_1 N_\infty/\bar{d}_p)(d_p/\bar{d}_p)^{-\gamma}, \qquad (5.27b)$$

where A_1 is a dimensionless factor that, with $\bar{d}_p$, may be a function of time and position. Friedlander (1977) has noted that when A is constant, Eq. (5.27) represents a special form of a self-preserving size distribution.

Substituting Eq. (5.27) into the expression for the scattering coefficient, (5.26), Friedlander has obtained

$$b_{sp} = \frac{\lambda^{3-\gamma} A N_\infty}{4\pi^{2-\gamma}} \left(\frac{6V}{\pi N_\infty} \right)^{(\gamma-1)/3} \int_{x_1}^{x_2} K_{scat}(x, m) x^{2-\gamma} \, dx, \qquad (5.28)$$

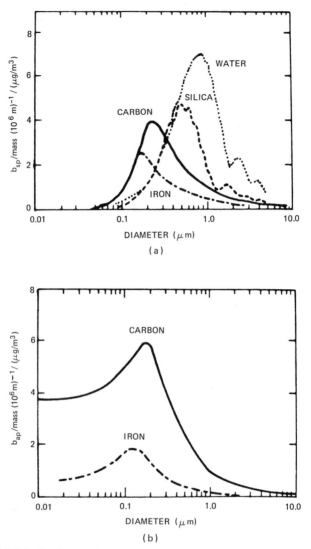

Fig. 5.17. (a) Calculated scattering cross section per unit mass at a wavelength of 55 μm for absorbing and nonabsorbing materials as a function of diameter for single-sized particles. The following refractive indices and densities (g/cm³) were used: carbon $m = 1.96\text{-}0.66i$, $\rho_p = 2.0$; iron $m = 1.33$, $\rho_p = 1.0$. (b) Calculated absorption cross section per unit mass at 0.55 μm for single-sized particles of carbon and iron. [Source: (a) Faxvog, (1975); (b) Faxvog and Roessler (1978); courtesy of the Optical Society of America.

where x_1 and x_2 correspond to the lower and upper size limits, respectively, over which the power-law form holds. For most aerosols studied, the lower limit of applicability of the power law is about 0.1 μm or somewhat less. This corresponds to $x_1 < 1$, and for this range K_{scat} is very small and so x_1 can be replaced by zero. The contribution to the integral for large values of x is small since γ is usually greater than 3 or 4 and K_{scat} approaches a constant, two. Hence, x_2 can be set equal to infinity. Carrying out the integration, the result is

$$b_{sp} = A_1 A_2 \lambda^{3-\gamma} N_\infty^{(4-\gamma)} V^{(\gamma-1)/3}, \qquad (5.29)$$

where A_2 is an empirical constant. The power-law dependence of the portion of the size spectrum contributing to light scattering determines γ via measurement of the wavelength dependence of the extinction coefficient.

Observations of aerosols in the earth's atmosphere have indicated that b_{sp} is proportional to the volume fraction (Covert et al., 1980), which gives a value of $\gamma = 4$ in the light-scattering range. It has been found experimentally that the light-scattering coefficient is sometimes proportional to the wavelength to the -1.3 power, corresponding to $\gamma = 4.3$. The inverse proportionality relationship between b_{sp} and λ indicates more light scattering in the short wavelength range (blue) than the longer wavelength (red) range. This suggests a greater range of vision in the red than in the blue for hazy conditions in the atmosphere.

5.5.3   Remote-Sensing Techniques.

The development of the relationships between scattered light and aerosols has stimulated the use of radiation transfer theory for remote sensing of particles in planetary atmospheres. Highly sophisticated experimental and theoretical techniques have emerged in recent years for interpretation of observations of sunlight and artificial light sources in the earth's atmosphere. A description of their application depends on further development of the concepts of radiant energy transfer.

In the general case of aerosol–light interactions in the atmosphere or within a confined space, the light is neither unidirectional nor monochromatic; each volume element is penetrated in all directions by radiation. This requires a refinement in the definition of the intensity of radiation. For an analysis of this case, an arbitrarily oriented small area ds is chosen with a normal vector $\mathbf{n}$ (Fig. 5.18). At an angle θ to the normal, we draw a line r, the axis of an elementary cone of solid angle $d\omega$; the resultant cross-sectional area perpendicular to S at the point P will be $ds \cos \theta$.

Let dE be the total quantity of energy passing in time dt through the area ds inside cone $d\omega$ in the wavelength interval λ to $\lambda + d\lambda$. For small ds and $d\omega$, the energy passing through ds inside $d\omega$ will be proportional to the product $ds \, d\omega$.

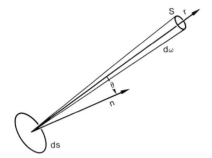

Fig. 5.18. Geometric factors determining specific intensity of radiation.

The specific intensity of radiation or simply the intensity I_λ is defined by the relation

$$I_\lambda = dE/(ds \cos \theta \, dt \, d\omega \, d\lambda). \tag{5.30}$$

The intensity is, in general, a function of the position in space of the point P, the directions $\mathbf{r}$, time t, and wavelength λ:

$$I_\lambda = I_\lambda(P, \mathbf{r}, t, \lambda).$$

If I_λ is not a function of direction, the intensity field is said to be *isotropic*. If I_λ is not a function of position, the field is said to be *homogeneous*. The total or integrated intensity of radiation is $I = \int_0^\infty I_\lambda \, d\lambda$. In the rest of this section, the subscript λ is dropped to simplify the notation.

Consider the radiant energy traversing the length dr along the direction in which the intensity is defined; a change in the intensity results from the combination of the effects of extinction (absorption and scattering) and emission:

$$dI(P, \mathbf{r}) = dI(\text{extinction}) + dI(\text{emission}).$$

The loss by extinction can be written as before in terms of the extinction coefficient b:

$$dI(\text{extinction}) = -b_{\text{ext}} I \, ds.$$

Emission by excited dissociated atoms and molecules in the air is usually small in visible radiation range compared with solar radiation. Thermal radiation is important at the far-infrared wavelengths of radiation but not in the visible. Hence, gaseous emissions can be disregarded for purposes of this discussion.

In general we may expect that only a part of the energy lost from an incident beam will appear as scattered radiation in other directions and that the remaining part will have been "truly" absorbed in the sense that it represents the transformation of radiation into other forms of energy.

Considering the case of scattering, a material is characterized by its scattering coefficient if from a beam of radiation (intensity I and frequency $\int$) incident on an element of mass dm is scattered from it at the rate

$$b_{scat}I \, dm \, d\int d\omega \text{ in all directions,} \tag{5.31}$$

where $b_{scat} = b_{sp} + b_{sg}$. To formulate and quantify the amount of scattering, the angular distribution of the scattered radiation in Eq. (5.31) needs to be calculated. This is done by introducing a phase function $p(\cos \theta)$ such that

$$b_{scat}I p(\cos \theta)(d\omega'/4\pi)dm \, d\int d\omega \tag{5.32}$$

gives the rate at which energy is being scattered into an element of solid angle $d\omega'$, and in a direction inclined at an angle θ to the direction of incidence of a beam of radiation on an element of mass dm. This is consistent with Eq. (5.31) if

$$\int p(\cos \theta) \frac{d\omega'}{4\pi} = 1.$$

Returning to the general case when both scattering and true absorption are present, we shall still write for the scattered energy the same expression (5.32). But in this case (in contrast to the case of scattering only) the total loss of energy from the incidental pencil must be less than that expressed by (5.31); accordingly,

$$\int p(\cos \theta) \frac{d\omega'}{4\pi} = \overline{\omega}_0 \leq 1. \tag{5.33}$$

Thus the general case differs from the case of pure scattering only by the fact that the phase function is not normalized to unity.

From our definitions, $\overline{\omega}_0$ represents the fraction of light lost from an incident beam due to scattering, while $1 - \overline{\omega}_0$ represents the remaining fraction which has been transformed into other forms of energy (or of radiation of other wavelengths). We shall refer to the parameter $\overline{\omega}_0$ as the *albedo for single scattering*. Moreover, when $\overline{\omega}_0 = 1$, a conservative case of perfect scattering occurs.

The simplest example of a phase function is

$$p(\cos \theta) = \text{const} = \overline{\omega}_0. \tag{5.34}$$

In an aerosol, a virtual emission exists because of rescattering in the r direction of radiation scattered from the surrounding volumes. The gain by emission can be written in the form of a source term.

$$dI(\text{emission}) = dJ \, dr \tag{5.35}$$

This equation defines the source function J.

Hence the energy balance over the path length dr takes the form

$$-\frac{dI}{b_{ext}\, dr} = I - J, \tag{5.36}$$

which is the equation of radiative transfer. This equation is useful in interpreting atmospheric visibility as discussed later in Chapter 8. Detailed applications require an expression for the source function J which can be derived in terms of the optical properties of the particles, and are beyond the scope of this book. For additional discussion, the reader should consult Chandrasekhar (1960) and Goody (1964).

A formal solution of the equation of transfer is obtained by integration along a given path from the point $r = 0$:

$$I(r) = I(0)\, \exp[-\tau(r, 0)] + \int_0^r J(r')\, \exp[-\tau(r, r')]b_{ext}\, dr', \tag{5.37}$$

where $\tau(r, r')$ is the *optical thickness* of the medium between the points r and r'. That is,

$$\tau(r, r') = \int_{r'}^{r} b_{ext}\, dr.$$

The source function $J(r')$ over the interval 0 to r must be known to evaluate the integral in Eq. (5.37). According to this equation, the intensity at r is equal to the sum of two terms. The first term on the right-hand side corresponds to Lambert's law, Eq. (5.22), often used for the attenuation of a light beam by a scattering medium. The second term represents the contributions to the intensity at s from each intervening radiating element between 0 and r, attenuated according to the optical thickness correction factor. For aerosol scattering, this term can result with an originally unidirectional beam in the r direction from light scattered by the surroundings into the particles along r and then rescattered by them in the r direction. This is the multiple-scattering term neglected in the derivation of Lambert's law. The criterion for neglecting this term can be expressed in terms of the optical depth. For $\tau_{scat} < 0.1$, the assumption of single scattering is acceptable, while for $0.1 < \tau_{scat} < 0.3$, it may be necessary to correct for double scattering. For $\tau_{scat} > 0.3$, multiple scattering must be taken into account.

When the medium extends to $-\infty$ in the s direction, and there are no sources along r, it starts at point 0 and continues indefinitely; or

$$I(r) = \int_{-\infty}^{r} J(r')\, \exp[-\tau(r, r')]b\, dr'.$$

Thus the intensity observed at r is the result of scattering by all of the particles along the line of sight, say through a path from the earth's surface outward through the atmosphere, or on a path tangential to the earth's surface. Basically, remote sensing then is concerned with measurement of the intensity $I(r)$ along such paths, for source functions which are solar light or artificial radiation such as a searchlight. Theory and independent information on aerosols are used to interpret the meaning of the observation determining optical thickness, and the extinction coefficient.

The measurements are made by the techniques listed in Table 5.4. Three basic groupings are given; (a) ground-based passive optical sensors, (b) airborne passive sensors (either aircraft, balloon, or artificial satellite), and (c) active sensor systems employing a controlled or man-made light source. The instrumentation designed and built for such purposes is diverse and ingenious. It has occupied the thoughts of astronomers, spectroscopists, meteorologists, and space engineers for many years. The references contain the details of the methodology and intercomparisons between remote sensors of different kinds and direct aerosol observations.

In general, the methods available are difficult to interpret quantitatively in terms of aerosol properties because of ambiguities in the size distribution–

TABLE 5.4

Some Remote-Sensing Methods for Atmospheric Aerosol Characterization

Method	Measurable	Light source	Reference
Ground based—passive			
Photometry and	Optical thickness;	Sun	Volz (1959)
radiometry	sky brightness	Sun	Shaw *et al.* (1973)
			Newkirk and Eddy (1964)
Polarimetry	Polarization of skylight	Sun/diffuse sky	Sekera (1967)
Polar nephelometry	Extinction coefficient	Sun/skylight	
Teleradiometry	Horizon brightness relative contrast	Reflected and scattered light	Malm and Walther (1980)
Airborne—passive			
Spectrophotometry	Albedo; optical thickness	Reflected sunlight	Hall and Riley (1978)
Limb occultation (satellite radiometry)	Optical thickness; polarization	Sunlight	Kondratyev (1975) Gille *et al.* (1977) MacCormick *et al.* (1981)
Active Sensing			
Transmission/backscatter	Optical thickness	Searchlight	Eltermann (1968)
Transmission/backscatter	Optical thickness	LIDAR[a]	Collis *et al.* (1964; 1970) Deluisi *et al.* (1976) Reagan *et al.* (1977)

[a] LIDAR: light detection and ranging.

concentration–distance profiles and variations in chemical properties contributing to the index of refraction. Two major intercomparison studies have been reported. The first was that of De Luisi *et al.* (1976), which involved LIDAR, multiwavelength radiometry and aircraft observations in the southwestern United States. A similar intercomparison was later reported by Reagan *et al.* (1977). An example of their results for the average size number–size distribution given in a power-law form (5.27a) with c the proportionality constant and γ the exponent for the light-scattering range is shown for three different measuring methods in Fig. 5.19. Although the distribution function lacks the detailed structure found in direct measurements, the agreement between techniques is not unreasonable considering the theoretical models applied.

5.6   MEASUREMENT OF PARTICULATE MASS CONCENTRATION

5.6.1   Mass Measurement.

Perhaps the moment of the particle size spectrum most commonly investigated is the product of the volume fraction and an average particle density. This yields, of course, the mass concentration of particles over a range of particle size relevant to the sampler. Mass concentration can be derived by several means including (a) gravimetry; (b) attenuation of low energy radiation (β radiation); (c) piezoelectric detection; and

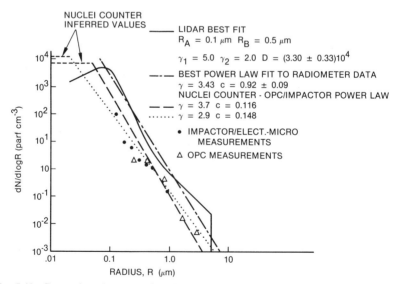

Fig. 5.19. Comparison between size distribution estimates from LIDAR soundings, solar radiometer data aircraft impactor observations as counted with an electron microscope, and an airborne optical counter (OPC). The LIDAR distribution corresponds to two subranges, the low range between R_A and R_B (γ_1), and $>R_B$ (γ_2). D is a scaling constant equivalent to c. [From Reagan *et al.* (1977); reprinted with permission of the American Meteorological Society.]

(d) indirectly, by integrating the third moment of the number–size distribution. The latter assumes an average density can be defined for a heterogeneous material.

The gravimetric method is the standard reference method used by most investigators. It involves taking a weight difference of the collecting substrate before and after sampling. The mass concentration collected over an interval in time can be calculated from the accumulation of mass if the surface area of the deposit or the collection efficiency and the flow rate are known. The errors in this method are reasonably well established, especially for porous or fibrous filters as collectors. The errors come from filter collection efficiency, from weighing, and from flow rate uncertainties. They also may involve variability in weights of substrates that absorb and desorb moisture at different humidities, or involve sample degradation in storage. Filters are generally weighed at humidities below 50% to ensure stability with moisture content in the environment; the influence of storage on stability of samples has been discussed for atmospheric samples by Smith *et al.* (1978). When sulfate, nitrate, and ammonium are present in particulate samples, the nitrate and ammonium can be lost preferentially during storage at room temperature as a result of evaporation. One would expect that certain volatile organics or other salts such as ammonium chloride also would be subject to change during storage. To minimize such effects, workers have attempted to stabilize samples by storage at low temperatures followed by early chemical analysis and weighing after sample collection. This method has not been tested extensively as yet, so it cannot be used with confidence.

Measurement of mass concentration by β radiation attenuation through a sample preconcentrated on a filter was first suggested a few years ago by Dresia *et al.* (1964). Continuous dust monitors built by commercial vendors appeared on the market in the early 1970s based on this principle. These devices appear to be useful for monitoring areas of high dust loading, such as mining operations. However, they have not proven to be a replacement for gravimetry in ambient air monitoring. Although they showed some promise for this application (Macias and Husar, 1976) and have been used in conjunction with the dichotomous sampler design (Loo *et al.*, 1970), their calibration for response versus mass concentration varies with material sampled. This variation is shown, for example, in Fig. 5.20. With such differences, it is questionable if the method can be useful for quantitative assessment of total suspended-particulate concentrations. Some of the β radiation monitors appear to be sensitive to humidity, too. This poses added problems in their use in the field as reliable monitoring devices.

Another method for continuous monitoring of mass concentration makes use of a piezoelectric effect. If a vibrating quartz crystal is used as an impaction plate, for example, or the collector for an electrostatic precipitator, detection of mass change can be achieved by measuring the shift in vibration frequency of the crystal with loading (Chuan, 1970). The piezoelec-

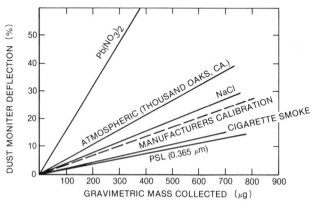

Fig. 5.20. β-gauge response versus gravimetric filter concentration for different materials. [After Landis (1975); courtesy of Pergamon Press.]

tric method has been used in commercial instruments, but has been found to yield uncertain results because the calibration varies with loading of the crystal, and the crystals are subject to adsorption of water vapor. Considerable care then must be used in applying such principles to mass concentration measurement.

5.6.2   Filtration Principles.   The collection on porous filter media is perhaps the most efficient means of particle removal available. Aerosol filtration is an effective means of air purification, while at the same time it has been widely used for sampling of airborne material for mass and chemical composition determination. There are a wide variety of filter media available, ranging from fibrous mats of relatively inert material to porous membranes. Fibrous mats and model filter arrays appear microscopically as stacks of overlaid cylinders, where the cylinders may be smooth or rough (Fig. 5.21). In contrast, the membrane media are plastic films with microscopic holes of near uniform size; e.g., nuclepore filters are produced of sheets of polyester, and the holes are introduced by neutron bombardment (e.g., Fig. 5.22).

Fibrous filters are the most economical and effective method for purification of air from suspended particles. This purification is achieved with a minimal loss of pumping energy associated with flow resistance, compared with other types of filters. The porosity of such materials is 85–99% and fiber diameter varies from 10^2 to 10^{-2} μm.

The membrane filters offer the advantage that particles do not become imbedded in the filter medium. Thus individual particles are readily identifiable and characterized microscopically on the filter surfaces. Furthermore, certain kinds of chemical analysis such as x-ray fluorescence analysis readily

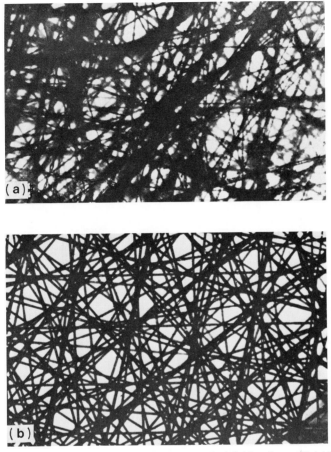

Fig. 5.21. Comparison between a fibrous filter mat on the left (a) and an artificial filter called a fan model made up of a cylindrical array of fibers (b). [From Kirsch and Stechkina (1978); with permission of Wiley.]

can be done in situ with minimal effects of filter interference on the membrane substrates.

Sampling devices range from simple filter holders to sequential configurations for automated routine air monitoring of many samples in series. Membrane filters can be obtained in different pore sizes, so that they have been used in series as particle size fractionators.

One of the best known filter samplers is the high-volume (hi-vol) sampler used for air monitoring. An example of this instrument is shown in Fig. 5.23. The device consists of a large filter holder oriented such that air passes downward through an 8 × 10 in. substrate with an electrically timed vacuum system. To minimize the collection of very large particles (larger than

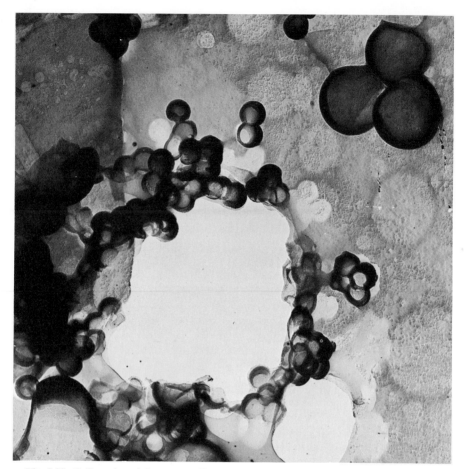

Fig. 5.22. Collected particles surrounding pores in a nuclepore membrane filter. Note that the particles are trapped at the edge of pores rather than in the blank areas of this filter.

roughly 30 μm diameter) a rectangular gabled roof is added. The mass of material accumulated on the filter has a somewhat arbitrary upper size limit which is dependent on winds and instrument height above the ground. The device nevertheless has been used as the standard method in the United States for monitoring total suspended-particle (TSP) concentrations (U.S. EPA, 1971).

A modernized version of the high-volume filter device is shown in Fig. 5.24. This instrument is an automated sequential filter sampler which has twelve 47-mm-diam filter holders arranged symmetrically around the sampling inlet duct to avoid a preferential sampling orientation (Fig. 5.24b). The filter holders are activated automatically to sample the inlet air in sequence over given intervals of time. An additional holder without flow capability is

Fig. 5.23. Two high-volume filter samplers in operational configuration. The setup is in compliance with siting requirements for baseline suspended-particle measurements at a rural site.

mounted in the center to obtain a sample of fallout. The device has a circular plenum with a sampling inlet covered by a shroud which has been designed to prevent particles larger than approximately 10 μm from entering the system (Fig. 5.24a). Addition of a cyclone on the inlet provides a means of size segregation at smaller diameters, say 3 μm. Thus this type of sampler can be used in pairs to obtain a size fraction less than 10 μm and a size fraction less than 3 μm in diameter. These fractions are of interest for projected air monitoring requirements in the 1980s (see also Section 9.3). The larger-diameter cut is being considered to characterize air pollution in the United States as distinct from blowing dust. The finer cut is of concern as an index of particles affecting visibility and representing a potential factor in respiratory disease. The sequential filter device can also be used with a wind direction sensor to automatically accumulate samples from a preferential direction.

Ambient filtration of air for monitoring purposes is only one important measurement application. Filtration has also been used for direct sampling of particles emitted in pipes, ducts, and stack gases. An example of a standard system currently used by the U.S. Environmental Protection Agency for flue gas sampling (e.g., U.S. EPA, 1977) is shown in Fig. 5.25. The sampling train first collects particles by filter in a heated vessel above 120°C

(a)

(b)

Fig. 5.24. A sequential filter unit containing several filter holders in a cylindrical plenum. The top photo (a) shows the unit with the plenum and inlet shroud mounted as in normal outdoor operation. Photo (b) at the bottom shows the symmetrically oriented filter units with the plenum removed. The center filter holder is used as a blank to obtain a fallout sample during operation. (Photos courtesy of J. Chow.)

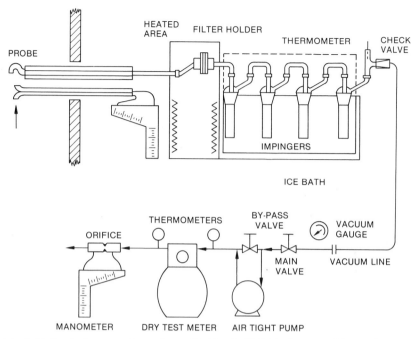

Fig. 5.25. Particulate sampling train specified by the U.S. Environmental Protection Agency Method 5 procedure. [U.S. EPA (1977).]

to minimize condensation of volatiles. Then the flue gases are passed through a series of chemical gas absorbers for emission characterization.

5.6.3   Idealized Filtration Theory.   The theory of filtration is a direct application of dynamical principles discussed previously in Chapter 2. The objective of the theory is to provide a framework for calculating the number of particles of a given size deposited per unit area or unit filter length as the sample depends on flow rate, porosity or filter diameter, temperature, pressure, presence of condensation or chemically reactive vapors, electrical fields, etc. The overall filter collection efficiency, combined with the pressure drop or flow resistance are crucial characterization parameters for selection of appropriate filters for air purification. Kirsch and Stechkina (1978) recently have reviewed in detail the classical theory of filtration on fibrous media without electrical effects. Their investigations form the basis for this section.

Particles are deposited from a gas layer adjacent to the substrate. Deposition takes place by convective diffusion and interception. Thus, the complex pattern of flow through a filter becomes a key to calculation of its efficiency. In principle, one could model the flow through a fibrous filter in terms of a superposition of flow around a cylindrical array (Fig. 5.21), taking into account the mutual interactions between fibers by using the packing density.

The character of flow through a fibrous net can be seen by examining the drag force $\mathcal{D}$ on a unit fiber length in terms of pressure drop ΔP across the filter:

$$\mathcal{D}^* \equiv \mathcal{D}/q_0\mu_g = \Delta P\pi a^2/\mu_g q_0 \xi H; \tag{5.38}$$

q_0 is the face velocity of the gas flow; a is the fiber radius; ξ is the packing density, or $(1 - \xi_v)$, with ξ_v the void fraction; and H is the filter thickness. A graph of this relation is given in Fig. 5.26 for flow regions where the fiber Reynolds number $\mathrm{Re_f} < 1$. As indicated in this figure, the filter resistance is not a single-valued function of the packing density. The absence of a single relationship is the result of the inhomogeneity of the filter structure, which comes from heterogeneities produced during manufacture. Although flow functions for fibrous filters are referred to flow around cylinders for $\mathrm{Re} \lesssim 1$, characterization is only qualitative.

The deposition on filters, estimated by a single fiber efficiency η, is then deduced in terms of semiempirical correlations based on dimensional arguments.

Thus in general,

$$\eta = \eta(\mathrm{Re_f}, \mathrm{Kn_f}, \mathrm{Pe_f}, I, \mathrm{Stk}, \mathrm{Gr}, \mathcal{D}_i^*, \xi, \mathrm{IH}), \tag{5.39}$$

where

$$\mathrm{Re_f} = \frac{2q_0 a}{\nu_g}, \qquad \mathrm{Pe_f} = \frac{2q_0 a}{D_p}, \qquad \mathrm{Kn_f} = \frac{\lambda}{a},$$

$$\mathrm{Stk} = \frac{2A(\mathrm{Kn})\rho_p R_p^2 q_0}{9\mu_g a}, \qquad \mathrm{Gr} = \frac{\tau_p g}{q_0}, \qquad \mathcal{D}_i^* = \frac{\mathcal{D}_i \tau_p}{m_p}, \qquad I = \frac{R}{a};$$

IH is the Inhomogeneity factor for the filter structure.

The total efficiency $\mathcal{E}$ is given in terms of the filter *penetration* N/N_0 by

$$\mathcal{E} = 1 - N/N_0 = 1 - \exp(-2\bar{a}\eta\xi H/\pi a^2), \tag{5.40}$$

where $\bar{a}$ is the mean fiber radius. Kirsch and Stechkina (1978) have reported single-fiber efficiency formulations for a "fan" model filter (Fig. 5.23) in the following form:

$$\eta_f = \eta_D + \eta_I + \eta_{DI}, \tag{5.41}$$

where η_D is the efficiency for convective diffusion, η_I that for interception, and η_{DI} a combination diffusion and interception term. From theoretical arguments,

$$\eta_D = 2.7\mathrm{Pe}^{-2/3}[1 + 0.39(K_H)^{1/3}\mathrm{Pe}^{1/3}\mathrm{Kn}] + 0.624\mathrm{Pe}^{-1}, \tag{5.42}$$

$$\eta_I = (2K_H)^{-1}[(1 + I)^{-1} - (1 + I) + 2(1 + I)\ln(1 + I)$$
$$+ 2.86\mathrm{Kn}(2 + I)I(1 + I)^{-1}], \tag{5.43}$$

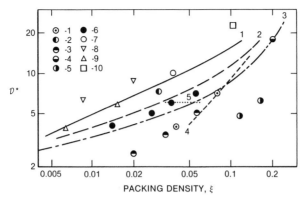

Fig. 5.26. Drag force for different filters versus packing density ξ. Curves: (1) data of Kirsch and Dvukhimenny (1975), (2) Chen (1955), (3) Langmuir (1942), (4) Ramskill *et al.* (1942), (5) Kimura and Iinoya (1959). Points: (1) data of Wong *et al.* (1956), (2) Whitby *et al.* (1961), (3) Blasewitz *et al.* (1951), (4, 5) Ramskill and Andersen (1951), (6) First (1951), (7) Humphrey and Gaden (1955), (8, 9) Thomas (1953), (10) Stern *et al.* (1960).

$$\eta_{DI} = 1.24(K_H)^{-1/2}Pe^{-1/2}I^{2/3}. \tag{5.44}$$

The hydrodynamic factor for cylindrical arrays is

$$K_H = 0.5 \ln \xi - 0.52 + 0.64\xi + 1.43(1 - \xi)Kn.$$

When Pe $\gg 1$, $I \ll 1$, Kn $\to 0$, these relations basically reduce to Friedlander's (1977) correlation, except for the multiplicative constants. The correlation for η_D versus Pe is well established, for example, for the fan model filter and different fibrous substrates (Kn $\to 0$). The correlation between η_0 and Pe is shown in Fig. 5.27 for several different experimental observations.

Kirsch and Stechkina (1978) recommend the following relationships to apply Eqs. (5.43) and (5.44) to commercial filter media:

$$\eta = \eta_f/IH,$$

where

$$IH = \mathscr{D}_f^*/\mathscr{D}_0^* = 4\pi(-0.5 \ln \xi - 0.52 + 0.64\xi)^{-1}/\mathscr{D}_0^*, \tag{5.45}$$

$$(\mathscr{D}_f^*)^{-1} = (\mathscr{D}_0^*)^{-1} + \frac{1.43(1 - \xi)IH^{1/2}Kn}{4\pi};$$

$\mathscr{D}_f^*$ and $\mathscr{D}_0^*$ are dimensionless drag forces on the filter and at Kn = 0, respectively. Thus calculation of filter efficiency reduces to estimation of η_f, $\mathscr{D}_f^*$, and IH.

The above theory has been extended to apply to ultrafine polydispersed fiber mats. Kirsch and Stechkina have calculated overall single-fiber collection efficiencies and compared them with experimental observations. An example of Dyment's (1969) different media is shown in Fig. 5.28. The

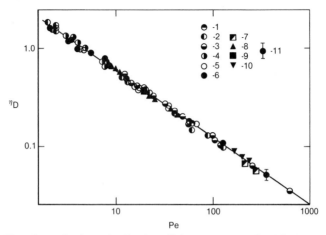

Fig. 5.27. Experimental values of collection efficiency versus Pe for diffusional deposition in (1–6) a fan model filter and (7–11) real filters, corrected for their inhomogeneity. The full curve is plotted according to an approximation equivalent to the first term in Eq. (5.41). [Experimental values and diagram from Kirsch and Stechkina (1978); diagram reprinted with permission of John Wiley & Sons.]

theory is in reasonably good agreement with observations of collection efficiency and particle radius (Fig. 5.28). It predicts a minimum in overall efficiency just below 0.1 μm particle radius. A similar comparability was found between observed efficiencies and face velocity.

Designs for filter systems, often require data on pressure drop. This is sometimes given in terms of the Kozeny–Carmen relationship. Pressure drop is discussed further in Section 10.4.4.

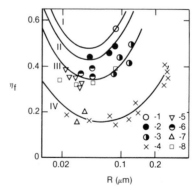

Fig. 5.28. Experimental values of collection efficiency at face velocity $q_0 = 19$ cm sec^{-1} for filters: Gelman GF/C (curve I, point 1); Gelman A (II, 2): Gelman GF/A (III, 3, 5, 6, 8); Gelman GF/D (IV, 4, 7). Particles: DES and DOP (points 1–5); selenium (6, 7); NaCl (8); the experimental data of Dyment with $q_0 = 20$ cm/sec. (See also Kirsch and Stechkina, 1978; courtesy of Wiley.)

5.6.4 Role of Electrical Charging. The superposition of electrostatic forces on particle behavior near a filter mat can have appreciable influence on filtration efficiency. The deposition patterns can take on significant treeing or branching of agglomerates on individual fibers. This aerodynamically distorts the cylindrical collector surface and branches the surface area, as well as distorting the electrical field around the collector. An example of charged-particle collection with treeing is shown on a cylindrical fiber in Fig. 5.29.

Basically the collection efficiency for an individual fiber with electrical forces can be estimated by models of Zebel (1968) and Kraemer and Johnstone (1955) when electrical forces are prevalent (see also Section 2.2). Pich (1978) has reviewed the theoretical models for different situations and examined the experiments available. As expected, electrostatic theory generally appears applicable for calculations where electrical forces are dominant.

5.6.5 Filter Adsorption and Other Chemical Interactions. Filtration has been used extensively as a preconcentrator for examining particulate matter. By the very nature of the filter, the filter substrate contains substantial surface area for chemical reaction, and may be made of potentially chemically active material. Therefore, care must be taken in the selection of a filter substrate which will be inert to passage of chemically reactive aerosol.

A practical example of potential interference in deriving total suspended-particulate concentrations in ambient air has appeared recently. Several years ago, certain glass fiber filter media were selected in the U.S. National Air Surveillance Network (NASN) for high collection efficiency and for relatively clean conditions relative to atmospheric particulate contaminants of interest (e.g., sulfate, nitrate, metals, and organic material). In 1966, Lee and Wagman noted that glass fiber filters were capable of adsorbing enough SO_2 to cause a positive error in observations of total suspended-particulate concentrations. More recent work, for example, of Spicer and Schumacher (1979) and Appel *et al.* (1980a), has indicated that a number of filter media used in the past for studies of aerosol particles in the presence of acidic gases are subject to the adsorption artifact. Examples are listed with the effect for NO_x in Table 5.5. In the case of NO_x, the artifact appears to be more severe than SO_2, perhaps as much as 8–10 $\mu g/m^3$ NO_x adsorption over a 24-hour sample period as opposed to 4 $\mu g/m^3$ or less for SO_2. The process involved is adsorption on the filter surface; this appears to be particularly severe with cellulose fibers which may be chemically reactive to acid gases, or to glass fibers which are alkaline in character unless carefully treated by acid washing just prior to use. Membrane filters which are chemically passive such as Teflon appear to be free of the artifact, but others such as cellulose- or polyester-based nuclepore or millipore are not. A Teflon-coated quartz, high-flow fibrous filter mat manufactured by Pallflex has been tested and

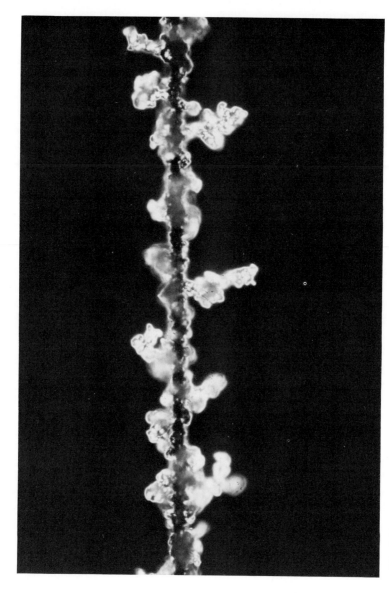

Fig. 5.29. Treelike branches of 1-μm-diam charged NaCl spheres deposited from an air stream on a circular tungsten cylinder. (Photo by D. Hochrainer.)

found to have minimal SO_2 and NO_x artifact but is subject to HNO_3 vapor adsorption (Mueller *et al.*, 1982).

No studies of organic vapor artifacts have been conducted on commercially available filters. However, from the experience with acidic gas adsorption one expects analogous effects with other reactive gases including some organic vapors.

A further complicating effect which probably influences collection of particles on filters (or other substrates) is the potential for chemical interactions between dissimilar particles on the substrate, and between reactive gases and particles on the substrate. For example, nitric acid is highly volatile, while sulfuric acid is not at normal temperature and pressure. Workers believe that sulfuric acid can be present in the condensed phase in ambient air, while nitrate probably exists only as an ammonium or other salt, or nitric acid dissolved in sulfuric acid. If ambient ammonia concentrations vary in the aerosol such that the particles on a substrate can acidify by sulfuric acid collection, release of ammonia and nitric acid can take place, giving rise to reduction in apparent values of suspended-particle concentration. Since the ammonium nitrate equilibrium vapor pressure in air is highly temperature dependent, anomalies in nitrate collection also may be found as a function of temperature alone (Stelson *et al.*, 1979). Such combinations of substrate variability have been suspected by investigators, but have not yet been studied or verified. However, investigators should bear these kinds of uncertainties in mind when comparing historical air monitoring data or environmental measurements of aerosols.

5.7 SIZE SEPARATION AND COLLECTION BY PARTICLE FORCES

5.7.1 Inertial Impaction.

In the methods discussed so far, continuous observations by particle size have been involved, giving detailed information on the particle concentration–size distribution, but limited detail on particle morphology. An important requirement for aerosol experimentation is the ability to sample and collect particles with size segregation. One method for such sampling uses the variation of inertial impaction with mass (or size). Devices which have been designed for this purpose are called *impactors*. They operate on the idea that a large particle will tend to collide with a surface when particle-laden air is directed to a surface, while small particles will follow the gas flowing around the collector. Typical impactor configurations are shown in Fig. 5.30. The air is forced through a converging nozzle and ejected onto an impaction plate oriented normal to the gas flow. The gas streamlines bend sharply aside, while particles with sufficient inertia will hit the plate. The basic design parameters of the impactor are the nozzle throat diameter or width and the distance from the nozzle exit plane to the plate.

TABLE 5.5　*Example Artifact Nitrate Results*

Experiment	Conditions	Approximate sample volume (m³)	Filter type→	Cellulose acetate	Poly-carbonate	Teflon (mitex)
1	Clean air	1.0		0	0	0
2	Clean air	0.95		0	0	0
3	2.6 ppm NO_2	1.0		0	0	0
4	2.0 ppm NO_2	0.95		0	0	0
5	1.8 ppm NO_2 (40% RH)	0.88		0	0	0
6	1.4 ppm $HONO_2$	0.75		30	0	0
7	1.5 ppm $HONO_2$	0.95		30	0	0
8	~8 ppm $HONO_2$	1.3		130	0	0
9	3.0 ppm $HONO_2$	1.3		40	0	0
10	3.3 ppm $HONO_2$ (30% RH)	1.1		540	0	0
11	5.5 ppm NH_3	1.1		0	0	0
12	0.3 ppm PAN	0.95		0	0	0
13	15 ppm N_2O	1.5		0	0	0
14	18 ppm N_2O (80% RH)	1.3		0	0	0
15	1.7 ppm NH_3 (40% RH)	1.4		0	0	0
16	30 ppm NO_2 (70% RH)	0.95		13	0	0
17	0.35 ppm $HONO_2$	4.1				
18	21 ppm NO_2	1.0				
19	~18 ppm $HONO_2$	1.3				
20	16.5 ppm NO_2 (17%)	1.3				
21	0.32 ppm NO_2 (25% RH)	1.2				
22	2.4 ppm NO_2 (25% RH)	1.2				
23	0.27 ppm $HONO_2$ (25% RH)	1.3				
24	3 ppm $HONO_2$ (25% RH)	1.5				

[a] Filter blanks have been subtracted.
[b] From Spicer and Schumacher (1979); reprinted with permission from Pergamon Press.

By operating several impactor stages at different flow conditions, the aerosol particles are classified into several size ranges from which the size distribution is determined. These single stages can be operated in a parallel or in a series arrangement. In the parallel flow arrangement, each of the stages classifies the airborne particles at different cutoff sizes, so that the difference in the amount of the deposit on any two stages gives the quantity of particles in the particular size interval defined by the respective cutoff sizes of the two stages. In the series arrangement, also known as the cascade impactor (Fig. 5.31), the aerosol stream is passed from stage to stage with continually increasing velocities and decreasing particle cutoff sizes. Thus each stage acts as a differential size classifier. Of the two flow systems, the cascade arrangement is by far the most popular, as evidenced by the many commercial cascade impactors currently available.

...from Gas–Filter Interaction Experiments[b] (NO_3^-, μg)

Glass AA	Glass A	Glass AE	Glass E	Glass–spectro	Nylon	Quartz (ADL)	Quartz (QAST)	Quartz (E-70)	Quartz (micro-quartz)	Glass (MSA)
	0		0	0	0					
	0		0	3	0					
	68		98	52	0					
	39		27	19	0					
	75		59	19	0					
	190		190	2,500	0					
	180		190	3,500	0					
	250		370	9,700	0					
	240		240	8,800	0					
	320		580	12,300	120					
	0		0	0	0					
	37		20	40	0					
	0		13	60	0					
	99		0	30	0					
	0		17	17	0					
	400		250	2,000	0					
		230				0	91			
117				11		0	37	45		
190				343		15	24	123		
240				46		0	19	58		
60		55		35		5	19		39	
56		95		30		23	23		35	
280		295				5	20		95	310
360		460				5	30		160	400

In the conventional impactor, the jet is formed in a nozzle (internal flow) and then impacts onto a plate. It is also possible to pass the impaction plate through the particle-laden air (external flow). The effectiveness of particle collection in the latter arrangement is comparable to that of conventional impactors. In operation, these impactors normally consist of impaction plates (or cylinders) mounted at the ends of rotating arms. As the arms are rotated through the air, particles are impacted onto the collection surface. The size of the particles collected depends upon the speed and width (or diameter) of the impaction surface as well as the size and density of the particles. These devices may be used to collect particles larger than 10–20 μm in diameter. Thus, for the collection of large particles, which may be difficult to sample efficiently in a conventional impactor, this type of impactor is a suitable alternative.

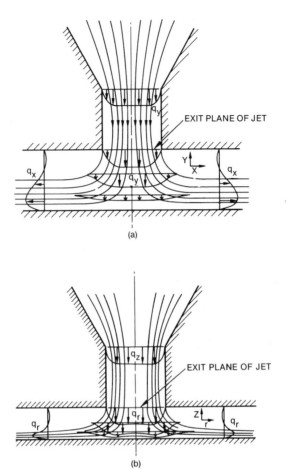

Fig. 5.30. Flow fields in rectangular and round impactors: W jet width or diameter, S distance from nozzle exit plane to impactor plate, L nozzle throat length. (a) Rectangular impactor ($H/W = 1$, $L/W = 1$, Re $= 3000$); (b) round impactor ($H/W = \frac{1}{2}$, $L/W = 1$, Re $= 3000$). [From Marple and Willeke (1976); reprinted courtesy of Academic Press.]

According to Marple and Willeke (1976), nearly ideal behavior of an impactor can be achieved if (a) in the region between the jet exit plane and the impaction plate, the axial y component of the air velocity is a function of y only, not the transverse direction x; and (b) the y component of the particles at the jet exit plane is uniform across the jet. These conditions are only approximately met in real impactors, as shown in Fig. 5.30. The key departures from ideality lie in the fluid boundary layers near the impactor walls. Particle–air flow behavior in these zones causes nonideal cutoff characteristics at large efficiencies in real devices.

Prior to 1970, theoretical studies of the mechanics of particle deposition in

Fig. 5.31. Example of a cascade impactor air sampling system. The two instruments here are Lundgren impactors mounted in parallel to a sampling inlet above. This instrument employs five rotating collector drums on which nozzles of each stage impinge. Foil surfaces are used for collecting; the drums rotate with constant velocity, permitting a strip of collected particles to be obtained, separated in time on the foil surface. The configuration shown also includes a total filter collector located directly below the top of the sampling inlet.

impactors relied on simple approximations to the gas flow field. Recently, Marple *et al.* (1974) have generated numerical solutions for the Navier–Stokes equations of viscous flow as applied to round and rectangular jet impactors. The computer solutions for the mechanical equations agree well with several experimental studies available for comparison. Their calculations provide useful and detailed design criteria for such devices that should eliminate much trial and error.

The theoretical efficiency curves derived by Marple and Liu (1974) are shown in Fig. 5.32. Several features of these curves are of interest. First, the curves generally are similar in shape except for extreme values of impactor (jet) Reynolds number. At very low Reynolds number, poor cutoff characteristics are the result of a thick fluid boundary layer in the jet of the impactor. At very high Reynolds number, the boundary layer adjacent to the plate becomes very thin, of a thickness approximately equal to a particle diame-

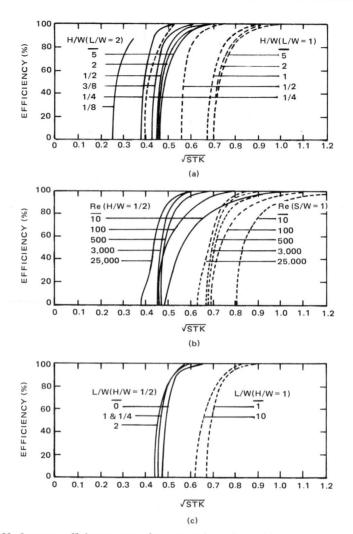

Fig. 5.32. Impactor efficiency curves for rectangular and round jet impactors, showing the effect of the ratio of jet to plate distance *h* and jet width *W*, Reynolds number, and throat length *T*. (a) Effect of jet to plate distance (Re = 3000); (b) effect of jet Reynolds number (*T/W* = 1); effect of throat length (Re = 3000). [From Marple and Liu (1975); reprinted courtesy of the American Chemical Society.]

ter. This evidently causes a knee in the efficiency curve, explained by smaller particles reaching the surface than for thicker boundary layers. Second, the efficiency curve is found to depend on Reynolds number and the ratio of the jet to plate distance *H* and the jet width or diameter *W*, except for large values of these parameters. Third, the round jet impactor will collect smaller particles than the rectangular jet configuration for similar values of *S/W* and Re.

The calculations of Marple *et al.* (1974) indicate that impactor designs should realize approximately ideal behavior for a jet with Re $\gtrsim$ 500. The influence of jet to plate distance is small above a value of 1.5 times the jet width or 1.0 times the jet diameter for rectangular or round impactors, respectively. Impactors should be designed for jet to plate distances larger than these values to minimize significant changes in cutoff efficiency with small changes in this distance. The key design parameter for impactors is the *plate collection efficiency* η, which gives the fraction of particles of a given size collected from the incident stream as a function of size. As found in Chapter 2, the collection efficiency is a function of a single dimensionless parameter, the ratio of the stopping distance of a particle to the half-width of the nozzle for the impactor, or the Stokes number to the half power. The Stokes number, for an impactor, is defined as

$$\text{Stk} \equiv [2\rho_p A(\text{Kn}) U_\infty R_p^2 / 9\mu_g]/(W/2),$$

where U_∞ is the mean nozzle throat velocity and W is the jet width or diameter. A well-designed impactor has a sharp cutoff at a Stk of approximately 0.5. Ideally the cutoff is a step change from an efficiency of zero to one centered at the cutoff diameter.

Impactors have been built in a variety of ways to achieve certain collection goals. In some cases, low-flow instruments have been used to sample materials for microscopic examination. In others, high-volume flow is required to collect large samples for gravimetric and chemical analysis. Instruments have been designed with single or multiple round or rectangular nozzles to spread samples on the collecting surface for examination. Impaction surfaces can involve smooth glass or plastic surfaces, or can be roughened with filter paper. Devices have been made with moving collection plates or rotating drums as impaction surfaces to obtain samples distributed over a time sequence.

A goal in size segregation by impaction is to collect particles as small as possible. Normal design of impactors with small incremental pressure differentials give a minimum size cutoff at approximately 0.5 μm in diameter. However, impactors can be designed for large pressure drops and achieve separation to as low as 0.05 μm (Hering *et al.*, 1978).

Actual jet impactor performance depends on several factors. The gas flow through the nozzle should be smooth without introduction or turbulence from burrs or nonparallel jet/impaction surface planes. Losses of particles at the inlet and between stages are often unpredictable, but may improve significant uncertainties when comparing material collected by stages versus a total collection measure such as a filter. Leaks can cause significant errors from uncertainties in flow.

Two of the important limitations of impactors for atmospheric particle collection have been elucidated recently. Attempts to use impactors to size-segregate particulate samples for chemical analysis showed anomalies in the distribution of some aerosol constituents which could only be explained by particle interstage losses and bounce-off from impactor substrates.

Particle losses in an impactor, generally referred to as wall losses or interstage losses, are the deposition of particles on surfaces other than the impaction plate. Currently, no theory exists to predict these losses, and thus they must be determined experimentally for different instruments.

Recently, three commercial cascade impactors have been investigated experimentally to determine the interstage losses (Marple and Willeke, 1976). In a detailed evaluation of the Lundgren impactor (rotating drum impaction plate) and of the Andersen viable aerosol sampler (2813 liter/min multiple round jets), and also in a similar evaluation of the Sierra high-volume cascade impactor (1133 liter/min, multiple rectangular jets), the interstage losses were determined as a function of aerodynamic particle size (i.e., equivalent diameter of unit-density spheres). They were expressed as the fraction of the total number of particles entering the impactor. This study showed that for the Lundgren and for the Andersen viable impactors, interstage losses are not a serious problem for particles smaller than about 5 μm, but increase rapidly for larger particles. The Sierra impactor, however, has appreciable losses for particles smaller than 5 μm. It was found that the interstage losses occurred primarily on both inner walls of the rectangular nozzle, apparently due to lateral impaction of the entering flow. Tapering of the nozzle inlets reduced these losses.

The limitations of impactors due to particles bouncing off the impaction plate or being blown off the impaction plate after collection are essentially the same: in both cases, particles which should have been collected are reentrained into the airstream. Thus, we shall use the term reentrainment for both blow-off and bounce-off. For single-stage impactors, this means that the concentration of particles collected on the impaction plate will be too small. For cascade impactors, the resulting size distribution will be shifted to smaller sizes, since reentrained particles will be collected on stages intended to collect smaller particle sizes.

The degree of particle reentrainment is a function of the type of particle and the nature of the impaction surface. Liquid particles will impact on any type of surface with very little reentrainment. The reentrainment of solid particles, however, is a strong function of the type of impaction surface used. The maximum amount of reentrainment is experienced with a dry, smooth surface and the minimum with a very sticky surface.

A detailed experimental study to determine the degree of reentrainment from several types of impaction surfaces was made using single-stage impactors (single round jets) designed according to theoretical criteria based on ideal collection efficiency by Stokes number. The results for one of the impactors are presented in Fig. 5.33. The losses occurred primarily for solid (methylene blue) particles larger than 2 μm diameter impacting on a smooth steel plate.

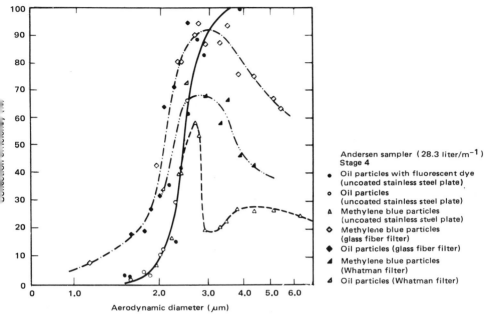

Fig. 5.33. Collection efficiency of a typical single-stage impactor for various collection surface media. [Data of Rao (1975); from Marple and Willeke (1976).]

With an oil-coated collector plate there was essentially no reentrainment, since the efficiency curve agreed well with the theoretical curve and attained nearly 100% efficiency. At the other extreme, the uncoated plate had severe reentrainment for solid particles larger than about $Stk^{1/2} = 0.47$, and had a maximum collection efficiency of less than 60%. When a filter paper was used as a collection surface, the efficiency curves were found to be between the two extreme cases of the oil-coated and the dry glass plates, and tended to have a characteristically poor sharpness of cut.

Reentrainment in an impactor with several stages (Andersen viable) was found to be similar to that in the single-stage impactor. The best collection efficiency was with liquid particles and the worst was with solid particles impacted onto an uncoated plate. The curve for collection on a glass fiber filter paper reached an efficiency of about 90%. It appears that most of the particles are collected in the void spaces between the fibers, while some always bounce off the top fibers, so 100% collection efficiency cannot be attained with glass filter paper. Also, the cutoff characteristics are generally poorer for the glass fiber paper than for a plate coated with oil or grease.

The collection of large particles ($\gtrsim 30$ μm diameter) in stagnant air is difficult by jet impactors because of the problems of particle entry into a tube associated with nonisokinetic sampling. To obtain high collection efficiency for such particles, Noll and Pilat (1971) devised a rotating blade impactor for atmospheric sampling. This device relies on the rapid rotation of a fan to

move air past an impactor plate for particle collection. The collection efficiency depends on the rotation speed. Since little interest has emerged in the
nature of very large particles in ambient air, this device has not seen widespread application.

5.7.2   Virtual Impaction—The Dichotomous Sampler.   A major potential
requirement in ambient air monitoring instrumentation has emerged with
United States air pollution legislation over the past fifteen years. The monitoring of airborne particles began in the 1940s and 1950s with dust-fall collectors. These devices were found to be a poor measure of airborne particle
concentration because only very large particles fall out at a significant rate.
Most of the pollution particles in urban air are smaller than 10 μm and fall
out only slowly. Thus, the United States government later adopted the high-
volume filter sampler, a simple inexpensive device for estimating total suspended-particulate (TSP) concentrations. Recently authorities have recognized that the monitoring of TSP alone is an incomplete measure of
particulate matter concentrations, which are relevant to human health effects (Miller *et al.,* 1979). Examination of such volume–size distribution as
that in Fig. 5.13 shows that coarse particles larger than a few micrometers in
diameter and fine particles in the submicrometer range contribute to mass
concentration designated as TSP. However, a few large particles from local
sources can heavily influence a sample which otherwise may be composed of
fine particles from a variety of sources. To distinguish between two (or
more) characteristic particle volume or mass modes in the air, size segregation at about 1 μm diameter has been suggested. An additional motivation
comes from the concern that respiratory disease, as well as other phenomena such as visibility impairment are influenced appreciably by finely divided
particles less than 2–3 μm in diameter. Since the chemistry and the sources
of atmospheric particles differ in the coarse and fine size ranges, revised air
monitoring techniques may be necessary to distinguish size classes.

For air monitoring purposes, gravimetric measures of total mass concentration from filters combined with chemical assessment generally require
relatively large amounts of sample. Also, as will be seen later, separate
samples free of the influence of chemical interactions during collection are of
interest. A device for monitoring applications has been developed which
may be used to supersede the high-volume sampler. The device is called a
dichotomous sampler. It collects particulate material in two size groups,
between 2 and 10 μm diameter, and less than 2 μm diameter. Segregation of
very large particles ($>$10 μm) is readily achieved by design of an inlet shroud
which restricts entry of particles larger than 10 μm diameter. Separation of
the coarse and finely divided particles is achieved by a method called *virtual
impaction.* In principle, this method avoids such difficulties as particle
bounce-off, reentrainment, and nonuniform deposition. In addition, it provides a separation of large and small particles such that they cannot chemi-

cally interact with one another after collection on a substrate. This sophistication in sampling may be important for characterizing chemically unstable particles in air.

Virtual impaction uses the principle of inertial separation, but the impaction plate is replaced by a zone of relatively stagnant air below the nozzle (Fig. 5.34). The virtual surface formed by deflecting streamlines gives sepa-

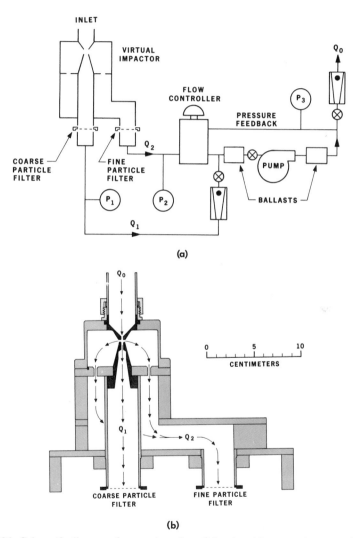

Fig. 5.34. Schematic diagram of a recent version of the virtual impactor (automated dichotomous sampler) designed by scientists at the Lawrence Berkeley Laboratory. (a) The flow system of the dichotomous sampler; (b) cross-sectional view of the nozzle configuration for the single-stage virtual impactor. [From Loo *et al.* (1979).]

ration conditions similar to those in conventional impactors. Large particles travel straight through into the low-flow region, while the small particles bend with the high speed flow as it moves radially around the receiving tube. The two size fractions then are deposited on separate filters (Fig. 5.34b). The keys to the performance of this device are the nozzle to receiving tube distance, the machining of the nozzle–gap region, and the stability of air flow in the system.

Loo *et al.* (1976) have designed and tested a dichotomous sampler system which features an automated substrate advancing and storage system (Fig. 5.34). The filter stacks can be transferred conveniently into a mass concentration measurement device using β radiation absorption, and an x-ray fluorescence spectrometer for elemental analysis.

5.7.3 Centrifugation. The deposition of particles can be achieved by introducing external forces normal to the flow of an aerosol. This is basically the principle of size separation devices employing centrifugal forces acting on the particles. Two types of particle samplers have emerged in this group. The first are cyclones, which are passive in nature, inducing spinning air motion and forcing particles to move outward to a collection surface. The second are centrifuges in which air is spun mechanically, causing particle migration and deposition on the outer walls of the device.

Cyclone Separators. These devices are often used in practice to remove particles larger than a few micrometers in diameter from gas streams. With careful design so that the gas enters in a uniform tangential flow and exists along the axis of rotation, sharp separation can be achieved. A well-designed cyclone has been used for sharp fractionation and is shown in Fig. 5.35a. For this design, the collection efficiency is given in Fig. 5.35b as compared with a Beckman dichotomous virtual impactor. This type of device has a performance which is comparable with well-designed impactors, but suffers from the disadvantage of not being capable of separating submicrometer particles with reasonable gas flow rates.

The theory for the cyclone is straightforward for particles in the Stokes range of particle behavior. The migration of the particle is calculated from a force balance in the radial direction. Assuming that the radial gas velocity is negligible, the radial particle velocity is

$$q_r = \frac{dr}{dt} = \frac{m_p r}{6\pi \mu_g R_p} \frac{d\theta^2}{dt}, \tag{5.46}$$

where r is the radial direction and θ is the azimuthal position relative to the axis of the cyclone. The particle trajectory is determined by rearranging this expression, or

$$\frac{dr}{d\theta} = \frac{m_p q_\theta}{6\pi \mu_g R_p}. \tag{5.47}$$

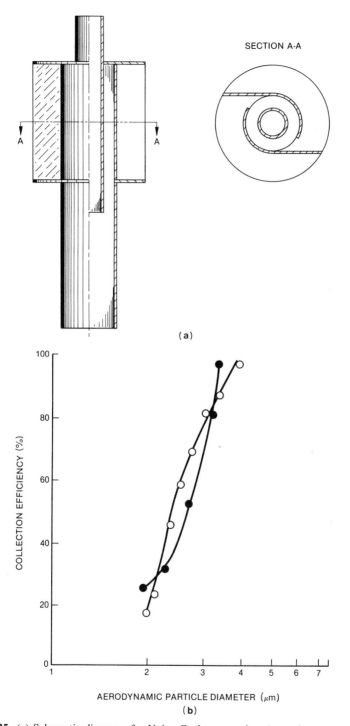

Fig. 5.35. (a) Schematic diagram of a Unico Cyclone sampler. Aerosol enters tangentially into the annulus at Section A-A and flows out of the inner tube. Particles are deposited on the outer walls of the annulus. (b) Comparison between the collection efficiency curves for the Unico Cyclone (○) operated at 10 liter/min and the Beckman dichotomous sampler (●). [From Mueller and Hidy (1983); courtesy of the Electric Power Research Institute.]

For motion in the θ direction, the particle velocity q_θ is taken as the gas velocity. If the gas velocity distribution is known, these particle trajectories can be calculated.

If the gas makes a defined number of rotations in the annular space of the cyclone, the number of turns N_t needed to remove particles of size R_p can be determined without knowledge of the flow field by using Eq. (5.47) and neglecting the particle radius compared with the spacing of the annulus:

$$N_t = \frac{\theta}{2\pi} = \frac{3\pi R_p}{m_p} \int_A^B \frac{dr}{q_\theta}. \tag{5.48}$$

By rearrangement, the radius of the smallest particle that can be removed in N_t turns is

$$R_{min} = \left(\frac{9\mu_g}{2\pi\rho_p N_t}\right)^{1/2} \left(\int_A^B \frac{dr}{q_\theta}\right)^{1/2}, \tag{5.49}$$

where A is the inner radius of the annulus and B is the outer radius of the annulus. Integration of Eq. (5.49) gives

$$R_{min} \simeq [\mu(B - A)/\rho_p N_t \bar{u}]^{1/2} \tag{5.50}$$

for design purposes, where $\bar{u}$ is the average gas velocity in the inlet tube.

Performance in terms of the number of rotations is normally determined empirically for a given design. Performance can be optimized for a fixed gas velocity by minimizing the diameter of the cyclone, which minimizes the radial distance the particle must move to collection. Small annuli also restrict the radial gas velocity components, which involve eddying of the aerosol.

Centrifuges.   Experimental investigations to determine the aerodynamic equivalent particle size for nonspherical particles led to the design of instruments to resolve individual submicrometer particle deposition by the influence of an external force. It was recognized in 1950 that centrifugation offers a useful prospect for such studies. The first aerosol particle size spectrometer actually providing a continuous size spectrum in terms of aerodynamic diameters was built by Sawyer and Walton(1950). Their centrifugal device, called a conifuge, deposited the particles according to their aerodynamic diameter in a size range between 0.5 and 30 μm on the outer wall of a rotating conical annular duct. The size separation was achieved by a laminar stream of clean air enveloping the aerosol entrance at the apex of the cone and flowing down toward the base. The aerosol particles, when entrained and leaving the apex, were subject to the centrifugal forces, and thus traversed the clean air layer in a radial direction at a velocity determined by their aerodynamic size. Because of this effect, a continuous size separation was obtained and particles of equal aerodynamic diameter were collected in concentric rings on the outer wall of the duct. At 3000 rpm, the instrument would permit an aerosol sampling rate of 25 cm³/min.

In spite of the promising aspects of the conifuge design concept, the instrument did not receive much recognition for more than 15 years, though Keith and Derrick (1960) extended the lower size limit for deposited particles to 0.05 μm and increased the sampling rate to 300 cm³/min. The lack of specific interest in the conifuge is surprising in view of the fact that gravitational versions of aerosol size spectrometers as introduced later by Timbrell (1954), for example, could not match the performance data of the conifuge. The smallest sizes deposited in the gravitational devices were, for example, not below 1.5 μm and the sampling rate was only 1 cm³/min.

In reviewing the situation of centrifugal aerosol size spectrometry and after assessing the limitations of a semidispersive cone-shaped helical-duct aerosol centrifuge (Goetz et al., 1960), Stöber and Zessack (1966) suggested that the performance of the conifuge-type size spectrometers could be improved by employing ring-slit aerosol entrances in modified designs featuring slender cones or cylindrical annular ducts. It was anticipated that ring-slit aerosol inlets would permit increased sampling rates as desired for many practical purposes. The cylindrical design would have the additional benefit of facilitating an exact theoretical performance evaluation. An actual instrument of the latter kind was subsequently built by Berner and Reichelt (1968) and showed that the experimental deposit patterns did, in fact, follow theoretical predictions.

During the following years, a variety of ring–slit centrifuges of the conifuge concept as well as the first spiral-duct centrifuge were built and tested (Stöber and Flachsbart, 1969). A comparison of the performance tests of these devices indicated that from almost all practical points of view the concept of the spinning spiral duct was superior to the other designs.

The theory for the cylindrical centrifuge is a straightforward application of force balance on particles in the annulus. If the centrifugal force acting on the particle is constant, the length from the entrance where a given size particle deposits is proportional to the aerosol flow rate, but inversely proportional to rotation speed and the square of the particle radius. These relationships are borne out by deposition experiments using particles of known radius and density.

The most important drawback of the ring-slit instruments was determined to be the difficulty of introducing the aerosol into the rotating duct without risking substantial particle losses or significant flow disturbances. In addition, a desirable increase of the aerosol sampling rate of a conifuge required a design of large instrumental dimensions. Maximum rates reported in such a case were 1.2 liter/min with a range of deposited sizes of about one order of magnitude (0.3–3.0 μm). Miniature designs suffered from greatly reduced sampling rates down to 12 cm³/min (Hochrainer and Brown, 1969) and the gain in size resolution was partially offset by increases in particle losses. In contrast, the spiral-duct centrifuge (Stöber and Flachsbart, 1969) permitted sampling rates of several liters per minute. Furthermore, the particles were deposited over a size range of almost two orders of magnitude, while losses could be kept low with suitable aerosol inlets.

The first spiral-duct centrifuge was built at the University of Rochester with a total duct length of about 180 cm. Although it was not a perfect instrument from the engineering point of view and comprised certain elements of overdesign, the new device could immediately be applied to the research problems it was intended for. Figure 5.36 shows a photograph of the instrument with the rotor lid removed. The disk-shaped rotor has a diameter of 26.2 cm. The essential part of it, the spiral duct, is of rectangular cross section and begins off-center. Then, with the inner wall touching the axis of rotation, the duct leads in a narrow semicircle from the center toward and parallel to the periphery of the rotor, from where it continues in wide curves for two and a half more turns.

In operation, the closed rotor spins at a standard speed of 3000 rpm in clockwise direction when viewed from above. Clean air is blown into the duct at the off-center inlet and passes first through an inserted laminator

Fig. 5.36. Spiral-duct centrifuge design with rotor housing open and rotor lid removed. The spiral channel through which the aerosol passes is readily seen. [From Stöber (1976); photo courtesy of T. Mercer and W. Stöber.]

which subdivides the flow by five thin parallel foils extending in the coaxial direction. In this way, the clean air is quickly stabilized and emerges as a laminar flow from the downstream end of the inserted section. The air then approaches an interchangeable aerosol inlet at the center of the rotor, where the aerosol enters coaxially through a nonrotating center bore. The aerosol is released into the duct as a thin layer parallel to the inner wall. Subsequently it is entrained into the laminar air flow.

The size separation in the spiral duct is a simple process. On leaving the center of rotation, the aerosol particles are subjected to centrifugal forces and begin moving in a radial direction across the air stream. Particle trajectories depend upon the operating conditions of the centrifuge and their aerodynamic size. Thus, while the air is drawn down the spiral duct toward the outlet, the particles are deposited, according to their size, in different locations on a foil or some other inserted collecting surface along the outer wall. The deposit represents a continuous size spectrum in terms of decreasing aerodynamic diameters beginning near the aerosol inlet and extending to the end of the spiral. To facilitate the sizing analysis, the collecting foil can easily be removed from the instrument and placed in a microscope mount.

Owing to the complexity of the geometry and the physical conditions in the spinning spiral duct, no quantitative theoretical model for the deposit pattern has been devised. Instead, empirical calibrations were made for different operating conditions. Typical results for the original design as obtained with monodisperse latex test aerosol particles are shown in Fig. 5.37. The curves indicate a maximum range of deposited particle diameters from less than 0.09 to larger than 5 μm, thus covering almost two orders of magnitude in size.

The graph also reveals another significant feature of the long-spiral-duct centrifuge: For small particle diameters of 0.1 μm, the deposit location of the particles is no longer strongly dependent upon the centrifugal forces acting on the particles. The upper two curves of Fig. 5.37 show this in their flattening of slope below 0.3 μm. They indicate that near the end of the spiral duct the motion of the particles toward the collecting surface is predominantly controlled by factors other than the centrifugal forces. Experimental evidence of Stöber and Flachsbart (1969) indicates that the deposition of particles at long distances down the duct is primarily due to an entrainment and transport of the particles by a slow secondary double vortex flow in the cross-sectional plane of the duct.

5.7.4   Precipitators or Migration Collectors.

In addition to the collection instruments based on inertial and centrifugal forces, workers have devised collectors which use thermal and electrical forces, as well as migration by Brownian diffusion.

Thermal deposition of dust particles was first studied by Aitken, who constructed an apparatus for ridding air of suspended particles by this

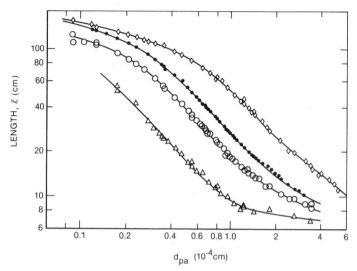

Fig. 5.37. Calibration curves of the original spiral-duct centrifuge for different operating conditions. 10 liter/min: ◇, 1500 rpm; ●, 3000 rpm. 5 liter/min: ○, 3000 rpm; △, 6000 rpm. [From Stöber (1976); reprinted courtesy of Academic Press.]

means, a line of work which was extended by Bancroft (1920). The application of this method to sample collection for size analysis of particles was first achieved by Whytlaw-Gray and Lomax in their hot-wire thermal precipitator, which was introduced in a modified form by Green and Watson (1935) and jointly patented by all four investigators (Whytlaw-Gray *et al.*, 1936). Air is drawn through a channel 0.51 mm wide past an electrically heated wire on either side of which is placed a cover glass. The cover glasses were located within the dust-free space surrounding the hot wire and are kept cool by the plugs in the massive metal body of the instrument. The velocity of the stream is controlled so that particles do not penetrate the dust-free space, but are repelled and brought by convection into contact with the cover glasses and deposited in a strip on either glass. No size sorting is attained, but there is a tendency for the larger particles to be deposited closer to the heating wire than the smaller ones. This is a fractionating effect which is to be expected from the theory of deposition in a thermal gradient. Unless the wire is very precisely centered, the deposits are of unequal density and counts must be made in both when an estimation of number of particles is required. Most of the sampling of aerosols has been done with a standard instrument, where the wire is maintained at approximately 100°C above the temperature of the ambient air and the rate of flow is about 7 cm³/min. Under these conditions, precipitation of particles from 5 μm diameter downward is virtually complete. Above 5 μm, errors occur owing to loss of particles falling onto the sides of the inlet. Prewett and Walton (1948) found that, for particles of unit density, the mean efficiency of four thermal precipitators of the same design was 96% for 2.5 μm particles, 91% for 5 μm particles, 97% for 10 μm particles (due to the sedimentation effect), and 75% for 20 μm particles.

The error in thermal precipitator sampling that arises from excess large particles falling into the sampling orifice has been considered theoretically by Davies (1947). His calculations referred to particles entering the top channel and not the actual sampling channel, where the flow rate is much higher. Walton (1947) pointed out that the greater part of this excess would sediment on the sloping walls of the top channel. The sample would be too large by the number of particles that would sediment into the narrow part of the sampling channel in the absence of a sampling flow.

Several modifications to the standard design have been introduced from time to time. Green and Lane (1964) reviewed these and noted that centering the wire heater enables equal deposits on the cover glasses to be obtained. Substitution of a ribbon for the heater wire also appears to give a more uniform deposit of particles.

According to Green and Lane (1964), the upper limit of number concentration that can be estimated with the thermal precipitator depends upon the accuracy of measurement of the volume of sample taken through the instrument. A volume displacement of 1 cm^3 would enable a smoke cloud of 10^6 particles/cm^3 to be estimated. To a certain extent, particle sizes also determine the upper limit because the presence of large particles enhances the overlap error in microscopic analysis.

For low concentrations the disadvantage of the thermal precipitator is the length of time taken to obtain a sample, so that rapid changes in particle concentration cannot be followed. A sample of 50 cm^3 volume of a dust cloud is sufficient to produce a deposit of suitable density; when there are about 1000 particles/cm^3, a volume of 500 cm^3 is necessary.

Particle collectors have also been designed based on the electrostatic precipitation of charged particles. Such devices have been built as offshoots of the mobility size classifiers discussed in Section 5.2. Particle precipitators of this type involve a corona discharge for electrification and a parallel-plate or cylindrical-plate geometry to generate an electrical field for collection. Examples of denuder or precipitator devices for sampling are discussed by Whitby (1976). Large, industrial air cleaning devices are discussed in Chapter 10.

Deposition by Brownian diffusion of particles makes use of knowledge of submicrometer particle removal in long tubes or channels, as discussed in Section 2.3. This method has been used for many years to examine aerosols by size segregation. The devices are multiple-channeled units. The diffusion method of determination of particle size, first applied by Townsend to the measurement of the gas ion mobility, proved to be very fruitful. For aerosols, it was first used by Nolan and Guerrini (1935) and Radushkevich (1939).

The method is based on measurement of the diffusional deposition of particles on the walls in the case of a laminar flow of aerosol through a channel of circular or square cross section. The *aerosol penetration,* defined as the ratio N/N_0 of concentrations at exit and entrance to the channel, can be accurately calculated theoretically, and from the penetration measurements it is possible to determine the particle diffusion coefficient through the relation between particle size. The equations for the diffusional deposition (and for a mathematically equivalent problem) have been obtained by a number of authors (see, for example, Fuchs and Sutugin, 1971). In practice, aerosol is often caused to flow through a system of parallel channels—a size-selective device which Rodebush (1943) proposed to call a *diffusion battery.*

When the aerosol flows through a channel, the smallest, most mobile particles are deposited first. Therefore, the apparent diffusion coefficient of a polydisperse aerosol, which can be calculated by means of the diffusion theory, increases with increase of parameter $y = xD_p/h^2\overline{q}$, where x is the axial distance from the inlet of a channel, $\overline{q}$ is the flow velocity, and h is the width of the channel. The diffusion measurements then are of value if they give the distribution of diffusion coefficients, or at least the value of the coefficient corresponding to a definite mean particle size.

The theory of diffusional deposition in channels derives from considerations discussed in Section 2.3. Solutions to the steady state diffusion equation have the form of an algebraic series. In the case of plane-parallel channels of width $2h$ (Fuchs *et al.*, 1962), the penetration of aerosol with a distribution function $n(R)$ is

$$\frac{N}{N_0} = 0.9149 \int_0^\infty n(R') \exp[-1.885y(R')] \, dR'$$

$$+ \, 0.059 \int_0^\infty n(R') \exp[-22.33y(R')] \, dR'$$

$$+ \, 0.0258 \int_0^\infty n(R') \exp[-151.8y(R')] \, dR' \tag{5.51}$$

where $y(R') = xD_p(R')/h^2\overline{q}$. For a monodisperse aerosol, this equation simplifies considerably in that the integral terms on the right-hand side degenerate to the exponential term since $\int_0^\infty n(R') \exp[\cdots] \, dR' = \exp[\cdots]$.

Different methods have been proposed for the determination of the parameters of the particle size distribution from experimental diffusion data. Pollak and Metnieks (1957) have suggested the so-called "exhaustion method." If the aerosol is considered a mixture of several monodisperse fractions characterized by increasing values of diffusion coefficients $D_1, D_2, \ldots, D_i$, and with particle concentration fractions $N_1/N, N_2/N, \ldots, N_i/N$, Fuchs and Sutugin (1971) note that

$$\exp\left(-\frac{KD'}{\overline{q}}\right) = \sum_i \frac{N_i}{N} \exp\left(-\frac{KD_i}{\overline{q}}\right), \tag{5.52}$$

where D' is an empirical value of the diffusion coefficient determined at the flow velocity $\bar{q}$ and K is a coefficient depending on the experimental configuration, such as parallel cylinders (Sinclair *et al.*, 1979) or plane-parallel channels (Gormley and Kennedy, 1938; DeMarcus and Thomas, 1952). In the limit $\bar{q} \to \infty$, the exponentials in Eq. (5.52) simplify to indicate that the empirical diffusion coefficient D' is the weighted-mean diffusivity for all classes of particles in the aerosol. As $\bar{q} \to 0$, terms in Eq. (5.52) can be approximated by the fraction of least diffusivity D_i. Therefore extrapolation of the experimentally determined $N/N_0 = f(y)$ to $\bar{q} \to 0$ will yield estimates D_1 and N_1/N. By subtracting N_1/N from the total number of particles, the form of $f(y)$ can be estimated for a particle collection without the first fraction. Extrapolating again to $\bar{q} \to 0$, one finds D_2 and N_2/N, etc. This method was adopted by Pollak and Metnieks using only the first term on the right-hand side in an algebraic series like Eq. (5.51); inclusion of higher-order terms leads to cumbersome expressions. Nolan and Scott (1965) evaluated this method, and found that at least nine fractions are required to obtain a reasonable approximation for a particle size distribution.

Fuchs and Sutugin (1971) do not recommend the above method because of inaccuracies developed in the iteration process. Instead they have used numerical solutions of Eq. (5.51) for a lognormal size distribution to generate families of curves of penetration versus $f(y)$. Sets of curves were obtained corresponding to a fixed geometric mean particle radius $\bar{R}_p$ in the range from 10 to 1000 Å, and for values of the logarithm of geometric standard deviation σ_g from 0 to 0.4. For the particle mobility, the Millikan formula, Eq. (2.6), was used. Figure 5.38 shows the $N/N_0 = f(y)$ curves taken from the work of Fuchs *et al.* (1962). In practice, this method compares experimental data plotted in the coordinates (N/N_0, log y) by superimposing the calculated curves plotted on the same scale on tracing paper on the curves in Fig. 5.38. The experimental curves are shifted along the abscissa until the experimen-

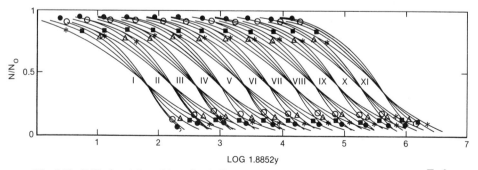

Fig. 5.38. Diffusional deposition of polydispersed aerosols in a plane-parallel channel. $\bar{R}$ (Å): I 10, II 16, III 25, IV 40, V 63, VI 100, VII 160, VIII 250, IX 400, X 630, XI 1000. Log σ_g: ○, 0; ●, 0.1; △, 0.2; ■, 0.3; *, 0.4. ●, the steepest curve identity; ○, next steepest curve; △, next steepest curve; ■, next steepest curve; *, shallowest curve.

tal points coincide most closely with one of the calculated curves. The σ_g is then estimated and the diffusivity D_p is found from the shift in curves along the log y axis.

A logical approach to analyzing diffusion battery data was used by Twomey (1963). Extrapolating solutions for diffusion of monodisperse aerosols to the case of polydisperse aerosols and obtaining an expression similar to Eq. (5.51), he showed that the experimental dependence $N/N_0 = f(y)$ can be presented as a converging series of functions of a form which is a Laplace transform of the size distribution function $n(R)$ he sought. Thus, this function can be found by means of the inverse Laplace transform. Theoretically, this method is applicable to aerosols with any size distribution. However, Twomey showed that if $f(y)$ is calculated from $n(R)$ and then, by means of an inverse Laplace transform, is changed back from $f(y)$ to $n(R)$, the latter can be significantly different from the original $n(R)$. The error due to this transformation increases greatly if, on account of imperfect measurements, the function $f(y)$ has been determined inaccurately. Only if particle penetration is determined over the whole range of measurements with error $\leq 0.5\%$ is it possible to obtain reasonable results by this method. Such accuracy being usually unattainable, Twomey's method seems to be applicable only for a crude estimation of the particle size distribution in very polydisperse systems, for example, in atmospheric aerosols.

In view of the uncertainties in applying the sizing technique with a diffusion battery, other methods such as light scattering or electrical mobility have been more popular alternatives for estimating size distributions.

5.8   CHEMICAL CHARACTERIZATION

In the previous sections, we have concentrated on reviewing the variety of methods for physical characterization of aerosol particles, primarily through the framework of the size distribution. In Fig. 5.1, the last stage of complexity in grouping of measurement devices adds chemical characterization. This can be accomplished either in terms of analysis of a whole sample corresponding to the total mass concentration, or it can be done on a size-fractionated basis. Chemical characterization becomes most important when considering a heterogeneous collection of aerosol particles such as those found in the ambient air or in the workplace. Chemical methods for characterization can be grouped in many different ways; we shall consider batch analyses of macroscopic samples first. Then batch analysis by microscopic, "near-single-particle analysis" will be surveyed. Finally, the prospects for continuous monitoring for certain components of an aerosol will be discussed.

5.8.1   Macroscopic Techniques.   Macroscopic methods for chemical analysis essentially take either all of the particulate matter sampled, or a significant portion of it for bulk analysis. The classical approach to this has been

through the application of standard microchemical techniques of wet chemistry. The unique analytical requirement for aerosol particle samples is the microgram quantities collected. The analytical methods adopted must be capable of detecting these quantities in a matrix of many different, often unknown components. New methods have been introduced over the past few years which involve spectroscopic examination by a variety of techniques including x-ray fluorescence, plasma emission spectroscopy, neutron activation analysis, photoelectron spectroscopy, and mass spectroscopy.

The wet chemical methods are summarized with references in Table 5.6. Basically this approach centers on the water-soluble extract obtained from filter substrates, impactor plates, or other collection surfaces. The extraction process has to be done with some care to ensure that all of the water-soluble material is removed. Standard extraction methods now involve the use of ultrasonic devices to maximize extraction efficiency. Once the extract is obtained, it may be subjected to a number of the methods listed in Table 5.6. Such that a detailed elemental breakdown by inorganic and (water-soluble) organic carbon can be accomplished. The methods listed yield either the concentrations of water-soluble ions measured in terms of certain oxidized states, or elements. For example, materials appearing as sulfate and nitrate

TABLE 5.6

Wet Chemical Methods for Particle Analysis

Species	Method	Reference
Cations		
Ammonium	Indol phenol blue colorimetry; ion chromatography	Katz (1977) Bouyouces (1977)
Anions		
Sulfate	Methylthymol blue colorimetry; ion chromatography	Colovos *et al.* (1976) Mueller *et al.* (1978)
Nitrate	Cadmium reduction; ion chromatography	Rand *et al.* (1978) Mueller *et al.* (1978)
Chloride	Ferric thiocyanate; ion chromatography	Rand *et al.* (1978) Mueller *et al.* (1978)
Elements		
Pb, Fe, Na, K, Ca, Cr, Ni, Zn, Mn, Si	Atomic absorption spectroscopy (AA)	Rand *et al.* (1978)
	Plasma emission spectroscopy	Fassel (1978)
Carbonaceous material		
Solvent-soluble organics	Extraction and carbon detection	Appel *et al.* (1980b) Mueller *et al.* (1981)

may include lower oxidation states, but the methods basically cannot distinguish these. The metal elements found are either soluble oxides or salts. The actual composition of the material is indeterminant, but workers have deduced the composition suspected to be present by an ion balance, combined with knowledge of the origins of the particulate material (see also Chapter 7).

Methods involving water-soluble extract chemistry have been discussed in several references. For specific application to aerosol analysis, the reader is referred to books and reports of the California Air and Industrial Hygiene Laboratory in Berkeley. Examples of modern analytical methodology and record keeping are provided in collected works such as Hidy and Mueller (1980) and Mueller and Hidy (1983).

5.8.2 Direct Particle Substrate Measurement.   With the emergence of modern chemical methods, new analytical approaches involving direct measurement of chemical components on portions of the collection medium are now available. The sampling substrate may be important in such applications. For example, porous, spongy filters like the glass fiber media and cellulose fiber media are difficult to analyze by direct techniques because of their thickness. Methods available largely concentrate on elemental analysis. X-ray fluorescence analysis or proton-excited x-ray emission spectroscopy (Giauque *et al.,* 1980; Flocchini *et al.,* 1976; Ahlberg *et al.,* 1978) are among the important developments accessible to the analyst. X-ray diffraction also has been used for particulate analysis (Jenkins, 1978). Neutron activation analysis offers a unique sensitivity for certain metals which can serve as a measurement comparison standard for others, such as lead (Ragaini *et al.,* 1980). The elements accessible by x-ray procedures and neutron activation analysis (NAA) are listed with detection limits in Tables 5.7, 5.8 and 5.9. The listings indicate that many elements are accessible by a combination of x-ray methods and NAA. These results have been compared with wet chemical determinations. Differences have not been fully explained, but they are believed to be related to substrate interference, the presence of non-water-soluble sulfur compounds, and error in the method. One determination reported for aerosol samples from St. Louis suggests that the bulk of the sulfur in ambient aerosol particles is sulfate (Dzubay and Rickel, 1978).

Considerable effort has been devoted to intercomparison of analytical methods by different techniques for elemental analysis. These are discussed, for example, by Ragaini *et al.* (1980) and Appel *et al.* (1980b). Differences have been reconciled, in part, by individual interpretation methods for the spectra obtained, and by differences in actual treatment of the samples in the analytical device.

Elemental analysis has been carried a step further recently by determination of valence state in filter-collected material. Novakov *et al.* (1972) and Appel *et al.* (1980b) have reported the application of photoelectron spectroscopy to filter-collected particulate samples. An example of the spectrum for

TABLE 5.7

Theoretical Limits of Detection for X-Ray Fluorescence Spectrometry[a,b]

Element and spectral line	Detection limit (ng/m^3)			
	Zr secondary target	Ag secondary target	Ni secondary target	Direct, Ag filter
Mg Kα	—	—	—	320
Al Kα	900	1400	7200	130
Si Kα	560	700	3500	70
S Kα	—	400	1400	40
Cl Kα	—	370	900	40
Ca Kα	—	—	200	—
Ti Kα	—	—	130	—
Fe Kα	—	—	120	—

[a] From Giauque *et al.* (1980); reprinted with permission of John Wiley & Sons.
[b] Medium Gelman GA-1, mass 5.0 mg/cm², air volume sampled 0.75 m³/cm², area analyzed 1 cm².

TABLE 5.8

Theoretical Limits of Detection for X-Ray Fluorescence Spectrometry, Tb Secondary Target[a,b]

Element and spectral line	Detection limit (ng/m³)
Sr Kα	17
Y Kα	16
Zr Kα	14
Nb Kα	12
Mo Kα	11
Pd Kα	10
Ag Kα	9
Cd Kα	10
In Kα	12
Sn Kα	13
Sb Kα	13
Te Kα	14
I Kα	17
Cs Kα	28
Ba Kα	38
La Kα	56

[a] From Giauque *et al.* (1980); reprinted with permission of John Wiley & Sons.
[b] Medium Whatman 41, mass 9.0 mg/cm³, air volume sampled 3.0 m³/cm², area analyzed 4 cm².

TABLE 5.9

Limits of Detection[a] for Trace Elements in Aerosols with Instrumental Neutron Activation[b] (ng/m^3)

Element	Gelman GA-1	Impactor film[c]
Na	300	200
Mg	100	10
Al	0.3	100
Cl	1	1
Ca	1	10
Ti	10	1
V		0.1
Mn	0.6	
Cu	0.1	1
Br		0.0003
Ln	0.002	0.0004
I		0.02
Ba	2	0.5

[a] Above blank.
[b] From Ragaini *et al.* (1980); reprinted with permission of John Wiley & Sons.
[c] Average results for 0.0006-cm Mylar, 0.0025-cm Teflon, and 0.0025-cm polyethylene covered with 0.004-cm sticky resin.

sulfur taken by this method is shown in Fig. 5.39 for a sample of particulate material taken near a major highway in Los Angeles. The spectrum shows several apparent valence states of sulfur in the sample which are identified with different compounds possibly present. The origins of these compounds are not known, but comparison with other samples away from a freeway indicate their absence. Therefore, sulfur compounds other than the oxidation state S(VI) (presumed to be sulfate) are suspected to be related to auto exhaust or diesel exhaust from vehicles present in the fleets of 1973 or earlier.

Recently, interest in the classification of carbon in particulate materials has led to the use of new carbon analyzers to estimate elemental carbon versus organic material. This can be done by thermal titration, as discussed by Mueller *et al.* (1982). The method indicates that often the particulate carbon found in urban ambient air is about one-half elemental carbon. Since this component is water insoluble, and generally organic solvent insoluble, it has been missed by other analyses. The presence of elemental carbon has been suspected from observations of light absorption as discussed by Rosen *et al.* (1979). The distinction between organic and elemental carbon in aero-

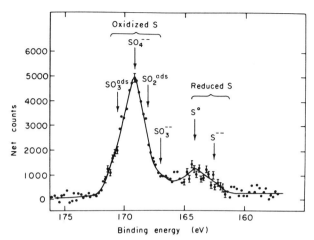

Fig. 5.39. Sulfur (2p) photoelectron spectrum of an ambient particulate sample. Expected positions of sulfate, sulfite, neutral sulfur, and sulfide are indicated. In addition, the expected line positions for adsorbed SO_3 and SO_2 are shown. [From Appel *et al.* (1980c); reprinted with permission of John Wiley & Sons.]

sol particles is important from the standpoint of the reaction chemistry of particulate material, as well as the optical effects of the material. These factors are discussed in more detail in Chapter 8.

Since organic compound identification is important in the assessment of certain health hazards of particulate material, continuing development of organic analysis at the microgram per cubic meter concentration level has taken place over the past few years. The use of solvent extraction in combination with fluorescence spectroscopy has characterized certain polynuclear aromatic compounds which may be carcinogens. These include benzopyrene, among others. The application of high resolution mass spectroscopy (e.g., Schuetzle, 1973; Cronn *et al.*, 1977) has revealed a wide variety of organic compounds potentially present in the condensed phase of particulate samples in air. Cautreels and Van Cauwenberghe (1978) have used gas chromatography in combination with mass spectroscopy to identify a large number of organic compounds in particles collected from urban air in Belgium. A listing of material found in the condensed phase from this work is given later in Table 7.9. A very large number of organic compounds can be identified by modern methods.

5.8.3 Microscopic Techniques. The hope of chemical characterization of individual particles or even the surface nature of individual particles has led investigators to apply new microtechniques to aerosol research early in their development. Early methods were developed which employed the electron microscope as an identification method, with particles being captured on a reactive substrate. The collection surface was selected to provide an indicator of a specific compound. Frank and Lodge (1967) found that sulfuric acid particles could be identified from a satellite ring structure after collection on a silicon surface. Later, Bigg et al. (1974) used specific chemical surface coatings to bring about colored chemical reactions which could be identified microscopically. These methods are generally semiquantitative in terms of mass concentration, but give a number density estimate and a rough particle size estimate, especially if they are used in conjunction with staged impactors.

Electron microscopy has been used for qualitative identification for some years (Nord, 1978). The scanning electron microscope has been used in conjunction with energy-disperse x rays (EDXs) for analysis of single particles (Lichtman, 1980). Microprobe techniques have been applied which include (a) Auger electron microscopy (AEM) (Lichtman, 1980); (b) photoelectron spectroscopy (ESCA) (Powell, 1978); (c) secondary-ion microscopy (SIMS) (Newbury, 1978); (d) electron microprobe analysis (EPMA) (Armstrong, 1978); (e) laser Raman microprobe analysis (MOLE) (Etz et al., 1977); and (f) laser microprobe mass analysis (LAMMA) (Wechsung et al., 1978). AEM, ESCA, and SIMS are surface-sensitive methods which provide knowledge of surface properties, but are difficult to extrapolate to heterogeneous materials. EPMA appears to give information related to bulk properties with a greater penetration depth; MOLE also may be useful in this respect. LAMMA has only recently become available as a research device, but will undoubtedly be used more extensively in fine-particle characterization.

In general, microprobe analysis methods have been considered semiquantitative in nature, but Armstrong (1978) has summarized some interesting new work using EPMA. He has investigated the feasibility of identification of individual particles by this method, sampling material in the Phoenix, Arizona area, for example. They have found that considerable care must be taken in transferring samples to the microprobe substrate with selection of conductive coating, and prevention of contamination. He determined that microprobe correction procedures accounting for electron scattering and x-ray production and absorption, as well as characteristic fluorescence patterns for specific size and shape are required to obtain useful information. With these factors taken into account, these investigators have applied EPMA as a microtechnique for fingerprinting individual sources in the Phoenix area. Identification of sources has been achieved by sample examination as far as 20 km away. The approach was used to distinguish between two secondary iron foundaries in the Phoenix area within 25 km and each other. Bulk sampling followed by macroscopic particle characterization failed to distinguish between the two similar operations. However, components of mixed metal oxide emissions were distinguishable in individual particles, and provided a basis for estimating the relative amounts of material emitted from the two industrial plants.

With further improvement and investigation of these microprobe techniques, it is expected that they will provide a useful series of techniques to characterize surface and bulk particle properties. In addition their application as source identification methods will enhance available techniques based on macroscopic sampling and chemical analysis (Cooper and Watson, 1980).

5.8.4 Continuous Methods. Continuous air monitoring for trace contaminants in ambient air has developed extensively over the past twenty years as a result of stimulation from new air pollution measurement requirements. Workers expect that similar needs will develop as certain chemical constituents of particulate material are identified as surrogates of public health effects. Techniques for the continuous chemical characterization of particulate matter are slow in coming because the amounts of material sampled are so small. In all cases examined so far, either a precollection method like filtration is required, or a special detector of high sensitivity is required.

Flame photometry has promise for measurement of sodium, lead, and potassium. An application to measurement of sodium and alkali metals has been reported (Crider, 1968). The continuous measurement of sulfur-containing particles has received considerable attention lately. The motivation for observation of sulfur-containing particles comes from the concern about sulfate in the atmosphere as a potential hazard. The continuous measurement of particulate sulfur in the ambient atmosphere requires that levels between 1 and 50 μg/m^3 be detected in the presence of sulfur dioxide gas and other sulfurous gases (e.g., Camp *et al.,* 1981). Under polluted air conditions, sulfur dioxide is by far the most predominant gas under normal conditions. Several designs for monitors have been proposed, and some have been tested to the feasibility stages. All of the instruments developed so far have employed the flame photometric detector as the sulfur measuring device. This detector measures essentially total sulfur contained in air passing through it so that a means must be devised to continuously remove either particulate sulfur or gaseous sulfur prior to detection. Both approaches have been used. The approach removing particulate sulfur involves detection of differences between total sulfur and gaseous sulfur to obtain the particulate component. An electrical collector with high removal efficiency and no gaseous removal capability has been employed as the particle cleaner. The method has been tested for feasibility, but appears to have been superseded by devices using the other principle.

The use of a continuous particulate sulfur analyzer with a gas denuder appears to have been thought of in 1975 (see, e.g., Mueller and Collins, 1980). Early tests showed that a silver-oxide-coated absorption tube could be added to the inlet of a total sulfur analyzer so that sulfurous gases could be removed at very high efficiency. Later devices also used lead-oxide-coated absorption tubes.

With present flame photometric detectors, the method appears to be limited to about 1 $\mu g/m^3$ sulfur. This approach has been extended by Huntzicker *et al.* (1978) to observe not only total particulate sulfur, but certain sulfur compounds, including sulfuric acid and ammonium sulfate.

A unique semicontinuous method for detecting sulfuric acid was reported by Charlson *et al.* (1974) using a nephelometer system which observed light scattering in air at different humidities. The device basically provides an automatic profile of water absorption for ambient aerosols which can be interpreted in terms of deliquescence profiles for certain compounds. Sulfuric acid in sufficient quantities in ambient particulate matter can be distinguished by this method.

Liquid water content in particulate matter can be measured semicontinuously in ambient aerosols. The method described by Ho *et al.* (1974) employs a microwave spectroscopic principle. The device uses a resonant cavity at approximately 10 cm wavelength, and compares the resonance with and without accumulated aerosol in a small filter located in the cavity. The device provides the liquid water present per unit volume of aerosol, which can be converted to a mass fraction with parallel filter measurements. The detection limit of this device appears to be in the range of 1 $\mu gH_2O/m^3$, depending on sampling interval and gas flow rate.

REFERENCES

Ahlberg, M. S., Leslie, A. C. D., and Winchester, J. W. (1978). *In* "Electron Microscopy and X-ray Applications to Environmental and Occupational Health Analysis" (P. Russell and A. Hutchings, eds.), Chap. 4. Ann Arbor Press, Ann Arbor, Michigan.

Ahlquist, N. C., and Charlson, R. J. (1967). *J. Air Pollut. Control Assoc.* **17,** 467.

Appel, B., Wall, S., Tokiwa, Y., and Haik, M., and Kothny, E. (1980a). *Atmos. Environ.* **14,** 549, 559.

Appel, B. R., Wesolowski, J., Alcocer, A., Wall, S., Twiss, S., and Giauque, R. (1980b). *In* "The Character and Origins of Smog Aerosols" (G. M. Hidy and P. K. Mueller, eds.), p. 69. Wiley (Interscience), New York.

Appel, B. R., Wesolowski, J., Hoffer, E., Twiss, S., Wall, S., Chang, S., and Novakov, T. (1980c). *In* "The Character and Origins of Smog Aerosols" (G. M. Hidy and P. K. Mueller, eds.), p. 199. Wiley (Interscience), New York.

Armstrong, J. T. (1978). *In* "Proceedings of Scanning Electron Microscopy" (O. Johari, ed.), Vol. 1, p. 455. Scanning Electron Microscopy, Inc., AMF/O'Hare, Illinois.

Bancroft, W. D. (1920). *J. Phys. Chem.* **24,** 421.

Berner, A., and Reichelt, H. (1969). *Staub* **29,** 92.

Bigg, E. K., Ono, A., and Williams, J. (1974). *Atmos. Environ.* **8,** 1.

Blasewitz, A. G., Judson, B. F., Katzer, M. F., Kurtz, E. F., Schmidt, W. C., and Weidenbaum, B. (1951). Rep. HW-20847, U.S. Atomic Energy Commission, Hanford, Washington.

Bouyouces, S. A. (1977). *Anal. Chem.* **49,** 401.

Bricard, J., Delattre, D., Madelaine, G., and Pourprix, M. (1976). *In* "Fine Particles: Aerosol Generation, Measurement, Sampling, and Analysis" (B. Y. H. Liu, ed.), p. 565. Academic Press, New York.

Buettell, R. G., and Brewer, A. W. (1949). *J. Sci. Instrum.* **26,** 357.

Camp, D. C., Stevens, R. K., Cogburn, W. G., Husar, R. B., Collins, J. F., Huntzicker, J. J., Husar, J. D., Jaklevic, J. M., McKenzie, R. L., *et al.* (1981). *Atmos. Environ.* **16,** 911.

Cautreels, W., and Van Cauwenberghe, K. (1978). *Atmos. Environ.* **12**, 1133.

Chandrasekhar, S. (1960). "Radiative Transfer," p. 527. Dover, New York.

Charlson, R. J., Vanderpol, A., Covert, D., Waggoner, A., and Ahlquist, N. (1974). *Atmos. Environ.* **8**, 1257.

Chaun, R. (1970). *J. Aerosol Sci.* **1**, 111.

Chen, C. Y. (1955). *Chem. Rev.* **55**, 595.

Cheng, L., and Soo, S. L. (1970). *J. Appl. Phys.* **41**, 585.

Collis, R. T. H. (1970). *Appl. Opt.* **9**, 1782.

Collis, R. T. H., Fernald, F. G., and Ligda, M. G. H. (1964). *Nature (London)* **203**, 1274.

Colovos, G., Panesar, M. R., and Parry, E. (1976). *Anal. Chem.* **48**, 1693.

Cooke, D. D., and Kerker, M. (1975). *Appl. Opt.* **14**, 734.

Cooper, J., and Watson, J. (1980). *J. Air Pollut. Control Assoc.* **30**, 1116.

Covert, D. S., Waggoner, A., Weiss, R., Ahlquist, N., and Charlson, R. (1980). *In* "The Character and Origins of Smog Aerosols" (G. M. Hidy and P. K. Mueller, eds.), p. 559. Wiley, New York.

Crider, W. L. (1968). *J. Sci. Instrum.* **39**, 2120.

Cronn, D. R., Charlson, R., Knights, R., Crittendon, A., and Appel, B. (1977). *Atmos. Environ.* **11**, 929.

Davies, C. N. (1947). *Trans. Instrum. Chem. Eng., Suppl.* **25**, 25.

Davies, C. N. (1976). "Sampling of Aerosols with Thin Walled Tube." Presented at the 12th Colloquium on Polluted Atmosphere, Paris,

De Luisi, J. J., Furukawa, P., Gillette, D., Schuster, B., Charlson, R., Purch, W., Fegley, R., Herman, B., Rabinoff, R., Twitty, J., and Weinman, J. (1976). *J. Appl. Meteorol.* **15**, 441, 455.

DeMarcus, W., and Thomas, J. (1952). Report ORN1-1413. U.S. Atomic Energy Commission, Washington, DC.

Dresia, H. P. (1964). *VDI-Z* **106**, 1191.

Dyment, J. (1969). Report AWRE 05/69. Atomic Weapons Research Establishment, United Kingdom.

Dzubay, T., and Rickel, D. (1978). *In* "Electron Microscopy and X-Ray Applications to Environmental and Occupational Health Analysis" (P. Russel and A. Huthings, eds.), Chap. 1. Ann Arbor Press, Ann Arbor, Michigan.

Eltermann, L. (1968). Environmental Research Papers, No. 285, p. 49. Air Force Cambridge Res. Lab., Bedford, Massachusetts.

Etz, S. S., Rosasco, G. J., and Cunningham, W. C. (1977). "Environmental Analysis" (G. W. Ewing, ed.), pp. 295–340. Academic Press, New York.

Fassel, V. A. (1978). *Science* 202.

Faxvog, F. R. (1975). *Appl. Opt.* **14**, 269.

Faxvog, F. R., and Roessler, D. M. (1978). *Appl. Opt.* **17**, 2612.

First, M. (1951). "Air Cleaning Studies." School of Publ. Health, Harvard Univ. Press, Cambridge, Massachusetts.

Flocchini, R. G., Cahill, T., Shadoan, D., Lange, S., Eldred, R., Feeney, P., Wolfe, G., Simmeroth, D., and Suder, J. (1976). *Environ. Sci. Technol.* **10**, 76.

Frank, E., and Lodge, J. P., Jr. (1967). *J. Microsc.* **6**, 449.

Friedlander, S. K. (1977). "Smoke, Dust and Haze," p. 167. Wiley, New York.

Fuchs, N. A., and Sutugin, A. G. (1971). *Int. Rev. Aerosol Phys. Chem. No. 2, 1971* Chap. 1.

Fuchs, N. A., Stechkina, I., and Starosselskii, V. (1962). *Br. J. Appl. Phys.* **13**, 281.

Gebhart, J., Heyder, J., Roth, C., and Stahlhofer, W. (1976). *In* "Fine Particles: Aerosol Generation, Measurement, Sampling, and Analysis" (B. Y. H. Liu, ed.), p. 565. Academic Press, New York.

Giauque, R., Goda, L., and Garrett, R. (1980). *In* "The Character and Origins of Smog Aerosols" (G. M. Hidy and P. K. Mueller, eds.), p. 147. Wiley (Interscience), New York.

Gille, J., Bailey, P., Craig, R., and House, F. (1977). *In* "Radiation in the Atmosphere" (H. S. Bolla, ed.), p. 305. Science Press, Princeton, New York.

Goetz, A., Stevenson, H. J. R., and Preining, O. (1960). *J. Air Pollut. Control Assoc.* **10,** 378.

Goody, R. (1964). "Atmospheric Radiation." Oxford Press, Oxford, England.

Gormley, P., and Kennedy, M. (1938). *Proc. R. Ir. Acad. Sci., Sect A* **45A,** 59.

Green, H. L., and Lane, W. R. (1964). "Particulate Clouds: Dusts, Smokes and Mists," 2nd ed. Van Nostrand-Reinhold, Princeton, New Jersey.

Green, H. L., and Watson, H. H. (1935). Nuclear Research Council Special Report Ser. No. 199. HM Stationery Office, London.

Hall, J., and Riley, L. A. (1978). *In* "Conference on Carbonaceous Particles in the Atmosphere" (T. Novakov, ed.), p. 29. Lawrence Livermore Lab., Berkeley, California.

Harper, W. R. (1967). "Contact and Frictional Electrification." Harper, New York.

Hering, S. V., Flagan, R., and Friedlander, S. (1978). *Environ. Sci. Technol.* **12,** 667.

Hidy, G. M. (principal investigator) (1973). Characterization of Aerosols in California. Interim Report, Phase I. For Calif. Air Resources Bd., Science Center, Rockwell Intl., 247–250.

Hidy, G. M., and Mueller, P. K., eds. (1980). "The Character and Origins of Smog Aerosols," Parts I and II. Wiley (Interscience), New York.

Ho, W. W., Hidy, G. M., and Govan, R. M. (1974). *J. Appl. Meteorol.* **4,** 871.

Hochrainer, D., and Brown, P. (1969). *Environ. Sci. Technol.* **3,** 830.

Humphrey, A., and Gaden, E. (1955). *Ind. Eng. Chem.* **47,** 924.

Huntzicker, J., Hoffman, R., and Ling, C. S. (1978). *Atoms. Environ.* **12,** 83.

Husar, R. B. (1973). Recent Developments in 'in situ' Size Spectrum Measurements of Submicron Aerosols. Presented at the Conf. on Instrument Monitoring in Ambient Air, Boulder, Colorado.

Jenkins, R. (1978). *In* "Electron Microscopy and X-ray Application to Environmental and Occupational Health Analysis" (P. Russell and A. Hutchings, eds.), Chap. 9. Ann Arbor Press, Ann Arbor, Michigan.

John, W. (1976). *In* "Fine Particles: Aerosol Generation, Measurement, Sampling, and Analysis" (B. Y. H. Liu, ed.), p. 650. Academic Press, New York.

Katz, M. (1977). "Methods of Air Sampling and Analysis," 2nd ed. p. 511. Am. Public Health Assoc., Washington, DC.

Keith, C. H., and Derrick, J. C. (1960). *J. Colloid Sci.* **15,** 340.

Kerker, M. (1969). "The Scattering of Light: And Other Electromagnetic Radiation." Academic Press, New York.

Kimura, N., and Iinoya, K. (1959). *Kogaku Kogaku, Chem. Eng. (Jpn.)* **23,** 792.

Kirsch, A., and Dvukhimenny, V. (1975). *Teoret. Osnovy Chem. Tekhnol.* **9,** 796.

Kirsch, A. A., and Stechkina, I. B. (1978). *In* "Fundamentals of Aerosol Science" (D. T. Shaw, ed.), Chap. 4. Wiley (Interscience), New York.

Knutson, E. O. (1971). Ph.D. Thesis, Univ. of Minnesota, Minneapolis, Minnesota.

Knutson, E. O., and Whitby, K. T. (1975). *J. Aerosol Sci.* **6,** 6.

Kondratyev, K. Ya (1975). *In* "Proceedings 4th Conference on CIAP" (T. Hard and A. Broderick, eds.) p. 254. U.S. Dept. of Transportation, Washington, DC.

Kraemer, H., and Johnstone, H. O. (1955). *Ind. Eng. Chem.* **47,** 2726.

Landis, D. (1975). *Atmos. Environ.* **9,** 1079–1082.

Langmuir, I. (1942), Office of Sic. Res. Dev. (OSRD). Rep. No. 865, p. 20.

Lee, R. E., Jr., and Wagman, J. (1966). *Am. Ind. Hyg. Assoc. J.* **27,** 266.

Levin, L. M. (1957). *Isv. Akad. Nauk SSSR, Ser. Geofiz.* **7,** 914.

Lichtman, D. (1980). Compound Identification of Atmospheric Particles. Report EA-1595. Electric Power Research Inst., Palo Alto, California.

Liu, B. Y. H., and D. Y. H. Pui (1974). *J. Colloid Interface Sci.* **47,** 155.

Liu, B. Y. H., Berglund, R., and Agarwal, J. (1974a). *Atmos. Environ.* **8,** 717.

Liu, B. Y. H., Marple, V. A., Whitby, K. T. *et al.* (1974b). *Am. Ind. Hyg. Assoc. J.* **35,** 443.

Liu, B. Y. H., Whitby, K. T., and Pui, D. Y. H. (1974c). *J. Air Pollut. Control Assoc.* **24,** 1067.

Loo, B. W., Jaklevic, J., and Goulding, F. (1976). *In* "Fine Particles: Aerosol Generation,

Measurement, Sampling, and Analysis'' (B. Y. H. Liu, ed.). p. 535. Academic Press, New York.

Loo, B. W., Adachi, R. S., Cork, C. P., Goulding, F. S., Jaklevic, J. M., Landis, D. A., and Searles, W. L. (1979). A 2nd generation dichotomous sampler for large-scale monitoring of airborne particulate matter. *Lawrence Berkeley Lab. Rep.* **LBL-8725.**

Marcias, E., and Husar, R. B. (1976). *In* "Fine Particles: Aerosol Generation, Measurement, Sampling, and Analysis" (B. Y. H. Liu, ed.), p. 535. Academic Press, New York.

McCormick, P., Grams, G., Hamill, P., Herman, B., McMaster, L., Pepin, T., Russell, P., Steele, H., and Swisster, T. (1981). *Science* **214,** 328.

McDonald, T. E. (1962). *J. Appl. Meteorol.* **1,** 391.

Malm, W. C., and Walther, G. G. (1980). A Review of Instruments Measuring Visibility-Related Variables. Report EPA-600/4-80-016. U.S. Environmental Protection Agency, Las Vegas, Nevada.

Marple, V. A., and Liu, B. Y. H. (1974). *Environ. Sci. Technol.* **8,** 648.

Marple, V. A., and Willeke, K. (1976). *In* "Fine Particles: Aerosol Generation, Measurement, Sampling, and Analysis" (B. Y. H. Liu, ed.), p. 411. Academic Press, New York.

Marple, V. A., Liu, B. Y. H., and Whitby, K. T. (1974). *J. Aerosol Sci.* **5,** 1.

May, K. R. (1967). *In* "Airborne Microbes." (P. H. Gregory and J. L. Monteith, eds.), Cambridge Univ. Press, London and New York.

Miller, F. J., Gardner, D., Graham, J., Lee, R., Jr., Wilson, W., and Bachman, J. (1979). *J. Air Pollut. Control. Assoc.* **29,** 610.

Mueller, P. K., and Collins, J. (1980). Development of a Particulate Sulfate Analyzer. Report EA-1492. Electric Power Research Institute, Palo Alto, California.

Mueller, P. K., and Hidy, G. M., Principal Investigators (1982). The Sulfate Regional Experiment. Report of Findings. Report EA-1901, Chap. 2. Electric Power Research Institute, Palo Alto, California.

Mueller, P. K., Mendoza, B. V., Collins, J. C., and Wilgus, E. S. (1978). *In* "Ion Chromatographic Analysis of Environmental Pollutants" (E. Sawicki, J. D. Mulik, and E. Wittgenstain, eds.), p. 77. Ann Arbor Sci. Publ., Ann Arbor, Michigan.

Mueller, P. K., Fung, K., Heisler, S., Grosjean, D., and Hidy, G. M. (1982). *In* "Particulate Carbon: Atmospheric Life Cycle" (G. T. Wolff and R. L. Klimisch, eds.), p. 343. Plenum, New York.

Nader, J. S. (1976). *In* "Air Pollution. III. Measuring Monitoring and Surveillance of Air Pollution" (A. C. Stern, ed.), 3rd ed., pp. 589–645. Academic Press, New York.

Newbury, D. E. (1978). *Proc. 13th Ann. Meet. Microbeam Anal. Soc.,* p. 65A. Ann Arbor, Michigan.

Newkirk, G., Jr., and Eddy, J. A. (1964). *J. Atmos. Sci.* **21,** 35.

Nolan, P., and Guerrini, V. (1935). *Proc. R. Ir. Acad. Sci., Sect. A* **43,** 5.

Nolan, P., and Scott, J. (1965). *Proc. R. Ir. Acad. Sci., Sect. A* **65A,** 39.

Noll, K., and Pilat, M. (1971). *Atmos. Environ.* **5,** 527.

Nord, G. L., Jr. (1978). *In* "Electron Microscopy and X-ray Applications" (P. Russell and A. Hutchings, eds.), p. 133. Ann Arbor Sci. Publ., Ann Arbor, Michigan.

Novakov, T., Mueller, P. K., Alcocer, A. E., and Otvos, J. W. (1972). *J. Colloid Interface Sci.* **39,** 225.

Penndorf, R. B. (1957). *J. Opt. Soc. Am.* **47,** 1010.

Pich, J. (1978). *In* "Fundamentals of Aerosol Science" (D. T. Shaw, ed.), Chap. 6. Wiley (Interscience), New York.

Pollak, L., and Metnieks, A. (1957). *Geofis. Pura Appl.* **37,** 183.

Powell, C. F. (1978). *Proc. 13th Ann. Meet. Microbeam Anal. Soc.,* p. 64A. Ann Arbor, Michigan.

Prewett, W. G., and Walton, W. H. (1948). Unpublished Ministry of Supply report. U.K.

Prochazka, R. (1964). *Staub* **24,** 353.

Pui, Y. H. 1975. Ph.D. Thesis, Univ. of Minnesota, Minneapolis, Minnesota.

Radushkevich, L. (1939). *Acta Phys. Chem. URSS* **11**, 265.

Ragaini, R., Ralston, H., Garvis, D., and Kaifer, R. (1980). *In* "The Characterization and Origins of Smog Aerosols" (G. M. Hidy and P. K. Mueller, eds.), p. 169. Wiley (Interscience), New York.

Ramskill, E., and Anderson, W. (1951). *J. Colloid Sci.* **6**, 416.

Rand, M. C., Greenberg, A. E., Taras, M. J., and Franson, M. A. (1978). "Standard Methods for the Examination of Water and Wastewater," 14th ed., p. 144, 613, 620. Am. Public Health Assoc., Washington, DC.

Rao, A. K. (1975). Ph.D. Thesis, Univ. of Minnesota, Minneapolis, Minnesota.

Reagan, J. A., Spinhirne, J., Byrne, D., Thompson, D., dePena, R., and Mamane, Y. (1977). *J. Appl. Meteorol.* **16**, 911.

Rich, T. A., Pollak, L. W., and Metnieks, A. L. (1959). *Geofis. Pura Appl.* **44**, 233.

Rodebush, W. (1943). Report ORNI-1413, U.S. Atomic Energy Commission, Washington, DC.

Rohman, H. (1923). *Z. Phyz.* **17**, 253.

Rosen, H., Hansen, A. D. A., Gundel, L., and Novakov, T. (1979). *In* "Proceedings Carbonaceous Particles in the Atmosphere" (T. Novakov, ed.), p. 49. Lawrence Berkeley Lab., Berkeley, California.

Sawyer, K. F., and Walton, W. H. (1950). *J. Sci. Instr.* **27**, 272.

Schuetzle, D., Crittendon, A., and Charlson, R. (1973). *J. Air Pollut. Control Assoc.* **23**, 704.

Schütz, A. (1966). *Staub* **26**, 18.

Sekera, Z. (1967). *Icarus* **6**, 348.

Semb, G. J., Whitby, K. T., and Sverdrup, G. (1980). *In* "The Character and Origins of Smog Aerosols" (G. M. Hidy *et al.*, eds.), p. 55. Wiley, New York.

Shaw, G. E., Reagan, J., and Herman, B. (1973). *J. Appl. Meteorol.* **12**, 374.

Sinclair, D., Countess, R. J., Liu, B. Y. H., and Pui, D. Y. H. (1979). *In* "Aerosol Measurement" (D. A. Lundgren, ed.), p. 544. Univ. of Florida, Gainesville, Florida.

Silverman, L., Billings, C. E., and First, M. W. (1971). "Particle Size Analysis in Industrial Hygiene." Academic Press, New York.

Smith, J. P., Grosjean, D., and Pitts, J. N., Jr. (1978). *J. Air Pollut. Control Assoc.* **28**, 930.

Spicer, C., and Schumacher, P. (1979). *Atmos. Environ.* **13**, 543.

Squires, P. (1972). *J. Rech. Atmos.* **6**, 565.

Stelson, A. W., Friedlander, S., and Seinfeld, J. (1979). *Atmos. Environ.* **13**, 369.

Stern, S. C., Zeller, H., and Schekman, A. (1960). *J. Colloid Sci.* **15**, 546.

Stöber, W. (1976). *In* "Fine Particles: Aerosol Generation, Measurement, Sampling, and Analysis" (B. Y. H. Liu, ed.), p. 351. Academic Press, New York.

Stöber, W., and Flachsbart, A. (1969). *Environ. Sci. Technol.* **3**, 641.

Stöber, W., and Zessack, U. (1966). *Zentrablb. Biol. Aerosol Forsch.* **13**, 263.

Sullivan, R., and Hertel, K. (1942). *Adv. Colloid Sci.* **1**, 37.

Thomas, D. (1953). Ph.D. Thesis, Ohio State Univ., Columbus, Ohio.

Timbrell, V. (1954). *J. Appl. Phys., Suppl. 3,* **5**, 86.

Twomey, S. (1963). *J. Franklin Inst.* **275**, 163.

U.S. EPA (1971). *Fed. Regist.* **36**, 8186.

U.S. EPA (1977). *Fed. Regist.* **42**, 41776.

van de Hulst, H. C. (1957). "Light Scattering by Small Particles" Wiley, New York.

Volz, F. E. (1959). *Arch. Meteorol. Geophys. Bioklimatol. Ser. B* **10B**, 100.

Walton, W. H. (1947). *Trans. Inst. Chem. Eng., Suppl.* **25**, 64, 69, 136.

Washimi, M. (1973). *Kansai Denryoku K. K. Soken Hokoku* **12**, 210.

Wechsung, R., Hillenkamp, F., Kaufmann, R., Mitche, R., and Vogt, H. (1978). *In* "Proceedings of Scanning Electron Microscopy" (O. Johari ed.), Vol. I, p. 611. Scanning Microscopy, Inc., AMF O'Hare, Illinois.

Whitby, K. T. (1976). *In* "Fine Particles: Aerosol Generation, Measurement, Sampling, and Analysis" (B. H. Y. Liu, ed.), p. 584. Academic Press, New York.

Whitby, K. T., and Clark, W. E. (1966). *Tellus* **18,** 573.

Whitby, K. T., Husar, R. B., and Liu, B. Y. H. (1972). *In* "Aerosols and Air Chemistry" (G. M. Hidy, ed.), p. 237. Academic Press, New York.

Whitby, K. T., Lundgren, D. A., McFarland, A., and Jordan, R. C. (1961). *J. Air Pollut. Control Assn.* **11,** 503.

Whytlaw-Gray, R., Green, H. L., Lomax, R., and Watson, H. H. (1926). British Patent No. 44551.

Willeke, K., and Liu, B. Y. H. (1976). *In* "Fine Particles: Aerosol Generation, Measurement, Sampling, and Analysis" (B. Y. H. Liu, ed.), p. 698. Academic Press, New York.

Wong, J. B., Ranz, W. E., and Johnstone, H. F. (1956). *J. Appl. Phys.* **27,** 161.

Zebel, G. (1968). *J. Colloid Interface Sci.* **27,** 294.

Zebel, G. (1978). *In* "Recent Developments in Aerosol Science" (D. T. Shaw, ed.), Chap. 8. Wiley (Interscience), New York.

CHAPTER 6

APPLICATIONS TO TECHNOLOGY

Aerosol science has found its way into a wide variety of technological applications. Perhaps best known is the use of spray generation principles for manufacturing dispersable consumer products such as personal deodorants, household sprays and cleaners, and pesticides. The use of aerosol technology is widespread in agriculture for dispersal of pesticides. There is an extensive application in the field of fine-particle production, and their subsequent use for material surface coatings, reinforcement, and strengthening. An obvious application also enters into the development of modern engineering of fuel combustion.

One must be concerned with the safety aspects of particle behavior in certain industrial applications. In the electronics industry great precautions are taken to ensure that particles are removed from ambient air so that microdevices are not contaminated during manufacture. There is a large body of information on dust explosions and damage to structures involving dust accumulation as an undesirable by-product. Aerosols have been used for many years in medical treatment of respiratory disease as means of therapeutic treatment. Recently large aerosol generators have been engineered into environmental health laboratories for human health studies.

In this chapter, these topics are discussed to provide a perspective on the variety of practical applications that have emerged with basic knowledge of suspended-particle behavior. In the discussion, the technology of clean room design for manufacturing and handling is deferred to the later chapter on control technology.

6.1 PARTICLE DISPERSAL FOR CONSUMER PRODUCTS

The concise scientific definition of an aerosol refers specifically to a colloidal state of material suspended in a gas. However, the term has acquired an additional meaning in common household usage. In the packaging field, the term aerosol now is synonymous with pressurized products that are released in a dispersed form from a can or a bottle. The discharge ranges from coarse fogs and mists to finely divided liquid or powder dispersions.

Although the list of products that can be or have been dispersed by the aerosol method is extensive, they all have common characteristics. The materials are packaged under pressure and are released by pressing a simple valve. They contain an active ingredient and a propellant that provides the force for expelling and breaking up the product. In many cases, the carrier or solvent for the active ingredient is also included in the suspension to make a useful product formulation.

A typical aerosol package is shown schematically in Fig. 6.1. The diagram shows how the pressure of the propellant forces the liquid out of the spray nozzle. Basically the expulsion of the material in the can at high pressure into ambient air gives conditions for breakup of the bulk fluid into small particles. The propellant gas may be a compressed, sparingly soluble gas at high pressure or it may be a soluble gas in the liquid contained in the can. Perhaps most common is propellant that is liquid under pressure but vaporizes in expansion to a gas. The best-known example of such propellants is the freon gas series. With the recognition that these halogenated hydrocarbons dissociate in the earth's upper atmosphere, the aerosol packaging industry has steadied somewhat from its explosive growth in the 1960s. Current propellant gases include carbon dioxide and isobutane rather than the freons.

A liquefied gas propellant has four advantages: (a) a constant pressure is maintained during discharge, (b) the quality of the spray remains during discharge, (c) a minimum of formulation is left in the container, and (d) a desirable effect of spray quality by reducing mean particle size. The use of such propellants has three disadvantages (a) more propellant is generally required than for a compressed gas, (b) the propellant cost may be higher than for compressed gas, and (c) there is a greater change in pressure with temperature than for a compressed gas.

Basically the transient action of a liquefied gas expanding under pressure through a nozzle provides energy for breakup during expulsion from a small nozzle. The primary atomization takes place in violent disintegration by the explosive action of the vaporizing propellant. It continues with the action of the shearing forces on droplets as they are forced from the nozzle jet into stagnant air.

The effectiveness of the aerosol in accomplishing its intended purpose depends largely on the size distribution of particles formed. Relatively little

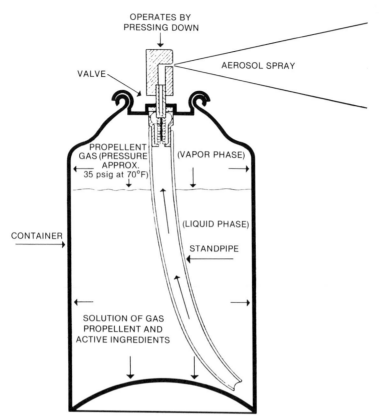

Fig. 6.1. Cross section of typical space or surface spray aerosol package. [From Downing (1961); reprinted with permission of Wiley.]

is known about the nature of the size distribution for submicrometer particles from pressurized spray cans. Liu (1967) has reported some exploratory data characterizing these atomization devices. Liu's measurements were made on several samples of cans containing 2% dioctyl phthalate and a liquefied freon propellant. The pressurized containers were standard units produced by the Johnson Company for an insecticide spray. He found the mean radii of sprays from several sample cans to be in the submicrometer range. The measured particle size of the aerosols generated is quite small compared with other disintegration methods. In these experiments, there appeared to be no systematic effect of pressure in the container. Liu's work indicated that the size distributions produced by spray cans were polydispersed. They could be characterized by a power-law distribution over a wide range of size, with the form $n(R) \sim R^{-4}$.

In a later work, Mokler et al. (1980) have reported additional data on particle size distributions from spray cans. Within a meter or two of the can, they found sprays with the size and concentration segregation listed in Table 6.1. A typical cumulative size distribution is shown in Fig. 6.2. The measured distributions are only approximately lognormal, but do not appear to vary appreciably from spray can to spray can. As aerosol generators, the spray cans may be convenient, but they produce a relatively coarse, polydispersed suspension with limited reproducibility from material to material.

6.2 AGRICULTURAL OR PEST CONTROL SPRAYING AND DUSTING

The use of devices to disperse quantities of pesticides for agricultural or public health applications has been widespread over the world. Their application ranges from individual household and domestic activity to very large-scale, systematic treatment as an integral part of agricultural practice.

In general, the control of pests or disease involves the distribution of a small amount of pesticide over a very large surface area. This may include surfaces of buildings, vegetation, or soil. The dispersal of pesticides is accomplished by suspending material in a liquid, usually water, then spraying, or by dusting using a finely divided powder. There are a variety of devices available for dispersing pesticides; techniques have been developed for different applications, involving ground-based aircraft operations. Good descriptive summaries are available from the World Health Organization (WHO) (1974), and the Food and Agriculture Organization (FAO) (Akesson and Yates, 1979).

6.2.1 Particle Size, Distribution, and Coverage.
The success of pesticide applications depends largely on control of the particle size range of the spray

TABLE 6.1

Simulated Aerosol Characteristics from Pressurized Spray Cans

Product type	Typical mass median aerodynamic diameter (μm)	Concentration of particles (mg/m^3)		
		<1 μm	1–3 μm	3–6 μm
Air freshener	6	1	4	9
Antiperspirant	6	7	42	74
Dusting aid	7	2	9	26
Fabric protector	3	1	3	2
Furniture wax	3	2	6	7
Hair spray	6	2	5	9
Paint	8	4	15	42
Wood panel wax	8	<1	1	3

[a] From Mokler *et al.* (1980); reprinted with permission from Ann Arbor Science Publishers.

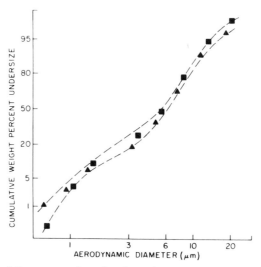

Fig. 6.2. Effect of the consumption of antiperspirant can contents on simulated breathing zone size distribution. The data from the middle third also fall within the dashed lines. Antiperspirant: ▲, first third; ■, final third. [From Mokler *et al.* (1980); reprinted with permission of Ann Arbor Science Publishers, Inc.]

or dry material, and on how this is affected by chemical, physical, and biological factors, some of which are described here:

1. The basic toxicity of the pesticide to the target pest. A small particle of a very toxic material may contain a lethal dose for the insect, whereas a larger drop or several small drops of a less toxic chemical may be required for lethal effect.

2. The physical characteristics of the chemical and its formulation, including density; fluidity of dusts; and vapor pressure, viscosity, and surface tension for liquids. For liquids, these characteristics affect the initial atomization process, and the aerial transport, evaporation, and deposition of the drops.

3. The collection efficiency of target surfaces, such as insects, buildings, and vegetation. This follows a collection efficiency as a function of Stk, where η is proportional to the drop diameter squared and its relative velocity, but inversely proportional to the width or frontal area of the object. Thus the collection efficiency of an object is increased rapidly by increased drop size and to a lesser degree by an increase in the relative velocity of the drop toward the object, but is decreased as the object size increases.

4. Location of the target insect, whether in the open or in a sheltered area, and in motion or at rest.

5. The local meteorological conditions. The presence and intensity of air turbulence, or the mixing and diffusion capacity of the air during application,

greatly affect the dispersion of pesticides, particularly the small aerosol particle concentrations.

6. The type and characteristics of the ground cover. Grassy savannah or low-growing agricultural crops permit relatively unrestricted downwind transport of finely atomized sprays; increased bush and tree cover increasingly filters particles out of the air.

7. The characteristics of the application equipment, which determines the size range or frequency distribution of drops as well as the spatial distribution of all solid and liquid materials directed to the target area.

Generally, some optimum droplet size range is recognized as most effective for each pesticide and for each formulation used for a specific control problem.[†] Maximum effective control of a target organism with minimum use of toxic materials and minimum adverse impact on the ecosystem is the objective. This simple statement covers a highly complex physical and biological phenomenon that occurs during and following an area application of pesticides. Research toward the objective has been conducted over many years. The earliest work with Paris green and toxic botanicals progressed through petrochemical products, culminating in the extensive use of the synthetic pesticides, DDT, and other organochlorines, as well as organophosphorus and carbamate materials.

Early observers found that although DDT and the other organochlorides are effective as direct contact applications, they are especially effective when used as residual applications. In contrast, many of the organophosphorus and carbamate compounds have a high contact and airborne toxicity to insects but degrade and lose effectiveness much more rapidly than the organochlorides, particularly in the presence of water.

In early laboratory and field studies researchers obtained data on the most efficient drop sizes to be used with given chemicals and on specified applications. Latta et al. (1947) calculated that the lethal dose (LD_{50})[‡] for *Aedes aegypti* was obtained with minimum dosage when drops of 22-μm diameter containing DDT were used with an air velocity of 2 mi per hr past the mosquito. At 3 mi per hr the drop size was found to be 18 μm, at 4 mi per hr 15 μm, and at 5 mi per hr 14.2 μm. Johnstone et al. (1949) came to the conclusion that the most effective drop size of a 10% DDT oil solution for control of resting and flying mosquitos considering both impaction rate and lethal dose would be 33 μm. This is below the minimum drop size (83 μm) containing a lethal dose of DDT; therefore, impingement of several such drops is required to kill an individual mosquito.

Yeomans et al. (1949) calculated from the collection efficiency relation-

---

[†] The targeting and delivery of pesticides to specific insects or vegetation is sometimes called *vector control*.

[‡] LD_{50} is a lethal dose to 50% of test animals in milligrams of chemical per kilogram of animal body weight; LC_{50} is the lethal concentration in milligrams of chemical per liter of air.

ship that a mosquito having a frontal width of 0.6 mm, at 2 mi per hr air velocity, would collect 15.8-μm drops most efficiently. This size is somewhat smaller than Latta's 22 μm for 2 mi per hr, because Latta used a smaller frontal width for his model mosquito. It has been shown in recent studies by Weidhass *et al.* (1970) that with the newer highly toxic organophosphate materials, a lethal dose for *Aedes taeniorhynchus* can be obtained from a 25-μm drop of Malathion, a 17 μm drop of Naled and a 20-μm drop of Fenthion. These sizes are much closer to the size of drops earlier observers found to be most efficiently collected by small insects; such drops did not contain lethal doses of organochlorides, however. Calculations show further that the minimum optimum drop size (the size of the drop most efficiently deposited) increases for larger insects. Thus houseflies showed increasing collection of drops up to 22.4 μm (David, 1946) and locusts up to 60 μm (McQuaig, 1962).

Drops carried by an airstream approaching a small object, such as a cylinder 3 mm in diameter (representing an insect body), are quite small in relation to the object (5–100 μm in relation to 0.32 cm, for example). Such drops tend to be diverted in two streams around the object. Thus those drops primarily directly in line with the center of the object will be deposited. As the size of drops or their velocity increases, those drops approaching in the projected frontal area of the object are less likely to be drawn around it by the airstream and are also deposited by impaction.

Earlier observers tried to evaluate application machines and techniques in relation to type of ground cover and weather in actual field trials. Johnstone *et al.* (1949) developed theoretical data based on atmospheric diffusion equations which indicated that if 10-μm drops are released at ground level, cumulative recovery (deposit) out to 6 miles would never exceed 60% under conditions of a small temperature inversion with 2 mi per hr velocity. At 5 mi per hr velocity with no atmospheric inversion, the recovery could be under 36%. Yeomans *et al.* (1949) showed from field studies that aerosol particles with a volume mean diameter (VMD) of 50 μm deposits up to 60% in 700 m under stable temperature inversion conditions, but less than 23% with more turbulent mixing conditions.

6.2.2   Application Volume.   Liquids may be applied in dilute or concentrated form. Dilute sprays are most frequently used for large-volume applications as large drops and with a wetting coverage. Concentrated liquids, generally those with very little or no diluting carrier, are applied as a fine spray, mist, or aerosol. Ultra-low-volume (ULV) is another name for the aerosol sprays from concentrates; ULV covers a wide range of volumes and dilutions of applied sprays, from a fraction of an ounce to a pint or more per acre.

The ULV treatment is a technique for applying a minimum amount of liquid per unit area compatible with the requirements for achieving control of a specific organism with a specific chemical. Whenever small volumes are applied, the liquid is finely atomized in order to maintain a desirable number of drops per unit of area or per unit of space volume. The number of drops available from a given volume of liquid is inversely related to the cube of the drop diameter.

If all drops are spread uniformly, the number of drops per unit of surface area per square inch will vary as the diameter cubed. Thus, for an application of 1 gal/acre with a 20-μm-diam drop size produces 9 drops/m^2, while a 50-μm drop size would give only 6 drops/m^2. This points out the large covering power of small particles, which permits thorough exposure of target organisms or other surfaces to the small volumes of applied aerosols. Akesson and Yates (1976) note, however, that deposition of aerosols also is limited by application conditions. Carrying this calculation one step further, consider the number of drops per cubic centimeter of air to a depth of 10 m from a 1 gal/acre application for uniformly dispersed drops of one size. Again, the radius-cubed relationship applies, and with 20-μm-diam drops, 0.08 drops/cm^3 would be found, while at 50-μm diameter only 0.005 drops/cm^3 would exist. Sprays directed specifically at small target organisms such as a mosquito must consider optimizing droplet concentration levels. This, in turn, suggests the use of droplet sprays less than 50 μm in diameter. No atomizer system produces drops of one size to vector control; the distribution of drops covers a wide range of size, from less than 1 up to 50 μm for a 20-μm VMD.

Ultra-low-volume techniques have applications for both airborne spray exposure (volumetric) as well as deposited spray. Thus the droplet size produced in such situations is in the micrometer size range except in special cases in which large drops can be used at a very low spatial distribution. A variety of sprayers using high liquid pressures, air shear such as that produced by high-speed aircraft or airstreams, and various two-fluid and spinning devices are available for producing the fine drops required for ULV applications.

Large scale, low-cost mosquito control programs have been adapted to ULV techniques, particularly for emergency treatment when entire cities or other large areas are treated. Increasing use is being made of ground-based generators for ULV-type treatments. Utilizing the "drift spraying" techniques, aerosol clouds of VMD below 50 μm can be carried by prevailing winds for downwind distribution in open areas from 300 m to a kilometer or more. However, the success of such applications is wholly dependent upon a temperature inversion condition in the air that will restrain any vertical diffusion of the spray-laden cloud. It must also be appreciated that evaporation of the released aerosol must be kept to a minimum by use of low-volatility spray formulations. Applications of aerosols under 50-μm VMD by aircraft have shown erratic results; only under unique weather conditions will the pesticides under 50 μm settle toward the ground in sufficient numbers to produce either a detectable deposit or provide an adequate number of drops for volumetric application in a target area.

Since exposure for adult insect control is largely by direct insect-droplet contact, the most effective droplet size is taken to be that small enough to impinge on the small surface presented while remaining airborne for a sufficient time for insect contact to take place.

6.2.3   Meteorological Factors.   The local meteorology is a significant factor controlling the success or failure of a vector control operation. The basic parameters are (a) vertical mixing as related to air stability; (b) air transport, which depends on wind velocity and wind velocity gradient with height and wind direction during drift spraying; and (c) relative humidity, particularly if water is the pesticide carrier. These factors affect the rate of dispersion and evaporation of pesticide materials released from either ground or aircraft equipment. The most significant of the factors listed is the air temperature gradient. When the air overhead is warmer than that at the ground under stable inversion conditions, any material released at the ground and transportable by air, such as aerosol particles smaller than 50 μm, will be carried along by the moving air at ground level and will not diffuse upward. The air speed under the inversion layer will control the mixing process in the area; higher speeds will cause more rapid ground-level dispersion. When temperature gradients are increasingly cooler overhead above a warm ground, the air becomes unstable and mixes vertically. Sprays are then diffused upward rapidly, and are dispersed and diluted by wind. Thermally stable air conditions and light winds normally occur in the early morning and in the evening.

Temperature inversions with low wind speed and weak velocity gradients provide the greatest vertical confinement of released sprays, and thus the best application conditions, particularly for fine sprays, or mists, when a large proportion of the released material is less than about 50 μm in diameter. This means that the time for application of aerosols and for optimum target exposure with sprays and mists should be early morning to midmorning, and late afternoon and evening under stable inversion conditions. Coarse sprays may be applied at any time during the day, the only limitation being the wind speed, which will displace an aerosol cloud generated by an aircraft significantly when wind exceeds 12 to 15 mi per hr. This range of wind speed also makes ground applications difficult to manage. Downwind concentration (essential for high exposure) is rapidly reduced by unstable temperature lapse rates and windy conditions; hence, this condition is favorable to weak downwind contamination in combination with a coarse spray.

Stable temperature inversions are produced by several different mechanisms; frequently more than one process may cause this effect. The most common is a radiation inversion involving heat loss or thermal radiation by the ground to cool air above. This occurs when the sun is low or below the horizon. Radiative heat loss cools the ground and air close to it mainly during the night. Another important cause for inversion is the influx over the land of a sea breeze along coastal areas. This cold air moves up valleys over the ground and undercuts warmer air away, forcing it upward, creating a stable temperature inversion condition. The sea breeze generally reaches its peak by mid-day. A third cause of temperature inversion is subsidence, a phenomenon in which air from a higher elevation is forced down into a lower level, such as a valley. The decrease in height ideally results in adiabatic compression, warming the air and establishing a warm layer over the ground, thus producing a stable inversion condition.

Because of the dominant effect of insolation, the inversion and lapse conditions follow a diurnal pattern, with lapse or well-mixed and neutral (no change in gradient with height) conditions prevailing during the day while the sun's effect is strong, and the stable inversion condition taking place when the sun is low during early morning and evening hours or at night. During cloudy overcast weather, the temperature gradient will vary from a neutrally stable to a stable inversion condition, depending on cloud cover and other conditions that may affect the temperature gradient.

Turbulence in the air is a normal daytime phenomenon which normally lessens under late afternoon temperature inversion conditions, and at night when the sun's heating of the ground is not contributing to vertical movement of the air. It is possible to have turbulence under a strong stable temperature inversion, but this mixing eventually decreases in the absence of energy sources for the turbulence. Stands of trees and forest canopy also act to dissipate turbulence near the ground. Here, the temperature will frequently remain the same (neutral with height) to the top of the forest cover. The wind velocity will be only a fraction of that above the forest cover. Thus both the thermal and frictional sources of turbulent energy are reduced in forests. Applying an aerosol by ground under the tree cover offers a realistic approach to efficient insect control. Normally, outside (or above) the forest cover, the daily changes in temperature and wind gradients will exist, with temperature inversions in early morning and late afternoon and daytime temperature lapse of turbulent mixing on most sunny days. During the night and under cloudy overcast, neutral conditions (neither strong lapse or inversion) would likely predominate.

Johnstone *et al.* (1949) determined both horizontal and vertical forest penetration distances, measured as the percentage of discharged aerosol that penetrated the forest to a stated distance. With injection from an aircraft, forest penetration varied with the density of the foliage cover. A very dense forest might have a vertical density twice its horizontal density, owing to the arrangement of leaves. Jonnstone *et al.* also showed that aerosols applied above the tree cover with little downwind speed do not penetrate but impinge on the foliage by horizontal wind motion. However, the aerosol droplets do penetrate horizontally if dispersed under the cover, as with injection from a ground aerosol generator. Penetration, however, would still vary with the density of the cover. For example, 15-μm aerosol particles released near the ground under inversion weather conditions and a 1–2-mi per hr wind speed gave deposits of DDT in the open (no cover) for a distance of 650 m. However, with light forest cover, this distance was reduced to 200 m; in dense jungle growth the distance was further reduced to 70 m of effective deposit. Their data also showed that increasing the droplet size to 200–300 μm increased vertical penetration of droplets through the forest canopy. Under these conditions, most of the penetrating droplets fall to the ground.

For mosquito larval control, sprays of 200–400-μm VMD are quite effective in reducing losses to aerial drift, and result in the highest deposit on the ground or in surface water. For example, experiments have shown that filtration by rice foliage up to 3 feet tall had very little effect on the penetration of a 200-μm-VMD spray. Bioassay of chemicals in paper cups that were placed at the top of the rice and also in the water beneath plants showed little difference, although both the recoveries were unexpectedly low, varying from 10 to 45% of the applied DDT spray (Akesson and Yates, 1976).

Brescia (1946) was one of the first researchers to try to evaluate the effects of downwind transport and droplet particle size on both adult and larval mosquito control. His tests involved thermal aerosols of 5- and 16-μm VMD. His results showed effective larva control of *Aedes taeniorhynchus,* also *A. sollicitans* and *Anopheles quadrimaculatus* at 0.001–0.002 lb of DDT per acre to distances of 650 m during strong stable inversions, and with grassy ground cover but no overhead canopy. Under a light forest canopy, this effective distance was reduced to 350 m; for a dense forest, to 100–150 m. For contact between airborne spray droplets and insects, an aerosol under 10-μm VMD, applied under strong inversion, with a positive low wind drift, was effective to about 1-km distance in the open areas, to around 0.5 km under light forest conditions. For conditions successively downwind, the number of airborne droplets and those deposited on the ground are increased by an increase in the applied dosage. Thus the foliage has a selective filtering capacity, but still permits a fraction of particles to pass no matter how dense the foliage may be.

Kruse *et al.* (1949) used thermal aerosol generators linked to the engine exhaust of an aircraft. With droplet sizes of 35–40-μm VMD, recovery on glass slides placed in the open in a 70-m recovery swath, in near calm conditions, was 9% of the discharged spray, and the peak was only 12% at the center of the swath. Kruse's foliage penetration data are not available, but he indicates that the dose required for a heavy, tall forest canopy would be ten times that for the open field, while for moderate low foliage or grass cover he suggested five times the open area dosage for LD_{90} control.

Since 1960, a wealth of information has been developed on field use of the organophosphorus insecticides applied as concentrates or diluted mixtures in nonvolatile petroleum and glycol solvents and diluents instead of volatile water-base emulsifiable concentrates of solutions. With near theoretical calculated concentrations of active ingredients of very high intrinsic toxicity, the phosphate and carbamate chemicals have made possible the reduction of liquid applied per unit of area to very low levels of 1 to 3 oz/acre, commonly referred to as LV (low volume) and ULV applications. However, it should also be pointed out that the use of these low application rates necessitates small droplet size (under 100 μm) to give an effective 1–3 drops/cm^2 on flat surface, or under 25-μm droplets to give air volume (up to 10 m height) dosage of 2–2.5 drops/cm^3.

The literature suggests that the effective downwind range for a ground aerosol applicator using organophosphorus chemicals for adult mosquito control can be as far as 3 km or more in open areas (Mount *et al.*, 1970) when drops of 10–15 μm are released. Dosages of nonvolatile phosphate chemicals for caged and natural adult mosquito mortality of 75–100% vary from 45 to 0.45 g/acre, depending on vector and chemical (Mount *et al.*, 1970). Recent data on the effect of dense and heavy forest or jungle growth on filtering of aerosols applied either by ground or aircraft indicate that for control of mosquitos in dense jungles an increase of three or more times the usual dosage per acre is required for effective control.

In the past, computation of the deposit expected from aerial spraying has relied on limited numbers of experiments and empirical factors. Recently, Miller (1980) has analyzed the deposition pattern of spray over a field using a well-stirred reactor model. The model was applied to a multiple crosswind swath pattern represented by Fig. 6.3a. The spray cloud is assumed to be well mixed over the air volume bounded by the flight pattern and height of aircraft flight. The retention of herbicide in experiments as deposits on a tract of land underneath in this case was 70–80%. The remainder drifted downwind, and decayed in mass deposited as indicated in Fig. 6.3b. Over the tract, the deposit from each swath adds to a cumulative maximum at the downwind edge of the volume. Miller's model assumes a well-mixed spray cloud whose change in droplet concentration with time is a function of the ratio of the product of the droplet concentration and the gravitational settling speed, to the height of aircraft release. The model is used to estimate the deposit downwind given the mean wind speed and a measure of atmospheric stability to mixing. With adjustable parameters of mass median droplet diameter and geometric standard deviation of the droplet distributions, the model estimates can be fit to several spray experiments for mean droplet diameters of approximately 250 μm, and a σ_g of 89 μm. This relatively simple mathematical approach shows promise for rationalizing a variety of available experimental data.

Regardless of the application, dispersion devices involve the breakup of a liquid or powder by pressurized flow of air through a nozzle as an atomizer. The pressurized flow basically supplies the energy for the breakup or disaggregation process. The principles of atomization have been described earlier in Chapter 4. Some nozzle configurations used in practice are shown in Fig.

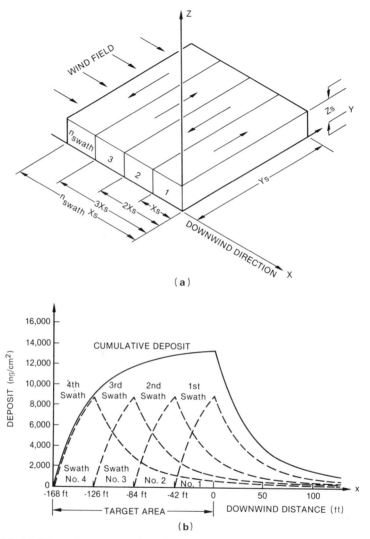

Fig. 6.3. (a) Schematic representation of a multiple swath pattern, χ_s is the length, y_s is the width, and z_s is the height of the swath. (b) Downwind target swath distribution from herbicide droplet deposit collected on Mylar strips. Each swath adds to a cumulative deposit which decays exponentially as the spray cloud drifts downwind. [Reprinted with permission from Miller (1980). Copyright 1980 American Chemical Society.]

6.4, which includes the basic geometry involved. In addition to atomizers employing nozzles, centrifugal devices such as spinning disks, cages, and brushes have been used. Sprays can be produced by breakup using a single-fluid device or two-fluid systems as indicated in the figure. The fluids are forced through the nozzle at high pressure; the nozzle geometry induces disintegration by various means.

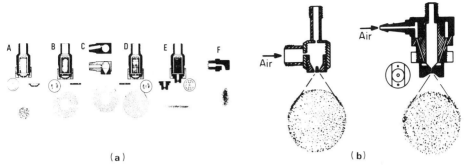

Fig. 6.4. Atomizer nozzles used for agricultural spray applications: (a) pressure type, producing a wide range of droplet sizes suitable for both aircraft and ground machine use; geometry of spray patterns are indicated; from left to right, jet, hollow cone showing whirl plate, centrifugal hollow cone, solid cone showing hole in whirl plate, fan and deflector fan. (b) Two-fluid atomizers for fine sprays; internal mixing on left and external mixing on right. [From Akesson and Yates (1976).]

Recently, a new spray nozzle has been reported which makes use of electrostatic dispersal of liquids (Coffee, 1971; Law, 1978). In principle, the technique is akin to the Millikan oil drop experiment in which the charge of electrons was first measured, and is similar in some respects to methods used in paint spraying. The special qualities are that Coffee's design is free of any moving parts and can be run on flashlight batteries. The droplets have curved trajectories so that when spraying the front of a plant, the chemical is also delivered to the back. The system supplies liquid to a high-voltage nozzle. The emergent charged liquid drops are subjected to an intensely divergent electric field which is sufficient to carry the droplets to the target, which is at ground potential. This field strength is about two orders of magnitudes greater than the gravitational force, making for very much reduced drift and wind influence. A unique quality of being able to deliver pesticides to the hidden surfaces of a plant when apparently spraying only one side arises from the curved flux lines of the electrically propelled droplets. As electrostatic theory predicts, the lines of electrical force are orthogonal to the equipotential contours so that the spray diverges from the high-voltage nozzle to converge on a target and thus to penetrate to surfaces which are not visible on a direct line of sight from the sprayer.

6.2.4 Equipment for Pesticide Application. The equipment for liquid dispersal using hydraulic nozzles includes many different types of sprayers, which are classified in terms of their mode of operation and mobility. They are either manually operated or power operated, and are stationary, hand-carried, or mounted on simple vehicles.

The intermittent sprayers are operated by hand using a simple pump, which may be of either solid-piston or plunger-piston type. Liquid is drawn from the container on the backward stroke of the pump and forced through the nozzle on the forward stroke.

Compression sprayers operate on the same principle as the well-known knapsack compression sprayers. The container is filled to approximately three-quarters of its capacity and the air in the remaining space is then compressed by means of a small built-in air pump of the plunger type. The container has to be sufficiently robust to withstand the pressure required to expel the liquid contents from the nozzle via a suitable trigger valve. Compression hand sprayers are useful for small-scale larviciding and other applications of limited extent or for emergency operations but they are not recommended for general use in control.

Large vehicle-mounted mist blowers are used extensively for mosquito control, for both exterior space spraying and larvicidal treatments. The size and capacity of these sprayers generally determine whether they are vehicle mounted or vehicle trailed. They can also be mounted on boats and amphibious or tracked vehicles.

Droplet size and rate of application can be varied to some extent by the choice of the nozzle size, the number of nozzles used, and the direction in which they are mounted in relation to the fan airstream direction. The velocity of the airstream also has an effect on the droplet size; in general, the higher the velocity, the smaller the resulting droplets. However, since most of these mist blowers operate with a high discharge velocity, only a comparatively small change in droplet size is possible.

Vaporizing pots are small electrically heated containers for the vaporization of pesticides into the atmosphere. Spirit burners are occasionally used as the heat source. The use of these devices in vector control is limited to applications in enclosed spaces, and in some countries their use is restricted or even prohibited.

In thermal fogging machines, the pesticide is dissolved in an oil of suitably high flash point which is vaporized by being injected into a high-velocity stream of hot gas. When discharged into the atmosphere, the oil carrying the pesticide condenses in the form of a fog. Two basic methods exist for the production of these fogs: In the first method, oil is injected into the exhaust gas of a pulse-jet internal combustion engine at a point where it will be completely vaporized and then immediately discharged. In the second method, gasoline vapor is burnt in a specially constructed chamber that is constantly supplied with a large volume of heated air at low pressure. The formulation is injected into a discharge tube through which the air is passing and is emitted as a dense fog. Such equipment is usually built into a frame complete with a power unit for mounting in a truck or utility vehicle. The discharge outlet is adjustable for direction and there is often provision for the control of droplet size.

The different types of equipment available for the application of pesticide powder (dusting) may be classified by manual and power-driven operation.

The essential parts of dusting equipment consist of some form of container or hopper to hold the dust, a means of feeding the dust at a constant rate to the discharge outlet, and the provision of an air blast or airstream to convey the dust to its desired location. Some means of agitation may be provided in the container to ensure evenness and continuity of flow during operation.

Hand-carried dusters may be improvised from readily available materials, such as a tin can with holes punched in the lid, or a loosely woven linen or fabric bag that may be shaken or struck with a stick. Air-operated types can be divided into two groups according to the method used to supply the air—the bellows type and the plunger air-pump type. The term "bellows" device describes any duster that is provided with rubber, leather, or plastic bellows, the squeezing of which produces a puff of air that expels the dust in a small cloud. The materials used for construction of the body may be wood, metal, plastic, or even cardboard (for disposable units).

The second type consists of rotary blowers where air is provided by a simple paddle-type fan, operated by rotating a crank with the hand. The cranking motion is transmitted through a gearbox to the blower and is also used to agitate the dust within a hopper and to operate the feeding mechanism.

Continuously operated dusting equipment is identical to the manually operated equipment, the only difference being that it is motor driven.

Vehicle-mounted dusters, for example, consist of a container from which dust is drawn into a Venturi device. Air flow through the Venturi device is generated by a motor-driven blower and the dust suspension is emitted through an outlet nozzle that may be directed at the target like the barrel of a gun.

Various mechanisms are used to feed the dust to a metering device but basically they take the form either of a worm-type feed or of a revolving or reciprocating brush. The metering device itself may be a sieve with fixed or adjustable holes or slots, or it may be a simple orifice of fixed or adjustable size. There is usually some variation in the discharge rate as the hopper empties. The dust from the metering device can be fed either into the blower intake or into the discharge side of the blower. The former method is preferable as it breaks the dust up before discharge.

6.2.5   Aircraft Applicators.   The use of aircraft for pest vector control dates back to the early 1920s—when applications were being made of chemicals such as Paris green and petroleum oils—but the technique was not widely used. Pesticide applications by aircraft became more common with the widespread use of DDT and have increased greatly in the last decade. This increase is due largely to the discovery of new insecticides that are highly effective at low dosages, to improvements and increased availability of application equipment, and to the development of new vector techniques. These developments have greatly reduced the costs of application by aircraft, particularly where treatments can be made over large areas. There has also been a growing appreciation of the potential value of aircraft for controlling certain vector-borne disease outbreaks and for preventing their occurrence in emergency situations arising from floods, earthquakes, hurricanes, and other disasters.

Aircraft are at present being utilized successfully for the control of mosquitos, blackflies, tsetse flies, and other disease vectors or pests. Information on the recommended use of aircraft for such control measures is given in a report of the WHO Expert Committee on Insecticides (WHO, 1970).

Aircraft in use can be classified in two groups—fixed wing and rotary wing. The principal advantage of the rotary-wing aircraft, or helicopter, is its ability to hover and maneuver within a confined space and to land in a small area—e.g., on a special platform built on top of a truck. This advantage is particularly useful in mosquito or blackfly control programs, since the machine may land at potential breeding sites for inspection and immediate spot treatment if required. A further advantage is that the downward draft from the rotor opens up dense foliage so that the pesticide droplets achieve a greater penetration. However, this effect can be produced only under specific conditions of height and forward speed of the helicopter; as height and forward speed increase, the downdraft effect is lessened. Fixed-wing aircraft can also produce a similar but less pronounced effect under specific conditions; air flowing over the top of the wing leaves the trailing edge in a downward direction so that if the aircraft is sufficiently close to the ground during the application the vegetation will be disturbed.

The purchase price of a helicopter is approximately three times that of a fixed-wing aircraft with the same load capacity. The maintenance costs are also correspondingly higher for the helicopter and the cruising speed usually lower. In certain circumstances, however, particularly when the airstrip of the fixed-wing aircraft is a long way from the area to be treated, the helicopter, being able to land in or near the area, can operate more efficiently than the fixed-wing aircraft. The versatility of the helicopter permits it to turn in a shorter distance, be loaded nearer the area to be treated, and to treat smaller, scattered areas more readily.

In addition to the many physical factors involved in the application of chemicals, aerial application requires knowledge of droplet size and behavior of aircraft construction and materials, and of the unique functions of low-flying aircraft. Although it has been demonstrated that the droplet size of pesticides has an important influence upon their biological effectiveness against adult insects, the VMD of pesticide particles dispersed from aircraft must, in practice, be greater than that considered desirable in applications at ground level, because the altitude at which the spray is discharged increases the difficulties of placing the pesticide in the target area.

Meteorological conditions that affect settling, evaporation, and deposition have a greater influence on the behavior of sprays dispersed above ground rather than at ground level. Aerosols having a spray droplet size of less than 50-μm VMD of concentrated (technical) pesticide are considered to be the most effective for direct contact with adult vectors of disease, but they are difficult to control when applied by aircraft, except under the most favorable weather conditions. On the other hand, larger droplets of concentrated pesticide containing more than the lethal dose per drop are wasteful as adult insecticide sprays. Thus, if larger drops are needed, the spray should be diluted with oils, glycols, or other solvents of low volatility. Coarse and medium sprays, which have minimal drift losses, can be used for larvicide applications on open water for blackfly or mosquito control. Larvicide application on irrigated pastures, rice fields, or other cultivated areas with tall, dense vegetation are somewhat more difficult. Where vegetation is dense, aerosols and fine spray applied in a moderate to strong wind are subject to high drift losses and also are more apt to impinge on the vegetation itself. Coarse sprays are less affected by wind and will give better vertical penetration of the vegetation, but the deposit on vegetation will always be high. The most effective aerial method against mosquito larva is the use of coarse sprays when wind speed is less than 3–6 mi per hr.

The formulations that most readily penetrate tall vegetation are free-flowing granules and pellets, but they are more expensive. Their application costs, too, are higher owing to their lower bulk density, which yields a lower aircraft payload. Fine sprays and mists when applied by aircraft, require temperature inversion conditions and wind speeds of not more than 6 mi per hr.

Typical spray patterns and their applications are summarized in Table 6.2 for different uses.

The equipment used for aerial applications is essentially the same as that used on ground equipment although certain modifications may be required. The most common type of equipment for the dispersal of liquid pesticides from aircraft is a boom and nozzle system. The number and location of the nozzles on the spray boom is determined by the volume and spray atomization desired, the type of spraying equipment available, and the aircraft on which it is to be used. The boom may be mounted behind the trailing edge of the wing, which in the case of a low-wing monoplane or a biplane, enables the pilot to see whether all the nozzles are functioning correctly, or it may be suspended below the wing, which may often be more convenient but increases the aerodynamic drag. Trials are usually carried out to ensure that the application rate, swath width, and distribution conform to the specific requirements.

Figure 6.5 shows the general layout of an aircraft spray system using a windmill to drive a pump. Liquid is pumped to a pair of linked valves controlled for spray release by the pilot.

Hollow Cone and flat fan hydraulic energy nozzles are the types most commonly used on aircraft and provide a range of droplet sizes of 100–500

TABLE 6.2

Spray Drop Size Range, Approximate Recovery Rate, and Recommended Use[a]

Spray size or type and description of spray system	Drop size range[b] (μm VMD)	Estimated deposit[c] in 1000 ft (%)	Use
"Coarse aerosols": cone and fan nozzles, and rotary atomizers	<125	<25	Aerosol applications, in vector control and control of forest insects and agricultural pathogens; used at low volume rates, primarily for adult insecticiding
"Fine sprays": cone and fan nozzles, and rotary atomizers	100–300	40–80	Forest pesticide chemicals, in large-area vector control, at low dosages of chemicals with low toxicity and rapid degradation; also useful for agricultural insect pathogens
"Medium sprays": cone and fan nozzles, and rotary atomizers	300–400	70–90	All low-toxicity agricultural chemicals where good coverage is necessary
"Coarse sprays": cone and fan nozzles, spray additives	400–600 with additives up to 2000	85–98	Toxic pesticides of restricted classification, when thorough plant coverage is not essential
"Sprays with minimum drift": jet nozzles and spray additives	800–1000 with additives up to 5000	95–98	All toxic, restricted—class herbicides such as phenoxy acids and others, within limitations such as growing season and location near susceptible crops
"Sprays with maximum drift control": low-turbulence nozzles	800–1000	≥99	Restricted nonvolatile herbicides, phenoxy acids, and others in the area of susceptible crops, subject to limitations of growing season and type of crop

[a] After Akesson and Yates (1976).
[b] Determined for water-based sprays; oils yield smaller droplets.
[c] 1000 ft downwind, wind speed 3–5 mi per hr neutral stability; material released under 10 ft height.

-μm VMD, depending on the size of the aircraft and the application rate and swath characteristics desired.

Aircraft also have been used to produce thermal fogs, by the injection of an oil-based pesticide formulation into the exhaust of the aircraft engine. However, thermal fogging has largely been replaced by mechanical droplet production.

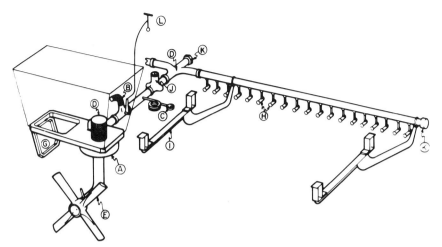

Fig. 6.5. This schematic diagram shows a basic aircraft spray unit. Included are a tank (outline), spray pump and propeller drive, and the control valve, which directs liquid to the boom and nozzle during spraying or back to the tank for recirculation. A, pump; B, control valve; C, pressure gauge; D, screen; E, propeller; G, dump gate; H, nozzles and check valves; I, boom mount; J, liquid; K, boom cleanout; L, valve level. [From Akesson and Yates (1976).]

Pesticides also may be applied from the air as dusts in which 80–90% of particles are smaller than 25 μm or in granular form (with particles of 1000–5000 μm). Granules can be formulated to meet a variety of application requirements. Hoppers must have a large opening at the top for easy filling and the lid must fit tightly. The flow of dry material is controlled by a gate that slides across a bottom outlet. A lever in the cockpit can be preset to open the gate the required amount to provide the correct dosage. Agitation is desirable for dust hoppers but is not generally required when granular materials are being dispersed.

Solid materials are distributed from fixed-wing aircraft using ram-air (Venturi) spreaders. Spinner spreaders are also commonly used on fixed-wing aircraft and give a wider, more uniform swath than the ram-air type, as well as reduced drag.

6.3 CONTROLLED LARGE-VOLUME GENERATOR APPLICATIONS

In some applications of aerosol technology, the capability of generating aerosols with large volumes of suspended material have been developed. Some situations dictate uncontrolled smoke dispersal, as in military applications (Green and Lane, 1964). However, others such as those mentioned in Section 6.2 can involve at least a degree of control of particle diameter, and the integrity of chemical composition of the aerial suspension. Another application of control requirements for sprays and mists is in the area of medical research and clinical practice. For example, aerosols have been used for therapeutic treatment of respiratory disease. Here the medication must be dispersed with a controlled particle size and volume for long periods of time under conditions when any chemical change in the suspended material is negligible. Recently, requirements for the study of the influence of air contaminants on respiratory disease have provided for the engineering of large exposure chambers with carefully controlled air properties. An example of such a design is discussed in detail below to illustrate the important factors involved. The controlled environment and experimental conditions could be applied equally well to clean room environments. The latter is an important adjunct to particle removal technology.

The design and operating principles of environmentally controlled chambers used within the United States for exposures of groups of human subjects to clean air and pollutant gases have been recently reviewed (Bell *et al.*, 1980). Three of these chambers have now been instrumented for controlled aerosol exposures of small groups of human subjects. A good example is the one at Rancho Los Amigos Hospital in Los Angeles (RLAH).

Design variables for the environmental control system which need to be considered and controlled in experimental aerosol exposures of human subject include the following:

(a) the concentration, size, chemical composition, and electric charge of airborne particles;

(b) the concentration of gases which may cause health effects or interact with particles;

(c) the temperature and humidity of the air; and

(d) the intensity and frequency of light that may affect subjects directly or indirectly through photochemical reactions of gases and particles.

Noise levels should be minimized to reduce subject stress and discomfort.

Air purification, temperature and relative humidity regulation systems developed for controlled exposures of humans to pollutant gases can be used for aerosol studies. Figure 6.6 shows a schematic diagram of the air purification and air conditioning systems for the RLAH human exposure chamber and high-capacity aerosol generator. The main chamber air supply is drawn from the outside through a stainless-steel duct at a flow rate of 14 m^3/min (500 ft^3/min). Large particles are removed by a prefilter, and a bed of activated carbon removes some pollutant gases. The air is passed through a high-efficiency particle (HEPA) filter, designed to remove 99.97% of the particles with diameters of 0.3 μm or greater, and is then heated. The air enters the first Hopcalite bed (mixture of manganese, copper, cobalt, and silver oxides on an activated charcoal support), where carbon monoxide and a large fraction of hydrocarbons are catalytically oxidized at about 250°C to water and carbon dioxide. The air is then cooled to room temperature before passing through one of two beds of Hopcalite, where additional gases are

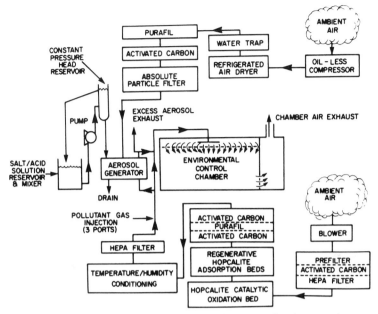

Fig. 6.6. Schematic of the air purification/conditioning system for the aerosol generator and chamber, the nebulizer solution feed, and the aerosol flow pattern for the RLAH human environmental exposure chamber. [From Bell *et al.* (1980); reprinted with permission of Ann Arbor Science Publishers.]

adsorbed. While one Hopcalite bed is adsorbing gases at 40°C, the other bed is being regenerated at 315°C. The partially purified air then passes through beds of activated carbon and beds of Purafil (potassium permanganate on alumina).

The purified air then passes through an air conditioning unit consisting of refrigerant coils, heaters, and purified steam injectors. The chamber instrumentation continuously monitors the chamber temperature and relative humidity linking the air conditioning unit to control automatically the temperature and relative humidity to within 1 and 5%, respectively.

Full capacity operation of the aerosol generator constitutes only 4% of the total chamber air flow, thus allowing a less complicated air purification system for the aerosol generator's compressed air supply than that for the main chamber system. An oil-less compressor is used to avoid contamination of the air with vapors which are difficult to remove. The dry compressed air supplies pressure to the nebulizers and is also used to dry rapidly the nebulized droplets when exposures are made with hygroscopic aerosols at low relative humidity levels. The activated charcoal, Purafil, and particle filters serve the same function as in the main chamber system.

The walls of the exposure chamber and all air ducts associated with the exposure facility are constructed of stainless steel to minimize problems of corrosion and reaction of gases and aerosols on the walls. As shown in Fig. 6.6, purified air or diluted aerosol is distributed in different directions above a false perforated ceiling inside the chamber. Air in this plenum is at a higher pressure than the chamber. Consequently, the air and particles are forced through the holes, causing good mixing of fine particles and pollutant gases. The air flow is vertically downward and uniform throughout the human subject exposure zone. Near the floor, air is directed toward the outlet grating in the bottom of the wall separating the main chamber from the entrance lock. The air is exhausted to a stack through the ceiling of the entrance lock.

The RLAH exposures required maximum sulfate or sulfuric acid concentrations of 1.0 mg/m^3 with mass median aerodynamic diameters (MMAD) of approximately 0.3 μm and geometric standard deviations (σ_g) between 2 and 4. Ammonium nitrate exposures required a maximum concentration of 5 mg/m^3 with a MMAD of about 1 μm and a σ_g between 2 and 3. These size distribution criteria were based on ambient air pollution measurements in the Los Angeles area, conditions for which this chamber was designed.

Other design criteria included

(a) production of an aerosol with a stable ($\pm 10\%$) size distribution and mass concentration over an 8-hr operating period;

(b) operation of the particle generator should not affect the regulation of temperature or relative humidity in the chamber;

(c) production of particles with low electric charge levels to prevent increased lung deposition resulting from electrostatic forces; and

(d) minimization of cost, space, and noise.

To meet these specifications for the RLAH chamber, a high-output generator system was built, as a generator system with high volume output. It was constructed mainly of stainless steel and consists of 20 nebulizers, a drying chamber, delivery duct, and several dampers. An aerosol mist is initially produced by nebulizing dilute aqueous salt or acid solutions. Clean air (2–3 m^3/min) withdrawn by a blower from the chamber purified air intake duct (Fig. 6.6) dilutes and dries the aerosol mist. The dried aerosol is then injected into the purified air for further dilution before entering the plenum of the exposure chamber.

The aerosol mass output of the generator is regulated by varying the number of nebulizers used. The MMAD and σ_g of the chamber aerosol are adjusted by varying the nebulizer solution concentration and the baffle position. Additional control of the chamber aerosol mass concentration is accomplished by adjusting the positions of both the output and bypass dampers.

Measurements were made of the electrostatic charge on ammonium sulfate aerosols at the position of the outlet damper of the generator and inside the chamber. Charge levels at the outlet were about 40% of the levels predicted by nebulizer charging theory and levels in the chamber decreased to about one-half of those at the generator outlet. Thus the initial charge of the generated aerosol is partially neutralized during the 2-sec passage through the stainless-steel delivery ducts to the chamber.

Based on the results of studies of Melandri *et al.* (1977), the electrostatic charge levels on the aerosol would be expected to increase the respiratory tract deposition in human subjects by only a few percent. Consequently, electric charge neutralizers were not added to this aerosol generator system. If for some other generation systems the electrostatic charge levels are high enough, compared to an ambient aerosol having an equilibrium charge distribution, to alter human lung deposition significantly, then Kr-85 charge neutralization systems should be included. Continuous radioactivity monitoring and frequent cleaning of the outside of the sealed Kr-85 gas tubes would be required to detect and prevent leaks and inadvertent exposure of human subjects to the Kr-85 gas.

Continuous real-time analyses are needed to assure the safety of the human subjects. An electrical aerosol size analyzer (EAA) and an optical particle counter (OPC) can be used to continuously monitor the aerosol mass concentration and size distribution over the 0.01–10-μm range during the exposure period. Alternatively, a light-scattering nephelometer can be used to continuously estimate the mass concentration of particles in the 0.1–1-μm range. A condensation nuclei counter may be used to continuously monitor the total number of particles smaller than 1 μm in diameter. Low-pressure, multistage cascade impactors that use piezoelectric crystals as collection stages have recently become available for continuously monitoring the mass concentration and size distribution of aerosols in the 0.05–1.5-μm range.

The RLAH chamber aerosols are continuously monitored by using an optical particle counter and electrical aerosol size analyzer with a minicomputer system interface. The EAA detects particles in the 0.013–0.42-μm size range and the OPC covers the 0.6–10-μm range. On command from the computer, the EAA is automatically stepped through the series of preset collector rod voltages and the value of the analyzer current is recorded after it has stabilized. A signal processor has been interfaced with the OPC to group the output pulses by pulse height into 256 channels. The computer then groups the OPC and EAA data into 24 discrete size intervals and calculates the number, aerosol surface, and volume concentration within each size group. The particle volume in the size range (0.42–0.6 μm) between the EAA and OPC is determined through linear interpolation of the volume data. The total mass concentration is determined from the particle density and the sum of the aerosol volume over all the size groups. The aerodynamic median and mean diameters; geometric median and mean diameters; and geometric standard deviation with respect to number, surface area, volume, and mass are calculated. Concentration and size distribution data may also be averaged and plotted off-line following an exposure with the RLAH system. The data can be displayed so that the chamber operators can monitor the safety of the atmosphere. If the concentration or size distribution deviates from acceptable limits, the operator may manually adjust the generator. A more advanced system, installed at the U.S. Environmental Protection Agency at Chapel Hill, NC, chamber, can automatically adjust the generator performance to control the chamber aerosol.

The aerosol generation and real-time monitoring system at RLAH has been used for exposure of human volunteer subjects for 2 hr/day on 2 or 3 successive days to nominal concentrations of 100 μg/m^3 of ammonium sulfate, 85 μg/m^3 of ammonium bisulfate, and 75 μg/m^3 of sulfuric acid aerosols (MMAD of approximately 0.3 μm and σ_g of 2.5 to 3) at 31°C and 40% relative humidity, simulating conditions observed during worst-case air pollution episodes in the Los Angeles area. Total filter and cascade impactor samples were collected during each exposure for gravimetric and chemical analyses to determine the particle mass concentration and size distribution as well as the ammonium, sulfate, and hydrogen ion concentrations. The EAA/OPC monitoring system was calibrated before each aerosol study against the data obtained from total filter and cascade impactor samples. The aerosol mass concentration and MMAD varied less than 10% over the 4–6 hr exposure period of a group of 5–7 subjects. Typically for sulfuric acid particles, the MMAD averaged 0.55 μm and repeated measurements taken at 5-min intervals over a 4-hr period had a standard deviation of less than 4%. The particle mass concentration averaged 252 μg/m^3 (at 40% RH the aerosol would contain about 50% water by weight) and repeated determinations had a standard deviation of less than 6%.

6.4   COMBUSTION OF SPRAYS AND DUSTS

The dispersal of material by spraying or by dispersal of solid particles plays an important role in combustion technology. Large industrial boilers or furnaces employ oil fuel or pulverized coal injection to provide an inlet stream for efficient combustion. Diesel engines, turbines, and certain kinds of rocket engines use fuel spray injection. The process of combustion concerns the steps of transport of fuel and oxidizer to the reaction or flame zone. Because of its technological importance, considerable research has been done on the burning of finely divided particles over the past 50 years. As one might expect, the combustion process for liquid droplets and solid particles is similar in some ways, but differs in details. In this section, our discussion will begin with sprays, and follow with pulverized coal.

The main objectives of liquid fuel firing are usually

(a)   a high combustion intensity,

(b)   a high combustion efficiency so that as little unreacted fuel leaves the combustion chamber as possible,

(c) a stable flame,

(d) the minimum deposition of soot or solid on the combustion chamber walls, and

(e) the maximum rate of heat transfer from the flame to the heat sink.

The most common method of firing liquid fuels is to atomize the liquid before combustion. The fuel is introduced into the combustion chamber in the form of a spray of droplets which has a controlled size and a velocity distribution. The main purpose of atomization is to increase the surface area of the liquid to intensify vaporization and to obtain good distribution of the fuel in the chamber and to ensure easier access of the oxidant to the vapor.

Following atomization, combustion takes place through a series of complex processes, most of which are only incompletely understood. The most important of these processes are

(a) Mixing of the droplets with air and hot combustion products—a process usually occurring under turbulent conditions.

(b) Transfer of heat to the droplets by convection from the preheated oxidant and recycled combustion gases and by radiation from the flame and the chamber walls.

(c) Evaporation of the droplets, often accompanied by cracking of vapor.

(d) Mixing of the vapor with air and combustion gases to form an inflammable mixture.

(e) Ignition of the gaseous mixture depending on the mixing conditions, a diffusion flame may either occur at the oxygen-rich boundaries of eddies containing many vaporizing droplets or individual droplets may be wholly or partly surrounded by their own small flames. Transition from one mechanism to the other can occur along the flame path.

(f) Formation of soot, and cenospheres[†] with residual fuels.

(g) Combustion of soot and of cenospheres, which are relatively slow processes.

In practice, these processes often occur simultaneously, resulting in an extremely complex aerosol system. Owing to this complexity, research has been centered on those areas considered to be most important. These have been reviewed by several workers including Essenhigh and Fells (1960), Williams (1968, 1973), and Hedley et al. (1971).

A combustion spray or aerosol differs from a premixed, combustible gaseous system in that it is not uniform in composition. The fuel is present in the form of discrete liquid droplets which may have a range of sizes and they may move in different directions with different velocities from that of the

---

[†] *Cenospheres* are fused hollow spheres or agglomerates made up of carbonized and ash material. They range from tens to hundreds of micrometers in diameter.

main stream of gas. This lack of uniformity in the unburnt mixture results in irregularities in the propagation of the flame through the spray, and thus the combustion zone is geometrically poorly defined.

The process of spray combustion is illustrated by means of the simplest case, that of a one-dimensional laminar flame moving at low velocity. The flame can be considered to be essentially a flowing reaction system in which the time scale of the usual reaction rate expression is replaced by a distance scale. This is shown in Fig. 6.7. As the unburnt spray approaches the flame front, it first passes through a region of preheating during which some vaporization occurs. As the flame zone is reached, the temperature rapidly rises and the drops burn. The flame zone can thus be considered to be a localized reaction zone sandwiched between, on one side, a cold mixture of fuel and oxidant and, on the other, the hot burnt gases. If the gas flow through the flame is one dimensional, then the flame front is planar. The nature of the burnt products depend on the properties of the spray in its unburnt state. If the droplets are large, then combustion may not be complete in the main reaction zone and unburnt fuel penetrates well into the burnt gas region. If the droplets are small, then a state of affairs exists that approximates very closely the combustion of premixed gaseous flame. Here the droplets are vaporized in the preheat zone and reaction after that is between the reactants in their gaseous states. The other factors that determine the time to reach complete combustion (that is, the length of the combustion zone) are the volatility of the liquid fuel, the ratio of fuel to oxidant in the unburnt mixture, and the uniformity of the mixture.

In most practical systems such as a furnace or a rocket engine, the combustion process is much more complicated because of two important factors. First, to a large extent the mixing of the fuel and oxidant takes place in the combustion chamber and thus the mechanics of the mixing process plays an important role. Second, the flow patterns are complicated by turbulence or recirculation and frequently cannot be represented by simplified models.

To obtain an understanding of the processes involved in spray combustion, it is necessary to have complete knowledge of (a) the mechanism of combustion of the individual droplets that make up the spray, (b) any inter-

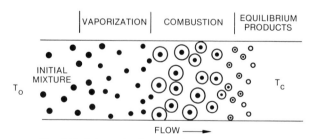

Fig. 6.7. Schematic diagram of spray combustion.

action between the individual droplets when they undergo combustion in the spray, and (c) the statistical description of the droplets that make up the spray with regard to size and spatial distribution.

Because the combustion of sprays is such a complex process involving simultaneous heat and mass transfer as well as chemical reaction, it is difficult to obtain detailed information about the burning rates or mechanisms by means of direct studies of spray combustion. A number of experimental techniques have been employed to study the combustion of isolated single droplets under controlled conditions. Theoretical analysis and mathematical models also are facilitated if only single isolated droplets are considered. Thus a considerable amount of effort has been devoted to the study of single droplets.

The next step is to apply these data to the multidroplet problem of spray combustion. However, the application of the results obtained for single droplets to obtain the characteristics and parameters of the combustion of sprays is not an easy matter and remains an incomplete solution to the combustion analyses.

Two types of droplet combustion can be identified. In the case most commonly considered the liquid droplet is a combustible liquid (the fuel) and it burns in a surrounding oxidizing atmosphere. The droplet acts as a source of vapor and since oxidant and fuel are initially separated, then the flame surrounding the droplet is a diffusion flame with mixing taking place in the flame reaction zone. The other type of droplet combustion involves monopropellants, e.g., hydrazine, which can sustain a self-decomposition flame without any additional oxidant. In this case, the flame is not controlled by diffusive mixing and has much in common with premixed spherical flames.

6.4.1   Droplet Combustion Theory.   The main requirement of a theoretical treatment of droplet combustion is that it gives an insight into the dependence of combustion on the chemical or physical processes that occur and also it must be in quantitative agreement with the experimental results. These criteria have been satisfied during the last thirty years or so by a number of theoretical interpretations of the combustion of isolated single droplets in an oxidizing atmosphere. The principal investigations have been made by Godsave (1952), Spalding (1952), Goldsmith and Penner (1954), Agafanova *et al.* (1957), and Lorell *et al.* (1958) although many other papers have been published dealing with particular aspects of droplet combustion (Williams, 1968).

The basis of most droplet combustion theories is the spherically symmetric model developed by Spalding and by Godsave; subsequent work has produced only minor refinements. This model is shown diagramatically in Fig. 6.8a. Here, the droplet is represented by a sphere of radius R_1 surrounded by a concentric flame zone of radius R_c. Concentric with the droplet and outside the flame zone lies another outer boundary which is taken to

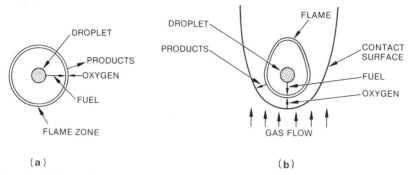

Fig. 6.8. (a) Single-droplet combustion with spherical flame zone. (b) Single-droplet combustion with gas flow causing weak convection and distortion of the flame zone.

be at infinite distance from the droplet. The composition of the gas at this outer boundary is taken to be solely that of the ambient atmosphere and is not influenced by the presence of the burning droplet with weak convection. The flame zone then becomes distorted into an ellipsoid (Fig. 6.8b). The flame is supported by the exothermic reaction of fuel and oxygen in the flame zone, the oxidant diffusing in from the outer boundary to the flame zone while the fuel vapor diffuses from the droplet surface. Heat is transferred from the flame zone to the droplet to provide the latent heat of vaporization of the liquid fuel.

Other than spherical symmetry, the classical model for droplet combustion requires the simplifying assumptions that

(a) the fuel droplet is considered a pure, single-component liquid;

(b) combustion occurs under quasi-steady state conditions;

(c) the temperature at the droplet surface is taken as the boiling point of the liquid;

(d) radiation and thermal diffusion effects are neglected unless extremely rapid burning occurs; and

(e) flame propagation is characterized by low velocity, and the flame zone has negligible viscosity so that constant pressure exists throughout the system

In all practical systems, the droplet must first undergo ignition and then settle down to steady state conditions. Once these are reached, the quasi-steady state is maintained only if the fuel is consumed in the reaction zone at the same rate as it evaporates from the droplet. Furthermore, the applicability of assuming a fixed droplet radius means that the velocity of the fuel vapor flowing away from the droplet surface is much greater than the regression velocity of the surface. These approximations lead to a small (usually negligibly small) error. However, the quasi-steady state theory ceases to be applicable when the liquid is above the critical point (Williams, 1973). Inherently connected with the quasi-steady state theory is the assumption that the temperature in the interior of the droplet does not vary with time, and it has been found that the temperature in the center of the drop may closely approach the surface temperature of the liquid. Williams (1968) reported that unless the heat of reaction is extremely low, or the fuel is highly nonvolatile, the heat flux to the droplet is large enough to lead to conditions under which the droplet surface temperature is only slightly below the boiling point of the fluid.

The major objective of the theoretical analyses is the prediction of the burning rate of the droplet together with elucidation of the gross structure, including flame size, temperature distribution, and other phenomology of the flames surrounding the droplet.

The mass burning rate $\dot{m}_p$ of a droplet is related to the rate of decrease in droplet size by

$$\frac{dm_p}{dt} = \dot{m}_p = -\frac{d}{dt}\left(\frac{4}{3}\pi R_1^3 \rho_l\right), \qquad (6.1)$$

where $\dot{m}_p$ is the mass of the droplet, ρ_l is the density of the droplet at the appropriate temperature, and R_1 is the droplet radius. This equation can be rewritten in a form which indicates that the square of the droplet diameter (or radius) should be proportional to time, or

$$dR_1^2/dt = -\dot{m}_p/2\pi\rho_l R_1. \qquad (6.1')$$

This relation has been shown many times experimentally, as illustrated in Fig. 6.9, under burning conditions; the square of the droplet diameter is a function of the burning time, being related by a proportionality constant K which has been termed the burning rate coefficient or evaporation constant. It is a convenient parameter since it is readily obtained experimentally and is related to the mass burning rate by

$$K = 2\dot{m}_p/\pi\rho_l R_1. \qquad (6.2)$$

This proportionality between $\dot{m}_p$ and R_1 has been known for a century, for large low-temperature evaporating droplets, based on diffusion theory. Yet workers thought for a considerable time that the rate of burning of drops was proportional to the surface area until Probert (1946) noted the theoretical distinction. Equation (6.2) subsequently has been verified experimentally by many investigators.

The species conservation equations are in an exact form but because of the boundary conditions imposed and the exponential form of the temperature dependence of the chemical reaction rate τ, it is not possible to obtain an analytical solution to give the burning rate.

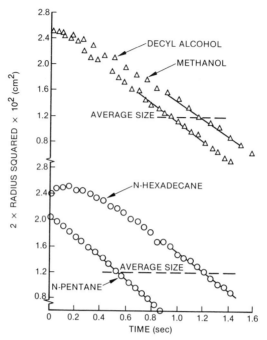

Fig. 6.9. Typical plot of the droplet diameter squared against time for stationary droplets burning in flame gases. [After Faeth and Lazar (1971); copyright American Institute of Aeronautics and Astronautics.]

A simplified analysis can be obtained based on the restriction that the chemical reaction is

$$\text{Fuel} + \text{Oxygen} \xrightarrow{\kappa} \text{Products (CO}_2\text{, H}_2\text{O, etc.)}$$

and that this reaction is "infinitely rapid"; that is, the chemical kinetics are not rate determining. Analyses based on this assumption have given the classical equations of droplet burning but subsequent workers have questioned the validity of this assumption. Brzustowski (1965) has indicated the general regimes (in the case of combustion of *n*-eicosane) in which the simplifying regime are applicable; these are indicated in Fig. 6.10.

With the assumption that the reaction rate is fast ($\kappa \to \infty$), the flame zone is assumed to be of infinitesimal thickness and can be represented by a surface rather than an extended reaction zone. It is also necessary to assume that the fuel and oxidant diffuse into the flame zone in the proportions required for stoichiometric combustion. As a consequence of the infinite reaction-rate assumption, the weight fractions of both of these species become equal to zero at the reaction surface.

As a result of these approximations, the equations are solvable analyti-

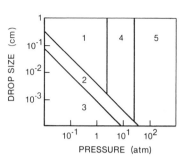

Fig. 6.10. Regimes of droplet combustion (*n*-eicosane/air): (1) diffusion limiting, (2) chemical kinetics limiting, (3) chemical reaction rate and evaporation slower than diffusion, (4) non-steady state effects important, (5) supercritical effects dominate. [After Brzustowski (1965).]

cally. The form of the expression for mass burning rate that is obtained depends upon the particular method chosen and the further approximations made, for example, whether thermal conductivities and specific heats are assumed to be temperature dependent or whether an average value is taken. However, the following equations given by Wise and Agoston (1958) are typical of the form of the relations. The mass burning rate is given by

$$\dot{m}_p = 4\pi R_1 \bar{\rho}_g \overline{D} \ln(1 + \mathcal{B}), \tag{6.3}$$

where $\bar{\rho}_g$ is the average density of the gas and $\overline{D}$ is the average diffusivity in the flame. $\mathcal{B}$ has been termed the transfer number and is given by

$$\mathcal{B} = (1/\lambda)[\bar{c}_p(T_\infty - T_1) + \Delta H y_{0,\infty}/\gamma], \tag{6.4}$$

where $\bar{c}_p$ is the average specific heat at constant pressure for the gas mixture, T_∞ is the temperature of the ambient atmosphere, T_1 is the temperature at the surface of the droplet, ΔH is the heat of reaction, $y_{0,\infty}$ is the mass fraction of oxidant in the surrounding atmosphere, λ is the heat of vaporization per unit mass of fuel evaporating, and γ is the stoichiometric mixture ratio of oxidant and fuel vapor (F), y_0/y_F.

The quantity $\bar{\rho}_g \overline{D}$ can be replaced by $\bar{k}_g/\bar{c}_p$ if the Lewis number is assumed to be unity. Here $\bar{k}_g$ is the average thermal conductivity. By substitution and using Eqs. (6.1) and (6.2), we get for the burning rate coefficient

$$K = (8\bar{k}_g/\bar{c}_p \bar{\rho}_g) \ln(1 + \mathcal{B}). \tag{6.5}$$

This is essentially the form of the equations first derived by Godsave (1952) and by Spalding (1952). An analysis by Goldsmith and Penner (1954) produced a somewhat more complicated expression allowing for the temperature dependence of $\bar{k}_g$ and $\bar{c}_p$.

Expressions have also been derived for the flame temperature T_c, the mass fraction of the fuel vapor at the droplet surface y_1, and the ratio of the flame radius R_c/R_L to droplet radius (Wise and Agoston, 1958):

$$T_c - T_1 = \frac{\Delta H - \lambda}{c_p} \left[\frac{1}{1 + \gamma/y_{0,\infty}} \right] + \left[\frac{T_\infty - T_L}{1 + y_{0,\infty}/\gamma} \right], \tag{6.6a}$$

$$y_1 = 1 - \frac{1 + y_{0,\infty}/\gamma}{1 + \mathcal{B}}, \tag{6.6b}$$

$$\frac{R_c}{R_1} = \frac{\ln(1 + \mathcal{B})}{\ln(1 + y_{0,\infty}/\gamma)}. \tag{6.6c}$$

These results indicate that (a) the mass burning rate is proportional to the droplet radius, that is, the square of the droplet radius is a linear function of time; (b) the ratio R_c/R_1 does not depend upon the droplet radius; (c) the mass burning rate is essentially unaffected by changes in pressure since $\dot{m}_p$ is proportional to $\bar{p}\bar{D}$, which is pressure insensitive; (d) the flame temperature is identical to the adiabatic flame temperature; (e) the mass fraction of the fuel in the vapor state at the liquid surface is less than unity, and thus the temperature at the liquid surface T_1 is less than the boiling point of the liquid.

By use of the conservation equations and Eqs. (6.6), it is possible to deduce the temperature and composition profiles in the flame surrounding a burning droplet. This has been achieved by Goldsmith and Penner (1954) and Wise et al. (1954). Typical profiles of a droplet burning in air are shown in Fig. 6.11.

The theory for droplet combustion with weak convection (Fig. 6.8) involves more geometric complexity than the case of spherical symmetry. For weak gas flow conditions, a model has been developed based on Stokes or Oseen flow. This calculation provides a perturbed-sphere model, and is described by Brzustowski and Natarajan (1966). The (small) convection for the

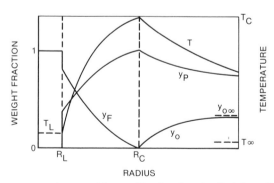

Fig. 6.11. Temperature and composition profiles for a burning fuel drop: fuel composition F, oxidizer composition o, and product compositions P. [From Williams (1968).]

ratio R_c/R_l [Eq. (6.6c)] is written for example in terms of a function of the Lewis, Grashof, and Schmidt numbers for the system.

Solution of the Conservation Equations for "Finite" Kinetics. Several investigations have considered the solution of the conservation equations without resort to the approximation of an infinite rate of the chemical reactions. This has established the accuracy of the infinite reaction rate approximation and has shown its limitations.

Since an exact analytical solution of the conservation equations is not possible if realistic chemical kinetic expressions are used, then numerical integration of the differential equations or other approximations are necessary to obtain the values of $\dot{m}_p$. This approach was taken by Lorell *et al.* (1958), who considered the combustion of a drop of ethanol of a fixed radius of 0.02 cm burning in air and in oxygen. They assumed the reaction rate to be given by a bimolecular Arrhenius expression of the form

$$\text{Rate} = (\text{const})[\text{Fuel}][\text{Oxygen}] \exp(-\Delta\hat{G}/\mathfrak{R}T),$$

and considered the effect of activation energies $\Delta\hat{G}$ of 0, 25, and 50 kcal/mol. As the activation energy is reduced, the increase in the reaction rate causes the thickness of the reaction zone to decrease until the system attains the flame structure observed in the limit for infinite reaction rate. In the cases studied the burning rates obtained assuming finite chemical kinetics did not differ much from that obtained by assuming infinite reaction rates.

An alternative approach has been that of Williams (1961). He used the Shvab–Zeldovich procedure to obtain a burning rate expression which is not independent of the chemical reaction rate. This expression is similar to Eq. (6.3); $\dot{m}_p$ is given by

$$\dot{m}_p = [4\pi R_l(k_g/\bar{c}_p) \ln (1 + \mathcal{B})] - 4\pi R_l \Delta H y_{0,1}/\bar{c}_p \lambda, \qquad (6.7)$$

but it contains an extra term allowing for the mass fraction of oxidizer $y_{0,1}$ at the liquid surface. This of course reduces to the form obtained assuming infinite kinetics if $y_{0,1} = 0$. Tarifa *et al.* (1962) investigated a more general case of droplet combustion by means of an approximate analytical integration method and considered effects of chemical kinetics and diffusion of species of different molecular weight. For an infinite reaction rate an asymptotic solution of the problem is obtained that is only strictly applicable when either the droplet radius or the pressure tends to an infinite value. In general, results tend rapidly towards their asymptotic value. However, for small droplets or very low pressures errors may be introduced by assuming that the reaction rate is infinitely fast.

Liquid fuels of commercial interest are usually mixtures of hydrocarbons with different boiling points which during droplet combustion undergo changes in the composition of the liquid phase, due to fractional distillation and also as a consequence of decomposition. Wood *et al.* (1960) have derived an expression for the burning rate coefficient for a binary mixture;

$$
K_{mix} = \frac{8k_g}{\bar{c}_p(y_A p_A + y_B p_B)} \ln \left[1 + \frac{1}{y_A \lambda_A + y_B \lambda_B} \right.
$$

$$
\left. \times \ \bar{c}_p(T_\infty - T_1) + y_{0,\infty} \left(\frac{\Delta H_A y_A + \Delta H_B y_B}{\gamma} \right) \right], \tag{6.8}
$$

where the subscripts refer to the two fuel components A and B. Equation (6.8) holds approximately for most mixtures because of the similarity of burning rate coefficients. If the boiling points of the components differ greatly then disruptive boiling occurs in which bubbles of the volatile component expand and burst in the droplet. The "d_p^2 law" holds except when the bubbles burst since atomization of the surface film occurs, enhancing, at that instant, the rate of combustion.

Experimental work carried out by a number of workers indicates that although the d_p^2 law is fairly well obeyed throughout the combustion of a droplet, the heat and mass transfer in the field around (and in) the burning droplets are in a transient state during the major part of the droplet lifetime. This is manifested by the change in R_c/R_1 during the droplet burning.

Kumagai and Isoda (1956) first tried to solve the time-dependent heat equation; additional analysis includes those of Kotake and Okazaki (1969). Some studies have also been performed exclusively investigating the temperature variation within the droplet during the course of combustion. Thus Parks *et al.* (1966) have investigated the temperature distribution within the droplet, while Shyu *et al.* (1972) have looked at the liquid-phase cracking in a heavy fuel oil during combustion. Essentially in these analyses the conservation equations for mass, and energy are solved while retaining the time-dependent terms.

Various methods of solution have been used but the most complete to date by Kotake and Okazaki (1969) involves numerical integration of the equations. In this analysis a non-steady state model was used but was still based on the major assumptions of the steady state analyses previously described, including spherical symmetry. The equations of concentration, velocity, and temperatures of the gas surrounding the drop and within the droplet were written as differential equations with respect to time and space and integrated.

These results showed that d_p^2 versus time gives approximately a linear relationship throughout the course of combustion, but the values of the flame temperatures and the ratio R_c/R_1 are much smaller than those of the quasi-steady state model and only approach their values towards the end of combustion.

Raghunandan and Makunda (1977) reviewed earlier investigations of non-steady state combustion. They have extended other work to include the effect of nonsteady state conditions in the condensed phase associated with droplet heat transfer. Their work indicated that this effect only lasts for 20–25% of the total burning time. Consequently, it cannot account for experimental anamolies in burning rate and flame temperature supported by other studies; nor can the non-steady state nature of the gas phase or the quasi-steady state theory with constant gas properties explain such differences.

Raghunandan and Mukunda (1977) then investigated the significance of variable fluid properties on the predictions of flame behavior from the quasi-steady state theory. Their model calculates values of the burning constant, flame position, and temperature somewhat better than the constant property case.

The Lewis number is an important basic parameter in spray combustion. By the definition of Le $(= k_g/\rho_g D_{AB} c_p)$, the variation of Le_A implies changes in the diffusion coefficient of species A in the (binary) mixture of gases. The results of Mukunda et al. (1971) show that the Lewis numbers based on binary diffusion coefficients are different from unity in the major part of the field in a propane–air diffusion flame. Typically, Le_A is in the range 1.0–3.0 for species A and Le_B around 1.1.

The effect of increasing Le_B is to increase K and T_c and to reduce R_c/R_l. All three parameters have a greater sensitivity to Le_B than to Le_A. Although the increase of Le_B produces favorable effects on K and R_c/R_l, it increases the value of T_c, which results in a greater deviation from experimental observations. Since the realistic values of Le_A and Le_B are approximately 1.5 and 1.1, it is thought that T_c must be lowered by including kinetic effects if all three characteristics are to be realistic.

Some of the seemingly realistic values of R_c/R_l have been obtained by the choice of low stoichiometric ratio by earlier investigators (Kassoy and Williams, 1968). A few experimental observations suggest that complete oxidation of the fuel may not take place at the flame zone.

The soot formation during combustion of droplets and the presence of significant amounts of carbon monoxide even beyond the diffusion flame zone are definite indications of oxygen to fuel ratio O/F being lower than stoichiometric at the flame zone.

The introduction of O/F lower than the stoichiometric value is contradictory to the theory using a single-step reaction with thin-flame kinetics. It is sometimes used essentially to approximate the kinetic effects. The use of lower O/F calls for changes in thermal properties of the system.

Table 6.3 summarizes the reaction schemes and results of calculations made using O/F slightly lower than stoichiometric. Raghunandan and Mukunda (1977) note that T_c approaches the observed flame temperatures with simultaneous improvement in the prediction of R_c/R_l.

6.4.2 Experimental Studies of Droplet Combustion.

Three techniques have been used to investigate the rate of combustion of single droplets: (a)

TABLE 6.3

*Effect of Oxygen to Fuel Ratio at the Flame on Combustion Parameters for Fuel Species A,
$Le_A = 1.5$, and Species B, $Le_B = 1.1$*

O/F	3.08	2.67	2.46
Reaction products per mole of C_6H_6	$6 CO_2 + 3 H_2O$	$4 CO_2 + 2 CO + 3 H_2O$	$3 CO_2 + 3 CO + 3 H_2O$
ΔH (cal/g)	9,700	7,980	7,110
K (mm²/sec)	0.7061	0.6873	0.6685
R_c/R_L	13.082	11.5685	10.636
T_c (°K)	2,622	2,460	2,373

[a] From Raghunandan and Mukunda (1977).

the captive (or suspended) drop method, (b) the supporting sphere technique, and (c) the free drop technique.

The *captive drop technique* may be used to obtain the rate of change of
droplet diameter or size as a function of time; examples of this are shown in
Fig. 6.9. Typically a single droplet (or array of droplets) is suspended from a
silica fiber, ignited, and the rate of combustion observed photographically or
by means of a photoelectric shadowgraph technique (Monaghan *et al.*, 1968;
Wood *et al.*, 1960). Experiments such as these have been used to look for the
d_p^2 versus time relationship. After an initial period in which the plots are
nonlinear and are generally attributed to the establishment of steady state
conditions, the plots are linear, giving fairly accurate values for the burning
rate coefficient. In this manner a considerable body of information has now
become available on the burning rate coefficients for a fairly wide range of
fuels and various ambient conditions. A number of complications are present
during most studies. First, droplets are not spherical due to the presence of
the supporting fiber, and an "equivalent diameter" must be used. Second,
unless the experiments are conducted under zero-gravity conditions the
flame shape is distorted and the mass burning rate is enhanced by the influence of natural convection. Third, the supporting fiber influences the results
to a small extent, this becoming of significance with small-diameter droplets,
thus limiting the range of droplet sizes that may be studied in this way.
Measurements of flame radius or ratio of R_c to droplet radius R_l may also be
obtained by these techniques. These reveal the fact that R_c/R_l varies during
combustion even under zero-gravity conditions and is also a function of the
initial droplet diameter. This implies that combustion actually is occurring
under non-steady state conditions; nevertheless the experimental values of
the burning rate coefficients under these conditions are sufficiently accurate
for practical purposes.

The *supporting sphere technique* provides a method of studying the
steady state combustion of simulated droplet burning. Here the diameter is

kept constant during combustion by supplying fuel to the surface of a supporting inert sphere at a rate equal to the rate of its combustion. The technique was first used by Khudyakov (1949). More recently the technique has involved the use of a ceramic porous sphere into which the fuel is fed by means of a small-diameter stainless-steel tube. The method is convenient in that different-diameter support spheres may be used for a variety of experiments involving the steady state combustion of droplets. In this way it has been used for measurements of flame shape, flame structure, various aerodynamic measurements, and for measurements of mass burning rates. In the latter experiments care must be taken in heat shielding or cooling the fuel supply lines so that no heat transfer to the unburned fuel can occur.

The *free drop technique* involves the investigation of moving droplets. In the free droplet technique, a single droplet or low-density cloud of droplets is produced by a suitable generator, such as an electrostatic or ultrasonic atomizer, a vibrating steel tube, a spinning disk atomizer, or a simple orifice. The droplet of controlled size so produced is allowed to fall under gravitational forces or is projected into a suitable hot environment generated by means of a flame or furnace. In the latter case self-ignition occurs or is caused by a pilot flame. A typical arrangement is reported by Hieftje and Malmstadt (1967). In some instances a geometrically simple self-sustained flame may be used, usually involving an annular flame or a stream of droplets. Apparently no analog of the stationary, flat one-dimensional flames used for premixed flame studies has been successfully developed for spray studies.

For the investigation of the effects of very high relative gas velocity a shock tube may be used, the droplets usually being first dispersed inside the shock tube by some atomizing device or single droplets injected or suspended inside the shock tube (Jaarsma and Derkson, 1970).

Whichever technique is employed to generate the droplets or gas flow in which they are studied, the main diagnostic instrument for the investigation of moving droplets is photographic, usually direct, but also may involve Schlieren shadowgraph studies of some kind.

Few measurements have been made of mass burning rates of droplets moving at velocities relative to a moving gas stream. For flow condition, the Frössling equation [Eq (2.26)] can be used to correct results for enhancement of mass transfer over the stagnant gas use.

A number of investigations have been concerned with the shock-wave ignition, combustion of liquid fuel droplets, and also the influence of a shock wave on burning droplets (Jaarsma and Derkson, 1970). Under certain conditions the burning rate of liquid droplets may be increased by many orders of magnitude by interaction with a shock wave. In the case of burning droplets the shock wave shatters the droplet into microdroplets, which, if the flame remains attached, results in a substantial increase in burning rate.

However, with low oxidizer concentrations the flame could in fact be blown off by the shock wave (Williams, 1973).

A considerable amount of data has been reported for single droplets burning suspended on quartz fibers; these have been summarized by Williams (1968). They are briefly mentioned here and are compared with the theoretical predictions.

In general, almost all experimental data must be considered as approximate values only, because they have been interpreted in terms of steady state combustion and without regard to the initial droplet size, which may affect the interpretation (Monaghan et al., 1968).

Despite the fact that the idealized, spherically symmetrical flame only applies to convection-free conditions which can only be achieved under zero-gravity conditions, little work has been carried out in this limiting situation. Kumagai and co-workers (1970) have found that the values of burning rate coefficient obtained are much less than those under natural convection conditions. They found that a free droplet (not suspended) has an intermediate value and that in both cases the initial droplet diameter has a small effect on the results. Typical values are (for a 1-mm-initial-diameter droplet) 7.5×10^{-3} cm^2/sec for a suspended droplet and 7.9×10^{-3} cm^2/sec for a free droplet of n-heptane. Presumably the difference can be attributed to heat conduction effects.

Most reported values of burning rate coefficients refer to suspended droplets burning under laboratory conditions of temperature and pressure and under the effects of natural convection. Few measurements have been reported for droplets burning with oxygen or at reduced pressures. Published data on burning rate coefficients for the n-alkanes in air and oxygen are summarized in Fig. 6.12. It appears that the lower alkanes have higher burning rate coefficients, the values tending to a constant value at high molecular weights.

Published results for other alkanes, alkenes, aromatics, alcohols, and a number of other compounds in air are given by Williams (1973). They range between 5×10^{-3} and 14×10^{-3} cm^2/sec. The alkane isomers, alkenes, and distillate fuels are similar to the n-alkanes; the aromatics and alcohols have somewhat lower values. A few measurements have been made with varying oxygen concentrations, the lower limit at room temperature is, however, restricted to about 18% O_2. Generally the burning rate constants increase approximately linearly with oxygen content, the values for air being increased by a factor of 2 for 100% oxygen.

With residual fuels, solid deposits are formed after the combustion of the volatile constituents; consequently plots of d_p^2 against time change abruptly when the distillate components are consumed. Masdin and Thring (1962) found that the cenospheres produced burn at between 0.3 and 0.1 of the rate of the volatile components.

A number of investigations have been made of mass burning rates using

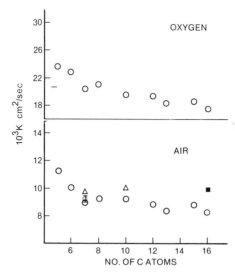

Fig. 6.12. Burning rate coefficient of *n*-alkanes burning in air and oxygen at room temperature as a function of molecular weight. Data based on different investigators as reported by Williams (1973).

the porous sphere technique. These can be converted to equivalent burning rate coefficients by means of Eq. (6.2). Values obtained in this way are not strictly equivalent to burning rate coefficients obtained from suspended-drop techniques because the heat conducted to the surface must also preheat the fuel to the surface temperature in addition to the heat of vaporization. This preheating requirement is not present during the steady state burning of a suspended droplet since this was accomplished during the droplet heatup period. In addition the differences in liquid density and heat transfer to the fuel feed pipe may be significant. Nevertheless, according to Aldred *et al.* (1971) values obtained with the porous sphere method are consistent with those obtained by the suspended-droplet technique.

Data have also been reported on the burning rate coefficients of various rocket propellant combinations; these range from 6×10^{-3} to 35×10^{-3} cm²/sec; temperatures increase by increasing T_c, k_g/c_p, and radiant heat transfer, but the increase is small because of the logarithmic relation in the burning rate expression [Eq. (6.3)].

Very few investigations have been made of the burning rates of droplets in air at elevated temperatures, principally by Kobayasi (1954) and Masdin and Thring (1962) for hydrocarbons and Tarifa *et al.* (1962) for propellants. The results for hydrocarbons all refer to droplets burning under the effects of natural convection. These indicate roughly an increasing burning rate coefficient with temperature. The higher-molecular-weight fuels have a higher temperature dependence presumably due to enhanced radiant heat transfer to liquid-phase cracking which can lead to carbon formation. Because of the extensive cracking of high-molecular-weight compounds, particularly with commercial fuels, the d_p^2 law is not obeyed at the higher temperatures because of cenosphere formation.

Burning rate coefficients also depend on pressure. The simple quasi-steady state theories predict little variation of burning rates with pressure within the regimes in which the d_p^2 law is obeyed. In actual fact, the burning rate coefficients vary markedly with pressure owing to natural convection. The variability is expressed as an exponential form proportional to pressure raised to the 0.4–1.0 power (Williams, 1973). The observed pressure dependencies vary somewhat with fuel and droplet size (and presumably also during the lifetime of a burning droplet).

A theoretical dependence of the burning rate of single droplets with pressure is due mainly to the pressure dependence of the latent heat of vaporization. This effect is small at low pressures, but at higher pressures, about one-tenth of the critical point of the fuel, this effect becomes more significant. As the liquid approaches its critical point the theory becomes complicated, and has been the subject of a number of investigations of near-critical-point evaporation and combustion (Savery *et al.,* 1971; Spalding, 1959a).

These effects may occur for combustion in high speed diesel engines, in liquid propellant rocket motors, and in gas turbines.

As soon as the temperature everywhere exceeds the critical temperature at supercritical pressures the droplet becomes a puff of gas, this then burning as a diffusion flame. Spalding (1959a) considered this type of combustion theoretically by approximating the droplet vapor as an instantaneous point source of fuel and then considering the time taken for the vapor ball to burn. This type of analysis has been further developed by Rosner (1967) who considered the puff of gas both as a point and a distributed source. The general effects of burning time indicate little effect until the mixture approaches the critical pressure; then a sharp decrease in burning time occurs. In the supercritical regime, the burning time increases.

Experimentally, supercritical combustion has been observed and no unusual change in the combustion zone was detected as the droplet passed through its critical temperature (Canada and Faeth, 1974). Droplet temperature measurements showed that in the critical regime the liquid temperature rises continuously from ignition to burnout; this is in contrast to the behavior in normal combustion when the liquid temperature reaches and remains at (or near) the boiling point for an appreciable period of the droplet lifetime. It was also shown that dissolved gas and other real gas effects must be taken into account and that for combustion in air it can be assumed that the gas at the droplet surface is a binary fuel/nitrogen mixture.

6.4.3   Ignition of Droplets.

The ignition, as well as the extinction of droplets, is a transient phenomenon involving the rapid transition from evaporation and chemically controlled processes to diffusional controlled near steady state combustion. Two modes of droplet ignition may be identified, as indicated in Fig. 6.13. These are "thermal" ignition and propagating or network ignition; the former involves the processes that occur when a cold droplet is exposed to a hot oxidizing environment and the latter occurs when a droplet is exposed to a combined influence of heat and radicals from an adjacent burning droplet. Most of the attention has been focused on the thermal ignition case. Any complete theory of droplet ignition must be able to accurately predict the duration of the time between the introduction of the droplet in the hot oxidizing atmosphere and the onset of flame, that is, the ignition delay time (or induction period). In addition, theory must be able to accommodate the fact that only droplets greater than a certain critical size are capable of igniting, those smaller only evaporate. The ignition of a droplet may be considered as consisting of essentially two stages, a preheat stage in which the droplet is heated without significant evaporation and a second stage in which ignition commences in the gas phase between the inflammable vapor and air. The first preheat stage is insensitive to composition and temperature and is essentially a transient heat transfer situation readily amenable to mathematical analyses, at least for pure liquids or simple mixtures. The second stage, involving gas-phase ignition of a hydrocarbon with oxy-

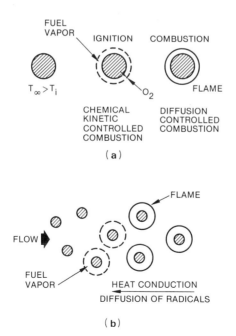

Fig. 6.13. Modes of droplet ignition: (a) spontaneous ignition and (b) network mechanism of ignition. [From Williams (1973).]

gen in a concentration field, is very dependent on molecular composition of the fuel and on temperature, and is not capable of detailed analysis at the present time and can only be treated in terms of macroscopic chemical kinetics. Essentially ignition involves a transition from a kinetic controlled situation to a diffusional control in the burning droplet; the reverse situation has been identified with droplet extinction.

El-Wakil and Abdou (1966) have carried out an extensive analysis of the factors controlling the magnitude of the ignition delay period. The analysis was semiempirical but led to expressions for the estimation of the physical and chemical delay times as well as for the total ignition delay time.

More detailed mathematical models have been developed to include the two limiting cases, ignition and extinction. Fendell (1965) numerically computed the ignition and extinction limits for an ethanol/air mixture. Peskin and Wise (1966) developed a linearized flame surface model which permits the inclusion of finite kinetics. In this way the combustion process can be studied as it changes with kinetic rate, and the correct mass burning rates for both the evaporative limit (zero reaction rate) and the infinite reaction rate limit (the classical droplet burning case) can be predicted. It was shown that transition from kinetic to diffusional control is rapid and is identified with an ignition temperature T_i, which is a function of activation energy for the reaction, rate coefficient, and mass fraction of fuel. In further analyses, Peskin et al. (1967) have developed the theory and obtained ignition and extinction conditions for various reaction rates, reaction orders, and mixture strengths. A typical plot of droplet radius against ambient temperature is given in Fig. 6.14. Data on ignition temperatures as a function of oxygen concentrations, as tests of the predictive methods, are given by Polymeropoulos and Peskin (1969) and LeMott et al. (1971).

An alternative approach to ignition and extinction problems has been made by Rosser and Rajapakse (1969) in which they considered the thermal stability of a reacting spherical shell. In this they showed that the overall temperature dependence of the critical droplet is approximately given by 0.5 $\Delta \hat{G} - \mathcal{R} T_\infty$.

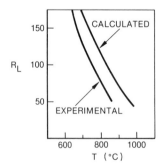

Fig. 6.14. Comparison of theoretical and experimental critical droplet radius (in μm) for ignition of No. 2 fuel oil. Theoretical curve from Peskin and Wise (1966); experimental curve from Wood et al. (1960).

Most ignition measurements have been concerned with ignition delay times based on the onset of the emission of light from a flame, although some studies have been directed at the time–temperature history of the droplet.

A number of investigators, primarily Kobayasi (1954), Nishiwaki (1954), Masdin and Thring (1962), and El-Wakil and Abdou (1966) have made measurements of the ignition delay t_i of suspended droplets of a number of fuels, both pure and commercial, in air or similar atmosphere under a variety of conditions. A number of investigations have also been made of fuels igniting in an oxygen atmosphere, of special propellant combinations, and of the influence of additives (Williams, 1973).

Masdin and Thring (1962) presented their data in the form

$$t_i \sim \exp(B/T),$$

and obtained the following values for B: kerosene 8.3 × 10³, diesel oil 8.8 × 10³, heavy fuel oil 7.9 × 10³, pitch creosote 11.3 × 10³, and benzene 22.8 × 10³. El-Wakil and Abdou (1966) correlated their results for a number of hydrocarbons by the expression

$$t_i = 10.3 d_{p0}^{8.5} \exp[a(T) + C_n],$$

where $a(T)$ is a function of the temperature, C_n is the number of carbon atoms in the hydrocarbon, and d_{p0} is the initial droplet diameter.

Other investigations have been concerned with ignition of fuel droplets in shock-heated gases (Kaufman and Nicholls, 1971). In this case, it has been shown that ignition can be adequately represented by an Arrhenius-law form and is governed by the formation of a combustible mixture in the wake of the shattered droplet. In the case of moving droplets recent studies show that ignition occurs in the wake of the droplet.

6.4.4 Combustion of Collections of Droplets. The convention that each fuel droplet is surrounded by its own envelope flame does not necessarily apply to sprays of relatively high number density. In the extreme, one can envisage two extremes in spray combustion behavior. The first is shown in Fig. 6.15a, representing an external sheath flame, burning outside the spray cloud; the droplets pool their vapor and burn as a group as if the cloud were a single droplet. The other extreme is the model of internal combustion when each droplet is surrounded by a flame (Fig. 6.15b). The external burning situation conceives a single diffusion flame surrounding the cloud with a flame separated some distance away from the cloud interface. This condition differs significantly from the conventional single-droplet combustion. It produces a rich nonflammable vapor mixture impeding thermal penetration in the gas phase, inhibiting ignition in the vicinity of individual droplets. Recent experiments probing burning sprays suggest that individual droplet flames are rarely encountered in the conventional spray combustion.

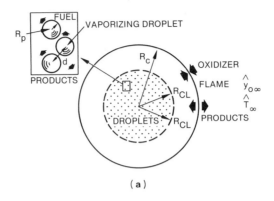

(a)

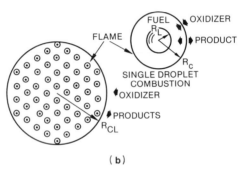

(b)

Fig. 6.15. (a) Schematic representation of external group combustion model. (b) Schematic representation of internal group combustion model. [From Chiu and Liu (1977).]

Quantitative criteria for the group-droplet burning conditions have been developed by Suzuki and Chiu (1971), and Labowsky and Rosner (1976), with recent extension by Chiu and Liu (1977). Basically, the formulation criteria require consideration of simultaneous chemical reaction and diffusion in an equivalent porous media. Labowsky and Rosner (1976) have made such an analogy, deriving a parameter for group combustion, called a *Thiele modulus,* proportional to the square root of the fuel volume fraction, and the ratio of the spray cloud radius R_{cl} and the spray droplet radius R_1.

Alternatively, Chiu and Liu (1977) have developed a droplet flame model in terms of a group combustion member G defined as

$$G = 3(1 + 0.276 \, Re^{1/2} \, Sc^{1/3})Le_A N \, R_1/R_{cl}, \qquad (6.9)$$

where the Re is droplet Reynolds number and Sc is the vapor Schmidt number. Thus the classification of burning is basically proportional to the spray cloud number density N, the Lewis number for the fuel vapor Le_A, and the average droplet to cloud radius ratio. This leads to two important observations. First, a spray classification into "dense" and "dilute" appears insufficient to provide a qualitative prediction of spray burning characteristics. The second observation concerns the transition from external combustion to internal group combustion as a result of significant droplet size reductions and cloud burnout, which reduce the group combustion number below the critical value.

In the first case, the Chiu and Liu model suggests that classification on the basis of group combustion number G provides a useful quantification of the combustible spray. For example, "high-G spray" may be defined as a spray with a group combustion number smaller than unity. High-G spray, accordingly, exhibits combustion characteristics external to the spray cloud, whereas a low-G spray burns with flame or flames located inside the two-phase zone containing droplets, fuel vapor, and oxidizer.

In order to provide practical spray characteristics, the range of group combustion number in industrial and jet engine combustors is estimated by taking the Lewis number and Nusselt number to be unity. The results are shown in Table 6.4. It is interesting to note that for a typical spray with $N = 10^3$ droplets/cm^3, cloud diameters of 10 cm, and $R_1 = 10$ μm, the group combustion number is 300. The spray is classified as high-G spray and the flame is a single envelope which surrounds the spray; i.e., external group combustion mode. Additionally, Chiu and Liu note that the group combustion numbers of dilute spray, say, $N = 10$–10^2 droplets/cm^3 with a droplet diameter of 10–100 μm are still larger than unity, and thus sprays fall in the category of intermediate- to high-G sprays.

The second observation concerns the transition from external combustion

TABLE 6.4

Group Combustion Number and Spray Characteristics[a]

		Group combustion number G[b]	
N (drops/cm)^3	R_{Cl} (cm)	$R_1 = 10$ μm	$R_1 = \mu$m
1	1	0.03	0.3
1	10	0.3	3
10	10	3	3×10
10^2	10	3×10	3×10^2
10^3	10	3×10^2	3×10^3
10^4	10	3×10^3	3×10^4
10^5	10	3×10^4	3×10^5

[a] After Chiu and Liu (1977).
[b] $G < 1$ is a low-G spray; $G \approx 1$ is an intermediate-G spray; $G > 10$ is a high-G spray.

to internal group combustion as a result of significant droplet size reductions and cloud burnout, which reduce the group combustion number below the critical value.

In a high-G droplet spray involving little or no droplet motion, the transition occurs as the droplet layer burnout causes a reduction in the cloud size. A characteristic time of the transition is of the same order of magnitude as the lifetime of the cloud. For an intermediate-G spray, the transition could occur by a combination of droplet size reductions and layer burnout. In a droplet cloud combustion accompanied by significant radial expansion, transition could occur without substantial burnout or size reduction if the droplet lifetime were much larger than a characteristic relaxation time. For fuel droplets with low volatility, the transition could occur without substantial burnout. This is likely to take place in the high speed injection of low grade oil droplets into a burner. A high-G spray with exceedingly high group combustion number may consume a large fraction of the fuel through external group combustion processes. Estimates of the relative heat generation by external and internal burning modes in high-G or intermediate-G sprays are complicated. The methodology for such estimates requires a detailed knowledge of the flow fields that are essential to an accurate estimation of the local G value and also to subsequent determination of combustion mode.

6.4.5   Application of Droplet Theories to Combustion Systems   There has been considerable interest in applying droplet combustion theories to a wide variety of spray combustion systems in order to increase the understanding of such systems so as to minimize pollutant emissions, or to describe them in quantitative detail. The application of theory is surveyed briefly below. More details can be obtained in the studies of F. A. Williams (1965), A. Williams (1968), or in Beer and Chigier (1972).

A mixture of a combustible spray and a suitable oxidant can support a flame in the same way as a premixed flame. That is, it may be supported as a stationary flame on a burner port or it may burn freely as a propagating flame; in addition, it will exhibit flammability limits.[†] The nature of the burnt products depends upon the properties of the spray in its unburnt state. If the droplets are large, then combustion may not be complete in the main reaction zone and unburnt fuel may penetrate well into the burnt gas region. If the droplets are small, then a state of affairs exists that approximates very closely the combustion of a premixed gaseous flame. Here the droplets are vaporized in the preheat zone and reaction after that is between the two reactants in their gaseous state. The other factors that determine the time to reach complete combustion, that is, the length of the combustion zone, are the volatility of the liquid fuel, the ratio of fuel to oxidant in the unburnt mixture, and the uniformity of mixture distribution.

[†] For mixtures of fuels and oxygen or air, there are limits of composition within which flame propagation may occur, and outside which a flame cannot be self-sustaining.

Theories for laminar flames supported by sprays have been reported, for example, by Williams (1965). These analyses have been restricted to dilute sprays in which the droplet velocity equals the gas velocity. Essentially the theories are thermal models in which heat is conducted to the oncoming droplets from the reaction zone in which conventional droplet combustion is assumed. Ignition of the droplets downstream occurs when the droplet reaches an ignition temperature. Account also can be taken of ignition delay time linking the flame propagation velocity with the ignition delay time and thus should be sensitive to variations in molecular structure of the fuel.

There is little experimental evidence available to test the theories. Most studies have been concerned with tetralin–air mixtures and there is now extensive evidence for both flame propagation velocities and inflammability limits; little information is available for other fuels but that which is available suggests that most hydrocarbon fuels behave in a similar way (Williams, 1968).

A second type of idealized combustor amenable to analysis based on single-droplet combustion theory is the heterogeneous well-stirred reactor, although, in fact, few studies have been undertaken. Courtney (1960) has considered combustion of monopropellants which are assumed to burn according to the d_p^2 law. It was shown that for a monodisperse spray, the combustion intensity I_c is given by

$$I_c \simeq Q_0/(1 + R_0^2 Q_0/5K),$$

where Q_0 is the rate of flow of gas from the chamber, R_0 the initial droplet radius, and K the burning rate coefficient appropriate to the turbulent conditions. This analysis is also applicable to a reactor in which conventional droplet combustion with a gaseous oxidizer occurs.

In practical systems such as a furnace or a rocket engine, the combustion process is more complicated because of two important factors. First, to a large extent the mixing of the fuel and oxidant takes place in the combustion chamber and thus the mechanics of the mixing process play an important role. Second, the flow patterns are complicated by turbulence or recirculation. The droplets that make up the spray must be adequately described with regard to size and spatial distribution and droplet collision processes must be taken into account. Nevertheless, despite the complexity of spray combustion, much valuable information has been obtained by the use of simplified analyses in which an initially uniformly distributed spray (in space) is assumed to burn in a plug-flow reactor and in which the d_p^2 law holds.

An early theoretical model was developed by Probert (1946), who calculated the mean drop diameter from known size distributions and calculated the combustion rate by assuming that the initial spray consisted of a constant number of drops of this mean diameter. The analysis was directed to turbojet studies, and this type of analysis gave a calculated performance lower than experimental values.

Tanasawa and Tesima (1958) applied a more general procedure applicable to sprays with a variety of droplet size distribution functions. They concluded that the rate of combustion was only slightly affected by the size distribution of the drops but is strongly affected by the mean diameter.

Williams (1973) has cited work showing that the evaporation and burning characteristics of polydisperse sprays in a turbojet cannot be described by those of the mean droplet size. Thus detailed calculations as a function of droplet size are required for adequate descriptions of combustion efficiency in practice.

Spalding (1959b) has considered spray combustion in a liquid-fuel rocket engine, and by using various approximations has developed analytical expressions to obtain either combustion efficiency or the minimum chamber length required for complete combustion. Using such a general theoretical treatment, Spalding (1959b) investigated the effects of droplet size, droplet injection velocity, final gas velocity, and the properties of the fuels and products. It was possible to obtain an analytical expression for the length of the combustion chamber necessary to obtain complete vaporization of a monodisperse spray. It was concluded that the combustion chamber necessary for complete vaporization is proportional to the square of the initial droplet radius, that increasing the injection velocity will increase the length required for complete evaporation, and that fuel preheat is advantageous. Similar conclusions have been reached by Williams (1965), who found that to obtain a short chamber length with a given mass flow, the droplet radius must be small, the vaporization rate should be large, and the injection velocity should be small. It was also shown that monodisperse sprays gave a higher combustion efficiency than a spray of a nonuniform droplet size distribution with the same average droplet radius. Williams et al. (1959) considered spray combustion in a diverging reactor for two types of sprays, a mono-sized spray and one with a Rosin–Rammler distribution of droplet size. In both cases, expressions were obtained for the reaction rate, which could be related to experimentally measured pressure profiles in an appropriate reactor.

Detailed models of spray combustion have been developed which allow for the role played by injection and atomization, and which permit the use of complex droplet size distributions and realistic droplet ballistics. The solution of the equations derived for such models is only possible by means of numerical methods using a high speed digital computer. Priem and Heidmann (1960) considered a vaporization-controlled mechanism and calculated the vaporization rate and the droplet histories to show the effect of different propellants, droplet size distributions, and ballistics. The results were correlated with an equation for the generalized chamber length. The results showed that as the range of droplet size in the spray increases, the combustion efficiency for a given length of chamber is determined by the proportion of large droplets present.

A spray combustion model has been used by Burstein and Naphtali (1960) which includes the process of droplet breakup. It was concluded from computer studies that this breakup is probably the most important effect in high velocity spray combustion. As a result of droplet breakup the combustion zone tends to be narrow and relatively insensitive to mass droplet size, size distribution, velocity, or temperature of the liquid.

In some circumstances, the injection of the spray and the formation of a combustible mixture may dominate the combustion characteristics of the system. Thus in the combustion of dense fuel sprays, the assumption that a burning droplet is immersed in an infinite quantity of gaseous oxidizer may no longer be valid and in that case may be controlled by a different mechanism. In particular, such cases are found when pressure jet atomizers are employed, for example, in internal combustion engines and turbojets. In addition, the injection velocities of droplets may be of importance. For example, in a recent analysis of droplet combustion in a jet combustion chamber Heywood *et al.* (1971) showed that as a consequence of the deceleration processes caused by injection and by turbulence that the droplet must be completely vaporized before a diffusion flame can be established. This causes a very fuel-rich region to be formed near the injector face, where droplet evaporation is taking place, but where there is little combustion due to lack of air.

In the past decade, considerable attention has been directed to modeling the combustion processes in a diesel combustion chamber so as to predict the influence of operating variables on heat release rates and on the formation of soot and NO_x. In compression ignition engines, it has been established that combustion occurs in the periphery of the dense injected sprays, and this situation has been difficult to analyze. However, Khan and coworkers (1972) have had success in modeling spray combustion in direct injection chambers in which the injected spray is considered to behave as a gaseous jet for estimation of air entrainment and local stoichiometries.

As far as industrial combustion chambers and furnaces are concerned, no comparable analyses have been undertaken although in many instances an analogous situation to that in compression–ignition engines can occur. Thus, McCreath and Chigier (1973) have shown experimentally that peripheral combustion occurs with conventional industrial pressure jets and have shown that the droplets in the inner zone move more slowly than those in the outer zone.

Before more detailed models of such combustion systems can be established, additional information is required, particularly in respect to entrainment of air and recirculated gases by sprays and of droplet collision effects, although some work has been carried out on the instability of spray combustion in rocket motors and associated supporting single-droplet combustion analyses (Strahle, 1964). These studies are of importance in their own right, and are also significant to combustion instability and combustion noise problems in industrial situations and internal combustion engines. Little direct work has been carried out in this area although one study has been reported using an idealized system (Hedley *et al.*, 1970).

6.4.6   Combustion of Solid Particles—Pulverized Coal.   A number of dusts are capable of sustaining flames; the number exceeds more than a hundred. However, the one of principal technological importance remains coal dust. The last quarter of the twentieth century probably will see a major resurgence of coal as a staple energy source in many nations. Because of the large capital investment in coal-fired systems, the pressures of air pollution emission control, and interest in coal-based synthetic fuels, research on coal combustion has surged ahead in the last few years. As with fuel sprays, a modern description of the particle burning process is complicated by the interactions of diffusion and chemical kinetics.

The process of particle combustion depends on the physical and chemical nature of the solid as it heats and burns. Coal is a complex material of volatile and nonvolatile components which becomes increasingly porous during volatilization of low-boiling-point constituents in burning. The crucial practical questions for boiler design concern whether pulverized fuel combustion is controlled by oxidizer diffusion or chemical kinetics. Combustion of pulverized coal is discussed below, first by examining the nature of coal, then by considering the chemical processes.

Coal may be described as a black, heterogeneous rock that is friable to hard, of variable vegetable origin with extraneous mineral additions of equally variable composition and quantity. Elemental analysis by weight of the organic constituents of different coals shows anything from 65 to 95% carbon, 2–7% hydrogen, up to 25% oxygen, up to 10% sulfur, and typically 1–2% nitrogen. Mineral matter, which becomes ash on firing, can vary in composition or quantity, with some recorded instances up to 50% by weight even in commercial coals although 30% is usually considered high. The same is true of moisture, with 70% recorded for some soft coals. "Proximate analysis," a method of subjecting the coal to thermal decomposition (or pyrolysis) under standard conditions, yields up to 40 or 50% "volatile matter," with the residue being ash and "fixed carbon."

Coals are transparent to translucent in thin section, and petrographic analysis shows variable maceral composition[†] that is a prime factor in the coking and caking properties of coals. In coking, coals may exhibit both thermoplastic and thermosetting behavior. Porosimeter and adsorption measurements reveal macropore and micropore structures that, respectively, control supply of reactive gas into the interior of a coal particle, and provide variable internal surface reactions. Accessible surface areas of different coals range from 100 to 500 m^2/g. On heating, many coals swell and subsequently con-

---

[†] Fragments of metamorphosed plant debris in coal.

tract as they pyrolyze, with internal surface area continually altering in consequence in both value and accessibility, and "intrinsic" reactivity changing as graphitization proceeds. With this variability, it is accepted that no two samples, or particles from a given source are alike. Further variability is introduced by crushing or grinding because of irregular fracture generating a range of particle shapes, with additional size, shape, maceral distribution, and ash variations even in a narrow sieve cut.

The search for relations governing the combustion behavior of fuels, and from which the performance of practical systems can be predicted, is complex enough when the fuel is well or reasonably well characterized, as in the case of gases or oils, involving as this does the need to combine the fundamentals of reaction kinetics, combustion aerodynamics, and heat transfer. With coal dust, however, its combustion characteristics are highly variable. Consequently, we are not yet able to design with total reliability from one set of reactivity conditions to another in many cases for a single coal, let alone from one coal to another given the same reactivity conditions.

The problems in coal combustion remain as they have been for many decades: first, the principal mechanisms governing reactions in the case of a coal from a single source require elucidation; and second, we need to know how to interpolate or extrapolate with confidence from a group of well-studied coals to predict behavior of new, uninvestigated coals. Without this latter ability, engineers are faced with the need to run detailed experimental tests on coals for every specific use, which is an extremely laborious task. A central requirement for addressing the second problem and to a lesser extent the first is the need to develop some working concepts of coal constitution for engineering applications.

Appreciation of this need for an understanding of coal constitution goes back many decades, perhaps being most forcibly expressed by Wheeler (1913). It was under Wheeler's direction that Marie Stopes developed an understanding and terminology of the banded ingredients in coals that provides today's basis for maceral component descriptions. This and related early work provided Essenhigh (1976) the basis for his studies of dust explosions.

A knowledge of coal constitution is desirable for several reasons. First, it is a means of characterizing a coal. Second, it is a means of relating the properties of one coal to another. Third, it provides a basis for evaluating and predicting reactivity behavior. There are several standard texts and papers on coal constitution and properties (Frances, 1961). A useful summary of later work is given by Raj (1976), and Elliott (1981).

Many years of study have provided certain generalizations of interest for characterizing coal. The carbon in coal is not graphite. However, coal properties and constitution are closely related to the geological processes of coal formation. The first stage is thought to be formation of humic substances or coal precursors by diagenesis from vegetable matter. In the subsequent coalification, or metamorphism, there is evidently a two stage process of "decarboxylation" in the first stage, and "demethanation" in the second. The result is carbon enrichment in the first stage at almost constant hydrogen content followed by a period of almost linear drop in hydrogen with further rise in carbon. The end point of coalification may be graphite (100% C), or it may be something slightly less than 100% C at a "coalification pole" corresponding to roughly $(C_{200}H_{40}O)_n$ (Essenhigh and Howard, 1971). According to Essenhigh (1976), plotting the H/C ratio against the O/C ratio for all coals generates the so-called "coal band" whose center line is given approximately by the empirical equation

$$H/C = 80(O/C)/[1 + 90(O/C)]. \qquad (6.10)$$

The boundaries of the band containing most coals are given by a range of ± 5 on the constant in the numerator. If the specific volume of the coals is v_s, then the equation obtained by the substitution of v_s/C for H/C in Eq. (6.10) describes also the variation of specific volume, although with slightly different values for the proportionality constants. This suggests that, within the limits of the scatter of the coal band, there is more commonality between coals than might otherwise be expected from their diverse geological origins. In addition, carboniferous coals are derived from primitive gymnosperm, while those of the Cretaceous and Tertiary periods are derived from a modern type of gymnosperm and from angiosperms.

There are some interesting differences between coals at different points in the coal band. Oxygen in the lignites tends to be high in carboxylic acid groups but with the H frequently replaced by metals through ion exchange from ground water. These metal ions are believed to be responsible for catalytic effects that contribute significantly to the higher reactivity of these coals. As carbon percentage or coal rank increases, reactivity generally tends to fall. Reactivity is often thought of as a joint function of the degree of swelling, the macropore and micropore structure of the coals under similar reactivity conditions, and some "intrinsic" reactivity (Essenhigh, 1976).

Reactivity should be influenced by structure and, from that point of view, the findings of x-ray analyses are significant (Hirsch, 1955; Diamond, 1956). These show coals to have a considerable fraction of material organized as a condensed aromatic ring structure in layers or lamellae, containing 15–20 carbon atoms per layer (sometimes interpreted as 2–3 rings per layer) in coals less than 90% carbon, with the number of atoms or rings per layer rising rapidly in coals above 90% carbon to 10 or more rings and 40 or more C atoms per layer. Coincidentally, this 90% C point corresponds with some overlap to the point at which the decarboxylation process is terminating and the demethanation process is starting. It also approximately corresponds to the point of peak specific volume (minimum density), and to minimum internal surface. Evidence of this sort provided the basis for Horton's (1952) suggestion that coalification was a two-stage process with bearing, as discussed more fully elsewhere, on Clark and Wheeler's (1913) two-component hypothesis of coal constitution, which they developed to explain combustion phenomena.

If coalification is a chemical reaction, then the coal precursor or parent material, presumably the ordered lamellae, is progressively turned into coal. It is then consistent with observed pyrolysis behavior that the parent material can be identified as Clark and Wheeler's "more easily evolved" component. The residue would then be the ordered lamellae of carbon and hydrogen that require stronger heating to decompose and that would generate mainly free hydrogen on doing so. The coal precursor or parent material is designated as component I, and the lamellae (presumed to be formed from component I) as component II.

Table 6.5 summarizes Essenhigh's (1959, 1976) interpretation of the relationship between the process of coalification, the resulting constitution, and the behavior on pyrolysis. In spite of the simplifications of Table 6.5, some evidence for this constitution model was recently obtained by Nsakala (1976). In temperature-programed furnace experiments, the loss in volatile material with temperature was found to segregate into two parts, consistent with the assumption of two components undergoing independent decomposition at different rates, and yielding two different sets of volatile products and one char.

TABLE 6.5

Constitution, Pyrolysis and Formation of Coals According to the Two-Component Hypothesis[a]

	(A) Constitution and pyrolysis	
Material decomposing	Temperature range (°C)	Product
Material volatilized out	Up to decomposition (T)	Moisture; absorbed and occluded gases: hydrocarbons
Coal precursor (I)	(Ia) T–550 (Ib) 550–650	Gases, vapors, tars; smoke forming volatiles generally, containing the volatile carbon
	(IIa) 650–1,000	Mainly H_2 gas
	(IIb) above 1,000	Fixed carbon
	(B) Coalification (coal formation)	
Stage 1	Up to 92.5% C; coalification reaction–transformation of component I [parent material into component II (lamellae)]	
Transition	85.0–92.5% C; Stage 1 goes to completion	
Stage 2	85.0–100% C; graphitization–condensation and coalescing of lamellae (component II); end point, graphite; the "liquid" structure breaks up and the intermediate structures formed are less perfect	

[a] After Essenhigh (1959).

The behavior of coal on pyrolysis is a highly complex chemical process the details of which are not well established. Phenomenologically, the main phenomenon in pyrolysis is the loss of weight by heating, with a roughly exponential decay in weight with time to a stable or equilibrium value that is temperature dependent. The percentage of volatile matter (VM) is measured under standardized time and temperature conditions; the measurement is often done by heating a charge in a crucible. The coke button from this test contains about 1% hydrogen and nitrogen, with these residual values effectively independent of the coal origin. In determining the VM, corrections must also be made for mineral matter decomposition. This can sometimes be substantial according to Essenhigh (1976).

Under nonstandard conditions, the VM yield can be changed by small to significant amounts. The most dramatic differences between techniques have been obtained in the so-called rapid heating experiments (Kimbert and Gray, 1967). Essenhigh (1976) notes that there is no clear evidence as yet that the rapid heating is the cause of an observed increased VM yield in such experiments. The explanation for this difference is not yet apparent.

Workers have speculated about the products and yields of coal pyrolysis. Essenhigh (1976) has hypothesized that effective theories that will enable prediction of combustion behavior from the characteristics of products will be based on a knowledge of coal constitution. At present this is not the case, and evidently represents a major deficiency of current theories.

6.4.7   Combustion and Flame Propagation.

Investigation of flame propagation in coal dust clouds has had a varied history, with initial interest stemming from explosion and safety research led by the famous work of Faraday and Lyell (1845), with interest only later being transferred to industrial conditions. One example of this transfer was the persistent use of widely quoted flame speed data that were ultimately identified as a set of curves published by de Grey (1922) without attribution. They are now believed to be based on about a dozen flame speeds[†] obtained by Taffanel and Durr (1912) for a single coal using a conical combustor, whereas the de Grey curves are for a set of coals and with different ash content. These curves were important as they showed flame speed values in the region of 10 m/sec, with maximum speeds at or about the volatile stoichiometric ratio. As such, they dominated hypotheses on propagation mechanisms and furnace design, with the proposition that anthracites must be fired fuel rich for stable flames in industrial boilers. This design assumption has been found to be incorrect. The matter of the equivalence ratio for the peak flame speed has been an open question ever since. Considerably more data are now available for design applications, particularly if a wider range of dusts than coal with some furnace studies and gallery explosion tests are included (Essenhigh,

---

[†] The speed at which a flame propagates through a combustible gas mixture.

1976). Ironically, workers now argue that flame speed has little relevance to practical situations, including furnace design, because industrial flame systems contain recirculating flows that modify the flame speed concept. In a one-dimensional or plug flow or nonrecirculating system, the flame speed can be thought of as a basic property of the combustible mixture. In a recirculating system, the flame speed is a function of the system as well as of the aerosol mixture. A flame front may exist, but the flame speed then is dependent, not independent, of combustion conditions. This would also seem to be true of systems in which a stabilizing gas flame, a heated stabilizing ring, or similar device such as an electric ignitor is used. In the limit of the perfectly stirred reactor, no flame front exists so no flame speed can be defined. Boiler flames are closer to this limit than they are to the plug flow limit.

The value in predicting or measuring flame speed therefore appears to be only as a means of checking the fundamental propagation and combustion theories. Essenhigh (1976) identifies at least four different modes of flame propagation, determined by the system but with the flame speed as a property of the mixture in the system of concern. These are summarized in Table 6.6.

The oldest flame propagation mechanism is that proposed by Faraday and Lyell (1845), which has been called the volatiles or touch paper theory. According to this model, particles would heat up until they pyrolyze, and ignition would then start in the volatiles. This view reasonably may be used to describe the behavior of coal in a fixed bed, for undoubtedly the coal has pyrolyzed substantially before the char enters the main bed combustion (and gasification) region. If the differences between overfeed and underfeed conditions, where diffusion limits the former, the volatiles model may be only partially applicable. Some more recent propagation studies in model coal beds show that the principal mechanism of flame propagation is radiation (Stumbar et al., 1970).

The Faraday and Lyell mechanism has otherwise been identified in only one other situation. In a reanalysis of Csaba's (1962) results, Essenhigh (1976) found that the reciprocal of flame speed was proportional to the dust cloud concentration and that the relevant constants of ignition temperature, rate of heating, and so forth could be extracted. For a (measured) flame temperature of 1400°C, the heating rate was found to be about 600°C/sec and the ignition temperature was about 375°C, which is about the temperature of pyrolysis of the coal used in the measurements.

About 70 years after Faraday and Lyell, Wheeler (1913) wrote that there seemed to be two types of explosion flames, a slow type, termed *inflammation,* and a fast type, termed *explosive combustion.* Gas analyses from the flames seemed to indicate that there was preferential combustion of the volatiles (Faraday mechanism) in the slow flames, but that particles in the fast flames were probably burning heterogeneously without time for pyroly-

TABLE 6.6

Approximate Flame Speeds and Burning Times for Different Ignition and Flame Propagation Mechanisms at Different Heating Rates[a]

Rate of heating (°C/sec)	System	Reference	Flame speed (m/sec)	Burning time (sec)		Remarks
				Volatiles	Coal to char	
10–10^2	Fixed bed	Faraday and Lyell (1845)	10^{-5}	10^2	10^3	Large particles used (run-of-mine or crushed)
$<10^3$	Plane flame	Faraday and Lyell (1845)	0.5–1.5	0.1	1	Ignition temperature about 370°C at pyrolysis temperature; flame length of meters
10^4	Plane flame	Smoot, Horton, and Williams (1976)	≤ 0.3	0.1	—	Diffusion flame formed by volatiles generated downstream of flame: mainly volatiles burning; flame length of centimeters
10^4	Plane flame	Howard and Essenhigh (1967)	0.3–0.5	0.1	1	Heterogeneous ignition followed by pyrolysis then by char burnout; flame length of meters
$>10^5$	Explosion flame	Wheeler (1913)	10–10^3	(?—total coal)	0.1	Heterogeneous ignition and combustion at limit of very fast flame speed

[a] After Essenhigh (1976); reprinted with permission of The Combustion Institute.

sis. There is still only indirect support for this view based mainly on estimated burning times of about 0.1 sec from flame lengths in explosion galleries. Some more direct support was provided about 20 years ago by the discovery that some flames were igniting heterogeneously even though followed by pyrolysis a short but distinct time after ignition [although there have been several criticisms of this result, based on misinterpretation of the findings (Essenhigh, 1976)]. These results are in qualitative agreement with the investigations of Juntgen and van Heek (1968) with pyrolysis delayed by rapid heating.

Another mechanism has been proposed by Smoot et al. (1976) that would account for phenomena in very small flames with low flame speeds or burning velocities. Based on a computer analysis of their experimental findings, Smoot et al. concluded that the particles were being heated inside such systems to pyrolyze some distance downstream from the flame front, with the flame then formed by upstream diffusion of the volatiles into the oncoming oxygen. As would be expected from this model, the peak flame speed is in the fuel-rich region but close to the volatile stoichiometric value. The stoichiometry in relation to volatiles remains poorly defined.

6.4.8 Diffusional Processes and Carbon Oxidation.

The process of combustion of a coal particle basically depends on the rate at which oxygen reaches the carbon surface. For nonporous particles, the burning rate for solid particles follows the same d_p^2 law as for liquid droplet, with a correction factor that is a function of excess air (Nusselt, 1924). For porous particles, account must be taken of the internal surface of the burning particles. This has been described in terms of adsorption and diffusion at the porous particle surface (Essenhigh and Fells, 1960; Mulcahy and Smith, 1969).

The regimes of combustion in particles can be grouped in terms of rate-controlling processes. These are often illustrated by an Arrhenius diagram (e.g., Fig. 6.16). Here the overall reaction rate v_T is plotted against the reciprocal absolute temperature. If the combustion reaction is controlled by slow reactions occurring by internal pore combustion, the rate should be dependent on temperature as indicated in zone I. At the opposite extreme, a gas-diffusion-limited fast reaction will be nearly independent of temperature as in zone III. Here the reaction occurs at the particle surface and the pore structure is not involved. An intermediate zone (II) also can be defined when both diffusion to the surface and internal pore reactions are equally important. Here a temperature dependence is expected which will be intermediate between the diffusion- and reaction-controlled extremes. The combustion process can be quite complex in theory; it may involve not only carbon char and volatilized but potential catalytic effects of noncarbonaceous components. However, it ideally begins with smooth, nonporous carbon spheres suspended in a gas.

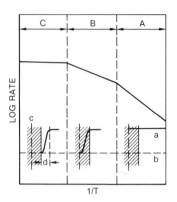

Fig. 6.16. Arrhenius diagram showing rate-controlling regimes in gas–porous solid reaction: (A) zone I, rate control by chemical reaction; (B) zone II, rate control by pore diffusion and chemical reaction; (C) zone III, rate control by diffusion to outer surface. (a) bulk oxidant concentration; (b) zero oxidant concentration, shaded portions solid; (c) center line; (d) boundary layer. [From Mulcahy and Smith (1969).]

Although the reactions of particulate carbon suspended in gases have been under investigation for 100 years or more, the first major advance in the quantitative analysis of the system was provided by Langmuir. The Langmuir (1915–1916) adsorption isotherm was originally derived specifically for the carbon oxidation reaction, and the specific rate τ_s may be written

$$\tau_s = k_1 k_2 \rho_s / (k_1 \rho_s + k_2), \qquad (6.11)$$

where k_1 and k_2 are the velocity constants for adsorption and for the oxygen monolayer desorption, respectively, and ρ_s is the surface oxygen. From this equation, it has been predicted that at low temperatures the reaction order with respect to oxygen should be zero; at high temperatures, it should be unity, so that at flame temperatures the reaction rate can be written as adsorption limiting, or

$$\tau_s \approx k_1 \rho_s. \qquad (6.12)$$

These predictions have been confirmed by experiment (Arthur and Bleach, 1952). At intermediate temperatures, a transition region exists in which the reaction order is fractional and is rising from zero to unity.

If a diffusion film of material exists above the solid surface then we also have the condition

$$\tau_s = k_0 (\rho_\infty - \rho_s),$$

where k_0 is the velocity constant (or the mass transfer coefficient) for oxygen transport and ρ_∞ is the mainstream oxygen concentration. Eliminating ρ_s we obtain

$$\frac{1}{\tau_s} = \frac{1}{k_0 \rho_\infty} + \frac{1}{k_1 \rho_\infty} = W_0 + W_1. \tag{6.13}$$

Because of the form of this equation the two quantities W_0 and W_1 have long been known as the diffusion and chemical resistances.

Applying this analysis to carbon combustion is a solid aerosol system, the assumption stated above (that at flame temperatures, $W_0 \gg W_1$) leads to

$$\tau_s = k_0 \rho_\infty, \tag{6.14}$$

which implies that ρ_s is effectively zero.

The velocity constant k_0 can be related to the oxygen diffusion coefficient D_0 such that

$$\tau_s = D_0 \rho_\infty / R_p. \tag{6.15}$$

This is equivalent to Nusselt's (1924) result that, in the quiescent system, the Nusselt number is 2 and the effective boundary layer thickness is equal to the particle radius. By integration and appropriate substitution the following relation for the burning rate coefficient is obtained:

$$K = \frac{8M_o D_0 p_\infty}{\rho_p \mathcal{R} T}, \tag{6.16}$$

where M_o is the molecular weight of oxygen. A form of this equation has been used to predict the value of K for the carbon residues of a set of coal particles. Essenhigh and Fells (1960) compared, calculated, and measured values of the burning rate coefficient for several English coals. The calculated values are about 1.5 times higher than measured values; the discrepancies are attributed to swelling, and photographic measurements have now established a swelling factor of the particles during heating.

The derivation of Eq. (6.15) is based on the assumption that reaction ideally takes place at a smooth exterior surface (zone III). In practice, experiments have shown that pore diffusion takes place with consequent internal reaction (zone II). This has been treated theoretically, and it has been shown that, for a first-order reaction, Eq. (6.15) becomes

$$\frac{1}{\tau_s} = \frac{1}{k_0 \rho_\infty} + \frac{1}{[k_1 + (k_1 D_0 s)^{1/2}] \rho_\infty} \tag{6.17}$$

where s is the specific internal surface (cm^2/cm^3). This in effect is substituting k_1' for k_1 in Eq. (6.13) where

$$k_1' = k_1 + (k_1 D_0 s)^{1/2} = k_1[1 + (D_0 s/k_1)^{1/2}].$$

When the oxidation takes place within the coal particle (zone II), the oxidant concentration varies from zero at the center of the particle to the surface concentration. When internal diffusion controls, Mulcahy and Smith (1969) give the rate expression

$$\imath = 2\mathfrak{N}_p \pi r_0 [r_0 D_0 \imath_s \rho_\infty]^{1/2}, \tag{6.18}$$

where $\mathfrak{N}_p$ is the number of pores of radius r_0 per unit external surface area. For small porous particles, the apparent reaction rate ($\imath$) is zero in oxidant. For thick, nonporous particles, diffusion is limiting, and the apparent reaction rate becomes half order in oxidant concentration (Blyholder and Eyring, 1957).

Other modifying factors so far omitted from this discussion include the influence of back reactions, such as readsorption of CO and CO_2; other reactions including reaction with water vapor; and gas-phase reactions, such as oxidation of CO. Studies of the carbon plus CO_2 and carbon plus steam reactions with readsorption have shown that the simple Langmuir isotherm is modified as would be expected and takes the form (Jolley and Poll, 1953)

$$\imath_s = k_1 k_2 \rho_{s1} / (k_1 \rho_{s1} + k_2 + k_3 \rho_{s2}), \tag{6.19}$$

where k_3 and ρ_{s2} are the velocity constant and surface concentration for the readsorbed components. There remains controversy concerning the detailed sequence of the steps in the mechanism (Walker et al., 1959).

Other carbonaceous dusts burn in a manner analogous to coal aerosols; they first lose volatile components which burn as though evaporated from liquid drops, and then the carbon residue, which behave in a similar manner. Of other dusts, many change phase and have more in common with liquid fuels so far as the controlling reaction mechanism is concerned. Some, such as sulfur, become liquid, then evaporate and burn to gaseous combustion products; others, mainly such metal dusts as aluminum, magnesium, titanium, and zinc, become liquid and evaporate, but then burn to give solid combustion products.

In aerosol reactions the important parameter for theoretical purposes is the nature of the rate-controlling step in the reaction sequence since this will determine the value of the burning rate coefficient. In liquid aerosols, theory and experiment are in accord with the assumption of rate control by heat transfer to the drop by conduction from the drop flame; the chemical processes, however, still determine the combustion limits. Determination of these is not easily predictable from theory, but Oehley's empirical argument seems to give good results according to Essenhigh and Fells (1960). Burning rates are influenced, at least to a second order of magnitude, by microturbulence and by radiation. The latter can materially affect the temperature distribution in the drop. It has been observed, for example, that radiation can initiate boiling well below the liquid surface, and this clearly will affect evaporation rates. For solid aerosols, the position is less clear. Values of the burning constant show the reasonable agreement between theory and experiment for the residues of single coal particles in an infinite atmosphere; as a further check, the burning times closely obeyed the d_p^2 rule. However, the assumption of diffusional rate control under all flame conditions has been questioned, the alternative proposal being that, in flames, the rate control is the chemisorption step. The basis of the argument is essentially that certain data may have been misinterpreted owing to confusion between the role of the velocity constants k_1 and k_2 (Essenhigh, 1955). Essenhigh and Fells (1960) as an example, explained that Tu et al. (1934) derived the "resistance" Eq. (6.13) without clearly specifying that it held only at high temperature because of the approximation of Eq. (6.12). They then chose an incorrect activation energy for k_1 and then interpreted their experimental data on spheres in terms of chemical control ($k_1 \gg k_0$) at high temperature ($>1000°C$). The source of this confusion is evident, but the Tu et al. argument has been generally adopted by subsequent theorists. One of the arguments used to favor the diffusion control theory is that the temperature coefficient of the carbon reaction at flame temperatures is too low to fit an Arrhenius exponential, but it can be represented by a T^2 dependence, which is to be expected from the diffusion theory. Unfortunately, however, if the activation energy is only 4 kcal, the Arrhenius exponential can also be represented to a good approximation by a T^2 dependence in the flame temperature range (1000–2000°C). This means that the two resistances W_0 and W_1 of Eq. (6.13) are indistinguishable by this test; so also is the distinction between external reaction (governed as k_1) and internal reaction [governed as $(k_1 D_0 s)^{1/2}$]. What does change, however, is the law governing the burnout time of individual particles since k_0 is inversely proportional to radius, but k_1 is independent of it. Neglecting pore diffusion we then have, for $k_1 \ll k_0$,

$$dR_p/dt = -k_1 \rho_0/\rho_1.$$

With integration this yields a burning time

$$t_b = d_{p\infty}/K_c \qquad (6.20)$$

for a single particle of initial diameter $d_{p\infty}$ in a free atmosphere, where K_c (cm/sec) is given from kinetic theory by

$$K_c = \frac{4\rho_p(2\pi \Re T/\overline{M})^{1/2}}{3p_\infty \exp(-\Delta \hat{G}/\Re T)}$$

where $\overline{M}$ is the mean molecular weight of the gas mixture. In a flame or finite atmosphere,

$$t_b = \xi_c d_{p\infty}/K_c, \qquad (6.20')$$

where ξ_c is a factor equivalent to a correction for excess air fraction.

Calculation of K_c gives values of the order of 1 cm/sec, so for isolated particles of 1 cm diameter we should expect diffusion to control the reaction. As the particle size drops and the air supply is increasingly restricted, we should expect a reversal of the respective roles as indicated in Fig. 6.17.This has been put to experimental test by comparing predicted and measured degrees of combustion on carbon particles derived from anthracite in a reactor. The comparison is shown in Fig. 6.18, which indicates that the chemical control prediction fits the measured values better than does the diffusional control prediction. While this is not conclusive, it supports the result implied by Hottel and Stewart (1940), who wrote the equation for the specific reaction rate in the form

$$\varkappa_s = (\text{const}) d_p^{-n}, \tag{6.21}$$

where d_p in the particle diameter and n a constant having a value between zero (for chemical control) and unity (for external diffusional control). This equation was then fitted to a collection of experimental data on flames from a variety of sources and a best fit obtained where, contrary to expectation, n had a value of zero.

The potential for distinction between oxidation mechanisms also can be deduced, in principle, from a plot of reaction rate normalized by surface area or volume. An example taken from Mulcahy and Smith (1969) uses data for the controlled burning of anthracite of Beer $et\ al.$ (1964). Anthracite is a hard, nonporous material before combustion which may undergo increase in porosity during burning. An Arrhenius diagram for the rate expression assuming (a) an impervious sphere model and (b) a porous sphere model is shown in Fig. 6.19. The plots are given for different reaction order n between 0 and 1. The comparison indicates that on the basis of kinetics data alone there is little to distinguish between the original Beer $et\ al.$ (1964) interpreta-

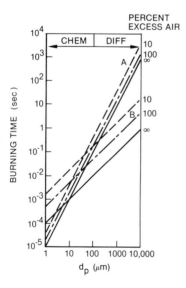

Fig. 6.17. Comparison of calculated burning times in a flame at 1500°C using hypotheses for diffusional and chemical reaction control (hypothetical carbon flame). Case A: physical control at the external surface by excess air; case B: chemical control after Eq. (6.20). [From Essenhigh and Fells (1960).]

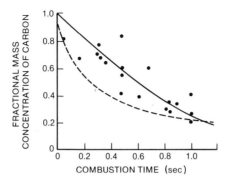

Fig. 6.18. Decay of carbon concentration in a dust flame with time. Comparison of experimental data (circles) with calculation assuming (i) diffusion rate control (dotted line), (ii) chemical rate control (solid line). [From Essenhigh and Fells (1960).]

tion of a diffusion-limited reaction ($n = 1$) as contrasted with reaction control ($n = 0$), or an internal porous sphere reaction of order one-half (Eq. 6.18).

The process of combustion may be influenced by convective effects characterized by diffusional transfer to the particle surface. This effect can in-

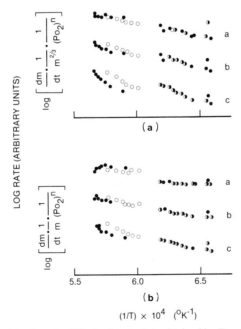

Fig. 6.19. Different treatments of the basic rate data obtained by Beer $et\ al.$: (a) impervious-sphere model where rate depends on surface area $m^{2/3}$; (b) completely penetrated porous-sphere model, where rate depends on particle mass m, Beer $et\ al.$ (1964): a, $n = 1$; b, $n = 0.5$; c, $n = 0$. [Graphs from Mulcahy and Smith (1969).]

volve microturbulence of very small eddies affecting the apparent reaction kinetics, and macroturbulence or forced convection related to the scale of mixing in the combustion chamber as a whole, influencing the ignition and stabilization of the flames. The possible role of microturbulence in promoting and then hindering the reaction of the liquid particles is reviewed by Essenhigh and Fells (1960). It is reasonable to expect that a promotional effect will operate in solid particles. Microturbulence is encouraged in solid aerosol systems by strong differential heating from localized burning or by small-scale differences in fluid-particle motion. As a result, the oxidant diffusion film around particles may be broken up during the combustion process. Evidence that this occurs has been cited from experiments showing the increase in burning rate with exposure of particle suspensions to low-frequency sound waves (Porter, 1948).

In allowing quantitatively for microturbulence, Spalding (1954) has applied Frössling's equation and obtained a multiplicative factor for the combustion rate equation for enhanced transport to particle surfaces.

In contrast with microturbulent factors, macroturbulence has a greater effect. In an enclosed jet flame, for example, the forced recirculation patterns have characteristic magnitudes approaching the combustion chamber dimensions. In dust flames such patterns feed back hot combustion products to the root of the jet, and have a major influence on the ignition and stabilization of the flame. Measurements on quiescent flames and stationary flames without recirculation give flame speeds of the order of 1 m/sec or less, in accordance generally with the simple radiation theories of flame propagation (Ledinegg, 1952). However, in horizontal jet–flame systems the horizontal convective motion of the dust exceeds 10 m/sec; to remain in suspension, flame speeds of 1 m/sec are too low for flame stabilization. Such systems are stabilized as a consequence of the extra energy supplied by the recirculating gases which increases the flame speed by a factor of 10 or more. Calculations suggest that this increase can be achieved by as little as 20% recirculation volume.

The importance of turbulence has also been demonstrated in moving flames. Flame speeds in aluminum dust clouds, contained in tubes and ignited at the closed end, were found to be increased if the flame passed through a turbulence grating (Cassel et al., 1949). This promotes an intense forced convection involving turbulent exchange and is regarded as an important factor in determining the intensity and magnitude of a dust explosion; it is possibly the counterpart of recirculation in stationary flames.

6.4.9 Design of Large Combustors. An important application of coal combustion is in large industrial boilers used for steam generation. A typical pulverized-coal system is drawn in Fig. 6.20. The combustion heat recovery is approximately 85–90% in a large, modern furnace–boiler system of this type. Most of the loss goes up the stack, as both latent and sensible heat in

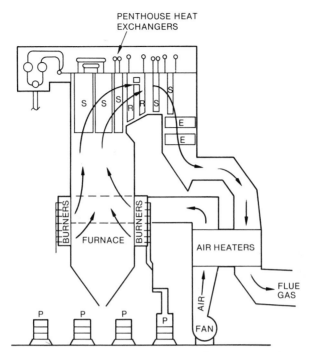

Fig. 6.20. Pulverized-coal-fired boiler. P = pulverizers; S = steam superheater; R = reheater; E = economizer. [From Breen (1976), reprinted with permission of The Combustion Institute.]

the flue gases. Condensation of the water (and possibly weak sulfuric acid) in the flue gases inside the system is avoided to minimize corrosion, so the stack gas temperature is generally maintained above 110°C.

In a typical large boiler about half the heat absorbed goes into the water-cooled walls of the combustion chamber (generally termed the furnace walls, or radiant water walls). Convective heat transfer is dominant downstream of this chamber, where the water is vaporized and superheated. Farther downstream is an "economizer" section for preheating the feed water before it enters the radiant water walls. Finally, a rotating drum with huge internal surface area is used to extract more heat from the cooled flue gases and transfer it to preheat the slightly cooler inlet air.

On the furnace's firing walls, up to 64 burners are mounted in one or more plenums called *windboxes*. Windbox pressures are typically 0.3 psi (2000 N/m²) or less above furnace pressure, which differs by only one psi (6900 N/m²) or less from atmospheric pressure. The steam pressure, on the other hand, may be as high as 3800 psi (26 × 10⁶ N/m²) and the steam temperature is often as high as 570°C (1050°F). This temperature, and the temperature of ambient air or water available for heat rejection in the condenser, results in an electrical generator cycle efficiency between 38 and 42%. Each 55°C rise in steam superheat increases station efficiency about approximately 1.5%. However, further improvements in efficiency by an increase in steam temperature are presently hampered by materials problems: loss of strength of alloy steels at higher temperatures, and corrosion due to dissociation of the steam molecules with subsequent hydrogen diffusion and hydrogen embrittlement or to formation of liquid-phase ash eutectics such as sodium–iron sulfates or potassium–aluminum sulfates on the hot gas side.

The combustion products leaving the furnace and entering the convective passes must be completely burned to avoid quenching at cool tube surfaces but still hot enough to bring the steam to design temperature at all steam flow rates. For a solid fuel, the furnace is sized first to provide sufficient volume for complete combustion of the fuel so that high temperatures and slag do not carry over into the convective superheater section, and furnace wall temperatures are maintained within allowable limits.

A large amount of the airstream is inert nitrogen (79%) and thus the air fan economics dictate the most important characteristic of burner design; i.e., that the burner pressure drop must be low (0.3 psi, 2000 N/m², or less). This results in nonintense turbulently mixed flames held in a low-flow air supply which are susceptible to flow instability and combustion-induced oscillations.

The burner configuration for pulverized-coal injection typically takes the form shown in Fig. 6.21. Fuel is injected through a series of nozzles for solid fuels. Primary air carries the entrained fuel in pulverized-coal firing, as shown in Fig. 6.21; its velocity is kept below 25 m/sec to reduce duct and burner erosion.

Owing to a basic low-pressure drop constraint, the oil and coal flames tend to form into a fire ball or cloud along the center line of the burner. In this case the fuel vaporization or volatilization inside the fire ball is faster than the turbulent mixing on the edge and the air mixes inward to form a stoichiometric flame boundary defining the fire ball. In small equipment where pressure drop is not such a critical economic consideration, more intense swirl and mixing can lead to air penetration completely through the cloud and thus to individual particle combustion mechanisms.

In a large boiler, the primary zone of raw air combustion exists axially for five to fifteen feet; within this zone adiabatic flame temperatures occur. Adiabaticity and recirculation within this raw air primary zone lead to flame holding and stability but also can lead to high nitric oxide production. Within approximately five feet of the flame, the mass of relatively cool product gases entrained off the firing wall becomes greater than the mass of raw air in the original burner stream. The temperature and quantity of the entrained gases depend on burner-to-burner spacing, so that this is an important primary design variable for nitric oxide control in addition to the overall heat transfer area.

Carbon burnout occurs downstream of the primary raw air zone by indi-

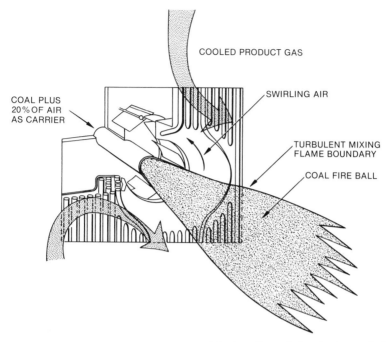

Fig. 6.21. Burner for pulverized-coal firing. Particle size: 60 μm; fuel and primary air characteristics: 70°C, 24 m/sec; main air flow: 316°C, 37 m/sec. [From Breen (1976); reprinted with permission of The Combustion Institute.]

vidual particle reactions. Liquid and solid fuel molecules can crack and form carbonaceous and/or ash deposits that will grow on burner parts if the burner is not designed to avoid such deposition. Once such carbonaceous deposits begin to form, the feedback leads to deteriorating conditions and finally to shutdown.

6.5   PRODUCTION OF SOOT IN FLAMES

An important by-product of the combustion of liquid or vapors is the formation of soot particles. Carbonaceous particle production is a serious limitation in the use of diesel engine technology for transportation, because of air pollution. However, the limitation also is a ''benefit'' in another industry. The production of carbon black is a major industry, with widespread use of the product for binders, dyes, etc. (see also Section 6.7).

Soot generated in combustion processes is not a uniquely defined substance. It normally looks a lusterless black but differs from graphite. The main constituents are carbon atoms; soot contains up to 10 mole % hydrogen, and even more if it is freshly produced, as well as traces of other elements. Much of the hydrogen is extractable in organic solvents, where it is associated in condensed aromatic ring compounds. Sometimes, materials are emitted which, when cooled, look like tar or glassy substance, either black, brown, or even yellow. The inspection of soot coming, for example, from a premixed flame or the exhaust valves of a diesel engine shows under the electron microscope that the building bricks of that soot are spherical or near-spherical elementary soot particles with mean diameters around 20–30 nm, corresponding to about 10^6 carbon atoms. These elementary particles adhere to one another to form straight or branched chains. These chains agglomerate and form the visible macroscopic soot flocculates, generally as a fluffy substance.

For the investigation of the internal structure of the elementary particles, x-ray studies have been very useful. Phase-contrast electron micrographs show that carbon black particles can be considered as an arrangement of bent carbon layers which follow the shape of the particle surface. Many dislocations and lattice defects are present. The density of these particles seems to be less than 2 g/cm³, owing to large interplanar spacing. Electron diffraction indicates the presence of single C—C bonds in soot. Soot particles collected during their growth show much stronger electron spin resonance signals than fully developed particles. Heat treatment improves the degree of ordering and reduces the hydrogen content (Wagner, 1981).

The time available for soot formation in combustion processes is of the order of milliseconds. During this time a solid phase of soot particles is formed from the fuel molecules via their oxidation and/or pyrolysis products. The possible paths for soot forming processes are indicated in Fig. 6.22. Here the logarithm of the molar mass (g/mole) of a species is plotted against its hydrogen content (as mole fraction in the molecule); soot formation increases with decreasing mole fraction of hydrogen in the fuel. Various fuel molecules are shown in the lower left corner, with $x_H > 0.5$; the soot particles are at the right top corner. Also indicated is the growth line of polyacetylenes. Soot is not formed along the growth line of the acetylenes, or the polycyclic aromatic compounds alone. In order for the fuel molecules to evolve to soot, there must take place a condensation of species with a particular hydrogen content or a condensation of species with higher hydrogen content and consecutive dehydrogenation, or a combination of these two limiting pathways from the lower left to the upper right corner in the figure (Wagner, 1981).

After a sufficient time, pure hydrocarbons exposed to high temperature in the absence of oxygen will reach an equilibrium state, which is graphite, and hydrogen with some stable hydrocarbons like acetylenes, large polycyclic aromatics in the gas phase.

For premixed hydrocarbon/air systems soot is expected to appear when there is insufficient oxygen to transform the hydrocarbon into CO and H_2, or when C/O approaches unity. Wagner (1981) notes that experimentally determined limits of soot formation, however, do not occur at C/O = 1 but usually

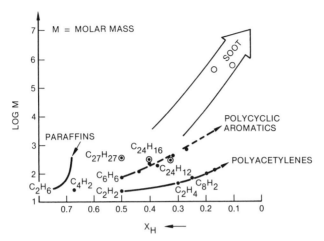

Fig. 6.22. Paths for soot formation. Molar mass plotted as a function of their hydrogen mole fraction x_H. [From Wagner (1981).]

closer to C/O $\approx$ 0.5. The critical C/O ratio is only weakly dependent on pressure or dilution with inert gas at constant temperature. Increasing temperature generally allows richer mixtures to be burned without the onset of sooting, as long as the flame front remains smooth.

A picture of the processes of soot formation in flames has come from the study of low-pressure acetylene oxygen flames. In a system where C/O = 1.4, at a pressure of 20 Torr, and an unburned gas velocity of 50 cm/sec, the downstream profile of particulate properties in the flame is shown in Fig. 6.23.

Under such conditions, some acetylene is left over and stays in the burned gases. Polyacetylenes accumulate and pass through a maximum around 15 mm above the burner. A large variety of polycyclic compounds such as pyrene, acenaphthylene, and many others seem to originate in this zone. A great many other hydrocarbons including various ring systems with and without side chains, mainly those having masses >250, begin to appear and disappear in the hot gases behind the oxidation zone. At 3.5–4 cm above the burner, their concentration passed below the limit of detectability of the mass spectrometer used, which had a mass range up to 500, so that particles larger than 500 mass units could not be measured (Wagner, 1981).

The formation of the higher-molecular-weight hydrocarbons proceeds via fast reactions, which are difficult to follow experimentally because so many different species are involved. In C_2H_2 flames their formation begins with reactions of acetylenes and radicals by way of a fast-radical polymerization. When the polymerized species become sufficiently large, they show branching and will undergo cyclization via unimolecular or bimolecular reactions.

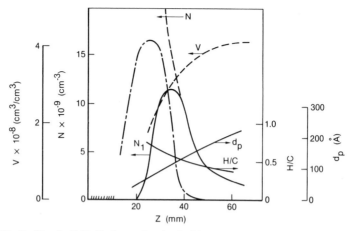

Fig. 6.23. Profile of a C_2H_2/O_2 flame showing buildup of polycyclic material followed by soot particle production. The flame zone is at the left. Volume fraction V in cm^3/cm^3 particle; diameter d_p (in Å), particle number density N (in cm^{-3}) and H/C ratio. The broken curve gives a measure of absorbing polycyclic material in the flame zone. [From Wagner (1978).]

Acetylenes and other highly unsaturated species are known to be capable of undergoing cyclization, for example, by Diels–Alder reactions. An important feature of these early steps is that, once the molecules are sufficiently large and sufficiently compact, radical addition complexes will have long lifetimes, thus preserving the radical character of the large species.

Measurements of newly formed soot particles are associated with a high H/C ratio (see Fig. 6.23) and strong radical character of the particles. The soot embryos are surrounded in the gas phase by a variety of polyacetylenes and ring compounds, including polyacyclic aromatics. A maximum of the polymeric material concentration appears at about 20 mm above the burner, which is higher in concentration than the final amount of soot.

Ionic or charged species are also formed in these flames and may play a role in soot formation (Olson and Calcote, 1981). Ion–molecule reactions can be extremely fast. Unsaturated species are formed during oxidation; therefore the formation of larger species could proceed via ionic polymerization. Whether this will be a significant contribution depends on the ion concentration.

For conditions similar to those of the C/O = 1.4 flame, Homann (1979) found at the end of the oxidation zone two maxima in the mass distribution of the ions. One belongs to the ions, the concentration of which increases towards the fresh gas; they grow very little in the oxidation zone. The other stronger maximum belongs to ions whose mass increases away from the burner to positively charged soot particles (up to about 6000 mass units, closing the gap between mass-spectrometric and electron-microscopic results). This nonequilibrium ionization is essentially concentrated in the region where the volume fraction V grows. Homann attributes this to an ionization mechanism related to the mass growth of soot particles. Wagner (1981) notes that the absolute concentrations of the ions in these flames are still the subject of discussion. Two independent groups who determined absolute ion concentrations came to the conclusion that the ion concentrations are well below the soot particle concentration in the main soot forming region.

The growth of soot volume fraction V and the decay of particle number concentration N in the standard flame are shown in Fig. 6.23. The N and V curves exhibit the typical behavior discussed above. Two curves for the particle number are shown. If one counts as soot particles those above a certain size, say 50 Å diameter, their number density (N_1) will first increase and then decrease. For another size limit, the curve will appear different on the burner side. This apparent difference is the result to the experimental methods. During the early phase of soot formation the carbonaceous embryos coagulate and become spherical. In the later phase, when the growth of V is slow, the soot particles continue to coagulate, no longer forming spheres, but particle chains. Optical methods identify these particle chains as a whole but electron-microscopic study yields the embryo or unit particles.

Early in the soot formation profile heavy hydrocarbon species are detected (Fig. 6.23) whose volume fraction is similar to the soot level. The differences in the soot formation tendencies and soot formation rates of various fuels such as C_2H_2, C_2H_4, paraffins, and C_6H_6 evidently are most important in this region owing to the different size and reactivity of the polymerized species formed. Wagner (1981) indicates that the observations in low-pressure flames of C_2H_2 and O_2 are to a certain extent typical, and propane or ethylene flames will be rather similar. However, benzene flames form soot more readily than acetylene flames. Mass-spectrometric analysis of the main reaction zone shows a composition rather different from that found for aliphatic fuels. Acetylenes are still major products, but are in reduced amounts. The concentration of aromatic species is about two orders of magnitude higher in C_6H_6 flames than for the C_2H_2 flames.

A very detailed investigation of intermediate products of a near-sooting benzene–oxygen flame has been performed by Bittner and Howard (1981). This shows, on the one hand, that the chemistry of these systems is complex indeed; on the other hand, it is obvious that there are many more comparatively large molecules with intact rings which may act as "embryos" for a more rapid growth of the soot volume fraction than in C_2H_2 flames. This suggests that in sooting low-pressure benzene–oxygen flames there are no positively charged soot particles with masses >1400 mass units; V reaches its final values while the particles are still comparatively small.

The mechanism of soot production provides a unique application for the theory of nuclei production and growth during vapor combustion. Until the 1960s, it was generally supposed that nucleation mechanisms for production of particles were dependent in some manner on the degree of saturation of a condensable vapor. However, since the work of Wagner and co-workers, a new mechanism of particle growth, termed "chemical" nucleation, has been elucidated which is independent of considerations of saturation of a condensable vapor (Homann and Wagner, 1965; Bonne and Wagner, 1965).

Chemical nucleation has received study (Homann, 1967; Wagner, 1981) in the important case of soot formation in combustion of carbonaceous fuel. It seems likely that in the future additional examples of chemical nucleation will be identified in processes in which particle production is at present attributed to homogeneous or heterogeneous physical nucleation mechanisms. One of the clearest and best-documented examples of chemical nucleation available is the acetylene flame.

A theory for the evolution of particles by polyacetylene formation can be derived from an approximate form of the growth equation [Eq. (3.11b)]. Brock (1972) has undertaken calculations from such a model, and has found the results to agree with the limited observations of soot particle size distributions in premixed acetylene flames of Bonne et al. (1964).

The mechanism of soot formation in combustion of liquid droplets is far more complicated than in a hydrocarbon vapor, for example, because of cenosphere production. Study of soot production from spray combustion remains largely descriptive, relying on accumulation of empirically based information.

Practical considerations in engine performance focus on fuel composition, including aromatic hydrocarbon content, and on the hydrocarbon to oxidant ratio (Longwell, 1976). In combustion engines, for example, the percentage aromatics specification has been set to avoid a series of combustion-related problems that were encountered with fuels containing more than the specified amounts. In jet turbine engines, for example, these problems are

(a) smoke formation under high power conditions,
(b) overheating of the combustion chamber from flame radiation, and
(c) formation of solid carbonaceous deposits, which causes fuel spray distortion and turbine damage when deposits become detached.

These effects have correlated quite well with hydrogen to carbon ratio, as expected from Fig. 6.22. The correlations, of course, are a function of the particular combustion system design, which must be empirically determined since adequate predictive techniques for the complex set of phenomena involved do not exist at present.

From the point of view of relating fuel composition to performance, the success of hydrogen to carbon ratio in correlating results is a significant simplification. It is, of course, related to fuel composition. It can be expected that differences among the structural variations possible within these compound types will occur that are not accounted for by simple hydrogen to carbon ratio; however, hydrogen to carbon ratio appears to be a good first-order measure of burning quality as well as of energy efficiency and cost in manufacturing fuels. The lower this H/C ratio, the more difficult it is for the fuel to burn.

The aviation gas turbine combustor and its fuel requirements can be considered as an example of the type of system that would be required for general transportation gas turbines and for the combustion fuel system needed for a Stirling Cycle engine in that it represents a continuous combustion system that must operate over a wide range of heat output and atmospheric conditions. With the possible exception of stringent pollutant emissions control and problems due to more frequent changes of output, the requirements for ground transportation appear less demanding than aircraft in that the range of atmospheric pressure and temperature is more limited and the penalties for flameout or inability to reignite are less severe. The new types of ground transportation systems seem to offer an excellent opportunity to design the combustors to operate satisfactorily over a wide range of fuel composition and volatility.

With the exception of the spark ignition Otto Cycle engine, none of the major liquid-fuel combustion systems vaporize and premix the fuel and air. Instead, the fuel is dispersed in the form of small drops, and a droplet–air mixture is burned. A luminous flame invariably results, since even an evaporated drop will leave a fuel-rich zone, and if the burning zone envelops the liquid drop, a very smoky diffusion flame can result, as discussed by Sjogren (1972).

When mixing of the total combustion air with fuel is completed, soot will oxidize to CO_2 if sufficient time at high temperatures is available. In large boilers and industrial furnaces the cooling rate of the combustion products is such that with combustion of heavy aromatic tars rather coarsely atomized, with substantial soot formation, soot depletion by burnout is attainable. In transportation systems, however, the need for compactness results in very rapid cooling rates, and soot burnout is more difficult to achieve. Longwell (1976) notes that current aviation gas turbines quench the combustion gases so rapidly that it is difficult, even when aromatic content of fuel is restricted to 20%, to achieve low smoke combustion. Similar, though less severe, problems can be expected in designing burners for automotive gas turbines. For intermittent fuel injection combustion systems like the diesel and fuel-injected stratified charge engines, fuel-rich combustion zones and very short times at high temperature are inherent in their design, so that some soot particles are emitted. One would expect that the low speed diesels used in ships or rail locomotives would show up in their ability to successfully use less refined grades of diesel fuel. It also should be noted that allowing a longer time at high temperature, while excellent for destruction of soot, CO, and unburned hydrocarbons, will generally result in increased production of nitric oxide.

The Otto Cycle engine which in principle vaporizes its fuel before ignition is not faced with the same problems; however, if vaporization is incomplete and large fuel drops are present, soot is formed. This requirement for vaporization before ignition limits the boiling range of the fuel and prevents the use of fuels with very high aromatic content because of their high boiling points, thereby limiting the fuel octane number and the compression ratio of the engine.

In all of these propulsion systems, fuel vaporization or very fine atomization and premixing with air seem called for. Fuel vaporization is, in fact, of considerable current interest in gas turbine combustion systems, although the practical advantages of droplet combustion systems capable of handling low hydrogen to carbon ratio fuels are still attractive.

The closely coupled tradeoffs between formation of soot and nitrogen oxides, combustion operating range, fuel composition, and many other factors makes optimized design an exceedingly complex problem for which much of the needed basic data on the chemistry of carbon formation, carbon burnout kinetics, droplet combustion for small drops, and turbulent flow are lacking, particularly for heavier, low hydrogen to carbon ratio fuels.

6.6 SAFETY HAZARDS OF DUST EXPLOSIONS AND FIRES

An important adjunct to the combustion of dusts are the industrial hazards associated with detonations and explosions of dust clouds. Although dust explosions have been recorded for over a century, appreciation of their true character was relatively slow in developing. Particularly great difficulty was experienced in accepting that an explosion could be caused by the dust alone, and that the presence of a flammable gas to support the explosion was not necessary. For instance, Morozzo (1795) gave an account of a flour dust explosion in Turin in 1785 which was published shortly afterwards in which it was postulated that ignition first occurred in flammable gas given off by the disturbed dust, although it was recognized that the dust itself contributed to the subsequent explosion. Over 100 years elapsed before it was generally accepted that dust explosions in coal mines could be initiated and sustained by coal dust alone, without needing support from flammable mine gas. When the position was finally clarified early in the present century, explosions in numerous other clouds of industrial dusts had occurred, and the way was open for a more rational assessment of the explosion hazard and the need for adequate precautions.

A review of the position in 1922 published by Price and Brown (1922) gave details of numerous explosions in various industrial dusts, and estimates of human and economic losses. A record of important dust explosions in the United States and Canada since 1860, excluding those in coal mining, gave details of the causes of explosions, the extent of damage and casualties, and numerous photographs of damage to industrial plants and structures (NFPA, 1957). Comparable statistics have not been published for the number of explosions and loss of life and property in other countries. The review of Brown and James (1962), covering data up to 1959, gave general references, particularly to dust explosions in factories.

6.6.1 Characteristics of Dust Explosions. A dust explosion may be envisaged as combustion of a dust cloud which results either in a rapid buildup of pressure or in uncontrolled expansion. The gas in which the dust is suspended takes part in the combustion, and hence in considering the properties of dust explosions both the nature of the dust (or the fuel) and the gas, the oxidant, are important. It is the expansion effect, or the pressure rise if expansion is restricted, which presents one of the main hazards in dust explosions. In some industrial plants involving pulverized fuel furnaces, dust clouds are continuously burned under controlled conditions so that in normal working the pressure is not permitted to rise dangerously. This process is not regarded as a dust explosion, although there may be similarities in the mechanism of the combustion.

The expansion effects in dust explosions arise principally because of the heat developed in the combustion and, in some cases, to gases being evolved from the dust because of the high temperatures to which it has been exposed. If the gases are not consumed in the explosion they would remain afterward and would contribute to the final gas volume, and hence would augment the expansion effects. The heat generated in a dust explosion is eventually lost to the surroundings and so the expansion and pressure effects are transient. However, it is a characteristic of explosions that flame speeds are high, whereas the time required for gas temperatures after the explosion to revert to the level of the surroundings could be many seconds. In the industrial environment, the heating rate may greatly exceed the cooling rate while the explosion is developing. Thus expansion and pressure effects take place. Indeed, in some cases, under controlled laboratory conditions, the rate at which the explosion propagates is so rapid that maximum pressures are obtained which are close to the theoretical values calculated on the basis of no heat loss while the explosion is developing. Heat loss to the surroundings is thus not likely to be the basis for adequate protection of industrial plants against dust explosions (Palmer, 1973).

In dust explosions the rate of propagation of flame through the dust cloud is fast compared with rates of flame spread in fires. For explosions in small enclosures, flame speeds range from hundreds of meters per second to the order of one meter per second. Rapidly burning dust fires are considered below this speed range. The type of explosion usually occurring under industrial conditions is termed a *deflagration*. An important property of deflagration is that the flame speed is less than the velocity of sound in the gaseous products of combustion. There is another mode of combustion, termed *detonation,* in which the flame speed is equal to the velocity of sound in the gaseous products and is accompanied by a shock wave. This velocity is of the order 1000 m/sec. Whether or not a true detonation can occur in a dust explosion in industry has not yet been established. Because of the energy released during the explosion, detonations should be possible and, indeed, explosions having velocities similar to those of detonations have been reported. The majority of evidence relates to explosions in coal mine galleries at least 100 m in length, but the explosions were initiated by powerful sources which in themselves were capable of generating strong shock waves. Whether or not detonations can develop from relatively small ignition sources has not yet been proven, but in any case a detonation would initially start as a deflagration according to Palmer (1973). It is general practice for protection against dust explosion hazards to assume that deflagrations rather than detonations occur.

Two basic parametrizations are often used to characterize industrial explosion hazards. These are the flame speed and the ignition conditions of a mixture. The flame speed in a dust explosion is not constant but depends on a number of factors. Two of the principal factors are the chemical composition of the dust and the oxidant. With the present state of knowledge the flame speed cannot be related simply to composition, including moisture content, or to heat of combustion. The flame speed depends upon the particle size of the dust: the smaller the particles the higher the flame speed, providing that with fine particles the dust is evenly dispersed in suspension. The flame speed also depends on the turbulence of the gas in which the dust is dispersed; broadly, increased turbulence leads to higher flame speed. The relationships between flame speeds and properties of the dusts including particle size and the turbulence of the gas have not yet been derived by theory, but rely on empirical results described, for example, by Palmer (1973).

For an explosion to occur, the dust cloud must be ignited. Apart from a few special instances where the act of dispersion of dust may ignite it, a separate source of ignition must be present and the concentration of dust in the cloud must be favorable. The main stimulus for the measurement of dust explosion properties has been the need to provide information on the safe use and handling of dust in industry, laboratory tests of ignitability of dust have been designed to incorporate sources of ignition likely to occur in practice. Two common sources would be a hot surface and a spark. Consequently, the minimum ignition temperature and the minimum ignition energy are the ignition properties most commonly measured in the routine testing of dusts. Examples of values for different dust mixtures, with over-pressures expected are given in Table 6.7.

A frequent source of ignition in industrial dust clouds is a spark. There are a number of ways by which these can be produced, including electrification, friction, impact, and hot cutting or welding. A characteristic of all these forms of sparking is that a small particle or volume of gas at a high temperature is produced for a short interval of time. Although the temperatures of sparks are well above the minimum ignition temperatures listed for dust clouds, the size of the spark is much smaller than that of the hot surface used for ignition temperature measurement. Because of the smallness of the source and its short duration, much higher temperatures are necessary to transfer the energy necessary to ignite dust clouds. It is much easier for experimental purposes to measure the energy delivered in an electric spark than that delivered by friction or other thermal processes. Consequently the routine method of measurement of minimum ignition energies of dust clouds is to pass an electric spark of known energy through the cloud and to observe whether ignition occurs. As with other explosion properties the minimum ignition energy of a dust cloud depends on the dust concentration, particle size, and moisture content. In addition, there is evidence that the design of the electrical circuit producing a spark can also affect the minimum ignition energy.

If a flame is to propagate through a dust cloud, the concentration of dust must lie within the lower and upper explosibility limits. The lower explosibility limit may be defined as the minimum concentration of dust in a cloud necessary for sustained flame propagation. It is a fairly well-defined quantity, and can be determined reliably in small-scale tests as discussed by Palmer (1973). Some examples are given in Table 6.7.

When the concentration of dust is raised above the lower explosibility limit and past the stoichiometric value, the flame speeds and vigor of the explosion increase. As the dust concentration is further increased, however, the quenching effect of the surplus dust on the explosion becomes more marked and eventually a concentration is reached at which flame propagation no longer occurs. This concentration is the upper explosibility limit and may be defined as the dust concentration above which flame propagation does not occur. The upper explosibility limit is not easy to measure in practice, mainly because of the difficulty of ensuring that a dust is uniformly dispersed in the cloud and that regions of low concentrations do not form by chance movement of the dust particles. Because of difficulties in measurement relatively few values have been determined and these are much higher than would be expected by analogy from combustible gases. There are also difficulties in ensuring that dust clouds could be maintained at concentrations above the upper explosibility limit under industrial conditions. Interest has been much less in the upper limit than in the lower explosibility limit, which in turn has only limited applications in the safe handling of dusts, for reasons given below.

The lower and upper explosibility limits depend not only on the composition of the dust, but also on properties such as particle size, and moisture content as well as on the type of ignition source used in the test. The particle size and shape also affect the dispersability of the dust; if this is poor owing to cohesive forces or to entanglement of fibrous particles, then the explosibility limits are also modified. Many of these factors do not arise in comparable experiments with flammable gases, for which the flammability limits are

TABLE 6.7

Typical Values of Key Parameters Describing Dust Explosion Potential[a]

Dust cloud material	Minimum ignition temperature (°C)	Minimum explosive concentration (g/liter)	Minimum igniting energy (mJ)	Maximum pressure (psig)	Maximum rate of pressure rise (psi/sec)	Comments
Charcoal, hardwood	530	0.140	20	100	1,800	0.5 g/liter; 27.7% volatiles, 62.5% fixed carbon
Pitch, coal tar	520	0.080	50	—	—	48.5% volatiles, 51.3% fixed carbon
Carbon black, acetylene	900	—	—	—	—	0.7% volatiles, 99.3% fixed carbon
Oil shale, CO; coal, IL	440 600	— 0.040	— 50	— 84	— 1,800	0.5 g/liter; 48.6% volatiles, 45.4% fixed carbon, 6.0% ash
Coal, Brookside, CO	530	0.045	60	88	3,200	0.5 g/liter; 38.7% volatiles, 48.8% fixed carbon, 12.5% ash
Coal, Cranston, RI	800	0.24	128	—	—	3% volatiles, 64.1% fixed carbon, 32.7% ash

	Lignite, ND	440	0.045	10	89	2,800	0.5 g/liter; 44.9% volatiles, 49.9% fixed carbon, 5.2% ash
Adipic acid		550	0.035	60	95	4,000	
Alfalfa		460	0.100	320	88	1,110	
Aluminum		650	0.045	50	84	>20,000	
Aspirin		550	0.015	16	87	7,700	
Calcium carbide		555	—	—	13	—	
Coffee		360	0.085	160	38	150	
Cornstarch		390	0.040	30	145	9,500	
Cotton linters		520	0.50	192	73	400	
3,5 Dinitrobenzamide		500	0.040	45	163	6,500	
Ferromanganese		450	0.130	80	62	5,000	
Malt barley		400	0.055	35	95	4,400	
Nylon		500	0.030	20	95	4,000	
Polystyrene		500	0.020	15	100	7,000	
Rice		440	0.050	50	105	2,700	
Soap		430	0.085	100	77	2,800	
Sulfur		190	0.035	15	78	4,700	
Vanadium		500	0.220	60	57	1,000	
Wheat flour		380	0.050	50	109	3,700	
Yeast		520	0.050	50	123	3,500	
Zinc		680	0.500	960	70	1,800	

[a] From U.S. Bureau of Mines test data and Palmer (1973).

well defined, and with dusts the difficulties have contributed to the lack of knowledge of explosion processes.

When a dust explosion occurs in an industrial environment, extensive damage may result if the explosion is initially completely confined within the structure, which, in turn, is ultimately too weak to withstand the full force of the explosion. Two of the factors influencing the violence of the explosion are the maximum explosion pressure and the maximum rate of pressure rise. The maximum values obtained with dusts in small closed vessels which are sufficiently strong not to burst or leak during the explosion may be as high as 1000 kN/m^2 (150 lb/in.^{2}) for maximum pressures and 100 MN/m^2 sec (15,000 lb/in.^{2} sec) for maximum rates of pressure rise. The values obtained under standard conditions depend on the explosion properties of the dust. The values cannot be predicted from knowledge of the composition and thermal properties of the dust, but must be obtained by measurement. They may be compared with each other for different dusts (e.g., Table 6.7), and relative explosion severities can be predicted. Maximum explosion pressures are used as indications of the likely severity of an explosion. Where it is intended to contain the explosion within the plant, the structural strength of the plant must be so designed that it is quite capable of withstanding the maximum pressure expected. The maximum rate of pressure rise, or alternatively the average rate of pressure rise, are often used in connection with the design of explosion relief venting. Where rates of pressure rise are high, products of combustion are being generated rapidly.

Dust explosions have the characteristic that the pressure waves transmitted through the atmosphere during the initial stages of the explosion can cause dust which is deposited to be resuspended. This disturbance may either arise because of air movement over the dust deposit causing the dust to be blown into suspension, or to vibration of structures so that dust deposits are shaken off and fall into the air. With either mechanism, or both, additional dust clouds are formed which can be ignited by the explosion flames already present. The initial stage of the process during which disturbance of the dust occurs is customarily termed the *primary explosion,* and the subsequent stages when the newly disturbed dust becomes involved with the flame is termed the *secondary explosion.*

In many dust explosions, the sequence has been as follows (Palmer, 1973): a dust explosion has developed inside a part of a plant (the primary explosion) and, because of inadequate explosion protection, flame has been discharged from that part to other areas. This discharge has occurred either because the equipment or building ruptured or because relief venting has been badly installed. In either event the result is that a flame has discharged from the plant and is accompanied by a disturbance of the air into additional areas. Dust settled in work areas then becomes dispersed and subsequently ignited by flames from adjacent areas. This subsequent explosion is the secondary explosion and its destructive effects are frequently much greater than those of the primary explosion. A greater amount of dust can become involved in the secondary explosion, and so the total amount of energy released may be greater. The results of secondary explosions may well be collapse of buildings, which can lead to extended damage and to casualties.

Although the occurrence of explosions has been recognized from years of experience, their hazards are not necessarily accounted for in recent trends in plant design in modern industry. The increased size of many new industrial operations means that the volume of the plant giving rise to the primary explosion may be factors of 10, or more, greater than was customary a few years ago. Hence the energy release in a primary explosion can be much greater and the protection requirements of the plant itself become a serious problem not only to the industrial operation, but to the surrounding community. The difficulty is aggravated because increase in size of plants tends to involve light structures and collapsible buildings. Provision of mechanical strength is expensive and it may not be economical to provide more than the minimum for safety. On the credit side, there is a tendency for large modern operations not to be situated in enclosed areas but to be built in the open away from residential areas. The risk is reduced of secondary explosions arising from the disturbance of dust external to the plant and causing damage to surrounding property. However, the total amount of energy which may be released during a single explosion in such a large building may be comparable to that obtained in the past as the sum of primary and secondary explosions. Information available at present on the safe methods of handling explosions on a large scale is very limited, and a need for such information is likely to increase rapidly in future for industrial and community safety.

The flames and pressure effects arising from dust explosions are a principal source of casualties, but there are also dangers arising from the combustion products of the explosions. These are deficient in oxygen and also contain toxic gases, particularly carbon monoxide. The concentration of toxic gases can be sufficiently high for a few breaths to be fatal, so that momentary unconsciousness due to the blast can result in death.

The toxic nature of the combustion products is particularly hazardous where the explosion has occurred in the working space with operations present, as occurs in secondary dust explosions or in coal mine explosions. In coal mines, because of the necessary mechanical ventilation systems, toxic gases from a relatively minor explosion may circulate over a wide area and cause casualties in regions where the blast and heat effects from the explosion were negligible. The hazards from toxic gases are additional reasons for not permitting dust explosions to occur in closed working spaces, and emphasize the desirability of venting industrial plants to atmosphere outside the work room.

Toxic gases are also likely to be present inside mechanical equipment after a dust explosion. Equipment must not be entered by operators or fire fighters until either the atmosphere has, by test, been shown to be safe, or breathing apparatus is worn by personnel entering.

Explosions which can occur from dispersed dusts can be spectacular and cause extensive damage, but only a minority of ignitions of dust result in explosions. The majority of ignitions occur in dust layers, deposits, or heaps, where the dust is not disturbed and a dust fire rather than an explosion is the result. There are several important aspects of dust fires which should be considered. These include (a) the dust which is ignited may in itself be a valuable material and that destruction of it by fire may cause substantial loss; and (b) even if the dust is of relatively little value, and it may even be a waste material, it can cause fire to spread to other neighboring materials, or buildings. There may also be a potential explosion risk, which could become reality if the burning dust were violently disturbed or otherwise dispersed into the air.

Past experience in industry or in the laboratory testing of dusts has shown that a wide range of dusts can give rise to explosions. Not all materials that will burn in air can cause dust explosions, even if finely divided and dry; that is, not all combustible dusts are explosive. All explosive dusts must be combustible. The reason why some combustible dusts are not explosive has not been definitely established, but it is clear that it is not related directly to the heat of combustion or calorific value of the dust. Some dusts with relatively high heat of combustion, such as graphite and some anthracite coals, are not explosive whereas other materials of lower heats of combustion, such as wood sawdust, are readily explodable. Dusts which are not explosive in air at atmospheric temperatures may be capable of explosions in a heated chamber, such as a furnace or oven. When combustible dusts are present, it is necessary to either refer to previous records or to carry out laboratory tests to decide whether the dust is explosive.

Over the years a considerable amount of information on the explosibility of dusts has been accumulated both from practical experience and from laboratory tests, and this information is of great assistance in assessing dust explosion hazards. Clearly where new materials are involved, or mixtures of old materials, or old materials made by new processes where their characteristics may be different, recourse must be made to further laboratory testing.

6.6.2 Protective Measures. The industries concerned in the manufacture or handling of explosible dusts are numerous. Some of the principal industries are (a) agriculture, (b) chemicals; including dyestuffs, (c) coal, mining and utilization, (d) foodstuffs, human and animal, (e) metals, (f) pharmaceuticals, (g) plastics, and (h) woodworking. This list gives a few examples and consists of those in which the principal products present dust explosion risks. There also is a wide range of industries which produce explosible materials in their processes, although dusts and powders may not be principal products as they are, for instance, in flour and sugar manufacture. Examples of industries where explosible dusts are present, and have in fact caused explosions, are rubber dust in the footwear industry, exparto grass dust in the paper industry, and aluminum dust in the manufacture of refrigerators.

Although the principal product of an industry may not be a dust and presents no dust explosion hazard, consideration must also be given to intermediate materials used in the manufacture of the final product, and to the processes involved in the manufacturing. These processes may give rise to by-products or waste materials which are explosible, and which in the factory may present as great a hazard as dusts in other industries where the principal product is explosible.

In attempting to assess the dust explosion risk, and the precautions which must be taken, due account should be taken of the size and nature of the plant and operations in which the explosible dust is involved. For example, any process in which explosible dusts are handled on the tonnage scale and where dispersion of this dust in air may occur would need detailed consideration to ensure that adequate explosion precautions have been taken. On a small scale, where perhaps only a few kilograms per hour of dust are involved, but where this dust is in suspension in air and operators are nearby (as in the polishing of metals accompanied by a dust extraction system), a dust explosion hazard is again present. Explosion protection is then advisable. Where the same quantity of dust is being manufactured under closely controlled conditions, and dispersion in air is most unlikely and ignition sources are not present as part of the process, there may be scope for the relaxation of explosion protection requirements. Such a situation might arise for instance where a pharmaceutical product is being synthesized and the process requires great care in the handling of materials, and where to prevent contamination the workroom is maintained scrupulously clean. Another example is where the dust has a high moisture content or is handled in a wet state.

Although it is good practice to eliminate as far as possible the formation of explosible dust clouds, and of sources of ignition, these steps give only partial protection. The reasons are plain; industrial operations demand the transport of explosible dusts, and the formation of suspensions in air is inevitable. The most severe precautions against the introduction or generation of sources of ignition may be taken, but because of unforeseen mechanical or human failures, complete elimination of ignition sources cannot be relied upon, particularly where powered machinery is involved. Sooner or later the conditions may occur for a primary dust explosion, namely generation of a suitable cloud of an explosible dust in the presence of a source of ignition. Palmer (1973) suggests that the minimizing of hazards can be achieved if reliance should be placed on the adequate functioning of explosion protection provided in advance. He suggests that basic precautions be taken for prevention of ignition, suppression or containment of the explosion flame, and allowance for an explosion to take its full course, but to ensure that it does so safely.

The method of protection selected depends on several factors including the design of the industrial process, the operational costs, the economics of alternative protection methods, the extent to which an explosion and its consequences can be foreseen, and regulatory requirements applicable.

Design of the buildings in which explosible dusts are handled must also be considered. Where the mechanical functions are inside a building, and explosion suppression equipment or ducted vents have been installed to prevent burning dust from being discharged from the plant; e.g., for bagging, the presence of dust within the building must be expected on occasions even with the most stringent precautions. Again, if a building is used for the storage of dusty material in sacks or drums, breakages and spillages can be expected. In all these cases there is a risk of dust being present within the room and the probability of a dust explosion must again be considered.

As a basic technical principle, the explosion hazard in a given plant with given materials increases as the amount of energy released by the explosion increases. The damage caused by the explosion of a large volume of combustible dust suspension may be more than proportionately greater than the damage from a smaller volume because larger structures are often relatively weaker. The escalation does not always occur because in a large volume there is always the possibility that parts of it may be at unfavorable dust concentrations for the explosion to propagate and this introduces an uncontrolled but positive safety factor. However, it is unwise to rely on this factor and in designing against an explosion hazard the safest method is to assume the worst conditions.

In trying to minimize the energy developed by an explosion, the most attractive method of protection is to prevent ignition of the dust cloud. As well as eliminating sources of ignition, an additional method which has been used with success is to reduce or eliminate the presence of the oxidant gas (usually the oxygen of the air). This may be done by introducing into the dust handling process an inert gas which either replaces the air originally present or at least dilutes it so that the level of oxygen is below that at which flames can be supported. This concentration level depends on the dust concerned, and also the inert gas used for diluting the air, and has to be determined in each case by an appropriate test.

Details for providing safety factors for explosion hazards are readily available in the industrial hygiene or safety literature cited in Palmer (1973).

6.7   INDUSTRIAL USE OF FINE POWDERS

The use of fine powders for industrial applications has become an increasingly important factor in aerosol technology. Finely divided powders now are used for reinforcement of materials; surface coatings; and for laminated, polycomponent materials. As a final example of applications for aerosol technology, aspects of this materials technology are surveyed in this section.

6.7.1   Production of Useful Powders.   The requirement for the production of a fine powder is initially for the formation of a large number of nuclei, the subsequent growth of which is controlled. To produce many nuclei a high degree of supersaturation is necessary and, as a result, nucleation will tend to be homogeneous. Under these conditions the size of the critical nucleus is approximately inversely proportional to the degree of supersaturation.

Having produced nucleation of particles, growth must be restricted or controlled. In condensation from the gas phase, growth is very rapid and can be controlled only by the separation of the fine powder and the provision of very large numbers of nuclei. Both these criteria are fulfilled by rapid quenching. The practical difficulties of preparing a fine powder depend considerably on the scale of operation. Colloidal suspensions and aerosols are not fine powders, but they contain fine particles: however, the amount of material available for particle growth may be limited and growth is consequently less difficult to control. In an industrial aerosol generator, nucleation is probably heterogeneous under normal industrial conditions without gas cleanup, but is often made so purposely by the introduction of foreign nuclei obtained, in their turn, by evaporation and subsequent condensation as discussed in Chapter 4. This gives rise to a reproducible aerosol with a narrow size range. Crystal structure can be modified by slight changes in the composition of the gas stream. Thus, when a number of metals are evaporated into a low-pressure (10 Torr) current of argon, the presence of about 0.5% by volume of oxygen causes changes in the shape of the crystals which become irregular and roughened, apparently without gross change in their chemical composition.

The application of aerosol particle generation offers certain special opportunities for powder production. Some examples are discussed below.

Oxides, metals, and carbides are among the materials which have been purified. For these, high temperatures are required, but rapid cooling is easily achieved.

In order to achieve the high temperatures required, one popular heat source which has the advantage of being usable in controlled atmospheres, is the high-intensity arc (Scheer and Korman, 1956). In this, the feed material, mixed with graphite if it is nonconducting, is made into the anode. Temperatures may reach 7,000°K in the tail flame as the vapor jet streams from the anode. Once out of the arc flame, the temperature of the vapor falls rapidly and the resulting high degree of supersaturation leads to rapid condensation. The use of the high-intensity arc in the production of a variety of fine powders has been reported by Holmgren *et al.* (1963) and many of these are available commercially (see also Table 6.8). Mean particle diameters usually lie in the range 0.005–0.1 μm. With most materials the particles are spheres. Polycrystalline particles are thought to result from vapor–solid condensation, while spheres are formed when there is a liquid intermediate.

Chemical reaction may occur almost incidentally in the course of the

TABLE 6.8

Some Fine Powders Prepared Using the High-Intensity Arc[a]

(A) Single components			

Oxides (in air)

SiO_2	Al_2O_3	Fe_2O_3	ThO_2
MnO_2	Nb_2O_5	NiO	Y_2O_3
UO_2	MoO_3	ZrO_2	MgO
WO_3			

Elements (in an inert atmosphere)

C	Al	Li	Ni
Fe	W	Mo	

Carbides (in an inert atmosphere)

ThC	TiC	B_4C	UC
TaC	SiC		

(B) Polycomponents		

Oxides

(Zn, Mn, Cu, Fe)O	$Al_2O_3 \cdot SiO_2 \cdot Fe_2O_3$
$LiAl(SiO_3)_2$	$MnO \cdot SiO_2$
$FeO \cdot Cr_2O_3$	

Carbides

$B_4C \cdot SiC$	$UC \cdot NbC$
$UC \cdot ThC$	

[a] From Veale (1972); copyright British Crown. Reproduced with permission of the Controller of Her Britannic Majesty's Stationery Office.

evaporation–condensation process, or it may be necessary to make special provisions for such a reaction. Examples of the first case are to be found in Amick and Turkevich's (1963) work, or in the use of the high-intensity arc in hydrogen atmosphere to prepare fine nickel, iron, or tungsten from the corresponding oxides (Holmgren *et al.,* 1963). Examples of special provisions are to be found in the control of the degree of surface hydroxylation of fine powders by control of the water content of the atmosphere in which condensation occurs (Everest *et al.,* 1971) or in the provision of a specific reactive atmosphere in which vapor–vapor reaction may take place. By means of this last procedure, an exceedingly wide range of materials, such as metals, metalloids, oxides, carbides, borides, nitrides, silicides, sulfides, and many others, can be prepared.

The heat source should be appropriate to the temperature required. For high-temperature reactions, arcs, chemical flames (Lamprey and Ripley, 1962), and direct or induction plasmas may be used (Neuenschwander, 1966;

Barry *et al.,* 1968). Of these, the latter, having no electrodes, may be operated in either corrosive or highly oxidizing environments, and with a range of different torch designs. The addition of facilities for the injection of quench gas results in a highly flexible assembly capable of exerting close control over all facets of any vapor–vapor reaction and of working over a very wide range of vapor concentrations. Some of the various reactions that have yielded fine powders are given in Table 6.9.

The theory of the nucleation and growth processes for a thermal decomposition (specifically of salt hydrate crystals) has been considered. If the thermal decomposition takes the form

$$A_{solid} \rightarrow B_{solid} + C_{gas},$$

the formation of the new phase B results from local structural fluctuations in

TABLE 6.9

Examples of Vapor-Phase Reactions which Have Produced Submicrometer Powders[a]

(A) Single components			
Oxides (in air)			
SiO_2	Al_2O_3	Fe_2O_3	ThO_2
MnO_2	Nb_2O_5	NiO	Y_2O_3
UO_2	MoO_3	ZrO_2	MgO
WO_3			
Elements (in an inert atmosphere)			
C	Al	Li	Ni
Fe	W	Mo	
Carbides (in an inert atmosphere)			
ThC	TiC	B_4C	UC
TaC	SiC		

(B) Polycomponents	
Oxides	
(Zn, Mn, Cu, Fe)O	$Al_2O_3 \cdot SiO_2 \cdot Fe_2O_3$
$LiAl(SiO_3)_2$	$MnO \cdot SiO_2$
$FeO \cdot Cr_2O_3$	
Carbides	
$B_3C \cdot SiC$	$UC \cdot NbC$
$UC \cdot ThC$	

[a] From Veale (1972), with references; copyright British Crown. Reproduced with permission of the Controller of Her Britannic Majesty's Stationery Office.

the lattice of A which produce conditions favorable to the formation of a nucleus of B. Such sites are situated at regions of disorder, e.g., vacancy, interstitial, or impurity clusters, meeting points of high-angle grain boundaries, etc.

In most decompositions the molecular volume of the product is less than that of the reactants. Because of this volume change, both reactant and product crystals become strained and this strain cannot be relieved until the critical shear stress in the neighborhood of the nucleus is exceeded. Thus, there is a strain energy involved in the growth of a nucleus up to, and especially beyond, its critical radius. Typically, the interfacial strain energy is of the order of 1 kcal/mol, and this is sufficient to account for the slow growth of small nuclei, which leads to the formation of a fine powder (Young, 1966). This is done conveniently by atomizing a solution or suspension of the material to be decomposed into a flame. A wide variety of single and double oxides has been prepared by this route (Table 6.10).

6.7.2 Uses of Fine Particles. Fine powders are finding increasing application for special purposes in materials technology. Perhaps best known are their use in surface coatings such as paints and inks. However, there are important new applications emerging including reinforcement and strengthening.

Fine powders are used as pigments in surface coatings and make definite contributions to the properties of such coatings as paints, enamels, and inks. Thus the chemical properties of the pigment assist in establishing the suitability of paints for exterior use, while physical properties have their place in determining sheen, texture, consistency, and thickness of the coating. The pigment field is very large, with its own textbooks (Nylen and Sunderland,

TABLE 6.10

Some Finely Divided Oxides Prepared by Atomizing Solutions into a Flame[a]

	Products	Starting solution
Single oxides	δ-Al_2O_3 Co_3O_4 Cr_2O_3 CuO α-Fe_2O_3 γ-Fe_2O_3 HfO_2 p-MnO_2 Mn_2O_3 NiO SnO_2 ThO_2 TiO_2 ZrO_2	Sulfate or chloride
	Mn_3O_4	Acetate
Double oxides	$CoFe_2O_4$ $MgFe_2O_4$ $MnFe_2O_4$ $ZnFe_2O_4$ $BaO6Fe_2O_3$	Chlorides
	$BaTiO_3$	Acetates or lactates
	(Ni, Zn) Fe_2O_4 $PbCrO_4$ $Cu_2Cr_2O_4$	Nitrates
	$CoAl_2O_4$	Sulfates
Mixed oxides	Al_2O_3–Cr_2O_3 Al_2O_3–CuO Al_2O_3–Fe_2O_3 Al_2O_3–NiO	Sulfates
	Al_2O_3–NiO Cr_2O_3–Fe_2O_3 Fe_2O_3–$CoFe_2O_4$	Chlorides

[a] From Veale (1972), with references; copyright British Crown. Reproduced with permission of the Controller of Her Britannic Majesty's Stationery Office.

1965) from which further information may be sought. We are concerned here with the part played in coatings by the fine-particle powders and with the specific properties of these powders that make them suitable for use in such coatings.

A paint film is typically about 20 μm thick, so all pigment particles must be smaller than this, but the actual size may vary quite widely from perhaps 10 μm down to perhaps 0.01 μm. In letterpress or lithographic printing, the ink film is less than 3 μm thick so the upper limit on particle size is more restrictive. This upper limit applies not to the size of the ultimate particle, but to any agglomerate which may resist the forces of application and persist in the final coating. Thus, although many pigments used in surface coatings are not less than 1 μm in diameter, and therefore are not strictly within the scope of our consideration, it is true to say that pigments represent one of the two largest uses of submicrometer powders.

A high gloss exists when a beam of light falling on a surface is reflected directionally. A mirror may reflect directionally more than 90% of incident light at high angles of incidence. Surface coatings do not usually reach this figure and it is typical of this type of reflection that, as the angle of incidence decreases from 90°, the amount of light reflected falls off. Gloss is imparted by extending a smooth surface over an area much greater than the wavelength of light. In a surface coating, it is largely a property of the air–binder or air–medium interface, with the pigment only showing up when the particles cause irregularity in the reflecting surface. The pigment particles must therefore be small and must not be present in too high a concentration.

A matte surface, on the other hand, is, on the microscale, rough, and multiple reflection of an incident beam occurs, producing a much more even distribution of reflected light. This is usually the result of a much higher pigment concentration in the medium. Large particles produce uneven effects, but particles can be somewhat larger than those in a gloss finish.

Pigments for surface coatings may be inorganic or organic. The more stable colors now available in the latter category are leading to increased use of organic colored pigments, as compared with inorganic coatings. However, since the latter are often less expensive, such materials as the different forms of iron oxide are likely to continue to be used. For white pigments inorganics predominate; titanium oxide (titania) is used in very large and increasing tonnage, whereas zinc oxide, antimony oxide, and lead sulfate- and barium sulfate-based pigments now find less use. For black coloration, some form of carbon black is usually used.

As regards particle size, zinc and antimony oxides, which are usually produced by burning the corresponding elements, and titania, together with carbon black, are usually produced in submicrometer particle size. An extra step is necessary to increase the size of titania particles, produced by the process of oxidation of titanium chloride to the optimum for light scattering and to avoid the bluish tint. Synthetic iron oxides usually produced by calcination of ferrous sulfate can take on a wide range of colors from pink to dark, bluish black. Differences in color intensity and also slight differences in shade for the α-ferric oxide pigment can be correlated with particle size (Hund, 1966).

Pigments are added to plastics and rubber to mask unattractive natural colors, to appeal to the eye, and sometimes for color coding. Although the general principles are similar to those for surface coatings, the pigment can now be dispersed through a much greater thickness. Organic dyes are much used, but for thermosetting plastics inorganic pigments predominate, because at the temperatures needed for curing (up to 300°C) organic pigments are not stable (Scott, 1961). The white reinforcing pigments much used are titania, silica, and zinc oxide, whereas carbon black is used more often as a reinforcement than a pigment.

Although the coating of paper has some similarity, in principle, with such surface coatings as paints, inks, etc., considered above (especially with regard to coloring), much paper is used as a surface for printing and as a recipient of ink coatings and other special coatings. Pigments are introduced into paper at two stages. The first is in the production of sheet paper where the pigment acts more as a filler, and the second when a true coating is applied to the sheet. Much of the pigmentation of paper and its opacity can be produced by the same submicrometer size materials added at either stage.

With few exceptions, colored pigment particles now in use are not strictly fine powders. Apart from carbon black and iron oxide, most colored pigments used in paints and inks are organic. This is due not to the inherent inexpensiveness of organic pigments, but to the intensity of the colors that may be obtained and to the recently developed light fastness. In plastics, inorganic colored pigments will probably continue to be needed for thermal stability. Here the increasing cost of some (e.g., nickel titanate) should spur the search for new inorganic pigments. While it is possible to predict, with some degree of assurance, the general color of an organic compound of a given structure, the connection between color and structure for an inorganic compound is frequently much more complicated. Explanation of an observed color is usually possible, but prediction is rarely possible at the present time. Future developments, in the long run, will change this state of affairs.

At the moment, the particle size of colored pigments is not usually closely controlled. In the future, it will probably become possible to separate the scattering and absorption functions with respect to particle size, and to optimize one or the other function by controlling particle size (see, for example, Fig. 5.17).

White pigments are generally inorganic in nature. The high refractive index of titania provides increased scattering power on the interface between titania and the medium. However, titania needs to be coated in order to reduce photochemical reaction with the organic medium which leads to discoloration or chalking. A nonchalking white pigment of equivalent scattering power would offer advantages. Silicon carbide has a refractive index similar to titania, and can occur as plate-shaped particles that would give good abrasion resistance to the coating. However, silicon carbide is a semiconductor and color control may depend critically on the concentration of impurities so that a reproducible white color may not be obtainable.

Much work has already been done on the dispersion of pigment particles in a medium. Recently, electron microscopy has been used to examine the dispersion of pigment particles in a paint film, and attempts will probably be made to correlate this final dispersion with that in the coating as applied and thus with the surface chemistry of the pigment particles themselves. Developments such as these may be expected to develop a firmer scientific basis for the pigment field, which is already technologically well advanced.

Dispersions of fine particles as a second phase in a matrix of a first phase are quite widely used to strengthen the first phase. The nature of the dispersed phase varies with that of the matrix; combinations which may increase strength are shown in Table 6.11. Since different types of matrices fail mechanically in different ways, it is to be expected that the mechanisms of strengthening will vary, and this is certainly true. Many details of the proposed mechanisms are, at the moment, subjects of considerable discussion.

Of the types of strengthening listed in Table 6.11, precipitation strengthening is unique in that, although the strengthening component consists of fine particles, these are generated in situ and do not have an existence separate from the bulk metallic matrix. Thus precipitation strengthening is not relevant to use of fine powders, and is not discussed further.

In dispersion-strengthened metals the dispersed phase is a hard ceramic. As a result, methods of preparation and fabrication are quite different from those for precipitation strengthening. The metallic matrix cannot now be fused without the risk of destroying the dispersion, and consequently fabrication and joining become more difficult.

The prototype of modern dispersion-strengthened metals (sintered aluminum powder with oxide dispersion) was announced in 1950. Since that time,

TABLE 6.11

Matrix-Dispersed Phase Combinations Leading to Strengthening

Nature of matrix phase	Nature of dispersed reinforcing phase	Type of strengthening
Metallic	Metallic	Precipitation strengthening
Metallic	Hard ceramic	Dispersion strengthening
Ceramic	Metallic	Dispersion strengthening
Elastomer	Inorganic or carbon	Reinforcement
Metallic	Hard ceramic	Cemented ceramic or cermet (higher proportion of refractory phase than in dispersion hardening)

a very wide range of metallic matrices and of finely divided hard ceramic particles—usually oxides—has been studied. In spite of this wealth of data, there is no completely accepted view of the detailed mechanism of strengthening by hard ceramic particle dispersions. A critical examination of theories for the yield strength, work hardening, and creep has shown that, while a number of the theories are of value, it is probable that no one theory will successfully describe the behavior of all dispersion-strengthened alloys (Ansell, 1968). In general, dispersion-strengthened metals are weaker than conventional alloys at lower temperatures, but more stable and stronger at higher temperatures.

Fine molybdenum dispersed in an alumina ceramic matrix inhibits grain growth. Increased strength in the ceramic matrix results from the finer grain size (Veale, 1972). Starting with submicrometer alumina and molybdenum, a matrix grain size less than 2 μm and a 98% relative mass density can be obtained after consolidation. According to McHugh *et al.* (1966), fracture energy is about 50% greater than that of alumina of the same grain size. Cutting tools of such strengthened alumina are expected to have lifetimes up to five times as long as normal alumina tips. This method of controlling grain growth is similar in action to the use of nonmetallic grain growth inhibitors, which are quite widely used in sintering ceramics and which appear to operate by deposition in, and pinning of, grain boundaries (Cable and Burke, 1963).

A dispersion of fine tungsten in ceramic uranium and plutonium carbides controls fission gas swelling. Ervin (1971) reports that the tungsten is precipitated by annealing at 1,400°C to exsolve it from the carbide into which it had been introduced by zone refining.

Dispersion strengthening would appear to provide a means of retaining a fine grain size in a ceramic without the need for the complexities and restrictions inherent in hot pressing. Whether the presence of the dispersion can be tolerated must depend on the use envisaged for the product.

Elastomers are composed of a tangled mass of kinked and intertwined chainlike molecules which, above the glass transition temperature are in a state of constant thermal motion. Elasticity arises from resistance to forces tending to distort the normal pattern within the solid. The tangled chains are free to slide past one other except where they are tied together by cross links. Such links limit the amount of stretch and increase the elasticity, and are formed by reaction of the chains with certain additives (curing agents or accelerators) mixed into the mass. Payne (1966) finds that heating also increases the extent and rate of cross-linking, which is termed vulcanization. Natural and synthetic elastomers vary greatly in their properties but, in general, are not able, as they stand alone, to meet the more stringent demands placed upon them, and need to be reinforced.

Depending on the function being fulfilled, elastomers deteriorate in two main ways. When used for energy absorption (as in cushioning, silencing, etc.) repeated deformations, small relative to the ultimate breaking strength, lead to a form of fatigue and a deterioration in the hysteresis behavior which is the main method of dissipating energy. In other types of usage (e.g., road tires) failure occurs by abrasion in which the surface of the rubber is torn away. There have, therefore, been two approaches to the study of the strength of rubber, one via the hysteresis, the other via abrasion due to cut growth, explained by the concept of tearing energy; Veale (1972) has noted that attempts have been made to find common ground in these two approaches. For amorphous rubbers there is a relationship between strain at break and hysteresis at break so tensile strength is often used as a measure of usable strength; like tear failure, tensile failure is initiated by some stress raising discontinuity. Tensile strength is always higher than tear strength as the volumes subjected to stress concentrations which initiate failure are smaller in tensile failure. However, under certain conditions, when the tear initiates at a notch, tear strength and tensile strength may be widely different from a uniform sheet. To resist abrasion, stiffness and mechanical hysteresis are needed, as well as good tear strength. Ervin (1971) indicates that stiffness prevents excessive elongation, while mechanical hysteresis leads to energy loss as heat, and consequent reduced wear.

It is difficult to define reinforcement exactly in terms of a single property of the elastomer system. For example, reinforcement is often considered to be represented by an increase in tensile strength, but, in some natural rubbers which already possess good tensile strength, such increase may be quite marginal. Payne (1966) indicates that it is probably preferable to consider as reinforcing a filler which causes increased stiffness, elastic modulus, and mechanical hysteresis, though usually at the expense of resilience and elasticity.

Within this definition and within reason it is true to say that any hard, finely divided solid of about 20–40 nm particle size will reinforce rubber. A number of recent reviews cover, in much greater depth than is justified here, many aspects, theoretical and practical, of the reinforcement of elastomers (Payne and Whittaker, 1971).

It has been found that a general classification into strongly and weakly reinforcing filters can be obtained by plotting the extension of a filled elastomer against the so-called *Mullins softening*—the reduction in elastic modulus caused by successive stretching. Appreciable softening, related to mechanical hysteresis, occurs only when there is true reinforcement (Buche, 1960).

To compare fillers, it is necessary that they exist in the same elastomer matrix and cross-linked to approximately the same degree. While the particle size or surface area of interaction between filler and elastomer are the prime factors, in this particular matrix it is not the only factor of interest. The bulk chemical nature of the filler may be involved, or the nature of its surface, or the degree of dispersion, or other factors. Careful work using carbon black of constant specific surface and surface chemical characteristics but differing in the degree of "structure" shows (Kraus, 1971) that tensile strength and rupture energy are functions of the carbon black concentration and a structure-dependent factor which correlates with the oil absorption of the black.

Carbon black is by far the most utilized reinforcement for rubbers. Millions of tons are used annually. A very wide variety of blacks is available, classified by method of manufacture or so-called structure. The available types and the progression of new types have been reviewed (Dannenberg, 1971).

Carbon black interacts with the elastomer at two stages in the processing, first when added to the raw rubber during milling, and second, during the vulcanization process. Addition of reinforcer to the raw rubber leads to the formation of "bound rubber"—a gel containing both rubber and reinforcer that is insoluble in the usual solvents for the elastomer. Bound rubber is usually pictured as hard islands (with carbon cores) set in a matrix of unchanged elastomer. The greater apparent degree of cross-linking within these islands includes a contribution from polymer–filler bonds formed during milling. The later increase in cross-linking during vulcanization involves interaction of functional groups, surface active sites, etc., with one other and with additives under the influence of increased temperature.

Apart from carbon blacks, the only fillers classified as fully reinforcing are specially prepared fine particle silicas and silicates of aluminum and calcium. Other oxides such as alumina and zinc oxide have been used experimentally. Such finely divided fillers are usually prepared by precipitation or condensation routes. The total produced is less than one-tenth of that of carbon black, and about two-thirds of this is silica (Smith, 1968).

The reinforcing action of non-carbon black fillers is even less well characterized and understood than that of carbon blacks. There are many points of similarity with blacks (e.g., formation of chemically bound rubber and modification of viscous properties), but also many differences. White fillers such as silicas and aluminum or calcium silicates are not capable of providing the wide range of surface groupings found on carbon black. Only silanol and siloxane groups are found on silica particles. The surfaces of silica and silicates are more polar than those of carbon black and are often considerably affected by the presence of small amounts of adsorbed moisture. To date, there appears to be no clear-cut evidence for chemical interaction between a white filler and an elastomer whether carbon based or not, but it must be recognized that different mechanisms of reinforcement may predominate at different stresses or in reinforcement–elastomer combinations of the same type but different composition.

There is relatively little information on fine-particle reinforcement in either thermosetting or thermoplastic matrices. In thermosetting materials, the degree of cross-linking is usually high and such materials fail by brittle fracture. The most important reinforcers are fibrous, but particulate fillers are used for special purposes (e.g., alumina and silicon carbide impart abrasion resistance and graphite imparts electrical conductivity) or in order to cheapen the product but, for these purposes, fine powders are not required according to Loewenstein (1966). Fine silica is added to control viscosity and impart thixotropy, but, as far as is known, there is no evidence that it exerts much reinforcing action (Schue, 1969). On the other hand, Galperin *et al.* (1965) show that fine titania increases tensile strength in a cross-linked epoxy resin, at least at certain relative humidities. No attempt appears to have been made to ensure good wetting or other bonding of filler to matrix, and this area may well benefit from study.

The effect of the use of fillers in linear thermoplastics is much less than when the polymer is cross-linked. Thus, although a reinforcing carbon black, in combination with coarser powders, increases the stiffness of polyethylene; this is accompanied by a decrease, rather than an increase, in tensile strength (Boonstra, 1965). On the other hand, Galperin and Kwei (1966) indicate that fine titania appears to somewhat increase tensile strength in poly(vinyl acetate), while an attempt to correlate changes in the tensile strength of polyethylene containing different fine fillers with the particle size of the filler showed a slight increase in strength with the finest filler according to Alter (1965).

Cermets and cemented carbides contain much higher proportions of ceramic phase than dispersion-strengthened metals (Schwartzkopf and Kieffer, 1960). Cermets normally contain 50–85% ceramic, while cemented ceramics contain more than 85% ceramic, although the distinction is quite arbitrary. Morral (1966) reports that industrially cemented carbides are the more important of these compounds with a cobalt-tungsten carbide or cobalt-mixed carbides of tungsten, titanium, tantalum, etc., predominating. In terms of practical usefulness, these materials represent an attempt to use the hardness of some ceramics while increasing somewhat the impact resistance by setting the ceramics in a metallic matrix. There are considerable differences in properties, depending on whether or not the metallic phase wets the ceramic. In general, oxides are not wetted by metals (this, in itself, leads to the use of metal-oxide cermets as containers for molten metals), whereas carbides are so wetted by a number of metals as indicated by Ramquist (1965). Consequently, carbide–metal bonding is stronger than that between oxide and metal.

The examples given above indicate the range and diversity of applications for fine particles. It appears that this material technology is still in its infancy and a variety of developments can be expected in the future. These will make increasing use of the unique properties of fine particles, and the expanding techniques for generating them by application of aerosol technology.

REFERENCES

Agafanova, F. A., Gurevich, M. A., and Paleev, I. I. (1957). *Sov. Phys.—Tech. Phys. (Engl. Transl.)* **2**, 1689.

Akesson, N. B., and Yates, W. E. (1976). *U.S. For. Serv., Res. Pap. PSW* No. 151, p. 4.

Akesson, N. B., and Yates, W. E. (1979). "Pesticide Application Equipment and Techniques," Bull No. 38. Food Agric. Organ. U.N., Rome.

Aldred, J. W., Patel, J. C., and Williams, A. (1971). *Combust. Flame* **17**, 139.

Alter, H. (1965). *J. Appl. Polym. Sci.* **9**, 1525.

Amick, J., and Turkevich, J. (1963). *In* "Ultrafine Particles" (W. E. Kuhn, ed.), p. 146. Wiley, New York.

Ansell, G. S. (1968). *In* "Oxide Dispersion Strengthening" (G. S. Ansell, T. D. Cooper, and F. V. Lenel, eds.), p. 61. Gordon & Breach, New York.

Arthur, J. R., and Bleach, J. A. (1952). *Ind. Eng. Chem.* **44**, 1058.

Barry, T. L., Bayliss, R. K., and Lay, L. A. (1968). *J. Mater. Sci.* **3**, 229.

Beer, J. M., and Chigier, N. (1972). "Combustion Aerodynamics." Appl. Sci., London.

Beer, J. M., Lee, K. B., Marsden, C., and Thring, M. W. "Flamme de Carbon Pulverise a Mélange Contrôle," p. 141. Inst. Fr. Combust. Energ., Paris.

Bell, K. A., Avol, E. L., Bailey, R. M., Kleinman, M. T., Landis, D. A., and Heisler, S. L. (1980). *In* "Generation of Aerosols and Facilities for Exposure Experiments" (K. Willeke, ed.), pp. 475. Ann Arbor Sci. Publ., Ann Arbor, Michigan.

Bittner, J. D., and Howard, J. B. (1981). "Particulate Carbon-Formation During Combustion" (D. C. Siegla and G. W. Smith, eds.), p. 109. Plenum, New York.

Blyholder, G. D., and Eyring, H. (1957). *J. Phys. Chem.* **61**, 682.

Bonne, U., and Wagner, H. (1965). *Ber. Bunsenges. Phys. Chem.* **69**, 35.

Bonne, U., Homann, K., and Wagner, H. (1964). *Symp. (Int.) Comb. [Proc.]* **10**, 503.

Boonstra, B. B. (1965). *In* "Reinforcement of Elastomers" (G. A. Kraus, ed.), p. 529. Wiley (Interscience), New York.

Breen, B. P. (1976). *Symp. (Int.) Combust. [Proc.]* **16**, 19.

Brescia, F. (1946). *J. Econ. Entomol.* **39**, 698.

Brock, J. R. (1972). *In* "Assessment of Airborne Particles" (T. T. Mercer, P. E. Morrow, and W. Stöber, eds.), Chap. 7. Thomas, Springfield, Illinois.

Brown, K. C., and James, G. J. (1962). *SMRE Res. Rep.* No. 201.

Brzustowski, T. A. (1965). *Can. J. Chem. Eng.* **43**, 10.

Brzustowski, T. A., and Natarajan, R. (1966). *Can. J. Chem. Eng.* **44**, 194.

Buche, F. (1960). *J. Appl. Polym. Sci.* **4**, 107.

Burstein, S. Z., and Naphtali, L. M. (1960). "Kinetics Equilibria and Performance of High Temperature Systems." Butterworth, London.

Canada, G. S., and Faeth, G. M. (1974). *Symp. (Int.) Combust. [Proc.]* **15**, 419.

Cassel, H. M., Das Gupta, A. K., and Guruswamy, S. (1948). *Symp. (Int.) Combust. [Proc.]* **3**, 185.

Chiu, H. H., and Liu, T. M. (1977). *Combust. Sci. Technol.* **17**, 127–142.

Clark, A., and Wheeler, R. V. (1913). *J. Chem. Soc.* **103**, 1704.

Coble, R. L., and Burke, J. E. (1963). *Prog. Ceram. Sci.* **3**, 197.

Coffee, R. H. (1971). *J. Agric. Eng. Res.* **16**, 98.

Courtney, W. G. (1960). *ARS J.* **30**, 356.

Csaba, J. (1962). Ph.D. Thesis, Dep. Fuel Technol. Chem. Eng., Univ. of Sheffield, Sheffield, England.

Dannenberg, E. M. (1971). *J. Inst. Rubber Ind.* **5**, 190.

David, W. A. L. (1946). *Bull. Entomol. Res.* **36**, Part I, 373; **37**, Part II, 1; **37**, Part III, 177; **37**, Part IV, 393.

de Grey, A. (1922). *Rev. Metall. Memoires* **19**, 645.

Diamond, R. (1956). Ph.D. Thesis, Univ. of Cambridge.

Downing, R. C. (1961). *In* "Aerosol Science and Technology" (H. R. Shepherd, ed.), p. 17. Wiley (Interscience), New York.

Elliott, M. A., ed. (1981). "Chemistry of Coal Utilization," 2nd Suppl. Vol., Chaps. 1, 8, and 19. Wiley, New York.

El-Wakil, M. M., and Abdou, M. I. (1966). *Fuel* **45**, 177.

Ervin, G. (1971). *J. Am. Ceram. Soc.* **54**, 46.

Essenhigh, R. H. (1955). *Sheffield Univ. Fuel Soc. J.* **6**, 15.

Essenhigh, R. H. (1959). *Proc.—Resid. Conf. Sci. Use Coal, Sheffield, Engl., 1958* p. A74.

Essenhigh, R. H. (1976). *Symp. (Int.) Combust. [Proc.]* **16**, 358.

Essenhigh, R. H., and Fells, I. (1960). *Discuss. Faraday Soc.* **30**, 208, 211.

Essenhigh, R. H., and Howard, J. B. (1971). *Pa State Univ. Stud.* No. 31.

Everest, D. A., Sayce, I., and Selton, B. (1971). *J. Mater. Sci.* **6**, 218.

Faeth, G. M., and Lazar, R. S. (1971). *AIAA J.* **9**, 2165.

Faraday, M., and Lyell, C. (1845). *Philos. Mag.* **26**, 16.

Fendell, F. E. (1965). *J. Fluid Mech.* **21**, 281.

Frances, W. F. (1961). "Coal," 1st ed. Arnold, London.

Galperin, I., and Kwei, T. K. (1966). *J. Appl. Polym. Sci.* **10**, 681.

Galperin, I., Arnheim, W., and Kwei, T. K. (1965). *J. Appl. Polym. Sci.* **9**, 3215.

Godsave, G. A. E. (1952). *Symp. (Int.) Combust. [Proc.]* **4**, 818.

Goldsmith, N., and Penner, S. S. (1954). *Jet Propul.* **24**, 245.

Green, H. L., and Lane, W. R. (1964). "Particulate Coulds: Dusts, Smokes and Mists," 2nd ed., p. 443. Van Nostrand, Princeton, New Jersey.

Hedley, A. B., Nuruzzaman, A. S. M., and Martin, G. F. (1970). *Nature (London)* **227**, 367.

Hedley, A. B., Nuruzzaman, A. S., and Martin, G. F. (1971). *J. Inst. Fuels* **44**, 38.

Heywood, J. B., Fay, J. A., and Linden, L. H. (1971). *AIAA J.* **9**, 841.

Hieftje, G. M., and Malmstadt, H. V. (1967). *Anal. Chem.* **40**, 1860.

Hirsch, P. B. (1955). *Proc. R. Soc. London, Ser. A* **226**, 143.

Holmgren, J. D., Gibson, J. D., and Scheer, C. (1963). *In* "Ultrafine Particles" (W. E. Huhn, ed.), p. 129. Wiley, New York.

Homann, K. H. (1967). *Combust. Flame* **11**, 265.

Homann, K. H. (1979). *Ber. Bunsenges. Phys. Chem.* **83**, 738.

Homann, K. H., and Wagner, H. (1965). *Ber. Bunsenges. Phys. Chem.* **69**, 20.

Horton, L. (1952). *Fuel* **31**, 341.

Hottel, H. C., and Stewart, I. M. (1940). *Ind. Eng. Chem.* **32**, 719.

Howard, J. B., and Essenhigh, R. H. (1967). *Symp. Int. Combust. [Proc.]* **11**, 399.

Hund, F. (1966). *Chem.-Ing.-Tech.* **38**, 423.

Jaarsma, F., and Derkson, W. (1970). *New Exp. Techn. Propulsion Energetics, Res. Advisory Group for Aerospace and Development (AGARD) Ser.* **38**, 123.

Johnstone, H. F., Winsche, W. E., and Smith, L. W. (1949). *Chem. Rev.* **44**, 353.

Jolley, L. J., and Poll, A. (1953). *J. Inst. Fuel* **26**, 33.

Juntgen, V. H., and van Heek, N. H. (1968). *Fuel* **47**, 103.

Kassoy, D., and Williams, F. A. (1968). *AIAA J.* **6**, 1961.

Kaufman, C. W., and Nicholls, J. A. (1971). *AIAA J.* **9**, 880.

Khan, I. M. *et al.* (1972). *Air Pollut. Control Transp. Engines Symp.* p. 205.

Khudyakov, J. N. (1949). *Izv. Akad. Nauk SSSR Otd. Tekh. Nauk* No. 4, 508.

Kimbert, G. M., and Gray, M. D. (1967). *Combust. Flame* **11**, 360.

Kobayasi, K. (1954). *Symp. (Int.) Combust. [Proc.]* **5**, 141.

Kotake, S., and Okazaki, T. (1969). *Int. J. Heat Mass Transfer* **12**, 575.

Kraus, G. (1971). *J. Appl. Polym. Sci.* **15**, 1679.

Kruse, C. W., Hess, A. D., and Ludvik, G. F. (1949). *J. Natl. Malar. Soc.* **8**, 312.

Kumagai, S., and Isoda, H. (1956). *Symp. (Int.) Combust. [Proc.]* **6**, 726.

Kumagai, S., Sakai, T., and Okajimi, S. (1970). *Symp. (Int.) Combust. [Proc.]* **13**, 779.

Labowsky, M., and Rosner, D. E. (1976). *Adv. Chem. Ser.* No. 168, p. 63.

Lamprey, H., and Ripley, R. L. (1962). *J. Electrochem. Soc.* **107,** 713.

Latta, L. R., Anderson, L. V., Rogers, E. E., LeMer, V. K., Hochberg, S., Lauterback, H., and Johnson, I. (1947). *J. Wash. Acad. Sci.* **37,** 397.

Law, S. E. (1978). *Trans. ASAE* **21,** 1096.

Ledinegg, M. (1952). "Dampferzeugung," p. 238. Springer-Verlag, Berlin and New York.

LeMott, S. R., Peskin, R. L., and Levine, D. G. (1971). *Combust. Flame* **16,** 17.

Liu, B. Y. H. (1967). Rep. No. 112. Part. Technol. Lab., Univ. of Minnesota, Minneapolis.

Loewenstein, K. L. (1966). *In* "Composited Materials" (L. Holliday, ed.), p. 129. Elsevier, Amsterdam.

Longwell, J. P. (1976). *Symp. (Int.) Combust. [Proc.]* **16,** 1.

Lorell, J., Wise, H., and Carr, R. E. (1958). *J. Chem. Phys.* **25,** 325.

McCreath, C. E., and Chigier, N. A. (1972). *Symp. (Int.) Combust. [Proc.]* **14,** 1355.

McHugh, C. O., Whalen, T., and Humenik, M. (1966). *J. Am. Ceram. Soc.* **49,** 486.

McQuaig, R. D. (1962). *Bull. Entomol. Res.* **53,** 111.

Masdin, E. G., and Thring, M. W. (1962). *J. Inst. Fuel* **35,** 261.

Melandri, C., *et al.* (1977). *In* "Inhaled Particles" (W. H. Walton, ed.), Vol. 4, p. 193. Pergamon, Oxford.

Miller, C. O. M. (1980). *Environ. Sci. Technol.* **14,** 824.

Mokler, B. V., Damon, E. G., Henderson, T. R., and Jones, R. K. (1980). *In* "Generation of Aerosols and Facilities for Exposure Experiments" (K. Willeke, ed.), p. 379. Ann Arbor Sci. Publ., Ann Arbor, Michigan.

Monaghan, M. T., Siddal, R. G., and Thring, M. W. (1968). *Combust. Flame* **12,** 45.

Morozzo, Count (1795). *Repert. Arts Manuf.* **2,** 416.

Morral, F. R. (1966). *In* "Fundamentals of Refractory Compounds" (H. Hansmer and M. Bouman, eds.), p. 229. Plenum, New York.

Mount, G. A., Lofgren, C. S., Baldwin, K. F., and Pierce, N. W. (1970). *Mosq. News* **30,** 331.

Mukunda, H. S., Marathe, A. G., and Ramani, N. (1971). *Proc. Natl. Heat Mass Transfer Conf. 1st, Indian Inst. Technol., Madras* Rep. HMT-45-71.

Mulcahy, M. F. R., and Smith, I. W. (1969). *Rev. Pure Appl. Chem.* **19,** 81.

National Fire Protection Assn. (NFPA) (1957). "Report of Important Dust Explosions." NFPA, Boston, Massachusetts.

Neuenschwander, E. (1966). *J. Less-Common Met.* **11,** 365.

Nishiwaki, N. (1954). *Symp. (Int.) Combust. [Proc.]* **5,** 148.

Nsakala, N. (1976). Ph.D. Thesis, Pennsylvania State Univ., State College.

Nusselt, W. (1924). *VDI-Z.* **68,** 124.

Nylen, P., and Sunderland, E. (1965). "Modern Surface Coatings." Wiley (Interscience), New York.

Olson, D. B., and Calcote, H. F. (1981). *In* "Particulate Carbon-Formation During Combustion" (D. C. Siegla and G. W. Smith, eds.), p. 177. Plenum, New York.

Palmer, K. N. (1973). "Dust Explosions and Fires," p. 372. Chapman & Hall, London.

Parks, T. M., Ablow, C. M., and Wise, H. (1966). *AIAA J.* **4,** 1032.

Payne, A. R. (1966). *In* "Elastomer Systems in Composite Materials" (L. Holliday, ed.), p. 290. Elsevier, Amsterdam.

Payne, A. R., and Whittaker, R. E. (1971). *J. Appl. Polym. Sci.* **15,** 1941.

Peskin, R. L., and Wise, H. (1966). *AIAA J.* **4,** 1646.

Peskin, R. L., Polymeropoulos, C., and Yeh, P. (1967). *AIAA J.* **5,** 2173.

Polymeropoulos, C. E., and Peskin, R. L. (1969). *Combust. Flame* **13,** 166.

Porter, R. C. (1948). *Chem. Eng. (N.Y.)* **55,** 100.

Priem, R. J., and Heidmann, M. F. (1960). *NASA Tech. Rep.* **NASA TR TR-R-67.**

Price, D. J., and Brown, H. H. (1922). "Dust Explosions." Natl. Fire Prot. Assoc., Boston, Massachusetts.

Probert, R. P. (1946). *Philos. Mag.* **34,** 94.

Raghunandan, B. N., and Mukunda, H. S. (1977). *Combust. Flame* **30,** 71.

Raj, S. (1976). Ph.D. Thesis, Fuel Sci., Pennsylvania State Univ., State College.

Ramquist, L. (1965). *Int. J. Powder Metall.* **1,** 2.

Rosner, D. E. (1967). *AIAA J.* **5,** 163.

Rosser, W. A., and Rajapakse, Y. (1969). *Combust. Flame* **13,** 311.

Savery, C. W., Juded, D. L., and Borman, G. L. (1971). *Ind. Eng. Chem., Fundam.* **10,** 543.

Scheer, C., and Korman, S. (1956). *In* "Arcs in Inert Atmospheres and Vacuum" (W. E. Kuhn, ed.), p. 169. Wiley, New York.

Schue, G. K. (1969). *SPE J.* **25,** 40.

Schwartzkopf, P., and Kieffer, R. (1960). "Cemented Carbides." Macmillan, New York.

Scott, J. R. (1961). "Powders in Industry," p. 195. Soc. Chem. Ind., London.

Shyu, R. R., Chen, C. S., Gondie, G. O., and El-Wakil, M. M. (1972). *Fuel* **51,** 135.

Sjogren, A. (1972). *Symp. (Int.) Combust. [Proc.]* **14,** 919.

Smith D. A. (1968). *Rubber J.* **150,** 54.

Smoot, L. S., Horton, M. D., and Williams, G. A. (1976). *West. States Sect., Combust. Inst. [Pap.]* **WSS/CI-76-3.**

Spalding, D. B. (1952). *Symp. (Int.) Combust. [Proc.]* **4,** 847.

Spalding, D. B. (1954). *Proc.—Inst. Mech. Eng.* **168,** 545.

Spalding, D. B. (1959a). *ARS J.* **29,** 828.

Spalding, D. B. (1959b). *Aeronaut. Q.* **10,** 1.

Strahle, W. C. (1964). *Symp. (Int.) Combust. [Proc.]* **10,** 1315.

Stumbar, J. P., Kuwata, M., Kuo, T., and Essenhigh, R. H. (1970). *Proc. Nat. Incinerator Conf., 4th* p. 288.

Suzuki, T., and Chiu, H. H. (1971). *Proc. Int. Symp. Space Technol. Sci., 9th, Tokyo* p. 145.

Taffanel, J., and Durr, A. (1912). *Am. Mines Mem. Sect.* **11**(2), 167.

Tanasawa, Y., and Tesima, T. (1958). *Bull. JSME* No. 1, 36.

Tarifa, C. S., Perez del Notario, P., and Moreno, F. G. (1962). *Symp. (Int.) Combust. [Proc.]* **8,** 1035.

Tu, C. M., Davis, H., and Hottel, H. C. (1934). *Ind. Eng. Chem.* **26,** 749.

Veale, C. R. (1972). "Fine Powders: Preparation, Properties and Uses." Wiley (Halsted Press), New York.

Wagner, H. (1978). *Symp. (Int.) Combust. [Proc.]* **17,** 3.

Wagner, H. (1981). *In* "Particulate Carbon-Formation During Combustion" (D. C. Siejla and G. W. Smith, eds.), p. 1. Plenum, New York.

Walker, P. L., Rusinko, F., and Austin, L. G. (1959). *Adv. Catal.* **11,** 134.

Weidhass, D. E., Bowman, M. C., Mount, G. A., Lofgren, C. S., and Ford, H. R. (1970). *Mosq. News* **30,** 195.

Wheeler, R. V. (1913). "Report, Explosions in Mines Committee," p. 9. HM Stationery Off., London.

Williams, A. (1968). *Oxid. Combust. Rev.* **3,** 3.

Williams, A. (1973). *Combust. Flame* **21,** 1.

Williams, F. A., Penner, S. S., Gill, G., and Eckel, E. F. (1959). *Combust. Flame* **3,** 355.

Williams, F. A. (1961). *Combust. Flame* **5,** 207.

Williams, F. A. (1965). "Combustion Theory." Addison-Wesley, Reading, Massachusetts.

Wise, H., and Agoston, G. A. (1958). *Adv. Chem. Ser.* No. 20, 116.

Wise, H., Lorell, J., and Wood, B. J. (1954). *Symp. (Int.) Combust. [Proc.]* **5,** 132.

Wood, B. J., Wise, H., and Inami, S. H. (1960). *Combust. Flame* **4,** 235.

World Health Organization (WHO) (1970). *W.H.O. Tech. Rep. Ser.* No. 443, Annex 18.

World Health Organization (WHO) (1974). "Equipment for Vector Control," 2nd ed. World Health Organ., Geneva.

Yeomans, A. H., Rogers, E. E., and Ball, W. H. (1949). *J. Econ. Entomol.* **42,** 591.

Young, D. A. (1966). "Decomposition of Solids," p. 8. Pergamon, Oxford.

CHAPTER 7

ATMOSPHERIC AEROSOLS

7.1 INTRODUCTION

In the last chapter, the variety in modern technology of aerosols was discussed. Now let us turn to the world of fine particles in the natural environment. The atmospheres of planets of the solar system are rich in suspended particulate matter, as is interplanetary and interstellar space. The richness of visual experience in observing the planets depends on gases and particles concentrated in their atmosphere. A summary of knowledge of aerosol behavior of different planets is given in Table 7.1. From this list, one can readily appreciate why astronomers have theorized so much about the presence of particles in planetary atmospheres. Best known of the planetary aerosols for obvious reasons are those of the earth. The earth's atmosphere is rich in particles. Their presence has been observed and reported for centuries in the literature. Only in the past twenty-five years have the scientific tools become available to characterize atmospheric aerosols in detail comparable with our laboratory knowledge. The knowledge of airborne particles through the late 1950s was reviewed extensively by Junge (1963a). Since this standard work was published, studies and the associated literature on atmospheric aerosols have exploded.

Airborne particles probably were recognized first in relation to sea spray drift, or to dramatic events associated with volcanic eruptions and forest or brush fires. However, the hazes associated with blowing soil or pollen dusts also contribute large quantities of particulate material to the atmosphere. Only in recent years has the significance of the contribution to the earth's air

TABLE 7.1

Summary of Suspended Particles Identified in Planetary or Lunar Atmospheres

Planet	Description	Estimated height of layers	Particle size (μm)	Composition	Reference
Mercury	Virtually no atmosphere—embedded in interplanetary dust	—	—	—	Dagani (1981)
Venus	Multilayers of cloud and haze	Upper 60–90 km lower 30–50 km	0.2–70, multimodal distribution	H_2SO_4 and crystalline sulfur	Knollenberg and Hunten (1980)
Earth	Multilayers of cloud and haze	Mainly below ~25 km	<0.05–>50; multimodal distribution	Sulfate, nitrate, soil organics, and mineral dust	See text
Mars	Multilayers of thin clouds and haze	Ground to 30 km and above	≲2	Soil dust and ferromagnetic minerals; some CO_2-H_2O-dust crystals (?)	Pollack et al. (1977)
Jupiter	Clouds and haze	(?)	(?)	Photochemical (?)—e.g., red phosphorus (?) ammonia, ammonium sulfide, polymers of methane, HC derivatives; interplanetary dust	Michaux (1967); Dagani (1981)
Saturn	Clouds and haze	(?)	(?)	Interplanetary dust, ammonia ice	Dagani (1981)
Titan	Haze	(?)	~0.2	Methane polymers (?)	Pollack et al. (1980)
Uranus	(?)	—	—	Methane ice (?), interplanetary dust	Dagani (1981)
Neptune	(?)	—	—	Methane ice (?), argon ice, interplanetary dust	Dagani (1981)
Pluto	(?)	—	—	Methane ice (?), interplanetary dust	Dagani (1981)

burden of extraterrestrial dust and the in situ production of particles by atmospheric chemical reactions become known. The latter is of particular interest in that the oxidation products of sulfurous and nitrogenous gases, and certain hydrocarbon vapors are prolific producers of small particles. Thus the "breathing" of traces of gases from natural biological chemistry in soils such as hydrogen sulfide or ammonia, and from vegetation as pinene or similar vapors actually contributes substantially to the atmospheric aerosol content. The direct transfer of particles to the air is often called *primary emissions*. The materials produced from atmospheric chemical processes are termed *secondary* contributions.

Added to the natural aerosol forming processes are the emissions from man's activities. With industrialization and urbanization of increasing large geographic areas, man emits substantial quantities of particulate matter. The expansion of agriculture also has enhanced the suspension of dust either directly by cultivation or indirectly by deforestation and temporary overpro-

Fig. 7.1. Examples of particulate matter collected from the atmosphere. (a) Particles sampled from Sheffield, England showing single spheres, smoke aggregates, and droplet residues; magnification as indicated by 1 μm mark. (b) A collection of sulfuric acid (spattered hemispheres with satellites), ammonium sulfate (rectangular images with satellites), and polycrystalline material from central Panama (22,400 $\times$ magnification)(photo by E. Frank).

duction resulting in soil erosion. Pollutant gases, including sulfur dioxide, nitrogen oxides, and certain reactive hydrocarbon vapors, also represent substantial potential for particle production in the air.

Illustrative photomicrographs of particles sampled in a remote tropical area and urban Sheffield are shown in Fig. 7.1 for comparison. The mixture found in remote areas is heavily influenced by sulfate species and crystalline solids. In contrast, the urban mixture contains sulfate mixed with a variety of crystalline debris and aggregates of spheres identified with combustion.

The estimated rate of particle injection into the air, which characterizes the global aerosol burden, is given in Table 7.2. This table represents a compilation from investigators in the early 1970s, who tried to estimate the relative contributions to the atmospheric aerosol. From this survey, the

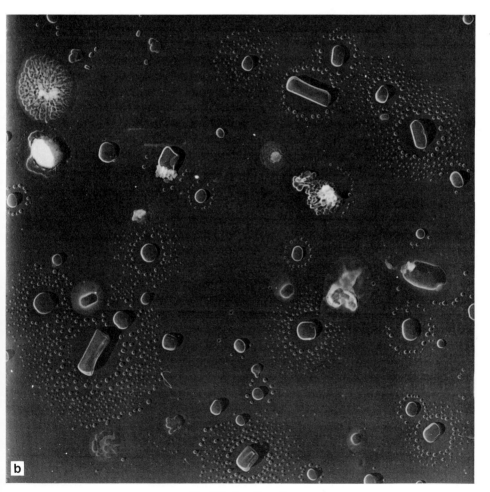

Fig. 7.1 (*continued*)

TABLE 7.2

*Rate at which Aerosol Particles of Radius Less than 20 μm
Are Produced in, or Emitted into, the Atmosphere*[a]

	Rate (10^9 kg/yr)
Natural particles	
Soil and rock debris	100–500
Forest fires and slash-burning debris	3–150
Sea salt	300
Volcanic debris	25–150
Gas to particle conversion	
Sulfate from natural sulfur gases	130–200
Ammonium salts from NH_3	80–270
Nitrate from N_xO_y	60–430
Hydrocarbons from plant exhalations	75–200
Subtotal	773–2200
Anthropogenic particles	
Particles by direct emissions	10–90
Gas to particle conversion	
Sulfate from SO_2	130–200
Nitrate from NO_x	30–35
Hydrocarbons	15–90
Subtotal	185–415
Total	958–2615
Extraterrestrial	0.5–50

[a] Reprinted from *Inadvertent Climate Modification* (SMIC
Report), 1971, by permission of MIT Press, Cambridge, MA.

natural contributions far exceed emissions from man's activities on a global
basis, but locally this is undoubtedly reversed, especially in parts of North
America and Europe. From the table, between 7 and 43% originate with
man, while the remainder is assigned to natural sources. The importance of
particles from atmospheric chemical reactions of gases also is shown from
data in the table. More than 50% of the estimated particle burden may come
from these secondary processes. Production of aerosols by chemical reactions represents an important loss mechanism for certain trace gases in the
atmosphere. Also it readily can be seen that the secondary material should
be dominated by particulate sulfur, present as sulfate, based on these estimates. Indeed sulfate is a ubiquitous constituent of atmospheric particle
populations.

The enormous quantities of particles injected into the earth's atmosphere
are mixed and aged by processes in the air to create a very diverse and
complicated mixture. The mix varies greatly with geographical region and
with altitude, but also has some remarkably common physical and chemical

features. Airborne particles in the atmosphere are intimately linked to its chemical processing, and contribute to visibility changes, as well as climate and weather effects. Discussion of these topics is deferred to Chapter 8. We begin with some details of the physical and chemical properties of the aerosol; then the origins of particles are covered in depth. The chapter ends with consideration of the dynamics of atmospheric aerosol particle distributions.

7.2 PHYSICAL CHARACTERIZATION

7.2.1 Vertical Structure. Before going into the physical and chemical character of atmospheric aerosols, let us introduce a framework for this discussion in relation to the structure of the atmosphere. The atmosphere has been classified in broad regions in accordance with vertical temperature structure, chemistry, and air motion. A conventional description is illustrated in the temperature profile shown in Fig. 7.2. The lowest part of the atmosphere is the *troposphere,* which generally is confined below heights of 15 km. At the lowest part of this region is the *planetary boundary layer* in which most of the natural and anthropogenic particles are injected or are

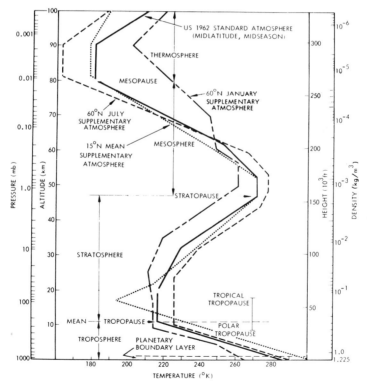

Fig. 7.2. Temperature–height profile of the atmosphere. [From U.S. Standard Atmosphere (1962).]

produced. This zone nominally is confined to 2 km height and below. The troposphere is defined in terms of a decreasing temperature with height, which falls off according to an adiabatic cooling on the average. About half the atmosphere is contained below 10 km. In this region, virtually all of the weather associated with clouds and precipitation is generated. Above the troposphere is an isothermal zone followed by a region with increasing temperature, defining the *stratosphere*. This region extends from 12–15 to 50 km altitude. The stratosphere is a zone of strong winds but with little vertical mixing associated with turbulence. Occasional veils of thin cirrus clouds appear in the lower stratosphere, but rarely above because of its dryness. Even higher, the temperature decreases again in the *mesosphere* to about 90 km. The mesosphere is the beginning of the influence of the rarified conditions approaching extraterrestrial space. Here chemistry takes a prominence along with electromagnetic phenomena that affect the dynamics of the air. This is a regime without water vapor or clouds, but with very small concentrations of suspended particles. Above the mesosphere is the *thermosphere,* a rarified gas condition that is the basic interface with the extraterrestrial environment.

A characteristic chemical measure of atmospheric chemistry that parallels the thermal structure is the atmospheric ozone structure. Ozone concentrations in the atmosphere show layering in structure on any given day. In any case, the ozone in the atmosphere is found to be maximum in a region near 20 km altitude. Here the net inorganic ozone forming processes are maximum, and give rise to its maximum concentration (Heicklen, 1976). This layer is widely known for its absorption of ultraviolet light, protecting the earth's surface from this part of the incoming solar radiation. As we shall see, ozone is a surrogate for photochemically related reactions which are important to aerosol behavior.

The distribution of water vapor with height is another important characteristic of the atmosphere. Typical observations show a rapid decrease in water vapor concentrations above the troposphere. Because of the dryness in the stratosphere and above, particle chemical reactions involving water vapor and interaction with water clouds are virtually confined to the troposphere, whereas only dry processes are involved at greater heights.

The vertical distribution of aerosol particles in the atmosphere has been measured by indirect means applying remote sensing of scattered light beams, and direct means employing balloon ascents carrying nuclei counters, optical counters, or filter–impactor samplers. Early profiles taken during balloon ascents are shown in Fig. 7.3. The data were obtained using a small expansion cloud chamber which is believed to activate and count all particles in the air larger than approximately 0.01 μm diameter. Thus these observations are believed to be representative of the total number N of particles in the air. As shown in Fig. 7.3, the distribution of particles tends to decrease rapidly from a range of 10^3–10^4 cm^{-3} at the ground to approach a

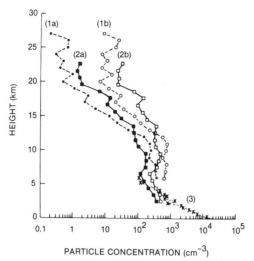

PARTICLE CONCENTRATION (cm^{-3})

Fig. 7.3. Variation of the average number concentration (#/cm^3) of Aitken particles with height over various locations: (1a) Average of seven flights over Sioux Falls, SD, 44° N, June 1959 to July 1960 (ambient); data of Junge. (1b) Same as (1a) but concentration at STP. (2a) Flights over Hyderabad, India, 17° N, March to April 1961 (ambient); data of Junge. (2b) Same as (2a) but concentration at STP. (3) Data of Weickmann (1956) based on flights over Germany. [From Junge (1963b); by courtesy of *Journal de Recherches Atmospheriques* and the author.]

low value of 0.1–10 cm^{-3} above 25 km. In the range between 7 and 12 km, there is a small increase in concentrations in some of the profiles.

As will be seen later, the distribution of particle concentration with size is dominated by particles of diameter 0.01–0.1 μm. This is sometimes called the *Aitken* range, named for an early observer (Aitken, 1923).

In contrast with the behavior of the small Aitken particles, the particles in a range larger than about 0.3 μm diameter show a somewhat different vertical structure, as illustrated in Fig. 7.4. Here we see that large (0.3–1.0 μm) particles observed with an optical counter show a decrease with height to the tropopause followed by a substantial increase in large particles between 12 and 18 km. This layer enriched in large particles was observed earlier by Junge and others (Changnon and Junge, 1961) from balloon ascents data. The maximum in large particles aloft is found between 12 and 25 km altitude depending on latitude. This layer of large particles is sometimes referred to as the *Junge layer* in the stratosphere. This zone of relatively high particle concentrations roughly coincides with the zone of maximum ozone concentration, but actually lies below it in altitude.

The distribution of particle concentration with height takes on broader interest considering the processes of injection and removal into different layers of the atmosphere. The net lifetime is some function of chemical reactivity and removal processes. The latter depend on the meteorology of

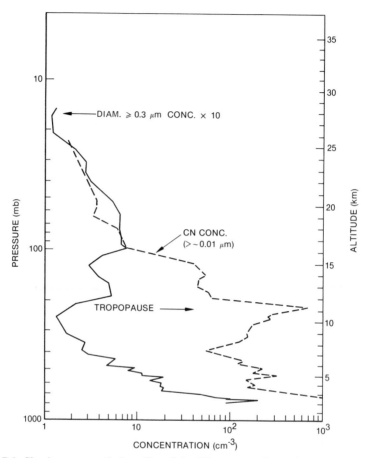

Fig. 7.4. Simultaneous vertical profiles of the CN concentration and aerosol concentration using the Wyoming CN counter and the Wyoming dustsonde over Laramie, 19 December 1973. [From Rosen (1974).]

the wind field for transport, and mixing by turbulence. For aerosol particles, the measure of a lifetime in the atmosphere is complicated by processes leading to a continuous change in particle size and composition. Particles tend to lose their identity by coagulation or accumulation of condensables. Thus, the term residence time or lifetime cannot apply to individual particles, but represents a parameter characterizing suspended particulate matter as a whole, or at least in classes by certain size intervals.

The overall lifetime in the atmosphere probably remains one of the most important time scales in the geochemistry of trace constituents. Its definition is given in terms of the ratio of the particles present to that initially present at a given time. The lifetime τ is ordinarily given in terms of a simple model corresponding to loss of particles at a rate proportional to particle concentration. Thus, the time τ required for the material to be reduced to 1/2.718 of its original concentration is its lifetime or residence time. This implies that a residual concentration of material may exist for very long times in the atmosphere. Thus, a minor but potentially significant fraction of emissions may be present for many months from a pollution event despite the fact that its original concentration has decreased rapidly to a lower level in much shorter time. This is a basic issue associated with decay of radioactive material long after an atmospheric injection from a nuclear explosion.

The residence times of particulate matter in different parts of the atmosphere as estimated prior to 1970 have been summarized by SCEP (Study of Critical Environmental Problems, 1970). Later results based on isotope behavior have been summarized by Pruppacher and Klett (1978), and are listed in Table 7.3. The more recent data tend to lower values than earlier estimates. However, both sets of data show a systematic increase in residence time with height in the atmosphere. The residence time ranges from a few days or less in the troposphere, where removal processes are active, to several weeks or months or even years as one traverses the stratosphere to the mesosphere. Above the tropopause the mixing and removal of particles by scavenging in clouds is negligible. Although a photophoretic force associated with solar radiation induces some downward flux at high altitude, this migration is small compared with gravitational settling (see, for example, Hidy and Brock, 1967). In any case, it can be seen readily from the table that volcanic eruptions or other disturbances which inject material into the stratosphere can be observed for months after injections.

7.2.2 Particle Size Distribution and Its Moments.

Because of the years of focus on relating particle properties to idealized spheres, much of the histori-

TABLE 7.3

Residence Time of Atmospheric Aerosol Particles at Various Levels in the Atmosphere[a]

Level in the atmosphere	Based on evidence prior to 1970	Based on evidence after 1970
Below about 1.5 km	—	0.5 to 2 days
Lower troposphere	6 days to 2 weeks	2 days to 1 week
Middle and upper troposphere	2 weeks to 1 month	1 to 2 weeks
Tropopause level	—	3 weeks to 1 month
Lower stratosphere	6 months to 2 years	1 to 2 months
Upper stratosphere	2 years to 5 years	1 to 2 years
Lower mesosphere	5 to 10 years	4 to 20 years

[a] From Pruppacher and Klett (1978); reprinted with permission of Reidel Publishing Co. and the authors.

cal effort investigating physical properties has been oriented in this conventional way. With the interest in the particle size distribution, physical properties are often described in terms of the size distribution parameters. The investigation of the size distribution of atmospheric particles and moments of the size distribution has been active for many years. The moments of greatest interest include the number concentration N, the surface area per unit volume S, the volume fraction V, and the closely related mass concentration M. Some typical recent measurements of these parameters for different ambient conditions are listed in Table 7.4. Tropospheric levels range over wide values; local areas have number concentrations ranging from 100 to 1000 cm^{-3}, while urban areas generally are much larger, as high as 10^6 cm^{-3}. Volume concentrations generally range from 10–50 μm^3/m^3 rurally to 100 μm^3/m^3 or more in cities. Mass concentrations have been found to range from tens of micrograms per cubic meter in remote areas to hundreds in cities. Typically dust storms can achieve mass concentrations in the milligram per cubic meter range; Eisenbud (1980) has cited early data taken in 1915 in Chicago indicating particulate loadings also in excess of 1 mg/m^3. However, concentrations in modern cities of the United States generally are well below the milligram per cubic meter level.

Surface area and density are also important physical parameters that can be associated with the size distribution. The nominal surface area S derived from particle number versus size measurements are included in Table 7.4. These are in the range of 30–3000 μm^2/cm^3. Actual surface areas may be much larger for porous particles. Surface areas for particles suspended in contaminated air are quite uncertain because of their capability for adsorbing gases. Corn and Reitz (1968), for example, reported an increase of a factor of 2 in the specific surface of particles from Pittsburgh air after degassing collected samples at 200°C.

TABLE 7.4

Typical Observed Values of the Particle Number Concentration, Surface Area, Volume Fraction, and Mass Concentration for Atmospheric Aerosol Suspensions[a]

Location	N (no./cm^3) $\times$ 10^{-3}	S (μm^2/cm^3)	V (μm^3/cm^3)	M (μg/m^3)
Marine background	0.4	30	12.1	—
Pt. Arguello, CA	12	325	102	—
San Nicolas Is. (1970)	2.4	—	—	30
Clean continental background	2	42	6	21
Goldstone, CA	30	200	12.7	50
Urban average	140	1130	70	100
Pomona, CA	100	1120	82	—
Urban and freeway mix	2200	3200	90	120–220
Labadie (MO) powerplant plume	130	576	36	—

[a] Data from Whitby (1978); Hidy (1975); Hidy and Mueller (1980).

The mass density of particles nominally can be estimated from the ratio of V and the mass concentration. Typically, this calculation yields densities in the range of 2–3 g/cm³. This range has been measured directly using a gas pyronometer technique, according to Jaenicke (1976).

Early work prior to the 1960s indicated that a size–number distribution with surprising uniform shape was to be found. Although the range below particle diameters of 0.5 μm was variable, the larger size range could be represented by a power-law form first associated with C. Junge. The form is

$$\frac{dN}{dd_p} = \text{const} \times V d_p^{-\gamma} \qquad (d_p > 0.5 \ \mu\text{m}). \qquad (7.1)$$

The exponent γ can vary from 3 to 5, but is often taken as 4; for this value, the constant in Eq. (7.1) is 0.003. An example of such a distribution is shown

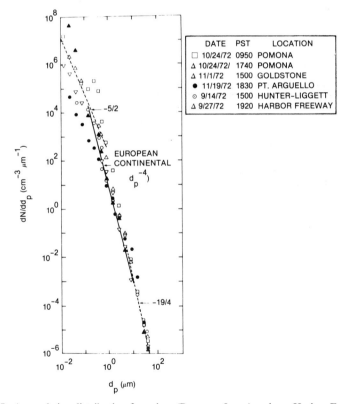

Fig. 7.5. Aerosol size distribution for urban (Pomona, Los Angeles—Harbor Freeway) and nonurban locations in southern California (Goldstone—Desert Continental background, Pt. Arguello—marine–surf, Hunter–Liggett Reservation—rural–arid). The regimes of slope −5/2 and −19/4 are, respectively, those for coagulation dominance and sedimentation dominance of Friedlander (1960). [From Hidy (1975).]

in Fig. 7.5, which illustrates the number–size distribution for the so-called large particle range (0.1–1.0 μm radius) and for the giant particles (>1.0 μm diameter). The power-law fit is often useful as an empirical form over at least two decades of diameter. However, power laws of different slopes can fit the small particle extreme and giant particle range as indicated in Fig. 7.5. These slopes can be rationalized on theoretical grounds as indicated in Section 7.6. These data may be considered reasonably representative of conditions near the earth's surface. Note the remarkable range in particle size of the tropospheric aerosol, from <0.01 to >50 μm in diameter. The range where the power law is a reasonable fit of the broad features of the data exceeds nine decades in concentration, as shown in Fig. 7.5.

Examples of rural data in the Aitken nuclei range, $R_p \lesssim 0.1$ μm, are shown in Fig. 7.6. These observations indicate large differences between rural continental stations (e.g., Sverdrup, 1977; Tanner and Marlow, 1977). Prahm *et al.* (1976) for marine air over the Atlantic Ocean are in contrast with Jaenicke's (1978) results for the central Atlantic Ocean region. In general, workers have found that the Aitken particles are depleted in remote areas, particularly in marine conditions where the aerosol tends to age without

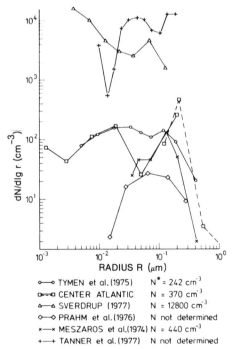

Fig. 7.6. Aitken nuclei and large particle size distributions from various authors. N is the total particle concentration if determined independently from the size distribution measurements, N^* the particle concentration within the covered radius range only. [From Jaenicke (1978).]

influence of terrestrial sources. Polluted air is frequently most heavily populated with the very small particles.

Observations of particle size–number distributions are limited above the
planetary boundary layer. Examples of changing distributions with height
are shown in Fig. 7.7 for the area over Nebraska. Here, the concentration of
most particles tends to decrease with height, except for the range around 1
μm radius, which remains roughly constant to 9 km. At higher altitudes,
there is a marked depletion in the largest size fraction. Presumably this is the
result of gravitational settling exceeding the mixing of large particles upward
by convection.

Conditions in the stratosphere reported by investigators are shown in Fig.
7.8. These results show a range of distributions, which have been taken by
different impactor methods using balloon ascents or aircraft. Perhaps the
most recent collection are reported by Farlow *et al.* (1979). Although the
earlier work shows a power-law decay in the large particle concentration of
smaller exponent than the troposphere, the Farlow *et al.* results fall closely
within the tropospheric pattern (compare Fig. 7.8 with Figs. 7.5 and 7.7).

The results reported by Farlow *et al.* were obtained using an impactor on a

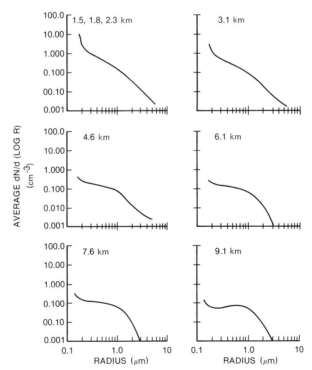

Fig. 7.7. Average particle size distributions at altitudes from 1.5 to 9.1 km over Nebraska
based on aircraft impactor data. Radii in micrometers (μm). [From Blifford and Ringer (1969).]

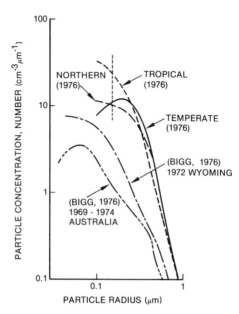

Fig. 7.8. General size distribution for particles in the lower stratosphere at different latitudes, 1976–1977, compared to earlier measurements. [After Farlow *et al.* (1979); reprinted courtesy of the American Geophysical Union.]

U-2 aircraft. Their results are similar to those reported by Bigg (1976) for radii greater than 0.3 μm. However, Farlow's data suggest enhancement of particles in the range $\leq$0.2 μm over the tropics. These workers hypothesized that the tropics are a region for injection of aerosol precursors, but this has not been followed up as yet.

Based on early evidence, Friend (1966) hypothesized that the stratospheric aerosol was composed of two distinct size populations, a large size range, with a mean radius of approximately 0.3 μm, and a nuclei range with mean radius approximately 0.05 μm. The small particle component was believed to be a highly variable component associated with new particle formation under certain circumstances. In recent measurements, no direct evidence of the small nuclei population has been reported to confirm its existence. The connection between Friend's hypothesis and Farlow's observations in the tropics has not been discussed.

For a number of years, workers concentrated on the number–size distributions and their apparent uniformity, with less interest in the surface area or volume distributions. With the advent of new methods for measuring the size distribution over a wide range by combinations of instruments, certain regularities in the surface area and volume–size distributions have been found. These tend to be obscured in the number–size distribution plots.

The work of Whitby and colleagues (Whitby, 1978) emphasized the form of the surface and volume distribution functions. Some typical examples for the atmospheric aerosol are shown in Fig. 7.9. Five different ways of plotting atmospheric particle size distributions are shown. The first (Fig. 7.9a) is the conventional number distribution method (see also Fig. 7.5). The second (Fig. 7.9b) is a cumulative distribution fit to a lognormal form with DGV representing the volumetric mean diameter, σ_g the (geometric) standard deviation, and χ_N^2 the normalized chi square function for particle number, defined as

$$\chi_N^2 = (1/N)(f_i - f_e)^2/f_e, \qquad (7.2)$$

where f_i is the experimental frequency of occurrence in size range and f_e is the expected frequency calculated from additive lognormal distributions. This measure is used to obtain goodness of fit of data to different lognormal forms and superimposed distributions.

The measured distributions can be fit by additive lognormal functions. The number distribution fit in terms of two functions is shown in Fig. 7.9c. The surface distribution is fit by three functions (Fig. 7.9d), and the volume

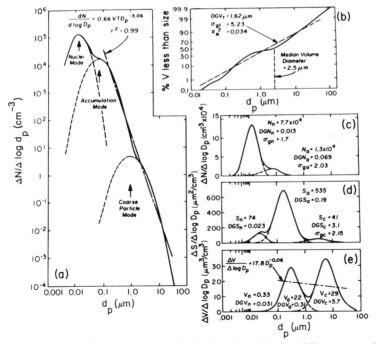

Fig. 7.9. Average model urban aerosol distribution plotted in five different ways. In (a) a power function has been fitted to the number distribution over the size range 0.1–32 μm. In (b) a cumulative lognormal distribution has been fitted to the range 0.1–32 μm. Methods (c), (d), and (e) fit the observations to superimposed lognormal forms covering major modes of the distributions, number, surface area, and volume–size distributions for nuclei (n), accumulation (a) and coarse (c) modes. [After Whitby (1978).]

distribution in this case is fit to three functions, the smallest particle component being too small to appear on the graph (Fig. 7.9e). The nomenclature in each graph refers to geometric mean diameters (DG) based on number, surface, and volume; the geometric standard deviation for each distribution function (σ_g); and the total number, surface area, and volume fraction in units given in Table 7.4.

The power-law fit for the number–size distribution equivalent to Eq. (7.1) is shown in Fig. 7.9a. The equivalent volume distribution for the power-law form is shown in Fig. 7.9e. Comparison of the two illustrates that apparent minor deviations from the power-law form of the number distribution are very large in the volume distribution, and are actually lost in the analysis if the power-law form is adopted.

The modes of the volume distribution show distinct breaks in the size ranges less than 0.1 μm diameter, 0.1–1.0 μm diameter, and greater than 1.0 μm diameter. Actual measurements taken in many different locations have shown that these modes are often distinguished from one another by the sources of particles as well as their chemical origins in the air. In rural or remote sites, where the background aerosol has aged substantially, one finds a situation with a single small submicrometer mode. If local sources of aeolian dust or sea spray are present the supermicrometer mode of large particles can be enlarged. Near combustion sources, such as highways, large numbers of very small particles are emitted. Thus, in source-enriched areas, one expects the new particles in the Aitken nuclei mode to be increased as implied in Fig. 7.9c.

Whitby *et al.* (Whitby, 1978) have named the submicrometer fraction between 0.1 and 1.0 μm diameter the *accumulation (a) mode*. Here, particle growth by condensation and coagulation of Aitken nuclei tends to accumulate since this range has a minimum in seasonal processes, including combined diffusional and inertial deposition or scavenging, turbulent surface deposition, and electrical phoresis. The nuclei (n) mode is designated as particles less than 0.1 μm diameter, and the coarse particle (c) mode is identified with particles larger than 1 μm diameter. This segregation corresponds basically to Junge's designation of nuclei, large, and giant particles. Observations of Whitby *et al.* (Whitby and Sverdrup, 1980; Gartrell *et al.*, 1980), for example, have demonstrated that the accumulation range can grow and shrink with processes essentially independent of the other modes. Observations indicate that the three modes can originate and behave roughly independently. For example, aged aerosol will have low volume concentration, but will have a dominant accumulation mode, unless disturbed by a large coarse particle contribution of blowing dust or sea salt. Similarly, freshly emitted combustion particles will influence the nuclei mode. This is particularly pronounced where growth by production of condensable species by chemical reaction takes place in the atmosphere.

The tropospheric particle size distributions are now frequently reported and interpreted in terms of the multimodal or trimodal distribution functions.

However, the power-law model also remains a factor in the literature, particularly in view of its susceptibility to rationalization from aerosol dynamic models (see also Section 7.6).

7.3   CHEMICAL PROPERTIES

Particles in the air can have a wide range of chemical composition depending on their origins or their exposure to chemically reactive gases. The gaseous components of the atmospheric aerosol are contaminants carried with the mixture of major constituents, oxygen, nitrogen, water vapor, and carbon dioxide. Of these, water vapor is most crucial to the behavior of the suspended particles since it interacts intimately with other (soluble) contaminants. Water content is one of the main components distinguishing the tropospheric aerosol from samples at higher altitude.

7.3.1   Tropospheric Aerosols.   With each year, the list of identified trace constituents in the troposphere becomes longer and more diverse. As knowledge of chemical processes develops and methods of detection improve, there seems to be no end to the identification of compounds that may be present in the gas–particle mixture. As an example, several gaseous sulfur and nitrogen constituents are listed in Table 7.5. The most common sulfurous gas found in the urban or near-urban atmosphere is SO_2. In some rural areas, however, hydrogen sulfide is found at locally comparable concentrations. Of the nitrogen oxides, nitrous oxide evidently is the most common gas in rural areas. However, nitric oxide and nitrogen dioxide are more abundant in urban air. Ammonia concentrations are generally low, in the parts per billion (ppb) to parts per trillion (ppt) range because of the chemical reactivity of this gas. There are several nitrogen oxide species found in the atmosphere that are products of chemical reactions, including nitric acid and peroxyacylnitrates.

Parallelling the variety of sulfurous and nitrogenous gases are the organic vapors, which are listed generically in Table 7.6. Although methane is by far the most common carbonaceous gas in the air, it is of least interest in its environmental impact. The olefinic and aromatic compounds along with the oxygenated materials represent potential sources of reactive species and some are believed to adversely influence human health. For example, aromatics, such as benzene, are believed to be carcinogens. Although the polycyclic aromatic compounds are generally very low in ambient concentration, they have been of concern for many years as potential carcinogens.

Suspended in this complex mixture of gases are the aerosol particles, whose chemical makeup is equally diverse and perhaps more poorly characterized. Typical estimated compositions of collected particulate material in North American cities and rural areas are listed in Table 7.7. The components of tropospheric particulate matter are characterized as major and minor by mass concentration. One readily sees in this list that the most com-

TABLE 7.5

Sulfurous and Nitrogenous Gases Known to Be in the Lower Atmosphere

Constituent		Concentration range (ppb)		Origin
		Urban	Rural	
Sulfur				
Sulfur dioxide	SO_2	5–500	1–50	Fossil fuel combustion
Hydrogen sulfide	H_2S	1–50	0.1–50	Biogenic, geothermal
Carbonyl sulfide	COS	0.1	0.1	Biogenic
Dimethyl sulfide	$(CH_3)_2S$	0.1	0.1	Biogenic
Carbon disulfide	CS_2	0.1	0.1	Biogenic
Methyl mercaptan	CH_3SH	—	~0.1	Biogenic
Nitrogen				
Nitrous oxide	N_2O	300	300	Biogenic
Nitric oxide	NO	10–1000	0.1–100	Fossil fuel combustion
Nitrogen dioxide	NO_2	1–500	0.1–100	Fossil fuel combustion atmospheric oxidation
Nitric acid	HNO_3	0.1–20	0.02–0.3	Atmospheric reactions of NO_x
Ammonia	NH_3	1–80	0.1–10	Biogenic, fuel combustion
Peroxyacetyl nitrate (PAN)	CH_3COONO_3	0.1–60	0.1–1	Atmospheric reaction product

mon constituents are identified as ammonium sulfate and nitrate, lead halides, and elemental and organic carbon material. Additional components include materials such as metal and halides as well as ubiquitous quantities of water.

Because of the universal prevalence of sulfur as sulfate in particulate matter in urban air and its implicated environmental effects, this component has received considerable attention. A generic description for the mixture of diverse chemical contributors, the *sulfur oxide particulate complex* (*SPC*), was coined a few years ago by Nelson and his colleagues (Nelson, 1975). The polluting tropospheric aerosols include sulfur dioxide (SO_2), particulate sulfate (SO_4^{2-}), and soot associated with the aerosols generated by fuel burning. Sulfur dioxide concentration, combined with total suspended-particulate matter (TSP), is the best index of industrial air pollution in a historical sense. The early documents supporting U.S. National Ambient Air Quality Standards (NAAQS) for SO_2 and TSP take note of the suspected health effects and other influences of these pollutants (U.S. PHS 1969). By the mid-1960s, information was available to show that ambient SO_2 concentrations and urban exposures in themselves were not sufficient to explain adverse health effects attributed to air pollution. However, in combination with particulate matter, there was evidence that the SPC was implicated in community health

TABLE 7.6

Organic Vapors Identified in Ambient Air

Constituent	Concentration range (ppb)		Origin
	Urban	Rural	
Methane	20,000	1000	Biogenic, geogenic
Alkanes (ethane, propane, · · ·)	100	10	Gasoline, fuels, solvents, etc.; partially burned hydrocarbon
Alkene· (olefins)	100	10	
Cyclic alkanes	1	0.1	Gasoline, fuels
Aromatics (benzene, toluene, · · ·)	1	0.1	Gasoline, fuels
Polycyclic aromatics (e.g., benzopyrene)	<0.1	<0.01	Combustion of fuel
Terpenes	0.1	0.1	Biogenic
Aldehydes (formaldehyde, · · ·)	1–10	0.1–1.0	Biogenic; atmospheric reaction products
Other oxygenates (ketones, alcohols, acids)	<0.1–1.0	<0.1(?)	
Alkynes (acetylene)	1–10	0.1–1.0	Partially burned fuel
Cyclic alkenes	0.1	0.01	Gasoline, fuels

deterioration from air contamination. The historic pollution incidents of Donora, PA in 1948 and London in 1952, involving SO_2 and "black" particulate matter, are the dramatic evidence perhaps most often cited in the literature (Williamson, 1973).

If sulfites and nitrites are present, they are generally in very small quantities; they are likely to be unstable as ions, but may exist as organic materials or metal-ion complexes. The sulfate and nitrate components are associated with the atmospheric end product for sulfur- and nitrogen-containing gases. Sulfate is widely present in quantities above 0.5–2.0 $\mu g/m^3$. There are large variations from place to place in nitrate concentration, ranging from <0.1 to >10 $\mu g/m^3$. Part of this variability appears to be related to unresolved sampling problems created by the volatility and instability of nitrates, and losses after filter collection and storage. Ammonium nitrate, for example, shows a strong temperature dependence on vapor pressure, and a susceptibility to decomposition which creates special problems for sampling (Stelson *et al.*, 1979). High concentrations of particulate nitrate are generally found in areas of elevated atmospheric ammonia concentrations, and cold air. Volatile nitric acid vapor then can be absorbed at low temperatures in particles and neutralized to form the ammonium salt. In warm air, nitrates appear to be

TABLE 7.7

Chemical Composition of Particulate Matter Sampled in Different North American Locations (Concentrations in $\mu g/m^3$)

Constituent	New York City[a] (Aug. 10–16, 1976)	Montague, MA[b] (Oct. 21–22, 1977)	Duncan Falls OH[b] (Oct. 21–23, 1977)	Allegheny Mountain[c] (Jul. 24–Aug. 11, 1977)	Pulaski, TN[b] (Oct. 21–23, 1977)	Denver, CO[d] (Dec. 18, 1978)	Los Angeles[e] (Aug. 14, 1977)
Secondary anthropogenic							
Ammonium	3.5	2.2	2.3	2.3	1.5	10.1	11.2
Sulfate	22.4	12(8.0)[f]	13(8.4)[f]	14.1	11(9.8)[f]	15	24.3
Nitrate	8.6	0.05	0.37	0.53	0.40	14	12.7
Carbon/organic							
Elemental C	—	—	—	—	—	33.2	—
Organic C	—	—	—	—	—	39.0	—
Total noncarbonate C	14	5.7	1.5	0.78	1.6	—	27.1
Automotive							
Lead and halide (Pb + Br + Cl)	1.6[g]	0.41	0.39	0.2	0.52	13.3	4.3[g]
Light metals							
Sodium	0.93	0.09	0.36	0.2	0.22	—	1.25
Magnesium	0.31	0.07	0.27	0.07	0.17	—	0.61
Aluminum	0.82	0.25	0.91	0.7	0.87	8.7	2.75
Silicon	—	0.50	3.0	1.0	3.3	25.4	10.5
Potassium	0.31	0.11	0.43	0.07	0.27	3.14	0.942
Calcium	—	0.16	0.77	0.33	2.2	—	1.55

Heavy metals

Titanium	0.084	<0.05	0.09	0.04	0.09	0.48	0.161
Vanadium	0.046	<0.05	<0.05	0.003	<0.05	0.06	0.074
Chromium	0.032	<0.05	0.05	—	0.05	0.054	—
Manganese	0.034	<0.05	0.05	0.009	0.05	0.041	0.076
Iron	1.48	0.23	0.82	0.32	0.71	7.0	2.74
Nickel	0.018	—	—	—	—	0.05	0.048
Copper	—	<0.03	<0.25	—	<0.25	0.164	0.042
Zinc	0.44	0.09	0.07	0.02	0.03	0.453	0.375
Arsenic	3.9×10^{-3}	—	—	—	—	4×10^{-3}	—
Selenium	7.8×10^{-3}	—	—	0.003	—	—	—
Total mass concentration	90	30	50	—	61	290	170

[a] Bernstein and Rahn (1979); Lioy et al. (1979); Daisey et al. (1979); Kleinman et al. (1979).

[b] Sulfate Regional Experiment, Mueller et al. (1983); metal data J. Winchester (personal communication).

[c] From Pierson et al. (1980).

[d] Denver Haze Study, Heisler et al. (1980a).

[e] Los Angeles Aerosol Study, Heisler et al. (1980b), <15 μm diameter, 0800–1200 PST.

[f] Parentheses refer to <2.5 μm diam data.

[g] Coastal cities may have significant fraction of Cl from sea salt.

less likely to form from nitric acid vapor or to remain stable in the condensed phase. Thus, nitrate in particles often appears as a winter maximum concentration, while sulfate has a summer maximum with an SO_2 minimum and a winter maximum accompanying an SO_2 maximum. Where large quantities of ammonia are available to neutralize acid, as in Los Angeles, Denver, and the San Joaquin Valley of California, high levels of nitrate have been observed (Heisler *et al.*, 1980b; Heisler and Baskett, 1981).

The acidity of the SPC is an important issue in relation to the chemical nature of the mixture of contaminants. Work initiated by Brosset (1978) and others (Charlson *et al.*, 1978; Stevens *et al.*, 1978) has indicated that a significant concentration of hydrogen ions is present in some particulate samples. Stoichiometric arguments have been applied to assign the acidity of the particulate material. Such calculations indicate that aerosol particle mixtures can be accounted for as ammonium salts of sulfate and nitrate. If particulate nitrate is present as a neutralized salt and chloride concentrations are small, the sulfur oxide mixtures frequently correspond to ammonium bisulfate and ammonium sulfate. Some examples of conditions where hydrogen ion content is in excess of ammonium bisulfate have been reported (Mueller and Hidy, 1983; Pierson *et al.*, 1980). Generally, insufficient observations of alkaline metal concentrations have been available to determine if sulfuric acid or other acids are actually present in appreciable quantities. However, the observations of Pierson *et al.* (1980) indicate that the acidity in western Pennsylvania as measured by infrared spectroscopy, Gran titration, and stoichiometry suggests that a mixture of ammonium sulfates and free sulfuric acid can coexist in some heavy pollution conditions. Data taken in the Allegheny Mountains near Pittsburgh during July and August 1977 indicate a high $[H^+]$ to $[SO_4^{2-}]$ correlation, with molar ratios approaching 2. Acidity was especially strong under conditions of high sulfate levels in this case. During conditions of regional pollution events in the eastern United States during July 1978, the calculated mass ratio of sulfate to ammonium of aerosol samples varied over periods of hours from 2.6 to more than 6, suggesting the presence of free acids as well as neutralized ammonium salts (Mueller and Hidy, 1983). In contrast with the Allegheny Mountain results, these investigations found elevated acidity with low sulfate concentrations.

Recently, Lazrus *et al.*, (1981) have proposed potentially useful parameters for characterizing the total acidity of air. The acidity of air can be defined as

$$\text{Acidity} \equiv \text{Equivalents of protonic acid} - \text{equivalents of base.} \quad (7.1)$$

These workers have attempted to estimate the acidity and potential acidity from samples taken during the first phase of the Acid Precipitation Experiment (APEX). For conditions in summer 1978 over the Ohio River Valley and Atlantic Coast areas, the acidity or potential acidity of air corresponded to calculated pH values from an ion balance roughly in the range between 2.5 and 4.

The presence of water in aerosol particles has been measured directly only in California (e.g., Hidy, 1975). It is known, however, that atmospheric particles are hygroscopic in nature, and so acquire water to equilibrate to a given humidity. Even at relative humidities less than 50%, particles are expected to contain about 5–10% water as noted by Meyer *et al.* (1973). The hygroscopic nature of particulate matter has been investigated by several workers (e.g., Winkler, 1969; Charlson *et al.*, 1978). An example of direct measurement of water content in particle samples by a microwave device is shown in Fig. 7.10. Here a marked dependence of humidity on water content is found.

The common crustal elements contained in particles are probably alkaline metal oxides or salts found either in soil dust, or contained in fly ash from emissions of fossil fuel combustion. There are traces of certain toxic elements in fly ash which are of concern in conjunction with the sulfate and nitrate species. Other possible sources of metals include petroleum, metal refining or smelting operations, and a variety of metal finishing operations.

Lead, said to be in the form of oxide, sulfate, or bromochloride, also is

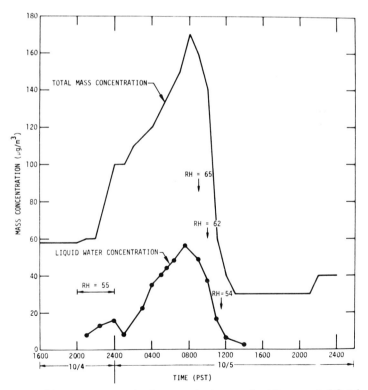

Fig. 7.10. Liquid water concentration in atmospheric aerosols at Pomona, 4–5 October 1972. [From Ho *et al.* (1974); reprinted courtesy of the American Meteorological Society.]

found in aerosol particles. The origin of this material is most likely automobile exhaust, except in the vicinity of smelting operations.

The carbonaceous material found in aerosol particles is made up of soot including elemental carbon and organics (Table 7.7). These may come from natural sources, incomplete combustion of fossil fuels, vegetative debris, or atmospheric chemical reactions. The organic fraction generally has been poorly characterized, but the other extractable portion of polar fraction has been reported for characteristic conditions as indicated in Table 7.8. The global burden is estimated by Hahn (1980) to be similar in marine and continental air, with concentrations roughly similar to that of sulfate in remote areas. Urban areas may have much larger carbon concentrations, as indicated in Table 7.7. Only recently have simple measurements segregating total elemental and/or organic carbon become available from different locations (e.g., Mueller *et al.*, 1982). The diversity of organic material found in urban particles is illustrated in Tables 7.9 and 7.10. The variety of compounds in Table 7.9 exemplify those expected from primary emissions including fuel combustion and chemical manufacturing. The distribution factor shown is the ratio of material found in the particles *P* to that found in the gas phase *G*. One notes that considerable quantities of low-molecular-weight, volatile materials exist in the vapor phase; some were observed only in the gas phase (Cautreels and Van Cauwenberghe, 1978). Roughly speaking, the lower the vapor pressure, the more material in the condensed phase. Table 7.10 gives a list of organic compounds identified by high resolution mass spectroscopy in particles sampled from photochemical smog in the Los Angeles area [Committee on Medical and Biologic Effects of Environmental Pollutants (CMBEEP), 1977]. These compounds are suggested as condensable products of atmospheric reactions with different hydrocarbon precur-

TABLE 7.8

Ether-Extractable Portion (EEOM) of Tropospheric Particulate Organic Matter[a]

Area	Particle size range	Scale height (m)	Ground-level concentration (μg/m³ STP)	EEOM mass (Tg)[b]
Marine	$r > 0.2\ \mu$m	1500	0.30 (0.15–0.60)	0.17 (0.08–0.33)
	$r < 0.2\ \mu$m	8000	0.30 (0.15–0.60)	0.86 (0.43–1.73)
	All particles	—	0.60 (0.30–1.20)	1.03 (0.51–2.06)
Continental	$r > 0.2\ \mu$m	1500	0.45 (0.23–0.90)	0.10 (0.05–0.21)
(nonurban)	$r < 0.2\ \mu$m	2000	0.45 (0.23–0.90)	0.14 (0.07–0.27)
	$r < 0.2\ \mu$m	8000	0.30 (0.15–0.60)	0.37 (0.18–0.74)
	All particles	—	1.20 (0.60–2.40)	0.61 (0.30–1.22)
Global	—	—	—	1.6 (0.8–3.2)

[a] From Hahn (1980); with permission from the New York Academy of Sciences.
[b] One teragram is 10^{12} g.

TABLE 7.9

Concentrations of Organic Compounds Formed in Air Samples Taken in Belgium[a]

	Concentration (ng/m³)		
Compound	Particle samples P	Gas-phase samples G	Distribution factor[b] P/G
Aliphatic hydrocarbons			
n-nonadecane	0.80	15.1	0.053
n-eicosane	0.85	7.55	0.113
n-heneicosane	1.08	4.12	0.262
n-docosane	2.33	4.23	0.551
n-triosane	4.75	3.38	1.41
n-tetracosane	8.15	4.63	1.76
n-pentacosane	9.50	5.74	1.66
n-hexacosane	9.73	8.70	1.12
n-heptacosane	11.1	9.03	1.39
n-octasane	8.10	7.80	1.04
n-nonacosane	15.8	7.32	2.41
n-tricontane	5.75	4.87	1.18
n-hentriacontane	11.2	3.99	2.81
Polyaromatic hydrocarbons			
Phenanthrene and anthracene	1.21	44.7	0.027
Methylphenanthrene and methyl-anthracene	0.90	10.2	0.088
Fluoranthene	2.22	8.52	0.261
Pyrene	3.17	3.36	0.488
Benzofluorenes	2.33	1.87	1.246
Methylpyrene	0.93	—	P
Benz(a)anthracene and chrysene	12.2	3.87	3.15
Benzo(k)fluoranthene and benzo(b)fluoranthene	23.1	2.01	11.5
Benzo(a)pyrene, benzo(e) pyrene and perylene	20.1	2.69	7.47
Phthalic acid esters			
di-isobutylphthalate	1.73	32.8	0.053
di-n-butylphthalate	101	353	0.286
di-2-ethylhexylphthalate	54.1	127	0.426
Miscellaneous			
Anthraquinone	1.59	5.66	0.281
Fatty acid esters			
Lauric acid	0.01	30.3	0.0003
Myristic acid	1.39	7.58	0.183
Pentadecanoic acid	3.60	5.35	0.673
Palmitic acid	29.0	4.77	6.08
Heptadecanoic acid	2.84	5.71	0.497
Oleic acid	2.06	—	P
Stearic acid	35.7	2.27	15.73
Nonadecanoic acid	1.91	1.02	1.87
Eicosanoic acid	9.04	3.00	3.01

(*continued*)

TABLE 7.9 (*continued*)

	Concentration (ng/m^3)		
Compound	Particle samples P	Gas-phase samples G	Distribution factor[b] P/G
Heneicosanoic acid	2.56	1.65	1.55
Docosanoic acid	13.7	—	P
Tricosanoic acid	3.23	—	P
Tetracosanoic acid	10.7	—	P
Pentacosanoic acid	2.55	—	P
Hexacosanoic acid	9.12	—	P
Aromatic acids			
Pentachlorophenol	2.43	—	P
Basic compounds			
Acridine, phenantridine, and benzoquinolines	0.94	—	P
Benzacridines	0.85	—	P

[a] From Cautreels and Van Cauwenberghe (1978). Low-molecular-weight organics found only in the gas phase were not included in this extract of the author's original table.

[b] P listed in the last column indicates that essentially all of the material was in particle form.

sors. The precursors are listed for each of the groups of compounds in the table.

Another dimension of tropospheric aerosol particles is their range of size. The chemical composition of aerosol particles tends to separate into distinct parts according to the origin of the material. Thus, the soil dust elements formed by comminution processes, for example, are concentrated in the coarse particles. On the other hand, the combustion-related components and secondary particles are in the finely divided material where condensation of material is involved during formation.

Examples of mass size distributions for several elements are shown in Fig. 7.11. In Fig. 7.11a are shown mass size distributions for elements that are associated with soil dust or comminution sources, including Al, Fe, and Na. The data are for West Covina, CA and Rubidoux, CA on the east side of the Los Angeles basin. Al and Fe are heavily weight in the coarse particles for these samples, while Na is somewhat more evenly distributed, particularly at West Covina. Distributions for a pair of elements associated with auto exhaust (Pb) and with fuel oil combustion (Ni) are shown in Fig. 7.11b. Here the enhancement of both elements in the fine particles is suggested.

Depending on the chemical nature of the particle environment, the size distribution of secondary products may display somewhat different mass median diameter. In general, a wide range of studies has shown sulfate to be

TABLE 7.10

Organic Constituents of Aerosol Particles Sampled from Photochemical Smog Attributed to Atmospheric Chemical Reactions[a]

Compounds identified	Possible gas-phase hydrocarbon precursors
Aliphatic multifunctional compounds	

1. $X—(CH_2)_n—Y$ $(n = 3, 4, 5)$ 1. Cyclic olefins

X	Y
COOH	CH_2OH
COOH	COH
COOH	COOH 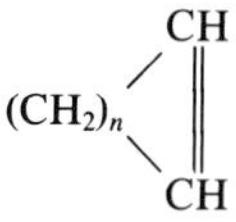
COOH	CH_2ONO
or (a) COH	CH_2ONO
COH	CH_2OH
COH	COH
COOH	COONO
or (a) COH	$COONO_2$
COH	COONO
COOH	$COONO_2$
COOH	CH_2ONO_2

and/or diolefins

$$>C=CH—(CH_2)_n—CH=C<$$

2. Others   2. Not known, possibly from aromatic ring cleavage

$CH_2OH—CH=C(COOH)—CHO$
$CH_2OH—CH_2—CH=C(COOH)—CHO$
$CHO—CH=CH—CH(CH_3)CHO$
$CH_2OH—CH=CH—CH=C(CH_3)CHO$
$C_5H_8O_3$ isomers[b]
Nitrocresols
$C_6H_6O_2$ isomers[b]

Aromatic monofunctional compounds

3. $C_6H_5—(CH_2)_n—COOH$ $(n = 0, 1, 2, 3)$ 3. Alkenylbenzenes
$C_6H_5—(CH_2)_n—CH+CHR$;
also toluene for C_6H_5COOH

4. $C_6H_5—CH_2OH$ 4. Toluene, styrene, other
C_6H_5CHO monoalkylbenzenes?
Hydroxynitrobenzyl alcohol

Terpene-derived oxygenates

5. Pinonic acid   5. α-pinene
Pinic acid
Norpinonic acid

6. Isomers of pinonic acid   6. Other terpenes?
$C_9H_{14}O_2$ isomers
$C_{10}H_{14}O_3$ isomers
$C_{10}H_{14}O_2$ isomers

[a] CMBEEP (1977).
[b] Isomers not resolved by mass spectrometry.

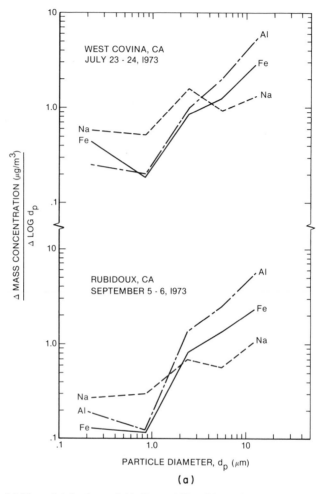

Fig. 7.11. (a) Mass distributions of Al, Fe, and Na with particle size. [From Hidy (1975).] (b) Mass distributions of Ni and Pb with particle size. When the Ni concentration was not detectable, the limit of detectability was plotted with a downward arrow. [From Hidy (1975).]

concentrated in the fine particles less than 1 μm in diameter. However, recent observations suggest a striking difference in classes of sulfate size distributions which depend on sulfate and ozone concentration, and relative humidity. An example from low pressure impactor measurements taken at Pasadena, CA is shown in Fig. 7.12. Hering and Friedlander (1982) attribute type I sulfur distribution to conditions favorable to particle growth by sulfate accumulation on existing particles in the air. Type II distributions are identified with new particle formation by heteromolecular nucleation of sulfuric acid in the air under conditions of high chemical reactivity and low concentrations of suspended particles available for growth.

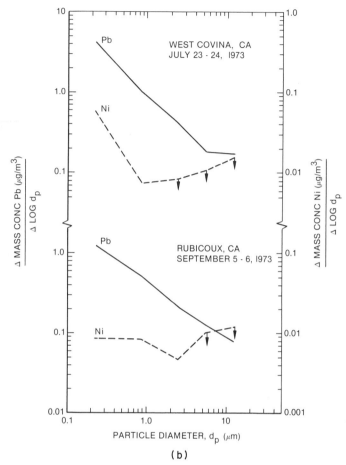

Figure 7.11 (*continued*)

7.3.2 Stratospheric Particles. Observations have become available for particle composition in the stratosphere, from occasional aircraft or balloon sampling. These results indicate that sulfate also is present in appreciable concentrations above the troposphere.

A "model" composition for the stratosphere aerosol is shown in Table 7.11. These data, based on the work of several investigators, including Junge *et al.* (1961), Castleman *et al.* (1974), and Lazrus *et al.* (1971), show that sulfate is by far the major component of aerosol at altitudes of about 20 km. Metals or metal compounds are very much lower in concentration than sulfate particles despite the potential presence of extraterrestrial material, or of dust from the earth's surface rising to high altitude.

In contrast to the tropospheric material, organics, as presently reported, are essentially absent from the stratospheric particles and nitrates are very

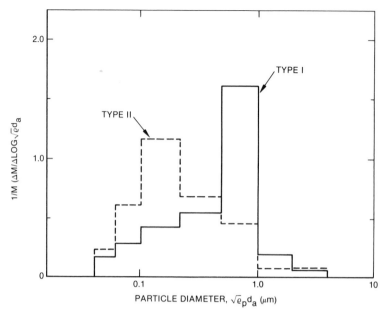

Fig. 7.12. Two distinct urban sulfur distributions from composites of several low-pressure impactor measurements in Pasadena, CA. The distributions are given in terms of the aerodynamic particle diameter corrected for atmospheric pressure, $\rho_p^{1/2}d_a$. Type I conditions include high relative humidity and high sulfate and ozone concentrations. Type II conditions involve lower relative humidity and low sulfate and ozone concentrations. [From Hering and Friedlander (1982).]

low in concentration. This is not surprising because it is expected that condensable organic vapors would have to be high in molecular weight and probably would not survive substantial degradation above the tropopause. Although nitric acid is present in the stratosphere, the absence of nitrate in the aerosol particles evidently is related to the high vapor pressure of HNO_3 relative to that of H_2SO_4; NH_4^+ or cations other than H^+ are not expected to be present in the large quantities required for the formation of nitrates of sufficiently low vapor pressure to persist at high altitude.

7.3.3   Geographical Distributions and Occurrence. An aggregated listing of the average composition of particle samples represents only one dimension of the chemical characterization of atmospheric aerosols. In recent years, sampling and analysis programs have provided information about the geographical distributions of particulate matter and their temporal variations. With the institution of air pollution control and regulation, much of the initial observational effort was focused on conditions in cities where human populations are centered. Later increasing concern has emerged for the large-scale influence of pollution, ranging from regional scale to global ex-

TABLE 7.11

Model Chemical Composition of Stratospheric Particles at 20 km Altitude during the Period 1969–1973[a]

Chemical component	Model concentration in air ($\mu g/m^3$)	Observed concentration range ($\mu g/m^3$)
Sulfate	0.6[b]	0.01–4 (Agung peak)[b]
Basalt[c]	0.05	0–0.7
NH_4^+	0.005	0–0.01
NO_3^-[d]	0	0
NO_2^-	0	0
Na	0.01	0.001–0.05
Cl	0.04	0.002–0.09
Br	0.002	0–0.003
Total	0.71	0.01–1

[a] From Hidy *et al.* (1974).
[b] For 8-yr averaged data during the period after the Mt. Agung eruption at 18 km; from Castleman *et al.* (1974).
[c] This includes all other components of the basalt; for example, Al, Ca, Mg, based on ratio to Si.
[d] Particulate NO_3^- as contrasted with HNO_3 vapor.

tent. This has resulted in extended measurement programs covering virtually every type of climate or air mass condition.

A perspective on geographical features of aerosol concentration distributions is derived from considering the "scales" of behavior. These help define the extent of atmospheric phenomena and are given in Table 7.12. Localized or microscale events take place over distances of a few kilometers and a day or less. These are represented by visual effects of a smoke plume, or a transient dust devil. The urban scale extends further into tens of kilometers over periods of at least a day. Examples of this kind of activity include haze buildup in large metropolitan areas like Los Angeles, London, or Tokyo. On even larger spatial scales, regional phenomena take place, which have been recognized in western Europe [Organization for Economic Cooperation and Development (OECD), 1977] and in North America (Hidy *et al.*, 1978a). These events evolve over distances of 100–1000 km, and remain active for up to 10 days or more. These have involved the accumulation of sulfate, for example, from large industrial areas, and dust storms from major arid areas such as the Sahara Desert. At the largest extreme are the planetary or global scale events, which may be characterized in terms of long-term changes over weeks to months, encompassing hemispheric properties. These are not necessarily dramatic disturbances, though volcanic eruptions penetrating to the lower stratosphere have such a character.

TABLE 7.12

Spatial and Temporal Scales for Trace Contaminant Behavior in the Atmosphere[a]

Pollution class	Meteorological class	Horizontal spatial range (km)	Temporal range (days)	Comments
Local	Microscale	0–10	≤1	"Classical" pollution impact— basis for nonreactive pollutant reactions
Urban	Mesoscale	10–100	1–2	Identified in the 1950s largely with reactive pollutant products such as photochemical oxidant
Regional	Synoptic scale	1000	1–5	Recognition of problems related to haze and visibility degradation, acid rain, sulfate
Global	Planetary scale	>1000	>5	Extensively surveyed in relation to dispersal of radio nuclides from bomb tests

[a] After Hidy *et al.* (1978a).

Knowledge of the combination of topographical features, emission distribution, prevailing winds and height of the surface mixed layer of air, and air chemistry generally are sufficient to explain at least qualitatively the spatial features of particle distributions. Quantification of the relation between sources and ambient conditions has occupied the efforts of mathematicians and engineers for a number of years. Progress in modeling technology has been made, but the reliability of methods for regulatory applications remains restricted (see also Chapter 10).

Perhaps the largest historical air monitoring program in the United States has involved TSP sampling in the National Air Surveillance Network (NASN). This system has provided data for over twenty years in some cities. Observations range from a few micrograms per cubic meter in remote areas to milligrams per cubic meter over 24-hr intervals in dusty or heavily industrialized areas. Over a period of approximately nineteen years between 1958 and 1977, national trends have shown a steady decrease in TSP levels. From 1962 to 1972, a nationwide improvement of 2% per year has been reported by the U.S. Environmental Protection Agency (U.S. EPA). During the same time period, particulate sulfate concentrations have either remained constant or increased somewhat in North America (e.g., U.S. EPA, 1978).

Effort has been expended on tracing the geographical distribution and trends in metal components of particles. The available data have concentrated on fuel-related elements, lead, vanadium, nickel, and titanium, and heavy-industry-related materials, cadmium, chromium, copper, iron, and manganese. These metals have been monitored by the U.S. National Air Surveillance Network in the United States, in some cases since the early 1960s. The results of a regional trend analysis (e.g., Faoro and McMullen, 1977) suggest a distribution relating to specific source categories and urban areas. Basically, trends from the mid-1960s to 1975 appear to be consistent with changes in particulate emission patterns associated with fuel use and industrial control.

The vertical distribution of suspended particulate matter has been poorly characterized relative to the geographical distribution at ground level. Some observations of vertical profiles of sulfate with sulfurous gases were reported by Georgii (1970) for the lower troposphere. Examples of elemental profiles below 10 km height for the midwest, coastal Pacific, and central California are shown in Fig. 7.13. These observations reflect a general decrease with

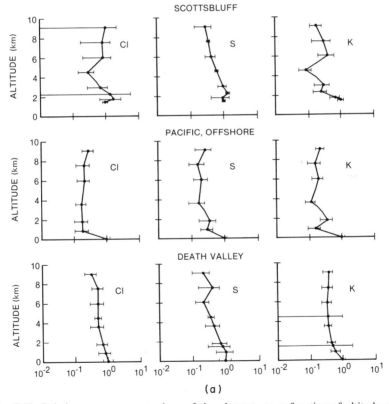

Fig. 7.13. Relative mean concentrations of the elements as a function of altitude at the various measurement sites normalized to ground level with range of variation in samples. Figure continues on next page. [From Gillette and Blifford (1971); reprinted with permission of the American Meteorological Society.]

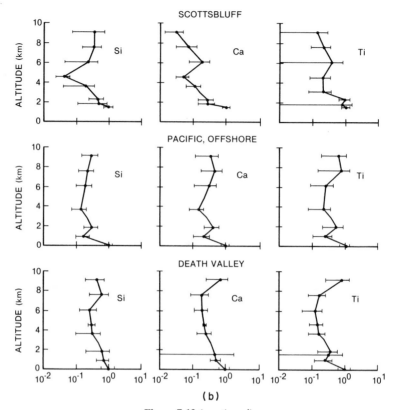

Figure 7.13 (*continued*)

height in sulfur and potassium. However, there is some evidence of increased concentrations of chloride and silicon over Scottsbluff, Nebraska, and increased Si, Ca, and Ti over Death Valley, CA in the samples taken (Fig. 7.13).

The spatial distributions of particulate matter in the stratosphere have been measured very sporadically by high-flying research aircraft. In the 1950s, special attention was given to radioactive material collection. Later in the 1960s, interest broadened to investigate the global distribution of particles, with emphasis on sulfate and its maximum in the Junge layer (e.g., Cadle, 1973). From aircraft sampling, Lazrus and Gandrud (1974) have reported the estimated latitudinal variation in particulate sulfate and nitric acid vapor. These are shown in Fig. 7.14 for the altitude range between 12 and 20 km for the spring of 1971. The patterns of distribution of these two constituents of the stratospheric aerosol are generally similar. There is an asymmetric tongue of low concentration extending upward at the equator. In the middle and high latitudes, higher concentrations are observed. The layer of

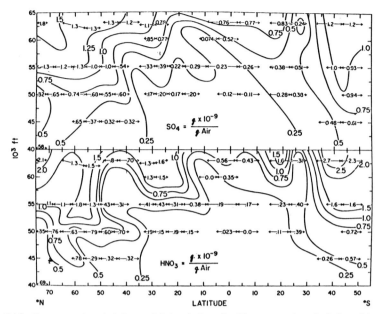

Fig. 7.14. Concentration (mixing ratio) isopleths of sulfate aerosol and nitric acid vapor, Northern Hemisphere, spring 1971. [From Lazrus and Gandrud, (1974); reprinted with permission of the American Geophysical Union.]

high concentration for both sulfate and nitrate begins at approximately 15 km in the high latitudes, but appears higher at 19–20 km in the tropics. The patterns shown are qualitatively consistent with a meridional circulation which involves the uplifting of relatively clean tropical air from equatorial heating and subsidence from cooling air near the poles.

7.3.4 Statistical Methods of Aerometric Analysis. When sufficient aerometric data are accessible, it is possible to analyze for the interrelationships between particle chemical components and other parameters believed to be characteristic of air chemistry, precursor gas emissions, or meteorology. Several investigators have attempted to achieve this goal by applying standard statistical regression techniques to existing sulfate data. Altshuller (1973) and Sandburg *et al.* (1976) have sought a relationship between long-term average, ambient sulfate concentrations and ambient levels and emissions of sulfur dioxide, while White *et al.* (1978) have attempted to establish a link between daily precursor emissions and daily sulfate levels. These studies have found that results are inconclusive for relationships involving emissions of sulfur dioxide or ambient SO_2 concentrations and sulfate levels, after applying straightforward statistical procedures. More sophisticated methods appear to be required to obtain defensible results.

The past application of statistics to the analysis of sulfate concentration patterns has attempted to find the "most important" variables which "best" predict sulfate variability. The type of analysis used in this work involves dependent or *predictant* variables, for example, sulfate, and independent or *predictor* variables (SO_2, NO_2, wind speed, mixing height, temperature, humidity, etc.). These independent variables are actually highly intercorrelated. This is typically the case with meteorological and air quality data, but there is no simple, objective way to find a "best possible" regression formula by application of stepwise calculations. Major uncertainties result from the instability of regression equations caused by the unavailability of truly independent variables in atmospheric data. Statisticians have realized for some time the potential dangers of multiple regression on intercorrelated variables. Swindel (1974) gives a graphic illustration of the instability of the regression equations determined under such conditions. Draper and Smith (1966) also make the same point. Methods are accessible which give results unaffected by intercorrelations of the variables. One example, a form of factor analysis called *principal component analysis* (PCA), has been used in the intercorrelations of the observable variables to uncover some of the physical and chemical processes basic to sulfate formation (Henry and Hidy, 1979, 1982; Mueller and Hidy, 1983). This method also can be used to establish a variety of elemental relationships in particulate matter (see also Moyers *et al.,* 1977). The technique is a useful way to examine large quantities of observational data for common aerometric relationships which can be considered complementary and independent.

Principal component analysis, analysis by empirical orthogonal functions, natural functions, or latent root analysis all refer to a method of multivariate linear statistical analysis or factor analysis. References to the statistical methods are to be found in Essenwanger (1976) and Harmon (1976). PCA is a special case of factor analysis but the two terms are used interchangeably by some authors.

PCA replaces the set of intercorrelated observables with a set of new, uncorrelated variables which are linear combinations of the original variables. The first principal component is that linear combination of variables which explains or accounts for a maximum of the variability of the original variables. The second principal component is the linear combination, uncorrelated with the first principal component, that explains a maximum of the total variability not already accounted for by the first component. The third-, fourth-, and higher-order principal components are defined similarly. The principal components are convenient to work with because they are statistically independent and, typically, the first few components explain almost all the variability of the whole data set. In cases reported for sulfate analysis (Henry and Hidy, 1979, 1982; Mueller and Hidy, 1983), 20 or 30 intercorrelated variables are replaced by 5 or 6 principal components which contain virtually all the information of the original variables.

The (linear) combinations of observables which characterize the data are derived from the data collection by mathematical manipulation of matrices. The linear combination of the variables Z_{ik} that accounts for a maximum of the total variability of the data set is written as a normalized variable:

$$Z_{ik} \equiv \frac{X_{ik} - m_i}{\sigma_i} = \sum_{j=1}^{n} A_{ij} P_{jk}, \qquad i = 1, \ldots, n \qquad (7.3)$$

where X_{ik} is the kth value of the ith observable such as wind speed or all SO_2 concentration values of X_{ik}, m_i is the mean of X_i, and σ_i the standard deviation of X_i; P_{jk} is the kth value of the jth (principal) component; $k = 1, \ldots$, M, is the number of cases (or time intervals); A_{ij} is called the loading of the ith variable on the jth principal component.

The mathematical basis for PCA is straightforward. By means of orthogonal transformations or rotations coordinates assigned to the observables chosen for analysis, the set of intercorrelated observables is transformed into a set of independent, uncorrelated variables P_{jk}. Values of P_{jk} are obtained with A_{ij} by using a uniqueness theorem from matrix algebra which states that a single matrix can be decomposed into a product of two matrices, one of which is the matrix whose elements are the correlation between pairs of all observations in the data set (the correlation matrix). If the variables are independent, then the correlation matrix is diagonal (actually an identity matrix). Mathematically, the procedure is to diagonalize the correlation matrix. This is done by finding the eigenvalues and eigenvectors of this matrix. It is a mathematical theorem that a correlation matrix will always have at least one nonzero eigenvalue and all the nonzero eigenvalues will be positive. The matrix of the diagonalizing transformation is composed of the eigenvectors as columns (or rows, depending on whether the original vector of variables was a row or column vector). The principal component associated with the largest eigenvalue is the first principal component, and so on.

The principal components are obtained from the inversion of Eq. (7.3) with the result (Harmon, 1976)

$$P_{jk} = \sum_{i=1}^{n} \frac{A_{ij}}{\lambda_j} Z_{ik}, \qquad (7.4)$$

where λ_j is the eigenvalue associated with P_j.

Given the estimation of principal components, one can then use a linear expression technique to deduce the contributions of the principal components to the variability of one of the parameters in the set. The values of A_{ij} range from $< \pm 0.2$ to $\geq \pm 0.9$. The former indicates essentially no correlation of variable i and component j; the latter indicates a strong positive or negative correlation of variable i and component j.

Once the principal components have been obtained, they can be used in a multivariate regression analysis to determine the strength of contributions of each component to account for the variability of the selected dependent observable.

The regression of Y, the dependent observable, on the P_i gives the formula

$$Y = \overline{Y} + \sum_{i=1}^{n} b_i P_i,$$ (7.5)

where $\overline{Y}$ denotes the mean value of Y. Because the principal components are independent, the least-squares estimates of the regression coefficients are given by the formula

$$b_i = \frac{1}{M-1} \sum_{k=1}^{M} Y_k P_{ik},$$ (7.6)

where M is the number of cases and Y_k is the kth value of the dependent variable. Application of this approach avoids the perceived ambiguities in a regression analysis using the actual aerometric observables as "independent" variables.

In addition to numerical stability, there are several advantages to this approach over traditional multiple-regression techniques. First, the principal components are mathematically independent. As seen in Eq. (7.6), b_i depends only on values of Y_k and P_i. Thus, any number of terms may be dropped from the regression while retaining its validity. Second, usually only a small number of components made of a much larger number of physical variables need be considered to explain most of the variability in the dependent variable. Third, eigenvectors with very small eigenvalues can be used to examine the exact nature of the intercorrelations present in the data set. Application of these properties can make the interpretation of the results useful, particularly if the components can be understood in terms of the physical and chemical processes under investigation.

Using the PCA model, Henry and Hidy (1979, 1982) have investigated interrelationships between particulate sulfate and a variety of more than fifteen other aerometric variables in several case studies. These variables included complementary aerosol parameters, including TSP, SO_2, NO_x, and O_3, and meteorological parameters, including wind speed and direction, absolute and relative humidity, temperature, and mixing height. The data sets were collected for a year or more in the 1970s in four different cities: Los Angeles, New York, St. Louis, and Salt Lake City, as well as nine nonurban sites covering the greater northeastern United States. This work has produced a variety of important interrelationships between variables that characterize aerosol behavior. One interesting result is illustrated in the breakdown of contributions to two major principal components for data from different sites in New York City (Table 7.13). These two components explained most of the variability in sulfate observed.

The analysis indicated that (a) certain key parameters influence all four

TABLE 7.13

Major Components of New York Area Data Sets (Values of $A_{ij} \times 1000$); Blocked Rows are those of Consistently Highest Values for All Sites[a]

Variable/site	Photochemistry				Local combustion sources/dispersion			
	Riverhead	Queens	Brooklyn	Bronx	Riverhead	Queens	Brooklyn	Bronx
Nitrogen dioxide	−309	−109	034	569	785	789	701	575
Sulfur dioxide	−506	−511	−539	−301	541	623	392	710
Ultraviolet intensity	244	612	642	597		−116	−317	−144
Relative humidity		030	−080	−254	−183	−181	326	−371
Maximum temperature	943	934	926	889	−009	−006	−032	−291
Minimum temperature	930	927	919	893	077	014	029	242
Wind speed, noon	−220	−162	−220	−347	−606	−511	−603	−381
Wind speed, midnight	−250	−275	−382	−429	−583	−551	−468	−373
Morning mixing height	−050	−065	−010	−115	−248	−289	−452	−425
Afternoon mixing height	268	359	321	366	−004	102	−320	−047
Daily average oxidant (O₃)	713	851	850	809	−047	−037	−082	−243
TSPM[b]	404	209	272	410	605	622	627	686
IRH[b]	169	−021	−092	−289	−200	−211	026	−323
Absolute humidity	946	824	786	672	−045	−043	112	−430
VENT^{-1}[b]	071	−010	022	085	140	165	431	290
% of total variance	26.4	25.4	26.1	27.2	15.5	14.5	15.2	15.9
R^2 for sulfate	0.17	0.062	0.034	0.130	0.107	0.203	0.224	0.095
Sulfate regression coefficient	1.94	1.62	0.923	2.26	1.53	2.93	2.38	1.93

[a] From Henry and Hidy (1979); reprinted with permission from Pergamon.

[b] TSPM is the mass concentration less sulfate and nitrate; IRH $\equiv (1 - \text{RH}/100)^{-1}$; VENT is a ventilation factor equivalent to the product of an average wind speed and mixing height.

New York sites; (b) a major fraction of sulfate variability can be assigned to a component that is dominated by temperature ozone and absolute humidity; (c) another major fraction is associated with SO_2 and NO_x gases, wind speed, and suspended-particulate concentration. These results are interpreted to mean that the sulfate behavior over New York is common to the entire metropolitan area, including a distant location at the eastern extreme of Long Island. Much of the sulfate behavior can be explained by photochemical oxidation chemistry as measured by the surrogates of temperature, ozone, and absolute humidity. Another fraction is linked with patterns associated with major emissions of reactant gases SO_2 and NO_x,[†] as well as primary particulate sources signified by TSP less sulfate and nitrate (TSPM).

The analysis of data from Los Angeles and St. Louis showed results similar to those for New York. However, Los Angeles sites showed a stronger influence of photochemical indicators than New York. In addition, a component strong in the influence of relative humidity appeared, which was interpreted to be a possible indicator of chemical reactivity in wet aerosols or in aqueous droplets. Since relative humidity is generally high in St. Louis and New York, this "factor" did not emerge from data taken in these cities. The Salt Lake City results were distinct from those for the other cities in showing a strong meteorological association with sulfate variability. High sulfate concentration was observed with low mixing heights and air stagnation conducive to pollution accumulation. The influence of relative humidity was found here too, but only a negligible influence of photochemical effects on sulfate emerged in the Salt Lake City area.

The examination by PCA of the nonurban data in the greater Northeast showed the dominance of seasonal and diurnal variations, along with winds and mixing conditions associated with regional-scale pollution behavior (Mueller and Hidy, 1983). However, the link between sulfate variability and its precursor SO_2 emerged distinctly in this data set, as compared with the four city results.

7.4 SOURCES OF PARTICULATE MATTER

Given the broad features of the phenomenological behavior of atmospheric aerosols, a major preoccupation of scientists over the past twenty-five years has focused on the sources and sinks of airborne particulate matter. With knowledge of the atmospheric content of particles, the specification of the sources and emission rates provides basic information for budgeting the suspended material, either on a global basis, or over specific geographical regions. In this section, we consider the sources of airborne particulate matter, including a discussion of primary and secondary material.

[†] In many cities the sources of these two pollutant gases are common, and are associated with combustion processes.

From the list in Table 7.2, the origins of atmospheric particulate matter are quite diverse. The sources also can vary appreciably with time. The global source distributions tend to deemphasize man's activities, and place perspective on the importance of natural emissions. The primary "sources" include the extraterrestrial background of interplanetary dust, sea salt particles, soil dust, volcanic debris, and forest or brush fires. On a regional and local basis, man's activities often overshadow the natural sources, and result from agricultural practice, fuel combustion, industrial processes, and construction. Perhaps the largest single source class of atmospheric particles is secondary in nature, resulting from atmospheric chemical reactions of gases. The precursors may come from natural or anthropogenic origins, and are grouped in terms of sulfurous, nitrogenous gases, and organic vapors. Most of the known aerosol-particle-producing reactions involve oxidation. However, one reaction involves the displacement of halogens in salts by acidic gas attack (e.g., Robbins *et al.,* 1959). Perhaps the best evidence for its importance comes from the depletion of particle chlorine relative to sodium in urban areas near the oceans (e.g., Miller *et al.,* 1972). Although this reaction should be recognized, it will not be covered further in this chapter.

7.4.1 Extraterrestrial Particles. Perhaps the most fundamental natural background level for the atmospheric aerosol consists of extraterrestrial dust which has been known for many years from astronomical observations (Öpik, 1958). The extraterrestrial component has its greatest influence in the high atmosphere, and may be best known through the explanation of the zodiacal light. This phenomenon is associated with the scattering of sunlight by particles at very high altitudes. The estimates of the contribution of extraterrestrial particles to atmospheric aerosols is small compared with sources listed in Table 7.2. However, this material undoubtedly makes up a much larger share of particles found above 30–40 km height.

The origins and composition of extraterrestrial dust have been reviewed recently by Millman (1979). Extraterrestrial or interplanetary dust refers to the solid particles present in space ranging in size from tenths of micrometers to meteors a few millimeters in diameter. The origin of this material is thought to be associated largely with debris from comets. The concentration of particles is largest along the plane of the ecliptic of the solar system.

The interplanetary dust is made up of clusters which appear as loose agglomerates of finely divided material which readily disaggregate on collision. The smaller submicrometer size particles are more dense, consisting of iron–nickel or iron–silicate material. Groupings of apparent density have been deduced, 8, 3, and 1 g/cm^3. The first two are high-density material with grain sizes of 0.2 to <0.01 μm. The larger particles are rough or fluffy aggregates, most of which are well compacted. This larger, fluffy material appears to be related to the absorption typical in producing the zodiacal light.

The size distribution of interplanetary dust is shown in Fig. 7.15, along

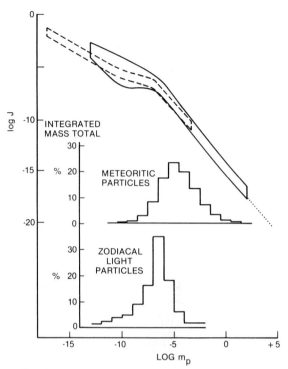

Fig. 7.15. Upper plot: (---) cumulative particle flux determined from microcrater counts on lunar rocks, general area of cumulative flux values found by numerous observational techniques; (···) a portion of the flux curve, estimated from results obtained with seismographs located on the lunar surface. Lower plots: frequency distributions for integrated mass totals of various particle masses; the zodiacal light particles represent the mass distribution of the particles that contribute to the luminosity of the zodiacal light in the wavelength range from the ultraviolet near 2000 Å to the near infrared at about 20 μm. J is the cumulative particle flux (m^{-2} sec^{-1}) [$(2\pi\text{sr})$]. m_p is the particle mass in grams. [From Millman (1979).]

with the cumulative particle flux. The flux to earth is of order 10^{-13} g/m² sec. The particle concentration in space tends to decay as the distance to the 1.0–1.5 power away from the sun. The size distributions deduced for meteoritic material and from optical observations of zodiacal light are shown in the figure. The median particle mass is of order 10^{-6}–10^{-7} g, which for spherical particles of unit density corresponds to a nominal mass median diameter of 0.4–1.0 μm.

Quantitative information on the chemical composition of interplanetary dust particles is now available from at least five independent observational techniques. Data for the average composition of the most common type of dust particle are summarized in Fig. 7.16. A convenient basis for comparison is Cameron's (1973) tabulation of elemental abundances in the solar system. In this context we are dealing with solid objects; the relative abundances

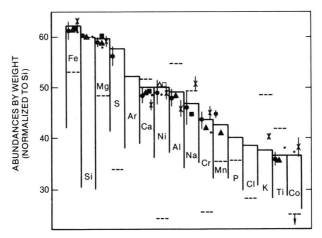

Fig. 7.16. Observational evidence for the relative abundances of the chemical elements in interplanetary dust. Cameron's 1973 values for the solar system are used as a basis for comparison. To illustrate strongly differentiated material, approximate average abundances for the earth's crust are also plotted. Large dots: 13 particles collected in the upper atmosphere, diameters 3–25 μm, the vertical lines representing root-mean-square deviations for a single observation; triangles: 19 deep-sea spherules from the mid-Pacific at depth 5000 m, diameters 300–400 μm, analyzed with the electron microprobe; squares: 16 meteor samples; small dots: meteorite residue in one impact crater from Skylab-IV, diameter 110 μm; crosses: abundances of upper-air positive ions, as measured at heights from 95 to 120 km above sea level, vertical lines indicating mean deviations of the measures among four groups of experimenters. Since neither meteor spectra nor upper-air ions provide reliable mean silicon values, normalization in these cases was to the average of iron and magnesium. The mean nickel values for meteor spectra and deep-sea spherules are less reliable than for the other elements, and this has been indicated by open symbols. ——, Cameron; ---, Ahrens. [From Millman (1979), with references.]

have been converted to weight and the relevant sequence of elements plotted on a logarithmic scale, normalized to Si at a value of 60. Ahrens's values (1978) for the earth's crust have also been plotted to illustrate material evolving in the solar system which is strongly differentiated from Cameron's index for the solar system. The particles collected from the upper atmosphere by Brownlee (1978) were all of this "chondritic aggregate" type, which accounts for more than half of the complete collection of more than 300 extraterrestrial particles. Other types include nickel-bearing iron sulfide and single-mineral grains of enstatite, olivine, or magnetite. Deep-sea silicate spherules from the mid-Pacific are of extraterrestrial origin on the basis of the abundances of certain trace elements (Ganapathy *et al.*, 1978). Iron spherules of interplanetary origin were also collected from the same red-clay core samples as the silicate spherules. The average abundances from the sixteen meteor spectra are for members of five well-known cometary streams. Residues from impacting particles can be detected inside microcraters, and the abundances plotted in Fig. 7.16 are for a relatively large crater produced by a meteoroid about 30 μm in diameter. Smaller craters, down to submicrometer sizes, have been analyzed for particle residues (Brownlee *et al.*, 1974) and in general show fewer elements than the larger craters, as would be expected for the impact of much smaller dust grains (Millman, 1979).

A strong enhancement of metallic upper atmosphere positive ion concentration is sometimes detected after the earth has passed through a meteor stream and these ions can be recorded by rocket-born mass spectrometers. This gives us a fifth method of measuring average abundances of the elements in interplanetary dust.

The evidence from Fig. 7.16 indicates that the overall average chemical composition of the interplanetary dust is very similar to that for the solar system, at least for the range of elements shown. It is completely unlike that of differentiated material such as the earth's crust. This must also be true to a reasonable extent for the various trace elements of mass higher than cobalt, as interplanetary material can be recognized in the lunar soils by a measure of these same trace elements. For the volatile elements more abundant by weight than iron we have no good average values. Both the aggregates collected by Brownlee and certain meteor spectra show evidence of C, H, and O, though the last element is probably also present as a contaminant from the atmosphere. The presence of atmospheric Na likely causes the high value of this element in the upper-air ions. In summary we see that among the abundant elements, interplanetary dust is most easily recognized by its high content of Fe, Mg, S, and Cr as compared with Si, Ca, Al, and Na.

7.4.2 Sea Salt.   The natural surface-level sources of particles are much larger than material captured from extraterrestrial scavenging. The oceans are known to be a prolific source of particulate material. One mechanism for production of particles from the sea is droplet breakup resulting from breaking waves and wind action on the wave crests. From spray formed at the sea surface, droplets are borne aloft by the winds to evaporate, producing sea salt particles. This process tends to form giant crystals which approach a 100 μm in size, and readily fall by gravity back to the surface.

A second mechanism was proposed several years ago by Blanchard and Woodcock (1957) which could produce small "satellite" droplets by bubbles of foam breaking on the water surface. The satellite droplets are much smaller than spray, and produce small salt crystals after evaporation (see also Fig. 4.13). It is hypothesized that the bubble ejection process accounts for small sea salt particles in the 1–10 μm size range which can remain suspended for extended periods.

Because the marine air generally has a shallow temperature inversion aloft, the sweeping of the wind fills the surface air layers with particles, but strongly inhibits mixing to heights above 1000 m. This is readily seen in Fig.

7.17 synthesized by Blanchard and Woodcock (1980). The sharp drop in sea salt concentration is seen at just above 1000 m height. The data in this figure also show the rapid increase in surface concentration of salt particles with wind speed. This is to be expected since it is well known that spray and foam (whitecapping) dramatically increase with wind speed.

The salt particle size distribution generated by the sea surface action follows a power-law distribution roughly proportional to $R^{-2.5}$ above 1 μm radius. The number of particles of different sizes increases nearly uniformly with wind speed, suggesting no strong differentiation in size of spray droplets with wind action on the waters.

The particle volume distribution of marine aerosols has a characteristic dominant mode in the giant or coarse ($\gtrsim 1$ μm) particle range, as expected from the particle generation mechanism involved. A volume–size distribution of particles sampled near breaking waves is shown in Fig. 7.18. For comparison, a volume distribution from a remote site in the California desert, north of Barstow, also is included, showing an arid soil dust distribution. This example contains two modes of roughly equal contribution to the volume, one a coarse particle mode and the other a large particle or accumulation mode in the 0.1–1.0 μm range. The coarse particles are mainly soil dust. Particle chemical analysis of this aerosol suggested that material in the accumulation mode was an aged component sharing urban particle character from coastal areas of California. For added comparison, an urban distribution, influenced by auto exhaust, is drawn in Fig. 7.18 to indicate a typical combustion aerosol particle distribution. In this last case, the modes are in

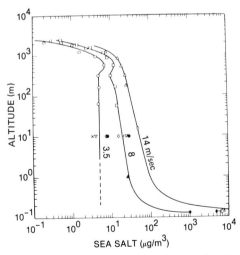

Fig. 7.17. Compilation of data on sea-salt concentration as a function of altitude and wind speed. The altitude ranges from <0.2 to >2000 m. [From Blanchard and Woodcock (1980), with references for observations; reprinted courtesy of the New York Academy of Sciences.]

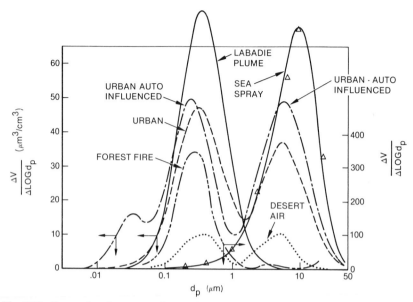

Fig. 7.18. Comparison between volume–size distribution for urban aerosol particles and for combustion-enriched samples including (a) urban auto-exhaust-influenced plume and a power plant plume in Missouri (Whitby and Sverdrup, 1980), (b) forest fire plume (Radke *et al.*, 1978), (c) desert background air (Goldstone, CA) (Hidy, 1975), (d) sea spray (Pt. Arguello, CA) (Hidy 1975).

the Aitken nuclei (0.01–0.1 μm) and large particle range (0.1–1.0 μm) rather than the giant particle range.

Far from land sources, one would expect that the chemical composition of marine aerosol particles would be dominated by sea salt. Sea salt composition is listed in Table 7.14. This composition varies little from place to place, except for volatile or biologically related species, ammonium and carbonaceous material. For comparison, composition of marine particulate matter taken offshore of California is shown in the table. We see that there is significant particulate enrichment in all species relative to sodium. Those elements or ions which are closest to sea salt are the halogens. These data illustrate well the complexity of chemical composition from the blending of particles from many sources, even in remote sites, supposedly dominated by a nearby natural source. For San Nicolas Island, as an example, sea salt evidently is only about 11% of the total suspended-particulate burden.

There also are very substantial differences in sulfate and nitrate content which are related to their production from oxidation of sulfur and nitrogen oxide gases in the marine air. Also, significant amounts of carbonaceous material are present, which is believed to be associated at least in part with production of particles from nebulization of fatty materials on the sea surface.

TABLE 7.14

Comparison between Sea Salt Composition (wt %) and Marine Aerosol Sampled in 1970 on San Nicolas Island (SNI), 90 Miles West of Los Angeles[a]

Element	Sea salt[b]	Sea salt $(elem/na)_{max}$	SNI aerosol[c]	Aerosol $(elem/Na)$[d]
Na	30.6	1.00	3.6	1.00
K	1.1	0.035	1.3	0.36
Ca	1.2	0.039	1.9	0.52
Si	$(1.4–94) \times 10^{-4}$	3.1×10^{-4}	5.0	1.39
Mn	$(2.5–25) \times 10^{-6}$	8.1×10^{-7}	0.04	0.011
Fe	$(5–50) \times 10^{-5}$	1.6×10^{-5}	1.6	0.44
Cu	1.8×10^{-3}	5.8×10^{-5}	0.6	0.17
Zn	2.8×10^{-6}	9.1×10^{-8}	0.2	0.056
Al	$5–50 \times 10^{-4}$	1.6×10^{-4}	2.8	0.78
C (noncarbonate)	8.7×10^{-3}	2.8×10^{-4}	3	0.83
Pb	1.3×10^{-5}	4.2×10^{-7}	0.4	0.11
Cl	55.0	1.79	9.9	2.75
Br	0.19	6.2×10^{-3}	0.2	0.056
NH_4^+	$(1.4–14) \times 10^{-6}$	4.6×10^{-7}	2.2	0.61
NO_3^-	2×10^{-3}	6.5×10^{-5}	5.7	1.58
SO_4^{2-}	7.68	0.25	17.5	4.86

[a] Aerosol data from Hidy *et al.* (1974).
[b] Based on dry weight composition from tabulations in Miller *et al.* (1972).
[c] Average of 13 samples taken about 50 m above the sea surface.
[d] Element or ion concentration normalized to Na concentration.

7.4.3 Suspension of Soil Dust. The action of the wind on loose soil areas can raise large quantities of dust from the earth's surface. Hazes from dust storms are well known in arid desert areas of the American southwest, the Sahara in Africa, and the loess storms in mainland China. As discussed in Section 4.3, the mechanism for the entrainment of dust depends on the aerodynamic forces induced by the air flow immediately over the particles. These forces must overcome the adhesive forces acting on the particles which retain them on the surface. The vertical flux of particles induced from turbulent air motion is deduced to be proportional to the kinetic energy transferred to the surface or ρu^{*3} where u^* is the friction velocity, equal to the square root of the shear stress at the surface divided by the air density (Gillette, 1980). The friction velocity is related to the drag coefficient of the surface, which in turn is roughly proportional to the square root of the wind speed near the surface.

The release and transport of soil particles requires an air flow above a threshold value. Gillette (1980) has described particle motion at the surface. Pictorially, the motions associated in the suspension are given in Fig. 7.19. Figure 7.19a shows the creeping or rolling motion of a giant particle close to

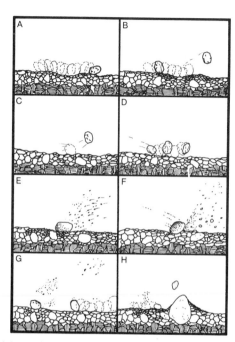

Fig. 7.19. Soil particle motions during wind erosion: (A) particle moving with wind speed slightly greater than threshold; (B) particle lifted by turbulent air fluctuations; (C) an airborne particle collides with the surface; (D) an airborne particle collides with the surface and rolls; (E) particle collision with breakoff of smaller particles encrusted on the colliding particle's surface; (F) a particle collision with splashing of soil; (G) combination of C, D, E, and F; (H) a nonerodible element shielding soil downwind. [Redrawn from Gillette (1980); reprinted courtesy of the N.Y. Academy of Sciences.]

threshold wind speed. In Fig. 7.19b the particle is lifted by air motion into the turbulent boundary layer, and in 7.19c the particle collides with the surface and bounces in the turbulent layer. In Fig. 7.19d a collision is followed by creeping and another lift. Figure 7.19e shows the disaggregation of smaller particles that were encrusted on the colliding particle's surface. In Fig. 7.19f, a coarse particle collision is followed by splashing of fine particles into the air. Taken together, parts (e) and (f) of Fig. 7.19 illustrate sandblasting of fine material by a much larger particle. In Fig. 7.19g, many of the movements in parts (a)–(f) are combined, and the small particles emitted are carried high into the air. Finally, Fig. 7.19h shows an example of the influence of a nonerodible element embedded in the surface. Rolling particles collide with the element potentially adding to the lifting; the nonerodible element also shields soil particles in its lee.

If the vertical flux of small particles is proportional to the flux of kinetic energy delivered to the soil surface, then the ratio of the vertical flux of fine material to the horizontal flux of sand should remain constant with wind speed. Physically, such a situation would reflect a soil surface that is uniformly susceptible to destruction by sandblasting by coarse particles. An example of such a surface would be a loose soil well mixed with fine and coarse particles. If the ratio of the vertical flux of fine material to the horizontal flux of sand increases with wind speed, the soil is probably being pulverized or disaggregated by energetic sandblasting, thereby producing a higher proportion of fine to coarse materials. If, on the other hand, the ratio of the vertical flux of fine material to the horizontal sand flux decreases with wind speed (that is, the proportion of fine material to coarse material in the eroding soil decreases), it may mean that fine particles from the surface mixture are being rapidly depleted at low wind speeds. An example of such a soil is one in which an unerodible layer is being reached but sandblasting continues from an upwind source of coarse material.

The wind speed at which the generation of dust begins is that for which aerodynamic forces are sufficient to overcome those forces holding individual particles in the soil. A number of theoretical and experimental studies have been made for idealized particle systems. These studies have not extensively considered natural soil surfaces, however, or man's impact upon soil surfaces and aggregate storage. Gillette measured threshold velocities on dry soils that were in both undisturbed and disturbed condition. The disturbance was a pulverization of the surface by the wheels of a truck. A list of threshold velocities versus soil types is given in Table 7.15.

The ordering of threshold velocities suggests that the threshold velocity is

TABLE 7.15

Wind Speed for Initiation of Dust Production versus Undisturbed or Disturbed Soil Type[a]

Approximate range of u^* (cm/sec)	Soil type
20–40	Disturbed soils having less than 50% clay and less than 20% pebble (<1-cm-diam) cover
25–40	Tilled bare sand soils
40–65	Disturbed pebbly soils
45–70	Bare clay soils that have been disaggregated by natural forces
40–150	Disturbed soils having a high salt content or more than 50% clay
140–200	Undisturbed sandy soils having a crust and soils covered with fine gravel
>150	Undisturbed soils having more than 50% clay and surface crusts and salt-crusted soils
>180	Soils covered by coarse (>5-cm) pebbles

[a] From Gillette (1980); with permission of the New York Academy of Sciences.

determined by the availability of loose, erodible, sand size particles at the surface. It also suggests that the threshold velocity is increased (erosion decreased) by the effect of aerodynamic partitioning of wind stress by non-erodible elements, such as pebbles and larger objects, and by the cementation of the soil by clay and salts.

The suspension of soil dust over land areas undisturbed or disturbed by man's activities produces large quantities of suspended particles, which heavily influence continental aerosol conditions. Estimates of the dust burden and strength of sources have been reported by Junge (1979) and are shown in Table 7.16. Substantial differences are seen between the two hemispheres; this is obviously related to the arid land area involved. A major contributor to the Northern hemisphere is believed to be the Sahara dust plume. The transport of dust from the Sahara has been found to extend over great distances across the Atlantic Ocean with the trade winds, and is a major example of natural long-range transport. The aeolian dust transport through long-distance air motion is estimated to exceed that from sand dune migration and the sediments by the River Niger. If other arid areas have a soil dust production close to that of the Sahara, we can expect such sources to have a major contribution to the variability of suspended particulate matter as a natural baseline.

The suspension of soil dust produces a particle size distribution which reflects the distribution of soils and rock weathered during the course of erosion. This comminution process should yield material concentrated in large to giant particles. Particle size distributions of soils and weathered rocks have been reported by Lerman (1979), who cited unpublished data of

TABLE 7.16

Estimates of the Global Tropospheric Dust Cycle[a]

Part of troposphere	Dust burden 10^6 metric tons	Source strength 10^6 ton/yr^{-1}
Northern hemisphere[b]	3.0[c]	150[d]
Southern hemisphere[b]	1.0[c]	50[d]
Whole troposphere[b]	4.0[c]	200[d]
Sahara plume	1.2–4.0	60–200[e]
Total troposphere (plus Sahara)[f]	3.2–12.0	130–800

[a] From Junge (1979); with permission of Wiley.

[b] Estimates disregarding any special production in deserts, particularly the Sahara area and its plume in the north Atlantic trade winds.

[c] Estimated uncertainty factor about ±2.

[d] The source strength was calculated from the dust burden assuming an aerosol residence time of 1 week. Estimated uncertainty factor about ±3.

[e] Range given by Prospero and Carlson (1972) and Jaenicke and Schültz (1978).

[f] After application of the uncertainty factors b and c.

Dapples. These are shown in Fig. 7.20. The size distribution of suspendable material takes a power-law form (Fig. 7.20a) analogous to that observed in the atmospheric aerosol particles.

The difference between weathered rock and residual soil in a mass distri-

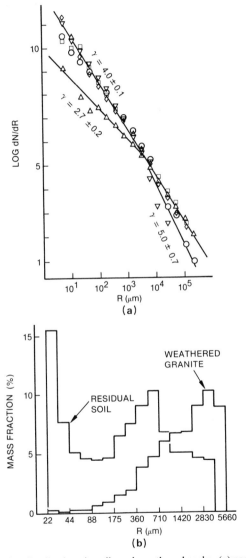

Fig. 7.20. Particle-size distributions in soils and weathered rocks. (a) particle-number spectra of size; *b* is the logarithmic slope of the straight lines drawn through the data points: (△) weathered granite; (◇) residual soil on granite; (□) soil from leucogranite; (○) soil from weathered syenite; (▽) residual soil from mica schist. (b) particle-mass histogram of a weathered granite and residual soil. Note shift to the abundance of smaller particles in the residual soil. [Data of Dapples as given by Lerman (1979); reprinted courtesy of Wiley.]

bution by size is illustrated in Fig. 7.20b. Residual soil has a bimodal distribution with a small fraction near 20 μm radius, and larger fraction at 500 μm diameter.

For aerosol suspension, the gravitational sedimentation process will strongly deplete particles larger than 10 μm. Thus the suspension of giant particles should be reduced rapidly as air travels away from a source area. Constraining the fine particle regime for soil is the breakup or comminution process, which becomes very inefficient below 10 μm because of the increase in cohesive and adhesive forces at this size. Some observations suggest nevertheless that more firmly divided material can be produced in soil breakup processes, and appear in the submission fraction of particulate samples. Lerman's (1979) reported observations lead to the suspicion that the weathering process may provide a mechanism for size reduction beyond what can be achieved by breakup alone. In particular, size decrease in soil material may take place by dissolution of soluble components of minerals. Thus small amounts of mineral-based material segregated into the submicrometer size range may be found in atmospheric particulate samples even though the bulk is expected in the giant particles. The processes of separation of the small particles are not known, but may be affected by water drop condensation and evaporation in clouds.

The segregation of suspended particles from soil origins with distance from the source or geochemical weathering can be extrapolated from the representative mass distributions drawn in Fig. 7.21. Here the maximum in the size distribution shifts to smaller sizes of suspended material such that the soil-dust-dominated aerosol from the Sahara realizes a mass median radius less than 10 μm after passing over the Cape Verde Islands, 850 km to the west of the African coast. The giant particles evidently have fallen out of the air by gravitational settling, leaving a remainder of smaller particles as the dust is transported westward with the Atlantic trade winds.

The chemical composition of aerosol dominated by wind-blown dust shows basically a close relation to material from the earth's crust. However, there is enrichment of certain elements, as indicated in Fig. 7.22. This can be shown readily by introducing *an enrichment factor,* which ratios elements to the average composition of crustal material. The definition of the enrichment factor based on iron is given in the caption of Fig. 7.22. Here the elemental composition of the Sahara aerosol, which was estimated to be more than 75% soil dust (by mass), is shown, normalized to iron in the crustal material.

The constancy of composition for the Sahara aerosol makes its chemical character a somewhat more fundamental quantity than it otherwise might be. Figure 7.22 shows this composition expressed in terms of a ratio to iron. The vertical bars are within one standard deviation of the enrichment factor over the seven samples, not the observed scatter directly. There are a number of features of this plot that are of interest. First, the majority of the elements have enrichment factors of essentially unity; that is, they are in the propor-

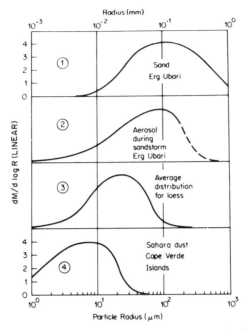

Fig. 7.21. Junge's (1979) comparison of different idealized mass distributions based on the following sources: (1) Schütz and Jaenicke (1974), sand from the Libyan desert, Erg Ubari; (2) same source and location as (1) but aerosol during sand storm; (3) Fütchtbauer and Müller (1970), average of 8 loess distributions from various continents; (4) Jaenicke and Schütz (1978), average mass distribution over the Cape Verde Islands. (Reprinted with permission of Wiley.)

tions of average crustal rock. At least twenty-five elements fit into the category of crustal rock, while only five elements have enrichment factors greater than ten in the Sahara dust samples. This situation is directly contrasted with an average of more than 100 aerosol particle samples taken at remote sites over the world. In this "global" average, nearly one-half the elements have enrichment factors of roughly ten or more. Thus from the point of view of enrichment factors, the Sahara dust particles are relatively straightforward chemically, with most elements at or near crustal proportions. In nearly all cases, the Saharan enrichments are depressed relative to the global average particulate values.

There appears to be some ordering to the enrichments of the Sahara dust. The light group IA and IIA elements, Na, K, Ca, and Mg, have low enrichment factors less than unity. Transition metals (e.g., groups IB and IIB, Au, Ag, and Hg) and metalloids (e.g., groups IIIA, V, and VI, Sb, As, and In) have enrichment factors between 5 and 100. Except for Ba, all the elements of groups IA and IIA are generally substantially enriched in tropospheric particle samples.

The rare-earth elements (REEs) are a particularly interesting case. With

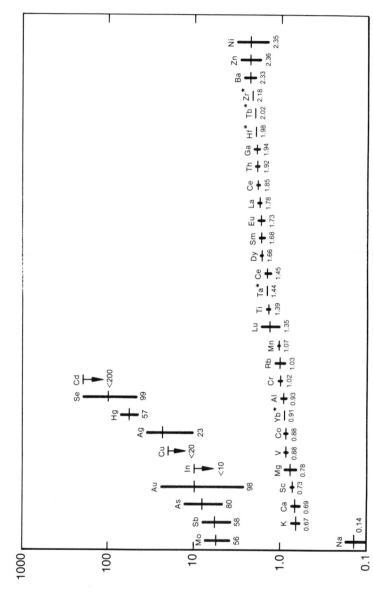

Fig. 7.22. Enrichment factors of the Sahara aerosol particles. The mean value is the horizontal bar; the range of the vertical bar is ± standard deviation. Enrichment factor = $(X/Fe)_{aerosol}/(X/Fe)_{crust}$. * = estimated value only. [From Rahn *et al.* (1979).]

the exception of Lu, they have enrichment factors between 1.6 and 2. These elements are most interesting because their enrichments have just about these same values in the aerosol from the rest of the world. No major pollution sources are known for the rare earths. Thus they are of crustal origin in the world aerosol but all have enrichments markedly higher than unity according to Rahn *et al.* (1979).

The enrichment factor has been used as an indicator for geochemical changes which may take place during exchange of condensed material between the earth's surface and the air. However, the finely divided particles have a much greater enrichment of certain elements than larger particles. The extreme of the Aitken particle range (≤ 0.1 μm diameter) has been summarized by Schütz *et al.* (1978) based on sampling at remote stations in the northern and southern hemispheres.

Figure 7.23 shows aerosol average particle–crust enrichment factors for

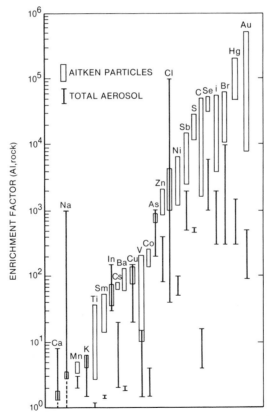

Fig. 7.23. Aerosol-crust enrichment factors relative to Al for Aitken particles and total aerosol of remote continental and marine regions. Al is approximately 4 ng/m³ in the nuclei samples. [From Schütz *et al.* (1978).]

elements in the Aitken nuclei range, compared to mean enrichment factors for total aerosol particle samples in remote continental and marine areas. The enrichment factors in individual samples can vary by roughly ten around the average values. The aerosol particle–crust enrichment factor was calculated relative to aluminum, as the crustal reference element. The mean crustal rock was used as crustal reference material. It can be seen that most elements are more enriched in the Aitken fraction than in the total aerosol. This is reasonable because the reference element Al is found primarily in particles larger than the Aitken size.

Sulfur and carbon were the most abundant elements by quite a wide margin, but the small particles contain much larger amounts of C and S relative to other elements than the total sample (less than 30 μm diameter). To a first approximation, the chemistry of the Aitken range seems to be primarily that of sulfur and carbon. This is reasonable, because both these elements have major gas-phase precursors to particle formation in the atmosphere. It is believed that most of particulate sulfur in the Aitken range has been produced by gas to particle conversion. However, this cannot necessarily be true for carbon, since combustion of fuel, including forest and brush fires, and man's activities generate large quantities of tiny carbon-containing particles.

The other elements have relative concentrations at least one order of magnitude lower than sulfur. Major crustal elements such as Al, Ti, Ca, and Ba, and major marine elements such as Na and Cl have concentration ranges of 1–10 ng/m^3, which is considerably lower than their abundances in the total suspended-particle samples. This is consistent with size distribution studies by Jaenicke and Schütz (1978), which showed that the number of Aitken particles and their size distribution were unaffected by mineral dust and sea salt particles.

Heavy metals such as Ni, V, Co, Cu, and Mn are also present within the Aitken particles. This enrichment may follow from injections of heavy metals associated with combustion products.

Other volatile elements such as Hg, Br, Se, As, and I, which have relatively high vapor pressures under atmospheric conditions, are also present in Aitken particles. Like sulfur and carbon, these elements may have a gas-phase origin. Their sources are not yet well characterized, however. Various processes, such as combustion of fossil fuels, volcanoes, biological methylation, and direct volatilization from the crust, have been proposed for these elements, but there is no general agreement as to which are the most important sources.

From the results cited in this section, we conclude that there is a more or less common chemical distinction between segments of a "baseline" natural tropospheric aerosol particle distribution which reflects the origins of the material. Large and giant particles contain the soil and sea salt elements, while the finely divided fraction is dominated by material from atmospheric sulfur chemical reactions, by absorption of volatile elements, and by finely divided carbonaceous material from combustion and from chemical reactions.

7.4.4   Volcanic Eruptions. Volcanism may intermittently inject large quantities of particulate material into the atmosphere (for example, Fig. 1.1). In most cases, this debris, sometimes known as *tephra,* is found in the troposphere and falls out relatively rapidly because of its large particle size. However, in the intense eruptions of Mt. Agung and Bali (in the early 1960s) or more recently of Mt. St. Helens in Washington state (1979–1982), Soufriere, St. Vincent (1979), and El Chichon in Mexico (1982), particles of ash reached the upper troposphere and stratosphere, where they experienced transport and dispersal over continents or even hemispheres. Indeed, aerosol emissions from the Bali eruption in 1963 are believed to have affected the particle burden and radiant energy in the stratosphere for several years after the event (Flowers *et al.,* 1969).

The production of dust and gases from volcanic activity is difficult to assess because of insufficient data. Nevertheless, it is possible to obtain a range of expected emanations based on observations of volcanoes, particularly those in Hawaii. For example, Stearns (1966) has given the magma eruptions from Kilauea volcano in Hawaii over many years ranging from 1.9×10^4 to 2.2×10^8 m^3. In 1965, a moderate size eruption took place emitting 1.1×10^7 m^3 of magma. This is adopted here for a typical "significant" eruption. The ratio of ash to magma produced varies from 0.1 for basaltic volcanoes to 10 for pyroclastic volcanoes (Wedepohl, 1969). An average may be a ratio of unity. If we assume that 50% of the ash from an eruption is suspended as a dust cloud, an annual eruption of 1.1×10^7 m^3 would yield an equivalent aerosol rate of $\sim 10^4$ ton/day. A major eruption could yield an amount of dust an order of magnitude larger or more at an equivalent daily rate.

In addition to emission of primary particles in the form of ash, volcanic activity produces significant quantities of "reactive" gases such as H_2S, SO_2, or HCl. For example, Rankama and Sahama (1950) estimated that the Valley of Ten Thousand Smokes in Alaska produces 3×10^5 tons annually of H_2S by fumarole activity. Similar activity on a worldwide basis might be an order of magnitude larger. If 60% of the H_2S is oxidized, fumaroles alone could produce globally $\sim 10^3$ ton/day of sulfate in the atmosphere. Suppose it is assumed that all of the sulfur contained in a molten lava flow is volatilized to sulfate aerosol. Taking the typical fraction of igneous rock as 0.05% S, a single emanation like the 1965 Kilauea eruption would produce an equivalent rate $\sim 10^2$ ton/day of SO_4^{2-}.

Recently, Sigurdsson (1982) has reported estimates of sulfur release from the Mt. St. Helens explosive eruption as compared with the Laki, Iceland crater-row eruption in 1783. The Laki eruption produced over roughly 8 months about 10^7 tons of sulfur, about 100 times that of Mt. St. Helens. The Laki eruption, though nonexplosive, produced an equivalent amount of airborne sulfuric acid of 10^5 tons SO_4/day. As a result of the Laki eruption, a haze was observed in Europe and in China.

Benjamin Franklin speculated that an unusually severe winter in Europe between 1783 and 1784 was related to disturbance of the earth's heat balance from the suspended volcanic debris.

In the absence of better information, we can assign a minimum intermittent rate of global sulfate production to an equivalent of 10^2–10^3 tons SO_4^{2-}/ day as a tentative emission rate. It should be recognized, however, that the composition of fumarole and volcanic clouds varies extremely widely. Therefore, this value may be quite inaccurate if applied to any given year.

The quantities of particulate material and gases injected into the atmosphere by volcanism are equivalent to that from large cities. Although such eruptions are relatively infrequent, their impact may be significant during the period of activity. A good example of such potential for effects of fallout and atmospheric influence recently comes from Mt. St. Helens. This volcano has distributed layers of ash amounting to centimeters in thickness on at least two occasions as far away as 200 km downwind of the mountain. Its tropospheric influence has been observed across the North American continent to the Atlantic coast; the plume has been identified in Europe (Meixner et al., 1981).

The size distribution of ash erupted from volcanoes varies with intensity and crustal disruption. Some examples of size distributions of ash measured in the Mt. St. Helens plume are shown in Fig. 7.24. The material collected in the troposphere is mainly in the form of large particles >1 μm in diameter. The stratospheric intrusion of the Mt. St. Helens plume included submicrometer particles of ash and acidic material as ammonium sulfate (Farlow et al., 1981). The material in the lower troposphere consisted of suspended particles which were primary in origin, believed to be heavily dominated by breakup or disaggregation with disintegration of the mountain top. Other less explosive eruptions involved production of particles in the smaller sizes, with condensation of vapors or fumes from emissions.

The ash from volcanic injections can have a varied composition, depending on the fraction of condensables entrained in the plume. The cloud of ash emerging from the Mt. St. Helens eruption contained large quantities of disaggregated crustal material as indicated in Table 7.17. However, the suspended particles also were enhanced in volatiles including sulfur, chlorine, arsenic, and lead. The sulfur and chlorine in particular probably come from condensation of sulfuric acid and volatile chlorides.

7.4.5 Forest and Brush Fires.   The last category of natural particle emissions involves combustion. Forest and brush fires have occurred from time to time across the globe. They are prolific sources of smoke as indicated in

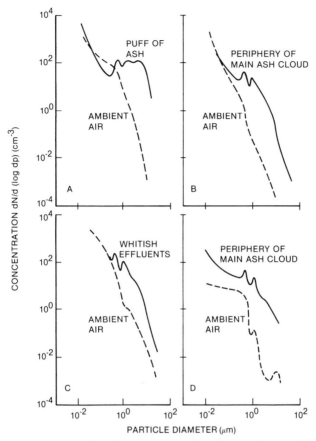

Fig. 7.24. Size spectra of small particles: (A) in a puff of ash at an altitude of 3 km and 13 km downwind of the volcano at 2239 h on 28 March 1980; (B) at the periphery of the dark ash cloud at an altitude of 3 km and 9.3 km downwind of the volcano at 2010 h on 18 May; (C) near the middle of the whitish-looking clouds at an altitude of 3.2 km and 130 km downwind of the volcano at 0230 h on 19 May; (D) at the periphery of the main ash cloud at an altitude of 2.7 km and 22 km downwind of the volcano at 2140 h on 19 May. [From Hobbs *et al.* (1981); reprinted with permission of the American Association for the Advancement of Science.]

the example, Fig. 7.25. Once considered a natural disaster, such fires now are thought to be an integral part of the ecological evolution of certain areas with rapidly growing vegetation. In many parts of the world, fires are set by man either accidently or deliberately as part of range or forest management practice. In this respect they are often said to be quasi-natural particle sources, as in the case of dust production from agricultural filling. Forest fires produce large quantities of finely divided particles in the submicrometer size range which may rise by heating of air near the fire to heights in excess of 5 km. Cadle (1966), for example, has quoted a figure of 2×10^{22} particles

TABLE 7.17

*Composition of a Particulate Sample Collected in
the Mt. St. Helens Plume on 19 May 1980; Average
Ash Composition is Shown for Comparison[a]*

Constituent	Particulate sample[b]	Average ash
Major elements (%)		
SiO_2[b]	≡65.0	65.0 ± 3.3
Fe_2O_3	6.7 ± 0.4	4.81 ± 1.1
CaO	3.0 ± 0.2	4.94 ± 0.92
K_2O	2.0 ± 0.1	1.47 ± 0.28
TiO_2	0.42 ± 0.03	0.69 ± 0.13
MnO	0.054 ± 0.004	0.077 ± 0.020
Trace elements (ppm)		
S	3220 ± 800	940 ± 380
Cl	1190 ± 400	660 ± 180
Ni	<20	15 ± 8
Cu	61 ± 10	36 ± 5
Zn	34 ± 7	53 ± 7
Ga	<8	18 ± 1
As	22 ± 4	≈2.8
Se	<7	<1
Br	<8	≈1
Rb	<17	32 ± 8
Sr	285 ± 7	460 ± 110
Zr	142 ± 7	170 ± 90
Pb	36 ± 11	8.7 ± 2.9

[a] From Fruchter *et al.* (1980); copyright 1980 by the American Association for the Advancement of Science.

[b] Collected on a traverse from 0 to 50 km from the volcano. All composition data for particulates are normalized to the average silica composition of ash samples. Analysis was by energy-dispersive XRF.

for every acre of burning forests. In the United States alone the U.S. Forest Service has estimated that 4×10^6 acres were burned in 1968. Of this, the majority was originated by man, either accidently or deliberately.

To calculate the production rate from forest fires, it is assumed that the particles are very small, with average diameter 0.1 μm, and a mass of 1.3×10^{-6} g. In the United States, then, the source strength would be equivalent to about 3×10^5 tons per day. On a worldwide basis, this value could easily be an order of magnitude larger. Taking the production rate as uniform over the area of the United States and over the world, for example, the production of aerosols could be 6×10^6 ton/day or even higher (Table 7.2).

Fig. 7.25. Smoke from forest and brush fires: (A) 1977 marble cone fire in Las Padres National Forest, CA; (B) horse fire break prescribed burning in Mt. Baldy, CA area. (Photos courtesy of W. Sniegowski, U.S. Forest Service.)

The uncontrolled combustion of brush and forest debris has not been studied extensively for the nature of the particles produced. However Radke *et al.* (1978) have reported comparative observations of the size distributions from the airborne particle baseline and those found in a forest fire plume in central Washington state. An example of a volume distribution taken at 1.7 km height 13 km downwind of a fire is shown in Fig. 7.18 with other examples of distributions from combustion emissions. The volume distribution is similar to the accumulation subrange in the urban aerosol, and coincides with this range in auto exhaust and power plant examples. Unlike the power plant case and forest fires, auto exhaust has a substantial nuclei mode and a coarse particle mode. For the forest fire example, the mean particle diameter is approximately 0.3 μm; there also are a few large particles present above 10 μm diameter. The contribution of material in the submicrometer size range corresponds well to the other combustion sources shown.

Examples of the elemental composition for slash fires and simulated grass fires is given in Table 7.18. The composition of smoke from fires varies over a wide range, producing mostly volatile and nonvolatile carbon material, with significant amounts of sulfate and nitrate and chlorine. The smoke from grass burns and trees is similar in makeup, with a possible distinction in nitrate content. Based on these data and other analyses, high K/Fe ratios may be useful indicators of smoke from burning vegetation. This indicator should be used with caution, however, considering earlier data of Shum and Loveland (1974) showing much lower K/Fe values than reported in Table 7.18.

TABLE 7.18

Chemical Characteristics of Fine Particles in Smoke from Slash and Field Burning in Weight Percent[a]

Component	Slash burn[b]		Field burn[c]	
	Average	Range	Average	Range
Na	0.56	0.02–0.91	0.23	0–0.86
Al	1.31	0–3.02	0.25	0.01–0.64
Si	0.42	0–2.29	0.15	0–0.32
S	0.14	0–0.64	1.19	0–4.51
Cl	4.93	0.74–7.17	9.92	4.45–20.1
K	0.60	0.57–0.61	8.2	1.5 –12.0
Ca	0.98	0.04–1.65	0.36	0–0.86
Fe	0.02	0–0.11	0.0	0–0.01
Br	0.05	0–0.12	0.03	0–0.05
Cu	0.03	0–0.12	0.0	—
Zn	0.0	—	—	—
Mn	0.14	0.01–0.35	—	—
F	1.23	0.10–1.80	0.30	0.18–0.63
NO_3^-	8.45	0.6 –11.65	0.89	0.24–2.49
SO_4^{2-}	2.46	0.12–3.30	5.03	0.15–17.3
Volatile C[d]	68.5	46.9–100	32.1	18.1–47.8
Nonvolatile C	3.9	0.6–8.1	9.6	0.2–30.5

[a] From Hester (1979).
[b] Average of 6 samples of different softwood fires in Oregon.
[c] Average of 8 cases for simulated burn of different field grasses (bent grass, blue grass, annual rye, wheat, or chard grass, perennial rye and tall fisene).
[d] Volatile $\leq 850°C$.

Another indicator for airborne material associated with vegetation burning is the enrichment of carbon-14 isotope relative to fossil carbon burns (Currie, 1982). The carbon isotope ratio technique is only now being applied for biological carbon identification.

7.4.6 Anthropogenic Sources. In parts of the world, man's domestic and industrial activities contribute a dominant share of the tropospheric particle burden. To illustrate their importance, a summary of estimated particulate emissions on a worldwide basis is given in Table 7.19. The production of aerosols by man's activities stems from a wide variety of sources. These have been reviewed in detail by Hammerle (1976) and by the U.S. Environmental Protection Agency in the recent documents on air quality criteria (U.S. EPA, 1982). Major categories are shown in Table 7.19 and include combustion of fuels and incineration (partially from transportation), industrial processing, and production.

Using a variety of sources of information, a tabulation has been made of potentially important sources for primary particle production, as shown in Table 7.19. The estimates are believed to be roughly representative of North

TABLE 7.19

Estimated North American Sources of Primary Particle Production[a]

	Annual consumption or production	Production rate (metric ton/ day)
Combustion of fuel and waste		
Coal	2.9×10^9 ton	6×10^4
Liquid hydrocarbons	1.3×10^9 ton	0.3×10^4
Natural gas[b]	1.9×10^7 ft^3	10^{-4}
Incinerators[b]	12×10^5 ton	2.5×10^2
Industry		
Steel	1.3×10^8 ton	0.18×10^4
Iron	9.1×10^7 ton	0.21×10^4
Smelting (Cu, Pb, Zn)	2×10^4 ton	10^2
Petroleum refineries	11.3×10^6 bbl	3×10^2
Chemical	—	$<10^3$ (?)
Pulp mills	25.7×10^6 ton	0.9×10^4
Portland cement	3.08×10^7 bbl	3.2×10^4
Bitumens cement	1.11×10^6 bbl	10^1
Food and feed	—	$<10^2$ (?)
		$\sim 1\text{--}3 \times 10^{5c}$

[a] From Hidy and Brock (1970).

[b] U.S. only.

[c] Range for projection of extrapolation to world industry.

America figures in the 1960s. They also may be reasonable qualitatively for the world if totals are compared with categories in Table 7.2. In Table 7.19, selected cases of combustion of fuels and industry are listed. Most of the estimates were made for the United States only, because information on production was lacking from other areas. Based on available emission factors and projections of production rates or combustion rates, it appears that coal burning in uncontrolled sources is by far the most significant single source of particles before 1970. Of the industrial emissions, it seems that Kraft paper pulp mills and Portland cement manufacturing are significant particle emitters along with the metal refining industry, for the United States.

The total production rate of particles by primary mechanisms of combustion processes and industry is in the range of 10^5 ton/day from data of the 1960s. Assuming that the bulk of the consumption of materials, and industrial activity remains centered in the United States, the value of $(1-3) \times 10^5$ ton/day is assigned for the sum of these sources worldwide.

An additional intermittent anthropogenic source of particles should be considered; that is, dust rise associated with agricultural activities. Taking the total acreage in the United States in farming, and assuming 10^4 $\mu g/m^2$ produced once a year during cultivation, this source could be in the range of 10^2–10^3 ton/day. As in the case of volcanic emissions, it may be more realistic to evaluate this source only during its seasonal peak. If cultivation takes place primarily over a period of 1–3 months, for example, the daily emission rate would increase by an order of magnitude during that period.

The material produced by man's activities ranges in particle size distributions from finely divided particles to large fragments of mineral material. In most modern industrial operations, fugitive dust and large particle emissions are controlled, so that emissions are concentrated in the fine particles. Mining activities or major construction may involve suspension of copious quantities of dust in the coarse particle range.

Examples of the typical size distributions for urban aerosols are shown in Fig. 7.10. An average urban volume–size distribution is compared with atmospheric aerosols enriched in combustion particles in Fig. 7.18. The urban particulate suspension evidently is heavily influenced by sources of fine particles between 0.1 and 1.0 μm. Fuel combustion is probably the largest contributor of such particles, with additions from industrial activities.

The compositions of particulate matter from transportation and stationary combustion sources are listed in Tables 7.20 and 7.21. Automobile exhaust for both leaded and unleaded gasoline is shown with diesel vehicle exhaust and residual oil using results summarized by Heisler et al. (1980a). Exhaust from cars with catalytic emission control using unleaded fuel contains only traces of metals, but is enriched in primary sulfate. Diesel vehicle exhaust is mainly carbonaceous, with a large component of elemental carbon. The residual oil composition listed in Table 7.20 is typical of Colorado crude oils,

TABLE 7.20

Chemical Composition of Fine-Particle Emissions (Percent ± Standard Deviation)[a]

	Source			
Component	Unleaded auto exhaust	Leaded auto exhaust	Diesel vehicle exhaust	Residual oil
SO_4^{2-}	13.3 ± 5.3	0	0	48.0 ± 12.0
NO_3^-	0	0	0	6.5 ± 0.4
NH_4^+	3.6 ± 1.5	1.8 ± 0.8	0	13.0 ± 4.0
Elemental C	43.3 ± 17.3	7.0 ± 4.3	54.0 ± 33.0	3.1 ± 2.5
Organic C	36.1 ± 14.4	21.6 ± 15.0	38.0 ± 22.0	7.0 ± 6.2
Al	0	0	0	0.53 ± 0.24
Si	0	0	0	0.96 ± 0.48
Cl	0	10.3 ± 4.1	0	0
K	0	0	0	0.28 ± 0.10
Ca	0	0	0	1.6 ± 0.6
Ti	0	0	0	1.1 ± 0.6
V	0	0	0	0.64 ± 0.15
Cr	0	0	0	0.047 ± 0.015
Mn	0	0	0	0.05 ± 0.01
Fe	0	0.4	0	3.0 ± 0.6
Ni	0	0	0	5.4 ± 1.2
Cu	0	0	0	0.08 ± 0.03
Zn	0	0.14	0	0.04 ± 0.20
As	0	0	0	0
Se	0	0	0	0
Br	0	15.0 ± 6.2	0	0.13 ± 0.02
Pb	0	40.0 ± 16.0	0	0.11 ± 0.06

[a] After Heisler *et al.* (1980a).

TABLE 7.21

Major Elements in Fly Ash, Coal, and Soil[a]

	Total concentration (wt %)			
	Fly ash		Coal (typical concentration)	Soil (range)
Element		Range		
Al	18.0	0.1–17.3	1.4	4–30
Si	10.0	1.41–28.6	2.6	25–33
Ca	10.3	0.11–22.2	0.54	0.7–50
Fe	3.3	1–29	1.6	0.7–55
Mg	1.7	0.04–7.6	0.12	0.06–0.6
Na	1.2	0.01–2.03	0.06	0.04–3.0
K	0.54	0.15–3.5	0.18	0.04–3.0
S	0.4	0.1–1.5	2.0	0.01–2.0
P	0.04	0.04–0.8	0.05	0.005–0.2
N	0.02	—	1.1	0.01–1
Ba	0.37	0.011–1.0	0.015	0.01–0.3
Sr	0.18	0.006–0.39	0.010	0.05–0.4

[a] From Pago *et al.* (1979).

since crude oils are quite variable in their trace of metals, but enriched in primary sulfate. Since crude oils are quite variable in their trace constituents, this composition should not necessarily be used as representative of different locations. For example, the vanadium content of fly ash from Indonesian oil may be an order of magnitude larger than that shown in Table 7.20.

The range of fly ash composition for fossil fuels is illustrated for coal in Tables 7.21 and 7.22. The minor and trace element contents of two mechanically sieved size fractions (>250 and >53 μm) of fly ash are shown in Table 7.22. Examination of data in Table 7.21 shows the fly ash matrix to be dominantly made up of Al, Si, Ca, Fe, Mg, Na, K, S, Ba, and Sr. These elements typically constitute about 50% of the dry weight of fly ash from coal combustion.

In their review, Page $et\ al.$ (1979) indicate a fly ash chemical composition $Si_{1.00}Al_{0.45}Ca_{0.051}Na_{0.047}Fe_{0.039}Mg_{0.020}K_{0.017}Ti_{0.011}$. This composition is consistent with fly ash derived from coal with intrusions of clay minerals, particularly kaolinite and lesser amounts of quartz, and also with intrusions of $CaCO_3$. The $CaCO_3$ content of the fly ash studied ranged from 2 to 4%, and that of gypsum was about 3.1%. Although the Ca concentration shown in Table 7.21 is within the range commonly found in fly ash, it is only comparable with Ca-rich fly ash. Examination of the chemical composition of coal samples of various coal ranks in the United States demonstrates that sub-bituminous and lignite coals contain considerably more Ca than anthracite or bituminous coals. The study of Natusch $et\ al.$ (1975) also showed fly ash derived from western lignite and sub-bituminous coals to be higher in Ca, Si, and trace elements and lower in S than those derived from eastern bituminous and anthracite coals.

Because most of the S in coal is discharged in the flue gas, the concentration of S in the fly ash (0.4%) is considerably less than the average concentration in coal (2.0%). Virtually all S in fly ash is in a soluble form (Page $et\ al.$, 1979). Ba and Sr concentrations in fly ash are substantially greater than their concentrations in coal. Concentrations of major elements in fly ash shown in Table 7.21 generally are comparable to their concentrations in soil.

Minor and trace elements in fly ash (Table 7.22) exhibit concentrations that are, in most cases, considerably higher than their concentrations in coal. The table shows that although substantial amounts of minor and trace elements could be mobilized into the environment from fly ash, their concentrations in fly ash and soil are, in general, comparable. Some exceptions, however, are B, Mo, and Se. Concentrations of these biologically toxic elements (B is toxic to plants and Mo and Se are toxic to animals) in fly ash greatly exceed their concentrations in soil. Note that the fossil fuel compositions listed have low K/Fe ratios compared with wood burning, again pointing to this ratio as an indicator distinguishing fossil and wood fly ash. Unfortunately, Small $et\ al.$ (1981) have also reported recently high K/Fe ratios in

TABLE 7.22

Minor and Trace Elements in Size Fractions of Fly Ash and in Coal and Soil[a]

	Total concentration ($\mu g/g$)					
	Fly ash				Coal	
Element	>250 μm	<53 μm	Bulk	Literature values	(typical concentration)	Soil (range)
As	10	17	14	2.3–6300	15	0.1–40
B	148	300	237	10–618	50	2–100
Cd	0.7	1.5	1.4	0.7–130	1.3	0.01–7.0
Ce	84	112	108	22–320	7.7	50
Co	10	15	13	7–520	7	1–40
Cr	52	65	64	10–1000	15	5–3000
Cs	4.1	4.6	4.9	1.5–18	0.4	—
Cu	45	70	50	14–2800	19	2–100
Eu	0.9	1.2	1.3	14.3	0.45	—
Ga	27	34	29	13–320	7	15–70
Hf	6	8.1	7.9	3.5–11	0.6	—
Hg	—	—	—	0.02–1.0	0.18	—
La	47	64	60	17–104	6.1	30
Lu	0.9	1.3	1.2	0.5–1.5	0.08	—
Mn	98	121	122	58–3000	100	100–4000
Mo	7	13	8.8	7–160	3	0.2–5.0
Nd	35	48	45	—	37	—
Ni	29	46	50	6.3–4300	15	10–1000
Pb	38	52	45	3.1–5000	16	2–100
Rb	44	51	53	36–300	2.9	30–600
Sb	2.9	4.4	3.8	0.8–202	1.1	0.6–10
Sc	12	17	16	3.7–141	3	10–25
Se	16	22	19	0.2–134	4.1	0.1–2.0
Sm	6.8	9.2	8.9	5.4–24	0.4	—
Tb	0.8	1.2	1.0	1.6	0.1	—
Th	16	21	21	13–68	1.9	—
U	6.3	7.7	7.4	0.8–19	1.6	20–250
V	—	—	11.9	50–5000	20	—
Yb	3.1	4.5	4.0	1.7–7.0	1	—
Zn	76	137	99	10–3500	39	10–300
Zr	140	187	183	50–1286	30	60–2000

[a] From Page *et al.* (1979).

copper smelter plumes, which would confound the use of this tracer for vegetation combustion.

The distinction between copper smelter particulate emissions and emission from power plants or incinerators is indicated by the enrichment ratios in relation to aluminum shown in Fig. 7.26. According to Small *et al.* (1981),

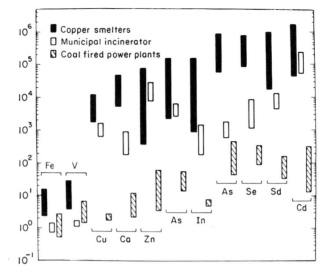

Fig. 7.26. Comparison of enrichment factors relative to aluminum from copper smelters, coalfired power plants, and municipal incinerators. [Reprinted with permission from Small *et al.* (1981). Copyright 1981 American Chemical Society.]

copper smelter emissions from samples in the southwestern United States indicate significant enrichment in the volatile elements As, Se, and Sb, and the chalcophilic elements (S, Cu, Zn, As, Se, Ag, Cd, In, Sb, W, and Au). The smelter particulate varied considerably among five different plants sampled. The composition apparently depends on the mineral ore charge as well as the process conditions. Municipal incinerator particles are strongly enriched in Cu, Zn, and Cd, and give opportunities for tracing this source in the absence of copper smelter interference.

From the examples shown above, man's activities provide a diversity of particulate material injected into the air. Many emissions may have a physical and chemical character which makes them difficult to distinguish from naturally emitted particles. However, the comparison of ambient particle samples with source-enriched samples offers avenues for estimation of source contributions in different locations. The use of chemical tracers for identification of source contributions to atmospheric particles has been explored extensively in the past ten years. This approach is sometimes called *receptor modeling;* it is discussed in Chapter 10 with source-based modeling for source attribution.

The segregation by size of the anthropogenic emissions from "condensation" of combustion products and from industrial processing leads to a chemical separation of soil-related elements from combustion-related elements. In a simplest "model" of urban air, one would expect that most of the particulate material from man's activities would be seen in the fine particles less than a few micrometers in diameter, while most of the natural sources would contribute to the larger particles, in the absence of fires. This picture appears to be borne out in cities if samples are taken well away from roadways or construction areas where coarse particles are suspended in the air.

7.4.7 Secondary-Particle Production. Examination of the chemical composition of samples of particles indicates that a large fraction of the suspended material is derived from atmospheric chemical reactions of reactive gases including sulfur dioxide and other sulfur gases, nitrogen oxides, and organic vapors, which form sulfate, nitrate, and organic particles. On a global scale, Table 7.1 suggests that natural emissions of sulfur and nitrogen or organic vapors dominate secondary-particle production. Additional perspective on the significance of secondary-particle contributions is presented in Table 7.23 for anthropogenic emissions of selected pollutants in the greater northeastern United States. In this region, the biogenic or other natural emissions of sulfur and nitrogen oxides is believed to be much smaller than man's contributions. If even a fourth of the sulfur oxides, nitrogen oxides, and reactive hydrocarbons are oxidized to condensable material in the air, the contributions to the airborne particle burden would be comparable with the primary particles.

Accumulation of condensed material as particles in the atmosphere can take place by two basic processes: first, by condensation of supersaturated vapor or by chemical reaction leading to spontaneous formation of new particles; and second, by condensation, absorption, or reaction on existing particles. In the latter case, the chemical reactions actually may occur on the surface of existing particles or within them.

With condensable precursors, particle formation may occur by heteromo-

TABLE 7.23

Summary of 1977 Emissions for the Greater Northeastern United States
(10^3 metric ton/day)[a]

Emission	Summer	Fall	Winter	Spring	Annual
Sulfur dioxide and primary sulfate[b]	81	83	94	81	85
Nitrogen oxides	33	32	33	32	32
Total emitted particles	41	41	42	40	41
Reactive hydrocarbons	42	49	69	48	52

[a] Data from EPRI, 1982; with permission from The Electric Power Research Institute.

[b] Primary sulfate emissions are based on estimates as 1% of the total sulfur oxides emitted for coal fired boilers, and 6% of the total sulfur oxides emitted for oil fired boilers (Mueller and Hidy, 1983).

lecular nucleation (homogeneous) or by heterogeneous nucleation. It is sometimes stated, but not proven, that heterogeneous processes are most likely in the atmosphere because of the large number of existing nuclei. One can readily see, however, that growth by condensation is limited by the rate of diffusion of vapor to the surfaces of nuclei. If conditions exist in which aerosol precursors evolve chemically at a rate exceeding diffusional transfer, supersaturation could build up to levels sufficiently high to permit heteromolecular (homogeneous) nucleation of new particles. It can be shown that H_2SO_4 can undergo heteromolecular nucleation at atmospheric concentrations in the absence of existing nuclei, but it is unlikely that HNO_3 can nucleate because of its relatively high vapor pressure. So little is known about the products of aerosol-forming organic reactions and their relevant physical properties that nothing can be said about the importance of heteromolecular nucleation in this case.

Growth by Condensation.   Regardless of the chemistry, certain physical constraints exist on aerosol particle–gas interactions. These restrictions relate mainly to (a) the thermodynamic stability of condensed material on particles of small diameter, (b) the kinetics of particle nucleation and condensation processes, and (c) reaction rates combined with absorption or adsorption on particles.

Particles must be close to or at equilibrium with respect to the surrounding vapor to exist in air for any length of time. Thus the partial pressure of condensed species on particles essentially must be less than or equal to the saturation vapor pressure at atmospheric temperature for stability. This presents no great problem for most inorganic salts or for sulfuric acid, even at parts per billion concentration in the gas phase. However, it places a severe constraint on the ability of HNO_3 or NH_4NO_3 to exist as pure compounds or as acids diluted in water. Nitric acid vapor pressure is sufficiently high to preclude its presence as a condensed material in pure form at atmospheric concentrations. Similarly NH_4NO_3 is a volatile salt whose vapor pressure is strongly temperature dependent. Its presence in air will be most likely at low temperature (see also Stelson *et al.*, 1979).

The requirement of low vapor pressure is particularly important to the stability of organic aerosol particles. The bulk of the organic vapors by mass concentration in the atmosphere that have been identified are in the range of carbon number less than six. Even if such materials reacted to form oxygenated materials, a survey of vapor pressure data suggests that only material with a carbon number much greater than C_6 would be thermodynamically stable in the condensed phase at concentrations of above 1 ppb (see Hidy and Burton, 1975; Grosjean and Friedlander, 1980).

Growth of particles by accumulation of material on existing particles can be classified by changes in the volume–size distribution. If the precursor is supersaturated, growth will occur at a vapor-diffusion-limited rate which depends on the supersaturation, the temperature, the particle size, and the accommodation coefficient at the surface. The proportionality to changes in particle size depends on the ratio of particle diameter and mean free path of the suspending gas. At one extreme the vapor-diffusion-controlled growth depends on volume to the two-thirds power; at the other, growth is proportional to volume to the one-third power, as summarized in Table 3.3. When the precursor is unsaturated, growth still may take place by absorption, or by chemical reactions in the particle. In this case the rate law should be proportional to the particle volume if the reaction is uniform throughout the particle.

The dominance of growth processes during chemical reactions in the atmosphere is difficult to determine since the rate of production of condensable material must be deduced indirectly.

Normally one expects that new particle formation from a supersaturated vapor could not take place in the troposphere because ample existing particles exist to act as nucleation centers. However, there may be special circumstances where nucleation may take place by both pathways in chemically reactive urban air (McMurry, 1977). For new particle formation, the most likely nucleating material is sulfuric acid, which may form new particles of approximately 100 Å diameter from gas-phase oxidation of SO_2. Recently, McMurry *et al.* (1981) have hypothesized that new particle production by heteromolecular nucleation appears to be dominant in plumes from large power plants with particulate emission control.

Heisler and Friedlander (1977) have investigated reactions to form organic particles in smoggy air contained in a large outdoor smog chamber. Polluted air containing hydrocarbon vapors and nitrogen oxides and particles was drawn into a large closed volume. The air was "doped" with cyclohexene, a particle forming organic vapor, and was allowed to react photochemically in the presence of sunlight while the production and growth of particles was observed. The results of these experiments indicated that particle growth was the dominant mechanism in the system. Growth was consistent with a vapor-diffusion-controlled accumulation of material on existing particles. This system is believed to be analogous to organic particle production in urban air.

Because the particle concentrations are very low in the stratosphere, the likelihood of new particle formation by nucleation should be greater than in the troposphere. Hidy *et al.* (1978b) have analyzed this situation for the case of sulfate production using production rates of sulfuric acid from gas-phase chemistry, and the theory of heteromolecular nucleation and condensation by vapor-phase-limited diffusion. They found that under normal conditions in the stratosphere, growth by accumulation of sulfuric acid on existing particles should predominate as in the troposphere. However, the theoretical estimates are very sensitive to water vapor concentration in the stratosphere. If the dry conditions in the stratosphere are disturbed by intrusions of tropospheric air, or from emissions of high-flying aircraft, nucleation of sulfuric acid could take place, producing locally large numbers of small particles.

The projection of potentially significant reaction mechanisms to form atmospheric aerosol has been a major challenge for the chemist. Basic laboratory experiments to determine rate constants and important reaction steps play a key role in eliminating many candidates. However, the complexities of the chemical interactions in atmospheric processes cannot be elucidated by fundamental experiments alone. Predictions of the behavior of reactive mixtures should be consistent with atmospheric processes or with atmospheric simulation experiments. Simulation of atmospheric phenomena using controlled laboratory prototypes is a well-known technique, particularly as applied to fluid dynamic processes. However, simulation of atmospheric chemical phenomena is less well established. With the possible exception of the stratospheric sulfate-related experiments of Friend *et al.* (1973) and the Castleman (1982) experiments of ion clusters, with application to the upper atmosphere, only simulations of the lower atmosphere, more specifically urban air, have been attempted. The principal method for tropospheric reactions used by many investigators is the static reactor or smog chamber approach, where air mixtures containing reactive contaminants are studied over a period of several hours. Such studies duplicate qualitatively, at least, many of the features of urban photochemistry to form ozone and other smog products. The principal deficiencies of simulators include the low volume to surface ratio compared with the atmosphere and the uncertainty regarding the role of reactions on the walls of the vessel.

A limited number of smog chamber experiments have been undertaken to investigate aerosol particle formation in smog. Aside from the early studies reviewed by Leighton (1961), these include the work of Groblicki and Nebel (1971), Wilson and colleagues (1972), and O'Brien *et al.* (1975). The results of these studies generally confirm that significant particle formation in mixtures of NO_x and reactive hydrocarbons requires an induction time until ozone begins to build up, reflecting the time required for the buildup of reactive intermediates such as HO, HO_2, RO_2, and $HONO_2$ from chain reactions involving olefinic hydrocarbons. Aerosol particles found in chamber experiments are composed of oxygenated organic material, including carboxylic acids and organonitrates. If SO_2 is present, suspended sulfate also is found. Much of the nitrate generated appears to accumulate on the walls of the chambers; this collection is believed to be associated with diffusional loss of HNO_3 or organic nitrates.

The presence of SO_2 in irradiated NO_x–hydrocarbon–air mixtures enhances particle formation, but does not significantly influence the maximum ozone level realized in the chamber. With a suitable choice of rate constants for reactions of HO and SO_2, and HSO_3 to form SO_4^{2-}, computer modeling can duplicate qualitatively many features of the Battelle chamber experiments (Miller, 1978; Atkinson *et al.*, 1982).

The smog chamber studies have been useful in classifying the reactivity of hydrocarbons for photochemical aerosol particle production. With SO_2 present, the experiments generally indicate that the terpenoid compounds such as α-pinene are highly prolific aerosol formers, followed in rough order by diolefins, cyclic olefins, high-molecular-weight terminal olefins, and aromatics. Without SO_2 present, the work of O'Brien et al. (1975) suggests, for example, that only diolefins and cyclic olefins of carbon number six or greater will generate significant quantities of aerosol at atmospheric pollutant concentrations for NO_x. Unfortunately such compounds have not been identified in air or in common hydrocarbon sources such as motor vehicle exhaust. The experiments of O'Brien et al. (1975) also suggest that NO_2 tends to suppress aerosol particle formation.

In another study using a smog chamber, Ripperton et al. (1972) have obtained evidence of the importance of ozone attack on an olefinic bond in aerosol particle formation, particularly in "natural" production from terpenoid compounds.

The laboratory experiments indicate that mixtures of gases found in the atmosphere are chemically active for aerosol particle production. Now let us examine chemical mechanisms of interest in such mixtures.

Sulfate Reactions. The oxidation reactions of SO_2 that potentially apply to the atmosphere have been reviewed by several investigators. Homogeneous reactions have been considered by Bufalini (1971), Calvert et al. (1978), and Calvert and Stockwell (1982). Heterogeneous reactions involving interactions with suspended particles have been examined by Hegg and Hobbs (1978), Beilke and Gravenhorst (1978) and Graedel and Weschler (1982). There are more than a dozen sulfate-forming reactions that may be relevant to atmospheric processes. These can be grouped conveniently in terms of homogeneous gas-phase reactions and heterogeneous reactions involving either particle growth by surface reactions or volume reactions. A summary of the two groups is given in Tables 7.24 and 7.25. The homogeneous reactions are broken down into subcategories whose end products are SO_3 or $RO_2SO_3^-$, and $ROSO_2^-$. Although the rates of reaction of these species with water or other species have not been reported, it is assumed that the reactions with water are fast to form H_2SO_4 and are not the rate-determining step in SO_4^{2-} production under atmospheric conditions. The heterogeneous reactions in Table 7.25 involve aqueous systems as well as solid nonaqueous particles.

Listed in Table 7.24 are three classes of homogeneous reactions. The first consists of inorganic oxidation mechanisms to form SO_3, while the second group involves organic radical oxidation agents generating SO_3. The third group of reactions forms $ROSO_2^-$ species. All of the reactions listed are exothermic and are favored thermodynamically. However, the first five reactions have been considered severely rate limited on the basis of available rate data (Calvert, 1974). The remaining example reactions listed appear to

TABLE 7.24

Examples of Theoretically Possible Homogeneous Removal Paths for SO_2 in the Troposphere; Estimated Rate of Oxidation is Based on Los Angeles Smog Conditions[a]

Reaction	$-\Delta H_{298}$ (kcal/mol)	Approximate rate constant (cm^3/mol sec)[b]
Inorganic reactions forming SO_3		
(1) $SO_2 + \frac{1}{2}O_2 + $ Sunlight $\rightarrow SO_3$	1.5–28	$4 \times 10^{-20} - 7 \times 10^{-16}$
(2) $O(^3P) + SO_2 + M \rightarrow SO_3 + M$	83	6×10^{-14}
(3) $O_3 + SO_2 \rightarrow SO_3 + O_2$	58	$<8 \times 10^{-24}$
(4) $NO_2 + SO_2 \rightarrow SO_3 + NO$	10	9×10^{-30}
(5) $NO_3 + SO_2 \rightarrow SO_3 + NO_2$	33	$<7 \times 10^{-21}$
(6) $N_2O_5 + SO_2 \rightarrow SO_3 + N_2O_4$	24	$<4 \times 10^{-23}$
Organic reactions forming SO_3		
(7) $\overset{\displaystyle O_3}{\overbrace{RCH - CHR}} + SO_2 \rightarrow SO_3 + 2RCHO$	~69	$\geqslant 1 \times 10^{-14}$ (?)
(8) $\cdot RCHOO\cdot + SO_2 \rightarrow SO_3 + 2RCHO$	~89	
(9) $HO_2 + SO_2 \rightarrow HO\cdot + SO_3$ (a)	17	$<1 \times 10^{-15}$
$\rightarrow HO_2SO_2$ (b)	~7	$<1 \times 10^{-18}$
(10a) $CH_3O_2 + SO_2 \rightarrow CH_3O\cdot + SO_3$	27	$<1 \times 10^{-18}$
Reactions forming $HOSO_2$ or $ROSO_2$ radical		
(10b) $CH_3O_2 + SO_2 \rightarrow CH_3O_2 + SO_2$	31	$~1 \times 10^{-14}$
(11) $HO\cdot + SO_2 \rightarrow HOSO_2$	~37	11×10^{-12}
(12) $CH_3O\cdot + SO_2 \rightarrow CH_3OSO_2$	~24	$~6 \times 10^{-13}$

[a] From Calvert (1974); revised with summaries of kinetic data of Calvert and Stockwell (1982) and Lloyd *et al.* (1982).

[b] Rate constants are given as second-order reactions at 1 atm pressure and 25°C.

TABLE 7.25

Types of Heterogeneous Reactions to Form Sulfate[a]

Aqueous

(13) $SO_2 + H_2O(l) \rightleftarrows H_2SO_3$
$H_2SO_3 \rightleftarrows H^+ + HSO_3^-$
$HSO_3^- \rightleftarrows H^+ + SO_3^{2-}$
$S(IV) \cdot O_2 = SO_2(aq) + HSO_3^- + SO_3^{2--}$
(13a) $(SIV) \cdot O_2 + O_2(aq) \rightarrow SO_4^{2-}$
(13b) $(SIV) \cdot O_2 + 2O_3(aq) \rightarrow SO_4^{2-} + 2O_2$
(13c) $(SIV) \cdot O_2 + 2H_2O_2(aq) \rightarrow SO_4^{2-} + 2H_2O$

Nonaqueous

(14) $\left.\begin{array}{l} SO_2 \text{ (ads)} \\ H_2O \text{ (ads)} \\ O_2 \quad \text{(ads)} \end{array}\right\}$ + carbon(s) or metal oxide(s) $\rightarrow H_2SO_4$ (ads)

[a] Here aq denotes dissolved species and ads species adsorbed on solids; $S(IV) \cdot O_2$ is the sum of dissolved SO_2, bisulfite, and sulfite ions in the presence of dissolved oxygen.

have rates that are sufficiently rapid to be of importance in the atmosphere, at least for polluted air with active photochemical processes. The free-radical reactions are driven photochemically by the reactive hydrocarbon–nitrogen oxide system, which is well known to be the coupled chemistry of smog, forming O_3 as one end product.

Reactions (7) and (8) in Table 7.24 correspond to the interpretation of Cox and Penkett's (1972) observations that SO_2 is oxidized at appreciable rates in the dark in ozone–olefin–air mixtures. Oxidation rates of a few tenths of a percent SO_2 change per hour may occur in smoggy air with reactive olefin species. Cox and Penkett suggested that either of two intermediates was involved in the SO_2 oxidation reaction: the ozonide illustrated in reaction (7) or the zwitterion has a diradical character and may be illustrated as reaction (8). Calvert's calculations of olefin–O_3 intermediates such as those in reactions (7) and (8) do not favor their importance as oxidizing agents. However, other radical species from the ozonide or zwitterion intermediates may be of interest, including those summarized by reactions (9)–(12). These classes of reactions may well account for SO_2 oxidation in the ozone–olefin mixtures.

Recently, the potential importance of radical addition reactions in the third class listed in Table 7.24 for SO_2 oxidation has become more fully appreciated. Such reactions are exemplified in the reaction series (9)–(12).

The rate of the $HO\cdot$ reaction was determined by Davis et al. (1973). With these experiments, the fractional rate of SO_2 disappearance may reach approximately a few tenths of a percent per hour in a moderate photochemical smog. Assuming that the rate of the $CH_3O_2^-$–SO_2 reaction is the same as that of the HO_2^-–SO_2 reaction, Calvert (1974) has estimated that the former will contribute a fractional oxidation rate of 0.2%/hr in moderate smog. Recent review of available literature has indicated this reaction pathway is slow in tropospheric conditions (Lloyd, 1982).

The $\cdot OH$ radical–SO_2 reaction, reaction (11), appears to be of particular importance in the troposphere and the lower stratosphere, where $\cdot OH$ concentrations are estimated to be high. Calvert (1974) has estimated the typical reaction of $\cdot OH$ or $CH_3O\cdot$ radical with SO_2 on the basis of analogies to reaction rates of $\cdot CH_3$, $\cdot C_2H_5$, and $\cdot CFH_2CH_2$ radicals with SO_2. The $\cdot OH$ and $CH_2O\cdot$ rates are listed as 0.23 and 0.5%/hr, respectively, at the low extreme. A later review of Lloyd (1982) has indicated that the $RO\cdot$ and RO_2^- radical reactions with SO_2 now have much less importance for SO_2 oxidation than $\cdot OH$ based on recent experimental studies.

Measurements of the rate constant for the $\cdot OH + SO_2 + M \rightarrow HOSO_2^- + M$ reaction are emerging from recent fundamental studies. At approximately 298%K, the measured rate constants are about 1×10^{-14} cm^3/molecule sec. This is sufficient to yield a theoretical conversion rate from a few tenths to $2–3\%SO_2$/hr in smoggy air. The $\cdot OH$ reaction now appears to be the dominant one for gas-phase SO_2 oxidation in the troposphere and stratosphere.

The radical addition products, such as $HOSO_2^-$, should react rapidly with other species to generate sulfuric acid, peroxysulfuric acid, alkysulfates, and mixed intermediates such as $HOSO_2ONO_2$. Any of these ultimately should lead to sulfate in the presence of water.

Summing all of the known homogeneous reactions for SO_2 oxidation, it is possible to rationalize a theoretical midday rate of sulfate production $>3\%SO_2/hr$ for moderate photochemical smog conditions (e.g., Lloyd, 1982). However, such rates are highly dependent on the presence of unstable intermediates at relatively high concentrations. High concentration of free-radical species is by no means a universal condition in nonurban air or in cities with minimal photochemical activity, as measured, for example, by ozone levels. However, calculations reported by Lloyd (1982) suggest that high SO_2–$\cdot OH$ reaction rates should be possible even at low reactant concentration with certain optimum hydrocarbon–NO_x ratios (approximately 5).

The oxidation reactions of naturally occurring sulfur gases such as hydrogen sulfide, carbon disulfide, and carbonyl sulfide also may be important to atmospheric sulfate production. On a global scale, the emissions of these and other sulfur gases are potentially major contributors to airborne sulfate. Cox and Sandalls (1974) suggested that the oxidation by $\cdot OH$ radicals is the main sink for H_2S in the troposphere. Through the free-radical $HS\cdot$, H_2S oxidation to sulfate is believed to pass through SO_2 as an intermediate. The $\cdot OH$ radical also can react with CS_2 to form COS and $\cdot SH$; the latter will oxidize further to SO_2. COS is oxidized by $\cdot OH$ further to CO_2 and $HS\cdot$. This series of reactions appears to be slow in the lower atmosphere, providing the potential for COS to contribute to sulfate production in the stratosphere (Crutzen, 1976; Torres *et al.*, 1980).

For conditions where photochemically induced homogeneous reactions cannot be important, the heterogeneous processes must be considered. Several important reactions are listed in Table 7.25. The class of reactions that has been used most frequently to explain high SO_2 rates in the presence of liquid-water-containing aerosol particles is the system involving SO_2 absorption in water followed by oxidation by dissolved O_2 to form sulfate. Catalysis of the oxidation in excess of $1\%SO_2/hr$ in clean water solutions (Johnstone and Coughanowr, 1968; Matteson *et al.*, 1969). Other metals and ions such as Mn^{2+} or Fe^{2+}, and suspended carbon also may be significant as catalysts for absorbed SO_2 oxidation in water (Hegg and Hobbs, 1978; Chang *et al.*, 1981). The absorption of SO_2 is suppressed by acidity, but can be promoted by the buffering effect of simultaneous absorption of ammonia. Scott and Hobbs (1967) have shown that the aqueous SO_2–O_2 reaction increases in rate significantly with the presence of ammonium ion.

It is well known that ozone is more soluble in water than oxygen. Therefore, one expects that ozone absorption with SO_2 would contribute to significant oxidation of SO_2. Experiments of Penkett (1972) have shown that oxidation of SO_2 in air at 7 ppb absorbed in water droplets with ozone present in surrounding air at 5 pphm, can be as large as $13\%SO_2/hr$. Thus, air containing ozone, fog, or closed droplets could be an important medium for SO_4^{2-} formation.

Recent experiments of Penkett *et al.* (1979) have shown that dissolved H_2O_2 provides a major pathway for SO_2 oxidation in water droplets. H_2O_2 is very soluble in water, and there appears to be sufficient H_2O_2 present in the air for this mechanism to dominate the O_2 and O_3 processes in some cases. The absorption of acid gases or the scavenging of acid particles tends to increase the acidity of water droplets in the air to pH of 4.5 or less. The aqueous H_2O_2 is only weakly dependent on acidity, but the O_2 and O_3 reactions are strongly suppressed with lowered pH in the water. With estimated atmospheric concentrations of H_2O_2, theoretical calculations indicate that heterogeneous oxidation rates can be large, in excess of a hundred percent per hour in fog or cloudy conditions, or in wet aerosol particle haze at high humidity.

The relative rates of SO_2 oxidation in water droplets by different reaction mechanisms are shown in Fig. 7.27 in relation to pH. The metal-catalyzed dissolved-oxygen reactions and the dissolved-ozone reaction decrease dra-

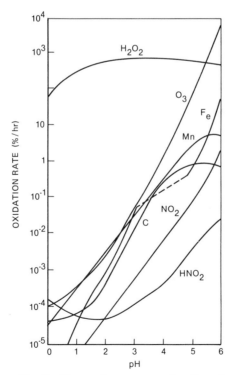

Fig. 7.27. Aqueous sulfur dioxide reaction rates as a function of acidity based on results reported by Martin (1982).

matically with pH. At pH above 5, these reactions are similar in rate, but are overshadowed by the ozone reaction at lower pH values. However, the H_2O_2 reaction theoretically could be the dominant mechanism below a droplet pH of approximately 4.5. Thus, the H_2O_2 reaction is favored by many investigators for a dominant aqueous mechanism in clouds and fog to produce sulfate.

From laboratory experiments, workers recently have speculated that additional dissolved oxidants such as NO_2, HNO_2, HNO_3, organic nitrates, and even free radicals such as ·OH may be of significance in aqueous SO_2 reactions. Martin (1982) has included reactions of HNO_2 in the oxidation process (Fig. 7.27). Aqueous free-radical chemistry has been discussed recently by Chameides and Davis (1982). The HNO_2 reaction appears to be minor in importance except at extreme pH. Free-radical chemistry has not been evaluated extensively yet, but preliminary studies show minimal significance.

The reported rates of SO_2 oxidation in clean water droplets must be considered maximum values. It is questionable whether they can ever be achieved in urban or polluted air since such aqueous reactions have been shown to be suppressed significantly by organic contaminants. The work of Fuller and Christ (1941) and later of Schroeter (1963) has indicated that the aqueous absorption of SO_2 and its subsequent oxidation are reduced by as much as an order of magnitude by dissolved organic acids or alcohols. Since this type of material is known to be present in the atmospheric aerosol sampled on the ground, one can expect that the aqueous oxidation will probably be most efficient in relatively clean conditions of clouds well away from the earth's surface. This significance of organic films on gas absorption has been demonstrated through the calculations of Graedel et al. (1983).

The heterogeneous mechanisms of SO_2 oxidation in the absence of liquid water are poorly understood. However, the recent work of Novakov et al. (1974) has shown that SO_4^{2-} can be produced rapidly on the surface of carbon particles suspended in the air. These workers have observed that significant amounts of SO_4^{2-} can be found on carbon particles generated by combustion of hydrocarbons in ppm-level SO_2-enriched air. Judeikis et al. (1978) have examined SO_2 adsorption on metal oxide particles followed by surface O_2 reactions for atmospheric applications. These appear to be of minimal significance in ambient air.

It is difficult to assess the significance of "dry" carbon or organic particles for SO_2 oxidation in the free atmosphere. There is little doubt that absorption of SO_2 on carbon particles freshly generated by combustion can provide a surface-catalyzed oxidation medium. The work of Yamamoto et al. (1973) further emphasizes that such a heterogeneous oxidation mechanism depends on a variety of factors, ranging from grain size of the carbon to temperature and to concentrations of SO_2, H_2O vapor, and oxygen as well as to the micropore structure of the particle surface. It would seem that oily, gummy, wet particles collected from the atmosphere would be poorly suited for nonaqueous reactions to form sulfate since their micropore structure would be minimal.

Comparison of the observations with the expectations of chemistry suggests qualitatively that several different mechanisms probably play a role in sulfate formation. Direct evidence from intermediates, aerometric conditions, or estimated rates does not provide for a distinction between mechanisms. However, the combination of laboratory experimentation and atmospheric observation leads to sufficient information on mechanisms to interpret sulfate behavior in the light of current knowledge.

The reactions that appear to be most significant for sulfate production on summer days or in urban air are consistent with homogeneous gas-phase processes driven by photochemistry. The identification of ozone and absolute humidity as closely related covariants of sulfate and absolute humidity emphasizes the role of the gas-phase reactions as contrasted with aqueous reactions, which should be associated with liquid water content. [Henry and Hidy (1979); Mueller and Hidy (1983)].

Other evidence, such as (a) the difference in sulfate size distributions observed in dry and humid conditions, and (b) the shift in sulfate variations to coincide with relative humidity by night or season, also indicate the conditions for heterogeneous or "mixed" reactions in the troposphere. Recent experiments in clouds reported by Hegg and Hobbs (1981) and Lazrus et al. (1981) give direct evidence of important SO_2 oxidation in cloud droplets analogous to the evidence for photochemical processes.

The accumulation of tropospheric observations provides a complex picture of sulfate chemistry in which a variety of different oxidation pathways of similar maximum rates can occur. These vary according to photochemical reactivity, water content, and temperature as well as nonsulfate particulate concentrations.

In the lower stratosphere the reactions of principal interest for sulfate formation appear to be photochemically induced. In particular, the homogeneous reaction of $\cdot OH$ and SO_2 or other sulfur-containing gases seems to be the leading candidate for SO_4^{2-} formation at this time. For estimated SO_2 concentration distributions in the northern hemisphere Harker (1975), for example, showed that the $\cdot OH$ radical oxidation process can account for a maximum in sulfate accumulation in the lower stratosphere in the region of the observed maximum. The gaseous compounds supporting SO_2 as an intermediate can include direct emissions, dominated by man's activities, or naturally occurring gases including H_2S, CS_2, and COS.

Nitrate-Forming Reactions.   As with the production of sulfate, nitrate can be formed in atmospheric particles, by a wide variety of homogeneous as well as heterogeneous reactions. The pathways of nitrate generation are not well understood, but it is likely that its reactions are interrelated with sulfate production at least in some circumstances.

Since both nitrous acid and nitric acid are much more volatile than sulfuric acid, it does not appear possible that they can exist in the atmosphere in pure condensed form as acids. Thus the presence of nitrate must involve formation of a condensable species such as NH_4NO_3, the absorption of organonitrates in particles, or the adsorption of a gaseous nitrogen oxide constituent by particles followed by stabilization through chemical reaction. Such heterogeneous processes, of course, can take place in an aqueous or nonaqueous medium.

The precursors for particulate nitrate formation are summarized in Table 7.26. These are classified in terms of (a) important nitrogen oxides, (b) volatile acids HONO and $HONO_2$, and (c) gaseous nitrates.

TABLE 7.26

Reactions Potentially Involved in Nitrate Formation

Species	$d\mathrm{NO_2}/dt$, or rate constant (ppm/min)[a]
Nitrogen oxides	
(15) $O_3 + NO \rightarrow NO_2 + O_2$	2.7×10^{-2}
(16) $O + M + NO \rightarrow NO_2 + M$	—
(17) $RO_2^* + NO \rightarrow NO_2 + RO^*$	2.5×10^{-3}
(18) $O_3 + NO_2 \rightarrow NO_3 + O_2$	4.0×10^{-4}
(19) $NO_3 + NO_2 \rightleftarrows N_2O_5$	1.0 to 23×10^{-4}
Volatile acids	
(20) $N_2O_5 + H_2O \rightarrow 2HONO_2$	2×10^{-5}
(21) $HO^* + NO_2 + M \rightarrow HONO_2 + M$	10^{-5} (M = 1 atm N_2)
(22) $NO + NO_2 + H_2O \rightarrow 2HONO$	—
(23) $HO_2^* + NO_2 \rightarrow HO_2NO_2$	
(24a) $HOSO_2O^* + NO \rightarrow HOSO_2^* ONO$	—
$+ H_2O \rightarrow H_2SO_4 + HONO$	
(24b) $HOSO_2O^* + NO_2 \rightarrow HOSO_2^* ONO_2$	—
$+ H_2O \rightarrow H_2SO_4 + HONO_2$	
Gaseous nitrates	
(25) $NH_3 + HONO_2 \rightarrow NH_4NO_3$	$\sim 10^{-6}$
(26) $RO_2^* + N_2O_5 \rightarrow$ R'C with O double bond, NO2 / ONO2	10^{-3}

$$ RO_2^* + N_2O_5 \rightarrow R'C\!\!\begin{array}{l}\overset{O}{\diagup\!\!/}\\ \diagdown ONO_2\end{array}\ \ NO_2 $$

$$ +\ \ R'C\!\!\begin{array}{l}\overset{O}{\diagup\!\!/}\\ \diagdown ONO\end{array} + \cdots $$

[a] Typical for smog reactant concentrations in the first hour of reaction (Calvert and McQuigg, 1975).

Because nitric oxide is relatively insoluble and nonreactive with water, the important nitrate-forming atmospheric oxides of nitrogen are believed to be NO_2, NO_3, and N_2O_5. These species are formed mainly in the atmosphere by the well-known "smog" reactions [(16)–(20)] and are not emitted primarily from material or anthropogenic sources.

With the nitrogen oxides coexisting with water vapor and sulfuric acid, the volatile nitrous and nitric acids can be formed via reactions (21)–(24). The mixed intermediates involving sulfuric acid are of interest as they link the NO_x and SO_x chemistries. Once the volatile acids are formed, they may react with ammonia in the gas phase to form, for example, NH_4NO_3 [reaction (25)]. The nitrogen oxides also react with radicals such as RO_2 to form organic nitrates and nitrites including peroxyacetylnitrate (PAN). Reaction (26) has been hypothesized by Calvert (1974) on the basis of an analogy to the NH_3–HCl reaction. For concentrations of NH_3 and $HONO_2$ of approximately 1 ppb, Calvert's work suggests that the ammonia reaction may represent a significant removal path of $HONO_2$ to form aerosol particles.

It is possible that condensable organic nitrates are formed via reaction (26) in the gas phase. Certainly the observation of such materials in smog chamber experiments (O'Brien et al., 1975) would provide some evidence for such a process. However, it is known that volatile organic nitrates such as PAN readily hydrolyze in an aqueous medium to form nitrite ion [reaction (31)]. Thus, the presence of such compounds resulting from gas-phase reactions could lead to particulate nitrate after stabilization with ammonium ion or another cation.

Of the gas-phase reactions involved in forming nitric acid vapor, reactions (20) and (21) are believed to be most important. Thus, it would appear that reaction (21) is of principal importance for $HONO_2$ formation in smog. The rate of conversion of NO_2 to $HONO_2$ by this reaction should be in the range of 2–8%NO_2/hr for the conditions used in the calculations in Table 7.24. With absorption of $HONO_2$ in wet particles and neutralization with ammonium ion, this value is not unreasonable for the upper limit of the estimated nitrate formation rate estimated in Los Angeles smog for the morning "peak" condition.

Once the nitrogen oxides are present or the acids begin to form, the interaction of these species with moist aerosols can take place. Some potentially important aqueous reactions of nitrogen oxides are given in Table 7.27. All of the nitrogen oxides contained in the atmosphere will react with liquid water to form traces of nitrate and nitrite. These reactions are generally reversible, however, so the anions must be stabilized by a cation such as NH_4^+. In the presence of dissolved oxygen or ozone, nitrite ion hypothetically can be oxidized to nitrate in analogy to the aqueous sulfate formation reactions.

None of the aqueous nitrate reactions has been studied with atmospheric applications in mind. However, there is a variety of information in the chem-

TABLE 7.27

Aqueous Reactions of Nitrogen Oxides of
Possible Application to Atmospheric Processes

(27)	$N_2O_5 + H_2O(l) \rightarrow 2H^+ + 2NO_3^-$
(28)	$NO + NO_2 + H_2O(l) \rightarrow 2H^+ + 2NO_2^-$
(29)	$2NO_2 + H_2O(l) \rightleftarrows H^+ + NO_3^- + HONO$
	$HONO + OH^- \rightarrow H_2O + NO_2^-$
(29a)	$2NO_2^- + O_2(aq) \rightarrow 2NO_3^-$
(29b)	$NO_2^- + O_3(aq) \rightarrow NO_3^- + O_2$
(30)	$2NO_2 + H_2SO_4 \rightleftarrows HNOSO_4 + HNO_3$
	$HNOSO_4 + H_2O(l) \rightleftarrows HNO_2 + H_2SO_4$
	$3HNO_2 \rightleftarrows HNO_3 + 2NO + H_2O$
(31)	$RONO_2 + H_2O(l) \rightarrow H^+ + NO_2^- + R'OH$

ical literature dealing with NO_x absorption in water. There is ample evidence, for example, from studies reported by Nash (1970) and Borok (1960) that NO_2 at trace levels in air is readily absorbed in aqueous solutions. The efficiency of absorption varies widely, however, depending on the acid–base content of the solution.

If significant quantities of concentrated sulfuric acid are formed in atmospheric aerosols, nitrate formation via absorption of NO_2 to form nitrosylsulfuric acid, $HNOSO_4$, may be of interest. The rate of absorption of NO_2 in sulfuric acid is fast, but the efficiency appears to be relatively low, according to Baranov *et al.* (1966). Again, it appears that absorption of ammonia or the presence of another basic cation has to be involved to drive the equilibria to nitrate production. Thus, in these aqueous reactions, nitrate formation may be limited by the concentration of ammonia rather than by any of the nitrogen oxide species.

The knowledge of nitrate-forming processes applicable to the atmosphere provides a less convincing basis for explaining the behavior of this ion than that of sulfate. The key difference in mechanisms evidently is centered around the volatility of nitrous acid and nitric acid and their equilibria in aqueous solution. The Los Angeles experience and the observations in the northeast described by Hidy (1975) and Mueller and Hidy (1983) emphasize the distinct difference between nitrate and sulfate evolution.

In analogy to sulfate, a survey of potential chemical reactions suggests that nitrate can be generated theoretically via homogeneous processes as well as by heterogeneous, aqueous processes. The complicated equilibria involved in nitrogen oxide ion solutions underscore the potential thermodynamic importance of basic cations in the presence of acidic sulfate, as well as temperature and moisture content (Stelson *et al.*, 1979; Lee and Schwartz, 1981). Based on current information, it is likely that little nitrate exists in particles relative to the gas phase, unless high concentrations of NH_3 are present to produce NH_4NO_3.

Interpretation of nitrate observations in the troposphere is difficult at present because of the paucity of reliable atmospheric aerosol data using multiple sampling techniques for gaseous nitrogen oxides, nitric acid, and ammonia, and particle collection on non-gas-adsorbing filter media. Confidence in the meaning of a variety of historical data obtained by different means will develop only after the sampling and analytical methods have evolved further and are used more widely to develop a body of data representing a range of atmospheric conditions.

The absence of NO_3^- in stratospheric aerosols appears to be related to the high volatility of HNO_3 since this species has been observed as a gas at altitudes above 15 km.

Organic Particle Formation.   Of the three major contributors to secondary aerosol production, the least is known about mechanisms for the organics. These processes have not yet been identified with any certainty. It is possible that organic materials can be polymerized in sulfuric acid solutions, for example. However, it is known that such reactions generally take place in very strong acid above 90% concentration. However H_2SO_4 equilibrium with water vapor in the lower atmosphere, should not exceed 40% concentration in water.

Two additional classes of organic reactions may be important in the atmosphere. They involve the attack of ozone or OH radical on olefins. There are sufficient quantities of olefinic material in the air to yield condensable organic particles in some tropospheric situations. The reactions of interest are molecular weight building or polymer forming, and involve oxygenation. The reactions probably involve initiation by O_3, $\cdot OH$, HO_2^- and NO_x interactions. They take the following form:

(32a) $O_3 + R'-C=CR \longrightarrow$ (Ozonide) $\longrightarrow$ (Alkoxy biradical) $\longrightarrow R'-\dot{C}HO\dot{O} + RCCHO$

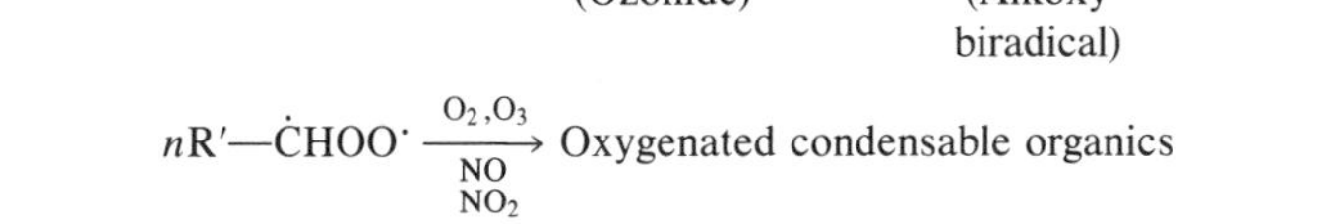

(32b) $nR'-\dot{C}HOO^{\cdot} \xrightarrow[\substack{NO \\ NO_2}]{O_2, O_3}$ Oxygenated condensable organics

(33) $OH + R'-C=CR \xrightarrow{O_2} (n)^{\cdot}OOC-C-R \xrightarrow[^{\cdot}OH]{O_2}$ Isomerization

(Alkoxy radical)

$\rightarrow$ Oxygenated condensable organics

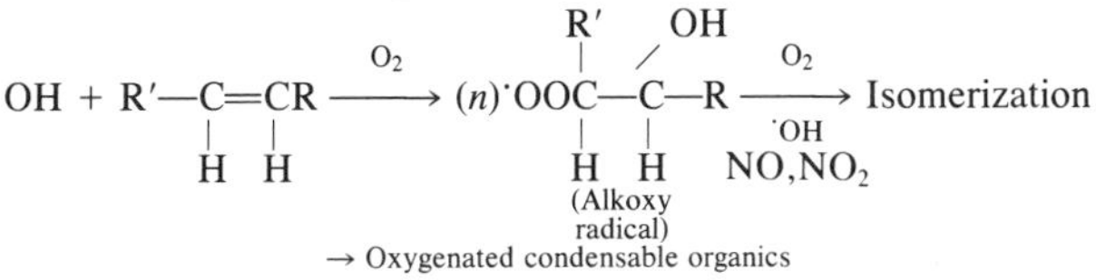

where R and R' are undesignated alkyl species (e.g., $H-$, CH_3-, C_2H_5-, . . .).

The ·OH abstraction and O_3 addition reactions probably have similar oxygenated end products. These may take the form of esters, alcohols, or carboxylic acids. Cyclic olefins can follow analogous pathways. These are discussed by Grosjean and Friedlander (1980). In the case of cyclopentene and cyclohexene, dicarboxylic acids appear to be an important condensable end product, and have been identified in urban air.

The empirical rates of production of aerosol from dry air ozone–olefin mixtures have been reported by Hidy and Burton (1975). Extrapolation of these data to olefins at the parts per billion level and ozone to atmospheric conditions suggests a production rate for organic material of tenths of micrograms per cubic meter per hour, which is an order of magnitude lower than projected from atmospheric observations in Los Angeles. Thus either the estimate of total precursor concentration is too low, or the ozone–linear olefin mechanism cannot explain the atmospheric processes. In contrast, Grosjean and Friedlander's (1980) work indicates that reactions of cyclic olefins are capable of producing organic particles with olefin concentrations in urban air at the parts per billion level. Much more knowledge is required about organic vapor reactions under atmospheric conditions to clarify the mechanisms of organic particle production. Currently there is little work of this kind being considered, except for recent smog chamber experiments of Grosjean and McMurry (1982).

7.5 SINKS—ATMOSPHERIC REMOVAL PROCESSES

Were it not for the fact that the removal of particles is relatively efficient, the earth's atmosphere would contain thick clouds of particulate matter, given the prolific sources at the planet's surface. Removal or loss of material from the troposphere can take place either by leakage to the upper atmosphere or by deposition at the ground. The loss to the upper atmosphere under average conditions is strongly suppressed by the thermally stable density structure defining the troposphere and stratosphere. However, there is vertical transport across the tropopause by deep penetration of large convective storms to heights above 15 km, and by interaction of a zone of persistent high winds called the *jet stream,* localized along the tropopause regime. The "tropospheric folding" around the jet stream offers a pathway for appreciable upward transport across the tropopause. The loss of material to the stratosphere has been estimated for typical conditions (Reiter, 1978). Although upward leakage of aerosol takes place, it is thought to be small relative to loss at the ground for particulate matter. The upward transport will not be considered further here.

Removal processes at the ground are classed in terms of dry processes in the absence of water clouds and precipitation, and wet processes when clouds and precipitation are involved. Dry deposition involves the scavenging of particles or the absorption of gases at the earth's surface on soil particles, stone or other solid surfaces, and vegetation. The removal is most effective for gases if the gases are water soluble or reactive with the collecting surface. Particulate matter is scavenged by gravitational fallout on surfaces, impaction, interception, or diffusional transfer to surfaces. External forces other than gravity such as electrical forces also may be involved. Wet removal processes involve (a) *rainout,* the absorption of gases or the assimilation of particles in cloud droplets or snow crystals, and (b) the *washout* of suspended material by collection during the fallout of hydrometeors[†] from clouds. Cloud and precipitation elements can transfer material from one height to another if condensation is followed by evaporation before the hydrometeors fall to the ground. Removal from the air is complete only if precipitation falls to the ground. Otherwise material is only transferred from one location to another in the air during the cloud formation and dissipation cycle.

Wet removal of gases takes place by absorption in hydrometeors. This is most efficient for soluble gases such as SO_2, ammonia, and nitrogen oxides. If the gases are reactive, they may be oxidized in hydrometeors prior to washout. Evaporation after earlier absorption and oxidation can result in new particle production, as exemplified in the series of heterogeneous reactions of SO_2 (Table 7.25) or of NO_x (Table 7.27).

The assimilation of aerosol particles into cloud droplets or ice particles may take place by addition of water to particles acting as nucleation centers for supersaturated water vapor. Other in-cloud scavenging processes of potential importance include collection by (a) particle impaction and interception on falling hydrometeors; (b) collection by electrical, thermophoretic, and diffusiophoretic forces; and (c) diffusional transfer to the falling particles. Beneath clouds, these processes also take place, but impaction, interception, and diffusional collection are believed to be most important.

7.5.1 Dry Deposition. The absorption and removal of the gaseous component of aerosols at natural surfaces has been investigated for some trace materials in the atmosphere. The knowledge of SO_2 behavior has been reviewed recently by Garland (1978); Sehmel (1980) has provided a more general review. The removal process is parametrized by the mass transfer coefficient to the surface, or the deposition velocity ℓ, as defined in Section 2.4,

$$\ell\left(\frac{cm}{sec}\right) \equiv \frac{\text{Rate of deposition}}{\text{Area}} \bigg/ \begin{array}{c}\text{Airborne contaminant concentration}\\ \text{near the surface}\end{array}$$

$$= \left(\frac{g}{cm^2\ sec}\right)\left(\frac{1}{g/cm^3}\right). \tag{7.7}$$

The estimation of deposition velocities depends on knowledge of the transboundary transfer rates. This is often expressed formally in terms of

---

[†] *Hydrometeor* is a generic term for ice and liquid water particles suspended in the atmosphere.

resistances to transfer of material at the ground (Wesely and Hicks, 1977). These include the aerodynamic resistance to diffusion near the surface, and the resistance to diffusion in the absorbing medium (water, soil, or vegetable material).

In vegetation, stomatal resistance also adds an important transport resistance. SO_2 absorption, for example, has been found to be a function of whether or not plant stomata are open. If the interfolial and stomatal resistance are negligible, gas absorption depends on the mass transfer process in the air near the surface, which is normally parametrized in terms of the properties of the surface layers, or the friction velocity u^*, and a surface roughness height z_0. The friction velocity in turn is proportional to air speed at some reference height, which the roughness is a property of the nature of the surface, as discussed for turbulent systems in Section 2.4. The deposition velocity for gases can vary over a considerable range of values depending on properties of the gas. An example of the range of values for SO_2 deposition over different surfaces is given in Fig. 7.28. For the case of SO_2, the deposition velocity found in the literature ranges from less than 0.1 cm/sec to nearly 10 cm/sec. Values reported depend on surface character, surface moisture, friction velocity, and atmospheric stability. For many cases, a

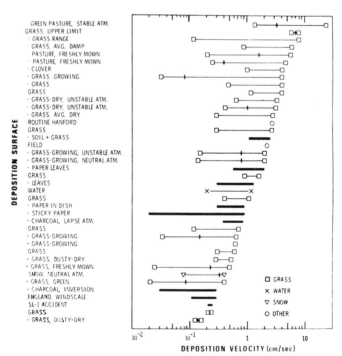

Fig. 7.28. SO_2 deposition velocity summary for different surfaces as summarized by Sehmel (1980) (reprinted with permission from Pergamon Press).

value of 1 cm/sec is assumed for absorbable gases in the absence of better information.

The dry deposition of particles is more complicated than for gases in relation to aerodynamic influences. From the discussion in Section 2.4, the deposition velocity will depend on particle size as well as density, in addition to the aerodynamic properties of the air boundary layer and the stickiness of the collector surface.

The wide range of variation of particle deposition velocity for different particle size and surface roughness is demonstrated in Fig. 2.11, for measurements in a wind tunnel. A comparison between a natural gas surface and artificial grass made of thin plastic strands illustrates further this behavior for particles. Slinn (1981a) has reported the curves shown in Fig. 7.29, which correlates Chamberlain's (1966, 1967) measurements using aerodynamic parameters. The deposition velocity is presented as an excess over gravitational settling normalized to the product of the square root of an overall

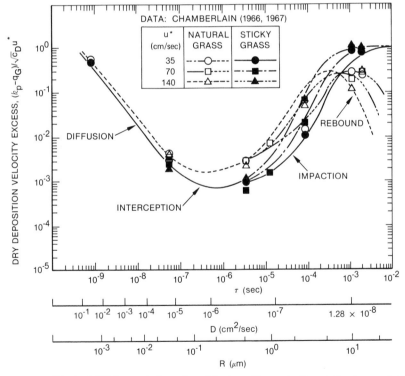

Fig. 7.29. Slinn's (1981a) correlation of particle deposition velocity relative to gravitational deposition rates for natural grass and sticky plastic artificial grass. Data of Chamberlain (1966, 1967) for grass height of 6–7.5 cm, corresponding to a roughness height of 0.6–1 cm, and a canopy drag efficiency of approximately 0.20.

surface drag coefficient C_D and the friction velocity u^*. The minimum in deposition occurs at approximately 0.2 μm radius, with increases with diffusion of smaller particles and impaction for larger particles. The results do not show an increase in deposition velocity for submicrometer particles for large roughness >0.1 cm, expected from Sehmel's results in Fig. 2.11. Some evidence also emerges in the data shown in Fig. 7.29 for large particle rebound from natural surfaces, as indicated by the departure from the "sticky" grass case above 15 μm radius. These results suggest the need for further experiments to clarify the behavior of particle deposition on naturally roughened surfaces.

The deposition of particles is further complicated by the fact that reentrainment may take place by winnowing or shaking of vegetation; Wesely *et al.* (1977) also have reported direct measurements of the vertical flux at the surface, which can reverse from downward to upward at different times. The reasons for these changes are not well understood, and will require additional study.

There is very little direct information about the particle removal rate expected as a result of deposition on trees, buildings, or other objects at the earth's surface. The process is complicated, too, by the possible influence of electrical charge and reentrainment of particles after collection. Neither of the processes can readily be accounted for by theory at present, though both mechanisms have interested investigators.

An interesting series of experiments was conducted in a wind tunnel on single conifer needles and conifer trees by Langer (1965) and Rosinski and Nagamoto (1965). This is one of the few studies in the literature that gives information on the relative importance of inertial effects, electrical charging, and reentrainment. For single needles or leaves, electrical charging of 2.4-μm-diam fluorescent ZnS dust with up to eight units of charge had no detectable effect at wind speeds of 1.2–1.6 m/sec. The average collection efficiency was found to be ~6% for edgewise cedar or fir needles, with broadside values an order of magnitude lower. Bounce-off after striking the collector was not detected, but reentrainment could take place above ~2m/sec wind speed. Tests on branches of cedar and fir by Rosinski and Nagamoto suggested results similar to those for single needles. The average scavenging efficiency was defined as the ratio of the number of particles deposited on the tree to the dose corresponding to the tree. This efficiency ran from 1 to 10% over the wind speed range 0.2–0.5 m/sec. Reentrainment was observed in early stages of a test, particularly on the upstream side of the trees, but it practically ceased after several hours of exposure. The average area coverage after several hours of exposure to 2.4-μm-diam ZnS particles was only about 0.4%.

Later wind tunnel experiments have been reported, including those of Little (1977) and Wedding *et al.* (1977). In Little's work, large differences in deposition velocities for 2.75–8.5 μm particles were observed between leaf laminae, petioles, and stems for different species (beech, white poplar, and nettles). Deposition velocities were proportional to wind speed and particle size. The Wedding *et al.* (1977) experiments confirmed Little's conclusion that rough leaves are about seven times better particle collectors than smooth leaves. In the Wedding *et al.* study, wind reentrainment of 6.77-μm-diam $PbCl_2$ particles was negligible, but simulated rainfall removed large amounts of particle deposit from the leaves.

One of the few studies of aerosol removal from the atmosphere by trees was reported a few years ago by Neuberger *et al.* (1967). They made measurements of ragweed pollen concentrations (large particles) and Aitken nuclei concentrations inside and outside a forested area. The data suggested that more than 80% of the pollen was removed in a dense coniferous forest. Parallel laboratory investigations showed that as much as 34% of the upstream concentration of Aitken nuclei would be removed by coniferous trees. Deciduous materials could account only for less than half as much removal of submicrometer particles.

7.5.2 Wet Removal and Deposition. The influence of rainfall on reduction of atmospheric contaminants was recognized in the mid-1950s when studies of sulfur oxides and nitrogen oxide behavior were undertaken in Europe. These early data suggested that there is a significant decrease in the concentrations of both gases and particulate matter after rainfall. The apparent decrease ranges from 24 to 73% over the city, which has an average rainfall comparable to some cities in the eastern United States and Europe.

Historical data are representative of ground-level conditions and are not necessarily representative of the situation aloft in clouds. The degree of apparent removal also is obscured by the fact that storm systems are advected across the terrain as they develop. Thus, a measurement of a decrease at a fixed ground station may reflect only the passage of clean air behind an advancing storm. Despite these reservations, the early observations in combination with the theory of scavenging provides useful evidence that rain clouds play a dramatic role in removal of soluble suspended materials.

The contribution of wet processes to the overall deposition rate of material will depend on the amount of precipitation as well as the chemistry of assimilation into rain. In two recent papers, Hales (1978) and later Slinn (1981b) have reviewed the state of knowledge of scavenging theory. Considerable effort has been devoted to the processes of scavenging, but their complexity combined with the limited usefulness of measurements in clouds has limited progress toward a unified, quantitative description of the processes.

The gas-phase components of atmospheric aerosols are removable by absorption if they are soluble. The rate of absorption depends on gas diffusion to the hydrometeor and chemical reactions within the hydrometeor. The gases do not react but only absorb; absorption is limited to thermodynamic equilibrium. The equilibrium condition is often described by Henry's law. The solubility coefficient and Henry's law coefficient $\mathcal{H}$ are related in the following way. For Henry's law, the partial pressure p_A of species A at equilibrium with a water solution is given by

$$p_A = \mathcal{H}x_A,$$

where x_A is the mole fraction of A in the solution. The solubility coefficient SC is given by

$$SC = \rho_1 \mathcal{R} T / M_1 \mathcal{H} \tag{7.8a}$$

or

$$SC = \rho_1 \mathcal{R} T x_A^* / M_1 p_A, \tag{7.8b}$$

where x_A^* is the mole ratio of A in the liquid at one atmospheric total pressure, and M_1 is the molecular weight of the liquid solvent (water).

The solubility coefficients typically decrease with temperature, as noted for CO_2 in Table 7.28. They also decrease typically in the presence of other constituents dissolved in water. For example, oxygen is about 20% less soluble in sea water than fresh water.

Some solubility coefficients applicable to trace gases in the atmosphere are listed in Table 7.28.

The collection of material in rainwater is sometimes described in terms of a *washout ratio* ξ, defined as

$$\xi \equiv \frac{\text{Concentration of material in ground-level precipitation}}{\text{Concentration of material in ground-level air samples}}.$$

For gases, $\xi \equiv \mathcal{H}$, the Henry's law constant if equilibrium is achieved between the hydrometeors and the air. The (wet) deposition rate at the ground, k_w, is $\xi \mathcal{P}$, where $\mathcal{P}$ is the precipitation rate.

Under nonequilibrium conditions, scavenging will depend on chemical reactions in the hydrometeor. Examples of reactions in water have been discussed above for SO_2 and NO_x.

The scavenging rate of particles in or below clouds is often expressed in terms of a collection efficiency $\eta(a, R)$ which depends on the radius a of the hydrometeors present and the aerosol particle radius R:

$$\eta(a, R) \equiv \frac{\text{Mass of particles collected during the hydrometeor's fall}}{\text{(Volume of air swept out by the hydrometeor)}} .$$
$$\times \text{(mass concentration of aerosol particles)}$$

For aerosol particles of a uniform radius $R \ll a$, the rate of mass accumulation on hydrometeors W then is given by

$$W = \pi M \int_0^\infty a^2 q(a) \eta(a, R) n_a(a)\, da \tag{7.9}$$

TABLE 7.28

Gas Solubility Coefficient (SC) in Water[a]

Noble gases (20–25°C)

He	9.4×10^{-3}
Ne	1.1×10^{-2}
Ar	3.6×10^{-2}
Kr	6.3×10^{-2}
Xe	1.4×10^{-1}
Rn	2.3×10^{-1}

Inorganic gases

CF_4	(25°C)	4.9×10^{-3}
SF_6	(25°C)	5.5×10^{-3}
N_2	(25°C)	1.6×10^{-2}
H_2	(25°C)	1.8×10^{-2}
CO	(25°C)	2.6×10^{-2}
O_2	(25°C)	3.1×10^{-2}
N_2O	(20°C)	6.7×10^{-1}
COS	(13.5°C)	6.7×10^{-1}
CO_2	(25°C)	8.3×10^{-1}
CO_2	(20°C)	9.5×10^{-1}
CO_2	(10°C)	1.2×10^{0}
CO_2	(0°C)	1.6×10^{0}
H_2S	(10°C)	3.4×10^{0}
Cl_2	(15°C)	3.3×10^{0}
SO_2	(18°C)	4.2×10^{1}
HI	(0°C)	4.9×10^{2}
HCl	(0°C)	5.1×10^{2}
HBr	(0°C)	6.1×10^{2}
NH_3	(25°C)	4.5×10^{2}
NH_3	(0°C)	1.2×10^{3}

Some organic gases

Methylamine, $MeNH_2$	(20°C)	1.26×10^{3}
Methylamine, $MeNH_2$	(10°C)	1.94×10^{3}
Methylamine, $MeNH_2$	(0°C)	4.17×10^{3}
Dimethylamine, Me_2NH	(20°C)	2.05×10^{3}
Trimethylamine, Me_3N	(10°C)	1.53×10^{3}
Dimethylether, Me_2O	(0°C)	3.7×10^{1}
Chloromethane, MeCl	(25°C)	2.6×10^{0}
Bromomethane, MeBr	(25°C)	3.9×10^{0}
Propylene, C_3H_6	(20°C)	1.5×10^{-1}
Propane, C_3H_8	(15°C)	7.7×10^{-2}

[a] Defined in Eq. (7.8); based on data from Gerrand (1976) and Slinn (1981b).

where M is the total mass of particles per unit volume, $q(a)$ is the vertical speed of the raindrops, and $n_a(a)$ is the raindrop size spectrum. Given the parameters η, q, and n_a, in principle, one can calculate the removal rate of particles for conditions where Eq. (7.9) is applicable.

A washout rate coefficient can be defined from Eq. (7.9) as

$$\Lambda = \pi \int_0^\infty a^2 q(a) \eta(a, R) n_a(a) \, da. \qquad (7.10)$$

The collection efficiency can be written in terms of the different mechanisms of interest. For cases where electrical, thermal, and diffusiophoretic forces are weak [for such special cases, see Wang *et al.* (1978)] Slinn (1977) has given expressions for mechanistic elements making up $\eta(a, R)$. These can be derived from appropriate expressions included in Chapters 2 and 3. Slinn (1977) has specified three of the collisional mechanisms and has suggested the following three formulas for computing the collision efficiency from the sum of the corresponding component efficiencies:

$$\eta_{impaction} = [(Stk - Stk_{cr})/(Stk + Stk_{cr} + \tfrac{2}{3})]^{3/2}, \qquad (7.11a)$$

$$\eta_{interception} = 3R/a, \qquad (7.11b)$$

$$\eta_{diffusion} = 4Sh/ReSc. \qquad (7.11c)$$

Here the Stokes number for the droplet and critical Stokes number are used as described for Eq. (2.43). The Sherwood number Sh can be calculated from the Frössling Eq. (2.26). More refined estimates of these component efficiencies are available in the more recent literature (Slinn, 1981b).

The corresponding numerical values of $\eta(a, R)$ obtained by summing the component efficiencies become large for both very large and very small particle sizes, and are small at intermediate sizes in the range of 0.1 μm (cf. Fig. 2.9). Since contributions of secondary mechanisms are disregarded in Eqs. (7.11), $\eta(a, R)$ values computed in this manner can be considered to be conservatively *low* estimates of actual behavior. An *upper-limit* estimate of $\eta(a, R)$ of unity can be used. Since this practice leads in some cases to efficiency values three orders of magnitude higher than those obtained from Eqs. (7.11), it is somewhat limited in value, at least in the present context where particle scavenging by nucleation is assumed unimportant. Because of this, Eqs. (7.11) is recommended for practical calculation under these conditions. Refinements for these expressions are available in recent literature, for example, Slinn (1981b).

When summed, Eqs. (7.11), give a collection efficiency which will be large for both large and small particles, with the minimum at intermediate sizes around 0.5 μm particle diameter (see also Fig. 2.9). The minimum, sometimes called the Greenfield gap for its discoverer, has been found in scavenging observations during precipitation. Three cases reported recently by Radke *et al.* (1980) are shown in Fig. 7.30. In these results, the minimum in scavenging is between 0.5- and 1.0-μm-diam particles. There also is observational evidence that the Brownian diffusion regime tends to decrease for particles smaller than 0.1 μm diameter. This result has not been explained as yet, but may be related to electrical charging effects.

The calculation of the washout rate coefficient requires estimation of the

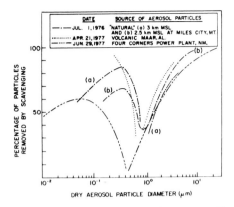

Fig. 7.30. Percentages of aerosol particles of various sizes removed by precipitation scaveng-ing. See Radke *et al.* (1980) for details of measurement conditions. (Permission to reprint from American Meteorological Society and authors.)

fall velocity of the hydrometeors. The fall velocity of raindrops, $q(a)$, de-pends on their shape as well as internal circulation (e.g., Pruppacher and Klett, 1978). For practical estimates, empirical fits to measurements are available. Those of Dingle and Lee (1972) are often applied. For $q(a)$ in cm/sec,

$$(0.05 \leq a \leq 0.7 \text{ mm}) \quad q(a) = 27.2692 - 1206.2884a + 348.0768a^2,$$

$$(7.12)$$

$$(0.7 \leq a \leq 2.9 \text{ mm}) \quad q(a) = 155.6745 - 613.4914a + 123.3392a^2.$$

The size spectrum of raindrops varies by cloud type. However, for practi-cal calculations, the Marshall–Palmer distribution is often chosen (cf. Prup-pacher and Klett,

$$n_a(a) = \frac{N_R}{\Delta_a} \exp(-\Delta_a a), \tag{7.13}$$

where $\Delta_a = 8.2 \mathscr{P}^{-0.21}$ mm^{-1} is a rainfall-dependent parameter ($\mathscr{P}$ is the rainfall rate in mm/hr). In this application, the raindrop concentration N_R should be approximately $1950 \mathscr{P}^{-0.21}$ drops/m^3. Typical cloud droplet distributions are given in Fig. 8.34.

If a polydispersed aerosol size distribution is involved, the collection effi-ciency is more complicated, taking the form

$$\Lambda = \pi \int_0^\infty \int_0^\infty a^2 q(a) \eta(a, R) n_a(a) n(R) \, da \, dR. \tag{7.14}$$

Examples of washout coefficient curves for various rain and aerosol size distributions have been reported by Dana and Hales (1976).

Irregular and varied geometries of snow or ice particles complicate the calculation of the washout coefficient because of ambiguities in size distributions, fall velocities, and scavenging efficiencies. Formulation of the relation for washout coefficient for snow particles is in the same form as for raindrops, but the drop radius is replaced by an equivalent diameter d_e and combined with a relation for the precipitation rate,

$$\Lambda \simeq \text{const} \times \mathcal{P}\eta(d_e, a)/d_m \qquad (7.15)$$

where the constant is of order unity and d_m is a characteristic cross-sectional length scale for the hydrometeor. These have been reported by Slinn (1981a), and are (a) graupel 0.014 cm, (b) plates and stellar dendrites 0.0027 cm, (c) powder snow and spatial dendrites 0.001 cm, (d) plane dendrites 3.8 $\times 10^{-4}$ cm, and (e) needles 1.9 $\times 10^{-3}$ cm. A semiempirical relation for η or Λ can be found in the work of Slinn (1981a,b) or Knutson and Stockham (1977).

The assimilation of suspended material during the cloud formation and decay process involves the transport of trace gases and particles into the cloud by air motion, transport inside the cloudy air, then microscopic processes of interphase transfer. Analysis of the microphysics of interphase transport suggests that contamination of cloud particles intimately involves the water nucleation process. The nucleation efficiency will depend on the degree of supercooling in the cloud, but condensation of water almost certainly depends on the hygroscopicity of the nucleating particles. Junge (1963a) has estimated that as much as 50–80% of the mass of aerosol may be active in the condensation process during a typical storm. The additional complexity of attachment by inertial forces or by diffusion along with collision and accretion of hydrometeors results in the net scavenging found in hydrometeors falling out of clouds as precipitation. The efficiency of both the nucleation process and the collisional processes in scavenging of material depends on the lifetime of the cloud and the air flowing through the cloud system. The detailed mathematical description incorporating all mechanisms has not been attempted. However, the microphysical basis for attempting such a calculation with useful approximation is accessible (Mason, 1971; Pruppacher and Klett, 1978). The limitation now lies in the scarcity of actual in-cloud and extra-cloud observations of air chemistry to confirm the validity of approximations required for a practical solution to such problems.

In the absence of detailed parametrization of rainout and washout processes, the use of collection rate theory accounting for the dynamic scavenging processes may be unwarranted. The scavenging of airborne material is often treated empirically in terms of material balances. These have been discussed in the literature using two analogous parameters, the washout ratio ξ mentioned earlier and the *rainout efficiency* ε_R after Junge (1963a).

For particles, the data available from nuclear bomb debris indicates that $\xi \approx 10^6$ (e.g., Gedeonov *et al.*, 1970). Slinn (1981b) has reviewed a variety of additional data for washout ratio; the data show considerable scatter but tend to decrease with precipitation rate and increase with particle size. An example of the latter dependence for elements found in particles is shown in Fig. 7.31. The range of washout ratios extends well over 10^4–10^6.

The rainout efficiency is calculated assuming that the mean concentration $\bar{\rho}_A$ in the air is uniform with height, and scavenging by washout is small compared with rainout. The rainout efficiency ε_R is defined by noting that the concentration of material A in rainwater χ_A is equal to the fraction in the air. In other words,

$$\chi_A = \rho_l \varepsilon_R \bar{\rho}_A / LWC, \qquad (7.16)$$

where LWC is the liquid water content in grams per cubic meter. For χ_A in milligrams per liter and ρ_l in grams per cubic meter $\bar{\rho}_A$ is in micrograms per cubic meter (STP), and $0 \leq \varepsilon_R \leq 1$.

If the concentration ratio $\chi_A/\bar{\rho}_A$ is taken at the ground, the rainout efficiency calculated on a volume basis is equivalent to the product of the washout ratio and LWC/ρ_l.

In many cases the liquid water content of precipitating clouds lies between 0.5 and 2.0 g/m³. For rural conditions where sources are absent, pollution becomes vertically well mixed to cloud base in a few hours. In such situa-

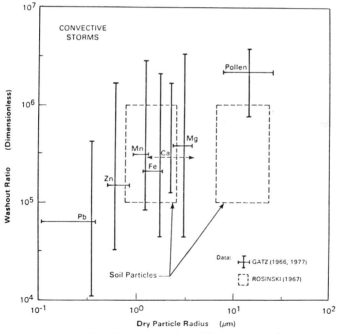

Fig. 7.31. Washout ratios for different aerosol particle constituents from convective storms. The horizontal variation in the data obtained by Gatz (1966, 1977) reflects different mass median diameters of material at different sites. The vertical range is for mean particle radii sampled in the St. Louis area. The range for soil particles found by Rosinski (1967) during different periods of a storm are indicated. [See also Slinn (1981b).]

tions, the rainout efficiency may be represented by ground-level conditions. Some values based on aerosol and rain data taken simultaneously estimating ε_R by precipitation event are shown in Table 7.29 for sulfate and nitrate. If particulate sulfate and nitrate only are used to calculate ε_R, values of ε_1 and $\varepsilon_3 > 1$ are found, which cannot occur by the definition of ε_R. The values for nitrate (ε_3) are much greater than unity, which is not possible. If the calculation includes sulfur dioxide, values of ε for sulfate (ε_2) are reduced compared with values based on particulate sulfate only (ε_1). There is a suggestion of segregation of events in cases where SO_2 absorption is important compared with particulate sulfate scavenging. Particulate nitrate concentrations are low at ground level, but were found to be substantially larger aloft in these experiments. Calculation of ε_3 based on value aloft provides lower values, which can be reduced further to less than unity by including the highly soluble gas, nitric acid. Estimates of nitric acid concentrations aloft at 300 m height are about 1 ppb; using this value in addition to particulate nitrate

TABLE 7.29

Calculated Values of the Rainout Efficiency Based on Data Taken during Summer 1977 and 1978 in the Northeastern United States[a,b]

Date	Aerosol composition (μmol/m³)			Rainwater composition (μmol/l)		Rain volume (cm³ hr)	Sulfate rainout efficiency		Nitrate rainout efficiency
	SO_2	SO_4^{2-}	NO_3^-	SO_4^{2-}	NO_3^-		ε_1	ε_2	ε_3
8/8/77	0.041	0.216	0.0032	35	25	222	0.16	0.14	7.8
8/9/77	0.041	0.080	0.0048	110	87	139	1.38	0.91[c]	18.0
8/11/77	0.123	0.213	0.0032	86	70	148	0.40	0.26	21.0
8/15/77	0.041	0.024	—	46	29	238	1.90	0.71[c]	—
8/18/77	0.041	0.021	0.0016	43	32	356	2.00	0.69[c]	20.0
8/25/77	0.082	0.047	0.0016	65	92	55	1.40	0.50[c]	57.0
8/27/77	1.68	0.134	0.0016	110	80	356	0.82	0.061	50.0
8/31/77	1.11	0.208	0.0016	64	65	255	0.31	0.048	41.0
7/4/78	0.082	0.040	0	25	13	32	0.62	0.20	Large
7/11/78	0.123	0.033	0.0064	69	77	196	2.09	0.43	12.0
7/15/78	0.082	0.224	0	95	68	46	0.42	0.31	Large
7/24/78	0.165	0.027	0.0016	42	20	1010	1.56	0.22	12.0
7/28/78	0.123	0.070	0.0016	81	47	270	1.20	0.42	29.0
7/30/78	0.123	0.053	0.0016	51	50	71	0.96	0.29	31.0

[a] From Hidy (1982). Copyright © 1982 by D. Reidel Publishing Company, Dordrecht, Holland.

[b] The rainout efficiency ε_1 is based on sulfate in rainwater and sulfate particle concentrations; ε_2 is based on rainwater sulfate and the sum of particulate sulfate and SO_2 (as sulfate) in the aerosol. The efficiency ε_3 is based on rainwater nitrate and particulate nitrate sampled on a Teflon-coated glass fiber filter.

[c] Hypothesized to be dominated by nucleation of water particle on a sulfate particle, and particulate sulfate scavenging.

concentrations, the resultant rainout efficiency then suggests that more of the nitrate found in rain may come from the gas phase rather than from particle scavenging per se. Intercomparisons of this kind by precipitation event and by constituent offer promise for elucidation of wet fallout processes from air monitoring data.

7.5.3 Precipitation Chemistry and Acidity. The process of aerosol scavenging from cloud action in the atmosphere is reflected well in the chemical composition of precipitation water, as noted in the previous discussion. The chemistry of major ions in precipitation was investigated many years ago by Junge (1963a). The deposition of scavenged material from polluted air has been a concern recently for potentially serious environmental effects. Recent studies in Scandinavia and in North America reviewing circumstantial evidence that the deposition of acidic precipitation water[†] may have widespread effects on vegetation and soils have not been verified in the natural environment (Hutchinson and Havas, 1980). However, fish kills and changes in aquatic biology have been observed in certain remote poorly buffered lakes and streams. These have been attributed by some investigators to the direct or indirect influence of acidity. Indirect effects appear to come from the accelerated leaching from soils of certain potentially toxic metals such as aluminum. The potential effects of rainwater acidity remain highly controversial, but they have awakened widespread interest in rainwater chemistry and its potential interaction with ecological systems.

To gain perspective on the chemical composition of precipitation water, consider the comparison between the composition of particulate matter and rainwater sampled in a similar geographical location (Table 7.30). Qualitatively this comparison shows that the enrichment of elemental material relative to sulfate is limited in the scavenging process. However, there is a dramatic difference between nitrate found in rain and in aerosol particles. This is a common observation from data in the eastern United States, but does not appear as pronounced in similar observations taken in southern California, for example, in Los Angeles and the San Joaquin Valley. The reason for the difference in the eastern air is not known, but must be related to particulate nitrate formation, which appears larger in the photochemical environment of California.

Acidity in precipitation water is often attributed to the concentration of sulfate and nitrate (Bowersox and de Pena, 1980). Calculations reported by Likens (1976) illustrate the significance of these anions in accounting for free acidity (at pH 4) of rainwater. The case shown in Table 7.31 is for a sample taken at Hubbard Brook, NH. The free acidity in this case is dominated by

---

[†] Acid rain in Europe was discussed in an early "forgotten" monograph of Smith (1872) in Europe. His work has gone largely unappreciated until recently, when his results were resurrected and found to be consistent with recent observations taken in the 1970s.

TABLE 7.30

Composition of Suspended Particulate Matter and Rainwater in Rural Pennsylvania in Summer[a]

Component	Particulate concentration Allegheny mountain (neq/m³)	Ratio to SO_4^{2-}	Rainwater concentration Kane, PA (μeq/l)	Ratio to SO_4^{2-}	Athens, PA (μeq/l)	Ratio to SO_4^{2-}
H^+	155	0.53	128	0.90	120	1.7
NH_4^+	128	0.44	100	0.70	19	0.27
NO_3^-	(8.5)[b]	0.029	60	0.42	54	0.76
SO_4^{2-}	292	1.00	142	1.00	71	1.00
Na^+	9	0.031	4.3	0.030	16	0.23
Mg^{2+}	6	0.020	8.2	0.058	2	0.029
Al^{3+}	78	0.27	—		—	
K^+	2	0.0068	5.3	0.037	1.2	0.016
Ca^{2+}	16	0.055	26.4	0.19	5	0.070
Anion sum-A	300	—	202	—	125	—
Cation sum-B	394	—	272	—	163	—
Cation sum less H^+ and NH_4^+-C	111	—	44.2	—	24.2	—
Ratio A/C	2.7	—	6.2	—	5.1	—
Ratio $A/(C + NH_4^+)$	1.3	—	1.4	—	2.9	—

[a] From Hidy (1982). Copyright © 1982 by D. Reidel Publishing Company, Dordrecht, Holland.

[b] Parentheses denote uncertain estimate of concentration.

TABLE 7.31

Sulfuric and Nitric Acids as Major Sources of Acidity in Precipitation, Based on a Sample of Rain Collected at Ithaca, NY, on 23 October 1975[a]

Substance	Concentration in precipitation (mg/l)	Contribution to free acidity[b] (meq/l)	Contribution to total acidity[c] (meq/l)
H_2CO_3	0.62[d]	0	20
NH_4^+	0.92	0	51
Al, dissolved	0.05[e]	0	5
Fe, dissolved	0.04[e]	0	2
Mn, dissolved	0.0005[e]	0	0.1
Total organic acids	0.34	2.4	4.7
HNO_3	4.40	39	39
H_2SO_4	5.10	57	57
Total		98	179

[a] Reprinted with permission from Likens (1976). Copyright 1976 American Chemical Society.

[b] At pH 4.01.

[c] In a titration to pH 9.0.

[d] Equilibrium concentration.

[e] Average value for several dates.

sulfate and nitrate ions. Neither chloride nor organic acids have a major role in acidity for this case. Note that with a titration to pH 9, carbonate and ammonium become important.

Interpretation of acidity from sulfate and nitrate concentrations oversimplifies the actual chemistry in rainwater. Acidity depends on the net balance between anions and cations present in the water. In the northeastern region, east of the Ohio–Pennsylvania state border, the rainwater acidity may well be dominated by sulfate and nitrate variations as, for example, at Hubbard Brook, NH and College Park, PA. However, further west the scavenging of alkaline soil dust is a significant factor in pH patterns. Junge's early measurements [reported by Stensland (1980)] and recent observations from the U.S. National Atmospheric Deposition Program (NADP) indicate the prevalence of the highest pH levels in midcontinent, with variable sulfate levels where alkaline dust contributes to cation content. On the other hand, the coastal areas show lower pH values with less soil dust influence.

Stensland (1980) has pointed out that the apparent trends in change in pH found between the 1950s and the 1970s in Illinois can be explained by major differences in the alkaline metal cations concentrations associated with drought rather than differences in sulfate and nitrate concentrations in rainwater. His comparison is summarized in Table 7.32 for observations taken at Champaign, IL.

Cogbill and Likens (1974) and Likens and Butler (1982) have speculated that significant trends toward increased acidity in rain have accompanied industrial expansion and urbanization of eastern North America after the mid-1950s. Taking into account Stensland's arguments on climatological variability, particularly in the Midwest, and a review of sampling and analytical methods used in the historical data, the apparent pH trends in acidity do not appear to be supported, except possibly in the Southeast (Hansen and

TABLE 7.32

Summary of Ion Concentrations (μeq/l) and pH for Precipitation Samples[a]

	[SO_4^{2-}]	[NO_3^-]	[$Ca^{2+} + Mg^{2+}$]	pH
1952[b]	60	20	82	6.05[c]
1977[b]	70	30	10	4.1

[a] From Stensland (1980).

[b] Median values for Champaign–Urbana, Illinois Airport.

[c] Includes an empirical correction for pH calculated from a balance of cations and anions. The correction is derived from Stensland's correlation between pH measurements and ion concentrations for a large number of observations. Without this correction, the value would be about 6.75.

Hidy, 1982). However, there remains little argument that rainfall over the Northeast is relatively acidic, and has been so for at least 30 years or more.

There remains the question of what is the actual background acidity or pH. Until recently, most tests considered the CO_2 at equilibrium with distilled water as representative of background pH levels. At 20°C, this equilibrium provides a pH $\simeq$ 5.6 for atmospheric CO_2 concentrations (Lerman, 1979). Observations of rainwater acidity from remote worldwide sites clearly show that the CO_2–distilled water baseline is rarely maintained (Table 7.33). Furthermore, the data listed in this table show the wide range of pH from less than four to greater than six experienced at such sites, some of which are located thousands of kilometers from any population centers. These observations also underscore the apparent distinction between rainwater at marine sites and those under continental influence. The former have substantially lower pH levels in precipitation water than the latter. Workers currently estimate that a background rainwater pH of about 5 is reasonable for the Northern hemisphere; values may be considerably higher with scavenged soil dust present.

7.6   DYNAMICS OF ATMOSPHERIC AEROSOL PARTICLES

An important consideration for characterizing aerosol behavior is the synthesis of dynamic processes to specify the evolution of the aerial suspension in time, taking into account sources and losses discussed in Sections 7.4 and 7.5. The dynamics of an atmospheric aerosol can be examined by macroscopic budgeting, or by microscopic analysis applying the rate expression for material balance of species. First, the budget concept is discussed as an application of analysis for geochemical cycling, then the microscopic ap-

TABLE 7.33

Average Value of pH from Remote Sites from 1972–1976[a]

Location	Average pH	pH range	Years	Number of values
Mauna Loa, HI	5.30	3.84–6.69	1973–1976	28
Pago Pago, Samoa	5.72	4.74–7.44	1973–1976	30
Prince Christian Sound, Greenland	5.73	4.70–6.81	1974–1976	13
Valentia Island, Ireland	5.43	4.2–6.8	1973–1976	48
Glacier Park, MT	5.78	2.60–7.10	1972–1976	33
Pendleton, OR	5.90	4.67–7.60	1972–1976	35
Ft. Simpson, NWT, Canada	6.27	5.22–7.14	1974–1976	14

[a] Data reported by the Environmental Data Service.

proach is summarized. Certain approximate schemes are introduced which apply to the size distribution function for an aerosol dominated by aging by coagulation and settling, and by condensation growth. The arguments for a quasi-steady state condition and for self-preservation in atmospheric aerosols are included as one theoretical rationale for observed uniformity in particle size distributions.

7.6.1 Budgeting. If the aerosol system is basically at steady state with respect to a given component, a macroscopic budget can be constructed. This material balance can account for inputs and losses, along with the suspended material retained in the air, given ambient concentrations, source and sink strengths, and residence times. Budgeting of this kind has been useful for elucidation of the geochemical cycling of certain elements (Garrels et al., 1975).

Global budgets for the flow of elements through the environment are instructive for exposure studies and ecological cycles. Semiquantitative budgets for components of the sulfur and nitrogen oxides have been reported in the literature. They have indicated the significance of man's activities in contrast to nature in many circumstances. Perhaps most extensively studied in the atmosphere is sulfur. Granat (1976) have prepared a regional atmospheric sulfur budget for northwestern Europe showing a balance in sulfur transport and deposition. A similar calculation for eastern North America has been prepared by Whelpdale and Galloway (1980). The method employs a synthesis of information on (a) emission rates of atmospheric sulfur, including anthropogenic (SO_2 and SO_4^{2-}) and natural sources from soil dust, sea salt, and biogenic emittors; (b) sinks including dry deposition and wet deposition; (c) transport into and out of the air reservoir from neighboring areas; and (d) retention in the air reservoir according to scaling by gas and particulate residence times.

The results of Whelpdale and Galloway are shown in Table 7.34. Unlike many atmospheric sulfur budgets, these calculations are based on "independent" estimates of each term. The inputs are derived for man-made and natural emissions. As expected, man's activities account for most of the atmospheric sulfur in this geographical region. According to this calculation, the average transboundary flow of sulfur from the United States to Canada is about the same as Canada's emissions. Total deposition of sulfur on eastern Canada is about the same as Canada's emissions. Total deposition of sulfur on eastern Canada is approximately the same as that on the eastern United States. Wet deposition and dry deposition are roughly equal over eastern North America, but wet deposition is larger over Canada. There also is a significant amount of sulfur estimated to be transported eastward over the Atlantic Ocean from the continent.

The calculations of Whelpdale and Galloway underscore the issues of international aspects of pollution involving sulfur, as has the situation in

TABLE 7.34

Magnitude of Sulfur Budget Terms for Eastern North America (Tg S/yr)[a,b]

Term	Eastern Canada	Eastern United States	Eastern North America
Inputs			
Man-made emissions	2.1	14	16.1
Natural emissions	0.32	0.44	0.76
Inflow from oceans	0.04	0.02	0.06
Inflow from west	0.1	0.4	0.5
Transboundary flow	2.0	0.7	—
Totals	4.6	15.6	17.4
Outputs			
Transboundary flow	0.7	2.0	—
Wet deposition	3.0	2.5	5.5
Dry deposition	1.2	3.3	4.5
Outflow to oceans	0.4	3.9	4.3
Totals	5.3	11.7	14.3

[a] From Whelpdale and Galloway (1980).
[b] 1 Tg = 10^6 metric ton.

Scandinavia. Both Sweden and Norway have estimated that the deposition of sulfur on their lands far exceeds that accounted for by their emissions. Through extensive observations and calculations, they have shown that their sulfur deposition can be accounted for by long-range transport of airborne pollutants from distant countries in Europe (OECD, 1977). With intensive industrialization in parts of the world, it appears inevitable the pollution components with atmospheric residence longer than a few days will have a multinational influence. New cases will arise where pollution will be transported across borders, imparting adverse effects to neighbors, who may have little "home-produced" pollution. The impact assessment technology for such situations has only recently begun to emerge.

Sulfur is essentially the only component of the atmospheric aerosol which has been examined for regional budgets. Similar analyses are needed for nitrogen oxides and eventually for carbon compounds. Taking into account differences in chemistry and deposition behavior, an improved perspective will be obtained in geographical areas which are environmentally or ecologically sensitive to large exposure to suspended-particulate deposition or ambient concentrations (see also Hutchinson and Havas, 1980).

7.6.2   Microscopic Dynamics—The Size Distribution.   A much more sophisticated approach to aerosol material balances makes use of an analysis of the size distribution function in Eq. (3.10). Consider an elemental volume of aerosol traveling in a turbulent medium with a mean wind. Then the "budget" for a particulate species in that volume is given by

$$\frac{d[Ng(v,\ n_i,\ \mathbf{r},\ t)dn_i\ dv]}{dt} =$$ A. Net gain by production from physical and chemical processes

+ B. Net gain or loss by collisions

+ C. Net gain or loss by condensation or absorption of gases

+ D. Loss by deposition or fallout

+ E. Loss by diffusional transport

− F. Loss by scavenging collisions by fallout

− G. Loss by raincloud processes.   (7.17)

The processes represented by D–G actually remove particulate material from the elemental volume. However, terms A–C essentially either generate new particles or particulate material, or transfer material to larger size ranges within the elemental volume. We shall consider material as "lost" or removed in this analysis from a given size and compositional class by either category of process.

It is impractical to deduce mathematically complete solutions to Eq. (7.17), so considerable simplification is sought for application to the atmosphere. Simplifications follow from estimation of the rate-limiting processes of importance. Suppose our discussion is restricted to the lower troposphere and to the ordinary size distribution function $n(v,\ \mathbf{r},\ t)$. Then we can estimate the rate-limiting terms by appropriate scale analysis using mathematical expressions for each of the terms. Consider first conditions where nucleation of new particles is negligible, and no sources are present. The net gain by collisional and scavenging processes is given by the coagulation equation taking into account Brownian motion, gravity settling, and turbulent collisions. Net gain by condensation and absorption of gas is assumed to be specified as a vapor-diffusion-limiting process. The fraction of loss of given size is given as approximately

$$\Phi_A n_i/\rho\bar{v},   (7.18)$$

where the vapor flux of condensable vapor A, Φ_A, is from Eq. (2.35); $\bar{v}$ is the average volume of particles over the size range considered. The actual loss in number/volume–time is the fractional loss times the number of i particles, n_i. The loss by sedimentation fallout is given by

$$-q_G\frac{\partial n_i}{\partial z};   (7.19)$$

the loss by diffusional transport is considered only in the vertical direction, proportional to a turbulent diffusivity D_t, or

$$-D_t \frac{\partial^2 n_i}{\partial z^2}. \tag{7.20}$$

The loss at the surface is estimated in terms of the deposition velocity ℓ for particles of a given size.

For purposes of calculation, let us use three ranges of particle size with mean radius centered at 0.05, 0.5, and 5 μm, whose concentrations are n_i. The concentration of the smallest particles $n_{0.05}$ will be identified with the Aitken nuclei counts ($N = 10^5/\text{cm}^3$), the middle range $n_{0.5}$ ($= 10^2/\text{cm}^3$) with an optical scattering instrument (roughly the visibility range), and the largest range $n_{5.0}$ ($= 1/\text{cm}^3$) with an impactor or total mass filter.

To estimate the significance of vapor condensation and absorption of gases, Eq. (7.18) is used. Consider a typical situation where the condensable gas is SO_2, with a concentration at the ground of about 10 μg/m^3 and at cloud base and above about 0.1 μg/m^3. It is assumed that about half of the gaseous SO_2 is removed by heterogeneous oxidation reaction to SO_4^{2-} involving "condensation" on particles.

From the discussion of Hidy (1973), the rate of loss by sedimentation and diffusion may be evaluated from Eqs. (7.19) and (7.20). By analogy to sedimentation, the loss in the first few meters of air to rough surface and vegetation is taken proportional to the concentration gradient at the surface, and the deposition velocity is the proportionality coefficient. Losses by various dry and wet collision processes are calculated from Eqs. (3.53), (3.54), and (3.55). The turbulent dissipation rates assumed are values taken from Lumley and Panofsky (1964). It is assumed that dry scavenging by sedimentation involves 10-μm particles moving through a cloud of smaller particles. Rainout is estimated assuming $a = 1$ mm and N_R is $10^{-3}/\text{cm}^3$. The droplet concentrations are taken as typical for cumulus and stratocumulus clouds as given by Fletcher (1966). In such an evaluation, essentially all of the mechanisms are assumed to operate continuously with varying degrees of efficiency except for mechanisms involving cloud development and precipitation. Cloud removal processes will be intermittent in nature, even though they represent an important factor in aerosol particle removal. There are estimates that, on the average, the earth is covered with clouds about half the time. Perhaps it rains (or precipitates) from cloud layers about 20% of the time. Thus, one may guess that about one-tenth to one-half the world aerosol particle population in the lower troposphere is influenced by washout and rainout at any one time. Washout, of course, will be of greatest interest near the ground, while rainout will be of importance at cloud base and above.

The results of the calculations are indicated in Tables 7.35–7.37. Bearing in mind the great deficiencies in the methods for making our estimations, establishing the relative importance of processes leads to some interesting

TABLE 7.35

Estimated Removal Rates of Aitken Nuclei in the Troposphere (particles lost/cm^3 sec)[a]

Process	Ground (urban) $N = 10^5$ cm^{-3}	Near cloud base (2 km) $N = 10^3$ cm^{-3}	In or above clouds (6 km) $N = 10^2$ cm^{-3}
Sedimentation	10^{-6}	10^{-8}	10^{-9}
Inertial and diffusional deposition on obstacles at the surface ($q_G = 0.1$ cm sec^{-1})	$\boxed{10^{-1}}$	—	—
Convective diffusion ($D_t = 10^4-10^3$ cm^2 sec^{-1})	$\boxed{10^{-1}-1}$	10^{-4}	10^{-6}
Condensation of vapors on particles	$\boxed{10^4}$	$\boxed{1}$	$\boxed{10^{-1}}$
Thermal coagulation	1	10^{-4}	10^{-6}
Scavenging by differential setting[b,c] ($R_2 = 10$ μm)	10^{-3}	10^{-7}	10^{-9}
Turbulent coagulation[d]	10^{-3}	10^{-9}	10^{-11}
Washout by 1-mm spherical hydrometeors ($N_R = 10^{-3}$ cm^{-3})	10^{-3}	10^{-5}	—
Rainout by cloud processes (nucleation + collisions) ($a = 10$ μm)[e]	—	$\boxed{10^{-2f}}$	$\boxed{10^{-4}}$

[a] After Hidy (1973); courtesy of Plenum, New York.

[b] Brownian diffusion to surface included.

[c] Calculated for $\rho_p = 1$ with 10 μm particle concentration, $N_2 = 10^{-1}$, 10^{-3}, and 10^{-4} cm^{-3}, respectively. N corresponds to the second particle in the collision.

[d] Calculated for turbulence dissipation rate $\varepsilon_t = 10^3$, 1, and 0.1 cm^{2}sec^3, respectively.

[e] Calculated for $N_R = 10^2$ cm^{-3}, cloud base; $N_R = 10$ cm^{-3} at 6 km.

[f] Aitken nuclei are assumed too small to be a factor in cloud droplet nucleation; Brownian diffusion is included in scavenging.

implications. Condensation of vapor or attachment of ions, as noted previously, is of primary significance in the smallest particle fraction, but appears to be important over all size ranges at all altitudes. Aside from condensation of vapors in the layers of urban air near the ground, turbulent coagulation, with washout removal by impaction and sedimentation, becomes significant for the largest particles. The vast quantities of condensable or absorbable gases present are significant in producing an effective "loss" by transfer of particles up the spectrum, and this process appears at least potentially as important if not more important than simultaneous processes of coagulation. It appears that convective diffusion in the atmosphere and thermal coagulation are more important contributors to removal for the small particles, but

TABLE 7.36

Estimated Removal Rates of Large Particles (R_p = 0.5 μm; $n_{0.5}$ = N) in the Troposphere (particles lost per cm³/sec)[a]

	Height		
Process	Ground (urban) $N = 10^2$ cm⁻³	Near cloud base (2 km) $N = 1$ cm⁻³	In or above clouds (6 km) $N = 10^{-1}$ cm⁻³
Sedimentation	10^{-6}	10^{-7}–10^{-8}	10^{-9}
Inertial and diffusional deposition on obstacles at the surface (q_G = 0.01 cm/sec)	10^{-5}	—	—
Convective diffusion (D_t = 10^5 cm²/sec)	10^{-3}	10^{-5}	—
Condensation of vapors on particles	10^{-1}	10^{-5}	10^{-6}
Thermal coagulation	10^{-4}	10^{-7}	10^{-9}
Scavenging by differential setting[b] (R_2 = 10 μm)	10^{-7}	10^{-11}	10^{-15}
Turbulent coagulation[b]	10^{-3}	10^{-9}	10^{-11}
Washout by 1-mm spherical hydrometeors (N_R = 10^{-3} cm⁻³)	10^{-4}	10^{-6}	—
Rainout by cloud process[b] (a = 10 μm)	—	$10^{-1\,c}$	$10^{-1\,c}$

[a] After Hidy (1973); courtesy of Plenum, New York.
[b] Same values of N_2, ε_t, and N_R as used in Table 7.35.
[c] Assumed 0.1 particle per cm³/sec nucleates.

essentially all mechanisms are weak for the larger particles. Even though turbulent coagulation is comparatively weak over all ranges of particle size, it may provide an essential mechanism to transport particles to larger particle sizes above ~0.5 μm radius. It is likely that condensation of vapors is also an important process in this range, however.

Depending on the altitude of cloud base and the frequency of cloud formation, the wet processes leading to precipitation should play leading roles with diffusion in removing larger aerosols away from the ground. Coagulation by Brownian motion will remain active in the loss of tiny particles only as long as their concentration remains high.

At altitudes exceeding 6 km, intermittent rainout will be the major factor in aerosol removal. However, one could expect that high in the troposphere above cloud tops, say at roughly 10 km, all mechanisms would be rather weak in removing particles compared with processes below 5 km altitude.

TABLE 7.37

Estimated Removal Rates of Giant Particles (R_p = 5.0 μm; $n_{5.0}$ = N) in the Troposphere (particles lost/cm³-sec)[a]

Process	Height Ground (urban) $N = 10^{-1}$ cm^{-3}	Height Near cloud base (2 km) $N = 10^{-3}$ cm^{-3}	Height In or above clouds (6 km) $N = 10^{-4}$ cm^{-3}
Sedimentation	10^{-7}	10^{-11}	—
Inertial and diffusional deposition on obstacles at the surface (q_G = 0.1 cm/sec)	10^{-6}	—	—
Convective diffusion (D_t = 10^5 cm²/sec)	10^{-6}	10^{-8}	—
Condensation of vapors on particles	10^{-6}	10^{-10}	10^{-11}
Thermal coagulation	10^{-6}	10^{-10}	10^{-12}
Scavenging by differential setting[b] (R_2 = 10 μm)	10^{-8}	10^{-12}	10^{-10}
Turbulent coagulation[b]	10^{-4}	10^{-10}	10^{-12}
Washout by 1-mm spherical hydrometeors (N_R = 10^{-3} cm^{-3})	10^{-4}	10^{-6}	—
Rainout by cloud processes (nucleation + collision)[b] (a = 10 μm)	—	$10^{-3\,c}$	$10^{-4\,c}$

[a] After Hidy (1973); courtesy of Plenum, New York.
[b] Same values of N_2, N_R, and ε_t as in Table 7.35.
[c] Assumed 0.1 particle per cm³/sec nucleates if nuclei are present.

From the considerations here, condensation of vapor and absorption of gases, including nitrogen oxides, sulfur-containing compounds, and high-molecular-weight hydrocarbons, emerge as an important factor in the effective growth and removal of particles as well as for the removal of the gases themselves. This has been recognized for some time, but still relatively few of the physical and chemical changes involved are understood at present. Thus, this aspect of aerosol dynamics in the atmosphere remains largely an open question at this time.

As can be seen readily from such crude considerations as introduced above, the problem of evaluating quantitatively the removal of aerosols is very complicated. Indeed, such characteristic scales as a residence time may have complex and diffuse definitions for aerosols.

7.6.3 Aerosol Aging without Vapor Condensation. For further analysis, consider a limiting case of an aging aerosol in the absence of new particle sources where the condensation and gas absorption are small. Many attempts have been made to explain the observed regularity in the shape of the aerosol distributions within these constraints. Some of the earlier theories relied only on Brownian coagulation and sedimentation to control the evolution of the spectrum [e.g., Junge (1963a) and later Friedlander (1960)]. However, Junge (1969) pointed out that these two mechanisms cannot be sufficient to account for tropospheric aerosol behavior. Other processes such as turbulent coagulation, washout by rain, and rainout by cloud droplets are important.

In trying to construct a useful model for the kinetics of the aerosol spectrum, Friedlander (1960, 1964) postulated that a steady state may be achieved in the atmosphere. In this model, the similarities in shape of the spectrum can be accounted for qualitatively by a balancing between particles entering the lower end of the spectrum by Brownian coagulation, and by removal from the upper end of the spectrum by sedimentation (see Fig. 7.5). Junge (1969) has termed the resultant of these two mechanisms a quasi-steady state or quasi-stationary distribution (QSD).

A hypothesis suggesting that the tropospheric aerosol particle size distribution take a self-preserving form (SPD) was advanced, based in part on an empirical correlation derived by Clark and Whitby (1967). They used the dimensionless variable defined in Eq. (3.63) to "scale" a large number of tropospheric aerosol distributions and obtained a universal correlation for such distributions, which supported the hypothesis that self-preservation of size distributions may be achieved in the atmosphere (Fig. 7.32).

To construct a theory for rationalizing the "universal" dimensionless form of tropospheric particle size distributions, a mathematical model can be adopted which incorporates the more important mechanisms and should apply to many cases for describing the average behavior of aerosols near the ground in the absence of condensation growth (Hidy, 1972). For an elemental volume of aerosol, the change in number distribution is given, following Eq. (3.11b), as

$$\frac{\partial n}{\partial t} = \frac{1}{2} \int_0^v b(\vartheta, v - \vartheta) n(\vartheta, z, t) n(v - \vartheta, z, t) d\vartheta$$

$$- \int_0^\infty b(v, \vartheta) n(v, z, t) n(\vartheta, z, t) d\vartheta + \hat{S} v^{2/3} \frac{\partial n}{\partial z} + D_t \frac{\partial^2 n}{\partial z^2}, \qquad z \leq 1 \text{ km};$$

$$\tag{7.21}$$

$$b(v, \vartheta) = \frac{2}{3} \frac{kT}{\mu_g} \left[(\vartheta^{1/3} + v^{1/3}) \left(\frac{1}{\vartheta^{1/3}} + \frac{1}{v^{1/3}} \right) \right]$$

$$+ 0.31 \left(\frac{\varepsilon_t}{\nu} \right)^{1/2} (\vartheta^{1/3} + v^{1/3})^3. \tag{7.22}$$

Here, the rate of change of particle concentration in the (volume) size range v, $v + dv$, is balanced by the gain by thermal coagulation of particles of volumes $\tilde{v}$ and $v - \tilde{v}$, by the loss of particles resulting from thermal coagulation of particles of all sizes $\tilde{v}$ colliding with v, by sedimentation, and by turbulent diffusion; $\hat{S} = 2\rho_p g/9\mu_g$. In the collision mechanism, electrical effects, chemical reactions, and noncontinuum effects have been disregarded as a practical first approximation.

Since there is a steady flow of particles entering the atmosphere at the ground, and these particles are readily mixed vertically into the air, it is likely that the aerosol in the surface layers of the atmosphere tends to achieve a quasi-steady state (QSD), or $\partial n/\partial t \approx 0$ in Eq. (7.21).

If the aerosol cloud retains a steady state at higher altitudes from the ground, a simpler model than Eq. (7.21) may be written for aerosols at cloud base and above based on the scaling in Tables 7.33–7.36. Above the ground, both the diffusional transport and sedimentation have to be present to achieve an approach to a steady state. Yet the aerosol spectrum should be dominated by the collision processes, particularly rainout. The sedimentation term remains a factor for removing particles continuously at very large particle radius ($R \gtrsim 50 \ \mu$m), but the dynamic term is still very small, as presumably is the convective diffusion term, compared with the collision terms, in contrast to the case near the ground. Then, for heights above 2 km, a range of the size spectrum will be governed by

$$\frac{1}{2} \int_0^v b(\tilde{v}, v - \tilde{v}) n(\tilde{v}, z) n(v - \tilde{v}, z) \, d\tilde{v} - \int_0^\infty b(v, \tilde{v}) n(v, z) n(\tilde{v}, z) \, d\tilde{v} \approx 0.$$

$$(7.23)$$

In Eq. (7.23) b is generalized to incorporate all significant collision processes, including those associated with cloud particle and aerosol particle processes. It is presumed that solutions to analytical models like Eqs. (7.21) and (7.23) will contain the Junge subrange, $n(R) \sim R^{-4}$.

Based on empirical observational evidence such as the results described by Clark and Whitby (1967), the aerosols in largely urban atmosphere tend to adjust in such a way near the ground as to be both self-preserving over a wide range of size and in a quasi-steady state. Further evidence of such a situation is seen in the averaged results, at least for rural aerosol distributions as a function of altitude, of Blifford and Ringer (1969). The bulk of their data can also be placed in the self-preserving framework, as indicated by Fig. 7.32. Then, from such data there exists the tendency to achieve or at least retain a "universal" distribution function over a wide range of particle size and number concentration in terms of the similarity variables

$$\Psi_R(\eta_R) = nV^{1/3}/N^{4/3} \tag{7.24}$$

and

$$\eta_R = (N/V)^{1/3} R,$$

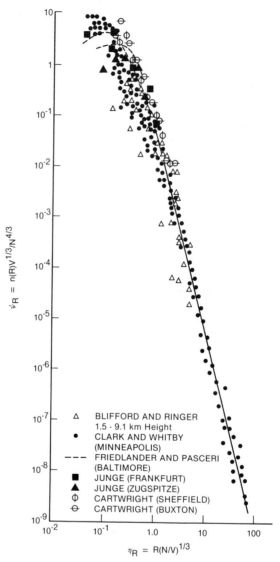

Fig. 7.32. Correlation of aerosol distributions observed primarily in polluted areas. Investigators are listed; references to original studies are found in Clark and Whitby (1967); aircraft data of Blifford and Ringer (1969) are also included.

which are equivalent to those defined in terms of volume, $\Psi = nV/N$ and $\eta = Nv/V$. Similarly the data of Clark and Whitby (1967), Peterson (1969), and Blifford and Ringer (1969) tend to confirm that the QSD model is applicable to correlation of the behavior of airborne particle distributions in the troposphere, purely on observational grounds.

The QSD constraints on the form of the particle size distribution, if Friedlander's (1960) arguments hold, have been extended by Jeffrey (1981). If the size distribution of atmospheric particles is maintained by a steady current of small particles coagulating to larger ones such that a small size range of the spectrum is dominated by coagulation and a coarse size range is dominated by sedimentation, then inspectional analysis gives (Jeffrey, 1981) a coagulation subrange (large particles):

$$n \propto I_0^{1/2} b^{-1/2} v^{-(\gamma+2)/3}, \tag{7.25}$$

where I_0 is taken as a (source or) current of small particles generated by condensation,

$$I_0 = \int_0^{v_1} v I(v) \, dv,$$

v_1 being the small particle limit of the coagulation subrange. The power-law form for n ($\propto R^{-\gamma}$) in this subrange then is found to be given by the exponent $\gamma_B = \frac{5}{2}$, $\gamma_S = 4$, or $\gamma_G = \frac{9}{2}$, depending on whether Brownian, shear, or gravitational collisions predominate.

The model of Jeffrey indicates that the sedimentation subrange for balancing the particle growth by coagulation with fallout is more complicated than Friedlander's form. A sedimentation subrange (giant particles) with coagulation by differential sedimentation gives

$$n \propto \left(\frac{\mu_g}{g\rho_p}\right)^{1/2} v^{-13/6} \left(K(I_0) - 0.04 \frac{g\rho_p}{\mu_g} v^{1/3} \frac{\partial n}{\partial z}\right)^{1/2}, \tag{7.26}$$

where $K(I_0)$ is a constant chosen such that the particle flow is maintained at I_0 at $v = v_2$, the particle volume at the lower end of the sedimentation subrange (equivalent to about 20 μm particles). For a condition where $\partial n / \partial z \to 0$, the exponent γ is 4.5 compared with Friedlander's estimate of 4.75. This suggests a similar falling off of the spectrum for Jeffrey's model in the limit of no-vertical-gradient coarse particles.

Self-Preservation and Altitude Dependence. Instead of defining the variables Ψ and η in terms of time as considered in earlier work, suppose a tendency toward self-preservation accompanies the tendency toward a steady state in the troposphere. If the variation in N and V can be attributed only to altitude, then Ψ and η depend only on the vertical coordinate z and not time. Hidy (1972) has reported an equation for a QSD form based on Eq. (7.21), but there are presently insufficiently detailed data for the number and mass distributions of aerosols as a function of altitude to determine if all the constraints leading to the QSD conditions are satisfied, or nearly satisfied, under atmospheric situations.

Even after extensive simplification of the dynamic equation for aerosols near the ground, the resulting integrodifferential equation is very complicated, and no complete solutions are currently available. However, it is possible to obtain approximate solutions for different extreme parts of the spectrum. Swartz (1968) has found an expression for the case where Brownian coagulation dominates turbulent collision processes. This may apply to tropospheric conditions near the ground, for particles less than ~0.5 μm radius, provided turbulence is not too intense. This is probably most applicable above the surface air layer ($z \gg 100$ m). Under such conditions, the self-preserving solution for Brownian coagulation with turbulent shear yields a distribution function of essentially the same shape as that for Brownian motion alone. Extrapolating this idea to atmospheric conditions should be useful as an approximation for the accumulation range of the atmospheric particle size distribution. The asymptotic solutions to the QSD equation given by Swartz should be applicable. As an approximation, Swartz found that

$$\Psi(\eta) \approx \text{const} \times \eta^{-\hat{A}/\hat{B}}, \qquad \eta \gg 1,$$

where

$$\hat{A} = \frac{V}{N^3}\left(\frac{V}{N}\right)^{2/3}\frac{d(N^2/V)}{dz}, \qquad \hat{B} = \frac{V}{N^2}\left(\frac{V}{N}\right)^{2/3}\frac{d(N/V)}{dz}. \qquad (7.27)$$

For the ratio of the "constants" $\hat{A}$ and $\hat{B}$ to be positive, $dN/dz > (N/V) \times dV/dz$ from the definitions in Eq. (7.27). The limited data of Blifford and Ringer (1969) and Peterson (1969) suggest that this is likely to be the case on the average, but may not be true on any one given sampling day.

Another interesting feature of the ratio $\hat{A}/\hat{B}$ is that it becomes identically 2 if $dV/dz \to 0$. In other words, the Junge subrange is found at very large η. The results of Blifford and Ringer indicated that $V(z) \to$ const for an average of the samples taken during their experiments, and Peterson's data indicate dV/dz varies much less than dN/dz with height. To ensure that the ratio $\hat{A}/\hat{B}$ must be greater than 2, it is implied that $V(z) \neq$ const exactly, or that the conditions of SPD and QSD may not be entirely satisfied. Jeffrey's (1981) result, Eq. (7.26), appears to be consistent with this conclusion.

From these theoretical results, the Junge subrange appears to be approached at high values of η under conditions observed near the ground, where V can be nearly constant. Unfortunately, as yet no solutions are available for small values of η to determine how wide a range of R^{-4} can be derived from the model.

At higher altitudes, Eq. (7.23) applied with collection in connection with rainout dominates, and the steady state solution of Liu and Whitby (1968) is appropriate. Their result indicates that Eq. (7.23) should predict a range of η^{-2} (the Junge subrange) over a portion of the spectrum, regardless of the form of the collision parameter b. As in the case for the near-ground model, however, it is not possible to establish over how wide a range this solution would be applicable.

7.6.4   Importance of Particle Source Distribution.   The explanation of regularities in the size distribution of tropospheric aerosols by means of the similarity theory remains incomplete. If one accepts Junge's (1969) arguments that the time available to achieve SPD and QSD solely by atmospheric processes is too long, then one must look to the relation of particle sources to the atmospheric aerosol spectrum. It is well known that many classes of particle distributions are nearly lognormal in shape, but some are fit well by a power-law form (Lerman, 1979). Such distributions are similar to the atmospheric aerosol spectrum over a range of sizes.

Measurements of size distributions from different sources often have a power-law form, and can be normalized to the SPD coordinates. Thus, the explanation for the remarkable similarity in atmospheric size distributions may be related to the statistical process of blending many similar distributions from different sources. The dynamic mechanisms in the atmosphere then would serve only to maintain the shape of a mixture of source-enriched distributions.

Friedlander (1970) has shown that if all source distributions are self-preserving, following Eq. (7.1) and the correlation in Fig. 7.32, the distribution resulting from mixing the sources together will also be self-preserving. His argument follows: Suppose that the aerosol mixture is composed of $\mathcal{V}_i$ volumes of aerosols from many different sources ($i = 1, 2, \ldots, k$), with values of $n_{R_i}(R)$ and V_i. Multiplying Eq. (7.1), rewritten for radius with the constant $= 0.05$, for each volume by $\mathcal{V}_i$ and summing over all volumes, we have

$$\sum_{i=1}^{k} n_{R_i}(R)\mathcal{V}_i = 0.05 \sum_{i=1}^{k} V_i\mathcal{V}_i R^{-4}.$$

However,

$$\sum_{i=1}^{k} n_{R_i}(R)\mathcal{V}_i = n_R(R)\mathcal{V},$$

where $\mathcal{V} = \sum_{i=1}^{k} \mathcal{V}i$; because of conservation of particle number in the size range $R, R + dR$. Further,

$$\sum_{i=1}^{k} V_i\mathcal{V}_i = V\mathcal{V}, \tag{7.28}$$

since the volume of material in the dispersed phase is conserved. Thus, the form of source size spectra is preserved in mixing for the final blend. In this respect, a slow rate of coagulation in the atmosphere need not prevent such processes from dominating the spectral shape if they are more active at the source of particle formation.

The question arises as to why the size spectra from various sources tend to achieve the shape displaying the R^{-4} dependence. By going through arguments equivalent to those used in establishing the dominant processes in the atmosphere, one finds that the processes of convective diffusion and collision by Brownian motion are dominant in the sources, particularly where intense turbulent mixing is involved. However, despite the much larger turbulent intensity and larger number of particles concentrated in source effluents, turbulent coagulation remains comparatively weak. This is readily illustrated for several sources by Hidy (1972). Consideration of the data characterizing collisional time scales and diffusional times suggests that they are not sufficient in most sources considered to achieve self-preservation.

The formation of spectra of a universal shape from different sources remains unexplained on the basis of the state of development of the theory of self-preserving spectra. However, some further clues may be derived from results of numerical experiments of Mockros *et al.* (1967) and Takahashi and Kasahara (1968). The results of these investigators indicate, without assuming self-preservation, that solutions to the kinetic equations including loss or gain terms other than collisional terms give asymptotic shapes in power-law form such as $R^{-\gamma}$. Such numerical calculations should be expanded as aids to better understanding of evolving aerosol particle distributions.

7.6.5 Self-Preservation and Vapor Condensation. The scaling of dynamic processes influencing atmospheric particle size distributions clearly demonstrates that the evolution can be influenced strongly by particle growth from condensation of vapors or by chemical reaction. Growth processes are particularly important in urban environments, or in situations where the reactivity of air is intense, as in photochemical smog. When growth takes place, there is a substantial accumulation of material in the range of fine particles in the range between 0.05 and 1 μm diameter. With growth of particles, the dynamic equation for the change in the size distribution function with time includes an additional term for the gradient of the particle current in v-space [c.f., Eq. (3.11b)].

For supersaturated vapor condensing in large particles, the current

$$I_0 = nv \propto \chi v^{1/3}, \tag{7.29}$$

where χ is defined in terms of the parameters in Eq. (3.65).

If the similarity parameter $\mathcal{B}(\mathcal{S} - 1)$ is constant, then one would expect self-preservation to occur in an aerosol cloud growing by vapor condensation and coagulation. Although Husar and Whitby's experiments indicated that such a process can occur in a closed vessel, no observations of such a system have been reported for the atmosphere.

It is possible that atmospheric conditions will involve three different domains for a growth-dominated aerosol. These have been identified by Friedlander (1978) in reference to aerosol formation in polluted air contained in a chamber irradiated with sunlight. The regimes are shown schematically in Fig. (3.15). In the first domain, new particles are formed by homogeneous or heteromolecular nucleation. In this domain, the number concentration of particles rapidly increases along with the volume fraction and total particle surface area. In the second domain, the coagulation rate begins to become rapid enough with large numbers of small particles to deplete the number of new particles formed that a maximum in N is realized. In this regime, nucleation continues, and, although coagulation actually depletes the total number of particles, the volume and surface area of particles continue to grow. The surface area eventually reaches a constant level, and finally a third domain is reached in which condensation continues on existing particles which act as nucleation centers, and coagulation continues to deplete numbers of small nuclei. Here the volume fraction continues by condensation growth, but the surface area approaches an asymptotic value after sufficiently long times. This is predicted in the SPD model of Pich *et al.* (1970). In the latter stage of evolution of aerosol-dominated condensation growth and coagulation, McMurry and Friedlander (1978) have shown that self-preservation of the size distribution and the surface area evolves into the form

$$S = 6.23 \times 10^3 V'^{3/5}, \tag{7.30}$$

where S is in reciprocal centimeters and V' is the rate of volume fraction growth and has the dimensions of cubic centimeters per cubic centimeters of air–second. This form comes directly from the self-preservation model for

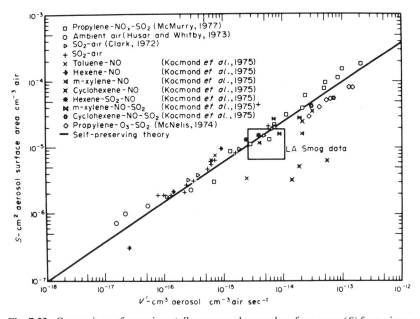

Fig. 7.33. Comparison of experimentally measured aerosol surface areas (S) for a given rate of aerosol formation I, with surface areas predicted from theory. A time of one hour was chosen for the theoretical line in accordance with laboratory time scales required to achieve constant surface areas. The range of values for Los Angeles smog was reported by Husar and Whitby (1973). [From Friedlander (1978); reprinted with permission of Pergamon.]

the size distribution. A correlation following Eq. (7.30) of smog chamber results for different reaction systems is shown in Fig. 7.33. A range of data is also included for Los Angeles aerosol conditions, reported by Husar and Whitby (1973). This result, along with the fact that the actual size distribution is normalized in dimensionless form, provides indirect circumstantial evidence for the applicability of a self-preserving description of the atmospheric aerosol in a reactive environment where growth of particles dominates the behavior of the accumulation range of the particle size distribution.

Jeffrey's (1981) model also is of interest in considering the QSD constraints on a condensation subrange for the atmospheric particle size spectrum. For the case where the particle volume is less than the lower size in the QSD coagulation subrange

$$n \propto I_0^{1/2}(\mu_g/kT)^{1/2}v^{-3/2} + \tfrac{9}{2}C(\mu_g/kT)v^{-5/4}, \tag{7.31}$$

where the coefficient C depends on the saturation vapor pressure of condensable species, p_s. The presence of a spectrum of this theoretical form has not been verified yet for atmospheric aerosols.

REFERENCES

Ahrens, L. H. (1978). "Origin and Distribution of Elements." Pergamon, Oxford.
Aitken, J. (1923). "Collection of Scientific Papers." Cambridge Univ. Press, London and New York.
Altshuller, A. P. (1973). *Environ. Sci. Technol.* **7**, 709.
Atkinson, R., Lloyd, A. C., and Winges, L. (1982). *Atmos. Environ.* **16**, 1341.
Baranov, A. V., Liberzu, E. A., and Popova, T. I. (1966). *Tr. Sib. Technol. Inst.* **38**, 77.
Beilke, S., and Gravenhorst, G. (1978). *Atmos. Environ.* **12**, 231.
Bernstein, D., and Rahn, K. (1979). *Ann. N.Y. Acad. Sci.* **322**, 87.
Bigg, E. K. (1976). *J. Atmos. Sci.* **33**, 1080.
Blanchard, D., and Woodcock, A. H. (1957). *Tellus* **9**, 145.
Blanchard, D., and Woodcock, A. H. (1980). *Ann. N.Y. Acad. Sci.* **338**, 330.
Blifford, I. H., Jr., and Ringer, L. (1969). *J. Atmos. Sci.* **26**, 716.
Borok, M. T. (1960). *An. Prikl. Khim.* **33**, 1761.
Bowersox, V. C., and de Pena, R. (1980). *J. Geophys. Res.* **85**, 5614.
Brosset, C. (1978). *Atmos. Environ.* **12**, 25.
Brownlee, D. E. (1978). *In* "Protostars and Planets" (T. Gehrels, ed.), p. 134. Univ. of Arizona Press, Tuscon.
Brownlee, D. E., Tomandl, D. A., Hodse, P. W., and Horz, F. (1974). *Nature (London)* **252**, 667.
Bufalini, M. J. (1971). *Environ. Sci. Technol.* **5**, 686.
Cadle, R. D. (1973). *In* "Chemistry of the Lower Atmosphere" (S. I. Rasool, ed.), p. 69. Plenum, New York.
Calvert, J. G. (1974). *Proc. Conf. Health Eff. Pollut., 1973*, Ser. No. 93-15, p. 19.
Calvert, J. G., and McQuigg, R. (1975). *Int. J. Chem. Kinet., Symp. No. 1*, p. 113.
Calvert, J. G., and Stockwell, W. (1982). *In* "Acid Precipitation: SO$_2$, NO, and NO$_2$ Oxidation Mechanisms: Atmospheric Considerations." Ann Arbor Sci. Publ., Ann Arbor, Michigan (to be published).
Calvert, J. G., Su, F., Bottenheim, J., and Strausz, O. P. (1978). *Atmos. Environ.* **12**, 197.
Cameron, A. G. W. (1973). *Space Sci. Rev.* **15**, 121.

Castleman, A. W., Jr. (1982). *J. Aerosol Sci.* **13,** 73.

Castleman, A. W., Jr., Munkewitz, H. R., and Manowitz, B. (1974). *Tellus* **26,** 222.

Cautreels, W., and Van Cauwenberghe, K. (1978). *Atmos. Environ.* **12,** 1133.

Chamberlain, A. C. (1966). *Proc. R. Soc. London Ser. A* **290A,** 236.

Chamberlain, A. C. (1967). *Proc. R. Soc. London Ser. A* **296A,** 45.

Chameides, W. L., and Davis, D. D. (1982). *J. Geophys. Res.* **87,** 4863.

Chang, S. G., Toosi, R., and Novakov, T. (1981). *Atmos. Environ.* **15,** 1287.

Changnon, C. W., and Junge, C. E. (1961). *J. Meteorol.* **18,** 746.

Clark, W. E. (1972). Ph.D. Dissertation, Dept. of Mech. Eng., Univ. of Minnesota, Minneapolis.

Clark, W., and Whitby, K. T. (1967). *J. Atmos. Sci.* **24,** 677.

Charlson, R., Covert, D., Larson, T., and Waggoner, A. (1978). *Atmos. Environ.* **12,** 39.

Cogbill, C. V., and Likens, G. E. (1974). *Water Resour. Res.* **10,** 1133.

CMBEEP (1977). "Ozone and Other Photochemical Oxidants," p. 51. Natl. Acad. Sci., Washington, D.C.

Corn, M., and Reitz, R. (1968). *Science* **159,** 1350.

Cox, R. A., and Penkett, S. A. (1972). *J. Chem. Soc., Faraday Trans.* **68,** 1735.

Cox, R. A., and Sandalls, F. J. (1974). *Atmos. Environ.* **8,** 1269.

Crutzen, P. J. (1976). *Geophys. Res. Lett.* **8,** 73.

Currie, L. A. (1982). *In* "Particulate Carbon Atmospheric Life Cycle" (G. T. Wolff and R. L. Klimisch, eds.), p. 245. Plenum, New York.

Dagani, R. (1981). *Chem. Eng. News* **29,** 25.

Daisey, J. M., Leyko, M. A., Kleinman, M. T., and Hoffman, E. (1979). *Ann. N.Y. Acad. Sci.* **322,** 125.

Dana, M. T., and Hales, J. M. (1976). *Atmos. Environ.* **10,** 45.

Davis, D. D., Payne, W. A., and Steif, L. J. (1973). *Science* **179,** 280.

Dingle, N., and Lee, Y. (1972). *J. Appl. Meteorol.* **11,** 877.

Draper, N. R., and Smith, H. (1966). "Applied Regression Analysis," pp. 147–150. Wiley, New York.

Eisenbud, M. (1980). *Ann. N.Y. Acad. Sci.* **338,** 599.

Electric Power Research Institute (1982). Rep. EA-2165-SY-LD, p. 24. Palo Alto, California.

Environmental Data Service (1972–1976). Atmospheric Turbidity and Precipitation Chemistry Data for the World; Global Monitoring of the Environment for Selected Atmospheric Constituents. U.S. National Climate Center, Asheville, North Carolina.

Essenwanger, O. M. (1976). "Applied Statistics in Atmospheric Science. Part A. Frequencies and Curve Fitting," pp. 252–288. Elsevier, Amsterdam.

Faoro, R., and McMullen, T. (1977). EPA-450/1-77-003. Office of Air Quality Planning and Standards. U.S. Environ. Prot. Agency, Research Triangle Park, North Carolina.

Farlow, N., Ferry, G., Lem, H. Y., and Hayes, D. M. (1979). *J. Geophys. Res.* **84,** 733.

Farlow, N., Oberbeck, V., Snetsinger, K., Ferry, G., Polkowski, G., and Hayes, D. (1981). *Science* **211,** 832.

Fletcher, N. H. (1966). "The Physics of Rainclouds." Cambridge Univ. Press, London and New York.

Flowers, E. C., McCormick, R. A., and Kurfis, K. R. (1969). *J. Appl. Meteorol.* **8,** 955.

Friedlander, S. K. (1960). *J. Meteorol.* **17,** 373 and 479.

Friedlander, S. K. (1964). *In* "Aerosols; Physical Chemistry and Applications" (K. Spurny, ed.), pp. 115–130. Czechoslovak Acad. of Sci., Prague.

Friedlander, S. K. (1970). The Characterization of Aerosols Distributed with Respect to Size and Chemical Composition." (Unpublished manuscript.)

Friedlander, S. K. (1978). *Atmos. Environ.* **12,** 187.

Friend, J. P. (1966). *Tellus* **18,** 465.

Friend, J. P., Leifer, R., and Trichan, M. (1973). *J. Atmos. Sci.* **30,** 465.

Fruchter, J. S., Robertson, D. E., Evans, J. C., Olsen, K. B., Lepel, E. A., Laul, J. C., Abel,

K. H., Sanders, R. W., Jackson, P. O., Wogman, N. S., and Perkins, R. W. (1980). *Science* **209**, 1116.

Füchtbauer, G., and Müller, T. (1970). "Sedimente und Sedimentgesteine," Fig. 408, p. 141. Schweizerbartsche Verlagbuchhandlung, Stuttgart.

Fuller, E. C., and Christ, R. H. (1941). *J. Am. Chem. Soc.* **63**, 1644.

Ganapathy, R., Brownlee, D. E., and Hodge, P. W. (1978). *Science* **201**, 1119.

Garrels, R. M. MacKenzie, F. T., and Hunt, C. (1975). "Chemical Cycles and the Global Environment." William Kaufman, Inc., Los Altos, California.

Gartrell, G., Jr., Heisler, S. L., and Friedlander, S. K. (1980). *In* "The Character and Origins of Smog Aerosols" (G. M. Hidy and P. K. Mueller, eds.), p. 665. Wiley, New York.

Gatz, D. (1966). Ph.D. Thesis, Univ. of Michigan, Ann Arbor.

Gatz, D. (1977). *ERDA Symp. Ser., No. 41.*

Gedeonov, L. I., Gritchenko, Z. G., Flegontov, F. M., and Zhilkana, M. I. (1970). *In* "Atmosphere Scavenging of Radioisotopes," Proc. Conf. in Patanga, USSR, translated from *Issled. Protsessov Samoochischeniya Atmos. Radioakt. Izot.* by Ldereman, NTIS TT-69-55099, Springfield, Virginia.

Georgii, H. W. (1960). *Geofis. Pura Appl.* **47**, 155.

Georgii, H. W. (1970). *J. Geophys. Res.* **75**, 2365.

Gerrard, W. (1976). "Solubility of Gases and Liquids." Plenum, New York.

Gillette, D. (1980). *Ann. N.Y. Acad. Sci.* **338**, 348.

Gillette, D., and Blifford, I. H., Jr. (1971). *J. Atmos. Sci.* **28**, 1199.

Graedel, T. E., and Weschler, C. J. (1981). *Rev. Geophys. Space Phys.* **19**, 505.

Graedel, T. E., Gill, P. S., and Weschler, C. J. (1983). *In* "Proceedings of the 4th International Conference on Precipitation Scavenging, Dry Deposition, and Resuspension" (H. Pruppacher, R. Semonin, and W. G. N. Slinn, eds.). Elsevier, Amsterdam (to be published).

Granat, L. (1976). *SCOPE Rep.* **7**, 87–134.

Groblicki, P., and Nebel, G. J. (1971). *In* "Chemical Reactions in Urban Atmospheres, Proceedings Symposium, 1969" p. 176. (C. S. Tuesday, ed.), Elsevier, North-Holland, New York.

Grosjean, D., and Friedlander, S. K. (1980). *In* "The Character and Origins of Smog Aerosols" (G. M. Hidy and P. K. Mueller, eds.). p. 435. Wiley, New York.

Grosjean, D., and McMurry, P. (1982). Report No. PA-098-8 (NTIS No. PB-82-262262). Coordinating Research Council, Environ. Res. Technol., Westlake, California.

Hahn, J. (1980). *Ann. N.Y. Acad. Sci.* **338**, 359.

Hales, J. (1978). *Atmos. Environ.* **12**, 389.

Hammerle, J. R. (1976). *In* "Air Pollution" (A. Stern, ed.), 3rd ed., Vol. 3. Academic Press, New York.

Hansen, D. A., and Hidy, G. M. (1982). *Atmos. Environ.* **16**, 2107.

Harker, A. (1975). *J. Geophys. Res.* **24**, 3399.

Harmon, H. H. (1976). "Modern Factor Analysis," 3rd ed. Revised, Univ. of Chicago Press, Chicago, Illinois.

Hegg, D. A., and Hobbs, P. V. (1978). *Atmos. Environ.* **12**, 241.

Hegg, D. A., and Hobbs, P. V. (1981). *Atmos. Environ.* **15**, 1597.

Heicklen, J. (1976). "Atmospheric Chemistry." Academic Press, New York.

Heisler, S. L., and Baskett, R. (1981). Rep. P-5381-701, for the Calif. Air Resources Bd., Environ. Res. Technol., Westlake, California.

Heisler, S. L., and Friedlander, S. K. (1977). *Atmos. Environ.* **11**, 157.

Heisler, S. L., Henry, R. C., Watson, J. G., and Hidy, G. M. (1980a). The Denver Winter Haze Study. Report No. P5417F to Motor Vehicle Manufacturers Assoc., Environ. Res. and Technol., Inc., Westlake Village, California.

Heisler, S. L., Henry, R. C., Mueller, P. K., Hidy, G. M., and Grosjean, D. (1980b). Aerosol Behavior Patterns in the South Coast Air Basin with Emphasis on Airborne Sulfate. Report No. P-A085, to Southern Calif. Edison, Co., Environ. Res. and Technol., Inc., Westlake Village, California.

Henry, R. C., and Hidy, G. M. (1979). *Atmos. Environ.* **13**, 1581.

Henry, R. C., and Hidy, G. M. (1982). *Atmos. Environ.* **16**, 177.

Hering, S. V., and Friedlander, S. K. (1982). *Atmos. Environ.* **16**, 2647.

Hester, N. E. (1979). Final Report AMC 58001.15FR to Oregon Dept. of Environ. Quality. Rockwell Int. Corp., Newbury Park, California.

Hidy, G. M. (1972). *In* "Assessment of Airborne Particles" (T. Mercer, P. Morrow, and W. Stöber, eds.), p. 81. Thomas, Springfield, Illinois.

Hidy, G. M. (1973). *In* "Chemistry of the Lower Atmosphere" (S. I. Rasool, ed.), p. 121. Plenum, New York.

Hidy, G. M., ed. (1974). "Aerosols from Engine Effluents. Monograph III. Climatic Impact Assessment Program," Chap. 6. U.S. Dept. of Transportation, Washington, D.C.

Hidy, G. M., Principal Investigator (1975). Final Report No. SC524.25FR, Contract 358. Calif. Air Resources Bd., NTIS No. PB-24947, Sacramento, California.

Hidy, G. M., Principal Investigator (1976). Report No. EC-125. Electric Power Res. Inst., Palo Alto, California.

Hidy, G. M. (1982). *Water, Air, Soil Pollut.* **18**, 181.

Hidy, G. M., and Brock, J. R. (1967). *J. Geophys. Res.* **72**, 455.

Hidy, G. M., and Brock, J. R. (1970). "Proceedings of the International Clean Air Congress, 2nd" (H. Englund and W. Berry, eds.), p. 1088. Academic Press, New York.

Hidy, G. M., and Burton, C. S. (1975). *Int. J. Chem. Kinet., Symp. No. 1*, pp. 509–541.

Hidy, G. M., and Mueller, P. K., ed. (1980). "The Character and Origins of Smog Aerosols," pp. 17–54. Wiley, New York.

Hidy, G. M., Tong, E. Y., and Mueller, P. K. (1978a). *Atmos. Environ.* **12**, 735.

Hidy, G. M., Katz, J. R., and Mirabel, P. (1978b). *Atmos. Environ.* **12**, 887.

Hidy, G. M., Mueller, P. K., Wang, H., Karney, J., Twiss, S., Imada, M., and Alcocer, W. (1974). *J. Appl. Meteorol.* **13**, 96.

Ho, W. W., Hidy, G. M., and Govan, R. M. (1974). *J. Appl. Meteorol.* **13**, 871.

Hobbs, P. V., Radke, L., Eltgroth, M., and Hegg, D. (1981). *Science* **211**, 816.

Hulburt, H., and Akiyama, T. (1969). *Ind. Eng. Chem. Fundam.* **8**, 319.

Husar, R. B., and Whitby, K. T. (1973). *Environ. Sci. Technol.* **1**, 241.

Hutchinson, T. C., and Havas, M. (1980). "Effects of Acid Precipitation on Terrestrial Ecosystems." Plenum, New York.

Jaenicke, R. (1976). *In* "Fine Particles: Aerosol Generation, Measurement, Sampling, and Analysis" (B. Y. H. Liu, ed.), p. 467. Academic Press, New York.

Jaenicke, R. (1978). "Meteor" *Forschungsergeb., Reihe* B No. 13, pp. 1–9.

Jaenicke, R., and Schütz, L. (1978). *J. Geophys. Res.* **83**, 3585.

Jeffrey, D. J. (1981). *J. Atmos. Sci.* **38**, 2440.

Johnstone, H. F., and Coughanowr, D. R. (1968). *Ind. Eng. Chem.* **50**, 1169.

Judeikis, H. S., Stewart, T., and Wren, A. (1978). *Atmos. Environ.* **12**, 1633.

Junge, C. E. (1961). *J. Meteorol.* **18**, 501.

Junge, C. E. (1963a). "Air Chemistry and Radioactivity," Chap. 2. Academic Press, New York.

Junge, C. E. (1963b). *J. Rech. Atmos.* **1**, 185.

Junge, C. E. (1969). *J. Atmos. Sci.* **26**, 603.

Junge, C. E. (1979). *In* "Saharan Dust" (C. Morales, ed.), Chap. 2. Wiley, New York.

Kleinman, M. T., Tomczyk, C., Leaderer, B. P., and Tanner, R. L. (1979). *Ann. N.Y. Acad. Sci.* **332**, 115.

Knollenberg, R. G., and Hunten, D. M. (1980). *J. Geophys. Res.* **85**, 8039.

Knutson, E. O., and Stockham, J. D. (1977). *ERDA Symp. Ser.,* No. 41.

Kocmond, W., Kittleson, D., Yang, J., and Dermerjian, K. (1975). Study of Aerosol Formation in Photochemical Air Pollution. Report EPA 650/3-75-007. U.S. EPA, Research Triangle Park, North Carolina.

Langer, G. (1965). *Kolloid. Z.* **204**, 119.

Lazrus, A. L., and Gandrud, B. W. (1974). *J. Geophys. Res.* **79**, 3049.

Lazrus, A. L., Gandrud, B., and Cadle, R. D. (1971). *J. Geophys. Res.* **76**, 8083.
Lazrus, A. L., Haagenson, P., Kok, G., Hulbert, B., Likens, G., Mohnen, V., Wilson, W., and Winchester, J. (1981). Progress Report No. 1 on APEX. Report U.S. Environ. Prot. Agency.
Lazrus, A. L., Haagenson, P., Kok, G., Hulbert, B., Kreitzberg, C., Likens, G., Mohnen, V., Wilson, W., and Winchester, J. (1983). *Atmos. Environ.* **17**, 581.
Lee, Y.-N., and Schwartz, S. E. (1981). *J. Geophys. Res.* **86**, 11971.
Leighton, P. (1961). "Photochemistry of Air Pollution," pp. 7–126. Academic Press, New York.
Lerman, A. (1979). "Geochemical Processes: Water and Sediment Environment." Wiley, New York.
Likens, G. E. (1976). *Chem. Eng. News* **54**, 19.
Likens, G. E., and Butler, T. J. (1982). *Atmos. Environ.* **15**, 1103.
Little, P. (1977). *Environ. Pollut.* **12**, 293.
Liu, B. Y. H., and Whitby, K. T. (1968). *J. Colloid Interface Sci.* **26**, 161.
Lloyd, A. C. (1982). Report EA-1907. Electric Power Research Inst., Palo Alto, California.
Lumley, J., and Panofsky, H. (1964). "The Structure of Atmospheric Turbulence." Wiley (Interscience), New York.
McMurry, P. H. (1977). Ph.D. Thesis, California Inst. of Technol., Pasadena.
McMurry, P. H., and Friedlander, S. K. (1978). *J. Colloid Interface Sci.* **64**, 248.
McMurry, P. H., Rader, D., and Smith, J. (1981). *Atmos. Environ.* **15**, 2315.
McNelis, D. N. (1974). Aerosol Formation from Gas-Phase Reactions of Ozone and Olefin in the Presence of SO_2. Report EPA 650/4-74-034. U.S. Environ. Prot. Agency, Research Triangle Park, North Carolina.
Martin, L. R. (1982). *In* "Acid Precipitation: SO_2, NO and NO_2 Oxidation Mechanisms: Atmospheric Considerations." Ann Arbor Sci. Publ., Ann Arbor, Michigan.
Mason, B. J. (1971). "The Physics of Clouds," 2nd ed. Oxford Univ. Press, London and New York.
Mastenbrook, N. J. (1968). *J. Atmos. Sci.* **25**, 299.
Matteson, M., Stöber, S., and Luther, H. (1969). *Ind. Eng. Chem. Fundam.* **8**, 677.
Meixner, F. X. *et al.* (1981). *Geophys. Res. Lett.* **8**, 163.
Meszaros, A., and Vissy, K. (1974). *J. Aerosol Sci.* **5**, 101.
Meyer, R., Hidy, G. M., and Davis, J. H. (1973). *Environ. Lett.* **4**, 9.
Michaux, C. M. (1967). "Handbook of the Physical Properties of the Planet Jupiter," p. 71. SP-3031, NASA, Washington, D.C.
Miller, D. F. (1978). *Atmos. Environ.* **12**, 273.
Miller, M. S., Friedlander, S. K., and Hidy, G. M. (1972). *In* "Aerosols and Atmospheric Chemistry" (G. M. Hidy, ed.), pp. 301–311. Academic Press, New York.
Millman, P. M. (1979). *Naturwissenschaften* **66**, 134.
Mockros, L., Quon, J., and Hjelmfelt, A. T., Jr. (1967). *J. Colloid Interface Sci.* **23**, 90.
Moyers, J. L., Ranweiler, L. E., Hopf, S. B., and Korte, N. E. (1977). *Environ. Sci. Technol.* **11**, 789.
Mueller, P. K., and Hidy, G. M., Principal Investigators (1983). The Sulfate Regional Experiment: Report of Findings. Report No. EA-1901, Chap. 6. Electric Power Research Inst., Palo Alto, California.
Mueller, P. K., Hidy, G., Warren, K., Lavery, T., and Baskett, R. (1980). *Ann. N.Y. Acad. Sci.* **338**, 453.
Mueller, P. K., Fung, K., Heisler, S., Grosjean, D., and Hidy, G. (1982). *In* "Particulate Carbon: Atmospheric Life Cycle" (G. Wolff and R. Klimisch, eds.). p. 343. Plenum, New York.
Nash, T. (1970). *J. Chem. Soc. A, No. 18A*, p. 3023.
Natusch, D. F. *et al.* (1975). *Proc. Int. Conf. Heavy Metals Environ., Toronto, Ontario* **2**, Part 2, 553.

Nelson, N., Chairman (1975). Report of Ad Hoc Sulfate Review Panel. U.S. EPA, Office of the Administration, Washington, D.C.

Neuberger, H., Hosler, C. L., and Kocmond, W. C. (1967). *In* "Biometeorology" (S. W. Tromp and W. H. Weike, eds.), Part 2, pp. 696–702. Pergamon Press, New York.

Novakov, T., Chang, S. G., and Harker, A. B. (1974). *Science* **186**, 259.

O'Brien, R., Holmes, J., and Bockian, A. (1975). *Environ. Sci. Technol.* **9**, 568.

OECD (1977). "Long Range Transport of Air Pollutants: Measurements and Findings." OECD, Paris.

Öpik, E. J. (1958). "Physics of Meteor Flight in the Atmosphere." Wiley (Interscience), New York.

Page, G. L., Elseewi, A. A., and Straughan, I. R. (1979). *Residue Rev.* **71**, 83.

Penkett, S. A. (1972). *Nature (London), Phys. Sci.* **240**, 105.

Penkett, S. A., Jones, B., Brice, K., and Eggleton, A. (1979). *Atmos. Environ.* **13**, 123.

Peterson, C. (1969). Ph.D. Dissertation, Univ. of Minnesota, Minneapolis.

Pich, J., Friedlander, S. K., and Lai, F. (1970). *J. Aerosol Sci.* **1**, 115.

Pierson, W., Brachaczek, W., Truex, T., Butler, J., and Korniski, T. (1980). *Ann. N.Y. Acad. Sci.* **338**, 145.

Pollack, J. B., Colburn, D., Kahn, R., Hunter, J., Van Camp, W., Carlston, C., and Wolf, M. (1977). *J. Geophys. Res.* **82**, 4479.

Pollack, J. B., Roges, K., Tom, O. B., and Yung, Y. L. (1980). *Geophys. Res. Lett.* **7**, 829.

Prahm, L. P., Torp, U., and Stern, R. M. (1976). *Tellus* **28**, 355.

Prospero, J. M., and Carlson, T. N., (1972). *J. Geophys. Res.* **77**, 5255.

Pruppacher, H. R., and Klett, J. D. (1978). "Microphysics of Clouds and Precipitation." Reidel Publ. Co., London.

Radke, L., Smith, J., Hegg, D., and Hobbs, P. (1978). *J. Air Poll. Control Assoc.* **28**, 30.

Radke, L., Eltgroth, M., and Hobbs, P. (1980). *J. Appl. Meteorol.* **19**, 715.

Rahn, K., Borys, R., Shaw, G., Shütz, L., and Jaenicke, R. (1979). *In* "Saharan Dusts" (C. Morales, ed.), Chap. 13. Wiley, New York.

Rankama, K., and Sahama, Th. (1950). "Geochemistry," p. 189. Univ. of Chicago Press, Chicago, Illinois.

Reiter, E. R. (1978). Atmospheric Transport Processes, Part 4, U.S. Dept. of Energy, Technol. Inf. Cent., Springfield, Virginia.

Ripperton, L. A., Jefferies, H. E., and White, O. (1972). *Adv. Chem.* **113**, 219.

Robbins, R. C., and Cadle, R. D., and Eckhardt, D. L. (1959). *J. Meteorol.* **16**, 53.

Rosen, J. M. (1974). An Airborne Condensation Nuclei Counter, Unpubl. Report, Univ. of Wyoming, Laramie.

Rosinski, J. (1967). *J. Appl. Meteorol.* **6**, 1066.

Rosinski, J., and Nagamoto, C. T. (1965). *Kolloid Z.* **204**, 111.

Sandburg, J. S., Levaggi, D. A., DeMandel, R. E., and Siu, W. (1976). *J. Air Pollut. Control Assoc.* **26**, 559.

SCEP (1970). "Man's Impact on the Global Environment," Report on the Study of Critical Environmental Problems. MIT Press, Cambridge, Massachusetts.

Schroeter, L. C. (1963). *J. Pharm. Sci.* **52**, 559.

Schütz, L., and Jaenicke, R. (1974). *J. Appl. Meteorol.* **13**, 863.

Schütz, L., Ketserides, G., and Rahn, K. (1978). *Ges. Aerosolforsch. Jahrestag., 6th, 1978*, pp. 179–183.

Scott, W. D., and Hobbs, P. V. (1967). *J. Atmos. Sci.* **24**, 54.

Sehmel, G. (1980). *Atmos. Environ.* **14**, 983.

Shum, W. Y., and Loveland, W. D. (1974). *Atmos. Environ.* **8**, 645.

Sigurdsson, H. (1982). *EOS, Trans. AGU,* **63**, 601.

Slinn, W. G. N. (1977). *Water, Air, Soil Pollut.* **7**, 513.

Slinn, W. G. N. (1981a). *Atmos. Environ.* **16**, 1785.

Slinn, W. G. N. (1981b). *In* "Atmosphere Science and Power Production" (D. Randerson, ed.), Chap. 11. U.S. Dept. of Energy, Technol. Inf. Cent.

Small, M., Germani, M., Small, A., Zoller, W., and Moyers, J. (1981). *Environ. Sci. Technol.* **15**, 293.

SMIC (1971). "Inadvertent Climate Modification" Report in the Study of Man's Impact on Climates (SMIC). MIT Press, Cambridge, Massachusetts.

Smith, R. A. (1872). "Air and Rain," p. 601. Longmans, Green, New York.

Stearns, H. T. (1966). "Geology of the State of Hawaii," p. 140. Pacific Books, Palo Alto, California.

Stelson, A. W., Friedlander, S. K., and Seinfeld, J. H. (1979). *Atmos. Environ.* **13**, 369.

Stensland, G. (1980). *In* "Polluted Rain" (T. Y. Toribara, M. Miller, and P. Morrow, eds.), p. 87. Plenum, New York.

Stevens, R. K., Dzubay, T., Russwurm, G., and Rickel, D. (1978). *Atmos. Environ.* **12**, 55.

Sverdrup, G. M. (1977). Ph.D. Dissertation, Univ. of Minnesota, Minneapolis.

Swartz, C. (1968). M.S. Essay, California Inst. of Technol., Pasadena.

Swindel, B. F. (1974). *Am. Statistn.* **28**, 63.

Takahashi, K., and Kasahara, M. (1968). *Atmos. Environ.* **2**, 441.

Tanner, R. L., and Marlow, W. H. (1977). *Atmos. Environ.* **11**, 1143.

Torres, A. J., Maroulis, P., Goldberg, A., and Bandy, A. (1980). *J. Geophys. Res.* **85**, 7357.

Tymen, G., Butor, J. F., Renoux, A., and Madelaine, G. (1975). *Chemosphere* **4**, 357.

U.S. EPA (1978). National Air Quality, Monitoring and Emission Trend Report, 1977. EPA-450/2-78-052. Research Triangle Park, North Carolina.

U.S. EPA (1982). Air Quality Criteria for Particulate Matter and Sulfur Oxides, Vol. II. Office of Research and Development, Research Triangle Park, North Carolina.

U.S. Public Health Service (USPHS) (1969). Air Quality Criteria for Particulate Matter. U.S. Dept. of Health, Education, and Welfare, Washington, D.C.

U.S. Standard Atmosphere (1962). "Handbook of Geophysics and Space Environments, 1965." (S. L. Valley, ed.), McGraw-Hill, New York.

Wang, P. K., Grover, S., and Pruppacher, H. (1978). *J. Atmos. Sci.* **37**, 1735.

Wedding, J. B., Carlson, R., Stukel, J., and Bazzaz, F. (1977). *Water, Air, Soil Pollut.* **7**, 545.

Wedepohl, K. R., ed. (1969). "Handbook of Geochemistry," Vol. 1, p. 245. Springer-Verlag, Berlin and New York.

Weickmann, H. (1956). *In* "Artificial Stimulation of Rain" (H. Weickmann and W. Smith, eds.), p. 81. Pergamon, New York.

Wesely, M. L., and Hicks, B. (1977). *J. Air Pollut. Control Assoc.* **27**, 1110.

Wesely, M. L., Hicks, B., Dannevik, W., Frisella, S., and Husar, R. (1977). *Atmos. Environ.* **11**, 561.

Whelpdale, D., and Galloway, J. (1980). *Atmos. Environ.* **14**, 409.

Whitby, K. T. (1978). *Atmos. Environ.* **12**, 135.

Whitby, K. T., and Sverdrup, G. M. (1980). *In* "The Character and Origins of Smog Aerosols" (G. M. Hidy and P. K. Mueller, eds.), p. 477. Wiley, New York.

White, W., Heisler, S. L., Henry, R. C., and Hidy, G. (1978). *Atmos. Environ.* **12**, 779.

Wilson, W. E., Jr., Levy, A., and Wimmer, D. B. (1972). *J. Air Pollut. Control Assoc.* **22**, 27.

Winkler, P. (1969). *Ann. Meteorol.* **4**, 134.

Yamamoto, K., Michiham, S., and Kunitaro, K. (1973). *Nippon Kagaku Kaiski* **7**, 1268.

CHAPTER 8

EFFECTS ON THE EARTH'S ATMOSPHERE

In Chapter 7, the rich diversity of atmospheric aerosol chemistry was surveyed. The presence of suspended particles in the earth's atmosphere provides for a variety of natural phenomena that are of interest and represent an important part of aerosol science. Particulate matter in the air exerts an influence on the transfer of electromagnetic radiation through the atmosphere. This manifests itself in changes in visibility and coloration as a result of light scattering and absorption. The wealth of sky color, shadow, and haziness that provide a varied and often beautiful setting for natural objects and for man's architecture is a direct result of the influence of suspended particles interacting with visible light.

Changes in transfer of radiation in different layers of the atmosphere are the crux of the atmospheric energy storage process. Aerosol particles also play a role in distributing solar energy throughout the atmosphere and consequently in affecting climate. A distinctly different role also is played by aerosol particles in the atmosphere, which involves the formation of clouds of condensed water. Suspended particles basically provide the nuclei for condensation of moisture, and for the nucleation of ice crystals in super-cooled clouds. Thus, in a sense, aerosols provide a skeleton through which are derived water vapor rain clouds and precipitation. The opportunity then presents itself for both weather and climate modification by injection of particulate matter into the air.

These three groups of phenomena represent important areas of study, and are discussed in detail in this chapter as examples of particle effects on atmospheric phenomena.

8.1 VISIBILITY

Most of the information obtained through our sense of sight depends on perception of differences in intensity or color among the parts of the field of view. For example, an object is recognized by color and brightness of differences over its surface, and in comparison with some background. The recognition of objects depends on the specific association of variation in brightness and color with the object.

Visibility in the atmosphere can be encapsulated in a phrase from Larcom's poem, *Black in Blue Sky:* "There is light in shadow and shadow in light, and black in the blue of the sky. . . ." As a dark object (or shadow) recedes from an observer, its brightness is dimmed by loss of light scattered or absorbed by the intervening atmosphere. The resultant effect is for objects to eventually blend into the horizon. In a descriptive sense, when an observer views an object near the horizon through a haze, he sees the object's outline but its texture (fine detail) may become obscured or its colors softened. The haze itself may be distinguishable as a dark or light layer, which takes on a gray to grayish-brown appearance. The effect of the haze is to soften the object, causing both the object and its background to approach the horizon brightness (Fig. 8.1). Both the object and its contrast with the background decrease, making the object more difficult to see. Hence the haze of the intervening air affects the observer by changing the brightness of viewed objects, their texture, and their colors. The ingredients of visibility then contain terms of human sensations, "brightness" and "color" which are psychophysical in character (Wyszecki and Stiles, 1967). These are parameters of human consciousness and cannot be measured by one instrument. The antecedents of brightness and color which can be measured physically are the amount of light energy entering the eye and its distribution among different wavelengths. Thus two separable phenomena are involved in visibility. First is the optical properties of the atmosphere as a semitransparent medium, and second is the psychophysics of the human eye–brain system.

A third element in visibility involves the psychology of a value judgment. Until recently, the practical application of visibility, as found in its definition,[†] involves considerations of distance of large objects and their implications as safety hazards to transportation. For many years, visibility records

---

[†] From Webster's *Third New International Dictionary:* ". . . the degree or extent to which something is visible (as by the degree of clearness of the atmosphere); the mean greatest distance prevailing over the range of more than half of the horizon at which a large object (as a building or ship) may be seen and identified depending upon its size, distance from the observer, the contrast between it and surrounding objects, glare, transparency and illumination of the atmosphere between the object and the observer, and the condition of the observer's eye unaided by special optical devices . . . a measure of the ability of radiant energy to evoke visual sensation: the luminous efficiency of light of a specified wavelength."

Fig. 8.1. Two examples of scenic vistas with aesthetic value. The presence of haze softens the features of landscape (a), blending the background and topography into the lake below. The sharp texture of mountain peaks impresses the viewer of atmospheric clarity at high altitude (b).

have been maintained at airports, for example, as an assistance to those responsible for monitoring air traffic safety.

Visibility impairment also has been recognized for sometime as a symptom of air pollution. The reduction of visibility resulting from the turbidity of polluted air has been experienced by nearly everyone. The inability to see great distances is considered to be an adverse effect and has caused numerous complaints to air pollution control authorities. The distance one can see is a parameter by which nearly everyone at one time or another gauges the severity of air pollution. Early links between suspended particulate matter and air pollution recognition were measured in cities, as indicated in Fig. 8.2. A survey conducted in St. Louis indicated that 70 and 75% of the survey population considered visible pollution a nuisance when TSP levels were 120 and 160 $\mu g/m^3$, respectively. None of the population appeared to be bothered at TSP levels below 50 $\mu g/m^3$. A definite empirical relationship was obtained which has a physical basis as will be seen later. The degradation in visual quality was observed to be proportional to suspended-particulate concentration in St. Louis. Similar relationships are at least implied if not well documented in other cities in the world.

The issue of aesthetics and visual perception has emerged at the national level in 1977. One of the amendments to the United States Clean Air Act legislation states that improvement in visibility associated with reduction in air pollution in certain pristine areas is a national goal (95th Congress, 1977). A simple statement of the ideal to preserve the wealth of scenic beauty in the United States, such as the Grand Canyon vistas, has created a variety of scientific questions about the measurement of changes in natural visual images. The psychology of placing human values on such changes thus becomes crucial to administration of public policy. Recent studies have begun to explore the methodology for relating human values. These have been discussed in a series of workshops, for example, Fox *et al.* (1979).

To follow the logic of visibility theory, let us consider first the atmospheric

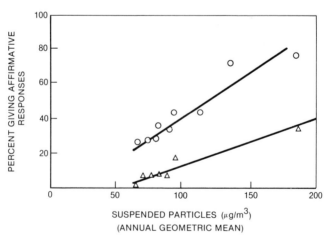

Fig. 8.2. Proportion of population in St. Louis stating that visible air pollution was present in their area of residence, and proportion of St. Louis population noticing air pollution has special connotation. [From USPHS (1968).]

optics, considering the human eye–brain system as a passive light detector.[†]
Then the discussion can be extended to include some elements of the psychophysics and psychology of seeing. The section on visibility is then completed with a brief survey of observations of visual effects as they are associated with aerosols.

An inventory of suspended atmospheric material reveals the presence of
diverse gases, liquid droplets, and solid particles. These contaminants affect
the transmission of light, and thereby the visibility. Perhaps one of the more
disagreeable effects is the brownish coloration of the atmosphere effected by
photochemical smog. This arises in some cases from the absorption of light
of short wavelengths by gaseous nitrogen dioxide (NO_2), if NO_2 concentrations are high (probably above 0.1 ppm) over at least a 10-km viewing path.
In other cases, the discoloration arises from absorption- and/or wavelength-
dependent scattering of light by suspended particles including elemental or
graphitic carbon particles or iron oxide particles. Perhaps most important is
the scattering of light by particles and droplets in rural areas.

Visibility is reduced because the atmosphere between an observer and a
distant object scatters light which comes from the sun and other parts of the
sky; some of this scattered light enters the eyes of the observer. When the
contrast between objects is reduced because of more scattered light, the
visibility is reduced. The essential features of the scattering of light by aerosol particles are complicated, because particles of different size have different characteristic scattering properties. Some of the details have been discussed in Section 5.3.1. From this earlier discussion, an important
parameter which characterizes the pattern of light scattered by a particle is
the ratio of the radius of the particle to the wavelength of light. Visible light
which affects the retina of the eye has wavelengths from about 0.40 to 0.70
μm. Thus we are concerned with how light of such wavelengths is affected
by particles of various sizes.

Large particles, those whose size exceeds several micrometers, will scatter light by three processes which are fundamental to optics. A portion of the
light incident on the particle will be reflected off the surface, and another
portion will be diffracted around the edges. Depending upon the composition
of the particle, some light may be refracted, pass through the interior, and
exit while again being refracted. These processes cause light to be deflected
from its original direction of travel (called the "forward direction") and
therefore are responsible for light scattering. We are all familiar with instances in which we have seen large particles in the air by means of the light
they reflect, and we have witnessed the phenomenon of a rainbow, which is
light refracted by water droplets. But diffracted light is much less familiar,

---

[†] This approach was treated in Middleton's (1952) classical book, which has served as a
standard for atmospheric visibility science for thirty years.

even though the amount of light diffracted around the edges of a large particle is similar to the amount reflected and refracted. If some of the incident
light of certain wavelengths will be absorbed within the particle, then the
light which is reflected or refracted may have a different color than the
incident light. Black smoke is an example of the case in which most of the
incident light is absorbed in the suspended particles. If there is no absorption, the scattered light has essentially the same color as the incident light.
Most of the light scattered by large particles has its direction only slightly
altered from the forward direction, and thus continues to travel in nearly the
same direction as it did originally. As a result, these particles cause the air
near the direction of the sun to appear white, nearly the same color as the
sun.

The aerosol particles most effective in scattering light are the smaller
ones, those of radius comparable to the wavelength of light. That is, a
greater proportion of the incident light will be scattered well away from the
forward direction when the wavelength of the light is about equal to the size
of the particle. Calculations show that most of the scattered light is deflected
by more than 1° of arc, but less than 45° from the forward direction. This
pattern is distinctly different from the predominantly forward scattering
from large particles. But the light scattered from intermediate size particles,
like that from large ones, is essentially the same color as the incident light.
As a result of scattering from intermediate size particles, the sky takes on a
hazy appearance if the atmosphere is sufficiently polluted with them. This
effect is particularly pronounced near the horizon, where one's view encom-
passes air with a greater concentration of aerosols and where the number of
particles in the line of sight is greater.

Light scattering from large and intermediate size particles is responsible
for the haze often found along seacoasts, over deserts, and within inland
valleys (Fig. 8.1a). Burning refuse releases great numbers of these particles
and can dramatically increase the effectiveness with which light is scattered
by the atmosphere (Fig. 7.25). Sharp differences in hazy, polluted air and
clean air may exist. A well-known example has been observed in California
when smog is blown inland with the sea breeze (Fig. 8.3).

The characteristics of scattering are quite different if the size of the particle is much smaller than the wavelength of light. Aerosols of a size less than
about 0.1 μm diameter scatter light of a particular wavelength equally well in
the forward and backward directions, and nearly as much intensity is scattered to the sides. This is in sharp contrast with the forward scattering from
larger particles. Another distinction is the fact that small particles scatter
light of short wavelengths more effectively than light of long wavelengths.
That is, a larger fraction of the incident blue light than of incident red light
will be scattered in all directions. This effect is responsible for the red color
of sunsets, which results because the blue component of sunlight has been
almost completely scattered out of the beam before it reaches the ground.

Fig. 8.3. Cool smoggy air moving inland with the sea breeze over Riverside, CA. The cool marine air from the Pacific Coast is undercutting, or flowing under warmer air inland after becoming polluted over Los Angeles. With the change in visual character of the polluted air is observed a sharp increase in pollutant concentration at the ground, for example, ozone. (Photo courtesy of E. Stephens.)

The reddish hue of the sun is more pronounced with a greater concentration of small aerosol particles in the atmosphere.

The theory of light scattering by small particles was first developed by Lord Rayleigh in 1871. However, the notion that the brightness of the day-time sky is due to the scattering of sunlight by particles suspended in the air is reported to have been already formulated in the early eleventh century by Alhazen of Basra, an Arabian physicist who carried out much of his work in Cairo (Kerker, 1969). Rayleigh added the suggestion that individual gas molecules could also scatter light and that this was actually responsible for the blue of the sky. We now know that this is not strictly correct; instead, Rayleigh scattering causing the blue color is due not to scattering from individual molecules but to scattering from groups of molecules in regions of the atmosphere where the concentration of molecules is momentarily greater than the average. Thus fluctuations in the density of air caused by the random motion of the gas molecules, increasing the concentration in some regions and decreasing it in others, give rise to scattering which accounts for the brightness of the sky, even if the atmosphere were to contain no particles. Because the fluctuations in the density of air are appreciable only over small volumes of atmosphere whose dimensions are much less than the wavelength of visible light, the scattering has all the characteristics of Rayleigh scattering from small particles and accounts for a portion of the blue color which we see.

8.1.1 Contrast and Visual Range. A physical basis for connecting visibility with optical changes in the air is the fact that man discerns objects by their *contrast* with the surroundings. The contrast may be in the color or brightness of an object compared with its background. Thus contrast is reduced when extraneous light is scattered toward the observer by particles in the intervening distance.

How light scattering produces a reduction in contrast can be formulated in terms of the theory of light transmission. The theory is linked with human observation in terms of a *visual range*. This is commonly defined as the greatest distance at which an observer can distinguish a contrast between an object and its background. Thus the visual range indicates how far we can see. It is unfortunate that "visibility" rather than "visual range" is often found in the literature describing the same concept. The nontechnical meaning of visibility is usually equated with contrast or the clarity with which objects stand out from their surroundings, with no reference to the distance at which the objects are perceived.

Another concept is important because it indicates the extent of the geographical area in which the visual range is restricted; this is the *prevailing visibility,* the greatest visual range which is attained or surpassed around at least half of the horizon, but not necessarily in continuous sectors. Roughly speaking, the prevailing visibility tells us how far out we have good visibility around half the horizon. It is a useful concept because it is a distance that can be determined by visual observations of the surrounding landmarks as viewed from one location, such as an airport control tower. As a result of its importance to aviation, a considerable body of data on prevailing visibility has been collected for locations around the world.

A theoretical formula for the visual range was first advanced by Koschmieder (1924). His derivation is based upon many assumptions about atmospheric conditions and knowledge of human perception, but three assumptions are fundamental. The first is that the object which we view (or the target) is black and therefore reflects no light, and that it is perceived against a white background. This condition approximates the common situation of a dark building seen against a light horizon sky. Consider an isolated object on the ground viewed from a distance (Fig. 8.4), and assume that multiple scattering does not occur, atmospheric refraction is unimportant, and the earth's curvature is disregarded. The contrast between the test object and the adjacent horizon sky is defined by the expression

$$C = (I_1 - I_2)/I_2 , \tag{8.1}$$

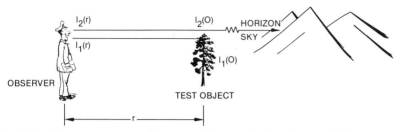

Fig. 8.4. Relative arrangements of observer, object, and horizon sky in definition of visibility. The angle between the lines of sight corresponding to I_1 and I_2 is very small.

where I_2 is the intensity of the background and I_1 is the intensity of the test object.

Expressions for the intensity can be obtained by integrating the equation of radiative transfer over the distance from the test object to the point of observation. If the extinction coefficient b_{ext} and source function J are not functions of distance through the light path r, the integration gives

$$I(s) = I(0) \exp(-b_{ext}r) + J[1 - \exp(-b_{ext}r)].$$

The location at which $r = 0$ corresponds to the location of the test object. Substituting in Eq. (8.1), the contrast as seen by an observer at a distance r from the object can be written as

$$C = [I_2(0)/I_2(r)]\,C(0)\,\exp(-b_{ext}r). \tag{8.2}$$

In viewing the horizon sky, the observer sees the virtual emission J resulting from the light from the sun and surroundings scattered in the direction of the observer by the atmosphere. This is sometimes referred to as the air light. By assumption, the air light is not a function of r. If 2 refers to the line of sight for the horizon sky, $I_2(r) = I_2(0) = J = \text{const}$ and Eq. (8.2) becomes

$$C = C(0)\,\exp(-b_{ext}r).$$

If the test object is perfectly black, $I_1(0) = 0$, $C(0) = -1$, then

$$C = -\exp(-b_{ext}r).$$

The minus sign in this expression results because the test object is darker than the background.

The visual range is defined as the distance at which the test object is just distinguishable from background. Hence the minimum contrast that the eye can distinguish must now be introduced into the analysis. This contrast is denoted by C^*. The visual range L_v for a black object whose contrast threshold is C^* is

$$L_v = -(1/b_{ext})\,\ln(-C^*). \tag{8.3}$$

Considerable effort has been invested in measurements of contrast thresholds for individuals and representative groups of people. The results of Blackwell (1946) are particularly noteworthy since they involved the analysis of almost half a million responses from a panel of 19 trained observers under controlled conditions. The results indicate that a contrast threshold can most precisely be determined if it is defined as the contrast which presented to an observer results in a 50% probability of its being perceived. With this definition, the threshold was found to be $C^* = (-)0.02$. This value is insensitive to the background intensity I_2 provided that the visual angle subtended by the target exceeds about $1°$ and daylight illumination prevails. This value for the threshold was in fact previously used by Koschmieder. Using the threshold contrast of 0.02, we obtain the idealized condition for the greatest distance a black target can be seen. This is sometimes given the name of the meteorological range. Taking the meteorological range as equivalent to the visual range, we obtain

$$L_v = -(1/b_{ext}) \ln 0.02 = 3.912/b_{ext} . \qquad (8.4)$$

Hence, as an approximation, the visual range is inversely proportional to the extinction coefficient. Since b_{ext} is a function of wavelength, the visual range defined in this way also depends on wavelength.

Because the contrast threshold appears logarithmically in this equation, the definition of the visual range is relatively insensitive to its exact value. Doubling C^* decreases the visual range by less than 20%. This is fortunate, for the relationship between the thresholds determined by controlled experiments and the thresholds for casual observers under urban conditions is not well established. Some evidence from controlled experiments suggests that the threshold contrast is 0.04 and may be more appropriate than 0.02. The reason for this is the fact that the 50% probability criterion previously mentioned left the observers with practically no confidence that they had really detected a contrast, whereas a threshold of 0.04 is found to correspond to a 90% probability of detecting a contrast, for which the trained observers felt considerably more confident in their identification.

Although the meteorological range is defined in terms of the extinction coefficient, Koschmieder's derivation emphasizes that it is not through the attenuation of light from the target that the contrast is degraded, for the target, being black, reflects no visible light. In fact, degradation results from extraneous light which impinges on aerosol particles in the line of sight between target and observer, with the subsequent scattering of a portion of this light toward the observer.

The total atmospheric extinction coefficient is the sum of contributions for the gaseous molecular scattering (b_{sg}) and absorption (b_{ag}), and aerosol particle scattering (b_{sp}) and absorption (b_{ap}) at certain wavelengths:

$$b_{ext} = b_{sg} + b_{ag} + b_{sp} + b_{ap} .$$

Molecular scattering coefficients for air have been tabulated (Table 8.1). For light of wavelength $\lambda = 0.5$ μm, the visual range calculated from Eq. (8.4) is about 220 km or 130 mi.

It is important to note that an effect not envisioned in Koschmieder's derivation is atmospheric "shimmer" or "twinkling." Particularly on sunny days, the path taken by a light beam is bent as it passes through air in which the density varies from place to place in irregular fashion. These density fluctuations result from the eddy motion of air in response to local heating by solar radiation, and occur on a much larger scale than the density fluctuations responsible for Rayleigh scattering by the atmosphere. As a consequence of their thermal origin, these fluctuations are known colloquially as heat waves. Such processes will obliterate fine details and distort the appearance of an object, perhaps leading to reduced visibility and a shorter visual range. The visual range may also be influenced by extraneous light sources within the field of vision, in the same way that the contrast threshold is known to depend upon such factors. Thus Koschmieder's relation is not expected to be valid at sunrise and sunset when the target is near the direction of the sun.

The realization of visual ranges where Rayleigh scattering dominates is relatively rare in populated areas because of the influence of air pollution. In

TABLE 8.1

Rayleigh Scattering Coefficient for Air at 0°C and 1 ATM[a,b]

λ (μm)	$b_{sg} \times 10^8$ (cm^{-1})
0.2	954.2
0.25	338.2
0.3	152.5
0.35	79.29
0.4	45.40
0.45	27.89
0.5	18.10
0.55	12.26
0.6	8.604
0.65	6.217
0.7	4.605
0.75	3.484
0.8	2.684

[a] From Penndorf (1957).
[b] To correct for the temperature, $b_{sg} = b_{sg=0°C} \times (273/T°K)$ at 1 atm. This approximate formula does not take into account the variation of refractive index with temperature.

many cities, visual ranges are reported with values less than 50 mi. This reduction is associated mainly with aerosol particle scattering and absorption, though gas absorption by NO_2 also may be a factor at times (Husar and White, 1976).

An example of the extinction coefficient contributions by wavelength of light for urban air is shown in Fig. 8.5. This case is taken from observations in Denver, CO during the winter of 1978. Direct observations, from which the extinction coefficient was calculated, were made using a transmissometer with a light path length of 5 km (Hall and Riley, 1976). The light-scattering coefficient was measured using an integrating nephelometer. The absorption of NO_2 was calculated from measured NO_2 concentrations at a point along the light path. The particle absorption coefficient was determined from

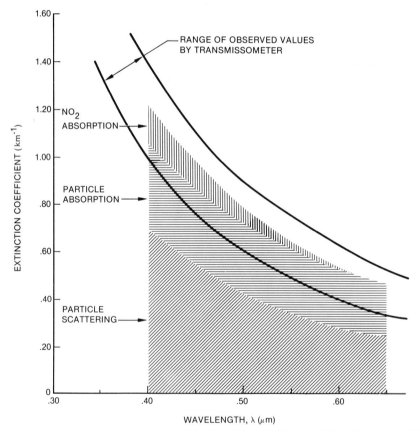

Fig. 8.5. Extinction budget for the period between 2020 and 2100 MST, 13 December 1978. Budget based on measurement of light extinction at night in Denver over a distance exceeding 5 km. Data for NO_2, b_{sp}, and b_{ap} are based on ground-level sampling at nearby measurement sites. [From Heisler *et al.* (1980).]

a filter method of Lin *et al.* (1973). The total extinction coefficient in this case was found to vary as the light wavelength to the negative 1.5 power. Extinction in the blue (short-wavelength part) is largest. The experiment indicates, within the range of observation, that the total extinction coefficient is made up of the sum of contributions from particle scattering and absorption and NO_2 absorption. About 60% of the total extinction accounted for was light scattering; the remainder was particle and NO_2 absorption. The contributions to light extinction will vary depending on location and the level of particle or gas concentrations.

In the absence of absorption from NO_2 and particles such as elemental carbon, extinction is dominated by light scattering, so Koschmieder's equation can be approximated with b_{sp} replacing b_{ext}. Experiments by Horvath and Noll (1969) were designed to examine this relationship by comparing the prevailing visibility with the scattering coefficient for ambient air as measured by a nephelometer. Figure 8.6 shows the observed reduction of prevailing visibility with increases in the scattering coefficient for a series of urban measurements from the top of a four-story building in Seattle, WA. The straight line represents Eq. (8.4), assumimg that the extinction coefficient arises from scattering alone, and assuming that the meteorological range L_v is identified with the prevailing visibility. This experiment and others indicate that Eq. (8.4) adequately describes prevailing urban visibility to within the precision of the observational data, generally less than $\pm 50\%$. However, more recent studies of Bajza *et al.* (1979) and Linak and Peterson (1981) have shown inconsistencies between instrumental measures of light extinction and perceived visibility in the southwestern United States, where visual ranges are often much larger than 10 miles. In the absence of better information, Koschmieder's relation remains in use for relating visual per-

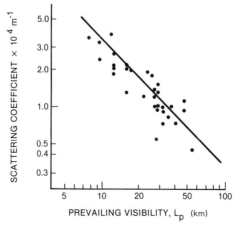

Fig. 8.6. Reduction of prevailing visibility with increasing scattering coefficient. [From Horvath and Noll (1969); with permission from Pergamon Press, Ltd.]

ception and the extinction coefficient. Available data support the notion that light scattering is a dominant cause of reduced visibility in polluted air taken in many locations, but particle absorption may be important in urban situations.

Light-scattering theory as discussed in Chapter 5 suggests that the relationship between the scattering coefficient and the composition and size of aerosols may be complex. Thus Koschmieder's relation does not provide us with a clear indication of how the visual range relates to the amount of aerosol in the air. It is useful to find a simple empirical relationship between the scattering coefficient and some easily measured parameter of the aerosol particle burden of ambient air, such as the concentration of suspended particles. In fact it is possible to find a simple relation: studies in a number of locations have revealed the existence of an inverse dependence between prevailing visibility L_p and mass concentration of the form

$$L_p = K_v/M, \tag{8.5a}$$

where K_v is a constant and M is the particle mass concentration determined from a high-volume filter sampler whose inlet excludes sampling particles $\gtrsim 10$ μm diameter. Another way of stating this relation is

$$b_{ext} \approx b_{sp} \approx \text{const} \times M \tag{8.5b}$$

for applications where b_{abs} is small. The constant K_v shows wide variations; the data from these and other measurements give an average value for K_v amounting to 1800 km μg/m^3 (or 1.8 g/m^2), but the scatter in the data is such that values of K_v between 900 and 3600 km μg/m^3 are possible. Thus for a particulate burden of $M = 100$ μg/m^3, the prevailing visibility is only 18 km (about 11 miles) on the average. These are convenient numbers to remember. In heavily polluted cities, the mass concentration often can amount to 300 μg/m^3, and the prevailing visibility is correspondingly reduced to 6 km.

The empirical relation, Eq. (8.5b) generally is not adequate to quantify relations between total suspended-particulate (TSP) matter concentration and prevailing visibility because the TSP measure is influenced by coarse particles. However, observations of the mass concentration of fine-particle diameter should show a strong relation with visibility. This is demonstrated indirectly in Fig. 8.7 for a data set taken in different parts of southern California. The light-scattering coefficient is found to be well correlated with the particle mass measured in the efficient light-scattering range of 0.1–1.0 μm diameter.

8.1.2 Perception of Visual Effects.   When the concept of visibility impairment is introduced to encompass the appearance of a complex scene, the distance over which *contrast detail* can be seen varies widely, as suggested in Table 8.2. The results indicate that if form, line detail, or texture are important to visual impressions (Fig. 8.1), the fine detail is lost much more

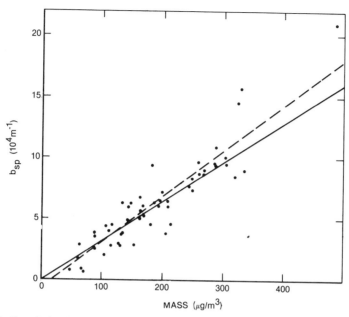

Fig. 8.7. Correlation between light-scattering coefficients b_{sp} with mass concentration of fine particles less than about 3 μm diameter. Data correspond to two km-averaged b_{sp} and mass concentrations determined on filters taken at 2 hr intervals in different parts of California (1972–1973). Solid line is the ratio of b_{sp}/M fit to be $b_{sp} = 0$, $M \rightarrow 0$. Dashed line shows linear regression fit between b_{sp} and M. [From Hidy (1975).]

TABLE 8.2

Visual Range of Contrast Detail Calculated from a Psychophysical Model Using Linear System Theory[a]

Detail of level	Characteristic object size at 10 km (m)	Examples for a scenic view at 10 km	Visual range (km)	
			100 km[b]	20 km[b]
Very coarse (form)	>100	Hills, valleys, ridgelines	79	16
Coarse (line)	50–100	Cliff faces, smaller valleys	76	15
Medium (texture)	25–50	Clumps of large vegetation, clearings on forested slopes	62	12
Fine (texture)	<25	Individual large trees, clumps of small vegetation	22	4

[a] From Henry (1979a).
[b] Assumed background visual range.

rapidly than the edge or broad features of large objects. The additional sophistication of the visibility concept introduces the requirement to consider such recent concepts as achromatic perception and the psychophysics of the eye–brain system.

A Linear System Analysis.   Henry (1977) has proposed a method to formulate quantitatively the psychophysical processes involved in visibility reduction as well as the physical processes. A great deal has been learned over the past fifteen years about visual acuity using the methods of linear system theory (for a basic presentation, see Cornsweet, 1970). This approach lends itself particularly well to the problem of visibility reduction of contrast detail by aerosols. Examples of contrast detail of interest are trees on a distant hillside or windows in a building. This section addresses the visibility of such contrast detail as predicted by linear analysis of the response of the eye–brain system.

According to linear system theory any system that is linear, isotropic, and homogeneous can be completely described by its sine-wave response, or transfer function; i.e., if $S(\omega)$ is the response of the system to a sine-wave input of frequency ω then the response to an arbitrary input $F(X)$ is given by

$$R(X) = \int_{-X}^{\infty} G(\omega - X)S(\omega) \, d\omega,$$

where

$$G(\omega) = \int_{-\infty}^{\infty} \exp(-\pi i \omega t)F(X) \, dX,$$

the Fourier transform of $F(X)$.

In general, this sine-wave response is a complex function. The modulus $| S(\omega)$, called the *modulation transfer function,* is used in this model. Ideally, complex scenes can be disaggregated into a series of bright and dark bands characterized by a spatial equivalent of a regular oscillation in time. A simple light–dark grating could be represented by a sine wave, for example.

The frequency of bands is spatial and is given in cycles per degree of arc rather than cycles per second. One defines the *modulation contrast* of a simple dark-linear-banded test pattern as

$$C_{\mathrm{M}} = (B_{\max} - B_{\min})/(B_{\max} + B_{\min}), \tag{8.6}$$

where $B_{\max}$ is the maximum luminance, and $B_{\min}$ the minimum luminance.

An experimental procedure is to reduce the contrast modulation of a sinusoidal test pattern until it is just barely visible; i.e., one measures the threshold contrast. In this way curves such as Fig. 8.8 are obtained. Note that the contrast decreases upward; thus the vertical axis is contrast sensitivity, the reciprocal of the contrast. It is important to note the logarithmic relationship of the threshold contrast and frequency and the maximum in the curve at about three cycles per degree. Many factors affect the exact loca-

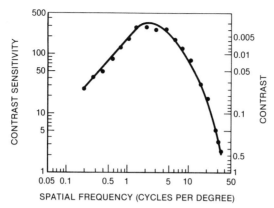

Fig. 8.8. Contrast sensitivity of human subjects for sine-wave grating peaks at three cycles per degree corresponding to a contrast threshold of 0.003 or 0.3%. [From Campbell and Maffei 1974 .]

tion of the curve: pupil size, mean retinal luminance, and age of the observer, to name only some of the most important; see Campbell and Green (1965) and Patel (1966). However, the basic shape of the curve is not changed by these factors. For purposes of discussion, let us assume an achromatic pattern, which is typical for many objects viewed through the atmosphere. The modulation transfer function of the eye–brain system is taken to be proportional to contrast sensitivity. For discussion here, the curve in Fig. 8.8 represents the modulation transfer function of the human eye.

The validity of the linear systems approach to the analysis of vision requires small visual angles; i.e., less than two degrees, contrasts near the threshold values, and high luminance levels. Even though the eye is not isotropic or homogeneous, its response to small perturbations is very nearly linear. If the above constraints are met, then it can be shown theoretically (Davidson, 1968) and experimentally (Campbell, 1966) that the eye–brain system response is indeed linear. These constraints are all satisfied in the case of calculating the visual range of windows in a skyscraper, for example, or for object textures concerned with the effect of aerosols on visibility. Consequently it is sufficient to consider only the visibility of sinusoidal contrast detail since the visibility of any pattern is determined by the amplitude of its largest Fourier component after multiplication by the modulation transfer function of the eye.

For a sinusoidal pattern of frequency ω structure observed through the atmosphere the visual range $L_v(\omega)$ of the sinusoid is

$$L_v(\omega) = b_{ext}^{-1} \log\left[\frac{1}{B'}\left(\frac{C_M(0)}{C_T(\omega)} - 1\right) + 1\right], \qquad (8.7)$$

where

$$B' = \frac{2}{(C_w + 1)(B_D/B_0 + B_B/B_0)};$$

C_w is the contrast of a white object against the horizon, B_0 is the luminance that a white object would have if it were in the same position as the background, and $B_D(r)$ and $B_B(r)$ correspond to the luminance at distance r for the maximum and minimum of the sinusoid relative to the horizon light (B_H). For example,

$$B_D(r) = B_D(0) \exp(-b_{ext}r) + B_H(0)[1 - \exp(-b_{ext}r)].$$

The contrast C_T is the threshold modulation contrast for the sinusoid. Equation (8.7) is closely related to Koschmieder's relationship, Eq. (8.4). The validity of the linear system analysis requires three assumptions in addition to the conventional ones for atmospheric visibility, namely:

(a)   Atmospheric luminance levels should be high, at least 10 foot-lamberts.

(b)   The contrast detail of interest should have a spatial frequency of at least 0.5 cycles per degree.

(c)   The contrasts in question should be close to threshold values.

The parameter B' in Eq. (8.7) is used to take into account the conditions of atmospheric illumination and solar angle. The value of C_w, the contrast of a white object under identical illumination, is low for an object in shadow and high for an object in direct sunlight. Equation (8.7) predicts that if all other factors are constant the detail on the object in shadow has a smaller visual range than in direct sunlight.

As written, the visual range is given in Eq. (8.7) as a function of ω, the spatial frequency. If the spatial frequency is fixed, then the visual range is inversely proportional to the attenuation coefficient as predicted by the classical theory. However, for a real object viewed through the atmosphere, the spatial frequency could never be constant. If b_{ext} decreases, then L_v would increase and the object, being further away, would appear smaller. Thus the spatial frequency of any contrast detail would increase. Similarly, an increase in b_{ext} would result in an apparent decrease in the spatial frequency. It is this interaction, due to the nature of the human eye, of distance, attenuation coefficient, and spatial frequency, that is basic to the theory developed by Henry.

An example of the use of Henry's theory calculates the minimum spatial frequency sinusoid visible at several distances as a function of b_{ext} for $C_M(0) = 1$ and $B' = 4$. A graph of the results appears in Fig. 8.9. Note the steepness of the curve for a distance of 40 km. This suggests that in clean air a small increase in b_{ext} would reduce minimum visible frequencies below the three to five cycles per degree level. This would essentially remove all

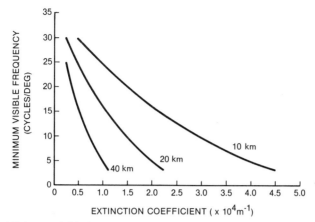

Fig. 8.9. Minimum visible frequency vs. extinction coefficient for objects at 10, 20, and 40 km. [From Henry (1977); with permission of Pergamon Press, Ltd.]

contrast detail from a hillside at 40 km. Residents of the area would notice that a hillside once plainly visible would be seen only in outline. On the other hand, the residents of a more polluted area might well judge nominal visibility in terms of a hillside 10 km distant. In that case, the curves show that a much larger increase in b_{ext} is needed to reduce that limiting frequency to under five cycles per degree.

As a final example, it is useful to examine available experimental evidence relating observed visual range to nephelometer readings, especially since the first example above gives reason to doubt the universality of a linear relationship between $1/b_{ext}$ and the visual range. However, it is difficult, if not impossible, to interpret these experiments in terms of the theory presented here. Detailed knowledge of the angular parameters and contrasts of targets used is necessary. Without these data the Fourier analysis cannot be performed. Even two-dimensional shape factors cannot be ignored, as indicated by Kelly and Magnuski (1975). Depending on the choice of targets, such experiments could either confirm or contradict the expected inverse relationship of visual range and the extinction coefficient.

From application of a linearized psychophysical model, visibility of contrast detail is determined by an interplay of the effects of distance, extinction coefficient, and response of the eye–brain system. A surprising result from this approach is that the visual range for contrast detail, in general, is not inversely proportional to the extinction coefficient, as classical theory would imply. Another result suggests that visibility, as measured by the minimum visible frequency, is much more susceptible to ambiguity in clean air areas than in polluted areas. As in the optical theory, a small absolute increase in the extinction coefficient in an unpolluted area would be a large percentage increase and would degrade visibility proportionately.

The linear system model can be used to estimate a threshold range of contrast compatible with the average response of the eye–brain system. This analysis comes from an application of a model for "just-noticeable differences" (JNDs) derived from Carlson and Cohen's (1979) experiments.

The model assumes that there are five channels in the human visual system relating to moderate to high frequency signals (>2 cycles per degree angle). Each channel responds only to a range or band of spatial frequencies. A signal detection scheme is used to determine the output of each channel and, finally, the outputs of all channels are summed to get the total output of the visual system. The results of the model are expressed as a number of JNDs (Henry, 1979b). A JND is defined as a change that would be correctly recognized 70% of the time. Also by definition, one JND is the threshold below which no change is perceptible. A change of two JNDs above threshold is defined by the following procedure: increase the signal to one JND, then increase it until a further change is just noticeable—this level is two JNDs above threshold. Thus, hypothetically, the results of Henry's analysis yield an estimated relationship between contrast reduction for a range of complex scenes. The estimates of JNDs suggest a novice observer would experience a readily noticeable change about four times as great as the threshold for an experienced observer. For contrast reduction less than 20%, the perceived degradation in a complex scene is approximately a linear function of contrast reduction; i.e., a contrast reduction of 10% causes twice the visibility degradation of a 5% reduction.

The visual range of texture in a complex scene is very sensitive to the angular size of the detail. As indicated in Table 8.2, fine detail is rapidly lost with distance. The degree to which this is true is determined by the inherent contrast of the texture, with high contrast texture being the more visible.

The perceptual "threshold" associated with the JND concept is distinct from that empirically used to obtain Koschmieder's relationship. The latter two-percent contrast threshold applies to a laboratory setting of controlled lighting conditions. The JND concept is based on a series of psychophysical experiments examining noticeable distinctions in *contrast detail* for different spatial wavelengths in relation to the observers' spatial frequency curve for contrast sensitivity (the modulation transfer function). From Henry's results, one would expect that a psychophysical response threshold to variations in contrast detail would be greater than 2%, perhaps as large as 5–10%. Experiments are needed under field conditions to verify this conclusion and to advance the application of this approach to the visual aesthetics of scenic vistas.

The theory of contrast detail offers an opportunity for alternate definitions of visibility indexes based on Fourier analysis of spatial features of scenery. Henry *et al.* (1981a) have proposed such an approach using photographs.

Digitization of color photographs allows the convenient inexpensive calculation of a number of spatially averaged visibility indices that currently cannot be determined in any other way. The usual visibility indices such as contrast, extinction coefficient, and visual range often do not necessarily accurately indicate the actual appearance of the scene. They also tend to be sensitive to irrelevant environmental conditions such as cloud cover and ground snow cover. One would like to have a measure of visibility which could quantitatively separate regional haze, for example, or quantitatively assess the amount of visible texture, an important clue in human perception of visual air quality. To accomplish these goals, visibility indices would have to be spatially averaged in some way using photographs. As a bonus, spatial averaging could be designed to decrease the sensitivity of the measures to changes in cloud cover and other extraneous environmental effects. Two candidates for new, spatially averaged visibility indices follow from modulation transfer theory.

The first uses as a measure of atmospheric clarity and regional haze the *modulation depth* of a view. This is the normalized standard deviation of the relative exposure over the area of interest in a photograph. It is defined as

$$\text{MD} = \frac{1}{(N-1)\overline{EX}} \left(\sum_{i=1}^{N} (EX_i - \overline{EX})^2 \right)^{1/2}, \tag{8.8}$$

where MD is the modulation depth, N the number of pixels (digitized picture elements), EX_i the relative exposure of pixel i, and $\overline{EX}$ the average relative exposure. The modulation depths for example clear day and hazy day photographs are given in Table 8.3. Since the modulation depth measures the variability of light, it is expected that a haze would lower the modulation depth by smoothing out or "luminating" some of the objects. General haze would decrease the modulation depth of the scene, whereas the presence of a distinct haze layer would increase the modulation depth. This is because these latter conditions add structural detail to the scene.

TABLE 8.3

Calculated Modulation Depth from Digitized
Color Photographs for Two Days in Southern
Wyoming[a]

	Modulation depth	
Color	Clear day	Hazy day
Red	0.299	0.210
Green	0.362	0.262
Blue	0.420	0.387
Achromatic	0.350	0.278

[a] From Henry *et al.* (1981a); courtesy of Pergamon Press, Ltd.

A second approach to unconventional measures of visibility is through Fourier analysis of the photographic scene using modulation transfer theory. Frequencies in the range of 0–3 cycles per degree contain most of the recognizable picture content of the scene. The response of the eye to the various spatial frequencies in the scene is determined by the eye–brain system. The modulation transfer function (MTF) of the human eye has a peak of about 3 cycles per degree (Fig. 8.8). Other spatial frequencies, both lower and higher, produce a smaller response in the eye. Spatial frequencies above 3 cycles per degree contain the finer detail, or texture of the picture. Given the MTF of the eye and the Fourier analysis of the input scene, the theory can be used to calculate how much degradation in the picture would produce a JND in the clarity of the photograph. In another sense, the amount of signal variability or the average relative photoexposure (power) in the higher frequencies of the scene is indicative of the amount of visible texture and could be used to quantify this property. Following Carlson and Cohen (1979), Fourier analysis of the average horizontal relative exposures was performed on the clear and hazy slides.

The relative amount of power in each frequency range can be calculated from digitized photos. Figures 8.10a and 8.10b are plots of the relative power in each frequency for the achromatic case for the clear and hazy day photographs. The general shape of the curves is typical of spectra for complex scenes, as discussed by Carlson and Cohen (1979). There is a broad distribution of the relative power; however, for higher frequencies, there is a decrease in spectral power that is proportional to $1/\omega^2$, where ω is the frequency. Theoretical calculations show that a single luminance edge gives a $1/\omega^2$ high frequency dependence. For complex scenes, the high frequencies are sums of many light and dark bands, or luminance edges.

Comparison of the achromatic power spectra of the clear and hazy day show distinctive characteristics. As expected, the power in the high frequencies in the hazy slide is reduced by a factor of 2–10 as compared with the clear day photograph. The power in some of the low frequencies for the hazy day is actually increased. This is caused by the haze layer itself, which appears as a broad low layer that is rich in low frequencies while reducing fine details. This type of result for a variety of cases involving photographic analysis shows promise for defining a psychophysically meaningful index for visibility impairment.

8.1.3 Perception of Color. In comparison with achromatic contrast the question of atmospheric discoloration, or chromatic contrast, is less well understood. The sensation of color from visual stimuli is a psychophysical concept for which no complete theory exists.

Helmholtz first proposed the three-color theory of color vision in 1866, yet evidence for three distinct photopigments in the retina was demonstrated only within the last twenty years. At least a dozen theories of color vision

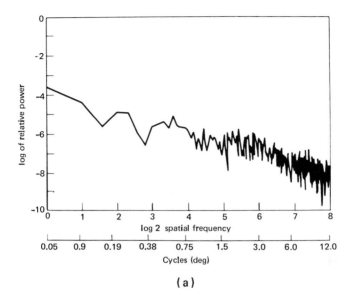

(a)

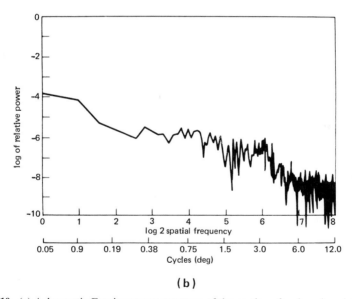

(b)

Fig. 8.10. (a) Achromatic Fourier power spectrum of the portion of a clear day picture near Rock Springs, Wyoming (11 June 1975, 08:50 MST). The ordinate is proportional to the average relative exposure of the slide. (b) Achromatic Fourier power spectrum of portion of the "hazy" day taken from Rock Springs, Wyoming airport (4 August 1975, 09:29 MST). The ordinate is proportional to the average relative exposure of the slide. [From Henry *et al.* (1981a); with permission of Pergamon Press, Ltd.]

exist today, but empirically color change remains judged by a chromaticity system developed many years ago.

Coloration is detected by the human eye–brain system through a comparison of the light intensities of various object features at different wavelengths. The colors associated with a given wavelength of visible light range from purple–blue at 0.4 μm, green at 0.51 μm, to red at 0.63 μm. The system adopted as a standard is that of the international *Commission Internationale de Eclairge* (C.I.E.).

Colors make up a chart of *chromaticities* which are plotted in terms of three tristimuli X, Y, and Z; these are defined in terms of spectral radiance $I(\lambda)$ over a given wavelength range, or

$$X = I(\lambda)\bar{x}\ d\lambda, \qquad Y = I(\lambda)\bar{y}\ d\lambda, \qquad Z = I(\lambda)\bar{z}\ d\lambda, \qquad (8.9)$$

where $\bar{x}$, $\bar{y}$, and $\bar{z}$ are the tristimulus weighting functions which have been determined by a large number of laboratory experiments. The chromaticity diagram is generally given in terms of coordinates $\bar{x}$, $\bar{y}$, and $\bar{z}$:

$$\bar{x} \equiv \frac{X}{X + Y + Z}, \qquad \bar{y} \equiv \frac{Y}{X + Y + Z}, \qquad \bar{z} \equiv \frac{Z}{X + Y + Z}. \qquad (8.10)$$

The sum of normalized coordinates $\bar{x}$, $\bar{y}$, and $\bar{z}$ is unity. A combination of coordinates $\bar{x}$ and $\bar{y}$ gives measures of color *hue* and *saturation*. As a color changes from a neutral gray or white, its color saturation is said to increase; pure colors are the most saturated.

The chromaticity coordinates do not specify color completely; a third parameter is needed to characterize brightness or *luminance*. The second tristimulus value (Y) is used to describe luminance. The brightness of a perfectly white object under given, fixed illumination conditions is arbitrarily assigned a luminance value of 100. The combination of chromaticity coordinates ($\bar{x}$, $\bar{y}$) and luminance (Y) is necessary and sufficient to specify the physical attributes of color under standardized laboratory illumination conditions.

From a large variety of observations, the apparent color of the sky has been estimated for a range of sunlight and cloud conditions (e.g., Wyszecki and Stiles, 1967). This range is shown in Fig. 8.11. The sky color covers a relatively small band of the color diagram. Interestingly, modest changes in NO_2 concentrations apparently vary indistinguishably in color within the natural band of sky color.

The C.I.E. system used to predict the colors of the atmosphere was not intended to be a theory of color vision. It is an engineering solution to the problem of setting color standards for world trade. As a method of defining a standard color system for paints, textiles, etc., the C.I.E. method works. However, its applicability has not been proven for judging small changes in colors of objects, plumes, or haze layers, under nonstandard and variable lighting conditions against varying colored surroundings, viewed through the

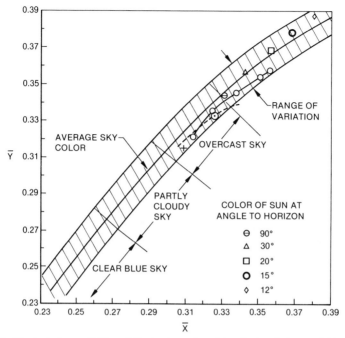

Fig. 8.11. Typical chromaticities of the horizon sky with calculated chromaticities of pollution haze layers: (+) standard light source C; (---) Los Angeles aerosol, 0 ppm NO_2 [scattering
angles 45°–135°, Husar and White, (1976)]; (—o—) Los Angeles aerosol, 0.1 ppm NO_2 [scatter-
ing angles 45°–135°, Husar and White (1976)]; (⊙) $b_{sp} = 6.2 \times 10^{-4}\,\mathrm{m}^{-1}$, 0.15 ppm NO_2 [Horvath
(1971)]. [Adapted from Wyszecki and Stiles (1967); from Henry (1979a); courtesy of the Air
Pollution Control Association.]

atmosphere. Even the variations in time of the apparent color of a plume
may be a factor of uncertainty that needs to be addressed. A partial list of
parameters of importance to perceived coloration other than the wavelength
distribution of the light includes (a) size of the object or area, (b) background
luminance level, (c) color of surrounding background, and (d) temporal variations. Each of these variables has been shown to have a potential effect on
perceived color as large as has been predicted for a polluted atmosphere. For
example, 21 colors are identical as judged by the C.I.E. 1931 standard observed with a 2° field of vision but are all different as perceived by the C.I.E.
1964 standard observer with a 10° field of view according to Wyszecki and
Stiles (1967). Obviously, the size of the field of view is important, since the
range of variation found is about as large as the color changes calculated for
nitrogen dioxide discoloration in Los Angeles smog by Husar and White
(1976). (See also Fig. 8.14.)

The work of Hunt (1953) shows very large changes in perceived colors on
the standard chromaticity diagram effected by changing only the luminance
level of the stimulus. More recently, Kelly (1974) has shown that the three

color mechanisms in the human retina all have different space and time sensitivities. Furthermore, Kelly showed that the visual effects of time and spatial variations are not independent. The effects noted by Kelly are not small and could influence the color threshold of a haze layer moving relative to the observer.

The effect of the color of the surroundings on the perceived color of an object is well known and is called *induced color chromatic adaptation*. In general, if the eye is adapted to a color, say blue sky, a nearby white area may take on the complementary color of the surrounding, in this case a light yellow–brown. Again, this effect may be large enough to explain some of the brown color of atmospheric haze. High clouds normally look white because they are partially illuminated by the blue light of the sky, eliminating the chromatic adaptation effect, yet dark low clouds do in fact often appear to be brownish.

The actual colors of the sky have been studied extensively. Figure 8.11, adapted from Wyszecki and Stiles (1967) and Judd *et al.* (1964), shows their results for the average range of sky color. This work was based on and compared with a large number of independent, direct visual colorimetric and spectrophotometric observations of unpolluted atmospheres. Because of chromatic adaptation, any color in the crosshatched area of Fig. 8.11 above an $\bar{x}$ value of about 0.30 to about 0.37 could be considered to be "white." Most reported calculations of the effects of pollution on sky color do not lie outside the crosshatched area. Thus, within the bounds of sensory perception there is a range of sensitivity to discoloration associated with aerosols in the natural environment which remains to be determined by experiment.

Some evidence of the phenomenon of induced color comes from observations of the winter haze over Denver, CO. This haze has often been characterized by a brownish-gray tinge. Some workers have identified the brown color with absorption of NO_2 or with wavelength-dependent scattering from aerosol particles (e.g., Fig. 8.5). During November and December of 1978, telespectroradiometer measurements were taken by R. Weiss, a typical example of which is shown in Fig. 8.12. The instrument measurements show that the haze is essentially gray, with no wavelength dependence of its radiance even though the haze was perceived visually to be brownish-gray. Given the fact that the objective measurement by the telespectroradiometer showed no significant difference between the haze and the gray card, why do most residents of Denver call it a "brown cloud?" The induced color phenomenon of the human eye–brain response may explain the apparent brown color where the instrumental measurement finds none. To adapt to a colored field of view, the eye may overcompensate for a nearby gray or white field, thus inducing the complementary color in the gray or white area. The color induced by sky blue is a yellow–brown to red–brown.

A systematic study of the magnitude and factors influencing induced color has been made by Kinney (1962). She studied the effects of size, purity of the inducing color, and luminance ratio of the inducing colored area to the

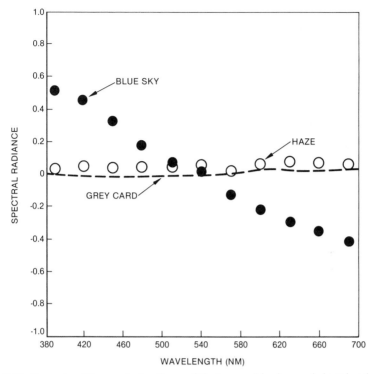

Fig. 8.12. Example of the spectral radiance of the haze and background sky taken by telespectroradiometer at Ruby Hill, CO, 1130 MST, 19 December 1978. [From Heisler *et al.* (1980).]

adjacent white area. This ratio was found to be very important; i.e., when the white area is twice the brightness of the colored area there is little induced color. However, when the situation is reversed, when the luminance of the gray area is one-half that of the adjacent colored area, then there is a significant amount of induced color. This is important with respect to the Denver winter haze, since the great amount of light absorption by elemental carbon in Denver air often causes the haze to be a gray band, dimmer than the deep blue sky next to it. These evidently are ideal conditions for induced color. Therefore, although the telespectroradiometer measurements showed the haze to be "optically" gray, it may appear brown to an observer.

Perceived Appearance of Plumes and Opacity. Obscuration of view often results from identifiable smoke plumes. These are readily observed in cities as well as in the rural environment. A major part of the effort to control air pollution in cities has been aimed at reduction of particulate emissions from stacks to eliminate visible plumes. The visual effects of plumes have been estimated from opacity, which basically relates to the transmission of light through the plume. People generally associate a stronger degradation in air quality with a "dirty," dark plume than with a white plume. Thus opacity historically has been estimated by comparison with a graduated gray scale, called a *Ringelmann scale* (e.g., Williamson, 1973). This measure is basically a psychophysical scale which relies on interpretation by an experienced observer.

The opacity of plumes can be estimated by calculation of the light transmittance through the plume for the extremes of scattering particles and observing (black) particles. The method described by Halow and Zeek (1973) takes into account different observer locations relative to the plume and the sun. The method uses Mie scattering theory to estimate the light extinction from particles. The calculations yield correlations with the Ringelmann scale as shown in Fig. 8.13.

The graph in Fig. 8.13a corresponds to a white plume, or a plume brighter than the background. The relationship is given in terms of the ratio of the plume luminance due to scattered light (B_{ss}) and the luminance of the background sky near the plume (B_0). This is related to the percentage of light transmitted through the plume (TR) by the expression

$$B_{ss}/B_0 = 100 + 100(B_p - B_0)/B_0 - TR, \qquad (8.11)$$

where B_p is the plume luminance. The theory correlates quite satisfactorily with observations reported by Connor and Hodkinson (1967) for white plumes. Roughly speaking, the Ringelmann number is linear in the luminance ratio for a range from 0 to B_{ss}/B_0 of 100.

The correlation for black plumes is shown in Fig. 8.13b. Here, relatively little sunlight or skylight will be scattered at most scattering angles, so that the evaluation of plume darkening will depend almost exclusively on light transmission. Transmission is given in terms of the ratio B_T/B_0, where B_T is the plume luminance due to transmitted light, calculated in terms of the extinction coefficient. The correlation shown in Fig. 8.13b yields a relation which reflects the observed threshold of a 2% contrast difference, but is roughly linear in the transmission range between 15 and 85%. The observations shown in Fig. 8.13b again come from Connor and Hodkinson for various times of day and plume viewing geometry. Halow and Zeek note that application of the correlations in Fig. 8.13 should be restricted to cases where the sky in the immediate vicinity of the plume is the same brightness as the sky behind the plume. Thus partly cloudy skies could cause an observer to report a variable Ringelmann number.

The discoloration of a plume or haze layer, in principle, can be estimated by estimating the relation between perceived color change and the contrast changes associated with a specific wavelength of light. This has been attempted by Latimer *et al.* (1978) and Williams *et al.* (1980). The calculation refers color change to a shift in chromaticity coordinates, measured by a ΔE parameter which is determined by changes in coordinates:

$$\Delta E = [(\Delta L^*)^2 + (\Delta U^*)^2 + (\Delta V^*)^2]^{1/2}, \qquad (8.12)$$

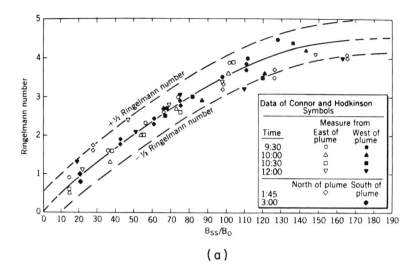

(a)

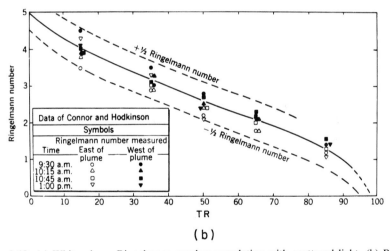

(b)

Fig. 8.13. (a) White plume Ringelmann number correlation with scattered light. (b) Black plume Ringelmann number correlation with transmittance TR. [From Halow and Zeek (1973); courtesy of Air Pollution Control Association.]

where

$$L^* = 116(Y/Y_0)^{1/3} - 16, \qquad U^* = 13W^*(U - U_0),$$

$$V^* = 13W^*(V - V_0), \qquad W^* = 25Y^{1/3} - 17$$

and

$$U = \frac{4X}{X + 15Y + 3Z}, \qquad V = \frac{6Y}{X + 15Y + 3Z};$$

U_0 and V_0 are defined in terms of reference to the tristimuli X_0, Y_0, and Z_0 which define the color of the nominally white object-color stimulus (Wyszecki and Stiles, 1967). The calculations are often normalized to $Y_0 = 100$.

Latimer *et al.* (1978) have estimated the terminal or threshold (just perceptible) value of ΔE with a contrast of 0.02 to be approximately 0.78. Calculations of this type probably overestimate the color change perceived by an observer for objects of narrow viewing angle. They do not take into account psychophysical research on the color appearance of small objects, such as signal lights and signs. An excellent, readable review of this field is given by Judd (1973). The most significant part of this review for atmospheric optics is Judd's discussion of the relative importance of brightness and color in determining the detection of visual signals. She states that "to make the C.I.E. 1964 space apply to dim targets of small angular subtense as well as to bright targets subtending more than 60 minutes of arc, we must generalize the color difference formula by introducing the factors k_u, k_v, and k_w, which indicate fractions of normal discrimination in the red–green, violet–green yellow and light–dark dimensions, respectively." Thus

$$\Delta E = [(k_u \Delta U^*)^2 + (k_v \Delta V^*)^2 + (k_w \Delta W^*)^2]^{1/2}.$$

Here ΔE, without the correction factors, is the color difference as used in Eq. (8.8) to assess plume discoloration. For small targets of $10'$–$30'$ of arc, the values of k_u and k_v are much smaller than k_w (see Fig. 8.14). A power plant plume 100 m wide seen from a distance of 10–35 km lies in this regime. Thus the detectability of the plume due to discoloration (the ΔU and ΔV parts of ΔE) is much less than the detectability caused by dark–light contrast, ΔW. It may be true that ΔE as calculated for visual appearance of haze layers seriously overestimates the importance of discoloration caused by NO_2. Such a conclusion requires further investigation for its application to the changes associated with haze layers.

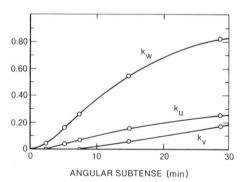

Fig. 8.14. Variation with angular subtense in the constants k_u, k_v, and k_w. [From Judd (1973).]

8.1.4   Psychological Factors.   The third element of visibility in the atmosphere concerns the human value judgment of perceived visual effects. This brings into play the psychology of interpretation of human impressions. This feature has largely been ignored in the literature of the atmospheric sciences until recently. However, the interface between science and visual psychological factors has been a part of philosophical literature since the writings of Leonardo da Vinci (e.g., Richter, 1833). There is good evidence that ancient sculptors knew of the human impression of contrast change in enhancing the aesthetic impression of brightly lit sculptures. [See, for example, the surface enhancement by shadowing of the famous horses of San Marcos, Vittori and Mestitz (1974).]

Objective measurements of psychological awareness of air pollution have been reported, and were reviewed in 1969 (U.S. PHS, 1969). The technology of enhancement of scenic impression has matured with photography, and with the television industry. It is well known, for example, that psychological tests of film usage indicate a preference for enhancement of certain color features over natural contrast. For example, the blueness of the sky is deepened in color film development based on such preferences. Studies showed that accurate reproduction of certain natural color contrasts of the original scene produced a photograph with low customer acceptance. The original investigations of this phenomenon by Bartleson and Bray (1962) revealed several facts. First, the preferred color of the blue sky in a reproduction is significantly different from the actual color of the sky. Second, the actual color of the blue sky is outside the range of psychologically preferred sky reproductions (Fig. 8.15). And third, the color remembered as the blue-sky color is significantly different from either the actual or preferred blue-sky color.

Figure 8.15 from Bartleson and Bray (1962) gives their results on the standard C.I.E. chromaticity diagram. These results were confirmed by another group of researchers more than a decade later using very sophisticated techniques (Hunt *et al.*, 1974). Thus there can be no doubt that the response of humans to photographic reproductions of natural scenes is significantly different from their response to the actual scenes with regard to the color of the sky. People tend to prefer the illusion of a sky blue that is a deeper, darker blue than the actual sky.

The implications of this work for the use of color photographs in visibility studies are significant for future development of aesthetic theories of visibility. Because of this tradition, the photographically produced sky colors are strongly distorted from psychophysical measurements linked to optical observations with a wavelength-sensitive detector. Thus the relationship of viewer preference for color reproductions to the response to a scene in a natural setting is uncertain and in doubt if one wishes to link wavelength-dependent optical measurement with human experience.

Recently, scientists have begun to orient some experiments of atmo-

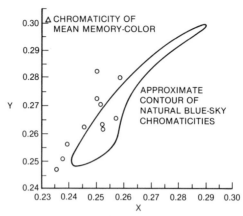

Fig. 8.15. Comparison between chromaticities corresponding to psychologically blue-sky reproductions (with respect to C.I.E. illuminant C), shown by date with the zone of natural chromaticities. Also indicated is the chromaticity of the mean memory color ($\triangle$) from several observers. [From Bartleson and Bray (1962).]

spheric vision to address the questions of psychological impressions (Malm *et al.*, 1981; Latimer *et al.*, 1981; Mumpower *et al.*, 1981). The first two are aimed at establishing objective methods for linking physical and chemical measures of air quality with visual impressions of scenic panoramas in parts of the United States where urbanization is absent. The third study concerns analogous questions for the urban environment.

The work of Malm *et al.* (1981) has concentrated on vistas in the southwestern United States. Their studies so far have affirmed the fact that visual impressions will be influenced by clouds, sunlight and angle relative to the observer, and the nature of the viewed topography. An attempt to investigate the observer threshold for estimating light contrast change has basically reaffirmed the range of discrimination found by earlier investigation—namely the 2–5% contrast difference range.

The work of Latimer *et al.* (1981) has used color photography of selected scenic vistas to survey several hundred groups of individuals with widely different demographic backgrounds. The motive of the survey was to explore relationships between (a) optical measures of haze, such as nephelometry and teleradiometry; (b) impressions of visual air quality (given in terms of an "arbitrary" visual air quality index, VAQI); and (c) a scenic beauty estimate (SBE) defined by work of Daniel *et al.* (1973). The VAQI is supposedly a measure of the influence of haziness on the panorama as distinct from its scenic value. The studies of Latimer and co-workers so far have indicated that the diverse observer groups surveyed were surprisingly well correlated in their reactions to perceived visual air quality and scenic beauty. Ratings of scenic beauty and visual air quality were found to be closely interrelated. However, visual air quality ratings were more sensitive to visual range changes than the scenic beauty ratings. The average correspondence between VAQI and SBE is shown in Fig. 8.16. The group shown a series of color photographs selected the vistas of Grand Canyon National Park as far superior to those from Mt. Lemmon or Kitt Peak near Tucson, AZ, or from the Smoky Mountains or Shenandoah National Parks of the eastern United States. However, even the reported visual impressions varied widely at the Grand Canyon depending on foreground features, cumulus clouds, or illumination shadowing.

The earlier conclusions about the blueness of the sky of Bartleson and Bray (1962) were verified again. The authors' most important conclusion was "that visibility impairment may have profoundly different effects on the human aesthetic experience . . . because of differences in landscape characteristics." This complicates the practice of interpretation of the measurements of atmospheric optical properties and aerosol behavior for regulatory purposes, and requires a sophistication in knowledge of the link to sensory perception which is basically absent at present.

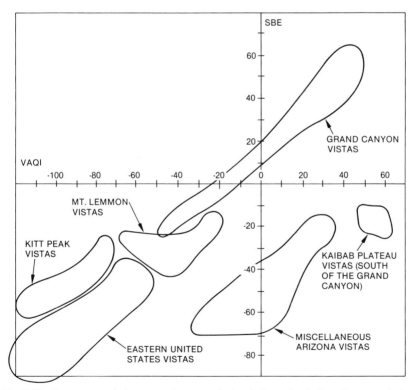

Fig. 8.16. Average scenic beauty estimates and visual air quality indices for several vistas. [From Latimer *et al.* (1981); courtesy of Pergamon Press, Ltd.]

The studies of Mumpower *et al.* (1981) employed certain cues of visual experience in the summer environment of Denver, CO to link physically measurable quantities of the air to visual impressions. At each observation point, two observers recorded at various times the weather at their observation site (clear, partly cloudy, overcast, hazy or foggy, rain or snow) and their overall rating of visual air quality (on a "subjective" scale). They then recorded observations in four directions (north, east, south, west) of (a) visual range, (b) the clarity of objects in the distance, (c) the color of the air, (d) the presence or absence of a border between colorless and discolored air, and (e) the suspected source of any degradation in visual air quality. The observers also took photographs of each scene for which they recorded data.

In addition, each observer collected data from at least one, and usually two (or more) individuals at or near their observation site. The respondents were asked to provide ratings of the present weather and the current overall visual air quality. They also responded to questions about the visual range, clarity of objects in distance, color of the air, presence or absence of a border, and imputed source of degradation, in one direction (of their choosing).

The procedures used in the Mumpower study differ from those in most studies of visibility in an important respect. First, most studies involve the collection of data from one (or a few) standardized observation sites. The advantage of such a procedure is generally argued to be that it provides the ability to make "precise measurements." For example, one can make precise measurements of visual range by establishing the physical distance of landmark objects. Measurements of visual range are considered free of the subjectivity inherent in scales used in human judgments. The severe disadvantage to such an approach is that it frequently leads to situation-specific results that cannot be generalized with confidence to all situations involving the phenomena under study. In order to understand human judgment of viewing, the processes of interest should be studied across the range of such combinations that individuals encounter. Otherwise, the results obtained will be peculiar to the particular combination of cues involved in the specific situation.

To try to prevent the results of the Mumpower study from being situation specific and, thus, to aim toward generalizations, observations were collected by individuals from throughout the Denver metropolitan region, at different locations and times. This approach was adopted to avoid the generation of results and conclusions that might be misleading because of undetermined effects of a standardization of protocol similar to that required for prevailing visibility.

The Mumpower study also elicited responses from both observers and their respondents in a less precise manner than is typical of studies of visibility. The study thus deliberately avoided a precise statement of "cues" for judgments of overall visual air quality. The reason for this strategy again was to avoid implicitly defining cues distinct from the manner in which individual observers might define and use them. As a possible consequence, the information acquired could mislead the analyst about the importance of protocol to observer judgments of visual air quality. The strategy of Mumpower *et al.* first tried to identify the categories or types of cues that observers relied upon in making judgments of overall visual air quality; attempts to develop more precise definitions or descriptions of these cues were reserved for future studies.

Given their basis, the Mumpower *et al.* results to date indicate (a) only a moderate degree of agreement in individual rating of visual air quality, (b) weak identification with perceptual measures of weather, visual range, clarity, color, presence of border, or inferred pollution source, and (c) weak empirical interrelationships of physical measures which account for only one-third to two-thirds of the variance in judgments of visual air quality. Their work indicates that clarity (of image) in viewed objects was the most important cue to visual air quality, but visual range per se was inadequate as a single measure of visual air quality. Finally, photographic displays did not appear to give the same human impression as elicited by real-time, on-site observations.

The results of the Mumpower study reinforce important questions raised by others about the subtleties of linking visual effects in the atmosphere to aesthetics as a complement to transportation-related safety considerations. The work does focus the direction of new work on objective relationships interpreted from imaging techniques, for example, by video camera (e.g., Viezee and Evans, 1980) rather than the nonimaging optical methods currently adopted in practice.

The three-stepped synthesis of atmospheric optics, psychophysics, and psychology remains in its infancy as part of a science describing atmospheric visual effects, but undoubtedly will be refined energetically as long as environmental regulatory legislation remains a stimulus.

8.1.5   Visibility and Weather Conditions.

With the interest in visibility as an indicator of pollution in the atmosphere, increasing efforts have been reported investigating the historical data on visual range. Although this parameter is an imperfect means of characterizing visual impressions, it nevertheless is the only description widely available at present. These data are derived primarily from airport observations of prevailing visibility. Such studies have been reported for certain cities and more recently for extended geographical regions. The latter work is motivated by concerns for regional-scale effects of widespread urbanization or industrialization, including the accumulation of pollution of large geographical areas under certain weather conditions.

Conditions for mean visibility for the United States are shown in Fig. 8.17. The pattern of isopleths indicates that there is a dramatic difference in prevailing visibility between the eastern and parts of the western United States.

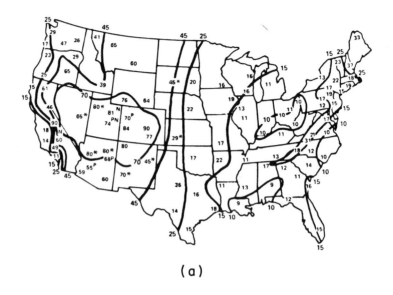

(a)

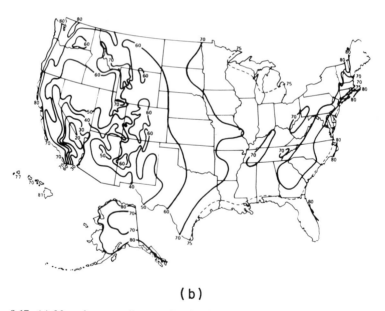

(b)

Fig. 8.17. (a) Map shows median yearly visual range (miles) and isopleths for suburban/
nonurban areas, 1974–1976. P, based on photographic photometry data; N, based on nephelometry data; *, based on uncertain extrapolation of visibility frequency distribution. [From
Trijonis and Shapland (1979).] (b) Annual mean relative humidity (%). [From National Oceanic
and Atmospheric Administration (NOAA), 1974.]

Relatively low visibilities of less than 20 miles generally are found east of the Mississippi River region, while much higher visibilities are reported in the West. The highest visibilities are reported in the arid, and unpopulated Southwest. The average distribution of visibility over the continent probably is related mainly to the influence of climatological factors, but it may involve air pollution, too. Though this pattern of distribution has been known for some time, large-scale climatological features associated with it have not been studied specifically, except for the role of humidity.

The pattern of visibility in Fig. 8.17a follows closely the distribution of relative humidity as indicated by comparison with Fig. 8.17b. The mean visibility patterns can also be associated with the differences of suspended particulate concentrations, as well as with water content of the air. Direct measurement of the aerosol liquid water content and the light-scattering coefficient have been reported for a range of relative humidities between 20 and 80%. A linear relationship between these parameters analogous to that of Eq. (8.5b) for aerosols sampled in urban air of Pasadena and Pomona, CA as well as dry desert air (Goldstone, CA) is shown in Fig. 8.18. A strong relationship between visibility and liquid water content (LWC) has been shown by many investigators in moist air near saturation; an example of correlated data is given in Fig. 8.19. This relation has the form $L_v = \text{const} \times (\text{LWC})^{-2/3}$. Here the visual range is empirically shown to be exponentially a function of liquid water content when a dense, moist haze or fog is present,

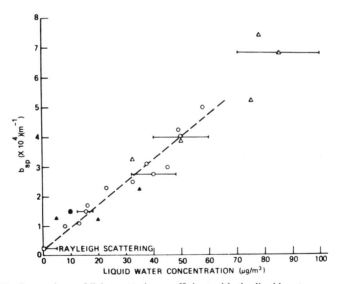

Fig. 8.18. Comparison of light-scattering coefficient with the liquid water concentration in atmospheric aerosols. $\bigcirc$, Pomona (October 4–9); $\blacktriangle$, Pasadena (September 20); $\triangle$, Pasadena (September 15); $\square$, Goldstone (November 1). [From Ho *et al.* (1974); courtesy of the American Meteorological Society.]

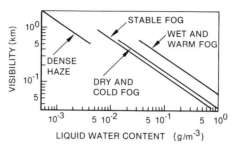

Fig. 8.19. Relationships between visibility and liquid water content. Curves 1, 2, and 3 correspond to the three average values of the proportionality constant 0.013 (dense haze), 0.034 ("dry and cold" fog), 0.060 ("wet and warm" fog). [From Tomasi and Tampieri (1976); courtesy of the Canadian Meteorological Society.]

with the proportionality constant differing somewhat for different fog or haze conditions.

The relationship between visual range and relative humidity can be readily appreciated considering the hygroscopic nature of suspended particles, especially those in the optically efficient light-scattering range (Covert *et al.*, 1980). The visual range basically is inversely proportional to mass concentration of particles. However, atmospheric particles are hygroscopic and tend to absorb moisture in a strongly nonlinear proportion to their (dry) mass, as suggested in Fig. 7.10. Thus, at high humidities, the actual mass concentration in the light-scattering range is much higher than the nominal mass concentration measured at relative humidities less than 50%.

8.1.6 Meteorological Factors versus Air Pollution. The role of air pollution in affecting visibility on a regional scale has been examined by several investigators. Early analyses for trends in decreased visibility with urbanization were reported by Munn (1972), and Husar and Patterson (1980) have analyzed historical visual range trends showing decreasing visibility in parts of the eastern United States. Trijonis and Yuan (1978) and Latimer *et al.* (1978) have conducted a similar analysis of historical airport visibility in the southwestern United States. Like the greater Northeast, the historical data suggest a deterioration in visibility over roughly 15 years from the 1960s to 1972. However, after 1972, there seems to have been improvement in visibility in the Southwest at least.

One explanation of trends in visibility is associated with man's increased consumption of raw materials containing sulfur compounds, which has resulted in increased SO_2 emissions over the years. As discussed in Chapter 7, the atmospheric oxidation of SO_2 produces particulate sulfate. Circumstantial evidence for relating light scattering in the troposphere to sulfate particles was collected by Waggoner *et al.* (1976) and others. Thus the trends in atmospheric sulfate were thought to be traceable to trends in coal and other fuel consumption for power generation or heating in the East, and the copper smelting industry in the West. Husar and Patterson (1980) showed a relation between coal consumption and the extinction coefficient derived from a number of airports reporting visibility between 1946 and 1978. This relationship is shown in Fig. 8.20. Though appealing for its direct explanation, their analysis did not account for potentially important climatological changes which could also influence visibility patterns. Suppose, for example, one takes the pattern for rainfall as a measure of the liquid water content trends in the troposphere, and rainfall is considered a climate variation indicator. A plot of United States rainfall data reported by Diaz and Quayle (1980) is added to the results in Fig. 8.20. The result suggests that the water content of the air also increased over the same period of coal consumption increase. Thus one must look more carefully into associations between visibility and air pollution to be sure of the cause and effect.

A similar argument has been reported about trends of visibility in the West associating particulate sulfate with changes in visual range. Sulfate in turn was related to patterns of SO_2 emissions from the nonferrous smelting industry. In the past, smelters in the West have reported very large SO_2 emissions, expecially at plants located in the southwestern United States. Perhaps most often cited is Trijonis' (1979) assertion that improved visibility was found during the mid-1960s when a smelter strike took place. This

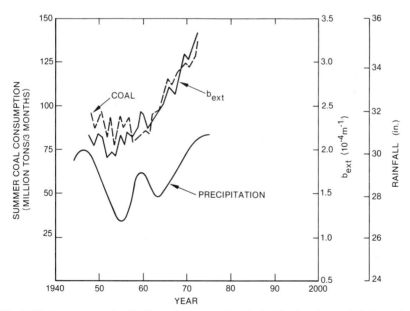

Fig. 8.20. Summer trends of U.S. coal consumption (dashed line) and annual U.S. precipitation compared with eastern U.S. average extinction coefficient, or haziness (solid line). [Adapted from Husar, and Patterson (1980); precipitation data taken from Diaz and Quayle (1980).]

conclusion was based on examination of airport visibility at several stations in the Southwest, which tended to improve during the strike period. An example for Phoenix, AZ is shown with estimated SO_x emission variations in Fig. 8.21a. The changes in visibility coincided with evidence of ambient particulate sulfate reduction based on available monitoring data at locations including Phoenix. Unfortunately, a simple interpretation of visibility data was confounded by observer factors and features of unusual weather.

Another evaluation of the Phoenix data has been reported by Henry *et al.* (1981b). They verified B. Herman's earlier conclusions about uncertainties in visibility observations at not only Tucson, but other stations, including Phoenix. Unlike Trijonis (1979), they concluded that the weather during the mid-1960s smelter strike was unusual. The apparent result of weather anomalies is indicated in Fig. 8.21b. Here the sequence of frequencies of occurrence of visual range between 40 and 45 miles reported by airport observers is shown, with a band of uncertainty given by three times the standard deviation σ of the data. The data sequence is partially adjusted statistically to remove weather anomalies during the period. Basically, the variability in frequency of occurrence in this visual range band generally falls within expected uncertainty before, during, and after the strike period. However, the improving trend begins ten months before the strike but begins to deteriorate again during the strike period. There was unusually hot and dry weather (good visibility at the maximum) followed by heavy rains and elevated moisture in the winter during the strike. Thus this example at Phoenix is an alternative interpretation of the Trijonis pollution hypothesis. Herman's[†] careful examination of the airport observations at Tucson also revealed important irregularities in observer reporting and target identification which confounded Trijonis' interpretation.

8.2 CLIMATE AND RADIATIVE ENERGY BUDGETING

Closely related to the processes that influence visibility are those that affect the transfer and retention of radiant energy in the earth's atmosphere. The radiant energy budget and its changes in turn are a direct link to climate. The sun is the main source of energy for atmospheric circulation, and for the earth's surface heat. The sun emits energy almost entirely by electromagnetic radiation, a portion of which is intercepted by the earth and its atmosphere. By comparison, the conduction of energy from the earth's core to its surface appears to contribute less than 1% of the energy the earth absorbs from solar radiation. If the earth absorbed solar radiation without loss, its temperature would continually increase. Since the mean global temperature is approximately constant, the planet must lose energy at the same rate as it is absorbed. The only way it can lose energy is by radiating into the void of

---

[†] Personal communication.

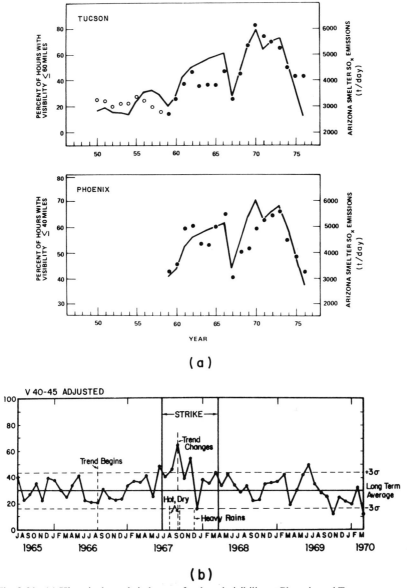

Fig. 8.21. (a) Historical trends in hours of reduced visibility at Phoenix and Tucson compared to trends in SO$_x$ emissions from Arizona copper smelters. The dots represent yearly percentage of hours with reduced visibility (measured on left axis). Note that the Tucson observation site moved in 1958; although this move did not produce a statistically significant change in reported visibilities, open dots are used to distinguish data prior to 1958. The lines represent yearly SO$_x$ emissions from Arizona copper smelters (measured on right axis). [Marians and Trijonis (1979).] (b) Monthly frequency of occurrence of Phoenix visibility better than 40–45 miles partially statistically corrected for meteorology. Dotted lines drawn for guiding eye, and show an uncertainty based in observations based on three times the standard deviation. [From Henry *et al.* (1981b).]

space. The balance between incoming and outgoing radiant energy is termed the energy or heat balance for the planet. Any major disruption of this balance would have serious consequences for life on the earth, with global warming or cooling. Indeed, there is speculation that major changes in living species have been caused by perturbation eons ago. Interpretation of prehistoric evidence suggests now that the relatively sudden demise of the great dinosaurs millions of years ago was brought about by climate change associated with aerosol particle clouds from eruptions of volcanoes, or from the impact of a giant meteorite on the earth's surface (Alvarez *et al.*, 1980).

The global energy budget takes account of gains and losses associated with absorption, reflection, and radiation of energy in different parts of the atmosphere and the earth's surface. Spatial differences in heating and cooling of the atmosphere basically create the driving forces for atmospheric circulation. The atmospheric radiation budget depends on energy absorption by water vapor and carbon dioxide, as well as by interaction with water clouds in the infrared parts of the energy spectrum. Ozone in the stratosphere has significance for absorption of incoming solar energy in the ultraviolet part of the energy spectrum. Aerosol particles also influence the radiation balance by scattering and absorbing light in the visible and infrared parts of the spectrum.

8.2.1   Role of Aerosol Particles.   The role of aerosol particles in the energy budget is shown schematically in Fig. 8.22. The incoming solar radiation enters through the upper atmosphere with a wavelength of approximately 0.6 μm at maximum energy. It is returned to space in the infrared with a wavelength at maximum energy of approximately 20 μm. Particles in the stratosphere, with a maximum concentration at approximately 20 km, scatter incoming sunlight and absorb outgoing infrared radiation. Tropospheric particles near the earth's surface also scatter and absorb incoming and outgoing radiation, reemitting radiation in the infrared. The effect on surface temperature depends on these complex interactions of energy absorption and transfer.

The possibility for aerosol particle interactions with the earth's radiation balance was recognized before the nineteenth century in Benjamin Franklin's work. Franklin was uncertain about whether the aerosols originated from comets or volcanoes, but it is clear that the idea that aerosols cause climatic cooling is at least two centuries old. Franklin's theory was well stated but not quantitative, and he did not give any data on the aerosols.

Early in this century, Humphreys performed a detailed quantitative analysis of the effect of volcanic aerosols, and revised his work in 1940. Considering the energy balance at the earth's surface, he estimated that volcanic aerosol injections, such as those by Krakatoa in 1883, could cool the surface by 7°K owing to their interception of solar radiation. He recognized the counteractive warming by the aerosols due to their "greenhouse" blanketing

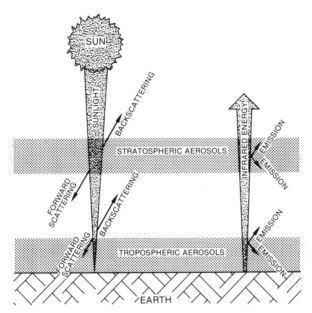

Fig. 8.22. This schematic drawing shows the interactions between aerosols and sunlight, and aerosols and infrared light. Aerosols in the stratosphere absorb and filter light from the sun, and absorb infrared light from below. Changes in the absorption and scattering with height theoretically influence the earth's heat balance. [After Pollack and Toon (1980).]

of thermal radiation from the earth's surface, but he argued that because of the small size of the aerosols their opacity to thermal radiation would lead to only a negligible warming of a few tenths of a degree. Humphreys overestimated the cooling effect of the aerosols, in part because he thought the aerosols must reflect most of the radiation they intercept back out of the atmosphere, while in reality they scatter most of it into the forward hemisphere. Also, Humphreys' simple assumption of an energy balance at the earth's surface must be replaced by a proper radiative transfer calculation, and the effect of the thermal inertia of the ocean in damping the radiative response of the earth must be included to yield a realistic magnitude for the temperature perturbation, since the thermal forcing due to the volcanic aerosols is too short lived to reach an equilibrium condition.

Rasool and Schneider (1971) focused attention on the aerosol/climate problem when they made radiative transfer calculations that indicated that an increase in atmospheric aerosols by a factor of 4 could cool the earth's surface temperature by about 4°K. Subsequently, a large number of further investigations of this problem have been made, and a fairly detailed understanding of the radiative aspects of the problem has been developed.

At the present time, Hansen *et al.* (1980) have noted that two major factors prevent reliable prediction of the future effect of aerosols on climate are (a) large uncertainties about aerosol optical properties, particularly anthropogenic aerosols, and lack of knowledge of how these will change in the future; and (b) the primitive state of the models for the complete climate system into which the aerosols' radiative perturbations must be inserted.

The physical properties of aerosol particles, such as size, shape, refractive index, and concentration in the atmosphere, control aerosol interaction with light according to a set of derived properties which are known as optical properties. Three fundamental properties, introduced in Chapter 5, are the optical depth τ, the measure of the size and number of particles present in a column of air; the single-scattering albedo, the fraction of light intercepted and scattered by a single particle; and the asymmetry parameter, an integrated measure denoting the portion of light scattered forward in the direction of the original propagation and the portion scattered backward toward the light source (see also Hansen and Travis, 1974).

A light ray traversing an aerosol-laden column of air is reduced by the effects of absorption and scattering as the exponent of the optical depth, which in turn is basically proportional to the product of the number of particles in a column of unit area and the cross-sectional area of a single particle.

A fraction of the light intercepted by a particle is scattered and a fraction is absorbed. The fraction scattered by any single particle is the single-scattering albedo $\tilde{\omega}_0$. Most atmospheric particles do not strongly absorb visible light, so $\tilde{\omega}_0$ for these particles is between 0.9 and 1.0. However, there are minor aerosol particle constituents such as soot that absorb visible light, and when small quantities of these constituents are present, $\tilde{\omega}_0$ can be reduced to about 0.5. The worldwide and even the local prevalence of absorbing compounds is poorly known. Most aerosols strongly absorb infrared light, and thus $\tilde{\omega}_0$ is less than 0.1 in the infrared. The product $\tilde{\omega}_0\tau_a$ is the scattering optical depth: a light ray traversing an aerosol-laden column will be reduced due to scattering as the exponent of $\tilde{\omega}_0\tau_a$. The absorption optical depth is $(1 - \tilde{\omega}_0)\tau_a$.

The light scattered by aerosols is not scattered uniformly in all directions but has a complicated angular distribution that can be a function of wavelength. A simple integrated measure of the angular scattering is the asymmetry parameter g, which varies from -1 to 1. If g were 1, all the light would be scattered into the hemisphere centered in the direction of the light beam's original propagation. If g were -1, all the light would be scattered backwards. For typical aerosols g is greater than zero. Owing to the small particle size relative to infrared wavelengths, g is about 0.5 or less for most aerosols in the infrared. At visible wavelengths, g tends to be close to 0.7. Fortunately, g depends only weakly upon the particle size and composition, and thus the observational uncertainty about g is less significant for climate than the uncertainty about $\tilde{\omega}_0$ and τ.

We have seen earlier that aerosols are not uniformly distributed over the earth, particularly in the lower atmosphere. At present, research into the effects of aerosol particles on climate is evolving from simple studies of global, time-averaged problems toward more complicated regional, temporally varying problems. However, a basic understanding of the dependence of climate on aerosol optical properties is most easily gained by reviewing global average calculations [see Ramanathan and Coakley (1978) for modeling assumptions]. In such calculations, atmospheric motions are generally ignored. The climate change is represented by a change in the global mean surface temperature, which is estimated by taking into account radiatively absorbing gases and particulate matter in a one-dimensional scheme.

Climatic changes such as droughts or a series of harsh winters are usually restricted to small areas of the earth. Often there will be compensating climatic changes from region to region: drought in one location may be nearly balanced by unusually high precipitation in another location. Periods of climatic extremes in numerous portions of the globe have been found to accompany small changes in global temperatures. We can calculate small changes in global temperature using global average climate models, but not local temperature and precipitation changes—the quantities of real climatic importance. The local changes must simply be inferred from the global average calculations on the basis of past climatic changes. Typical changes in the earth's global mean surface temperature over the last thousand years, during which the climate has varied considerably, have been about 1°C, whereas the difference between the present and the ice-age mean temperatures is only about 5°C. Calculated global temperature changes of even several tenths of a degree are therefore thought to be significant.

The direct radiative effect of aerosol particles on atmospheric temperatures can be accurately computed with available radiative transfer techniques. Such calculations have limited significance because they exclude important feedback effects in the climate system as well as effects of the aerosols on processes such as cloud formation. In fact, most of the estimates of particle impact on climate have been based on calculations of only the direct radiative effect, and it is logical to first try to understand this before proceeding to account for additional complications.

To illustrate the degree to which aerosol particles can affect global temperature through their direct radiative effect, including an indication of how the results depend upon various particle parameters, a recent example calculation is chosen. The one-dimensional radiative–convective model selected is described by Hansen et al. (1978). It is basically similar to the model first employed by Manabe and Wetherald (1967) more than a decade ago. The Hansen et al. model computes a global average temperature under the assumption that relative humidity, cloud cover and height, and carbon dioxide levels do not change in the climate perturbation experiment.

Figure 8.23 shows the global average temperature change that results from varying the amount of atmospheric particles while keeping everything else

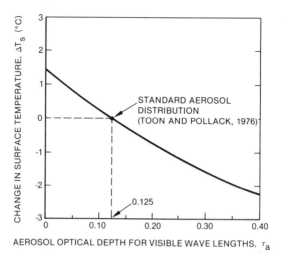

Fig. 8.23. Change in surface temperature as the optical depths increase with aerosol amount, compared to "standard" particle distribution. [From Hansen *et al.* (1980); reprinted with permission of the New York Academy of Sciences.]

constant. The global average aerosol distribution of Toon and Pollack (1976) is employed as a standard for comparison. This distribution has a total optical depth of 0.125; it contains sea salt, basalt, and sulfates in the lowest three kilometers of the atmosphere, where the volume-weighted mean particle radius of the size distribution is about 2 μm; it contains sulfates and basalt between 3 and 12 km, with a mean radius of about 0.5 μm; it has sulfuric acid particles in the stratosphere, with an optical depth of 0.005 and an effective particle radius of 0.2 μm. The important point to note is that, for these assumed "standard" aerosols, increasing the amount of particles by a factor of 2 would cause an estimated cooling of about 1.2°K. The effect is essentially linear within the incremental range of optical depths of interest.

Figure 8.24 illustrates how the surface temperature would change if the single-scattering albedo of the aerosols were changed, with all other parameters left unchanged. It shows that making the particles perfectly conservative (nonabsorbing) would result in a further cooling of only 0.1°K. On the other hand, by introducing substantial absorption in the particles, it is possible to cause a large heating at the ground. The graph may be somewhat misleading in the sense that most materials tend to fall close to the scattering range of the diagram, but it is appropriate from the theoretical standpoint, since the temperature effect theoretically is almost linear with a change in single-scattering albedo.

The strong dependence of surface temperature on the single-scattering albedo of the suspended particles is a major difficulty when it comes to estimating how additional aerosol particles will affect the temperature. This

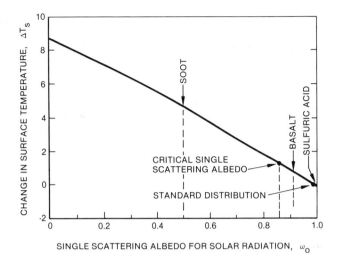

Fig. 8.24. Change in surface temperature as the single-scattering albedo of the aerosols is changed, compared to the standard aerosol distribution. The "critical albedo" is that value which would eliminate the cooling effect of the aerosols. [From Hansen *et al.* (1980); reprinted with permission of the New York Academy of Sciences.]

means that we need to know not only the principal composition of the particles, but also the composition of any impurities that may affect the single-scattering albedo. However, it is noted that the *critical albedo,* the single-scattering albedo for which a change in aerosol amount would neither heat nor cool the surface, is relatively small. Thus suspension of most aerosol particles tends to have a cooling effect. Yet there may be significant exceptions, as in cases where some desert and soil particles are suspended, or urban aerosols containing a substantial amount of carbon are present.

Figure 8.25 shows that the temperature impact of aerosols also depends upon the altitude at which they are suspended. In this case, Toon and Pollack's standard aerosol distribution has been employed, but an additional layer of aerosols has been added at the height designated in the graph (Fig. 8.25). The results are more complicated than in the previous calculations, because there are competing effects in the solar and thermal parts of the spectrum. The cooling effect due to the increase of the earth's albedo becomes somewhat stronger when the aerosol particles are at greater altitude, where they are above the other radiative constituents. On the other hand, the greenhouse warming in the thermal infrared also increases as the altitude increases within the troposphere, because the aerosols radiate at a colder temperature. Aerosol particles of the type described by Toon and Pollack's (1976) standard model cool the surface for particles at all altitudes, with maximum cooling for low-level aerosols. However, it should also be emphasized that if the aerosol particles are either (a) more absorbing or (b) larger

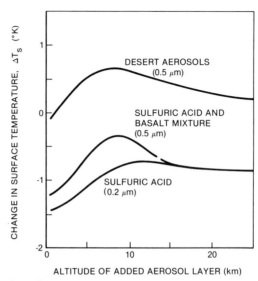

Fig. 8.25. Change in surface temperature as the altitude of a layer of aerosols is varied. The atmosphere contains the standard aerosol particle distribution of Toon and Pollack (1976), plus a layer of either (1) sulfuric acid; (2) 60% sulfuric acid, 40% basalt mixture; or (3) model desert particles. The optical thickness of the added layer is 0.1, except in the case of the desert aerosols, for which it is 0.025. [From Hansen *et al.* (1980); with permission of the New York Academy of Sciences.]

than the standard case, effect will tend to be more in the direction of heating. This is illustrated in Fig. 8.25 for desert aerosols with substantial absorption.

The optical thickness also depends on aerosols in the thermal infrared window region. The infrared optical thickness is sensitive to the largest particles in the size distribution; i.e., to the large particle tail of the distribution. Thus, accurate knowledge of the aerosol particle size distribution is required in order to reliably state the direct radiative effect of the aerosol cloud.

There are several other parameters that influence the radiative effect of the aerosols, in addition to those considered so far. One parameter that has received considerable attention is the ground albedo; however, it is constant within the range 5–25% over most of the earth, and thus apparently does not significantly affect the above results computed for a 10% ground albedo. Another is the asymmetry of the single-scattering phase function of the aerosol particles. Toon and Pollack note that this is included implicitly in their results through the Mie scattering computations for different size distributions. The parameters $\tilde{\omega}_0$ and τ_a are the principal variables that need to be known in order to relate tropospheric aerosols and climate. The major complication for the analysis is that aerosol particles are not uniformly distributed, either horizontally or vertically, throughout the troposphere, and therefore $\bar{\omega}_0$ and τ_a vary widely with location. The value of the critical albedo also varies with location because it depends upon the ground albedo. In addition, once the aerosols affect the radiation budget in one area, the atmosphere responds with complex changes in the wind and clouds. Because aerosols are heterogeneous, theorists cannot provide a single answer to the question of whether tropospheric aerosols cool or warm the earth: there is no single answer.

Most aerosol particles exist near the earth's surface, but the Junge layer of particles is found about 20 km above the surface, in the stratosphere. These aerosols normally have such a small τ that they have negligible effect on climate. However, after large volcanic eruptions, their optical thickness can be as large as that of the aerosols in the troposphere. The relation of climate to the optical properties of stratospheric aerosols is slightly different than for tropospheric aerosols, as has been demonstrated by several investigators (Pollack and Toon, 1980).

Figure 8.26 shows that the relation between the surface temperature and the optical depth for the stratospheric aerosols is quite close to that given in Fig. 8.22 for tropospheric aerosols. However, for stratospheric aerosol particles, the dependence of warming and cooling upon $\bar{\omega}_0$ and upon the variation of τ_a with wavelength is different than for tropospheric aerosols. The stratosphere is not strongly coupled to the earth's surface by atmospheric dynamics, and if the stratospheric aerosols absorb some incoming solar energy, the energy would not be conducted to the surface. Indeed, it is found both observationally and theoretically that the presence of stratospheric particles simultaneously causes the surface to cool and the stratosphere to warm. The aerosols warm the stratosphere by absorbing solar and infrared energy. The solar energy that the stratospheric aerosols and backscatter does not reach the surface, causing cooling there. Hence, stratospheric aerosol particles cool the surface for all values of $\bar{\omega}_0$.

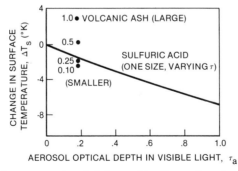

Fig. 8.26. The surface temperature is influenced by the number of sulfuric acid particles in the stratosphere and by the size of volcanic dust particles. Large particles tend to warm the surface whereas small ones cool it; the sulfuric acid particles are usually smaller than 0.1 μm, and thus they have a cooling influence. [After Pollack and Toon (1980).]

The size of the stratospheric aerosol particles is an important factor in determining whether the climate warms or cools. Small particles, as shown in Fig. 8.26 tend to cool the surface, whereas larger ones warm the surface. For stratospheric particles whose size is less than 0.1 μm, the ratio of the infrared τ_a to the visible τ_a is 0.1. For 0.25-μm-radius particles, however, the ratio is 0.5, and for 0.5-μm-radius particles the two optical depths are nearly equal. The reason infrared opacity is more important for stratospheric aerosol particles is that they are at high altitude and low temperature. The larger cold particles in the stratosphere are very effective at blocking infrared radiation coming up from the atmosphere and surface below, thereby warming the surface. Because tropospheric aerosols lie below much of the atmosphere and are almost as warm as the ground, they are not nearly as efficient at blocking the escape of infrared energy.

Globally distributed stratospheric aerosol particles a few tenths of a micrometer in radius will always tend to warm the stratosphere. However, the earth's surface will cool if most of the suspended particles are smaller than a critical size—near 0.5 μm, depending somewhat upon the aerosols' composition and the earth's albedo. Hence τ_a and the size of the stratospheric aerosol particles are the most important parameters needed to estimate the effect of the aerosols on the climate. The major complicating factor is that, after a volcanic eruption, the aerosol particles change in size and composition as they spread out over the earth. Although observational data are few, it seems that the size of the particles varies for several months; as the dust veil begins to form, the particles may be larger than the critical size of about 0.5 μm.

For calculation of climatological change as exhibited through the earth's surface temperature, the stratospheric and tropospheric optical depths need to be known, with the atmospheric particle size distribution and the tropospheric particle single-scattering albedo at visible wavelengths. Although information is yet to be compiled, considerable progress has been made toward either measuring directly these properties, or providing data for estimating them.

Experimental projects to determine $\bar{\omega}_0$ have used one of three approaches: direct measurement of $\bar{\omega}_0$ in the atmosphere (Herman *et al.*, 1975), laboratory measurement of the absorption by collections of atmospheric aerosols and calculation of $\bar{\omega}_0$ (Patterson *et al.*, 1977), or laboratory measurement of the absorption by pure materials known to compose aerosols and calculation of $\bar{\omega}_0$ (Palmer and Williams, 1975; Twitty and Weinman, 1971). The laboratory studies have shown that materials such as sulfates and sea salt are very transparent to visible light and have $\bar{\omega}_0$ very close to unity. Most soil particles are only moderately absorbing and have $\bar{\omega}_0$ above 0.9, although smaller values are occasionally found.

Accurate direct measurements of $\bar{\omega}_0$ are quite difficult to make, and the available values could be incorrect. Also, measurements have not been performed in enough locations to obtain a "typical" value. Most of the lowest $\bar{\omega}_0$ values have been found in urban regions, which make up only a small fraction of the earth's area. For these reasons, the available direct studies may be misleading for application to average global conditions. At the same time, it is also quite possible that the laboratory studies have overlooked minor, highly absorbing materials such as iron oxides or soot. If 10–20% of the atmospheric particles were composed of soot, the value of $\bar{\omega}_0$ would be below the critical value even if the remaining bulk of the material were completely transparent. If the absorbing particles were much smaller than the typical suspended particles in the air, an even smaller mass fraction could be very significant (Bergstrom, 1973). It has been found that a small amount of soot is responsible for the low values of $\bar{\omega}_0$ in urban areas (Rosen *et al.,* 1978).

8.2.2 Upper Atmospheric Effects of Volcanic Eruptions. The global radiation balance can be disturbed enough to be measured under some conditions. Disturbances have been documented both in the stratosphere and in the troposphere. Effects in the lower troposphere are suspected but are less certain. The earth's energy balance can be perturbed by a variety of possible processes. Robock (1978) has investigated both "internal" and "external" causes for such disturbances using a numerical model which includes radiation and convective heat transport. His experiments suggest that a variation in annual mean surface temperature can be forced internally by random turbulent eddies in the atmosphere. External causes may include (a) disturbances in the solar constant associated with sunspot activity, (b) volcanic eruptions, or (c) additions to the carbon dioxide and aerosol particle burden and direct heating by man's activities. Robock's experiments indicated changes in heating in itself were inadequate to cause climate changes observed in the past. The influences of man's activities are potentially large, but contributions from CO_2 heating appear to roughly cancel the cooling from aerosol particles. However, the effects of volcanic dust injections in the upper troposphere and stratosphere suggest a major detectable effect which should be observable in historic data.

A large volcanic eruption provides an excellent opportunity for a case study of the climatic response to a change in atmospheric aerosols. Volcanic aerosols injected into the stratosphere tend to be spread globally by stratospheric winds, so if their optical thickness is substantial, they should be capable of producing a measurable climatic response.

The effect of volcanic aerosols on climate has been considered by a number of different investigators (Lamb, 1970; Pollack *et al.,* 1976; Mass and Schneider, 1977). Lamb has conducted a comprehensive study, at least from an empirical point of view. He has listed a chronology of volcanic eruptions and a "dust veil index" for the period subsequent to 1500 A.D., and, in general, he has found some correlation of volcanic events with climate perturbations. Lamb's dust veil index is based in part on observed temperature data because of the lack of direct knowledge about the aerosols.

Pollack *et al.* (1976) examined observations from the past century, 1870–1970, for which it is possible to estimate the aerosol optical thickness from measurements of the transmission through the atmosphere for solar radiation. The measurements are only intermittent; they are made with several different techniques, and they exist for only a small number of locations. However, for periods of substantial volcanic contributions, the investigators consider the estimates are probably reliable within about a factor of 2. With the assumption that the measured opacity was from stratospheric particles composed of sulfuric acid, Pollack *et al.* calculated a radiative energy balance to obtain the Northern Hemisphere temperature perturbation, as shown in Fig. 8.27. They calculated an increase in temperature of about 0.5°K between 1890 and 1920, which is in surprisingly good agreement with observations. However, between 1940 and 1963, the observed temperature *declined* by 0.2–0.3°K, while the model showed an increase because of an absence of substantial volcanic eruptions and a continuing CO_2 buildup.

The numerical experiments suggest that there are other mechanisms besides volcanic aerosols and CO_2 affecting the mean global or hemispheric temperature. This makes it difficult to compare observations with model results over a long period and draw a firm conclusion about the role of aerosols in climate change. One way to try to minimize the effect of the superposition of different mechanisms is to make a statistical analysis of the temperature for a few years before and a few years after a large number of different volcanic eruptions. This has been done by Mass and Schneider (1977) who found a small, but possibly significant, cooling after volcanic events, with a magnitude on the order of 0.1°K.

A complementary approach has been taken by Hansen *et al.* (1978). They examined one of the largest and best-documented volcanic eruptions. The

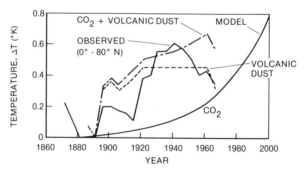

Fig. 8.27. Observed Northern Hemisphere surface temperature changes and model results of Pollack *et al.* (1976). [From Hansen *et al.* (1980); reprinted with permission of the American Association for the Advancement of Science.]

eruption of Mount Agung on the island of Bali in 1963 was of particular interest. This eruption occurred at a time when tropospheric temperatures were being measured routinely, accurate measurements of the aerosol optical depth were being made at several astronomical observatories, and there was direct sampling of the composition of stratospheric aerosol particles. Using the measured time history of the aerosol optical depths and a one-dimensional radiative–convective climate model, Hansen *et al.* (1978) computed the expected effect of the aerosols on stratospheric and tropospheric temperatures. In their model, they employed a thermal inertia at the surface for a uniform thermal mixed layer in the ocean of 70 m.

Figure 8.28 shows the computed and observed temperatures in the stratosphere. The observations were made over Port Hedland, Australia, and are smoothed with a 3-month running mean. At altitudes near 20 km (pressures ~ 60 mbars), the oscillation in the observed temperature increase may be due to an unaccounted for seasonal disturbance, but the theory appears to indicate that the aerosols did result in stratospheric warming of a few degrees, with a time scale for the increase on the order of several weeks.

Figure 8.29 shows the computed and observed averaged temperatures in the troposphere for the latitude range 30°N to 30°S. Hansen *et al.* report that after the Agung eruption, the average tropospheric temperatures evidently decreased by a few tenths of a degree with a time scale on the order of one year, in agreement with their theoretical result.

Based on the Hansen *et al.* (1978) computations, the physical explanation for the computed effects is straightforward. The overall effect of the added aerosol particles on the bulk of the atmosphere and the surface is cooling,

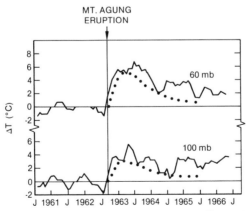

Fig. 8.28. Observed stratospheric temperatures over Australia at 60 mbars and 100 mm pressure altitude and computed temperatures after the eruption of Mount Agung, assuming that the added stratospheric aerosols are sulfuric acid and that the average depth of the thermally mixed layer of the ocean is 70 m. [From Hansen *et al.* (1978); reprinted with permission of the American Association for the Advancement of Science.]

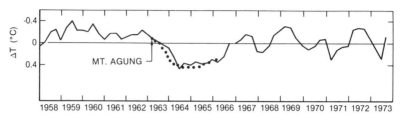

Fig. 8.29. Observed tropospheric temperatures between 30°N and 30°S and computed temperatures after the eruption of Mount Agung, assuming that the added stratospheric aerosols are sulfuric acid and that the average depth of the mixed layer of the ocean is 70 m. ——, observed, ···, computed. [From Hansen *et al.* (1978); reprinted with permission of the American Association for the Advancement of Science.]

because the sulfuric acid droplets are highly reflective to solar radiation and thus tend to decrease the amount of solar radiation received by the earth–atmosphere system. The aerosols also interact with the thermal radiation and warm the earth–atmosphere system through a blocking of radiation from the surface and troposphere. However, particles in the air have apparently too small an optical thickness in the infrared regime for the greenhouse effect to exceed the albedo effect. The effect of the aerosols on the local (stratospheric) temperature is heating, because they absorb thermal radiation from the warmer atmosphere and surface below more effectively than they release heat to space, and they absorb a small amount of solar radiation because they have broad absorption bands in the near infrared. The amount of energy involved in the stratospheric heating is small, but the local temperature is strongly influenced because of the low density of the atmosphere in that region.

From the theory, the thermal effect of the aerosol particles depends on their composition and size distribution, properties that are not very accurately known for the first few weeks after the volcanic eruption. At that time, the aerosols may have included a substantial fraction of absorbing silicate (basaltic) dust. To examine the potential effect of aerosol composition, computations were made for two extreme cases—pure basaltic material and sulfuric acid (75% concentration by weight in water). The results of Hansen *et al.* (1980) show that the basaltic dust results in a substantial increase in the stratospheric heating and some decrease in tropospheric cooling.

The greatest weakness in the model computations is probably the absence of interactions of the computed heating with the atmospheric dynamics. A related defect in the computations is the omission of potential cloud-cover feedbacks; the radiative perturbation due to aerosols could conceivably induce a change in the clouds that would counteract (or reinforce) the direct climatic effect of the aerosols. Advances in analysis will require a model in which the cloud cover is computed from first principles and is allowed to interact realistically with the radiation field and the atmospheric dynamics.

However, Hansen *et al.* (1980) note that the extent to which the observed climatic effect agrees with that obtained from a simple radiative model in fact provides evidence that such potential feedbacks do not overwhelm the direct radiative effect.

8.2.3 Tropospheric Effects and Man's Activities. The typical optical depths of tropospheric aerosols vary on shorter time and space scales than do those of stratospheric aerosols. A general trend shows much lower tropospheric optical depths at higher latitudes than at lower latitudes, and much lower optical depths over oceans than over continents (Toon and Pollack, 1976). There is little evidence for large global-scale changes in the tropospheric optical depth. The optical depth has been monitored at remote mountaintop sites for several decades, and no long-term trends have yet been detected (Roosen *et al.,* 1973; Ellis and Pueschel, 1971). However, at many rural sites near cities, industrialized areas, or agricultural areas, including parts of Japan, the eastern and southwestern United States, and the Soviet Union, there is evidence of a large upward trend in optical depth during the past several decades (Yamamoto *et al.,* 1971; Husar and Patterson, 1980; Machta, 1972). Tropospheric aerosol particles tend to be removed from the atmosphere relatively rapidly, within days. Hence, major changes are expected on regional but not global scales.

Of the major aerosol constituents—sea salt, soil, and sulfate—large optical depth changes due to variations in sea-salt concentration are the least likely, whereas changes due to soil particles are much more prevalent and perhaps best documented. Significant regional suspended-soil-particle enhancements result from dust storms near desert regions and during droughts in semiarid regions of the world. For example, the optical depth of the dust from the Sahara Desert often reaches 0.5 over the Atlantic Ocean and locally sometimes exceeds 2.0 (Carlson and Caverly, 1977). Also, large variations in the dust over the Atlantic Ocean may be partially caused by drought in Africa. A well-known example of drought-caused dust was the Dust Bowl era of the 1930s, which affected the plains of the south–central United States. No observations allow us to estimate exactly how much of an impact the soil particles have on the global average optical depth, but local anthropogenic enhancements are well known.

Knowledge of the optical depth associated with sulfate particles and their variability comes rather indirectly. Since particulate sulfate seems to constitute about a third to half the world's mass of fine-particle mass concentration, it may supply as much as half the global average aerosol optical depth. Indeed, since the optical depth is proportional to the surface area of the aerosols and the mass is proportional to their volume, a fixed mass of small sulfate particles has a much larger optical depth than the same mass of large soil dust or salt particles. Man's contribution to the sulfate supply may be as much as 25–40% of the total aerosol optical depth. Studies in New York City

(Leaderer *et al.*, 1978) suggest that sulfates are responsible for 50% of the light scattering there. Weiss and co-workers (1978) find that sulfate may dominate the light scattering in the midwestern and southern United States, while Waggoner and co-workers (1976) reach a similar conclusion about light scattering in Scandinavia. Several pollution episodes associated with greatly enhanced sulfate levels and reduced visibility over extensive regions of the eastern United States have been mapped by large-scale monitoring networks. Zones with high particulate sulfate concentrations do not necessarily coincide with zones of reduced visibility. However, particulate sulfate is a regionally important contributor to fine particle concentrations.

The identification of particulate sulfate with light scattering of fine particles also comes historically from the focus on this component of the aerosol. It should be borne in mind that the remaining half to two-thirds of the fine particles, including the organic fraction, remains largely uncharacterized as to composition and optical effects.

Annual averages of aerosol optical depths are much higher in central Europe and the eastern United States than in surrounding regions (Flowers *et al.*, 1969); the difference may be due to sulfate particles which are enhanced by man's activities through oxidation of SO_2 emissions.

There are basically two schools of thought about the significance of the influences of anthropogenic aerosols on surface temperature. Those expressing concern that tropospheric particle loadings are increasing and causing thermal changes at the ground are represented by the work of Budyko (1969) and Bryson (1972). Those who consider global climate effects unlikely include Ellsaesser (1975) and Landsberg (1975). A recent study of Hoyt *et al.* (1980) supported the latter view. Hoyt *et al.* examined trends in atmospheric transmission of sunlight in locations near Albuquerque, NM, Madison, WI, and Blue Hill, MA from 1940 to 1977. They found evidence for a weak effect of particulate matter resulting in urbanization around the sites, but dramatic evidence of the influence on the upper atmosphere of the Mt. Agung, and the Fuego volcanic eruptions. However, the trend in atmospheric transmission under normal conditions was found to be very small. Over the period of observation, they found that the vertical optical depth increased by about 0.041. Extrapolating the change observed to global conditions, Hoyt *et al.* (1980) estimated a depletion in radiation of approximately 0.05% associated with increases in anthropogenic aerosols between 1940 and 1977. Interestingly, this is in the range projected by Ellsaesser (1975) and Hidy and Brock (1971). At such a rate, it would take 1400 years to double the present global aerosol concentration in the troposphere. In the opinion of Hoyt *et al.*, such a slow growth rate of anthropogenic aerosols is too weak to be important on a global scale. However, regional and local disturbances may be of greater consequence.

Perhaps the most famous theory connecting tropospheric aerosols and climate is R. Bryson's "human volcano" hypothesis (1972) that the 0.3°C cooling that has been observed between 1940 and 1970 is due to particles injected into the atmosphere by humans. Theory suggests that if about 25% of the present aerosol optical depth were due to anthropogenic aerosols with a high value of $\tilde{\omega}_0$, they could be the cause of the observed cooling. But we do not even know if the globally averaged optical depth has changed by such an amount as 25%, although the optical depths in many regions are known to have changed significantly since about 1940 (Yamamoto *et al.*, 1971; Husar and Patterson, 1980; Machta, 1972).

Additional contributions to the optical depth could have been made by agricultural activities. Other possible causes of the temperature change since 1940 include surface albedo changes from alterations in human land use (Sagan *et al.*, 1980) and natural, random climate fluctuations (Robock, 1978). The human contribution to tropospheric particles is probably great enough to have increased the optical depth by 25%, though the timing of the change is not known.

The possible effects of tropospheric aerosols may be detectable on a regional scale. Bryson (1972) has suggested that the dusty Rajputana Desert of India, site of the ancient Indus Valley civilization, is an example of an anthropogenic aerosol desert. He believes that overgrazing of arid lands by the domesticated animals of the Rajputana inhabitants allows the wind to blow large quantities of dust into the atmosphere; the dust increases the infrared radiation from the lower atmosphere, causing the air to cool and subside, which suppresses rainfall (rain normally falls only in ascending air parcels). Harshvardhan and Cess (1978) have reexamined this problem with a better radiative transfer scheme, and they find that the dust contributes negligibly to the infrared cooling of the air. As yet no one has considered the impact of the dust on the solar radiation, and a solution as to whether the dust in the atmosphere over the Rajputana Desert is partly responsible for creating the desert remains to be determined by further study.

Another good candidate for empirical testing of the effects of tropospheric aerosols on climate is Sahara dust over the Atlantic Ocean. The radiative properties of this dust have been characterized by Carlson and Caverly (1977), who stated that the dust should lead to cooling at the surface and increased solar heating in the atmosphere above the ocean. The significance of these factors for the weather over the ocean has not yet been investigated.

Husar and co-workers (1980) have attempted to correlate for each season of the year observed increases in aerosol levels over the eastern United States with regional climate changes observed since 1948. Of the seasons and regions considered, they found that the aerosol levels increased most dramatically in the Smoky Mountain region during summertime. Simultaneously the same region also experienced the largest changes of any region in several climatological variables, including higher humidity and lower temperature after 1950, with noontime temperature decreasing by about 1°C.

One might imagine, because of the obvious importance of aerosols to visibility in urban areas, that the radiative effects of aerosols on the urban climate would be well known. But in addition to the large quantities of aerosol particles, cities also have large concentrations of gaseous absorbers. They release large quantities of heat into the environment, and their surface properties, such as ground heat capacity and reflectivity, differ substantially from those of surrounding rural areas. Theoretical studies (Ackerman, 1977) have shown that the radiative impact of aerosols on the urban climate is less important than the release of heat or the urban modification of surface properties. Although aerosols do modify the surface radiation energy budget, these modifications seem to be partly offset by changes in other heat transfer processes, such as latent and sensible heat transfer by atmospheric motion.

The most significant radiative impact of the aerosols in urban areas seems to be that they warm the atmosphere aloft and stabilize it against convection. Mixing processes that remove pollutants from urban areas are highly sensitive to the density stability of the atmosphere. Urban dwellers quickly learn that temperature inversions, which cause the atmosphere to be stable, lead to high-pollution episodes. Although aerosols tend to promote such inversions, they are primarily caused by large-scale meteorological processes. Aerosols seem capable of affecting the climate of cities, but because other processes have a stronger influence, it is necessary to make detailed empirical studies that can ascertain the aerosol–climate relations.

8.3 CLOUD FORMATION AND DYNAMICS

A major manifestation of atmospheric processes in the atmosphere is the development of clouds of condensed water. Clouds are virtually universal in their presence in the troposphere, and fallout of precipitation from them represents a critical part of the natural means for distributing water over the earth's surface. Clouds are a natural result of the process of supersaturation as warm air rises, cools, and expands, or mixes with cooler surrounding air.

Water vapor can supersaturate by three processes: (a) adiabatic cooling of ascending air; (b) cooling by mixing during entrainment of cool air into warm, moist air; and (c) isobaric cooling by radiative transfer. The first two are basically the same as those described for aerosol generation methods in Chapter 4. The last is difficult to achieve in the laboratory under normal circumstances.

Air can rise in the atmosphere by its buoyancy if it is warmer than its surroundings. During its ascent, the air will expand and cool. Although heat may be added to the parcel through the effects of radiation, frictional dissipation, and mixing with the environment, in many situations the resulting temperature changes are of secondary importance to that arising from the expansion process.

Although most cooling of moist air to saturation occurs via the expansion which accompanies lifting, there are also a variety of cooling mechanisms that can occur isobarically.

Dew or frost are often isobarically formed by the night-time radiational cooling of calm, moist air in ground contact. The same cooling process may also produce ground fogs. Isobaric cooling leading to fog or stratus cloud formation may also occur when a mass of moist air moves horizontally over a colder land or water surface or over a colder air mass.

From the discussion in Chapter 4, mixing between a warm, moist air mass and a cool, dry air mass can produce supersaturation conditions in the atmosphere as well as modify adiabatic cooling. This was recognized at the turn of the century by Taylor (1917), who noted certain fogs forming over the ocean in advected air by air mass advection occurring as a result of mixing.

The presence of suspended particles in the troposphere provides the microphysical basis for the condensation of water vapor to form water clouds. The particles serve as nuclei for droplet or ice crystal production when supersaturation occurs, or when supercooled condensed water is present. Once clouds form, they evolve to dissipate by evaporation, or they develop precipitation in the form of rain, snow, or hail. The crux of the latter route basically involves the growth of some initially nucleated particles by various mechanisms by a factor of a million or more from large aerosol size of tens of micrometers to hydrometeors of precipitation size in tenths of centimeters. Thus there is a direct conceptual link between aerosol dynamics in the atmosphere to cloud microscale processes. The water cloud broadly begins with the dynamic processes of large aerosol particles in terms of number and mass concentration in the air. Ultimately the theory is concerned with continuing growth of a very small fraction of the total number of wet particles present until precipitation takes place. Growth processes include (a) vapor diffusion to droplets or ice crystals, (b) selective growth of ice at the expense of vapor diffusion from coexisting supercooled droplets, and (c) collision and coalescence or accretion of droplets on ice particles.

The history of cloud physics is a long one as part of the science of meteorology. However, appreciation of the microphysics began in the late nineteenth century with studies of nucleation in expansion chambers (Coulier, 1875), and Wegener's ice-forming particles led to concepts of cloud initiation. The thermodynamic concept of adiabatic cooling was merged with microphysical concepts in the mid-nineteenth century through Kelvin's and Hertz's theories and Renou's interests in ice crystals. By the 1930s, Wegener and Bergeron introduced the notion of colloidal instability in clouds containing supercooled drops and ice crystals leading to rain formation. Bergeron envisioned that in such clouds, ice crystals will grow by diffusion of vapor from supercooled water drops until droplets are consumed or the ice crystals fall from the cloud.

A second colloidal growth concept of droplet or ice particle collision and coalescence arose from early ideas of eighteenth-century physics in Bonifas's thoughts on hailstone formation. The same basic idea for raindrops was suggested by Musschenbroek. The collisional growth mechanism was developed further in the nineteenth-century studies of Reynolds and Rayleigh.

The Wegener–Bergeron mechanism dominated conceptions of cloud particle growth until Simpson reviewed the collision mechanism in 1941 based on aircraft pilot reports of precipitating clouds over India with tops thought to be warmer than 0°C. The existence of warm cloud precipitation resurrected the collision-based mechanism.

After the Second World War, interest in cloud microphysics processes dramatically expanded, with stimulation from military operations as well as great interest in the 1946 experiments of Langmuir and Schaefer. These workers demonstrated the feasibility of cloud modification by introducing a colloidal instability in supercooled stratus clouds in a dry ice seeding, which led to a minisnowfall.

The quantitative theory for evolution of water clouds remains incomplete for bridging the microscopic fluid dynamic processes with microphysics. Yet progress toward detailed knowledge of water cloud processes including electrification phenomena has emerged through the 1960s, and has been sufficient to interpret descriptively many features of cloud behavior. Further, workers are now in a considerably improved position to appreciate the subtleties of experiments attempting to modify weather through clouds, or to look for inadvertent weather modification resulting from air pollution.

Discussion of the details of water cloud formation and precipitation processes is well beyond the scope of this book. For such descriptions, the reader is referred to texts such as Fletcher (1966), Mason (1971), Hobbs (1974) or Pruppacher and Klett (1978). In the context of bridging the technology of aerosols to atmospheric physics, certain aspects of water cloud processes are discussed in the following subsections. These include (a) a summary of cloud microstructural properties in relation to aerosol particles and nucleation of water and ice, (b) an introduction to hydrometeor growth processes, and (c) elements of attempts to modify cloud processes by microphysical change.

8.3.1 Cloud Microstructure and Nucleation.   There are basically two kinds of nucleation involved in cloud formation where aerosol particles play a role. These are in the water vapor condensation process (cloud condensation nuclei, CCN), and in the ice-forming process (ice-forming nuclei IFN). The physicochemical properties of these nuclei are quite different.

Condensation of Water Vapor.   The atmosphere is a stable gas–particle system so that it does not depart significantly from the thermodynamic equilibrium in its properties. Experimental studies of clouds in the atmosphere have shown that supersaturation of water vapor in air rarely exceeds one percent. Supersaturations well in excess of unity are required to achieve homogeneous nucleation of water vapor or nucleation on ions at atmospheric temperatures. Therefore, the nucleation of cloud droplets takes place exclusively by heterogeneous processes involving aerosol particles.

For the nucleation process, the critical radius and supersaturation can be estimated from Eq. (3.43).

A Kohler diagram (Fig. 8.30) for different natural aerosols as analogs to mixtures of ammonium sulfate, sodium chloride, and insoluble material illustrates the relation between the supersaturation ratio of water vapor and particle radius at equilibrium. The larger the aerosol particle, and the more hygroscopic its makeup, the lower the equilibrium supersaturation sustained over a droplet at particle radii below 1 μm. Above 1 μm radius, the equilibrium supersaturation ratio is basically the same.

If atmospheric particles are maintained at or near vapor–condensed phase equilibrium, they display a hygroscopic nature or they may deliquesce such that they will accumulate water with increasing humidity. This is readily illustrated from Winkler's (1967) studies of pure salts, as shown in Fig. 3.7. The increase in mass of natural aerosol particles is very dramatic for relative humidities above 50% (see also Fig. 8.31) such that the mass of wetted nuclei may be more than four times that at saturation compared with absolutely dry

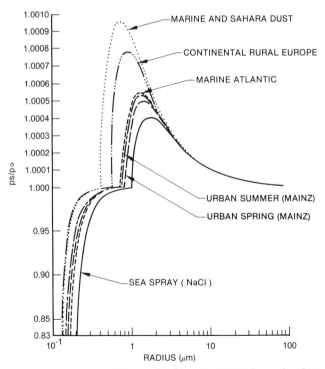

Fig. 8.30. Calculated variation of the equilibrium supersaturation ratio of an aqueous drop formed at 20°C on various natural aerosol particles. Calculations are based on the water uptake of particle deposits on the same substrate. [From Hänel (1976); courtesy of Academic Press.]

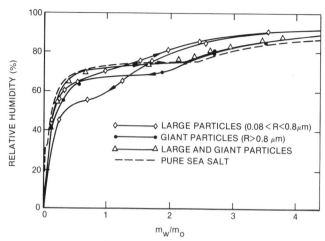

Fig. 8.31. Equilibrium growth of natural aerosol deposit; maritime aerosol, Helgoland Island, September 1966, west wind. [From Winkler (1967).]

conditions. Thus the process of cloud formation can begin gradually with a wet haze developing into a regime of low supersaturation. Atmospheric particles also exhibit the "hysteresis" effect such that particles retain more moisture as they dry out from 100% humidity conditions. This can lead to the persistence of a wet haze in the air well below 100% relative humidity, with accompanying reduced visibility.

The critical supersaturation of water vapor for activating aerosol particles composed of mixtures of sodium chloride and other material is shown in Fig. 8.32. The results in the figure indicate that water-soluble particles will act as most effective condensation nuclei. Sodium chloride particles would be active at 1% supersaturation to (dry) particle of 0.01 μm. Water-insoluble particles with a contact angle of 7° or less will be active as CCN at 1 μm radius, but contact angles of 12° or more effectively would substantially decrease the nucleation ability of particles at all sizes. The curves in the figure also show readily that the larger particles will nucleate water at progressingly lower supersaturations.

For the condensation of water vapor, we conclude that the preferential properties would (a) constitute relatively large particles to avoid the vapor pressure change associated with the Kelvin effect, and (b) favor hygroscopic surface properties to be compatible with vapor accumulation at low supersaturation.

The number of CCN present in the atmosphere has been measured in several locations using portable vapor diffusion cloud chambers analogous to the laboratory devices outlined in Chapter 4. These studies have shown that the number of aerosol particles activated as CCN at a few percent supersaturation is smaller than the total aerosol particle population. An

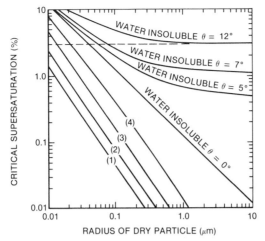

Fig. 8.32. Critical supersaturation for water nucleation ($J = 1$ particle/sec) on a spherical substrate particle with volume fraction (ε_v) of water-soluble component, and contact angle θ for water-insoluble particles, and as a function of the volume percentage of NaCl in mixed particles; for 0°C. Mixed particles: (1) $\varepsilon_v = 1.0$; (2) $\varepsilon_v = 0.4$; (3) $\varepsilon_v = 0.1$; (4) $\varepsilon_v = 0.01$. [From Pruppacher and Klett (1978); courtesy of Reidel Publ. Co.]

example of such data is given in Table 8.4. Even in remote areas and in marine air, the CCN concentration generally represents 10% or less of the total suspended-particle concentration.

Median worldwide concentrations of CCN from several studies are shown in Fig. 8.33 as a function of supersaturation required for activation. In general, the number of CCN in air over the continents is substantially larger than over the oceans. This is comparable with larger numbers of large and giant particles found in continental air. From these results, one would expect that cloud droplet populations would be larger in number in clouds in continental air as compared with marine conditions. This has been observed, as indicated in typical cloud droplet size spectra, such as those in Fig. 8.34.

Cloud droplets are characterized by size distributions which are similar in shape but broader than equivalent aerosol particle spectra. Clouds over the ocean tend to have larger but fewer droplets present than continental clouds. Given similar levels of liquid water content and time for growth, this would be expected with the differences in CCN concentrations observed. The more CCN activated, the less they will grow since they will tend to compete for condensable water in the cloud.

Fogs have droplet size distributions similar to those in clouds. Radiation fogs have been observed to have a mean radius of 2–14 μm (Tampieri and Tomasi, 1976); a typical size distribution is shown in Fig. 8.35, which is fitted to a modified gamma distribution.

Interestingly, the liquid water content in clouds is similar despite the form

TABLE 8.4

Comparison between Total Concentration of Aerosol Particles and Concentration of Cloud Condensation Nuclei Activated at 1% Supersaturation at Various Locations[a]

Location	Type of nuclei Number of Aitken particles ($cm^{-3} \times 10^{-3}$)	Number of CCN ($cm^{-3} \times 10^{-2}$)
Washington, DC	78	20
	68	20
	57	50
	50	70
Long Island (NY)	51	2.2
	18	1.1
	6.5	1.5
	5.7	0.30
Yellowstone National Park (WY)	1.0	0.15

[a] From Pruppacher and Klett (1978); courtesy of Reidel Publ. Co.

and graphical origin of clouds. Table 8.5 gives an indication of this conclusion. One can expect that the liquid water content will range between 0.1 and 10 g/m³ with a typical value of about 1 g/m³ in clouds.

Ice Formation. Ice represents an important component of water clouds. Its behavior is highly complex, involving growth from vapor diffusion from

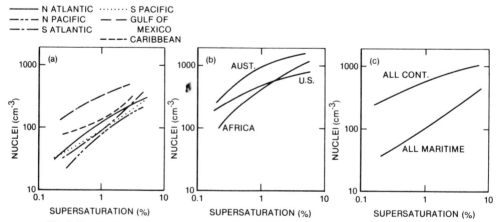

Fig. 8.33. Median world wide concentration of CCN as a function of supersaturation required for activation: (a) in air over oceans, (b) in air over continents, (c) all observations. [From Twomey and Wojciechowski (1969); courtesy of the American Meteorological Society and the authors.]

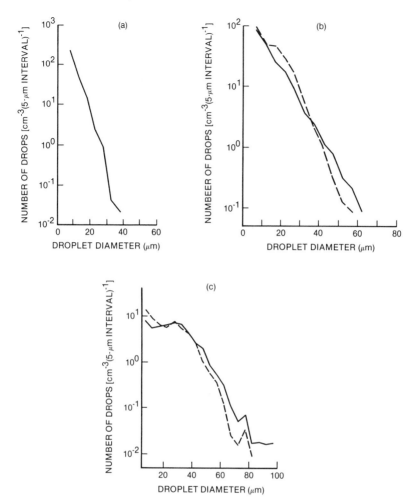

Fig. 8.34. Mean droplet size distributions in clouds (a) fair weather cumulus, average total drop concentration 293 cm^{-3}; (b) cumulus congestus over central United States, average drop concentration in 10 clouds with precipitation echoes by radar (dotted line) 188 cm^{-3}, average of 21 clouds without echoes 247 cm^{-3} (solid line); (c) tropical cumuli over Gulf of Mexico. Average total drop concentration of 11 clouds with echoes 52 cm^{-3} (dashed line), 26 clouds without echoes 58 cm^{-3} (solid line). [From Battan and Reitan (1957); courtesy of Pergamon Press, Ltd.]

accretion and riming, and disaggregation by splintering or fracturing. Like liquid water, the production of ice in the atmosphere proceeds through heterogeneous nucleation. Ice-forming nuclei (IFN) exhibit three basic modes of action. In the first, water vapor is adsorbed directly onto the nucleus surface; at sufficiently low temperature, the adsorbed vapor is converted to ice. This is called the *deposition mode.* In the second, *freezing mode,* the nucleus initiates the ice phase from inside a supercooled droplet. The nu-

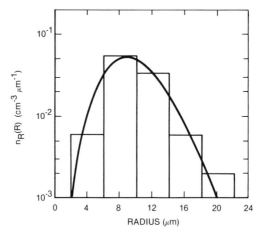

Fig. 8.35. Droplet size spectrum for (a) radiation fog in the Po Valley, Italy, fitted to a modified gamma distribution. [From Tampieri and Tomasi (1976).]

cleus enters the droplet by a variety of scavenging mechanisms. The third, *contact mode,* of action involves ice-phase formation at the moment the nucleus contacts a supercooled drop.

The measurement of IFN in the atmosphere is done by devices which simulate contact or freezing of supercooled droplets in a chamber, with the vapor deposition on particles maintained at very low temperatures in humid air (Allee, 1970; Langer *et al.*, 1967). None of the instruments is capable of allowing for the different modes of ice-forming action, nor can they simulate the time scale for temperature and supersaturation variation in natural clouds. Thus IFN concentrations reported in the literature should be interpreted with some caution.

TABLE 8.5

Average and Maximum Liquid Water Content in Clouds of Different Types[a]

Cloud type	Liquid water content (g/m^3)	
	Average	Maximum
Stratus and Stratocumulus	0.17–0.38	0.36–0.88
Altostratus and Altocumulus	0.11–0.16	0.24–0.70
Cumulus	0.19–1.0	0.21–3.00
Cumulus congestus	0.5 –3.9	2.10–6.50
Cumulonimbus	0.33–2.5	1.7–10.00

[a] Summarized from Fletcher (1966).

Comparison between concentrations of the two different classes of nuclei shows that the ice nuclei population is very small compared with the CCN measured at the same time. However, in some cases, the variation in CCN and IFN is correlated. When such correlation exists, the nuclei populations can be attributed to similar air mass origins. Correlation between the two populations is generally not expected since the two classes of aerosol particles play distinctly different roles, requiring complementary physical and chemical properties.

Observations of IFN have shown dramatic short-term variations from time to time over periods of days, while the total airborne particle population has remained reasonably stable. These variations occasionally involve intermittent "bursts" of large number of IFN, sometimes called "storms." The origins of the IFN storms remain uncertain, but there is some evidence leading to speculation that injections of extraterrestrial dust may be significant here (Bowen, 1953; Rosinski, 1979). Other explanations relate IFN to the long-range transport of dust from China and Mongolia, or from the Sahara in Africa.

The apparent effectiveness of ice nuclei is strongly temperature dependent as indicated in Fig. 8.36. The mean numbers of IFN concentration evidently exhibit no systematic geographical variation. Thus, far from local source, the atmospheric aerosol is reasonably uniform in its capability to initiate ice formation.

According to Huffman (1973), the effectiveness of IFN also depends on relative humidity and is logarithmically related to supersaturation over ice at a given temperature.

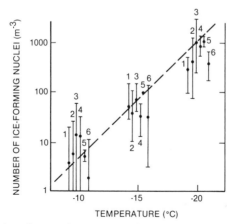

Fig. 8.36. Range of median number concentration of IFN as function of temperature for various geographic locations, 44 stations. The dashed line represents $N_{IFN} = 10^{-5} \exp(0.6 \, \Delta T)$. (1) Europe, (2) North America, (3) Asia, (4) Africa, (5) South America, (6) Australia. [From Bigg and Stevenson (1970); courtesy of *Journal de Recherches Atmospheriques*.]

Comparing concentrations of IFN and cloud ice particles at nearby locations, particularly at warm temperatures, workers (Pruppacher and Klett, 1978) have found that the concentration of ice particles may exceed by many orders of magnitude the concentration of IFN determined at the cloud-top temperature. Observations to this effect were made over the Cascade Mountains in the United States, in stable cap clouds over Wyoming, in cumulus clouds over Australia and Tasmania, and in cumulus clouds over Missouri. These observations show that the maximum enhancement ratio or enhancement factor defined as a ratio of the ice crystal concentration to the IFN concentration determined at the cloud-top temperature can be as large as 10^4–10^5 at temperatures between -5 and $-15°C$. This factor tends to decrease with decreasing cloud-top temperature, and reaches unity between -25 and $30°C$.

Heymsfield (1972) found an opposite situation prevailing in cirrus clouds. While at cirrus cloud level and $-30°C$, the IFN concentration was typically 10^3/liter, the maximum ice crystal concentration was only about 50/liter. This behavior was attributed to a dilution mechanism operating in cirrus clouds. The multiplication mechanisms appear to be associated with (a) mechanical fracture of ice crystal dendrites, needles, or other fragile bodies during collisions; (b) splintering during riming[†]; and (c) fragmentation of large cloud drops during freezing.

IFN can operate effectively over a wide range of particle size, but they tend to favor large particles. The sizes of natural IFN are summarized in Table 8.6. The effectiveness of nucleation at constant temperature below $-10°C$ tends to drop sharply below 50 μm diameter.

Some clues as to possible sources of IFN are provided by the chemical identification of aerosol particles found at the center of snow crystals (e.g., Pruppacher and Klett, 1978). Typically, one solid silicate particle, usually identified as clay, has been found in the central portion of a snow crystal. These findings suggest that desert and arid regions of the earth's surface are a major source of IFN.

The notion that surface soils act as an IFN source is also strongly supported by laboratory experiments. The ice-forming capability of silicate particles collected at various parts of the Northern Hemisphere has been tested by several investigators. When these particles were allowed to function as IFN in the deposition mode at water saturation, the threshold ice-forming temperature (which is conventionally taken as the temperature at which one particle in 10^4 produces an ice crystal) was found to range typically between -10 and $-20°C$. Clay particles such as kaolinite, anauxite, illite, and metabentonite have a threshold temperature as warm as $0°C$ (Mason, 1960).

Surprisingly, clays such as kaolinite often exhibit varying ice-nucleating ability, or "nucleability," depending on the location at which the clays are

---

[†] Riming occurs when water droplets collide and freeze on ice particles.

TABLE 8.6

Sizes of Natural Ice-Forming Nuclei Given by Listing of Rosinski
(1979) (Courtesy of Elsevier)

All particles d_p (μm)	Maximum frequency d_p (μm)	Temperature (°C)	Location
0.5–8	3.0	> −20	Japan
0.1–7			Japan
0.2–8		> −21	Michigan
1–80	5–15	> −20	France
0.1–13	0.5–1		Greenland
0.5–7 minerals[a]	2–3	−12	Colorado
0.5–5 organics[a]			
0.2–7 minerals[a]	2–3	−19	Colorado
0.5–3 organics[a]			
0.5–10 minerals	2–7	−12	Montana
1–5 organics			
0.5–10 minerals	2–5	−15	Montana
1–5 organics			
0.2–10 minerals	2	−20	Montana
1–5 organics			

[a] Ice-forming nuclei separated over an area previously exposed to cloud seeding; all organic particles contained silver (not AgI).

sampled. Similarly, soils from different parts of Australia are found to be generally less active than those from the Northern Hemisphere, the threshold temperature of Australian soils ranging between −18 and −22°C (Patterson and Spillane, 1967). The observations suggest that other substances admixed to soils in small quantities may importantly affect their ice nucleation behavior.

Experiments by Roberts and Hallett (1968) showed that clay particles, having been involved once in ice crystal formation, can exhibit a considerable improved nucleability. Aerosol particles which behave in such a manner are termed preactivated. Roberts and Hallett showed that the IFN activity spectrum for each of the tested clays, containing particles of diameters between 0.5 and 3 μm, shifted by more than 10°C toward warmer temperatures provided that the temperature of the air in which the clay particles were kept never rose above 0°C and its relative humidity never fell below 50%.

Clays also seem to exhibit significant differences in their ice-nucleating ability according to their mode of action. Roberts and Hallett (1968), for example, showed that kaolinite acting in the deposition mode has a typical threshold temperature of about −9°C, and reaches full activity (1 : 1 ice crystal production ratio) near −20°C, while montmorillonite has a threshold temperature of about $-25°C$, and reaches full activity near $-30°C$. In contrast, Hoffer (1961) found that the warmest freezing temperature of 50–60-μm-radius drops containing kaolinite and montmorillonite was $-13.5°C$, and that the median freezing temperatures were -24.0 and $-32.5°C$, respectively. Gokhale and Spengler (1972), on the other hand, found that drops of 2–3 μm radius, freely suspended in the air stream of a wind tunnel, froze at temperatures as warm as $-2.5°C$, with full activity between -6 and $-8°C$, when contacted by red soil, sand, or clay particles. While monitoring the IFN concentration in Japan, investigators detected a new source of IFN; Isono and Komabayasi (1954) and Isono (1959) observed that pronounced IFN storms arose following the eruption of volcanoes. In support of the notion that volcanoes act as IFN sources, Mason and Maybank (1958) and others found during laboratory experiments that volcanic ash and other volcanic materials were capable of serving as IFN with threshold temperatures as warm as $-7.5°C$. However, in contrast to these findings, Price and Pales (1963) found no significant increase of the IFN concentration during the eruption of volcanoes on the Hawaiian Islands, indicating that not all volcanic material has a high ice nucleability.

Recent laboratory and field studies demonstrate that IFN also have a biogenic source. Vali (1968) noted during experiments that soils with a relatively high content of organic matter exhibited a higher ice nucleability than pure clays or sand. This observation led him to suggest that decaying plant material contributes to the IFN content of the atmosphere. Subsequent detailed studies by investigators showed that some IFN from soils are produced by decomposition of naturally occurring vegetation such as tree leaves. Leaf-derived nuclei (LDN) were found to initiate ice by the freezing mode at temperatures typically between -4 and $-10°C$. The diameter of these particles ranged between 0.1 and 0.005 μm.

The global ubiquity of these IFN was established by testing plant litters collected at various geographic locations in different climatic zones (Vali and Schnell, 1973). Highly active IFN (active as freezing nuclei at $-1.3°C$) were found to be present during the early stages of decay of aspen leaves. Fresch (1973) demonstrated that the ice-forming capability of these decaying leaves is closely related to a single strain of aerobic bacteria (*Pseudomonas syringae*), which by themselves act as IFN. Such nuclei were termed bacteria-derived nuclei (BDN). Whether or not LDN and BDN are to be regarded as acting independently of each other has not, as yet, been established. Schnell (1974) also found that some organic material from the ocean surface, termed ocean-derived nuclei (ODN), can be quite effective IFN. His experiments showed that LDN, BDN, and ODN may act as IFN in both the freezing and deposition modes, being generally more efficient in the freezing mode. Unfortunately, no studies on the contact mode of these particles are available.

Rosinski (1979) has reviewed the significance of IFN capability of a large number of organic compounds. There are a variety of materials that have an activity threshold of $-1°C$. Except for a limited number of organic substances concentrated in the ocean surface, the world oceans are not a source of IFN. Field observations have demonstrated that maritime air masses are consistently deficient in IFN.

Two processes have been suggested for IFN deactivation. Georgii and coworkers showed that contaminant gases such as SO_2, NH_3, and NO_2 severely reduce the ability of atmospheric aerosol particles to serve as IFN. The higher the concentration of these gases, the stronger is the deactivation. An additional mechanism of deactivation was proposed by Georgii and Kleinjung (1967). They suggested that in urban areas, where the concentration of Aitken particles may reach $10^6/cm^3$ and higher, IFN become deactivated as a result of coagulation with these particles, which are generally found to be poor IFN.

Observations show that ice-forming nuclei have to meet a number of specific characteristics. The most important of these are summarized below from the discussion of Pruppacher and Klett (1978).

Insolubility Requirement. In general, IFN are highly water insoluble. The negative correlation that is observed between the concentrations of IFN and sea-salt particles gives some evidence of this fact. The obvious disadvantage of a soluble substrate is that its tendency to disintegrate under the action of water prevents it from providing the structural order needed for ice embryo formation. In addition, the presence of salt ions causes a lowering of the effective freezing temperature.

Size Requirement. Field studies generally have shown that airborne particles of the Aitken size range are considerably less efficient IFN than large particles. Although it is tempting to interpret such observations in terms of an IFN size effect alone, the measurements may partly reflect a dependence of particle chemistry on size (e.g., silicate particles, which are known to be good IFN, are mostly confined to the large size range).

However, there are other clear indications that IFN size is important. Since the solubility of a substance increases with decreasing particle size, we might expect, on this basis alone, to find that IFN occur predominantly in the larger size ranges of airborne particles. The existence of very small organic IFN is consistent with the fact that such particles are known to be highly water insoluble.

Chemical Bond Requirement. Numerous experimental studies show that the chemical bonds exhibited at its surface also affects its nucleation behavior. Considering the fact that an ice crystal lattice is held together by hydrogen bonds (O—H---O) of specific strength and polarity, it is quite reasonable to assume that an IFN must have similar hydrogen bonds available at its surface in order to exhibit good ice nucleability. While asymmetric molecules tend to point their active H-bonding groups inward to achieve minimum free energy at the solid surface, molecules with rotational symmetry cannot avoid exposing their active H-bonding groups, thus allowing maximum interaction with a water molecule.

In view of this bond requirement, it is not surprising that certain organic compounds such as those having —OH, —NH$_2$ or =O are potentially significant as IFN.

Crystallographic Requirement. Experiments also have shown that the geometrical arrangement of bonds at the substrate surface is often of equal or greater importance than their chemical nature. Since ice nucleation on a foreign substrate may be regarded as an oriented (or epitaxial) overgrowth of ice on this substrate, it is reasonable to assume that this overgrowth is facilitated by having the atoms, ions, or molecules that make up the crystallographic lattice of the substrate exhibit, in any exposed crystallographic face, a geometric arrangement which is as close as possible to that of the water molecules in some low-index plane of ice. In this manner, atomic matching across the interface between ice and the substrate particle may be achieved.

If there are but small crystallographic differences between ice and substrate, either or both the ice lattice and the substrate may elastically deform so that they may join coherently. Thus, strain considerations suggest that the solid substrate should have an elastic shear modulus which is as low as possible in order to minimize the elastic strain energy. If there are large crystallographic differences between ice and the substrate, dislocations at the ice–substrate interface will result, leaving some molecules unbonded across the interface and causing the ice embryo to be incoherently joined to the substrate. The interface may then be pictured as being made up of local regions of good fit bounded by line dislocations. These dislocations at the interface will raise the interface free energy. In addition, any elastic strain within the ice embryo will raise its bulk free energy. Both effects will reduce the ability of the aerosol particles to serve as an effective IFN.

An unequivocal proof of the necessity, although not of the sufficiency, for a substrate to meet certain crystallographic requirements for good ice nucleability was given by Evans (1965). From a study of the effectiveness with which AgI nucleates ice-I$_h$ and the high-pressure modification ice-III, he was able to show that even under conditions where ice-III is the stable phase, ice-I$_h$, which has a closer crystallographic fit to AgI than ice-III, was consistently nucleated by AgI. However, despite the obvious importance of the crystallographic properties of a substrate to its ice nucleability, there is no unique correlation which can be established between ice nucleation threshold and any of the crystallographic characteristics such as symmetry or misfit. The main reason for this irregular behavior evidently is the role played by surface-active sites (Pruppacher and Klett, 1978).

Active-Site Requirement.   Experiments have established that ice nucleation, like heterogeneous water nucleation, is a highly localized phenomenon in that it proceeds at distinct active sites on a substrate surface. Not surprisingly, it happens that sites which are capable of initiating ice nucleation are also generally active with respect to water vapor adsorption and water nucleation. Experimental evidence for the effectiveness of topographic surface features to initiate the ice phase from the vapor is taken from studies of the ice-nucleating properties of CuS, AgI, and PbI$_2$ in the depositional mode. Observations during these studies revealed that ice crystals appear preferentially at cleavage and growth steps, at cracks, in cavities, and at the edges of the substrate surface.

Adsorption studies by Gravenhorst and Corrin (1972) have established that particles from AgI samples containing impurity ions have a considerably higher ice nucleation efficiency than "pure" AgI. The lower nucleation efficiency of pure AgI has been attributed to the presence of relatively inactive physical adsorption sites at its surface, causing water molecules initially to be adsorbed in the form of extended water "patches" within which the water molecules exhibit strong lateral interaction. Only on approaching water saturation do multilayers develop with an adsorbate–vapor interface which assumes the energetic properties of a liquidlike surface prior to nucleation.

Experiments with AgI and numerous other substances show that the region surrounding active sites should have a hydrophobic character for the surface to exhibit a high ice nucleation efficiency. One reason for this is that it is easier for water molecules to join a disordered water cluster on a low energy (hydrophobic) surface than to enter an oriented array of water molecules on a polar (hydrophilic) surface. Similarly, it is easier for a water cluster than an oriented film to achieve an icelike structure. Also, the growth of an ice embryo is facilitated by surface diffusion of weakly adsorbed molecules near the active site.

An interpretation of the size variability in activity of IFN concerns the hypotheses that a given solid substance has a characteristic area density of ice nucleation active sites of varying quality. It also implies that experiments on nucleation thresholds should be interpreted with caution, since the outcome may well depend on the area of the substrate under study.

Two additional effects which demonstrate the importance of active sites for nucleation in the deposition mode are of interest. The first effect emerges under certain conditions in which IFN may become activated, and in this state exhibit a considerably improved ice nucleability (memory effect). To behave in this manner the IFN either must have been previously involved in an ice nucleation process and formed a macroscopic ice crystal, or they must have been exposed to temperatures below $-40°C$. The work of Roberts and Hallett (1968) has shown that the observations are interpreted in terms of the retention of patches of ordered, icelike layers of water molecules at the surface of a substrate, where each patch can be considered to represent the remnant of a macroscopic ice crystal which developed over an active site.

The second effect involves the previously mentioned observation that foreign gases or vapors such as NO_2, SO_2, and NH_3 strongly reduce the nucleability of IFN (Georgii and Kleinjung, 1967).

Activity in Freezing and Contact Modes.   So far the significance of active sites for ice nucleation in the deposition mode has been discussed; we now consider their significance in the freezing and contact modes. Unfortunately, no quantitative experiments are available to help elucidate these questions. One can speculate that the ice nucleation process proceeds somewhat differently in the freezing mode than it does in the deposition mode. Pruppacher and Klett (1978) envisage the following process. Once a water-insoluble particle becomes submerged in water, it is surrounded by an abundance of water molecules. At any one moment, a large number of these molecules are linked together into small structural units in which some of them tend to be tetrahedrally bonded, while other molecules seem to be uncoupled. Suppose now that the surface of the submerged solid particle is generally hydrophobic, but contains hydrophilic sites where water molecules are preferentially adsorbed. The molecules most likely to be adsorbed are those that exhibit uncoupled bonds. In this manner the already existing structural units become "anchored" to the solid surface, causing them to be less vulnerable to destruction by the thermal motion in the water. As the temperature of the water is lowered, more and more uncoupled bonds become connected to the particle's surface, thus allowing individual structural units to be connected together to form clusters in which individual water molecules have considerable freedom to move their dipoles into an orientation most favorable for a tetrahedral, icelike arrangement. Eventually, the anchored cluster may reach embryo size.

Pruppacher and Klett (1978) point out alternatively that if a particle exists with a surface which has a strong, uniform affinity to water owing to the presence of an array of strongly hydrating ions, polar groups, hydroxyl groups (OH—), or oxygen (O) atoms in the solid surface, water molecules will become adsorbed in a close array, with most of the dipoles of individual molecules oriented more or less alike. Such an arrangement is not conducive to ice nucleation, owing to the structural entropy penalty imposed on such an adsorbed layer. In this case, multiple adsorbed layers may be required before the freedom of orientation among the water molecules in the outermost adsorbed layer is sufficiently large for some of them to assume icelike orientations while others remain anchored.

Little is known quantitatively about the importance of active sites to ice formation in the contact mode. Observations have shown that dry particles of many compounds such as clays, sand, soil, CuS, and organic compounds are considerably better IFN when acting in the contact mode rather than in the freezing or deposition modes. In an attempt to explain this effect, Fletcher (1970), for example, pointed out that the observed differences between nucleation in the contact and freezing modes may be caused by the partial solubility of any solid, especially when in the form of small particles. Thus it is reasonable to assume that active sites at the surface of a particle are especially vulnerable to erosion by dissolution after a particle has become immersed in water. Although the erosion effect may account for some differences between the contact and freezing modes, it cannot explain the significant difference between the ice nucleability of some clays in the contact and deposition modes. In addition, the erosion effect is unable to account for the fact that in all three modes, AgI exhibits practically the same threshold temperature of $-4°C$ for particles of the same size.

A different explanation for contact nucleation was given by Evans (1970). He suggested that only those compounds initiate ice formation in the contact mode which exhibit a strong affinity for water, thus adsorbing water molecules from the liquid or vapor in a close array. In such a case, ice nucleation is hindered owing to a structural entropy penalty imposed on the adsorbed layer. However, during the initial brief moments of contact between a particle and a supercooled water drop, adsorption is incomplete and disordered despite the strong affinity. Thus, during this period the energy barrier to the formation of a more ordered, icelike arrangement may be considerably lower, and thus nucleation may be much more likely, than in an adsorbed and firmly attached oriented array. Although this explanation is attractive, it hinges on the assumption that the time required for building up an oriented water film is much longer than the time needed to form an ice embryo in the disordered adsorbed layer. Unfortunately, this assumption has not yet been justified.

A third explanation for contact nucleation has been given by Guenadiev (1972), who conjectured that an IFN acting in the contact mode must build up a critical ice embryo which is in equilibrium with the water of the supercooled drop, rather than with the surrounding water vapor. Since at any given temperature the former requirement is less stringent, an IFN may nucleate on contact with a supercooled drop even though the ice particle is of subembryo size with respect to ice formation from the vapor. Although this mechanism can account for clay particles being better IFN in the contact mode than in the deposition mode, it cannot explain, for example, why AgI exhibits the same nucleation threshold in both modes, nor can it explain why clay particles exhibit a better ice nucleability in the contact mode than in the freezing mode.

Given the complexity of the surface chemistry involved in the phase transition of atmospheric water, it is not surprising that the idealized theory of heterogeneous nucleation of ice is not readily applicable to natural water systems. The extension of the "capillary" theory of nucleation to ice formation is described by Fletcher (1970). It offers a useful qualitative model for ice nucleation, but is not sufficiently comprehensive to provide a quantitative framework for the microphysics of ice phase transition in the atmosphere. Such developments await further sophistication in dealing with properties of the solid–fluid interface.

8.3.2 Growth of Cloud Particles. Once cloud particles are formed, their evolution in supersaturated, moist surroundings ultimately can produce precipitation if a colloidal instability exists in the cloud. The heart of the precipitation process is the establishment of a mechanism for production of large hydrometeors from a large number of cloud particles over a limited cloud lifetime. The conditions within the cloud could be affected by local changes in droplet concentration associated with entrained CCN, or shifts in icing by IFN. The change in the particle spectrum also not only depends on growth processes but has to take into account evaporation and glaciation.[†] The latter is defined as the onset of appearance of large numbers of ice crystals in a cloud.

The lifetime of clouds varies from a few minutes for small cumulus (convective) systems to perhaps an hour for very large cumulonimbus storms rising from a cloud base of 2–4 km to 10–15 km altitude. Thus the cloud particle evolution process is highly time constrained. A comparative overview of the basic growth processes is seen in Figs. 8.37–8.41. The diffusional growth rate for individual aqueous drops at 1% supersaturation, a temperature of 20°C, and a condensation coefficient of 0.045 is shown in Fig. 8.37. The change in radius by a factor of 10 (mass by a factor of 2.2) takes approximately 100 sec. There is a tendency for small droplets containing less electrolyte to catch up with the size of more concentrated drops. This occurs because the curvature and solute effects rapidly decrease with growth. Calculated growth by collision and coalescence for two different model droplet distributions are shown in Fig. 8.38 for 30–50 μm droplets, as representing those reaching precipitation size. The growth of particle by collision is relatively slow unless times in excess of 1000 sec are available. Droplets present in concentrations of ten per liter are estimated to grow much faster in maritime clouds than in continental clouds, suggesting a greater possibility of precipitation from the former. Finally, the change in mass with time of a spherical ice particle of initial radius 10 μm falling through a maritime cloud and through a continental cloud of supercooled droplets is shown in Fig. 8.39. Here the dry growth by the vapor scavenging process yields a rapid growth for the ice particles relative to the other mechanisms. The growth here is faster in a maritime cloud due to a broader assumed droplet size distribution. Such calculations imply that the mixed ice–liquid droplet cloud offers the possibility for rapid growth processes. It is important to note that

---

[†] Glaciation occurs when water droplets freeze, turning the cloud into a suspension of ice crystals.

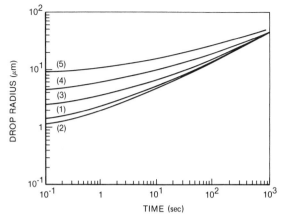

Fig. 8.37. Diffusional growth rate of individual aqueous solution drops as a function of time at 1% supersaturation and 20°C: $\alpha_c = 0.045$; the initial drop radius corresponds to that of a salt-saturated solution drop in equilibrium (salt grain mass of m_s) with the relative humidity at which the sale deliquesces. (1) NaCl: $m_s = 10^{-12}$ g; (2) $(NH_4)_2SO_4$: $m_s = 10^{-12}$g; (3) NaCl: $m_s = 10^{-11}$g; (4) NaCl: $m_s = 10^{-10}$g; (5) NaCl: $m_s = 10^{-9}$g. [Based on data of Low (1971); from Pruppacher and Klett (1978); courtesy of Reidel Publ. Co.]

the growth of ice particles by the Wegener–Bergeron mechanism is strongest near $-12°C$ since the vapor pressure difference between ice and supercooled water is a maximum here.

Perhaps more realistic is the calculation of growth of the entire droplet size spectrum in clouds since the rapid growth of only a few droplets among many is necessary for precipitation. The voluminous literature on such calculations is discussed by Pruppacher and Klett (1978). An illustration of results for vapor-diffusional growth in cumuloform clouds considering an air parcel steadily rising in a supersaturated environment is shown in Fig. 8.40. Here a typical model distribution and chemical composition is assumed for

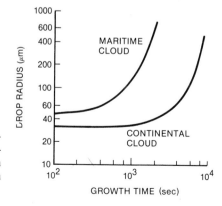

Fig. 8.38. Size variation of a drop growing by collision and coalescence in a maritime- and a continental-type cloud. [From Braham (1968); courtesy of the American Meteorological Society.]

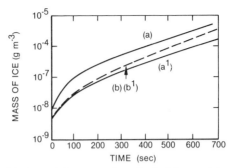

Fig. 8.39. Change with time of the mass of a spherical ice particle of initial radius 10 μm falling through a maritime cloud and a continental cloud. Curve (a) refers to the calculations using a characteristic marine drop size distribution (a) and only the average 10-μm drop radius of the drop spectrum in that cloud (a'). Curve (b) refers to the calculation using a continental drop size distribution and to the calculation using only a 10-μm average drop radius of the drop spectrum in that cloud (b'). [From Ryan (1972); courtesy of *Journal de Recherches Atmospheriques*.]

maritime and continental aerosol particles as nuclei (Lee and Pruppacher, 1977). The nuclei distribution was assumed to be at 99% RH, taking into account the deliquescence process. Initial vertical pressure, temperature, and humidity distributions were used which are thought to be typical of a macroscopic cloud environment. Allowing the air to rise with a constant updraft of 1 m/sec and permitting CCN entrainment yields droplet spectra as shown in Fig. 8.40 for two environments. Entrainment evidently assists in broadening the droplet distribution because it provides a continuous supply of small droplets, as observed. The droplet spectra are broadened in the nonentrainment case by diffusional growth. The droplet distribution tends to become bimodal for various height levels in the model cloud. In continental clouds, the estimated number of droplets larger than 20 μm diameter presumably results largely from activation of a broadened initial CCN distribution. The maritime case gives too low a small droplet population in relation to larger droplets compared with observation.

The deficiencies in the Lee and Pruppacher (1977) "single-pass" calculation for a cloud evidently can be rectified in part by considering that actual clouds develop in a sequence of lifting and sinking motions. Such apparent buoyant oscillation will involve a cycle of mixing, condensation, and evaporation influencing the cloud droplet spectrum. This cycle has been simulated in the calculations of Mason and Jonas (1974). They follow the rise and subsidence of a buoyantly unstable "thermal" parcel of air, which is followed by a second thermal air parcel picking up the residue of the first. Although most of the droplets in the first parcel evaporate, a few of the largest particles survive to experience further growth in the second rising parcel. Through this sequence, a few large drops can be produced without

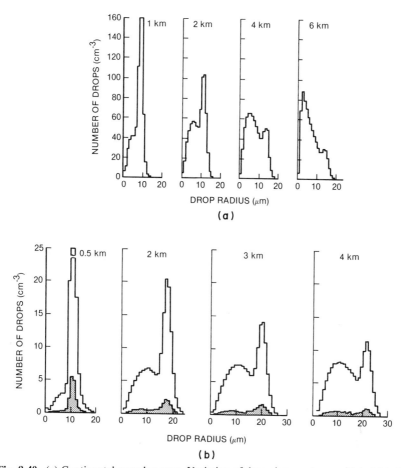

Fig. 8.40. (a) Continental cumulus case. Variation of drop size spectrum with height above cloud base. Entrainment model, aerosol particles entrained. (b) Maritime cumulus case. Variation of drop size spectrum with height above cloud base. Entrainment model: aerosol particles entrained. The shaded area of the histogram refers to NaCl particles, the unshaded area to $(NH_4)_2SO_4$ particles; numbers on top of histogram refer to number of drops in the histogram interval. [From Lee and Pruppacher (1977); courtesy of *Pure and Applied Geophysics* and the authors.]

requiring the presence of giant hygroscopic nuclei. The subsidence of the first parcel also boosts the relative concentration of small droplets to levels expected from observations.

Alternatively, the growth of hydrometeor size in clouds can be accounted for by extending the calculations for collision and coalescence with a stochastic, material balance for each droplet species, analogous to Eq. (3.18). In principle, the collision model can account for the details of the particle size distribution as it evolves, given theoretically realistic collision efficiencies for droplets outside the Stokes hydrodynamic range. A number of numerical simulations of the dynamic equations for droplet spectra applicable to cloud evolution have been attempted, and have been summarized by Pruppacher and Klett (1978). One comprehensive treatment was reported by Berry and Reinhardt (1974). In their model, these investigators separated the water droplet spectrum ($n_R \langle \ln a \rangle$) into two parts, one consisting of cloud droplets (S1), and the other mainly of large hydrometeors (S2), which contain most of the liquid water in a mature cloud. Included in the calculations are parametrization for the collision of cloud droplets with one another (S1–S1, or autoconversion), accretion collection (S1–S2 interaction), and large hydrometeor self-collection (S2–S2 interaction). The initial liquid water to feed the S2 spectrum is provided by autoconversion; this mode becomes weak in the early stages compared with S1–S2 collections. The S2–S2 interactions then become mainly responsible for forming the maximum and shape of the hydrometeor part of the spectrum. As an example of the graphical display of the evolution of the cloud particle spectrum for this model is shown in Fig. 8.41. As the system develops, the S2 spectrum matures and collects most of the cloud water droplets. The drop radii a_1 and a_2 are, respectively, those corresponding to the volumetric mean radius of the droplet spectrum (S1) and the mean radius of the distribution of water content (large drops, S2). The evolution shown in Fig. 8.41 is calculated by using an approximate form of the collision kernel b, taken as the sum of droplet and hydrometeor volumes.

By probablistic arguments, the Berry and Reichardt studies and related work indicate that a statistical description of the collection–accretion process can account for accumulation of significant amounts of liquid water in hydrometeors on a time scale compatible with natural clouds. This accumulation can be enhanced by modest levels of turbulence in clouds, as noted in the calculations of de Almeida (1975). For turbulence characterized by tur-

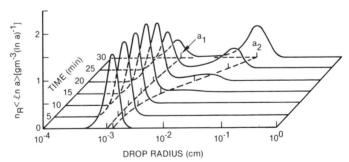

Fig. 8.41. Time evolution of the liquid water spectrum. Initial radius of droplet size distribution $a_1 = 12 \ \mu$m. The S1 spectrum is shown on the left. The S2 spectrum begins to evolve after about 20 min at the depletion of S1. [From Berry and Reinhardt (1974); courtesy of the American Meteorological Society.]

bulent energy dissipation rates of 1 cm²/sec³ droplet growth is enhanced, but beyond this level growth rates can decrease. Another feature of de Almeida's calculations indicates that the air turbulence has a tendency to suppress biomodal particle distribution formation. Pruppacher and Klett (1978) speculate that turbulence enhances the coagulation rate of small droplets among themselves and with larger drops, but has little effect on large drop collisions with each other. This emphasizes the autoconversion and accretion modes preferentially to hydrometeor self-collection, giving rise to faster transfer of water by the S1–S2 pathway.

The addition of condensation and breakup processes to the collision mechanism further complicates the evolution of the cloud particle spectrum. The work of Jonas and Mason (1974), for example, suggests that condensation tends to enhance the broadening of the droplet spectrum produced by collision processes.

The continual collision models tend to predict an unrealistic "flow" of water mass to larger and larger drops. Without a continuous source of small droplets, this prevents the evolution of a quasi-steady state distribution, which is believed to be present in many cloud systems. A means of maintaining a small droplet supply other than condensation on entrained particles is through breakup of large drops, splintering of ice crystals. Breakup may take place through collision, or by hydrodynamic instabilities. Ice splintering may take place during contact nucleation of supercooled drops.

Finally, there remains the question of the hydrometeor evolution when ice crystals and frozen drops emerging into graupel[†] are present. The calculations of Beheng (1978) are instructive here. His calculations extend those of Berry and Reinhardt (1974) to consider a situation in which platelike ice particles are grown by riming within a cloud droplet spectrum evolving by collision. The study indicates that the graupel formed by the riming process tend to accumulate water while suppressing the growth of the large drop portion of the droplet spectrum, obviously by depletion through the riming. The droplet depletion in turn tends to narrow the graupel size distribution, with a diminishing accumulation of frozen water mass. Thus, in the absence of other mechanisms, the combined droplet collision and ice accretion process tend to inhibit hailstone formation and large drop production.

8.4 MODIFICATION OF CLOUDS

With the background in Section 8.3, the stage is set to inquire about the possibilities for modification of cloud and precipitation processes employing atmospheric aerosol perturbations. With the development of modern cloud physics, the scientific basis for such modifications has become more sound. From the discussion, cloud process modification could be achieved in princi-

---

[†] Graupel consists of semispherical, soft, ice hydrometeors of 0.5–6 mm size, with densities of 0.1–0.9 g/cm³.

ple by addition of microscopic dispersed chemical substances to act as CCN or IFN at key periods during the evolution of the cloud system. The question that remains to be answered is: Can man consciously or inadvertently add sufficient amounts of suspended material at critical times in cloud evolution to catalytically (or otherwise) disturb the "colloidal" stability of natural clouds?

The history of attempts to modify clouds is a long one, which predates scientific knowledge of atmospheric processes. Rainmaking by appeal to the gods was well known to primitive man, and followed his civilization. By the nineteenth century, the firing of rockets into clouds for hail suppression or rainmaking became a part of the natural philosopher's repertory. The art of rainmaking is sometimes called "pluviculture."[†]

Attempts to develop a science of cloud seeding began in earnest after Schaefer and Langmuir's experiments in the late 1940s with a great practical interest in attempting to enhance precipitation over certain watersheds, and to suppress supercooled fog formation over highways and airports. The success of cloud seeding has been highly controversial based on experience to date (Kerr, 1982). However, its devotees remain convinced that effects can be detected in controlled studies. Cloud modification reached a stage of interest a few years ago where concern for legal aspects of redistribution of atmospheric water became a significant legal and potential regulatory issue (e.g., Dennis, 1980). Such questions of environmental precedent remain unresolved to this day.

A natural cloud of water droplets of small size is a system having a very large degree of colloidal stability. Attempts to control the precipitation of clouds must, therefore, be concerned with the means by which this stability can be upset. Water clouds differ from one another in two basic microscopic ways—in liquid water content and in droplet size spectrum. The liquid water content does not vary as much as expected from an adiabatic model, but seems to be limited to average values of the order of 1 g/m^3 or less by mixing processes whose vigor is related to the existing liquid water content. In droplet spectra, however, clouds differ very widely from one another, maritime clouds having small numbers of rather large droplets and continental clouds having large numbers of small droplets. Since stability in these warm clouds can only be upset by collision and coalescence processes between the droplets, and since collision efficiency is a very strong function of the droplet radii, details of the size spectrum evolution are of paramount importance in determining the intrinsic stability of the cloud microstructure. If the cloud becomes unstable and precipitation develops, then it presumably does so through the action of the small proportion of much larger than average droplets present in the spectrum. Thus it is the interaction of these droplets

[†] The reader who is interested in the empirical experience of the subject is referred to a fascinating book on the history of rainmaking in the United States published recently by Spence (1980).

with average droplet population which in general is considered most susceptible to be influenced.

For clouds that penetrate above the freezing level, the presence of ice crystals has to be considered. The liquid water content and droplet spectra are similar to those of warm clouds, with the exception that vigorous thunder clouds may have central liquid water contents substantially larger than 1 g/m^3. Such clouds, however, are so far from a state of stability that they normally are not considered susceptible to modification by microscale perturbation.

If the stability of clouds is to be modified via their microstructure, two accessible parameters are identified—the droplet size spectrum and the concentration of ice crystals present. The importance of these two quantities lies in the fact that only one droplet in a million is significantly larger than its fellows and one ice crystal per liter of cloud is important. The part of the microstructure important to stability is thus of a sufficiently small concentration that modifications can be contemplated without the expenditure of a prohibitive amount of energy. The technology of modification techniques essentially addresses this problem.

8.4.1 Approaches to Inducing Cloud Change. In attempting to modify the microstructure of warm clouds, the only known changeable quantity is the size spectrum of the cloud droplets, and the only hypothesized ways in which it seems possible to modify the spectrum are to add nuclei, to add droplets, or to increase the collision efficiency.

Maritime clouds containing perhaps 50 droplets/cm^3 are unstable once they grow to depths of more than about 1000 m, so that little cause would be served by trying to enhance their instability. On the other hand, continental clouds and particularly clouds in drought conditions consisting of large numbers of small droplets may reach considerable depths without precipitating. The time required for natural precipitation to develop may be so long that the cloud dissipates without having produced rain. It is upon such clouds that attention is often focused.

Since it is clearly impractical to attempt to deplete the air of its nuclei content in order to decrease the droplet concentration, any modification must consist of an addition to the nucleus population. The only addition of value would be one which increases the number of very large droplets. To accomplish this, rather large hygroscopic particles can be introduced into the air below cloud base in numbers sufficient to provide about one large drop per liter. Because of its solubility and innocuous properties, an obvious choice for the seeding material is sodium chloride. It seems possible that a sufficient concentration of large salt particles released below the base of a cloud of moderately continental character might stimulate precipitation if the vertical development of the cloud is several thousand meters and its lifetime an hour or more.

Several experiments were performed in the 1950s to test this seeding technique. For example, Davies (1954), in East Africa, sent up balloons carrying charges of finely ground salt which was dispersed explosively on reaching cloud level. A large proportion of the treated clouds were observed to rain after about 20–30 min. Augustin (1954), in Madagascar, dispersed salt particles from an aircraft and claimed substantial rainfall to result, though much of it not until the next day. Fournier d'Albe and co-workers (1955) used blowers to disperse NaCl particles 10^{-9}–10^{-8} g in mass from the ground in the Punjab and observed apparent increased rainfall in the region downwind from the generators. The results of these experiments were inconclusive.

A more direct method of modifying the droplet spectrum, suggested by Ludlam (1951) is by the actual introduction of large droplets from some sort of water spray. By using suitably equipped aircraft, it is possible to introduce the droplets at any level in the cloud, but the optimum position is presumed to be just above the cloud base.

In convective clouds with rising velocities and depth, an increase of a factor of a thousand in mass occurs for droplets injected at cloud base, but only a factor of fifty for droplets released near the cloud top. Therefore, it appears more economical to release droplets near cloud-base than near cloud-top level, and for this reason cloud-base injection was recognized by early workers.

Bowen (1952) carried out experiments to test the effectiveness of this seeding method. An aircraft fitted with a water tank and equipped with two horizontal spray bars flew through selected clouds about 300 m above cloud base. The spray gave droplets estimated to have a median radius of about 25 μm. In performing the experiments situations were chosen in which several similar nonprecipitating clouds were visible, and the behavior of the seeded cloud was compared with that of the others. In almost all cases, the seeded cloud behaved in a manner different from the others and precipitation often developed in it. The initial trials of the method, while not yielding statistically significant results, supported qualitatively the expectations of the theory.

Later, Braham and co-workers (1957) described experiments in which water was released at a rate of 450 gallons per mile into the bases of cumulus clouds. In this experiment, two similar clouds were chosen and one of the pair was seeded on a random basis. Varying results were obtained, but in many cases the treated cloud rained while the other remained stable. Statistical examination of the results showed that the likelihood of the effect being due to chance was less than 2%, giving further confirmation that this seeding produced positive results.

Another method of modification of the cloud droplet spectrum which has attracted some interest involves electrical effects to increase collection efficiency. Strong electric fields or charges on small droplets may sufficiently disturb the surfaces or the motion to enhance collision and coalescence. Thus the presence of such fields or of opposite electric charges on colliding droplets may enhance the collision rate within the cloud, building larger droplets and perhaps causing instability to develop.

Vonnegut and Moore (1958) made progress in this direction by investigating the way in which space charge can be introduced into the lower atmosphere. They supported fine wires 0.025 cm in diameter and about 7 km long on masts about 10 m above ground level, and by applying a potential of about 30 kV produced a corona along its length. The ions produced rapidly became attached to Aitken nuclei in the air stream and their mobility was reduced to such an extent that they were carried downwind in detectable quantities to distances as great as 8 km. In this experiment, a voltage potential was used so that the ions were all of one sign, and their presence was detected by measuring anomalies in the normal atmospheric potential gradient. If these ions were carried into clouds, they would tend to charge drops similarly with consequent inhibition of collisions. Vonnegut proposed a different role for the ions in developing cumulus clouds, but this work is considered controversial and generally has not been pursued actively.

When attention is turned to subfreezing clouds, modification associated with the Wegener–Bergeron mechanism becomes a possibility. Since a subfreezing cloud becomes unstable when ice crystals are present in concentrations of about one per liter, it is simply necessary to devise a means of producing them at this concentration in clouds not normally cold enough to contain natural ice crystals in order to markedly effect stability.

Two methods have been evolved to achieve this effect. The first, discovered by Schaefer (1946), consists of dropping into a supercooled cloud pellets of CO_2 (dry ice, $-78°C$) which chill the droplets formed by condensation to well below the temperature for homogeneous freezing and so produce a cloud of tiny ice crystals in the wake of the pellets. Schaefer found that, in fact, pellets of any material cooled to a temperature below $-40°C$ are active in this way, but dry ice is most convenient from a practical point of view.

The second method, discussed as early as 1938 by Findeisen (1938) involves the addition of quantities of ice-forming nuclei, which will be active at temperatures warmer than those characteristic of natural nuclei, to the air mass in which clouds are forming. This suggestion was not taken up practically until Vonnegut (1947) discovered the great efficiency of silver iodide as an ice crystal nucleus. Since the 1950s this method has become the major technique used in cloud modification.

The technique of seeding clouds with silver iodide particles was discovered, as was the dry ice method, during the course of research on Project Cirrus. Vonnegut (1947) found that when silver iodide was formed into a smoke by intense heating, the particles produced had a nucleation threshold near $-4°C$, while at $-15°C$, the yield of active nuclei was about 10^{15} per gram of silver iodide burned.

Vonnegut developed several methods of producing the silver iodide smoke. This was a nucleating agent sufficiently active that very small amounts could modify the ice nucleus content of the atmosphere by several orders of magnitude. Field trials immediately following this discovery showed some signs of success, though the results achieved were not as spectacular as had been obtained using dry ice. In more recent experiments, however, the technique has been amply demonstrated to be an effective one for cloud modification. Various means of distributing the smoke from burners on the ground or from aircraft have been developed (Fletcher, 1966). In cases where aircraft burners are used, the seeded clouds are more readily identified and their development studied.

It is difficult to obtain quantitative information on the overall effect of cloud seeding with silver iodide by studies of individual clouds. Braham (1960), on the basis of radar studies of a large number of clouds in Arizona, attempted to address these issues. If a suitable cloud is seeded at its base with silver iodide, it takes about 20 min for precipitation to develop. Braham found that, for the region studied, only about 30% of clouds judged suitable for seeding would persist for the required additional 20 min, and of those that did persist almost half rained by natural processes in any case. The statistical effect of seeding is thus small, and under most conditions the number of experiments required to give a statistically significant quantitative result is prohibitively large. For this reason, as well as because it is ultimately the rainfall reaching the ground that is important, most quantitative cloud seeding experiments are based on large-scale repetitive programs. A number of these have emerged in the 1960s, and have been recognized for their achievements and limitations (Elliot et al., 1978; Silverman, 1976; Grant, 1969).

If a supercooled cloud is seeded with dry ice or silver iodide, the results may be viewed in two ways. The ice crystals produced grow at the expense of the cloud droplets and fall with increasing rapidity, emerging from the cloud base as snowflakes or raindrops, emphasizing precipitation elements. If instead we focus on the cloud droplets, then the effect of seeding is to deplete their number and, as the precipitation falls, to decrease the liquid water content of the cloud. This aspect of seeding results is also important since the dissipation of clouds in the vicinity of airfields can aid materially in aircraft traffic flow. Conditions where such dissipation is necessary are mainly those in which a continuous cloud deck extends over a large area, so that seeding of such layer clouds must be considered rather than the cumulus clouds typical of rainmaking experiments.

Either dry ice or silver iodide may be used as the seeding agent for dissipation by depleting small droplets. Since cloud temperatures are often warmer than $-10°C$, dry ice tends to be preferred because of its greater efficiency IFN at these temperatures, and the ease with which comparatively large quantities may be dispersed. Since the cloud areas involved are usually not large, few problems arise in handling the quantities of material required. This technique has shown promise of being effective and economical for use in situations where supercooled stratus or fog is a major visibility problem. It has been used operationally for cleaning airports of fog in recent years. The restriction to supercooled clouds is, however, often a limiting factor in the usefulness of the method. For warm clouds entirely different approaches must be applied.

Hail and Lightning Suppression. As well as modifications of cloud structure simply designed to produce precipitation, attempts also have been made to seed clouds in an effort to suppress the formation of hail and to reduce the incidence of lightning strikes.

Hail develops when the rate at which an ice particle grows by collision while falling through a supercooled cloud becomes so great that some of the water is accreted in liquid form and later frozen. Hail is therefore associated with supercooled clouds of very high liquid water content, which will usually be compact clouds of great vertical development. While it has been shown that the natural IFN concentration in clouds in which hailstorms are forming is not significantly lower than normal (Soulage, 1958), it has been suggested that the introduction of large numbers of artificial nuclei into incipient hail clouds might cause sufficient glaciation at sufficiently high temperatures to reduce the rate of liquid water accretion on precipitation elements.

To this end, programs have been designed in Europe and America to study the effects of silver iodide seeding on hail suppression. Since hail is even more variable than rainfall, and a negative result is much more difficult to detect than a positive one in small sample experiments, any assessment of the effectiveness of the method must await the design and performance of large-scale, long-term statistical experiments. A survey of early experiments has been given by Frank (1958). Later programs have been initiated in the Soviet Union, with considerable success claimed. Air over five million hectares apparently is being treated for hail suppression techniques in recent Soviet programs (Sulakvelidze *et al.,* 1967; Battan, 1977; also Dennis, 1980).

A major study was initiated in the United States in 1969 to try to verify the apparent successes of the Russian programs. The National Hail Research Experiment (NHRE) operated in northeastern Colorado for a number of years, but produced inconclusive results with regard to the Russian methods [National Center for Atmospheric Research (NCAR), 1974].

Any attempt to reduce the incidence of lightning strikes would best be based on a detailed knowledge of the mechanism of electrification of clouds, and unfortunately such knowledge is not at present available. Partly because silver iodide seeding is the most readily available means of modifying cloud structure, and based upon the thought that glaciation may prevent the great vertical development characteristic of thunderclouds, several programs of cloud seeding and lightning strike evaluation have been initiated in the

United States. Of these, perhaps the best known is Project Skyfire (Barrows, 1958).

Although the results published for Project Skyfire contained no conclusions on the effectiveness of silver iodide seeding in reducing lightning strikes, other evidence now suggests that such seeding in fact may increase such strikes. This conclusion appears reasonable in light of an increasing acceptance of the view that the major part of cloud electrification is caused by interaction between ice crystals and supercooled water droplets. The seeding increases the number of ice crystals in the cloud and hence the rate of charge separation, leading to a result directly contrary to the aims of the program. Such conclusions are, however, as yet only tentative.

8.4.2 Verification of Technology.   Of all the methods of cloud modification discussed above, only one has received really broad scale trial in many different geographical areas. This is the seeding of subfreezing clouds with silver iodide smoke. Some of this seeding has been in an attempt to minimize hailstorms and modify hurricanes because of their extensive potential for damage, including effects on crops, or to reduce the number of lightning strikes, because of their potential for starting forest fires. However, the great majority of experiments still have been conducted in an effort to increase rainfall.

According to Fletcher (1966), the crux of the practical utilization of cloud modification centers on its success as a precipitation producer. Any measure of "increase" implies knowledge of the "normal" amount of rain which would have fallen in the absence of any seeding experiments. Most of the difficulty in assessing the results of cloud seeding experiments stems from a lack of knowledge of this normal amount of precipitation, and skillful statistical design is required if significant answers are to be obtained despite this gap in our basic knowledge.

There are two possible approaches to the problem of determining the effectiveness of seeding in a given situation. The first, which is the more appealing from a fundamental point of view, involves three phases: (a) a knowledge of the natural evolution of clouds and in particular the precipitation produced, (b) a knowledge of the way in which individual clouds of various types are affected by seeding, and (c) a detailed cloud census for the region in which seeding is contemplated. Supplied with this information, it should be possible to make a reliable estimate of the overall results of cloud seeding in an area, and this is one of the aims of the detailed applications of cloud physics studies. Unfortunately, our understanding is as yet too meager to predict confidently the evolution of a cloud with the precision necessary for a rainfall estimate. Similarly, the behavior of a seeded cloud is equally uncertain, and no cloud census programs appear to be sufficiently comprehensive to give the information necessary.

Constraints of Natural Variability.   Before the concept of an evaluation experiment is considered, it is useful to examine the natural influences that affect cloud evolution. With these in mind, it should be possible to devise an experiment which will yield a definite result with the least amount of effort. In practice, this reasoning has rarely been followed. Experiments have been performed over prolonged periods without well-thought-out design. Then when the results are analyzed, it is often found that any effects produced by seeding are masked by random fluctuations produced by natural causes.

The simplest type of experiment to design and perform is Fletcher's (1966) *serial experiment.* A geographical region is selected whose promise from a cloud seeding point of view is to be assessed, and the available historical data on its rainfall is examined. In many cases of interest, rainfall data are available for a period of more than fifty years and allow a reliable value of mean rainfall and standard deviation about this mean, for a selected time unit, to be determined. From this information, it is then possible to make certain predictions about the natural rainfall to be expected in the future. Cloud seeding operations are then initiated in the area and the actual rainfall compared with that expected on the basis of the historical data.

In order to evaluate the effectiveness of this method, an estimate of the accuracy of the predicted value is needed of the "unseeded" rainfall in the target area. Some idea of this quantity can be obtained by examining the historical rainfall records. Examination of available rainfall data normally indicates that an experimental duration of a few days is too short because the variability of rainfall from day to day is too great. On the other hand, a period of tens of years must be ruled out for practical reasons. While the longer the period chosen, the less the inherent uncertainty due to natural fluctuations, explicit figures must be examined if a suitable averaging period is to be found for a given region. Under typical continental conditions, rainfall varies in the mean between 1.5 and 12 cm/month. The standard deviation for such cases ranges between 1.0 and 6.5, with coefficient of variation between 1.0 and 2.0 (e.g., Fletcher, 1966). Though the rainfall totals for any single day are far from normally distributed, when sufficient days are added together into groups the total rainfall totals for the same calendar month of each year are normally distributed to a first approximation, and the approximation becomes satisfactory for yearly rainfall totals.

If we make this approximation for monthly totals, then it is possible to predict estimates of the expected precipitation for some future month and to assign confidence limits to this prediction.

In cloud seeding experiments, it is usual to regard a rainfall increase as significant if the probability of an increase as large or larger occurring by chance is less than 5%. Examination of tables of the normal distribution function shows that for an increase to be significant at this level, it must exceed at least 1.65 standard deviations. Typical natural rainfall data have a standard deviation of about 13–20% of the mean. Thus experiments would require precipitation increases of 85% or more above the mean to be significant statistically in a one-month study where the standard deviation is at the low of the range. With higher variability in rainfall, the change in the mean would have to be larger. Experience with cloud seeding experiments suggests that the increase in precipitation is probably only of the order of 10%, and certainly does not approach 100% as required for significance in an experiment of this kind. Fletcher concludes, therefore, that a month is too short an averaging time for a simple serial experiment of this type.

A year is the next convenient time unit to consider. For a one-year experiment, McDonald (1958) has examined data from several stations in the United States. He found that the coefficient of variation for each station is much less than for the monthly totals, but even in the most favorable case of Iowa City, an increase in rainfall of 23% is required for significance, while 40–50% increase is required in other areas. An experiment designed on such a simple basis is thus insensitive to the detection of small precipitation increases and does not represent a satisfactory approach to the evaluation of cloud seeding programs. The principal obstacle to the use of a simple serial experiment to determine the effects of seeding then is the great variability of natural precipitation even over periods as long as a year.

Alternatively, control-area experiments attempt to reduce the natural uncertainty involved in predicting the unseeded rainfall in a target area by using data from stations outside the target area to aid with the prediction. Thus the predicted rainfall $\mathcal{P}$ should have the form

$$\mathcal{P} = A + \sum_i \mathcal{B}_i X_i, \tag{8.13}$$

where A is the historical mean value, X_i are meteorological variables (rainfall, cloud cover, etc.) associated with one or more control areas during the seeded period, and the coefficients $\mathcal{B}_i$ are determined from analysis of the historical data. If the rainfall in the target area is normally closely related to that in the control areas, this procedure may be expected to increase greatly the accuracy of the predicted rainfall.

An analysis of commercial rainmaking experiments carried on in 1950–1954 in Oregon has been made by Decker et al. (1957) using this method. They employed eleven control variables X_i in their analysis and evaluated the coefficients $\mathcal{B}_i$ by a study of weather records for the previous four-year period. This reduced the coefficient of variation to the extent that a 15.5% increase in rainfall would have been regarded as significant. Since the observed increase was only 6%, no conclusions could be drawn.

Fletcher (1966) has noted that a defect inherent in all methods which rely upon historical records to evaluate the coefficients in Eq. (8.13) is that they neglect the effects of any long-term climatic change. While the coefficients could be made time dependent and extrapolated to new values for the test period, the amount of data available is not usually sufficient to make this feasible, so that this defect remains. A related defect is of a practical nature and relates to the measurement of precipitation. Since this measurement usually involves interpolation from sparse rain gauge data, it is not particularly precise, and it becomes necessary to ensure that the same rain gauges and evaluation procedures are used during the seeded period as were used for the historical records. This precludes the installation of special rain gauge networks and usually results in sparse and inadequate precipitation data for coverage of a test area.

Fletcher (1966) has suggested that a method which eliminates both the defects above is to seed selectively during only some of all possible opportunities and to compare the rainfall in the target area during seeded and unseeded periods. Since no historical rainfall data are involved, special dense networks of rain gauges can be installed over the target area and the measurements should be much more representative than those obtained from existing rain gauge stations. This is the approach that has become increasingly popular in recent experimental designs, where selection of clouds for seeding is done randomly.

It is necessary that no bias be introduced in the selection of seeded and unseeded situations, and several means of ensuring this have been adopted. Langmuir (1950) adopted a seeding schedule having a seven-day periodicity in experiments conducted in New Mexico, and sought a resulting seven-day periodicity in rainfall figures. Such a periodicity was in fact found, commencing at about the same time as the seeding operation, and was hailed as a substantial effect due to cloud seeding. However, Lewis (1951), for example, found similar seven-day periodicities in earlier weather records, and Brier (1954) showed that these could probably be explained by the orderly progress of circumpolar pressure patterns. Because such real or apparent weather periodicities are so common, seeding programs on a simple alternating basis are unsuitable, and a randomized method must be adopted. This method amounts to characterizing certain situations as seedable, and then after this has been done, deciding on a completely random basis whether or not seeding is to be carried out. Equivalent seedable situations may be simply successive weeks or other convenient short-time units; they may be individual storms possessing certain suitable characteristics, or they may be determined in a more complicated manner. The important thing is that the situation be characterized as seedable before a decision is made as to whether or not it is to be seeded. Randomized experiments have been undertaken since the 1950s with government funding because of an understandable reluctance on the part of commercial operators and their clients to let half of the seedable situations pass without any action. An early large-scale randomized experiment was commenced in the Snowy Mountains area in Australia in 1955 (Adderley and Twomey, 1958). Since that time, other experiments have been initiated in the United States, including the Sierra Nevada

Mountains of California, in Arizona and Illinois as well as in several other regions in Australia, and Israel.

The necessity for systematic statistical design in experiments of this type can hardly be overemphasized. The claimed results of any rainmaking program will rightly be subjected to searching evaluation by professional statisticians, and it is only by proper statistical design in its initial stages that an experimental program can hope to produce a convincing result.

Important examples of the controversy which can emerge from ambiguous design and treatment of data has emerged in the literature from the major experiments, the Skajit River orographic cloud program (Hastay and Gladwell, 1969), the Rocky Mountain–Colorado River Basin Study (Elliot *et al.*, 1978), and the Wolf Creek–Climax Study (Grant and Elliott, 1974; Grant and Mielke, 1967).

Hobbs and co-workers have criticized the conclusions of each of these projects on different grounds of apparent inadequacies in data analysis. For example, the Skajit River study analysis may not have taken complete account of anomalous runoff for the central river during the experimental year. This resulted in an apparent overstatement of the achieved increases in precipitation. In the case of the Climax and Wolf Creek experiments, the critics questioned the results because of the limitations of the perceived design. This results from lack of support for the designers' hypothesis that 500-mbar temperatures are good measures of cloud top temperatures in the Rocky Mountains (Hobbs and Rangno, 1978). The Colorado River Basin randomized experiment has been criticized on the basis that the statistical treatment of the data may have actually masked results which would have shown more conclusive results of seeding than actually reported (Hobbs and Rangno, 1979; Rangno and Hobbs, 1980).

Recently the cloud seeding experience of scientists in Israel has been evaluated (Kerr, 1982). Two Israeli series have been attempted, the first between 1961 and 1967, and the second between 1969 and 1975. The first was exploratory in nature, seeking to determine whether silver iodide seeding of winter clouds would increase rainfall in northern Israel. The initial experiments suggested increases of 15% in rainfall from seeding in target areas. The second series was confirmatory in design, with a rigid statistically randomized method. The seeding was undertaken in the watersheds feeding the Sea of Galilee. Using 388 experiment days (Gagin and Neumann, 1981), the seeding enhanced rainfall by 13% in the target area as a whole and 18% in a smaller, selected catchment area. According to the reported results, the probability that the increase resulted from a chance occurrence of particularly rainy days was 2.8% in the overall target area and 1.7% in the selected catchment area.

Although the analysis and results need to be verified independently by other statisticians, they appear to show a particularly favorable conclusion. The results are more impressive in the sense that they are physically consistent with expectations. Gagin and Neumann hypothesized that clouds would be most susceptible to seeding action at temperatures above $-20°C$ since at lower temperatures natural IFN are as effective as artificial stimuli. Above $-5°C$, AgI loses its efficiency as an IFN. Thus clouds with tops between -5 and $-20°C$ are shown to be optimum for seeding. They have reported that seeding had little effect at cloud-top temperatures warmer than $-10°C$ or colder than $-21°C$. However, with cloud-top temperatures between -15 and $-21°C$, rainfall increased 46%, with a significance level of 0.5%. If confirmed, these results provide strong evidence for the effectiveness of seeding in supercooled clouds, at least under northern Israel's winter conditions.

Regardless of the continuing scientific debate on the quantification of results of seeding, the technology is being used in different parts of the world. The extent of its usage has signaled a social and regulatory concern for the equitable geographical distribution of water resources and for public protection against accidental excesses in radically changing natural processes.

8.4.3 Inadvertent Weather Modification. Concern for the potential of inadvertent cloud modification emerged as one of the questions asked of the adverse effects of air pollution in 1960s. Visible signs of cloud formation associated with emissions from industrial sources, forest fires, or cities have been reported by several investigators (Warner, 1968; Hobbs *et al.*, 1970; Eagan *et al.*, 1974). The former observations were made of CCN produced in paper mill plumes in Washington state. The latter concerned the production of CCN from sulfur cane burning in Australia, which was associated with visible cloud formation. The production of nuclei from paper mill plumes led the investigators to speculate that they were responsible for enhanced precipitation from warm clouds in the area. However, subsequent calculations of Hindman *et al.* (1977) did not support this hypothesis.

A photograph containing possible circumstantial evidence for cloud formation and evolution from industrial origins in a cloud street near Lake Erie is shown in Fig. 8.42. In this region, rows of cumulus clouds are often observed. However, in this case, only a single cloud street was seen, originating on the left of the photo with an industrial area.

Rainfall anomalies near metropolitan areas were documented by Huff and Changnon (1973). A list of apparent effects in summer rainfall is given for nine midwestern cities in Table 8.7. Evidence for an urban effect on precipitation was suggested in data from St. Louis, Chicago, Cleveland, Detroit, and Washington, D.C. Evidence was weak or nonexistent for Indianapolis, Tulsa, and New Orleans. According to the authors, the effect appeared more pronounced in summer than in winter and tended to be maximum 15–45 km downwind of the cities.

Perhaps the most dramatic rainfall anomaly and the one that has been the longest center of controversy is that of LaPorte, IN, a steel industry commu-

Fig. 8.42. A cloud street near Lake Erie extending over a hundred miles. The chain of clouds may have been induced by aerosol particles injected into the air from industrial activity at the left side along the southern shores of the lake.

TABLE 8.7

Summary of Possible Urban Effects on Summer Rainfall for Nine Cities[a]

City	Observed effect	Maximum change (mm)	(%)	Approximate location
St. Louis	Increase	40	15	16–20 km downwind
Chicago	Increase	50	17	50–55 km downwind
Indianapolis[b]	Indeterminate	—	—	
Cleveland[c]	Increase	35	15	40–80 km downwind
Washington	Increase	28	9	Urban area
Baltimore[d]	Increase	43	15	Urban and northeastward
Houston[e]	Increase	33	17	Near urban center
New Orleans	Increase	45	10	NE side of city
Tulsa	None	—	—	

[a] From Huff and Changnon (1973); courtesy of the American Meteorological Society.
[b] Sampling density not adequate for reliable evaluation.
[c] Estimated orographic effect subtracted (maximum actually 27% greater than city).
[d] 50–65 km downwind of Washington—not included in original study.
[e] Urban effect refers only to air mass storms—apparently little or no effect in frontal storms.

nity near South Bend on the southern tip of Lake Michigan. Changnon first reported this anomaly in 1968. Subsequently, the controversy over interpretation of the anomaly has continued in a series of papers (Changnon, 1971; Holzman and Thom, 1970; Ogden, 1969; Machta *et al.*, 1977; Clark, 1979; Changnon, 1980). Changnon's (1980) latest review still maintains that urban effects are the main factor in causing the anomaly. Yet the results remain in question and are still being debated because of differences in observer methods and uncertainties of lack of supporting data from neighboring areas.

If urban conditions influence cloud formation, their explanation may lie in macroscopic processes associated with convective disturbances from spatial differences in heating or surface roughness. Or effects may be associated with CCN and IFN emissions from cities or individual sources.

There is ample evidence that cities and industrial sources can produce large numbers of CCN. These may include soluble secondary aerosol particles such as ammonium sulfate or nitrate. Unfortunately there does not appear to be unambiguous evidence from precipitation chemistry trends in the United States or Scandinavia that can be interpreted as a dramatic trend in increased sulfate in rainfall. Such a trend could be interpreted as evidence of the presence of contaminants in precipitation from pollution scavenging. Although trends in increasing nitrate concentration in precipitation have been reported, this component cannot be interpreted as a CCN trend because it is likely to be strongly influenced by scavenging of nitric acid vapor. No trends in soluble metals or insoluble elements in rainwater have been examined.

The question of man's influence on IFN is a complex one. Some of the particles at the center of ice crystals consist of insoluble combustion products. This finding suggests natural or anthropogenic combustion sources for IFN. In support of this notion, Hobbs and Locatelli (1969) observed a significant increase in the concentration of IFN downwind of a forest fire, and Pueschel and Langer (1973) observed increased IFN concentrations during sugar cane fires in Hawaii.

Evidence for the effectiveness of other sources of anthropogenic IFN has been provided by Soulage (1958, 1964), Telford (1960), Admirat (1962), and Langer *et al.* (1967), for example, who showed that certain industries, in particular, steel mills, aluminum works, sulfide works, and some power plants, release considerable amounts of IFN into the atmosphere. The particles emitted from electric steel furnaces in France were found to be of particulately high effectiveness as IFN, with a threshold temperature of about $-9°C$. Wirth (1966) and Georgii and Kleinjung (1968) showed that the IFN concentration is comparatively high throughout heavily industrialized Europe. For example, the mean IFN concentration observed during summer, 1964, at ten different locations in Europe ranged from 2.6 to 53.7 per liter at $-20°C$.

These high IFN concentrations can be understood in part from the fact that some of the particles emitted during industrial processes consist of metal oxides, most of which are known to have a high nucleability. Thus Fukuta (1958) and other investigators found that the oxides Ag_2O, Cu_2O, NiO, CoI, Al_2O_3, CdO, Mn_3O_4, and MgO exhibit threshold temperatures between -5 and $-12°C$, while oxides such as CuO, MnO_2, SnO, ZnO, and Fe_3O_4 have threshold temperatures between -12 and $-20°C$. Also, particles of Portland cement ($3CaO \cdot SiO_2$, $2CaO \cdot Si_2O$, $3CaO \cdot Al_2O_3$, $4CaO \cdot Al_2O_3 \cdot Fe_2O_3$) were found to act as IFN at temperatures as warm as $-5°C$ (Murty and Murty, 1972).

Although some specific anthropogenic sources may emit IFN, the anthropogenic emission from urban complexes as a whole is generally deficient in IFN. Thus, studies by Braham and Spyers-Duran (1975) carried out in the area of St. Louis show that, on the average, fewer IFN were found downwind in the city than upwind. Hidy et al. (1969) also reported low concentrations of IFN and CCN in Los Angeles air, which when transported downwind would represent negligible influences on regional nuclei populations in southern California. This result suggests that anthropogenic combustion products emitted into the air over urban areas are generally poor IFN and, in addition, may be capable of deactivating existing IFN.

The work on inadvertent weather modification so far would lead prudent investigators to be concerned for potential changes based on aerosol behavior. However, little concrete, definitive information exists today on which to project adverse effects on agriculture or suburban ecology. Weather changes outside of thermal radiation anomalies is the least understood of all the air pollution issues raised in the last decade.

REFERENCES

Ackerman, T. P. (1977). J. Atmos. Sci. **34,** 531.
Adderley, E. E., and Twomey, S. (1958). Tellus **10,** 275.
Admirat, P. (1962). Bull. Obs. Puy de Dome No. 2, p. 87.
Allee, P. A. (1970). "Proceedings Conference on Weather Modification," p. 244. Am. Meteorol. Soc., Boston, Massachusetts.
Alvarez, L. W., Alvarez, W., Asaro, F., and Michel, H. (1980). Science **208,** 1095.
Augustin, H. (1954). Rapport Technique Surles Operations, de Pluie Artificelle á Madagascar in 1953. Madagascar, Service Mét., Publ., 23.
Bajza, C. (1979). In "View on Visibility-Regulatory and Scientific" (Air Pollut. Control Assoc., ed.), p. 243. Pittsburgh, Pennsylvania.
Barrows, J. D. (1958). Project Skyfire, U.S. Advisory Committee on Weather Control. Rep. 2, p. 105. Washington, D.C.
Bartleson, C. J., and Bray, C. P. (1962). Photogr. Sci. Eng. **6,** 19.
Battan, L. J. (1977). Bull. Am. Meteorol. Soc. **58,** 4.
Battan, L. J., and Reitan, C. H. (1957). In "Artificial Simulation of Rain" (H. Weickmann and W. Smith, eds.), p. 184. Pergamon, Oxford.
Beheng, K. K. (1978). J. Atmos. Sci. **35,** 683.
Bergstrom, R. W. (1973). Contrib. Atmos. Phys. **46,** 223.
Berry, E. X., and Reinhardt, R. L. (1974). J. Atmos. Sci. **34,** 1814, 1825, 2188, and 2127.
Bigg, E. K., and Stevenson, C. (1970). J. Rech. Atmos. **4,** 41.

Blackwell, H. R. (1946). *J. Opt. Soc. Am.* **36**, 624.

Bowen, E. G. (1952). *Q. J. R. Meteorol. Soc.* **78**, 37.

Bowen, E. G. (1953). *Aust. J. Phys.* **6**, 490.

Braham, R. (1960). *Proc. Am. Soc. Civ. Eng.* **86** (IRI), 111.

Braham, R. (1968). *Bull. Am. Meteorol. Soc.* **49**, 343.

Braham, R., and Spyers-Duran, R. R. (1975). *J. Appl. Meteorol.* **13**, 940.

Braham, R., Battan, L. J., and Byers, H. R. (1957). *Meteorol. Monogr.* **11**, 47.

Brier, G. (1954). *Bull. Am. Meteorol. Soc.* **35**, 118.

Bryson, R. A. (1972). *In* "The Environmental Future" (N. Polunin, ed.), pp. 133–177. McMillan, New York.

Budyko, M. I. (1969). *Tellus* **21**, 611.

Campbell, F. W. (1966). *Proc. Symp. Inf. Process. Sight Sensory Syst. 1965* p. 177.

Campbell, F. W., and Green, D. G. (1965). *J. Physiol. (London)* **181**, 576.

Campbell, F. W., and Maffei, L. (1974). *Sci. Am.* **231**, 106.

Carlson, T. N., and Caverly, T. S. (1977). *J. Geophys. Res.* **82**, 3141.

Carlson, C. R., and Cohen, R. W. (1979). Image Descriptors for Displays: Visibility of Displayed Information. Report ONR, Contr. N00014-74-C-0184. RCA Lab., Princeton, New Jersey.

Changnon, S. A., Jr. (1968). *Bull. Am. Meteorol. Soc.* **58**, 1069.

Changnon, S. A., Jr. (1971). *Science* **172**, 987.

Changnon, S. A., Jr. (1980). *Bull. Am. Meteorol. Soc.* **61**, 702.

Clark, R. R. (1979). *Bull. Am. Meteorol. Soc.* **60**, 415.

Connor, W. D., and Hodkinson, J. R. (1967). Optical Properties and Visual Effects of Smoke-Stack Plumes. Rep. 999-AP-30. U.S. Public Health Service, Washington, D.C.

Cornsweet, T. N. (1970). "Visual Perception." Academic Press, New York.

Coulier, P. (1875). *Pharm. Chim. Paris* **22**, 165.

Covert, D., Waggoner, A., Weiss, R., Ahlquist, N., and Charlson, R. (1980). *In* "The Character and Origins of Smog Aerosols" (G. M. Hidy and P. K. Mueller, eds.), p. 574. Wiley, New York.

Daniel, T. C., Wheeler, L., Boster, R., and Best, P. (1973). *Man-Environ. Syst.* **3**, 330.

Davidson, M. (1968). *J. Opt. Soc. Am.* **58**, 1300.

Davies, D. A. (1954). *Nature (London)* **174**, 256.

de Almeida, F. C. (1975). Research Report 75-2. Dept. of Meteorol., Univ. of Wisconsin, Madison.

Decker, F. W., Lincoln, R. L., and Day, J. A. (1957). *Bull. Am. Meteorol. Soc.* **38**, 134.

Dennis, A. S. (1980). "Weather Modification by Cloud Seeding." Academic Press, New York.

Diaz, H. J., and Quayle, R. G. (1980). *Mon. Weather Rev.* **108**, 249.

Egan, R. C., Hobbs, P. V., and Radke, L. F. (1974). *J. Appl. Meteorol.* **13**, 535.

Elliot, R. D., Shaffer, R., Court, A., and Hannaford, J. (1978). *J. Appl. Meteorol.* **17**, 1298.

Ellis, H. T., and Pueschel, R. (1971). *Science* **172**, 845.

Ellsaesser, H. (1975). *In* "The Changing Global Environment" (S. F. Singer, ed.), pp. 235–269. Reidel, London.

Evans, L. F. (1965). *Nature (London)* **206**, 822.

Evans, L. F. (1970). *Proc. Conf. Cloud Phys.* p. 14.

Fletcher, N. H. (1966). "Physics of Rainclouds." Cambridge Univ. Press, London and New York.

Fletcher, N. H. (1970). *J. Atmos. Sci.* **27**, 1098.

Flowers, E. C., McCormick, R. A., and Kurfis, K. R. (1969). *J. Appl. Meteorol.* **8**, 955.

Findeisen, W. (1938). *Meteorol. Z.* **55**, 121.

Fournier d'Albe, E. M., LaTeef, A. M. A., Rasool, S. I., and Zaidi, I. H. (1955). *Q. J. R. Meteorol. Soc.* **81**, 574.

Fox, D., Loomis, R. J., and Green, T. C., eds. (1979). Proceedings of the Workshop in Visibility Values. Rep. WO-18. U.S. Dept. of Agric., For. Serv., Fort Collins, Colorado.

Frank, S. R. (1958). Survey and History of Hail-Suppression Operations in the U.S., U.S. Advisory Committee for Weather Control, Report, Vol. 2, p. 264. Washington, D.C.

Fresch, R. W. (1973). Res. Rep. AR-106. Dept. of Atmos. Res., Univ. of Wyoming, Laramie.

Fukuta, N. (1958). *J. Meteorol.* **15,** 17.

Gagin, A., and Neumann, J. (1981). *J. Appl. Meteorol.* **20,** 1301.

Georgii, H. W., and Kleinjung, E. (1967). *J. Rech. Atmos.* **3,** 145.

Georgii, H. W., and Kleinjung, E. (1968). *Pure Appl. Geophys.* **71,** 181.

Gokhale, N., and Spengler, J. I. (1972). *J. Appl. Meteorol.* **11,** 157.

Grant, L. O. (1969). An Operational Adaption Program of Weather Modification for the Colorado River Basin, Bureau of Reclamation Report, p. 98.

Grant, L. O., and Elliot, R. E. (1974). *J. Appl. Meteorol.* **13,** 355.

Grant, L. O., and Mielke, P. W., Jr. (1967). *In* "Proceedings of the Symposium on Mathematical Statistics and Probability" (L. M. LeCam and J. Neyman, eds.), Vol. 5, pp. 115–132. Univ. of California Press, Berkeley.

Gravenhorst, G., and Corrin, M. L. (1972). *J. Rech. Atmos.* **6,** 205.

Guenadiev, N. (1972). *Pure Appl. Geophys.* **99,** 251.

Hall, J. S., and Riley, L. A. (1976). *Prog. Astronaut. Aeronaut.* **49,** 205.

Halow. J. S., and Zeek, S. J. (1973). *J. Air Pollut. Controll Assoc.* **23,** 676.

Hänel, G. (1976). *Adv. Geophys.* **19,** 73.

Hansen, J. E., and Travis, J. D. (1974). *Space Sci. Rev.* **16,** 527.

Hansen, J. E., Wang, W., and Lacis, A. (1978). *Science* **199,** 1065.

Hansen, J. E., Lacis, A., Lee, P., and Wang, W. (1980). *Ann. N. Y. Acad. Sci.* **338,** 575.

Harshvardhan, H., and Cess, R. O. (1978). *J. Quant. Spectros. Radiat. Transfer* **19,** 621.

Hastay, M., and Gladwell, J. S. (1969). *J. Hydrol. (Amsterdam)* **9,** 117.

Heisler, S. L., Henry, R. C., and Hidy, G. M. (1980). The Denver Winter Haze 1978. Report P-5417 for the Motor Vehicle Manufacturers Assoc. of the U.S.. Environ. Res. and Technol., Inc., Westlake Village, California.

Henry, R. C. (1977). *Atmos. Environ.* **11,** 697.

Henry, R. C. (1979a). *In* "View on Visibility-Regulatory and Scientific" (Air Pollut. Control Assoc., ed.), p. 27. Pittsburgh, Pennsylvania.

Henry, R. C. (1979b). *In* "Proceedings of the Workshop in Visibility Values" (D. Fox, R. J. Loomis, and T. C. Green, technical coordinators), Fort Collins, Colorado, January 28–February 1. U.S.D.A., Forest Service, Washington, D.C.

Henry, R. C., Collins, J. C., and Hadley, D. (1981a). *Atmos. Environ.* **15,** 1859.

Henry, R. C., Hidy, G. M., and Collins, J. (1981b). Analysis of Historical Visibility Data in the Southwest. Rep. P-A849-201. Environ. Res. and Technol., Inc., Westlake Village, California.

Herman, B., Brouning, R. S., and DeLuisi, J. J. (1975). *J. Atmos. Sci.* **32,** 918.

Heymsfield, A. (1972). *J. Atmos. Sci.* **29,** 1348.

Hidy, G. M., Principal Investigator (1975). "Characterization of Aerosols in California (ACHEX)," Vol. IV. Final Report Contr. 358 to Calif. Air Resources Bd., Sci. Cent. Rockwell Int'., Thousand Oaks, California.

Hidy, G. M., and Brock, G. R. (1971). *In* "Proceedings, International Clean Air Congress, 2nd" (H. M. Englund and W. T. Berry, eds.), pp. 1088–1097. Academic Press, New York.

Hidy, G. M., Green, W., and Alkezweeny, A. (1969). *In* "Aerosols and Atmospheric Chemistry" (G. M. Hidy, ed.), pp. 339–344. Academic Press, New York.

Hindman, E. E., II, Taj, P., Silverman, B., and Hobbs, P. (1977). *J. Appl. Meteorol.* **16,** 753.

Ho, W. W., Hidy, G. M., and Govan, R. M. (1974). *J. Appl. Meteorol.* **13,** 871.

Hobbs, P. V. (1974). "Ice Physics." Oxford Univ. Press, Oxford, London and New York.

Hobbs, P. V., and Locatelli, J. D. (1969). *J. Appl. Meteorol.* **8,** 833.

Hobbs, P. V., and Rangno, A. L. (1979). *J. Appl. Meteorol.* **18,** 1233.

Hobbs, P. V., Radke, L. F., and Shumway, S. E. (1970). *J. Atmos. Sci.* **27,** 81.

Hoffer, T. (1961). *J. Meteorol.* **18,** 766.

Holzman, G. B., and Thom, H. C. S. (1970). *Bull. Am. Meteorol. Soc.* **51**, 335.

Horvath, H. (1971). *Atmos. Environ.* **5**, 333.

Horvath, H., and Noll, K. E. (1969). *Atmos. Environ.* **3**, 543.

Hoyt, D. V., Turner, C. P., and Evans, R. D. (1980). *Mont. Weather Rev.* **108**, 1430.

Huff, F. A., and Changnon, S. A., Jr. (1973). *Bull. Am. Meteorol. Soc.* **54**, 1220.

Huffman, P. J. (1973). *J. Appl. Meteorol.* **12**, 1080.

Humphreys, W. J. (1940). "Physics of the Air." McGraw-Hill, New York.

Hunt, R. W. G. (1953). *J. Opt. Soc. Am.* **43**, 479.

Hunt, R. W. G., Pitt, I. T., and Winter, L. M. (1974). *J. Photogr. Sci.* **22**, 144.

Husar, R. B., and Patterson, R. (1980). *Ann. N.Y. Acad. Sci.* **338**, 399.

Husar, R. B., and White, W. (1976). *Atmos. Environ.* **10**, 199.

Isono, K. (1959). *Nature (London)* **183**, 317.

Isono, K., and Komabayasi, M. (1954). *J. Meteorol. Soc. Jpn.* **32**, 29.

Jonas, P. R., and Mason, B. J. (1974). *Q. J. R. Meteorol. Soc.* **100**, 286.

Judd, D. (1973). *In* "Color Vision" pp. 65–73. Natl. Acad. Sci., Washington, D.C.

Judd, D. B., MacAdam, D. L., and Wyszecki, G. (1964). *J. Opt. Soc. Am.* **54**, 1031.

Kelly, D. H. (1974). *J. Opt. Soc. Am.* **64**, 983.

Kelly, D. H., and Magnuski, H. S. (1975). *Vision Res.* **15**, 911.

Kerker, M. (1969). "The Scattering of Light: And Other Electromagnetic Radiation." Academic Press, New York.

Kerr, R. A. (1982). *Science* **217**, 519.

Kinney, L. S. (1962). *Vision Res.* **10**, 503.

Koschmieder, H. (1924). *Contrib. Atmos. Phys.* **12**, 33, 171.

Lamb, H. H. (1970). *Philos. Trans. R. Soc. London, Ser. A* **266A**, 425.

Landsberg, H. (1975). *In* "The Changing Global Environment" (S. F. Singer, ed.), pp. 197–234. Reidel, London.

Langer, G., Rosinski, J., and Edwards, C. P. (1967). *J. Appl. Meteorol.* **6**, 114.

Langmuir, I. (1950). *Bull. Am. Meteorol. Soc.* **31**, 386.

Latimer, D. A., Bergstrom, R., Hayes, S., Liu, M., Seinfeld, J., Whitten, G., Wojcik, M., and Hillyer, M. (1978). Rep. EPA-450/3-78-110a, b, c. U.S. Environ. Prot. Agency, Research Triangle Park, North Carolina.

Latimer, D., Hogo, H., and Daniel, T. C. (1981). *Atmos. Environ.* **15**, 1865.

Leaderer, B. P. (1978). *J. Air Pollut. Control Assoc.* **28**, 321.

Lee, I. Y., and Pruppacher, H. R. (1977). *Pure Appl. Geophys.* **115**, 523.

Lewis, W. (1951). *Bull. Am. Meteorol. Soc.* **32**, 192.

Lin, C. I., Baker, M., and Charlson, R. J. (1973). *Appl. Opt.* **12**, 1356.

Linak, W. P., and Peterson, T. W. (1981). *Atmos. Environ.* **15**, 2421.

Low, R. D. (1971). Res. Develop. Rep. ECOM-5358. U.S. Army Electronics Command, Fort Monmouth, New Jersey.

Ludlam, F. H. (1951). *Nature (London)* **167**, 254.

McDonald, J. E. (1958). *Adv. Geophys.* **5**, 223.

Machta, L. (1972). *Bull. Am. Meteorol. Soc.* **53**, 402.

Machta, L., Angell, J., and Korshover, J. (1977). *Bull. Am. Meteorol. Soc.* **58**, 1068.

Malm, K., Kelly, K., Molenar, J., and Daniel, T. (1981). *Atmos. Environ.* **15**, 1875.

Manabe, S., and Wetherald, R. T. (1967). *J. Atmos. Sci.* **21**, 241.

Marians, M., and Trijonis, J. (1979). "Empirical Studies of the Relationship Between Emissions and Visibility in the Southwest." Report EPA Grant 802015. Tech. Serv. Corp., Santa Fe, New Mexico.

Mason, B. J. (1960). *Q. J. R. Meteorol. Soc.* **86**, 552.

Mason, B. J. (1971). "The Physics of Clouds," 2nd ed. Oxford Univ. Press, London and New York.

Mason, B. J., and Jonas, P. R. (1974). *J. Meteorol. Soc.* **100**, 23.

Mason, B. J., and Maybank, J. (1958). *Q. J. R. Meteorol. Soc.* **84**, 235.

Mass, C., and Schneider, S. H. (1977). *J. Atmos. Sci.* **34**, 1995.

Middleton, W. E. K. (1952). "Vision Through the Atmosphere," p. 122. Univ. of Toronto Press, Toronto, Canada.

Mumpower, J., Middleton, P., Dennis, R., Stewart, T., and Veirs, V. (1981). *Atmos. Environ.* **15**, 2433.

Munn, R. E. (1972). *Atmosphere* **11**, 156.

Murty, A. S. K., and Murty, B. V. R. (1972). *Tellus* **24**, 581.

95th Congress (1977). The Clean Air Act as Amended August 7, 1973, Section 169A, Committee Print No. 99-3020. U.S. Govt. Printing Office, Washington, D.C.

NCAR (1974). National Hail Research Experiment Project Plan, 1975–1980, p. 286. Boulder, Colorado.

NOAA (1974). "Climatic Atlas of the United States." U.S. Dept. of Commerce, Washington, D.C.

Ogden, T. L. (1969). *J. Appl. Meteorol.* **8**, 585.

Palmer, K. F., and Williams, D. (1975). *Appl. Opt.* **14**, 208.

Patel, A. S. (1966). *J. Opt. Soc. Am.* **54**, 689.

Patterson, M. P., and Spillane, K. T. (1967). *J. Atmos. Sci.* **24**, 50.

Patterson, E. M., Gillette, D. A., and Stockton, B. H. (1977). *J. Geophys. Res.* **82**, 3153.

Penndorf, R. (1957). *J. Opt. Soc. Am.* **47**, 147.

Pollack, J. B., and Toon, O. B. (1980). *Am. Sci.* **68**, 268.

Pollack, J. B., Toon, O., Sagan, C., Summers, A., Baldwin, B., and Van Camp, W. (1976). *J. Geophys. Res.* **81**, 1071.

Price, S., and Pales, J. (1963). *Arch. Meteorol., Geophys. Bioklimatol., Ser. A* **13A**, 398.

Pruppacher, H. R., and Klett, J. D. (1978). "Microphysics of Clouds and Precipitation." Reidel, London.

Pueschel, R., and Langer, G. (1973). *J. Appl. Meteorol.* **12**, 549.

Ramanthan, V., and Coakley, J. A. Jr. (1978). *Rev. Geophys. Space Phys.* **16**, 465.

Rangno, A. L., and Hobbs, P. V. (1978). *J. Appl. Meteorol.* **17**, 1661.

Rangno, A. L., and Hobbs, P. V. (1980). *J. Appl. Meteorol.* **19**, 347.

Rasool, S. I., and Schneider, S. H. (1971). *Science* **173**, 138.

Richter, J. P. (1883). "The Literary Work of Leonardo DaVinci," 3rd ed., 1970, Vol. I. Phaedon, London.

Roberts, P., and Hallett, J. (1968). *Q. J. R. Meteorol. Soc.* **94**, 25.

Robock, A. (1978). *J. Atmos. Sci.* **35**, 1111.

Roosen, G. I., Angione, R. J., and Klemke, C. (1973). *Bull. Am. Meteorol. Soc.* **54**, 307.

Rosen, H., Hansen, A. D. A., Gundel, L., and Novakov, T. (1978). *Appl. Opt.* **17**, 3859.

Rosinski, J. (1979). *Adv. Colloid Interface Sci.* **10**, 315.

Ryan, B. F. (1972). *J. Rech. Atmos.* **6**, 673.

Sagan, C., Tom, O. B., and Pollack, J. B. (1980). *Science* **206**, 1323.

Schaefer, V. J. (1946). *Science* **104**, 457.

Schnell, R. C. (1974). Res. Rep. AR-111. Dept. of Atmos. Res., Univ. of Wyoming, Laramie.

Silverman, B. (1976). *J. Weather Modif.* **8**, 107.

Soulage, G. (1958). Influence des Emissions d'iodure d'ayent de l'Association d'Etudes sur le pouvoir glocagne de l'air en periode greligene. Assoc. d'Etude de Moyens de lutte Contre les fleaux atmospheringe, Toulouse, Rept. No. 6, p. 24.

Soulage, G. (1964). *Nabila* **6**, 43.

Spence, C. C. (1980). "The Rainmakers." Univ. of Nebraska Press, Lincoln.

Sulakvelidze, G. K., Bibilashvili, S. H., and Lapcheva, V. F. (1967). "Formation of Precipitation and Modification of Hail Processes," p. 208. Israel Program for Scientific Translations, Jerusalem.

Tampieri, F., and Tomasi, C. (1976). *Tellus* **28**, 333.

Taylor, G. I. (1917). *Q. J. R. Meteorol. Soc.* **43**, 841.

Telford, J. (1960). *Bull. Am. Meteorol. Soc.* **16**, 676.

Tomasi, C., and Tampieri, F. (1976). *Atmosphere* **14**, 61.

Toon, O. B., and Pollack, J. B. (1976). *J. Appl. Meteorol.* **15**, 225.

Trijonis, J. (1979). *Atmos. Environ.* **13**, 833.

Trijonis, J., and Shapland, R. (1979). Rep. EPA 450/5-79-010. U.S. Environ. Prot. Agency, Research Triangle Park, North Carolina.

Trijonis, J., and Yuan, K. (1978). Rep. EPA-600/3/78/039. U.S. Environ. Prot. Agency, Research Triangle Park, North Carolina.

Twitty, J. T., and Weinman, J. A. (1971). *J. Appl. Meteorol.* **10**, 725.

Twomey, S., and Wojciechowski, T. A. (1969). *J. Atmos. Sci.* **26**, 684.

U.S. Public Health Service (USPHS) (1969). Air Quality Criteria for Particulate Matter, p. 100. Dept. of Health, Educ., and Welfare, Washington, D.C.

Vali, G. (1968). *Proc. Cloud Phys. Conf.* p. 232.

Vali, G., and Schnell, R. C. (1973). *Nature (London)* **246**, 212.

Viezee, W., and Evans, W. E. (1980). Rep. EA-1434. Electric Power Research Inst., Palo Alto, California.

Vittori, O., and Mestitz, A. (1974). "Four Golden Horses in the Sun." Int. Fund. for Monuments, New York.

Vonnegut, B. (1947). *J. Appl. Phys.* **18**, 593.

Vonnegut, B., and Moore, C. B. (1958). Preliminary Attempts to Influence Convective Electrification in Cumulus Clouds by the Introduction of Space Charge into the Lower Atmosphere, Conf. on Atmos. Electr., A. D. Little, Inc., Cambridge, Massachusetts.

Waggoner, A. P., Vanderpol, A., Charlson, R., Larsen, S., Granat, L., and Tragardh, C. (1976). *Nature (London)* **261**, 120.

Warner, J. (1968). *J. Appl. Meteorol.* **7**, 247.

Weiss, R. E., Waggoner, A., Charlson, R., Thorsell, D., Hall, J., and Riley, L. (1978). *In* "Proceedings Conference on Carbonaceous Particles in the Atmosphere" (T. Novakov, ed.). p. 257. Lawrence Berkeley Lab. Publ. 9037, Berkeley, California.

Williams, M., Treiman, E., and Wecksung, M. (1980). *J. Air Pollut. Control Assoc.* **30**, 131.

Williamson, S. (1973). "Fundamentals of Air Pollution." Addison-Wesley, Reading, Massachusetts.

Winkler, P. (1967). Diplom. Thesis, Meteorol. Inst., Univ. of Mainz, Mainz, Germany.

Wirth, E. (1966). *J. Rech. Atmos.* **2**, 1.

Wyszecki, G., and Stiles, W. S. (1967). "Color Science." Wiley, New York.

Yamamoto, G., Tanaka, M., and Arao, K. (1971). *J. Meteorol. Soc. Jpn.* **49**, 859.

CHAPTER 9

HEALTH EFFECTS OF INHALED AEROSOLS

In the previous chapter, the role of aerosol particles in atmospheric phenomena was discussed. Let us now return to processes which affect more directly human activity. Traditionally one of the greatest motivating forces for the study of aerosol behavior relates to public health. It has long been known that finely divided particulate matter is a contributor to respiratory disease. Indeed, a principal portal of entry of contaminants into the human body is through the respiratory system. Once materials are breathed into this system and retained in the body, they may contribute to a variety of diseases. The link between disorders of the human body and specific contaminants has been difficult to verify in most cases. However, diseases of pneumoconiosis, emphesyma, and lung cancer represent classes of results from suspected chemical assaults. Viral or bacteriological links of infections have been established with influenza, bronchitis, pneumonia, and tuberculosis.

To gain perspective on the potential influence of respiratory disorders relative to disability or death associated with all causes, consider statistics reported by the U.S. Public Health Service. Shown in Table 9.1 is a listing of the ten leading causes of death in the United States in 1900 and 1967. With certain qualifications (Spiegelman and Erhardt, 1974), there have been some major shifts, with reduction in death rates during that period of nearly 70 years. In particular, diseases of the heart and cancer and cerebral strokes account for more than 60% of deaths in 1967 in contrast to about 18% in 1900. Accidental deaths have been fairly constant, rising from 4.2 to 6.1% during the period, while respiratory diseases (tuberculosis, influenza, and

TABLE 9.1

The Ten Leading Causes of Death: United States, 1967 and 1900[a]

Rank	Cause of death	Rate per 100,000	Percentage of all deaths
1967			
	All causes	935.7	100.0
1	Diseases of heart	364.5	39.0
2	Malignant neoplasms	157.2	16.8
3	Cerebral hemorrhage[b]	102.2	10.9
4	Accidents, total	57.2	6.1
5	Influenza and pneumonia[c]	28.8	3.1
6	Certain diseases of early infancy[d]	24.4	2.6
7	General arteriosclerosis	19.0	2.0
8	Diabetes mellitus	17.7	1.9
9	Other diseases of circulatory system	15.1	1.6
10	Other bronchopulmonic diseases	14.8	1.6
1900			
	All causes	1,719.1	100.0
1	Influenza and pneumonia[c]	202.2	11.8
2	Tuberculosis	194.4	11.3
3	Gastroenteritis	142.7	8.3
4	Diseases of heart	137.4	8.0
5	Cerebral hemorrhage[b]	106.9	6.2
6	Chronic nephritis	81.0	4.7
7	Accidents, total	72.3	4.2
8	Malignant neoplasms	64.0	3.7
9	Certain diseases of early infancy[d]	62.6	3.6
10	Diphtheria	40.3	2.3

[a] From National Center for Health Statistics (1970).
[b] And other vascular lesions affecting central nervous system.
[c] Except pneumonia of newborn.
[d] Birth injuries, asphyxia, infections of newborn, ill-defined diseases, immaturity, etc.

pneumonia) have dropped from 23 to about 6%. Much of this change is in reduction of tuberculosis mortality. For modern man, heart disease, strokes, and cancer (including lung cancer) loom as major health factors. These diseases certainly have natural origins, but some investigators suspect that they have a relation to exposure to toxic materials in pollution (U.S. PHS, 1969).

The mortality statistics do not reveal the full story of the disability associated with respiratory disease. Additional perspective comes from *morbidity* data. Morbidity refers to a general term which is applied to measures of both incidence and prevalence of disease. A National Health Interview Survey

(NCHS, 1964) introduced the concept, stating that "Morbidity is basically a departure from a state of physical and mental well being resulting from disease or injury, of which the individual is aware." Awareness connotates a degree of measurable impact on the individual or his family in terms of restrictions and disabilities caused by the morbidity. It includes not only active or progressive disease but also impairments; that is, chronic or permanent defects that are static in nature, resulting from disease or injury, or congenital malformations. Given a certain subjectivity in morbidity data, the significance of respiratory disease is readily seen in data listed in Table 9.2. Clearly respiratory disorders far surpass all other categories shown in both incidence and bed disability. Thus we infer that the direct impact of respiratory disorders is in morbidity of humans rather than in mortality. However, the latter may well be heavily affected in an indirect manner through terminal diseases.

9.1 ENVIRONMENTAL EXPOSURES TO TOXIC MATERIALS

9.1.1 Toxic Chemicals. Toxic environmental factors may be found in the food and water humans consume, as well as the air humans breathe, and in materials contacting the skin. Some broad classes of physiologically active agents in the occupational environment are listed in Table 9.3. Materials which contribute to respiratory disease are listed in Table 9.4. These include a wide range of airborne material such as the community pollutants (SO_2, NO_2, and ozone); inorganic silicas and carbon; animal and vegetable debris; metallic and organic vapors or fumes and sprays; bacteria viruses and fungi; consumer aerosol products; and tobacco products. Considering the range of potential exposures of suspended particles, it is remarkable that respiratory disease measurably affects such a small portion of the population.

Of the many particulate materials and gases listed, a group which has received increasing attention in recent years is the chemical carcinogens. Some of these are listed in Table 9.5. They originate from a broad group of organic and inorganic compounds. Most of these are found in very small quantities in the industrial or community environment. Yet their presence represents the potential for public health hazard as a derivation of the affluence and well being of our modern world.

Tobacco Smoke. One of the most controversial aerosol contaminants involved in respiratory disease is tobacco smoke. Smoke has not been considered an environmental health problem like other air pollutants because it has been judged a matter of personal choice and therefore avoidable. Recently, however, public concern has been focused on involuntary smoke inhalation by people who must share the air with tobacco smokers, and some measures have been introduced to control smoking in public places. The

TABLE 9.2

Annual Incidence of Acute Conditions and Associated Disability Reported in Interviews per 100 Persons According to Specified Condition Groups: United States, Each Fiscal Year Ending June 1958 to June 1969 (Civilian Noninstitutional Population)[a]

Measure; fiscal year ending June	Condition group					
	Total	Infective and parasitic	Respiratory	Digestive	Injuries	All other
Incidence per 100 persons per year						
1969	207	23	122	10	24	28
1968	189	22	106	9	29	24
1967	190	24	104	9	28	25
1966	212	25	126	10	25	25
1965	213	28	116	11	30	28
1964	209	30	110	11	30	28
1963	219	24	127	11	28	28
1962	222	27	128	12	29	26
1961	202	28	110	13	28	23
1960	203	24	119	11	26	23
1959	215	26	126	12	29	22
1958	260	23	169	14	28	26
Days of bed-disability per 100 persons per year						
1969	419	47	250	21	47	55
1968	337	44	185	18	44	47
1967	297	45	147	16	45	44
1966	366	54	196	20	47	48
1965	349	55	170	18	45	61
1964	346	60	157	20	52	57
1963	380	54	206	20	43	57
1962	381	57	202	19	44	59
1961	332	61	150	19	47	55
1960	369	56	197	20	41	55
1959	360	53	190	17	49	51
1958	519	53	352	20	43	50

[a] From Cole (1974).

TABLE 9.3

Classes of Physiologically Active Agents Occurring in Occupational Situations: Their Origin and Distribution[a]

Class	Examples of material	Examples of origin	Distribution
Acids and alkalis	Sulfuric and nitric acid, sodium hydroxide	Industrial processing, agricultural	Airborne dust or spray
Chemical carcinogens	See Table 9.5	See Table 9.5	Vapor, dust
Drugs	Estrogens	Synthesis and purification	Airborne dust
Metal poisons	Arsenic	Ore smelting	Airborne dust
	Cadmium	Welding fume	Airborne dust
	Mercury	Caustic soda manufacture	Vapor
Microbiological agents	Proteolytic enzymes	Detergent formulation	Airborne dust
Mineral dusts and fibers	Quartz	Mines and foundries	Airborne dust
	Asbestos	Insulation operations	Airborne fibers
Noxious gases and fumes	Hydrogen sulfide	Petroleum refining	Gaseous diffusion
	Nitrogen oxides	Nitric acid manufacture	Gaseous diffusion
Pesticides	Parathion	Mfr., formulation and application	Vapor and dust
	Dieldrin	Mfr., formulation and application	Vapor and dust
Plants	Poison ivy	Brush clearing	Skin contact
	Cotton	Milling	Airborne dust
Plastics	Fillers, plasticizers, dyes, foaming agents, catalysts	Chemical production and blending	Vapor and airborne dust
Solvents	Trichlorethylene	Degreasing metal parts	Vapor
	Toluene	Paint formulation	Vapor

[a] After Kay (1977); courtesy of the American Physiological Society.

extent to which cigarette smoke constitutes a health hazard to the smoker himself and to the nonsmoker who is exposed is still a point of some debate. We shall outline the issues here, but the reader is referred to detailed reviews such as Falk (1977) for more information.

The more serious of the health hazards to the tobacco smoker are coronary heart disease, cerebrovascular disease, aortic aneurisms, peripheral vascular disease, pulmonary emphysema, chronic bronchitis, peptic ulcer, and allergic conditions; the most serious hazard is cancer—of the lung, the esophagus, the bladder, and the pancreas. Nicotine and carbon monoxide have been associated with several of these major diseases, but the caresatine

TABLE 9.4

Important Agents Producing Environmental Lung Diseases[a]

Type	Examples
Inorganic dusts	
Free silica	
Crystalline	Quartz, tridymite cristobalite
Amorphous	Diatomaceous earth, silica gel
Silicates	
Fibrous	Asbestos, sillimanite, talc, sericite
Other	Mica, Fuller's Earth, kaolin, cement dust
Carbon	Coal, graphite, carbon black, soot, charcoal
"Inert" metals	Iron, barium, tin
Biological "dusts"	
Vegetable	Moldy hay ("farmer's lung"), mushroom compost, bagasse (sugar cane), maple bark, *B. subtilis* enzyme (detergents), malt, grain weevil, cork, roof thatch, lycoperdon, cotton, flax, hemp, jute, sisal, thermiophilic actinomycetes
Animal	Pigeons, parrots, budgerigars, hens, pituitary snuff
Toxic chemicals	
Irritant gases	Oxides of nitrogen (silo filler's disease), sulfur dioxide, ammonia, chlorine, ozone, phosgene, carbon tetrachloride, hydrogen chloride, chlorinated camphrene, chloropierin, "smoke"
Metallic fumes, vapors, oxides, and salts	Beryllium, silicide, mercury, cadmium, platinum, manganese, pentoxide, nickel, osmium, copper
Plastics	Polytetrafluoroethylene, toluene diisocynate (TDI)
Organic oils	Mineral oil, "cutting" oils
Respiratory infections	
Bacteria	Tuberculosis (miners, nurses, pathologists), anthrax (wool, mohair, and alpaca workers), glanders (grooming of horses, mules)
Virus/rickettsiae	Psittacosis (pet shops, turkey dressings, pigeons), variola (smallpox handlers), Q fever (lab workers, abattoirs)
Fungi	Coccidiomycocis (irrigation, farm work, archeology), histoplasmosis (poultry, pigeons, starlings), cryptococcosis (pigeons)
Respiratory carcinogens	Arsenic, cobalt, nickel, hematite, uranium (pitch blends, "yellow cake"), isopropyl oil, gas retort fumes, chromates, asbestos, other silicates (talc), polynuclear aromatics
Consumer aerosol products	Hair sprays, deodorants, paints, pesticides
Tobacco products	Cigarettes, cigars and pipes, marijuana

[a] From Brain (1977).

agents for cancer have not been identified satisfactorily despite considerable efforts over the years.

In the following, we summarize a discussion of Falk (1977) which focuses on the chemical composition of smoke. A list of potentially toxic compounds

TABLE 9.5

Some Carcinogenic Chemicals[a]

Chemical entity	Target organs	Species activity	Industrial occurrence	Potentially exposed persons
Polycyclic aromatic hydrocarbons (alternate type)				
Benz[*a*]pyrene	Skin, lung	Hamsters, rabbits	Tar, petroleum, and derivatives	Countless
Aromatic amines and nitro compounds				
2-Acetylaminofluorene	Liver, bladder, and others	Rats	Research	Tens
4-Aminobiphenyl	Liver, bladder, mammary	Rats, mice, dogs, man	Research	Tens
Benzidine and salts	Bladder	Dogs, man	Dye intermediates	Hundreds
3,3'-Dichlorobenzidine	Bladder and others	Rats	Pigment, intermediate, curing	Hundreds
4-Dimethylaminoazo-benzene	Liver, bladder	Rats, dogs	Formerly dye uses, now research	Tens
α-Naphthylamine	Bladder	Man	Dyes and agriculture chemicals	Hundreds
β-Naphthylamine	Bladder	Man	No U.S. production, research	Hundreds
Auramine [tris(4-aminophenyl)-methane]	Liver, intestines, bladder	Rats, mice, man	Dye manufacture and use	Thousands
Rhodamine B (an aminoxanthene)	Injection site, s.c.	Rats	Dye manufacture and use	Thousands
4,4'-Diamindiphenyl-methane	Liver	Rats	Elastomers	Thousands
4,4'-Methylene-bis(2-chloroaniline)-3,3-dichloro-4,4-diamino-diphenylmethane	Liver, lung	Rats	In elastomers	10,000
Michler's ketone (4,4'-tetramethyl diamino-benzophenone)	Liver	Rats	Antioxidant in sol-ventless ink	Thousands
4-Nitrobiphenyl	Bladder	Man	Research	Tens
N-Nitroso compounds				
N-Nitrosodimethy-lamine	Liver, kidney	Rats	Research solvents, rocket fuel manufacture	Tens
α-Haloethers				
Bis(chloromethyl) ether	Lung, skin	Animal species	Intermediate research uses	Hundreds
Chloromethylmethyl-ether	Lung, skin	Man, animal species	Manufacture of ion exchange resins	Hundreds

TABLE 9.5 (*continued*)

Chemical entity	Target organs	Species activity	Industrial occurrence	Potentially exposed persons
Alkylating agents				
β-Propiolactone	Skin	Rodents	Acrylate production sterilizing agent	Hundreds
Ethyleneimine	Skin	Rodents	Paper, textiles, fumigant	Hundreds
Trimethylomelamine	Antitumor mutagenic	Mice	Flame retardance	Thousands
TEPA [2,4,7-tris(1-aziridinyl) phosphine oxide	Antitumor mutagenic	Drosophila, mice	Flame retardance and insect sterilization	Thousands
Organic Phosphates				
Dipterex	Liver and others	Rodents	Pesticide	Thousands
Dimethoate	Liver and others	Rodents	Pesticide	Thousands
Naturally occurring carcinogens				
Aflatoxins	Liver and others	Rats and others, man	Infected nuts	Thousands
Rotenone	Mammary	Rats	Pesticide	Thousands
Carbamates				
Urethane (ethyl carbamate)	Lung, stomach	Rodents	Plastics industry	Thousands
Ethylenethiourea	Thyroid, liver	Rodents		Thousands
Chlorinated hydrocarbons				
Vinyl chloride	Liver and others	Rodents	Poly(vinyl chloride) manufacture propellant	Thousands
Chlorinated hydrocarbon pesticides	Liver and others	Rodents	Manufacture formulation and application	Thousands
Trichloroethylene	Liver	Mice	Solvents	Thousands
Miscellaneous				
Aminotriazole	Thyroid, liver	Rodents, fish	Herbicide	Thousands
Benzene	Bone marrow	Man	Solvents	Many
Arsenic in trivalent form	Skin, liver, lung	Man, mice	Pesticides, drugs	Thousands
Asbestos, various forms	Lung and mesothelium	Man, rodents, rabbits	Insulation	Thousands
Beryllium and zinc beryllium compounds	Bone and lung	Rodents, rabbits, monkeys	Aerospace	Thousands
Various cadmium compounds	Administration site and testes	Rodents, fowl	Smelting, welding, pigments, and so forth	Thousands
Chromium and compounds	Administration site, lung	Man, rodents, rabbits	Smelting pigments	Thousands

(*continued*)

TABLE 9.5 (*continued*)

Chemical entity	Target organs	Species activity	Industrial occurrence	Potentially exposed persons
Cobalt, oxide, sulfide	Administration site	Rabbits, rats	Metalliferous industries	Hundreds
Iron-carbohydrates	Administration site	Rabbits, rodents	Drug industry	(?)
Lead, phosphate, acetate, tetraethyl	Kidney, testes, lymph	Rodents	Petroleum and other industries	Thousands
Nickel, metal, carbonyl and other compounds	Lung and nose	Man, rabbits, rodents, cats	Smelting and refining	Hundreds
Selenium compounds	Liver and thyroid	Rats	Glass, rubber, lubricants, pigments	Thousands
Zinc chloride	Testes	Rats, fowl	Not pertinent	Not pertinent

[a] From Kay (1977), with references; courtesy of the American Physiological Society.

in the tobacco smoke aerosol is shown in Table 9.6. Comparison of this list with those in Tables 9.4 and 9.5 indicates clearly the potentially severe exposure to airborne chemicals considered generators of respiratory disease. Of special significance is the potential for cancer.

In chemical analyses of different types of tobacco, the carcinogenic benzo[a]pyrene (BaP) has been singled out for comparison. In one case, smoke yielded amounts of BaP from 0.9 μg (cigarette) to 3.4 μg (cigar) and 8.5 μg (pipe) for each 100 g of tobacco consumed; for pyrene and anthracene, even greater differences have been observed. In another study, the high levels of BaP in pipe tobacco smoke were confirmed, but cigar smoke contained less BaP than cigarette smoke, based on the same amount of tobacco consumed. The total phenol content of these smokes paralleled the BaP concentrations: 69 mg (pipe), 25 mg (cigarette), and 10 mg (cigar) per 100 g of tobacco smoke consumed (Hoffmann *et al.*, 1963).

These findings were at first surprising because of the difference in incidence of lung cancer between cigarette smokers and cigar or pipe smokers. Initially the lower lung cancer risk of the pipe smoker was attributed to differences in smoke inhalation and other smoking habits. Recently, however, Dontenwill *et al.* (1970) suggested that the level of BaP in smoke is not an adequate indicator of carcinogenic potency of tobacco smoke.

Changes in concentration of smoke components as the cigarette becomes shorter have been described: pyrene, BaP, and benz[a]anthracene (BaA) were at higher concentrations in the smoke from shorter cigarettes. By following the decrease in cigarette length from 6 to 0.4 cm at regular intervals and analyzing the BaP content of the smoke, investigations have found

TABLE 9.6

Some Potentially Toxic Chemicals in Tobacco Smoke[a]

Species	Concentration (μg/cigarette)[b]
Carbon monoxide	$(14-20) \times 10^3$
Nitric oxide	45–580
Nitrogen dioxide	23–49
Nitrosoamines	
(*N*-nitrosodimethytamine/*N*-nitrosopyrolidine and others)	0.004
Aliphatic amines	2–22
Propionitrile	30
Acrylinitrile	10
Nicotine	800–3,000
Other alkaloids	1–90
Heterocyclic compounds	
Pyridine	220
Benzo[*a*]pyrene	3–23
Carbozole	2
Alcohols (methanol, C_{17}—C_{26})	15–180 (MeOH)
Phenols (phenol, cresols, . . .)	8–182
Aldehydes (formaldehyde, acetalydehyde, . . .)	8–65
Sulfur compounds (H_2S, thiophene)	2–12
Hydrocarbons (alkane, alkenes, simple aromatics)	1–500
Polycyclic aromatics (indane, indine, fluorene and derivatives)	0.4
Trace metallic elements	
Iron	230 ppm
Antimony	60 ppm
Zinc	35 ppm
Selenium	6 ppm
Organic pesticides	<3
Biological contaminants/toxic metabolics	
Fungi	0.2–200
Viruses (tobacco mosaic)	Detectable

[a] Based on data from Falk (1977); courtesy of the American Physiological Society.
[b] Unless otherwise stated.

an increase in quantity from 3 μg at the beginning, to 9 μg at 2-cm length, and finally to 23 μg at the shortest butt length, all expressed per cigarette smoked (Lindsey, 1962). These values are much higher than those obtained from smoke of cigarettes manufactured more recently. Total phenol and nicotine content of smoke was found by Seehofer *et al.* (1965) to increase in similar fashion as the cigarette became shorter. In the *mainstream,* the aerosol a smoker inhales, an increase in CO as the cigarette was smoked cannot be explained as a loss of filtering capacity as the tobacco is burnt, but it results from a lack of dilution by air passing through the cigarette paper as the cigarette becomes shorter (Mumpower *et al.*, 1962).

It is a generally accepted practice to divide smoke constituents into the particulate and gas-phase components. This separation distinguishes chemicals that can be removed by mechanical filtration from those that require activated carbon for their retention or are not retained at all. However, Falk (1977) has noted that the entire aerosol needs to be treated holistically because many of the vapor-phase components are adsorbed on particulates, which may cause their partial retention by tobacco or filter tips or their deposition in the lung with the particulates. Other gas-phase components with high vapor pressure may not remain associated with particulates and may behave like gases.

Smoke exposure has been described in terms of the mainstream flow to the smoker himself, and the *sidestream* component which is inhaled by the nearby nonsmoker. For evaluation of the sidestream, Hoegg (1972) has reviewed the mainstream and sidestream smoke compositions. The smoking of one cigarette, 20 sec, was sufficient to use 347 mg tobacco, producing about 10^{12} particulate in the mainstream, whereas 550 sec was spent between puffs, and the sidestream smoke produced in the same time contained about 3.5×10^{12} particulates, and resulted from combustion of 411 mg tobacco.

The ratio of sidestream to mainstream smoke concentrations of BaP, pyrene, cadmium, and oxides of nitrogen was found to be between 3 : 1 and 4 : 1. For the oxides of nitrogen it should be mentioned that the oxidation of nitric oxide (NO) to nitrogen dioxide occurs in air, increasing toxicity considerably. An unexplained observation in a comparison of the content of oxides of nitrogen in filter versus nonfilter cigarettes was that the filter cigarettes gave off more NO_x in sidestream smoke (77 μg) than the nonfilter ones (55 μg). No other constituent has shown such differences (Falk, 1977).

For phenols and nicotine the sidestream to mainstream ratio is low (2 : 1 to 3 : 1) depends on the humidity of the tobacco. The ratio for water vapor is about 40 : 1 (7.5 mg in mainstream and 300 in sidestream). With an increase in moisture content, the phenol yield declined by 20%; the amount in the mainstream decreases more than the amount in the sidestream increased. For nicotine there is no change in total smoke content, but the concentration in the mainstream decreased while that in the sidestream went up with an increase in moisture content. The moisture content of tobacco is therefore of importance for estimation of mainstream and sidestream contributions of many water-soluble components.

Perhaps the most concern about adverse health effects for the sidestream inhaler has been expressed in terms of carbon monoxide. Its distribution between the sidestream and mainstream gives a ratio of 5 : 1; i.e., 75 ml in the sidestream to 15 ml in the mainstream. The smoker would thus inhale 14–20 mg in the smoke of one cigarette. The highest CO concentration reached in a test chamber with the smoking of 24 cigarettes was found to be 70 ppm. The adverse effect on people due to CO while staying in a smoke-filled room can best be addressed from experimental data on human volunteers during their recovery after breathing 50 ppm of CO for several hours: after 1 hr exposure, the level of carboxyhemoglobin (HbCO) was 2.1% and it remained twice the normal level for 2 hr after the exposure period ended. Normal levels were reached after 5 hr time. When the exposure lasted 3 hr, the HbCO level before recovery was four times the normal level. After exposure to CO the HbCO level was still six times the normal after 2 hr recovery and still twice the normal level after 11 hr (Stewart *et al.*, 1970). These data suggest that the long recovery period required after prolonged exposure to CO will not allow complete disappearance of the induced HbCO level before returning to a smoke-filled environment the following day.

From the point of view of public welfare, it seems clear that the nearby nonsmoker actually suffers as much or more hazard than the smoker himself. Evidently the sidestream contains significantly higher concentrations of certain contaminants than the mainstream. This generally is not appreciated by the public in such considerations.

Potentially toxic compounds exemplified in Tables 9.5 and 9.6 react with the respiratory system in various ways. The system is a warm, highly moist environment. Thus such soluble gases as SO_2 and NH_3 are absorbed and can cause irritation in the upper respiratory tract. Such less soluble gases as ozone and nitrogen dioxide reach further down into the lung for their deposition. Particulate matter deposits in different parts of the respiratory tract according to size and hygroscopicity.

9.1.2 Association Between Particles and Disease.

Given the potential for enhancement of respiratory disease from environmental chemical factors, institution of public health measures requires proof of association. The accumulation of evidence is achieved considering three elements—epidemiological studies, toxicological experiments, and human clinical studies.

Epidemiology is the branch of medicine concerned with the conditions associated with the occurrence of widespread diseases. When dealing with the effects of air pollution, epidemiologists have endeavored to correlate environmental factors with the occurrence of disease, discomfort, or physiological change in body functions. Frequently such studies have sought to find associations between levels of specific pollutants and the rate of mortality or morbidity within an exposed population. Although an association between a specific pollutant or combination of pollutants and an accompanying or subsequent effect does not prove the existence of a causal relationship, the association may be so obvious as to provide sufficient basis for policy decisions by individuals or governments. As epidemiology has played a central role in assessing the effects of air pollution, we shall comment on some of its strengths and weaknesses.

The challenge for epidemiology is in gathering and processing the appropriate statistical information in a manner that will separate out effects arising from causes other than pollution. For example, to obtain a true measure of the correlation between the levels of air pollutants in a community and the incidence of lung cancer, the effects of an individual's smoking must be isolated. The correlation between smoking and lung cancer is strong enough that it may obscure in some cases a statistical relationship between air pollution and lung cancer. On the other hand, if a true measure of the effects of air pollution is to be obtained, the smokers of a population cannot be neglected. Owing to synergistic effects, air pollution may in fact have a more serious effect upon smokers than nonsmokers. A dramatic illustration is provided by the results of a study of the incidence of occupational lung cancer among a group of asbestos workers (Selikoff et al., 1968). Of a population of 87 nonsmoking workers, during the period from 1963 to 1967 there were no deaths from lung cancer; however, of 283 workers who did smoke, there were 24 deaths from lung cancer reported, compared with only 3 that had been expected on the basis of their smoking habits or occupations alone. Thus multiple causative factors can be very important.

One difficulty in obtaining reliable epidemiological data for the effect of air pollutants is the heterogeneous character of urban populations. The residents of a community at the time of a given study have a varied background with respect to previous exposure to contaminants, partly as a result of moving from one place to another, but also partly because the nature of the pollutants in some communities has changed during the past twenty years, as new industries are introduced and automobile traffic increases. Another difficulty facing epidemiological studies is the lack of reliable measurements of ambient levels of pollution, a consequence of an insufficient number of monitoring stations throughout most of our metropolitan regions. The reported air pollution levels in a city may in fact represent conditions for only one location, and this is probably not indicative of either the average levels or peak levels for other parts of the city. It has therefore been difficult to obtain reliable assessments of the long-term health effects of specific airborne contaminants.

By comparison, clinical studies of air pollution toxicology in laboratory experiments offer better control over the ambient conditions of exposure for human or animal subjects, and in the case of animal experiments, physiological change in the respiratory system can be studied systematically by autopsy for different degrees of pollutant exposure. Clinical studies on animals have shown that a variety of respiratory disease phenomena can be exacerbated by airborne toxic chemicals. Studies on human subjects also have shown a variety of effects on the respiratory system associated with pollutant inhalation. The principal difficulties for interpreting such laboratory studies lie in (a) the extrapolation of animal experiments to human, and (b) the short-term limits of human responses to exposures where long-term, cumulative effects are likely to be of primary interest. In spite of such limitations in medical science, the qualitative evidence for potential adverse health associated with inhaled aerosols is strong. Furthermore, physiological knowledge of the respiratory system is sufficient to identify the importance of aerosol interactions and their pathology.

Toxicological studies compared with epidemiological studies are also limited in another way (Table 9.7), the gross statistical insensitivity imposed by the small number of subjects normally tested. Thus experiments involving low levels of pollution and 50 animals would have little chance on a statistical basis alone of detecting diseases that might be produced at a rate of only 100 individuals per 100,000 similarly exposed population. And this is a statistically significant rate for many types of epidemiological studies.

Epidemiological comparisons of urban and rural populations in the 1950s and 1960s consistently showed a higher incidence of chronic lung disease for the former. Mortality rates from lung disease were also higher. This is true even when the effects of smoking are removed from the statistics. The remaining higher morbidity and mortality is known as the *urban factor*. Several examples of the urban factor for various studies of mortality from lung cancer are given in Table 9.7. For nonsmokers, the mortality rate from lung cancer is more than 2.3 times higher in an urban environment than rural. For smokers, the difference is generally not as pronounced, ranging from 1.26 to 2.23, but the mortality in both urban and rural environments is considerably greater. Although statistical evidence such as this is compelling, it is not necessary a priori to conclude that the cause of the urban factor is air pollution, for there are social and economic differences between urban and rural populations, including the social stress from crowded conditions, noise level, as well as diet and customary amounts of exercise (Reid, 1964).

TABLE 9.7

Lung Cancer Mortality Studies; Number of Deaths from Lung Cancer per 100,000 Population[a]

Smokers, standardized for age and smoking			Nonsmokers			
Urban	Rural	Urban/rural	Urban	Rural	Urban/rural	Study
101	80	1.26	36	11	3.27	California men—1967
52	39	1.33	15	0		American men—1958
189	85	2.23	50	22	2.27	England and Wales—1957
—	—	—	38	10	3.80	Northern Ireland—1966
149	69	2.15	23	29	0.79	England; no adjustment for smoking—1964
100	50	2.00	16	5	3.20	American men—1962

[a] From Lave and Seskin (1970); reprinted with permission of the American Association for the Advancement of Science.

One study by Ishikawa *et al.* (1969) may have avoided many of these complications and provided evidence that air pollution may cause or contribute to emphysema. A comparison was made of autopsied lung material from residents of two cities: Winnipeg, Manitoba, and St. Louis, MO. The Canadian city has a relatively low level of air pollution, whereas the American city characteristically has high levels of industrial contaminants. Emphysema was found to be seven times more common in St. Louis for ages 20–49 and twice as common for ages over 60. Smoking was significantly associated with the disease, but could not be isolated as a factor. The incidence of severe emphysema was found to be four times as high among cigarette smokers in St. Louis as among a comparable group of smokers in Winnipeg. These results are suggestive, but do not prove a causal link between air pollution and emphysema. A question remains as to whether the markedly different climatic conditions of the two cities may be a contributing factor.

As a result of difficulties such as this in making definitive statements from epidemiological studies on different populations, there is considerable interest in the temporal variation and effects of air pollution on the same population. The simultaneous occurrence of abnormally polluted air and excess morbidity and mortality may be so pronounced as to seem to imply a causal relationship; indeed, in the extreme pollution episodes that occurred in Donora, PA in 1948 and London in 1952, there can be little doubt about this. However, the more subtle episodes that occur several times a year in many regions of the world require more care in their analysis because the weather patterns that are largely responsible for the episode are frequently accompanied by characteristic temperature changes, and this in itself can give rise to an increase in respiratory illness. Abrupt temperature changes place a strain on the body and nervous system, while prolonged warm spells contribute to exhaustion. Furthermore, cold decreases mucus transport, thereby reducing the efficiency for removing airborne materials from the lung.

Hodgson (1971) found that mortality in New York City from respiratory and heart disease is quite significantly related to environmental conditions, especially air pollution (as indicated by the mass concentration of aerosol particles) and temperature (in terms of degree days). Some 73% of the variation in daily mortality from the heart and respiratory diseases listed in Table 9.8 is associated with concurrent variations of air pollution levels and temperature, with pollution showing a more significant correlation. The expected increase in average daily mortality for a moderate increase in particulate concentrations is 20–30 deaths, almost 20% of the average daily total number of deaths attributable to heart and respiratory diseases. Studies of this nature have provided evidence that air pollution is exacting a toll, and that its long-term influence is not limited to spectacular but rare episodes. This "proof" of pollution effects remains highly controversial to this day despite more than a decade of effort beyond the initial assessment (U.S. EPA, 1982).

TABLE 9.8

Categories of Respiratory and Heart Diseases Included in a Study of Mortality and Air Pollution Levels in New York City[a]

Respiratory diseases	Heart diseases
Tuberculosis of respiratory system	Arteriosclerotic heart disease, including coronary disease
Malignant neoplasm of respiratory system	
Asthma	Hypertensive heart disease
Influenza	Rheumatic fever and chronic rheumatic heart disease
Pneumonia	
Bronchitis	Other diseases of the heart, arteries, veins
Pneumonia of newborn	Certain types of nephritis and nephrosis of the kidneys

[a] Reprinted with permission from Hodgson (1971). Copyright 1971 American Chemical Society.

Before discussing deposition and pulmonary defense mechanisms, let us examine the anatomy of the respiratory system and its common diseases further for important definitions and terminology.

9.2 THE RESPIRATORY SYSTEM

Human beings cannot live without producing self-sustaining amounts of energy, which in turn is produced by oxidizing the organic molecules in food. Thus the consumption of oxygen and the production of carbon dioxide during food assimilation are indispensable to life. It follows that the human body must have an organ designed to exchange carbon dioxide and oxygen between the circulating blood and the atmosphere in sufficient volumes to sustain life. The respiratory tract is faced with conflicting demands. It must be an efficient organ of gas exchange, but it must also create a protective barrier which effectively excludes unwanted, harmful agents.

9.2.1 Basic Anatomy. The human respiratory system is illustrated in Fig. 9.1. It is often convenient to divide it into an upper system, including the nasal cavity, pharynx, and trachea, and a lower system of the air passages below the trachea, including the bronchi and lungs. The bifurcation at the lower end of the trachea from which the right and left bronchi lead is the first of a series that produces a treelike arrangement of successively smaller bronchi. The trachea and bronchi are kept open by transverse rings of cartilage, spaced at intervals along the length of the air passages. The narrowest bronchi, about 1–3 mm in diameter, occur after eight to thirteen generations. They lead to even finer passages called bronchioles, which are distinguished from the bronchi in that they have no cartilage. Branching continues for

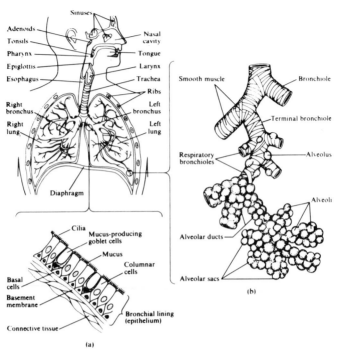

Fig. 9.1. (a) Air passages of the respiratory system of man. (b) Details of the terminal respiratory units of the lung. [From Williamson (1973); with permission of Addison & Wesley.]

another ten to fifteen generations, ending in terminal bronchioles approximately 0.6 mm in diameter. Bronchioles and terminal bronchioles are encircled by muscle which during spasms can block off the air passages.

The terminal bronchioles divide into three or more respiratory bronchioles, which are distinguished by the fact that they have occasional outpouchings called alveoli. In turn, the respiratory bronchioles branch into several passages known as alveolar ducts, whose walls contain five or six alcoves or sacs formed by groups of alveoli. The alveoli are the functional unit of the lung. It is across their thin membrane that oxygen transfers from the air in the lung to the pulmonary capillaries and carbon dioxide passes in the opposite direction. Although each alveolus is small, with a diameter of about 0.2 mm, efficient transfer of oxygen and carbon dioxide in respiration is ensured by their great number. It is estimated that as many as 400 million alveoli are contained in a healthy adult lung, for a total surface area of about 750 m², an area more than 25 times the area of the skin. Air can pass from one alveolar duct to another through pores between the alveoli. Thus, should a passage in one duct become blocked, air can be supplied from another.

The alveolar epithelial cells (type I pneumonocytes) completely line the alveolar spaces and are joined together by tight junctions. The capillary endothelium and the alveolar epithelium seem to share a common basal lamina whenever they are in close proximity. Type II pneumonocytes, which produce a surfactant, are present in niches between the type I cells. This surfactant is a detergentlike phospholipid–protein complex which keeps a film of fluid evenly covering the alveolar surface and allows uniform expansion of alveoli during respiration. *Alveolar macrophages* are the main *phagocytes* present in the alveolar spaces; they are cells capable of engulfing or digesting foreign particles harmful to the body.

Three processes are essential for the transfer of oxygen from outside air to the blood flowing through the lung: (1) ventilation, (2) perfusion, and (3) diffusion within the air spaces and across the air–blood barrier. *Ventilation* is the tidal process in which air is moved in bulk between the atmosphere and the lung. *Perfusion* is the action of the cardiovascular system in pumping blood throughout the lungs. *Diffusion* is the passive movement of gases down a concentration gradient between the alveolar gas and pulmonary capillary blood.

In order to deliver the inhaled gas to the alveolar surface uniformly, the lung has an elaborate distribution system. After moving through the nasal cavity or through the mouth, air reaches the throat or pharynx. Unless swallowing occurs, the epiglottis does not cover the larynx, and the gas passes through the larynx into the trachea. The trachea branches into left and right bronchi, which branch repeatedly until the alveoli are reached.

Movement of air into the lungs is caused by the action of the respiratory muscles, which expand the chest wall so that the lungs fill passively. At the end of a normal expiration, a large amount of gas remains in the lungs. Both the rate at which breathing occurs (respiratory frequency) and the amount of air coming in with each breath (tidal volume) are variable and can be altered to match metabolic demands. The ventilation rate is adjusted by the brain to keep the concentration of oxygen and carbon dioxide in the alveoli relatively constant. An important parameter for gauging the response of the respiratory system to certain air pollutants is the airway resistance, which, when multiplied by the rate at which air flows through the upper respiratory system, gives the difference in pressure between the ambient air and the air in the lung. High resistance indicates that there is an obstruction or constriction of the air passages, so that for a given pressure differential air flow is slower than in a normal system.

Although ventilation is essential, gas exchange does not occur unless blood is uniformly distributed throughout the lungs. The right side of the heart pumps blood in the veins which is high in carbon dioxide and low in oxygen. The pulmonary blood vessels are highly branched, and a thin film of blood is spread just beneath the alveolar surface.

The third requirement for gas exchange in the lungs is diffusion. Diffusion is very effective over small distances. In the alveoli, the blood is separated from the air by two specialized layers of cells, which together are less than 1 μm thick. The gas exchange capability of the lungs must be linked to the needs of the whole body by means of the circulatory system and by the unique properties of the blood.

During the last decade, it has become obvious that the lung also has many nonrespiratory functions: for example, the endocrine function of the pulmonary circulation has been emphasized. The lungs metabolize *vasoactive* (blood vessel active) substances and may be involved in such complex humoral processes as blood pressure *homeostasis* (physiological steady state). Brain (1977) also notes that the lung involves the uptake and release of pharmacological substances. Few studies of this phenomenon have been conducted on human beings, although there is substantial information from animal experiments.

9.2.2   Respiratory Diseases.   Two groups of respiratory disease are of interest in relation to inhalation of aerosols. The first are those that are likely to be initiated by natural allergens or are infectious in nature. The second are those caused more by airborne contaminants from man's activities than by natural factors. The first group includes asthma, influenza, bronchitis, and pneumonia. According to some physicians, chronic diseases such as bronchitis, emphysema, and lung cancer belong to the second category.

Asthma is a respiratory condition which owes its name to a clinical symptom of shortness of breath. If asthma is caused by obstructed airways, it is called bronchial asthma; if red blood cells fail to carry oxygen away from the lungs, it is cardiac asthma. Owing to airway obstruction in the respiratory system, oxygen transfer becomes deficient. This compels the asthmatic to force air into his lungs. As the pressure increases, the stretched alveoli are chocked with air. The added stress may inflame and tear the tender alveolar membranes and produce scar tissue. Eventually this process can produce emphysema. Asthma is believed to be associated with allergenic reactions in tissues, producing such materials as *histamine*. This chemical overrules the normal physiological balance which govern the muscles, blood vessels, and glandular secretions. Common allergies include pollens from vegetation, mold spores, debris from fur and feathers, and residential dust. Exposure to consumer aerosols and insecticide sprays also can induce bronchial asthma. Asthma is rarely a primary factor in mortality, but is more prevalent in morbidity. In contrast with asthma, *influenza* is a highly infectious respiratory disease caused by bacteria or a virus. The disease runs its course in epidemics, which can spread rapidly through 25–40% of the population. It is abrupt and brief in duration, but may result in complications of bronchitis or pneumonia. The disease is often transferred by airborne particles containing the virus. The virus attacks and destroys the cells lining the respiratory tract. Following cell injury, fluid from the blood is exuded, and the supporting tissue evidences acute inflammatory changes. Changes are generally limited to the nose and the larger respiratory passages.

Influenza epidemics have been documented since the mid-eighteenth century. One case described by Hippocrates is suspected as early as the fifth century B.C. The first clear evidence of an influence epidemic was in 1610. Many of the epidemics appear to originate in Russia and Asia. Although the majority of the population easily recovers from the flu, deaths have been reported from the disease or its complications.

Bronchitis is a term for an inflammation of all or part of the bronchial tree. In response to the irritation of the respiratory tract, mucus production is increased by cells in the bronchial wall and a cough reflex is initiated to assist in eliminating the secretions. Airway resistance increases as the mucus layer grows in thickness. Acute bronchitis is a short-lasting disease caused by one or more irritants such as a virus, fumes of chemical agents, or dust. Infection and fever is frequently, but not necessarily, present. The disease runs its course within a week; however, if untreated or if irritants persist, acute bronchitis can develop into chronic bronchitis. This is a long-standing inflammation of the bronchial lining, resulting in permanent destruction of cells and loss of cilia. Mucus glands are commonly enlarged and are excessively productive. Scar tissue narrows the air passages and produces rigidity and distortion. The condition is characterized by a persistent cough and production of sputum. Infection and swelling of the terminal bronchi in chronic bronchitis can cause a collapse of the walls with a result similar to pulmonary emphysema. In a significant number of cases, chronic bronchitis may develop into emphysema.

The disease *pneumonia* is an inflammation of the lung, and has been recognized since ancient times. Pneumonias are grouped as (a) primary, or those caused by microbial attack on the lung directly; (b) those that occur in specific forms during systemic infections and other diseases; and (c) those caused by a mixed infection and other diseases; and (c) those caused by a mixed infection from previous disease or injury, aspiration of foreign material, airway obstruction, or drug-impaired resistance and shock; (d) those from aspiration of oils, chemicals, or dusts; and (e) by allergic reaction.

The modern description of the disease began with Stermby's nineteenth century discovery of the pneumococcus bacillus and its association with pneumonia. Pneumococci settle and multiply in an area of the lung and cause inflammation, which spreads rapidly involving part or all of the lobe. First congestion occurs, then there is an accumulation of mucus, fibrai, and cocci in the alveolar spaces. The bacteria also can enter the blood and occasionally localize on the cardiac valves, in the bone joints, or peritoneum with serious consequences. Other pneumonias involve streptococci, staphylococci, and so-called gram-negative bacilli. Viral pneumonias also have been recognized since the turn of the century as well as systemic conditions with cytomegalovirosis, measles, and rheumatic fever. Inhalation of irritant chemicals can give rise to lipoid pneumonia and an allergic reaction such as Loeffler's syndrome or eosinophikic pneumonia.

Pulmonary emphysema (from the Greek, to inflate or blow into) is a condition of the lung in which a thinning and destruction of the alveolar walls results in an enlargement of the air sacs. The lung becomes hyperinflated owing to dilation and coalescence of the alveoli. The disease is progressive, with no known cure. It is characterized by shortness of breath, due to incapacitation of the lung. When the disease is well developed, any physical exertion is followed by great difficulty in breathing; simultaneously the heart is placed under stress as it attempts to increase blood circulation through the viable portions of the lung to meet the oxygen demands of the body. Because the alveolar surface area available for producing phagocytes is markedly reduced, foreign bodies cannot be eliminated efficiently, with the result that the lung is susceptible to further injury. Death finally occurs either from heart failure or from an acute episode of infection.

Emphysema is one of the fastest growing causes of death in the United States. The toll has increased seventeenfold since 1950 to an annual rate of 20,252 deaths in 1966, and it is popularly suspected that air pollution plays a causal role. Evidence is accumulating that susceptibility to emphysema may be enhanced by hereditary factors (Lieberman, 1969).

Lung cancer (bronchogenic carcinoma) is a condition in which cells of the linings of the air passages of the lung undergo a change that leads to their uncontrolled growth. Typically it develops in the epithelial lining of the bronchi or bronchioles. The mechanism by which it is caused is unknown. The resulting malignant tumors block air circulation and destroy pulmonary structure, thereby reducing the ability of the lung to function. Unless growth is arrested or the tumors removed at an early stage, the disease evolves to a condition where metastases occur, and the disease spreads through the lymphatic system to other parts of the body, resulting eventually in death. Other carcinomas may originate in the alveoli, but this is rare.

Pneumoconiosis, a generic term for a group of diseases of the lung caused by inhalation of specific dusts, is an example of occupational disease caused by inspiration of dust-laden air and may be the extreme of lung affliction in urban man. The different types include silicosis, asbestosis, and anthracosis. These diseases are most commonly found to afflict workers in the mineral industries.

Silicosis can result from exposure to high concentrations of dust-bearing free silicas (SiO_2), such as quartz and sand, which are used in foundries, sandblasting, and pottery manufacturing. The particles when lodged in the lung pass through the alveolar walls and collect in the lymphatic system. Damage to the alveolar membrane is caused by production of silicotic nodules and a fibrous structure in the lung wall. The condition may be irreversible when developed beyond a certain stage. Emphysema can result from the alveolar walls losing their elasticity and breaking down, thereby forming large cavities perhaps 3–4 mm in diameter.

The other forms of pneumoconiosis are also characterized by fibrosis of the lung. Asbestos—a combination of silicates of magnesium and iron—is also a potent lung irritant. This may be due partially to the shape of some forms of this material (chrysotile or crocidolite), which even in the most finely divided state maintains its needlelike form and thus creates a cytotoxic or fibrogenic reaction.

Perhaps the most widespread pneumoconiosis is anthracosis, commonly known as coal miner's "black lung." In some mines as many as one-third of the workers suffer from the disease (Tokuhata et al., 1970). Autopsies reveal that the lung in the advanced stage of the disease does indeed have a black color, although the color is not due simply to the dust that is retained. Physiological changes such as fibrosis contribute to the disease symptoms. Some 10–40% of the dust may be free silica. The observation that the lungs of city dwellers are blackened is probable evidence that some of the inhaled solid aerosols of urban air are retained in the lung. However, again we emphasize that a clear distinction should be maintained between occupational exposures to such things as coal dust and exposure to soot in polluted air. No disease has as yet been causally related to exposures to outdoor suspended particulate matter alone. The concentration of aerosols in urban air is much smaller than the concentrations found in mines and foundries, so the exposures are markedly different.

The inhalation of coal dust is usually not very incapacitating unless silica is present (Morgan and Lapp, 1976). The coal macule, an accumulation of carbon, macrophages, and fibroblasts, forms around the terminal bronchioles altering their structure. The adjacent respiratory bronchiole then dilates owing to the stresses of the respiratory cycle. The result is focal emphysema, but this is usually not disabling (Dannenberg, 1977). Occasionally, massive pulmonary fibrosis occurs containing amorphous black conglomerates. This severe form of coal miner's pneumoconiosis is frequently accompanied by a variety of autoimmune phenomena and by endarteritis that obliterates the lumen (passageway) in blood vessels.

9.2.3 Respiration of Toxic Chemicals. Chemical constituents of aerosols promote biochemical reactions leading to respiratory disease in different ways. The soluble and partially soluble gases are often irritants.

Table 9.5 lists some of the substances that produce cancer of the lung and other organs. Benzo[a]pyrene, which evidently is noncarcinogenic in itself, is oxidized by arylhydrocarbonhydroxylase (AHH) in the microsomes of cells to an epioxide which is carcinogenic (Thakker et al., 1976). Alterations by other enzymes and coupling to glucuronic acid reduce its carcinogenicity and aid its excretion (Thor, 1977). Alveolar macrophages are rich in AHH (Cautrell et al., 1973).

Benzo[a]pyrene is a combustion product of coal, oil, gasoline, and tobacco. It probably is not a major carcinogen, because men installing pitch roofs inhale large concentrations of benzo[a]pyrene with only a slight increase in their incidence of lung cancer after many years of exposure (Hammond *et al.,* 1976). Benzo[*a*]pyrene is discussed here because it is one of the many products of combustion which may act synergistically to cause cancer.

β-Naphthylamine is an example of a chemical which is on the list of fourteen compounds considered most dangerous by the U.S. Occupational Safety and Health Administration (OSHA, 1974). It is evidently hydroxylated and phosphorylated by the liver and excreted by the kidneys into the bladder, where the phosphate group is removed and a carcinogen created. Unprotected workers in chemical plants producing β-naphthylamine often developed bladder cancer (Baetjer, 1973a).

Vinyl chloride, the monomer from which vinyl plastics are made, causes angiosarcomas of the liver, a rare type of cancer. The problem does not seem to be widespread, because only 58 angiosarcomas in the world have been attributed to vinyl chloride prior to the mid 1970s. The vinyl plastics on the market today are probably noncarcinogenic because they are almost free of the monomer, but phthalate plasticizers added to give flexibility might cause other toxic effects.

Although various metals can cause cancer (Table 9.5), they also have other effects according to Seaton and Morgan (1975). Fume fever occurs several hours after the inhalation of zinc oxides during galvanizing operations. It is a severe flu-like syndrome with general malaise, fever, and dyspnea (painful breathing) that lasts about a day. Inhaling oxides of cadmium, nickel, manganese, and other metals may injure the alveolar walls and cause a pneumonitis. Certain metals are classic poisons, namely arsenic, lead, and mercury. Beryllium compounds are used in a variety of industries. When inhaled, they cause pulmonary granulomas similar to Boeck's sarcoid, often after a latent period of several years (Gee *et al.,* 1977).

The inhalation of asbestos fibers may cause pleural plaques, diffuse interstitial pulmonary fibrosis (asbestosis), mesothelioma of the pleura (or peritoneum), and bronchogenic carcinoma (Becklake, 1976). These have been found 15–30 yr after initial exposure. Contributory factors to the carcinogenicity of asbestos are cigarette smoking (Selikoff *et al.* 1968) and possibly the nickel, chromium, and polycyclic hydrocarbons associated with the asbestos fibers (Baetjer, 1973b). Asbestos bodies are ferruginous bodies with an asbestos core. They form when a fiber (too big to be ingested) is surrounded by partially fused macrophages, which cover it with a mucopolysaccharide–protein–hemosiderin-containing coating. This coating renders the fiber nonfibrogenic, but many fibers remain uncoated and coated fibers may lose their coatings. Ferruginous bodies also form around fiberglass particles and other substances. Talc often contains chrysotile asbestos (both being magnesium silicates). Seaton (1975) report that talc miners and mill workers inhaling high doses of talc have an increased incidence of pulmonary fibrosis and lung cancer.

Pneumoconioses are seldom simple. They usually result from the inhalation of mixtures of dusts (e.g., silica, coal, and cigarette smoke) and may be complicated by nonspecific infections, tuberculosis, and autoimmune phenomena.

9.3 DEPOSITION OF AEROSOL PARTICLES

The respiratory system has several defense mechanisms to protect it and counteract the effect of exposure to irritants. Normally, inspired air is warmed and moistened as it makes its turbulent passage through the nasal cavity and trachea. By the time it reaches the bronchi, it is completely saturated with water vapor, thus keeping the pulmonary membranes from drying out. Gaseous air pollutants of moderate to appreciable solubility, such as sulfur dioxide, are mainly absorbed by the moist lining of the upper respiratory tract and have little chance of entering the lung. For this reason, the irritating effects of SO_2 are localized when experienced in the relatively low concentrations normally found in ambient air. Insoluble gases will of course enter the lower lung system in the same concentration as they occur in the ambient air.

The upper respiratory system effectively filters larger particles from the airstream during inspiration. Hairs in the nasal passage form the first line of defense. Particles which succeed in avoiding entrapment may be caught by the second line of defense, the mucus covering over the lining of the nasal cavity and trachea. Some particles may also be caught by the fine hairlike *cilia* that line the walls of the upper respiratory system, the bronchi, and bronchioles. A continual wavelike motion of the cilia has the important effect of moving the mucus and entrapped particles to the pharynx, where they can be eliminated by swallowing or expectoration. Irritants such as tobacco smoke stimulate ciliary activity and improve mucus transport, but very high concentrations of some pollutants can have the opposite effect.

Particles smaller than about 3 μm in radius can escape the defense mechanisms of the upper respiratory system and enter the lung. A small fraction of these may impact on the bronchiolar membranes and become trapped. The slow movement of air in the lung, together with the large horizontal surface area of lung tissue, permit an appreciable fraction of the large particles to settle onto the lower surfaces of the bronchioles and alveoli. The major subdivisions within the respiratory tract differ markedly in structure, size, function, and response to deposited particles. They also have different mechanisms and rates of particle elimination. Thus determination of an effective tissue dose from inhaled aerosols depends on (a) regional deposition, (b) retention times at the deposition sites and along the elimination pathways, and (c) the biological effects of physical and chemical properties of the particles. This section centers on the deposition pattern of inhaled particles and the factors that determine it, including the influence of gaseous cocontaminants.

9.3.1   Deposition Mechanisms.   Particles deposit in the various regions of
the respiratory tract by a variety of physical mechanisms. Deposition efficiency in each region depends on the aerodynamic properties of the particles, the anatomy of the airways, and the geometric and temporal patterns of
flow through them. These processes have been reviewed in detail, for example, by Lippmann (1977), and Lippmann and Altshuler (1975). The discussion below follows largely from their two reviews and from two more recent
papers (Lippmann, *et al.*, 1980; Lippmann, 1983).

Five mechanisms whereby significant particle deposition can occur within
the respiratory tract are interception, impaction, sedimentation, diffusion,
and electrostatic precipitation. However, in most cases, only impaction,
sedimentation, and diffusion are of importance.

Interception.   Interception, while usually insignificant, can be important
in the case of fibrous particles. It takes place when the trajectory of a particle
brings it close enough to a surface so that an edge contacts the surface,
implying that the particle size is a significant fraction of the airway diameter.
The probability of deposition by impaction and sedimentation is determined
by the aerodynamic diameter d_a. Nonfibrous particles with >5 μm will
usually deposit in large airways by impaction or sedimentation. For asbestos
and other fibrous particles, Timbrell (1972a) found that d_a is about three
times the fiber diameter for fibers with a length/diameter ratio >10; thus a
fiber of 200 μm length and 1 μm diameter would have an aerodynamic
diameter of ~3 μm. With such fibers, inertial and settling mechanisms play a
minor role in the conductive airways compared to interception, which is a
function of the fiber length. Fibers as long as 200 μm have been found in
human lungs (Timbrell, 1972b). Straight fibers such as amphibole asbestos
penetrate more into the alveoli than the similar size but curly chrysotile
asbestos, because the straight fiber assumes a closer parallel orientation to
the flow streamlines.

Impaction.   Inhaled air follows a tortuous path through the nose or
mouth and branching airways in the lung. Each time the air changes direction, the momentum of particles tends to keep them in their prior directions,
which can carry them onto airway surfaces. Impaction probability at each
level and bifurcation depends on the stopping distance. The location of the
particle within the airway must also be considered. For example, in a bifur-
cated airway, the deposition probability during inhalation is higher for a
particle traveling along the center line of the parent tube than for particles
nearer to the walls, and the highest density of deposit is at or near the *carina*
of the bifurcation (the "leading edge" of the branch).

No adequate mathematical model exists for impaction deposition of the
bronchial tree. The air flow is periodic. In all but quiet breathing, it changes
from turbulent in the trachea to laminar in deeper airways. In the larger
airways, where impaction is important, the flow pattern is never fully devel-

oped since the Reynolds number is high and the length of each segment only about three times its diameter.

Sedimentation.   Gravitational sedimentation is an important mechanism for deposition in the smaller bronchi, the bronchioles, and the alveolar spaces where the airways are small and the air velocity low. Sedimentation becomes less effective than diffusion when the terminal settling velocity of the particles falls below ~0.001 cm/sec, which corresponds to a unit density sphere of about 0.5 μm diameter.

Diffusion.   Brownian motion increases with decreasing particle size and becomes a more effective mechanism of particle deposition in the lung as the root-mean-square displacement approaches the size of the air sacs in the alveoli. Diffusional deposition is important in small airways and alveoli and at airway bifurcations for particle diameters below about 0.5 μm. For radon and thoron daughters attached to particles that are nearly molecular in size, diffusional deposition efficiency will be high in the nasopharynx passage and in large airways such as the trachea.

Electrostatic Precipitation.   Particles with high electrical mobility can have an enhanced respiratory tract deposition even though no external field is applied across the chest. Deposition can result from the image charges induced on the surface of the airways by the charged particles. Test aerosols resulting from the evaporation of aqueous droplets can have substantial mobilities, and the results of some experimental deposition studies using such aerosols without charge neutralization are accordingly suspect. However, most ambient aerosols have reached charge equilibrium and have low charge levels. Thus the deposition due to charge is usually small in comparison to deposition by the preceding mechanisms.

Particle size is generally the most important parameter in regional deposition. There are a number of ways of expressing particle size as discussed previously. In this discussion, and most health-related literature, particle size is expressed in terms of actual or equivalent diameters.

Aerodynamic diameter d_a is the most appropriate parameter in considering particle deposition by impaction and sedimentation, which generally account for most of the deposition by mass. On the other hand, diffusional displacement, which is the dominant factor for $\leq$0.5-μm particles, depends only on particle size and not on density or shape. Interception also depends on the linear dimensions of the particle and its shape, since aerodynamic forces affect particle orientation within the airway.

The conversion of microscopic linear or projected area measurements of nonspherical particles to aerodynamic diameters requires assumptions about the aerodynamic shape factor, density, and the relation between projected area and volume. Though such conversions can be made accurately, in many cases they are unreliable. An alternative approach is to measure aerodynamic size directly using an aerosol centrifuge (see also Section 5.7.3).

A complicating factor for water-soluble particles is the change in size which may take place in humid atmospheres. Furthermore, dry aerosols of materials such as sodium chloride, sulfuric acid, and glycerol will take up water vapor and grow in size within the warm and nearly saturated atmosphere in the lungs (as might be expected from the results in Section 8.3.1). Since total and regional depositions are sensitive to particle size, changes in size due to droplet growth could cause significant changes in deposition pattern and efficiency.

Air velocity has already been mentioned as one of the important respiratory factors affecting deposition. Increasing velocity augments impaction deposition, but decreases sedimentation and diffusional deposition by decreasing residence time. The calculation of deposition at a constant velocity is relatively simple, but may provide little insight into the effect of velocity in normal breathing where the velocity is variable in time and reverses direction twice for each cycle. During exhalation, the flow profiles in the airways will differ from those during inhalation and will affect particle deposition, especially the contribution from impaction.

The velocity profiles in the larger conducting airways are quite different from those normally encountered in fluid conduits. The laryngeal jet can have a marked effect on flow patterns and on particle deposition. Sudlow *et al.* (1971) measured the flow profiles within a hollow bronchial cast without a larynx, and reported that the flow downstream of the tracheal bifurcation was highest near the inner walls of the daughter tubes. When similar measurements were made in a cast with an upstream larynx (Olson *et al.*, 1973), the flow downstream of the tracheal bifurcation was highest near the outer walls.

The average length of a human bronchial airway is only about three times its diameter, so that the flow never achieves a fully developed profile. Furthermore, each segment terminates in a bifurcation, so that the flow profile entering the daughter tubes is asymmetrical; i.e., the maximum velocity is close to the cardinal ridge. Such profiles have been demonstrated by Olson *et al.* (1973) and Pedley *et al.* (1971), who have also shown that the turning of the flow results in a secondary swirling motion in the bulk flow. Finally, the flow is periodic and with frequent reversals. At peak flow, there will be at least intermittent eddying in the trachea, but the Reynolds number decreases with increasing lung depth, so that in the smaller conducting airways it is always laminar, and in the alveolar region it is low in Reynolds number or approaching a flow dominated by viscous forces.

Since the flow is laminar in most of the anatomical dead space, the central core velocity is almost twice the average velocity. Even in very shallow breathing, a substantial fraction of the inhaled volume penetrates beyond the dead space, and particles greater than 2 μm in diameter, having appreciable sedimentation rates, or less than 0.1 μm in diameter, having large diffusional displacements, can deposit in substantial amounts in peripheral airways.

The *tidal volume* is an important respiratory parameter. The air inhaled at the start of each breath goes deeper into the lung and remains there longer than the air inhaled later in the breath. It follows that the deeper the air goes and the longer it stays, the greater its depletion of inhaled particles. Thus, for quiescent breathing, when the air velocity is low and mixing is minimal, and for tidal volumes of only two to three times the dead space volume, a large proportion of the inhaled particles can be exhaled. Conversely, during heavy exertion, when larger volumes are inhaled at higher velocities, impaction in the large airways, sedimentation and diffusion in the smaller airways and alveoli, and mixing with lung air will be greater.

Intrasubject variations in airway anatomy affect particle deposition in several ways: (1) the diameter of the airway determines the displacement required by the particle before it contacts the airway surface, (2) the cross section of the airway determines the flow velocity for a given volumetric flow rate (velocity affects particle deposition, as discussed under respiratory factors); and (3) the variations in diameter and branching patterns along the bronchial tree affect the mixing characteristics between the tidal and reserve air in the lungs. For particles with aerodynamic diameters below ~ 2 μm, such convective mixing can be the single most important factor determining deposition efficiency.

There are also large intersubject differences in respiratory tract anatomy. For example, the average alveolar air space dimension has a substantial coefficient of variation when measured either post mortem on lung sections or *in vivo*[†] by aerosol persistence during breath holding. In the former case, Matsuba and Thurlbeck (1971) reported a mean size and variation of 0.678 ± 0.236 mm. In the latter case, Lapp et al. (1975) found corresponding values of 0.535 ± 0.211 mm.

The effective diameters of the conductive airways for air flow are defined by the surface of the mucus layer. In normal subjects, where the mucus layer on the conductive airways is only about 5 μm thick (Dalhamn, 1956), the reduction in free cross section by the mucus is negligible. On the other hand, in individuals with bronchitis, the mucus layer can be much thicker, and in some locations it can accumulate and partially or completely occlude the airway. Air flowing through partially occluded airways will form jets, which will probably cause increased small airway particle deposition by impaction and turbulent diffusion.

One of the important functions of the conductive airways is to warm and humidify the inspired air. When the air is inhaled through the nose it is already saturated at body temperature before it reaches the pharynx. On the other hand, air inhaled through the mouth may not reach saturation before the major or segmental bronchi. The resulting loss of water by the mucus in the large airways may alter the character of the mucus and affect particle

---

[†] Within the living subject; *in vitro* refers to experiment in tissue outside the body.

clearance via the mucus, and conceivably may contribute to the progression of bronchitic symptoms.

There is relatively little information on how temperature and humidity affect particle deposition. Based on the preceding discussion, it might be assumed that low ambient humidity could dry the mucus. This does not appear to be the case, at least in normal nose-breathing humans. Mucus transport of particles in the nose is not affected by even extremely low humidity, such as 10% relative humidity at 23°C (Proctor *et al.*, 1973). However, ambient humidity can greatly affect the size of many pollutant particles and thereby their total and regional deposition efficiencies.

Many of the finely divided particles in the atmospheric aerosol contain water-soluble acids and/or salts which derive from gaseous precursors. Thus the sulfates, nitrates, and other hygroscopic salts will grow rapidly in the moist environment of the upper respiratory tract and deposit as if they were larger particles than their "dry" entry diameter.

Any discussion of the deposition of inhaled atmospheric aerosols would be incomplete without consideration of the combined influence of airborne contaminants on the lungs. Inhaled irritants can affect the fate and toxicity of the inhaled particles by altering airway caliber, respiratory function, clearance function and/or distribution function, and survival of the cells that line the airways.

A large portion of the total urban aerosol, and the dominant portion of the aerosol particles of submicrometer size, is formed within the atmosphere from gaseous precursors associated with combustion sources. Thus any atmosphere containing elevated sulfate and nitrate aerosol concentrations is likely to contain elevated SO_2 and NO_2 gas concentrations. Similarly, elevated gaseous oxidant concentrations would be associated with elevated hydrocarbon aerosol concentrations.

The fate and effects of sulfur dioxide have been studied to a greater extent than those of any of the other pollutant gases. Because of the solubility of SO_2 in tissue fluids, inhaled air is depleted of it and little penetrates to the deeper airways. SO_2 reaching the lungs diffuses rapidly to the bronchial walls and its concentration rapidly diminishes with depth. SO_2 can be delivered to the bronchioles and alveoli when adsorbed on inert particles which are small enough to reach and deposit in these airways, but there are no quantitative data on dosage delivered via this route.

SO_2 alters the mechanical function of the upper and lower airways in man and experimental animals. Nasal flow resistance may increase in human subjects exposed to 1 ppm of SO_2 for 1–3 hr (Andersen *et al.*, 1974); the response is accelerated at higher concentrations (Speizer and Frank, 1966). While acute SO_2 and NO_2 exposures increase pulmonary flow resistance $\mathfrak{R}_L$, the concentrations required to elicit this response are at least several times higher than those found in ambient air (Frank *et al.*, 1962; Stresemann and von Nieding, 1970). Animal tests indicate that $\mathfrak{R}_L$ can also be elevated by sulfate aerosols, and that the response to SO_2 can be potentiated by simultaneous exposure to hygroscopic nonirritant particles (Alarie, 1973).

Increased pulmonary flow resistance reflects the bronchoconstrictive effects of the pollutant. The reduction in airway cross section in the larger bronchial airways results in increased flow velocities and should therefore increase particle deposition by impaction. In tests at the New York University Institute of Environmental Medicine, increased bronchial deposition of inhaled particles was observed in tests in which two normal young men were preexposed to SO_2. Tests on two other normal subjects with 5- and 9-ppm concentrations of the gas did not produce any significant shift in regional deposition (Lippmann, 1977).

Other major pollutant gases, NO_2 and O_3, are less soluble than SO_2 and penetrate more effectively to the alveolar zone, where they exert their major toxic effects, as shown in controlled inhalation studies on small animals. There are virtually no data concerning their effects on pulmonary function in man at realistic air pollution levels, though Bates et $al.$ (1972) have shown that exercise potentiates the functional effects of O_3 on human volunteers.

While some apparently normal cigarette smokers have increased bronchial particle deposition, the increase is relatively small compared to that seen in individuals with clinically defined chronic bronchitis (Lippmann et $al.$, 1971). In the same study, greatly increased tracheobronchial particle deposition was also seen in some asthmatics.

9.3.2 Experimental Deposition Data. There have been relatively few attempts to measure regional particle deposition in humans. A much larger number of studies have explored total deposition. For particles between 0.1 and 2 μm in aerodynamic diameter, deposition in the conductive airways is generally small compared to deposition in the alveolar regions, and thus total deposition approximates alveolar deposition. Total deposition as a function of particle size and respiratory parameters has been measured experimentally by numerous investigators. Many reviews on deposition have called attention to very large differences in the reported results (Davies, 1964; Hatch and Gross, 1964; Lippmann, 1970b; Task Group on Lung Dynamics, 1966).

Differences in results between investigators can be attributed to uncontrolled experimental variables and questionable experimental techniques. The major sources of error have been described by Davies (1973). Figure 9.2 shows data from studies performed with improved precision. Tidal volumes varied from 0.5 to 1.5 liters. All appear to show the same trend with a minimum of deposition at about 0.5 μm diameter. In all the four studies using di-2-ethyl-hexyl sebacate (DES), somewhat lower absolute values were obtained by different investigators as reviewed by Lippmann (1977).

It is also apparent that in most studies involving more than one subject,

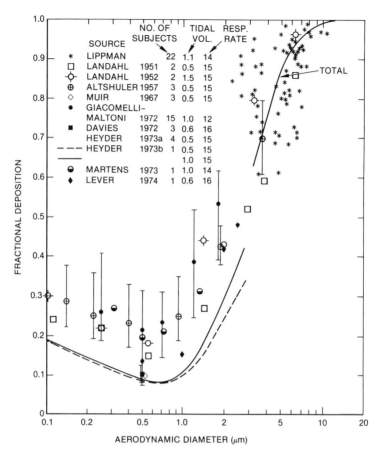

Fig. 9.2. Total fractional deposition in the respiratory tract during mouthpiece inhalations as a function of aerodynamic diameter except below 0.5 μm, where deposition is plotted versus linear diameter. Tidal volume in liters; respiration in min^{-1}. [From Lippmann and Altshuler, (1975), with references; with permission of Keter Publishing House.]

there was considerable individual variation among subjects. Davies *et al.* (1972) showed that some of this variation could be eliminated by standardizing the expiratory reserve volume (ERV), and that deposition decreases as ERV increases. This was confirmed by Heyder *et al.* (1973; Heyder, 1974), who reported that there was little intrasubject variation among six subjects when their deposition tests were performed at their normal ERVs.

The data of Heyder *et al.* (1973) appear to represent deposition minima for normal men. Their test protocols were carefully controlled. There were no respiratory pauses, no variations in respiratory flow rates, and no electrical charge on the particles. With more realistic aerosol and respiratory parameters, higher deposition efficiencies would be expected. The data of Landahl *et al.* (1951, 1952), Altshuler *et al.* (1957), Giacomelli-Maltoni *et al.* (1972), and Märtens and Jacobi (1973), which show higher fractional deposition, agree quite well with one another and provide the best available data for total deposition in normal humans.

The deposition data in Fig. 9.2 were obtained from inhaled and exhaled particle concentrations, except for the values of Lippmann *et al.* (1971), which are based on external *in vivo* measurements of γ-tagged particle retention. Most of the Lippmann *et al.* measurements were made with particles larger than 2.5 μm, while almost all of the data from the other studies refer to particles of smaller size. Despite the minimal overlap in size range, the two sets of data appear to be consistent. The large amount of scatter among the individual data points for the larger particles is due to a quite variable deposition in the head and tracheobronchial tree, as illustrated in Figs. 9.3, 9.4, and 9.5. The data in Fig. 9.3 are for nonsmokers only, breathing through a mouthpiece. Cigarette smokers have a similar median behavior for head deposition, but a larger scatter. Figure 9.5 shows that the median and upper limits of tracheobronchial deposition are higher for cigarette smokers than for nonsmokers, but the lower limit is about the same.

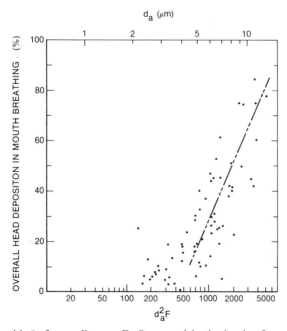

Fig. 9.3. Deposition of monodisperse Fe_2O_3 aerosol in the heads of normal nonsmoking human males during mouthpiece inhalations as a function of d_a^2F, where F is the average inspiratory flow in liters/min, and aerodynamic diameter at 30 liter/min inspiratory flow. An eye-fit line describes the median behavior for depositon between 10 and 80%. Total respiratory tract depositions in these tests are shown in Fig. 9.2. [From Lippmann and Altshuler (1975); with permission of Keter Publishing House.]

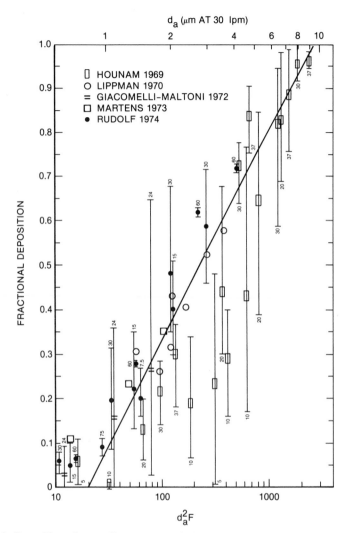

Fig. 9.4. Deposition of monodisperse aerosols in the head during inhalation via the nose versus $d_a^2 F$, where d_a is the aerodynamic diameter at 30 liter/min (μm). The heavy solid line is the ICRP Task Group, 1966 deposition model, which is based on the data of Pattle (1961). For the Hounam *et al.* (1969), Giacomelli-Maltoni *et al.* (1972), and Rudolf and Heyder (1974) data, the symbol shows the median value, and the bars show the range of the individual observations. The number at the end of the bar indicates the inspiratory flow rate. [From Lippmann and Altshuler (1975); with permission of Keter Publishing House.]

Some inhaled particles deposit within the air passages between the point of entry at the lips or nares and the larynx. The fraction depositing can be highly variable, depending on the route of entry, particle size, and flow rate. In most cases, the nasal route is a more efficient particle filter than the oral

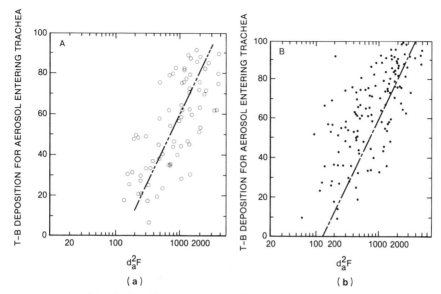

Fig. 9.5. Deposition in the ciliated tracheobronchial (TB) region during mouthpiece inhalations, in percentage of the aerosol entering the trachea. Left panel (a) shows data for normal nonsmoking human males. The eye-fit median line is reproduced on the right panel (b), which contains data for cigarette smokers. Data are for aerosol particles of aerodynamic diameter in μm for 30-liters/min inspiratory flow. [From Lippmann (1977); courtesy of the American Physiological Society.]

pathway, especially at low and moderate flow rates. Thus those people who normally breathe part or all of the time through the mouth may be expected to deposit more particles in their lungs than those who breathe entirely through the nose. During extreme exertion, the flow resistance of the nasal passages, a very large fraction of the deposition takes place within the anterior unciliated nares, where the particles are retained until they are removed mechanically; i.e., by nose blowing, wiping, sneezing, etc.

There are very few data on head deposition during mouth breathing. The data on nonsmoking normal humans (Fig. 9.3) were based on external scintillation detector measurements of γ-tagged particles deposited in the head immediately after their inhalation via a mouthpiece.

Head deposition during nasal inhalation has been studied using monodisperse aerosols by Pattle (1961), Hounam *et al.* (1969), Lippmann (1970a), Giacomelli-Maltoni *et al.* (1972), Märtens and Jacobi (1973), and Rudolf and Heyder (1974).

Figure 9.4 shows head deposition data for monodisperse aerosols during inhalation via nose masks, as a function of the controlling quantity $d_a^2 F$, where F is the flow rate in liters/min. The solid line is taken from the ICRP Task Group model (1966), based on Pattle's (1961) data on one subject. These values, and the Hounam *et al.* (1969) data on three subjects, were based on measurement of the reduction in airborne particle concentration as constant flows were drawn in through the nose and out of the mouth. The data of Lippmann (1970a), Giacomelli-Maltoni *et al.* (1972), Märtens and Jacobi (1973), and Rudolf and Heyder (1973) estimated nasal passage deposition on the basis of the total deposition difference in paired studies on the same subject; i.e., mouth in–mouth out versus nose in–mouth out. Since deposition in the oral passage is assumed to be negligible in this technique, these nasal passage deposition values may be exaggerated for particles larger than about 2 μm.

Pattle (1961) used methylene blue particles between 1 and 9 μm in aerodynamic diameter and constant flow rates of 10, 20, and 30 liter/min; Hounam *et al.* (1969) 1.8–8 μm particles and constant flows of 5–37 liter/min; Lippmann (1970a) 1.3–3.9 μm Fe_2O_3 particles and inspiratory flows averaging about 30 liter/min; Giacomelli-Maltoni *et al.* (1972) 0.25–1.8 μm wax particles and average flows of about 24 liter/min; Märtens and Jacobi (1973) 0.3–1.9 μm latex particles and an average flow rate of about 30 liter/min; and Rudolf and Heyder (1974) used 0.5–3 μm di-2-ethyl-hexyl sebacate particles and constant inspiratory flows of from 7.5 to 30 liter/min. Pattle (1961) found that the data from the three different flow rates could be normalized by plotting $d_a^2 F$ against retention. Hounam *et al.* (1969) found that the best straight-line fit to his data was obtained by plotting retention against $d_a^2 F$. In Fig. 9.4, his $d_a^2 F$ data at 37 and 30 liters/min lie close to the line, while those at 20, 10, and 5 liter/min fall increasingly below the line. The Rudolf and Heyder (1974) data show a similar but lesser trend, with the 60- and 30-liter/min data tending to lie above the line, while the 15- and 7.5-liter/min data straddle the line. A better straight-line fit can be obtained for the Rudolf and Heyder (1974) data by plotting $d_a^2 F$ vs. retention.

In the Hounam, Giacomelli-Maltoni, and Rudolf data, most of the scatter is due to variability between individuals. Hounam *et al.* (1969) also measured the pressure drop across the nose and mouth in each deposition study, and presented a plot of the data in terms of percent deposition vs. $d_a^2 \hat{R}$, where $\hat{R}$ is the resistance across the nose and mouth in mm H_2O. The same data on this plot exhibited less scatter than the $d_a^2 F$ plot, with the greatest improvement at large values of percent deposition. The fact that particle deposition correlates better with $d_a^2 \hat{R}$ than $d_a^2 F$ indicates that the air path dimensions are variable. Since the air paths are distensible, their dimensions can be expected to vary with flow rate in a given subject, perhaps accounting for the observation that a flow factor exponent higher than unity helps to unify the data.

Rudolf and Heyder (1974) also determined head deposition efficiencies for aerosols exhaled through the nose. The efficiency was essentially the same as that measured during inhalation for a flow rate of 7.5 liter/min. At 15 and 30 liter/min, the efficiency for 1–3 μm particles during exhalation was about

25% higher than during inhalation. However, at 60 liter/min, the exhalation efficiency was about 7% lower. It should be noted that while deposition efficiencies are similar on inhalation and exhalation, the amount deposited on exhalation will be lower because the concentration decreases on passage of the aerosol inward through the nasal passages and the lungs.

9.3.3 Deposition in the Tracheobronchial Zone. The only measurements of tracheobronchial (TB) deposition reported in the literature are those of Lippmann and Albert (1969) and Lippmann *et al.* (1971). The TB deposition values reported by them represent the difference between the initial thorax burden and that measured 20–24 hr later, after the completion of bronchial clearance; an updated version of these data is presented in Fig. 9.5. TB deposition for a given particle size varies greatly from subject to subject among nonsmokers, cigarette smokers, and patients with lung disease. The average TB deposition is slightly elevated in smokers, and greatly elevated in the subjects with lung disease. However, in normal patients and nonbronchitic smokers, each individual has a characteristic and reproducible relationship between particle size and deposition. TB deposition includes both deposition by impaction in the larger airways and deposition by sedimentation in the smaller airways. Impaction deposition predominates for large particles ($d_a > 3$ μm) and high inspirating flow rates ($F > 20$ liter/min), while sedimentation deposition becomes a larger fraction of a diminishing TB component for smaller particles and lower flows.

The influence of the laryngeal jet on particle deposition in the trachea and larger bronchi was studied using monodisperse γ-tagged aerosols and a hollow bronchial cast of the human airways used in previous tests by Schlesinger and Lippmann (1972). In these tests to explore the effect of the larynx, the monodisperse ferric oxide particles had an aerodynamic diameter of 9 μm and the inspiratory flow rate was 45 liter/min. Tracheal deposition was much greater when the larynx was present, but the deposition efficiency in the other large airways was reduced. This reduction is consistent with the shift of flow streamlines away from the bifurcations, as shown by Olson *et al.* (1973).

At the sites of concentrated deposition, the epithelial cells receive much higher than average doses from carcinogens in the deposited particles. Schlesinger and Lippmann (1972) reported that the average distribution of particle deposition within the lobar bronchi of the hollow bronchial cast was remarkably similar to the distribution of sites of primary bronchial cancer in these same airways. The deposition distribution for each particle size is illustrated in Fig. 9.6, in which the particle fraction within each of the five lobar bronchi is compared to the distribution of primary cancer sites. This presentation of the data shows that there is no consistent variation in deposition distribution with particle size, at least between 1.7 and 12.2 μm. Thus,

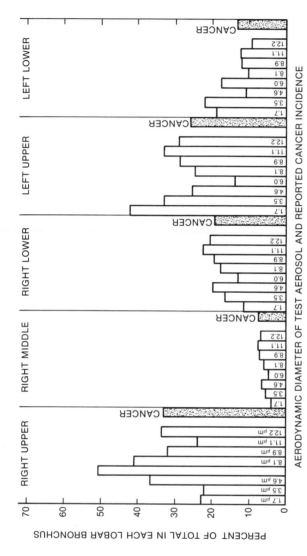

Fig. 9.6. Comparison of the distribution of particle deposition within the lobar bronchi of a hollow bronchial cast of the human lungs for a variety of particle sizes, and the reported distribution of primary bronchial cancer within lobar bronchi. [From Schlesinger and Lippmann (1972); courtesy of the American Industrial Hygiene Association.]

while increasing particle size increases the amount of deposition in the lobar bronchi, it does not greatly change its proportionate amount among the individual members.

Lippmann and colleagues have measured particle deposition in all bronchi with diameters of at least 0.5 cm. For branches with smaller diameters a statistical sampling technique was used to select those to be counted, so as to give a representative portrayal of the deposition profile along the cast. The total cast contained 583 bronchial segments and deposited activity is routinely measured in 83 of these. Separate analysis of bifurcations and lengths and separation of longer segments into more than one counting region involved a total of 160 separate counting sections.

The same hollow cast of the large airways was sacrificed in order to determine the degree to which the deposition is concentrated on the airway bifurcations. In this test, γ tagged particles with an aerodynamic diameter of 6 μm were drawn at 30 liter/min through the laryngeal and hollow bronchial casts. Measurements were made with a collimated scintillation detector of cast segments one centimeter in length in the axial direction. These were designated as either bifurcation or length segments, the former consisting of segments centered on the tracheal or a bronchial bifurcation and the latter, lengths between bifurcations. The bifurcation regions indicated in Table 9.9 were dissected from the cast. The dividing spurs (crescent-shaped area in the sensing zone) were separated from the peripheral surfaces and counted separately. The tracheal carina, the ridge separating the origins of the two first bronchi, had 55% of the activity of the total tracheal bifurcation segment. The average percentage of the deposition in each of the next three airway generations on the dividing spur was similar to that on the tracheal carina, but the lobar and segmental spurs had much more concentrated surface deposits. The main reason for the higher surface density in these airways is that they have higher deposition efficiencies.

These data reinforce the notion of an association between localized regions of selective deposition and primary cancer sites. The percentage of the deposition in a given airway segment of the first four branching levels which occurs in the region containing the bifurcation varied from 33 to 73%. As shown in Table 9.9, the average surface dose is relatively low on the bifurcations of the trachea and major bronchi, but increases markedly on the bifurcations of the next two branching levels. This pattern of relative dose is similar to reported patterns of cancer origin sites within these large airways.

On the average, for every ten primary lung cancers, only one arises in the major bronchi, six arise in the lobar and segmental bronchi, and three arise in the distal to the segmental bronchi (Veeze, 1968). Thus the primary sites of bronchogenic cancer are distributed in a similar fashion to the sites of highest particle deposition density within the tracheobronchial tree.

The 0.3-μm-median-diam aerosols used in the current study have size distributions similar to that reported for cigarette smoke (Keith and Derrick,

TABLE 9.9

Bifurcation Region Dose for 6-μm γ-Tagged Particles Drawn through a Hollow Cast of the Large Airways at a Flow Rate of 30 Liters/min[a]

Airway bifuraction region–daughter segments[b]	Deposition on dividing spur (%)	Spur surface area		Average surface dose to spur[c]
		(cm²)	(percent of region)	
11, 12	55	7.5	44.4	1.4
111, 112	61	1.6	25.0	2.9
121, 122	50	2.7	41.5	1.0
1111, 1112	57	3.1	43.1	1.0
1121, 1122	63	1.0	27.0	19.0
1221, 1222	58	1.0	44.1	20.0
1211, 1212	55	0.9	51.7	7.0
11211, 11212	32	0.8	26.6	14.0
11221, 11222	80	1.7	34.7	9.3
12211, 12212	65	1.0	50.0	6.5
12121, 12122	70	0.8	31.4	18.0

[a] From Lippmann and Altshuler (1975); with permission of Keter Publishing House.

[b] Binary code in which 1 indicates the major daughter tube and 2 the minor daughter. Thus 121 describes the larger daughter tube which branches from the smaller of the major bronchi.

[c] Normalized on the basis of 1.0 for the region with the lowest dose.

1960) and also to that for the accumulation mode of the air pollution aerosol: the small particles deposit preferentially at large airway bifurcations with almost as high concentrations as the larger particles. Thus the local dose at the bifurcations within the higher cancer risk bronchial generations may be due in part to direct particle deposition at these sites.

The studies of Lippmann and Albert (1969; Lippmann *et al.,* 1970) also provide *in vivo* data on alveolar deposition, based on the fractional thorax retention one day after test aerosol inhalation. Brown *et al.* (1950) and Altshuler *et al.* (1966) estimated alveolar deposition on the basis of measurements of inhaled and exhaled aerosol concentrations and assumptions about volume partitioning within the airways. Landahl *et al.* (1951, 1952) performed mouth breathing experiments with monodisperse aerosols of triphenyl phosphate with diameters ranging from 0.11 to 6.3 μm. They did not use the data obtained in these experiments to describe alveolar deposition directly; instead, they compared them to predicted values for the same breathing patterns obtained from previous calculations (Landahl, 1950). The agreement was sufficiently good for the authors to conclude that the predictions were verified.

Brown *et al.* (1950) studied regional deposition during nose breathing using china clay aerosols with a narrow particle size range ($\sigma_g = 1.25$); the count median diameters were between 0.9 and 6.5 μm. Brown et al. collected the exhaled air in seven sequential components, and used the CO_2 content of each fraction as a tracer to identify the region from which the exhaled air originated. The validity of these data depends upon the accuracy of the association between the various exhaled air fractions and their presumed sources. The Brown et al. data have been criticized (Altshuler et al., 1966) on the basis that the simple model employed was inadequate, and that rapid diffusion of CO_2 caused the measured CO_2 values to differ significantly from the corresponding CO_2 concentration in the alveolar spaces. The resulting error in the volume partitioning caused an underestimation of the alveolar deposition.

Altshuler et al. (1966) estimated regional deposition from mouth breathing experiments on three subjects in which measurements of both the concentration of a monodisperse triphenyl phosphate aerosol and the respiratory flow were made continuously during individual breaths (Altshuler et al., 1957). Using a tubular, continuous filter bed model as a theoretical analog for the respiratory tract, regional deposition in the upper and lower tract components was calculated for various values of anatomic dead space. The upper tract penetration during inspiration, pause, and expiration was derived from the expired aerosol concentration. The alveolar deposition estimates varied with the volume of anatomic dead space. The particle size for maximum alveolar deposition was estimated to be greater than 2 μm, but a more precise estimate could not be made since only one particle size was used in the experiments. The alveolar deposition values of Altshuler et al. (1966) for their best estimates of dead space are plotted in Fig. 9.7, which also shows alveolar deposition values obtained in mouth breathing inhalation tests on nonsmoking normal patients. These data are based on external measurements of the retention of γ-tagged particles after the completion of bronchial clearance. It can be seen that in the region of particle size overlap, the two sets of data are in good agreement.

The data of Fig. 9.7 also provide an estimate of the alveolar deposition which could be expected when the aerosol is inhaled via the nose. It is based on the difference in head retention during nose breathing and mouth breathing, calculated on the basis of the straight-line relations plotted in Figs. 9.3 and 9.4. It can be seen that for mouth breathing the particle size for maximum alveolar deposition is near 3 μm, and that about half of the inhaled aerosol of this size is deposited in the alveolar region. For nose breathing, a broader maximum of 25% is predicted at 2.5 μm. Indeed, with nose breathing alveolar deposition is nearly constant, about 20% for the size range from 0.1 to 4 μm.

9.3.4 Predictive Deposition Models.

Mathematical models for predicting the regional deposition of aerosols have been presented by Findeisen (1935),

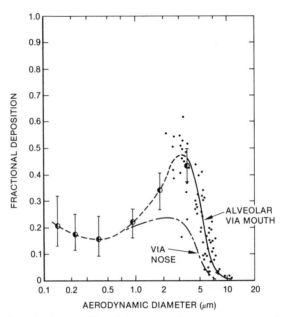

Fig. 9.7. Deposition in the nonciliated alveolar region, in percentage of aerosol particles entering a mouthpiece, as a function of aerodynamic diameter, except below 0.5 μm, where linear diameter was used. Individual data points and eye-fit solid line are for the same Fe_2O_3 aerosol tests plotted in Figs. 9.2, 9.3, and the left panel of Fig. 9.5. The upper dashed line and its extension to the left is an eye fit through the median best estimates of Altshuler *et al.* (1966) on three subjects whose range is shown by the vertical lines. The lower dashed curve is an estimate of alveolar deposition during nose breathing, and is based on the difference in head depositions shown in Figs. 9.3 and 9.4. [From Lippmann and Altshuler (1975); with permission of Keter Publishing House.]

Landahl (1950, 1963), and Beeckmans (1965). Findeisen's simplified anatomy, with nine sequential regions from the trachea to the alveoli, and his impaction and sedimentation deposition equations were used in the International Commission on Radiological Protection (ICRP) Task Group's (1966) model. According to Lipmann and Altshuler (1975), a comparison between the various predictions and the experimental data indicates that for total and alveolar deposition, Landahl's calculations come closest, but alveolar deposition for particles with aerodynamic diameters larger than 3.5 μm is overestimated.

The ICRP Task Group's (1966) model was adopted by the ICRP Committee II in 1973. Although unavailable in official publication, the 1966 Task Group report has been widely quoted and used within the health physics field. It subdivided the respiratory tract as follows: (a) nasopharynx, anterior nares to the larynx or epiglottis; (b) trachea and the bronchial tree down to and including the terminal bronchioles; and (c) pulmonary or functional area (exchange space) of the lungs.

Assuming suspended-particle clouds occur as lognormal distributions, the ICRP deposition calculations were made for three different ventilation states typified by tidal volumes of 750, 1,450, and 2,150 cm^3, respectively, at a respiratory frequency of 15 cycles/min. The lung deposition calculations were based on the anatomic model proposed by Findeisen (1935). The method of calculation was similar to that used by Findeisen, except for the deposition due to diffusion, where the Gormley and Kennedy (1949) formulation was used. Nasopharyngeal deposition was calculated using the empirical relation determined by Pattle (1961) in his experimental study. Experimental data of Dennis (1961), Brown et al. (1950), Van Wijk and Patterson (1950), and Wilson and La Mer (1948), as reinterpreted and adjusted by the Task Group, were compared to the results of the model calculation. The Task Group judged that "the calculated curves do not appear to be seriously at odds with the available experimental data, particularly in view of the inferential nature of the latter and the uncertainties of many of the particle size measurements."

One of the important conclusions of the Task Group study was that the regional deposition within the respiratory tract can be estimated using a single aerosol parameter, the mass median diameter. This is indicated in their work and is illustrated in Fig. 9.8. It can be seen that for a tidal volume of 1,450 ml, there are relatively small differences in estimated deposition over a wide range of geometric standard deviations ($1.2 < \sigma_g < 4.5$). A second important conclusion of the Task Group was the finding that the

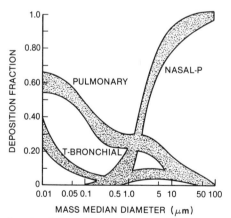

Fig. 9.8. Example of regional deposition predictions based on model proposed by ICRP Committee II Task Group on Lung Dynamics (1966), indicating effect of variations in σ_g: each of the shaded areas (envelopes) indicates the variable deposition for a given mass median (aerodynamic) diameter in each compartment when the distribution parameter σ_g varies from 1.2 to 4.5 and the tidal volume is 1,450 ml.

calculated total and compartmental deposition rates changed relatively little over a range of tidal volumes between 750 and 2,150 ml. Thus one expects that at least qualitative estimates of regional deposition patterns can be made for decision making, based on such a model.

The ICRP Task Group model does not agree well quantitatively with the best available experimental data on normal human beings. This is illustrated in Fig. 9.9, which shows a comparison between the *in vivo* alveolar deposition according to Fig. 9.5 and the Task Group prediction for polydisperse aerosols with the indicated mass median diameters. Furthermore, the Task Group model does not provide any indication of the large variability in deposition efficiencies among normal subjects, nor for the abnormalities produced by cigarette smoking or lung disease.

The Task Group model has had extensive application in inhalation hazard evaluations and many specialized air samplers have been designed to separate airborne particles into those deposited in one or more of the model's

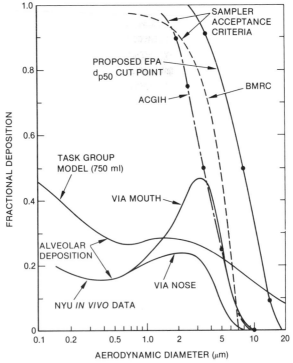

Fig. 9.9. Comparison of sampler acceptance curves of The British Medical Research Council (BMRC) and The American Conference of Governmental Industrial Hygienists (ACGIH) with alveolar deposition according to ICRP Task Group model and median human *in vivo* data from Lippmann (1977). The 1982 proposed EPA d_{p50} cut-point curve for new thoracic particle (TP) particulate standard also is included.

three regions. Two other sampler acceptance criteria have had extensive application in the field of occupational health. One is that of the British Medical Research Council (BMRC), and the other is that of the American Conference of Governmental Industrial Hygienists (ACGIH). Both were based primarily on the experimental deposition data of Brown *et al.* (1950) and both define "respirable" dust as that which penetrates a precollector whose particle retention characteristics simulate those of the human respiratory tract down to and including the terminal bronchioles. Their application is limited to those airborne dusts such as coal and quartz which are essentially insoluble in the lung. The sampler acceptance criteria of BMRC and ACGIH are shown in Fig. 9.9. The BMRC criteria can readily be met with horizontal elutriator collectors (settling chambers with fixed-flow inlets), while the ACGIH-AEC criteria are matched by cyclone precollectors. The background and application of the BMRC and ACGIH criteria have been reviewed in detail elsewhere (Lippmann, 1970b). The proposed 1982 EPA d_{p50} cut-curve for the TP standard is also shown in the figure. This curve is estimated with the 50% cut point for a sampler inlet at 10-μm aerodynamic diameter. Such a sampler will collect a larger mass concentration for apparent lung deposition than the other proposals.

9.3.5 Deposition of Atmospheric Particles. Recent data described in Chapter 7 indicate that the mass or volume distribution of the ambient aerosol as a function of particle size generally has at least two distinct modes. One mode results from the formation of small particles in the atmosphere and their growth by accumulation of condensed material and coagulation. These particles vary from approximately 0.01 to 2 μm in diameter. The second mode is created by particles injected into the atmosphere by mechanical processes. Most of the mass of these suspended particles is contributed by particles larger than approximately 3 μm. The upper size limit is quite variable, increasing with increasing atmospheric turbulence.

The nature of the size–mass variation of atmospheric aerosol particles, combined with the size-selective characteristics of the human respiratory tract, provides a basis for improving the characterization of the potential inhalation hazard from pollutant particles. Two important deposition and aerosol concentration factors are involved. First, particles contribute to inhalation hazard only when they are deposited in the respiratory tract. Exhaled particles do not contribute to the hazard. Second, the potential hazard from deposited particles depends greatly on where they are deposited. One major distinction is between the conductive airways and the alveolar zone, since the retention times for nonsoluble particles are usually very much greater in the alveolar zone. There may also be important distinctions for subregions of the conductive airways, but the influence of these differences on inhalation hazard is less clearly established at this time.

In the 1960s the U.S. National Ambient Air Quality Standard (NAAQS) was established on the basis of a so-called total suspended-particulate exposure, as determined by the high-volume filter sampler (Fig. 5.23). With the research on particulate matter in the atmosphere, and sampler performance, workers determined that the high-volume device collected material with an uncertain and variable upper size fractionation near 30 μm in diameter. Although particles of this size and larger can enter the human body, from Fig. 9.8, more than 90% are deposited in the nose. Since hazard from inhaled particles is said to be confined to the respiratory tract, a suitable size cutoff for airborne particles is approximately 10 μm diameter (Fig. 9.9). Particles of this and smaller diameter recently have been defined as the *thoracic particles* (TPs); this range is now being considered for revising the NAAQS for the 1980s. The TP levels will be an improved index for a human-health-related standard, and will clarify a variety of issues related to the control of blowing dust, which can heavily influence the levels of suspended-particle concentration in arid areas.

Consideration also has been given to further segregation of particulate matter for regulatory purposes, noting the qualitative association between finely divided particles ($\lesssim$3 μm) and alveolar deposition. By separating the surveillance of particles by size (and by chemical composition) some workers believe that an improved public-health-based policy for regulation of particulate sources can be achieved.

The deposition in the alveolar region of the human lung can be simulated by an air sampling device that mimics the collection characteristics of the conductive airways (Miller *et al.*, 1979). A proposed sampler would consist of two stages in series. Particles that would be deposited in the head and tracheobronchial regions would be collected in the first stage, and particles that would reach the alveolar region would be collected in the second stage.

In an advanced design for a sampler, the first stage would have an inlet that simulates the aspiration characteristics of the human nose; i.e., it would not permit the entry of particles which are too large to be inhaled. This stage could also incorporate a thermostatically controlled inlet and have a wetted wall entry section, so that hygroscopic particles would grow to the same equilibrium size that they achieve in the human respiratory tract.

The second stage would collect all the particles that penetrate the first. This requirement is easily met by a variety of filters. The second-stage sample collected would be much larger than the alveolar-zone deposition, since those particles small enough to reach the alveolar zone have a high probability of being exhaled without deposition. However, a reasonably constant factor can be assumed for the fraction deposited. Considering the collected material, convention suggests that the fraction less than 0.1 μm in diameter be disregarded because of its low mass contribution in ambient air, combined with its relatively low deposition efficiencies. Within the 0.1–3.5 μm range, there is a relatively low and reasonably constant alveolar-zone deposition of approximately 20% for nose breathing normal subjects. Thus, it would be reasonable to assume that 20% of the second-stage sample represents alveolar deposition. For cigarette smokers, and for other people with chronic bronchitis or asthma or both, this would be a conservative estimate of alveolar-zone deposition.

The particle size cutoff characteristics of the first-stage collector should be similar to those of the upper airways. Based on the data in Fig. 9.9, an inlet capable of sharply separating particles smaller than 10 μm in diameter from larger ones (say 10 μm at 50% efficiency) is a reasonable design parameter to assume collection of all particles to enter the respiratory system. The second-stage of the device has been proposed to a 50% cut point in the range between 1 and 3 μm since this is where the minimum occurs between the accumulation and coarse particle modes are observed in tropospheric particle distributions. Such a cut also would permit separate chemical determinations of two major types of pollution corresponding to combustion or secondary particles, as opposed to fugitive dusts and locally injected coarse material. On the other hand, the respirable dust separator characteristics with a 50% cut at 3.5 μm, which is widely used in occupational health work, is also consistent with the deposition and ambient aerosol size data, and may be preferred on the basis that it would provide a more conservative estimate of alveolar deposition, and a common response to a similar requirement.

Two sampling devices have been suggested for air monitoring applications. They are the dichotomous impactor sample (Fig. 5.34), described by Loo *et al.* (1978), and a stacked filter sampler reported by Flocchini *et al.* (1976). Most experimenters prefer the dichotomous impactor approach because the stacked filter sampler is believed to have ambiguities in the flow rate–size fractionation achieved using nucleopore filters of different pore diameter.

9.3.6 Summary. Inhaled particles deposit in the component regions of the respiratory tract according to (1) their size, density, and aerodynamic drag factor; (2) the average rate and time pattern of flow; and (3) the geometry of passageways and spaces. The controlling physiological quantities differ among individuals and vary in time. The particle parameters will also change during breathing if the particles act as condensation nuclei within the warm and humid airways. The processes have been far too complex for complete rigorous mathematical analysis, and the experimental difficulties have limited the extent and reliability of empirical data.

None of the available models provides reliable estimates of aerosol particle deposition in healthy normal adults, since their predictions for total and alveolar deposition efficiencies differ from the best available experimental data for normal adults. Furthermore, they do not give a reliable measure of the very large variability in deposition efficiencies among normal humans, nor of the changes produced by cigarette smoking and lung disease. However, there have been significant advances in the measurement of deposition in recent years, and effort continues to provide improved theoretical under-standing of the deposition process.

Although considerable work has been initiated to study the size-segre-gated character of tropospheric particles, and methods for their sampling, the recent scientific debate on revising the (U.S.) National Ambient Air Quality Standards (NAAQS) has failed to confirm a two-segregated standard with basis of health effects. However, there remains a recommendation to consider a (secondary) NAAQS for fine particles (<3 μm diameter) on the basis of visual aesthetics and visibility impairment.

9.4 DEFENSE MECHANISMS

Despite continuous exposure of the lung to particles characteristic of occupational and ambient environments, the surfaces of the respiratory tract are relatively free of foreign matter. Even in lungs seriously compromised by disease, the respiratory system is still surprisingly efficient at cleansing it-self. The diseased and blackened lungs of miners who succumb to coal worker's pneumoconiosis contain less than 10% of the dust originally depos-ited there. The healthy lung exhibits even greater potency in the pursuit of cleanliness. This is the result of a complex system of respiratory defense mechanisms which both prevents particle entry and removes particles that have been deposited. *Clearance* refers to the dynamic processes that remove particulates from the respiratory tract; it is the outflow of particles previ-ously deposited. Highly soluble particles and gases dissolve rapidly and are absorbed into the blood from the respiratory tract. Their metabolism and excretion resemble that of an intravenously injected dose of the same ma-terial.

Less soluble particles that are deposited on the mucus blanket covering pulmonary airways are moved toward the pharynx by the cilia. Also present in this moving carpet of mucus are cells and particles which have been transported from the nonciliated alveoli to the ciliated airways. Similarly, particles deposited on the ciliated mucous membranes of the nose are pro-pelled toward the pharynx. There, mucus, cells, and debris coming from the nasal cavities and the lungs meet, mix with salivary secretions, and enter the gastrointestinal tract after being swallowed. Since the particles are removed with half-lives of minutes to hours, there is little time for solubilization of slowly dissolving materials. In contrast, particles deposited in the nonci-liated compartments have much longer residence times; there, small differ-ences in *in vivo* solubility can have great significance.

A number of factors can affect the speed of mucus flow. They may be divided into two categories: those affecting the cilia themselves and those affecting the properties of the mucus. The following aspects of ciliary action may be affected by environmental insult: the number of ciliary strokes per minute, the amplitude of each stroke, the time course and form of each stroke, the length of the cilia, the ratio of ciliated to nonciliated area, and the susceptibility of the cilia to intrinsic and extrinsic agents that modify their rate and quality of motion. The characteristics of the mucus are frequently even more critical. The thickness of the mucus layer and its rheological properties may undergo wide variations.

According to Brain (1977), many studies reported in the literature do not characterize ciliary motility and the quantity of the mucus separately. In most cases the interaction of the two processes, mucus transport, is the variable that is measured. Most evidence suggests that mucus secretion is a more sensitive process than is ciliary activity. In many instances, the quantity and rheological characteristics of the polymeric gel which constitutes mucus may be affected independently of any change in the cilia.

9.4.1 Bronchial Clearance. The mucosa of the bronchial tree contains goblet cells, which produce mucus, and ciliated cells, which propel surface materials toward the throat where they are swallowed. Glands in the submucosa also discharge fluid onto the mucosa. These are mixed glands in that they contain both cells producing serous fluid and cells producing mucus. The surface of the mucosa consists of metachronal fields, each containing from a few to several hundred ciliated cells. The cilia in each field beat synchronously in waves that move in the general direction of the throat. Between the metachronal fields are areas of nonciliated globlet cells (secreting mucus) and, more rarely, areas of *squamous metaplasia*. Mucus on these nonciliated areas often remains in place until a patch of moving mucus makes contact, adheres (because of stickiness) and pulls the static mucus along. Mucus generally ascends the bronchial tree in one or several streams. It does not ascend as a single layer with uniform movement.

The ciliated bronchial escalator apparently does not transport inhaled foreign particles unless they are attached to mucus droplets, flakes, or plaques (Iravani and van As, 1972). The latter are conglomerations of droplets and flakes. The mucus layer, therefore, is not a continuous blanket covering the mucosa, but scattered minute to extensive patches with free space between them. Such patches are rare near the terminal bronchioles and are numerous in the larger bronchi and trachea, where most of the foreign particles are usually deposited.

The mucus patches rest on the tips of cilia, which, when fully extended, usually propel them towards the head. For the return stroke these cilia bend, sliding under the mucus. Then they extend again for the forward stroke.

The upper parts of the serous layer just beneath the mucus patches move with the mucus, but the lower parts move little and often not at all. The serous layer acts as a lubricant in which the cilia beat and upon which the mucus slides. Under physiological conditions its depth is rather constant. When a person is severely dehydrated, the serous layer becomes thin and the mucus thick. The cilia tend to get stuck in the mucus and their function is impaired.

When irritant gases such as SO_2 are inhaled, copious fluid is produced and the thickness of the serous layer is increased. In this case the tips of the cilia may not always reach the mucus layer. In addition, irritant gases may directly injure the ciliated cells. Thus, environmental irritants can alter the physiology of the mucosa of the bronchial tree by interfering with ciliary function and the amount and properties of the secretions. In chronic bronchitis, cilia may beat in disorganized fashion—sometimes in the wrong direction and occasionally in a swirling pattern. This results in whirlpools and eddies where particles may accumulate. There are also areas of ciliary inactivity and stasis. In chronic bronchitis the serous and mucus secretions are increased several fold (reducing the concentration of irritant present). These secretions are transported up the bronchial tree at a more rapid rate, apparently because there are areas of increased ciliary activity, more forceful expiration, and occasional coughing (Dannenburg, 1977).

When irritants cause chronic bronchitis, there are more secretions and more rapid transport of the mucus and irritating particles in streams. There are also more areas of stasis where the transport mechanism is ineffectual.

9.4.2 Pulmonary Lymphatic System. In the bronchial tree are "pulmonary tonsils or protrusions" known as BALT (bronchus-associated lymphoid tissue). Bronchus-associated lymphoid tissue has a pseudostratified epithelium devoid of goblet cells and heavily infiltrated with lymphocytes which seem to be migrating into the inner space of the bronchial tree. Dannenburg (1977) notes that granulocytes and monocytes may migrate similarly, but there is little information on this possibility. The lymphocytes in BALT enter the connective tissue of the mucous membrane *lamina propria* of both the bronchi and the gut, where they become cells that secrete immunoglobulin A (IgA) into the seromucus surface fluids. IgA can inhibit virus growth and agglutionate bacteria.

The phagocytic cells in the mucus blanket are of two origins: (a) the alveolar macrophages that ascend the bronchial tree from the alveoli, and (b) the granulocytes and monocytes (young macrophages) that enter from the blood as a result of chemically induced motion (chemotactic influences), usually associated with inflammation. Local irritation by environmental toxicants, cigarette smoke, and infectious agents increases the number of granulocytes and macrophages (of both types) in the mucus blanket.

Not all phagocytic cells in the bronchial tree reach the throat in a viable condition. According to Heppleston (1963), some die in route, releasing the particles they have ingested: the more toxic the particle, the sooner its release. Thus, small particles from the alveoli as well as large particles that impinge on the bronchial mucosa ascend the bronchial tree both within phagocytes and also in a free state.

The mechanism that attracts the alveolar macrophage into the terminal bronchiole is unknown.

Most inhaled particles are cleared via the bronchial escalator, but a small proportion of these particles accumulate in the lymphatic system of the lung. Some may be carried to the "pulmonary sumps" in a free state with the alveolar fluid. *Pulmonary sump* is a term introduced by Macklin (1955); it is a vaguely defined region located at the end of the terminal bronchiole forming the *hilus* (small recess or opening) of a primary pulmonary lobule. At this place there is an accumulation of lymphoid cells and the centrilobular (periarterial) lymphatics begin. (There are also sumps and lymphoid cells where the perilobular [perivenous] lymphatics begin.) Macklin (1955) postulated that a layer of fluid continuously flows over the alveolar surfaces and is absorbed by the lymphatics of the sump. Since only a small fraction of inhaled particles accumulates in the sumps and lymph nodes, the flow of fluid over the alveolar surface to the sumps must be small, except perhaps when pulmonary edema is present. The first site of accumulation seems to be in the lymphoid tissue at the terminal bronchiole near Macklin's sump, where the centrilobular lymphatic vessels begin. These vessels then follow the branches of the pulmonary artery along the bronchial tree to the hilus of the lung. En route they are interrupted by larger and larger lymph nodes, which filter out many of the particles in the lymph.

Each primary pulmonary lobule has a terminal bronchiole at its hilus. Groups of these primary lobules form a secondary lobule with a larger bronchiole at its hilus and are demarcated by connective tissue. The outer lobules have the pleural (lung covering membrane) surface at their base. Perilobular (perivenous or paraseptal) lymphatic vessels begin at the venous side of the capillaries and then accompany the veins that course along the septa that demarcate the secondary lobules. These lymphatics (which are also interrupted by lymph nodes) and the veins then follow the bronchial tree to the hilus of the lung. Lymph from the hilar lymph nodes flows (via the tracheobronchial and tracheal lymph nodes) into the right lymphatic duct and the thoracic duct (for the left upper lung in man) and then into the blood stream.

There are no lymphatics in the alveolar walls. Interstitial fluid in these walls circulates during respiration to the centrilobular and perilobular lymphatics, respectively, but these two lymphatic systems probably provide for intermingling of blood vessels en route to the pulmonary hilus.

Insoluble inhaled particles that reach the alveoli and do not ascend the bronchial tree are deposited in three sites: (a) interstitially around the terminal bronchiole, (b) interstitially in the septa and under the pleura, and (c) in the numerous lymph nodules and nodes interspersed along the periarterial and perivenous lymphatic vessels. This distribution is readily seen in the lungs of coal miners, for example, because of the large amount of carbon inhaled. Nonetheless, it seems that less than 1% of all inhaled particles is retained interstitially in the lung and lymph nodes. Over 99% ascends the bronchial tree and is swallowed and eliminated in fecal matter.

The flow of mucus over the surface of the bronchial epithelium is not uniform. There are holes where new bronchi enter and the mucus must flow around them. When two bronchi of equal size join, the flowing mucus at the carina must split, half entering the stream on the front and half entering the stream on the back of the junction. There are also areas of unciliated squamous metaplasia in the bronchial mucosa around which the stream of mucus must flow. In these three places, eddies occur in the mucus stream where particles may linger for hours. In these same places alveolar macrophages may die, be destroyed, and release the particles that they had ingested deep within the lung. If the particles are toxic, or irritant smoke is present in the bronchial tree, death of such phagocytes is hastened. Sometimes the phagocytes contain compounds like benzo[a]pyrene which they can convert to carcinogenic material. Thus one expects the incidence of respiratory cancers to be highest in the areas of minimal mucus or stagnating flow.

9.4.3   Alveolar Processes.   Macrophages are usually credited with keeping the alveolar surfaces clean and sterile. Alveolar macrophages are large, mononuclear, phagocytic cells found on the alveolar surface. They do not form part of the continuous epithelial cells (type I pneumonocytes) and great alveolar cells (type II pneumonocytes). Rather, the alveolar macrophages rest on this lining. These cells are largely responsible for the normal sterility of the lung (Green and Kass, 1964). It is the phagocytic and lytic (cell destroying) potential of the alveolar macrophages that provides most of the known bactericidal properties of the lungs (Brain, 1977). Like other phagocytes, alveolar macrophages are rich in *lysosomes* (cell dissolvers). The lysosomes attach themselves to the phagosomal membrane surrounding the ingested pathogen. Then the lysosomal membranes become continuous with the phagosomal membrane and the lytic enzymes kill and digest the bacteria. Among the hydrolases they are known to contain are proteases, deoxyribonuclease, ribonuclease, β-glucuronidase, and acid phosphatase. Although these enzymes constitute an important aspect of the lung's defensive posture, when kept in a chronically activated state, their digestive capacity may serve to damage pulmonary tissues. Release of lysosomal enzymes, particularly proteases, from activated macrophages and leukocytes may be involved in the development of emphysema. Release may occur as a consequence of cell death, or cell injury. Increased deposition of inert or infectious particles acts to recruit additional macrophages and thus the effect may be reinforced.

Hence, physiological concern is also directed toward the macrophages' ability to ingest nonliving, insoluble dust and debris. Rapid cellular destruction of insoluble particles prevents particle penetration through the alveolar epithelia and facilitates alveolar–bronchiolar transport. Schiller (1961) and later Sorokin and Brain (1975) found little evidence that macrophages laden with dusts can reenter the alveolar wall; only free particles appear to penetrate. Thus phagocytosis (cell abnormalities) plays an important role in the prevention of the entry of particles into the fixed tissues of the lung. Once particles leave the alveolar surface and penetrate the tissues subjacent to the air–liquid interface (type I and II cells, interstitial and lymphatic tissues), their removal is slowed. Particles remaining on the surface are cleared with a biological half-life estimated to be twenty-four hours in humans, while particles that have penetrated into "fixed" tissues are cleared with half-lives ranging from a few days to thousands of days. Therefore, the probability of particle penetration is critical in determining the clearance of particles from the unciliated regions of the lungs.

Brain (1977) notes that pathological processes such as fibrosis may also impair particle clearance from these compartments. The probability of particle entry into a fixed tissue in which it would have a long biological half-life is reduced if the particle is phagocytized by a free cell. Therefore *endocytosis* or internal cell disease emerges as a central theme of macrophage activity and the ability to quantify endocytosis and to study variations in the rate of endocytosis are essential to the understanding of respiratory defense mechanisms against inhaled particles. Only then can the effects of differing environmental conditions on endocytosis be estimated. Endocytosis has been much studied, but most techniques which are available measure ingestion or destruction of phagocytes (phagocytosis) *in vitro*. Attempts to quantitate phagocytosis *in vivo* are considerably less frequent. Hepatic macrophages have been studied as the major component of the reticuloendothelial[†] system and the rate of disappearance of various particulate materials from the circulation is well described. In contrast, the rate at which phagocytosis occurs in the living lung is not well established. Serial sacrifice and visual observation can give some clues, but other techniques to estimate endocytosis *in vivo* are needed for this research.

During the last few years, it has become increasingly obvious that not all pulmonary macrophages are alveolar macrophages. Another important subdivision of pulmonary macrophages is the airway macrophage. These mononuclear cells are present in conducting airways of both large and small caliber. They may be present as passengers on the mucus escalator, or they may be found beneath the mucus lining, apparently adhering to the bronchial epithelium. These airway macrophages probably represent the result of alveolar–bronchiolar transport of alveolar macrophages, although it has been suggested that they are the product of direct migration of cells through the bronchial epithelium.

The third subdivision of pulmonary macrophages is interstitial macrophages found in the various connective tissue compartments in the lung. These include alveolar walls, sinuses of the lymph nodes and nodules, and peribronchial and perivascular spaces. Connective tissue macrophages have

---

[†] Reticuloendothelium is reticular tissue associated with the squamous epithelial lining of the lung.

been considered in some detail by Sorokin and Brain (1975). All pulmonary macrophages are usually considered to be relatives of monocytes and macrophages throughout the body. This extended family includes the Kupffer cells of the liver, the free and fixed macrophages in the spleen, lymph nodes, bone marrow, the peritoneal macrophage of the serous cavity, and the osteoclasts of bone.

A major aspect of macrophage function of great importance to environmental health is the fate of the pulmonary macrophage. Differences exist among the varying classes of pulmonary macrophages and their fate must be considered separately. Although little is known about the latter stages of the life of the alveolar macrophage, the possibilities are limited and easily enumerated. They may be subject to alveolar–bronchiolar transport mechanisms, they may enter the lymphatics or connective tissue, or they may enter the circulation. Finally, it is also possible that some never leave the alveolar surface but persist for long periods of time, die there, and are then ingested and digested by younger, more vigorous siblings.

Brain (1977) speculates about the mechanisms responsible for alveolar–bronchiolar transport but little supporting evidence exists. He points out that most particles deposited in the alveoli are ingested by alveolar macrophages. Some of these cells find their way to the bronchioles, and are then carried to the pharynx by ciliary action. Migration through alveolar pores, or other collateral pathways between adjacent bronchial paths cannot be excluded. However, almost all of the macrophages are located on the surfaces of alveoli or bronchi. Thus, it seems unlikely that macrophages migrate to the bronchioles by penetrating between alveolar cells or by emerging from lymphatic pathways.

Since some macrophages find their way to the airways, we must ask how these cells move to the mucus escalator. It is possible that macrophages exhibit chemically directed locomotion because of a concentration gradient in a chemotactic factor. The phenomenon of chemotax or chemical stimulation has been studied *in vitro*. Little is known about the chemotactic behavior of alveolar macrophage movement in situ. There is also no evidence to suggest that other induced flows, such as gravitational stimulation, account for a directed migration of macrophages. Little experimental evidence suggests that the fluid lining of the alveolar region moves mouthward, but many investigators assume that it does.

The direct entry of alveolar macrophages into lymphatic pathways and connective tissue has often been suggested but rarely proved. The presence of particle-containing macrophages in these compartments is compelling evidence for many investigators. However, the entry of alveolar macrophages on the one hand and the entry of bare particles which are subsequently ingested by connective tissue macrophages already present on the other hand cannot be readily distinguished. During alveolar clearance, some non-ingested particles may follow lymphatic or vascular channels from alveoli into the peribronchial, perivascular, or subpleural outer coatings of the blood vessels and thus penetrate into the connective tissue of the lung. They are then stored by resident macrophages already present. This pathway may be more common when conditions favor increased lymphatic permeability (pulmonary edema). Then a greater number of particles might pass into these vessels through clefts between endothelial cells, to be carried along lymphatic drainage paths until filtered out by macrophages located farther along in lymphoid foci. Although inhaled particles can be found in connective tissue macrophages, Brain (1977) has noted there is little evidence that implicates movement of surface macrophages into connective tissue compartments. He does not exclude totally this possibility, however uncommon it may be. Yet such penetration may be of consequence to environmental immunological lung disease since it provides a pathway for antigens in or on alveolar macrophages to meet reactive lymphocytes in the connective tissue.

When the fate of airway macrophages is considered, it should not be assumed that the particle-containing mononuclear cells in the airways are necessarily the product of alveolar–bronchiolar transport. Some may be, but others may derive from blood monocytes which have migrated from the bronchial circulation directly to the airways. Alternatively, they may derive from as yet undescribed local monocytic cell renewal systems subjacent to the bronchial epithelium.

Although the pulmonary macrophages are essential to host defense, the normal activity and movement of pulmonary macrophages may also cause harm. Because the macrophages are actively phagocytic, inhaled toxic, radioactive, or carcinogenic particles become concentrated within pulmonary macrophages. What begins as a diffuse and relatively even exposure becomes highly localized and nonuniform. "Hot spots" are formed, and if thresholds for certain effects exist, these "hot spots" of intensified dosage are probably significant in the development of disease.

Similarly, adherence of some airway macrophages to the airway epithelium may increase airway exposure to inhaled toxic materials. More importantly, perhaps, this close association with the bronchial epithelium can lead to transbronchial transport of inhaled particles and subsequent reingestion by subepithelial connective tissue macrophages. These cells, like their relatives in the alveolar and airway compartments, also segregate, retain, and perhaps metabolize carcinogenic and other toxic particles. The metabolites probably represent critical but as yet unidentified links to disease.

Brain (1977) notes that additional research needs to be carried out on the structure and function of the pulmonary lymphatics. The timely removal of particles through these channels was often considered to be the major route for the clearance of material from the lower respiratory tract. The importance of lymphatic clearance may have been exaggerated for several reasons. Most of the early judgments were based on morphological studies.

Histological procedures, however, usually washed away the particles and cells being cleared via the airways, and so the importance of this route was underestimated. It is also difficult to draw quantitative conclusions about dynamic processes when only static observations are available. The importance of a clearance pathway is a function not only of the number of particles in the pathway, but also of the rate at which the particles are moving. Estimates of the percentage of particles cleared via the lymphatics has dropped based on recent work according to Brain. Yet appreciation of the importance of those particles entering the lymphatics has greatly increased. Because particles in lymphatics are cleared slowly, they attain significance in the pathogenesis of many lung diseases. When months and years have passed after exposure to particles, these connective tissue burdens may constitute the major reservoir of retained particles.

Little is known regarding the nature and properties of the air–blood barrier and the potential for particles to penetrate it. The extent to which the pulmonary capillary bed serves as a protective filter eliminating debris, pathogens, and cells from the circulating blood is not well described. The relative importance of these clearance pathways in health and disease is not well understood. This lack is not without good reason. Complicated experimental protocols would be necessary to accurately quantify fractions carried by the different clearance pathways. Theoretically, clearance by the various routes could be measured by monitoring the final common pathways for the airways, the blood vessels, and the lymphatics using various ingenious approaches. In the airways, a balance needs to be made for material deposited in greater detail if alveolar clearance is to be measured. In the lymphatic route, a single final pathway would have to be rationalized surgically for many collateral openings into the bloodstream. Examination of lung washings could be helpful in determining what part of the uncleared particle residues are lodged in alveolar macrophages and what part remain in the connective tissue. Further allowance must be made for the possibility of crossover from one clearance path to another, and it is important to separate the amount of material cleared by extracellular processes from that cleared by cellular mechanisms. As yet few studies have provided quantitative details sufficient to describe and evaluate the complexities of alveolar clearance. It is also likely that the extent to which each pathway is used depends on the size, solubility, and chemical properties of the deposited particles, as well as the time elapsed after the exposure and the degree of disease present.

Finally, it should also be remembered that deposition and clearance must be considered together. Researchers believe that the mass burden of the toxic substance in the lung is of primary importance. The actual amount of a substance in the respiratory tract at any time is called the *retention*. When the exposure is continuous, the equilibrium concentration (achieved when the clearance rate matches the deposition rate) is also the retention. Thus the relative rate constants of deposition and clearance determine the equilibrium level; it is the properties of the particle that are presumably related to the probability of a pathological response.

9.5 PULMONARY INJURY FACTORS

The lung is more than an important portal of entry for environmental contaminants. It may also be the critical organ most likely to be injured or compromised by inhaled particles or gases. The abundance of specific diagnostic labels for pulmonary disease does not alter the reality that universal mechanisms are operating. The lung responds to environmental insults in a variety of ways, but its repertory of responses is not unique. The same catalog of responses and mechanisms that characterizes the injured respiratory tract is also common to other organ systems. Thus a general pathology provides an essential backdrop for our understanding of the lung and environmental insults (Florey, 1970). Further progress in basic pathology is essential. Table 9.10 lists some of the changes that may take place in the lung. These changes are direct or indirect consequences of the physical and physiological changes in cells brought about by their contact with inhaled agents. The cell responses, in turn, cause the loss and distortion of alveoli, deposition of new alveolar tissue, alterations in capillary and airway diameter, and other anatomic alterations. These changes can occur in combination, and their severity is related to the duration and concentration of exposure.

TABLE 9.10

Mechanisms of Pulmonary Injury[a]

Proteolysis: Elastin and collagen destruction leading to
    emphysema
Fibrosis: Increased connective tissue and scarring
Pulmonary edema and altered alveolar stability
Immunologic responses: Asthma, hypersensitivity lung
    diseases, extrinsic allergic alveolitis
Inflammation: Irritation leading to mucosal edema,
    increased mucus production and bronchitis, enhanced
    cell renewal
Altered susceptibility to infection: Cytotoxic and
    competitive effects on macrophage function, altered
    mucociliary transport due to changes in cilia or the
    quantity of rheological character of mucus
Degenerative changes: Necrosis, calcification, and
    autolysis
Bacterial, viral, and fungal infection
Pulmonary carcinogenesis

---

[a] From Brain (1977).

9.5.1 Pathological Mechanisms.   Although pathological mechanisms operative in the lung are not exclusive to it, many arise from unusual interactions with environmental factors. Certain mechanisms seem especially worthy of study. Connective tissue proteins are an integral part of the lung. Albuminoid substances of connective tissue fiber and elastic tissue (collagen and elastin) help maintain alveolar, airway, and vascular stability; limit lung expansion; and contribute to lung recoil at all lung volumes. Because of their fundamental role in lung structure and function, collagen and elastin balance (synthesis and degradation) deserve additional attention.

Two groups of environmental lung disease are associated with aberrations of normal collagen and elastin balance: emphysematous and fibrotic disorders. To what extent this association involves changes in the rate and nature of synthesis, or in the role or nature of degradation of these proteins, remains speculative. Analysis of collagen composition, synthesis, and degradation rates would be extremely useful both for diagnosis and for assessment of therapy of these connective tissue lung diseases.

Unfortunately, the measurement of such parameters in the lung is difficult. Quantification of rates and composition analysis can be done by medical procedures, and can be performed on lung biopsy tissue. However, the large number of different cell types, and the heterogeneity, interactions, and insolubility of lung connective tissue have made accurate quantification difficult. Improved approaches are needed and *in vitro* models of lung biochemical function may be useful. Appropriate animal models of emphysema and fibrosis could be exploited to quantify the rates of elastin and collagen synthesis and degradation and to compare them to normal rates. In emphysema, further studies are needed to examine the release of protein-digesting enzymes from leukocytes and macrophages, their effect on lung cells, and the protection afforded by inhibiting protein digestion in the lung or sera. In fibrosis, investigations need to be pursued studying the role of lymphocytes, macrophages, and fibroblasts (large fiber forming cells), and the effect of fibrogenic agents on the collagen balance and on the replication and differentiation of each essential (parenchymal) cell type.

Attention to this kind of injury has increased because of recent new knowledge about pulmonary emphysema associated with inborn α_1-antitrypsinenzyme inhibitor deficiency in humans. Imbalances between protolytic activity and its control or inhibition have important implications as a generalized mechanism of lung injury in other pulmonary pathological states caused by air pollution or the inhalation of occupational dusts.

Brain (1977) points out that much more needs to be known regarding the nature of the particles that trigger the connective tissue diseases. Dust particles of appropriate size and shape may deposit on alveolar surfaces and stimulate production of excess collagen in the alveolar membrane. Particle size, shape, and resistance to chemical attack may be more important in fibrogenicity than chemical characteristics. Asbestos, glass, and other fibrous dusts all have been shown to stimulate collagen synthesis. Fibers over 5 μm in length are sometimes incompletely ingested by macrophages, and this may result in macrophage death. Growth of fibroblasts *in vitro* has been shown to require a solid supporting particle of minimum critical dimensions.

In addition, there is some evidence that fibrogenesis may also occur as a two-step process. This would especially apply to highly fibrogenic particles such as silica which are highly symmetrical in shape (5 μm in diameter) and which are engulfed by macrophages following deposition. Silica has not been shown to exert a direct stimulatory effect on fibroblasts. Rather, the interaction of a particle with a macrophage is thought to release factors which stimulate local production of collagen fibroblasts. It is unlikely that macrophages differentiate into collagen-synthesizing fibroblasts.

Silica and asbestos have the added disadvantage of being cytotoxic to alveolar macrophages. Within a few minutes they can cause dissolution or *lysis* of cells by direct interaction with the plasma membrane, or, if successfully ingested, in several hours cause rupture of secondary lysosomes, releasing lysosomal hydrolases into the cytoplasm. The resulting dead macrophages can become focal points for further fibrogenesis. In addition, the particles are released anew on the alveolar surface to cause more irritation.

The responses of the lung to inhaled antigens and allergens and the general area of environmental allergic respiratory disease is emerging as an important area of basic investigation. Advances in this area may help to identify hyperreactors to a number of industrially important materials, such as toluene diisocyanate (TDI), cotton dust, molds (Farmer's Lung and Bagassosis), and proteins causing a variety of forms of extrinsic allergic alveolitis. Immunological research addressed to the interaction of these inhaled materials with connective tissue, pulmonary cells, and the lymphatic and vascular systems in the lung holds promise. Little is known about the degradation of proteins deposited on the respiratory tract surfaces. No doubt in many instances macrophages and pulmonary clearance defend the body against excessive antigenic stimulation. However, there may also be circumstances when clearance pathways may cooperate with the immune system and preserve and present immunogenic molecules to the immune system. Thus the issue of how and when pulmonary macrophages either suppress or enhance the immunogenicity of antigens requires investigation.

It is known that several organic dusts can induce an immune response in bronchial or alveolar tissues. The type I and type III allergic reactions are most commonly seen according to Rose and Phills (1967). The type I response, often termed immediate hypersensitivity, occurs when an inhaled antigen reacts with cell-sensitizing antibody. Cells of the bronchial wall may be sensitized by this antibody and react to produce histamine, a slow-reacting substance associated with hypersensitivity and retarded movement (melancholia). In persons sensitized to a specific inhaled antigen, bronchoconstriction, mucosal edema, excessive secretion, and eosinophil infiltration can occur.

The type III, or *Arthus reaction,* is characterized by formation of a complex between an antigen and a precipitating antibody in the presence of complement. Tissue damage occurs from inflammatory responses caused by deposition of these complexes along capillary membranes. The complex is phagocytized by leukocytes and causes them to release their lysosomes. In addition, histamine may be released from most cells producing local edema. It is unknown whether antigens cross the alveolar and capillary membranes to react with circulating antibodies, or if lymphocytes and other reactive cells on the alveolar and bronchiolar surfaces are responsible for the initial antibody–antigen reaction. The site of action of TDI and the active components of cotton dust needs to be better described in this respect.

Particles inducing *hypersensitivity pneumonitis* tend to be derived from fungal, bacterial, or serum protein sources. Examples are the fungus *Thermoactinomytes vulgaris* in Farmer's Lung, *bacillus subtilis* in detergent worker's lung and bird fancier's disease (Pepys, 1969).

Neoplastic responses to inhaled particles is another area of major and growing concern. Although smoking has been determined to be the major cause of respiratory carcinogenesis, other factors, especially industrially produced agents, are gaining strong epidemiologic support. Selikoff *et al.* (1964) have shown that asbestosis has been correlated with bronchogenic carcinoma, as well as mesothelioma of the pleura and peritoneum. In addition, these workers (Selikoff *et al.,* 1968) indicate synergism between asbestos and cigarette smoking may cause a substantially increased risk. Other agents imposing an increased risk are the polycyclic aromatic hydrocarbons, radioisotopes, chromates, and compounds involved in nickel refining (Brain, 1977).

Bronchogenic cancer is many times more frequent in asbestos workers who smoke cigarettes than in nonsmoking asbestos workers. Thus a synergism exists between inhaled asbestos and smoke particles as far as this type of cancer is concerned. Cigarette smoke itself has many components that could act synergistically to cause bronchogenic cancer. It contains benzo[a]pyrene, tars, ^{210}Pb, ^{210}Po, and kaolinite (a clay particle somewhat toxic to macrophages), as well as a variety of other substances (Falk, 1977). Smoke bypasses the nasal mucosa, for it is inhaled through the mouth. The acids, aldehydes, phenols, and various gases (CO_2, NO_2, HCN) in cigarette smoke are directly irritating to the bronchi and impair the ciliary action of the bronchial mucosa. Chronic bronchitis and atypical mucosal cells (a type of carcinoma in situ) are more frequent in smokers.

The effects of cigarette smoke and other atmospheric contaminants on alveolar macrophages and on the immune reactivity of the lung have recently been reviewed by Holt and Keast (1977). In general, both bacterial clearance and immune function are impaired, but not always.

9.5.2 Detection and Quantification of Lung Injury. An essential aspect of continuing study of environmental lung diseases relates to improved determination of dose–response curves. Physicians need to know with precision what the respiratory tract is being exposed to as well as a measure of lung damage with similar precision and sensitivity.

It is now widely appreciated that it is essential to specify the mass, particle size, and solubility of inhaled particles. The ICRP lung model emphasized the importance of these parameters in determining burdens of particulates in the respiratory tract. It also provided a common framework which can be used in summarizing experimental data. But there are still major problems in quantifying the pulmonary dose. The relative merits of intratracheal and inhalation exposures have recently been discussed (Brain *et al.*, 1977). Switching from nose to mouth breathing and the effects of exercise are largely unexplored. Systematic descriptions of species differences are also essential. Although many mammal species have been used for studies in aerosol particles deposition, a systematic description of the differences in reactions observed among commonly used laboratory animals has not been reported as yet. It is difficult to abstract such a description from the literature because so many different kinds of animals and aerosol particles have been used in various combinations.

The lung responds to environmental injury in many ways. This diversity in response is matched by a diverse array of possible tests designed to detect and quantify lung damage. Table 9.11 lists some of the approaches commonly used. Almost all these approaches demand further refinement, and

TABLE 9.11

Methods of Measuring Lung Injury[a]

Properties	Parameters and methods
Mechanical properties (pulmonary function)	Compliance
	Resistance
	Lung volumes
	Flow volumes, partial flow volume curves
	Maximum breathing capacity
	Closing volumes
	Frequency dependence of compliance
Gas exchange	
Adequacy of ventilation	
Alveolar gas tensions	
Arterial pCO_2, pO_2.	
Distribution of ventilation	N_2 washout
	Xe^{133} scans
	Perfusion scans—particles or infused Xe^{133}
	Diffusing capacity

(*continued*)

TABLE 9.11 (*continued*)

Properties	Parameters and methods
Radiologic techniques	Atelectasis
	Fibrosis
	Degree of aeration
	Bronchography (tantalum)
	Lung volumes
	Focal lesions
Mucociliary transport	*In vitro*
	In vivo
	Nasal
	Airway
	Mucus studies
	Cilia studies
Lung lavage	Surfactant levels
	Cell numbers
	Cell differential counts
	Viability
	Biochemistry
	Histochemistry
	In vitro functional assays of macrophage activity
	Migration
	Phagocytosis
	Bactericidal activity
Histology on whole lung	Fixation techniques. Problems: lung volumes, cellular heterogeneity
	Sectioning and microscopic analysis of tissue
	Airway and alveolar dimensions: morphometric approaches
	Cell types: light and electron microscopy
	Histochemistry
	Vascular changes
Renewal of lung constituents	Metaphase counts—colchicine
	Uptake of tritiated thymidine
	Problems in identifying cell types
	Collagen and elastin breakdown and synthesis
Identifying pulmonary carcinogens	Experimental pulmonary carcinogenesis
	Ames assay
Microbicidal activity	Recognizable experimental pulmonary infections (morbidity and mortality studies)
	Bacterial aerosol models, *in vivo* models
	In vitro killing
	Phagocytosis
	In vitro
	In vivo

a From Brain (1977).

many have serious limitations. For example, one of the major problems with pulmonary function tests is that they are rarely specific. Measurements of airway resistance reflect not only direct alterations in the airway but may also reflect parenchymal changes. For example, if the parenchyma becomes weakened by emphysema, support for the airways is diminished. They become more collapsible and airway resistance may increase. Similarly, compliance measurements do not exclusively reflect changes in the pulmonary parenchyma. If airway obstruction occurs, then the parenchyma served by the airway will largely be excluded and the lung will appear to be stiffer.

Further refinements are needed for improved histologic, histochemical, and morphometric analysis of lung damage. Various fixation procedures have been used for the fixation of lung tissue. The main criteria are to preserve optimally the pulmonary parenchyma in accordance with the desired goal: examination of ultrastructure, investigation of acellular alveolar lining layer, application of histochemistry. Techniques exist which permit cutting sections of whole lungs for pathologic examination. These sections were originally intended for studying the pathology of coal worker's pneumoconiosis, and for comparison with radiological examinations during the individuals' lives. In addition to evaluating the lung sections qualitatively, quantitative morphometric methods have been developed to assay the degree of tissue damage.

Brain (1977) has suggested that it would be desirable to identify enzymes or other factors released from damaged lung into serum which could be sampled. Unfortunately, currently no biochemical parameters are known which indicate the presence of toxic lung damage. It thus becomes important to obtain more complete knowledge about the biochemical events leading to cell death or to repair and pulmonary cell proliferation. Only then will understanding of the mechanisms of response be increased, and perhaps more useful indicators of lung damage can be derived.

Techniques for sampling directly from the lung are also desirable. Improved biopsy techniques which can be used for repeated sampling from live animals need to be developed. Lung washing both in sacrificed animals and in isolated segments of living animals is also a valuable approach. For example, research on pulmonary macrophages often depends on our ability to isolate pure populations of macrophages from the lungs of animals and man. Alveolar macrophages represent a relatively homogeneous cell population accessible by lung lavage. However, success in isolating pulmonary macrophages depends on the particular subclasses involved, and, unfortunately, not all categories of pulmonary macrophages can be recovered with the same ease and purity. To date, investigators evidently have not been able to prepare a pure suspension of interstitial pulmonary macrophages. Techniques for dissolving the lung and liberating individual cells are being developed and intensive efforts are under way with regard to separation of isolated lung cells. However, interstitial macrophages are less numerous and little is known about the existence of unique physical or chemical properties which could be exploited in separating these cells. Conceivably, their tendency to attach themselves to glass or some specific surface receptor could be exploited. If a lung which has been washed to eliminate airway and alveolar macrophages has been dispersed, presumably those cells attaching to glass represent a relatively enriched population of interstitial macrophages.

Improved assays for identifying pulmonary carcinogens also need to be developed. One approach is to administer the suspected material to a laboratory animal by intratracheal instillation. Saffioti *et al.* (1968) have developed such an approach, and it has been frequently imitated. An alternative to experimental carcinogenesis has been implemented by Ames *et al.* (1973). This test operates under the working hypothesis that carcinogens are also mutagens. Using an *in vitro* bacterial test system (*Salmonella typhimurium*), Ames *et al.* have been able to detect known classes of mutagens at very low concentrations. The Ames assay is based on the use of controlled mutation of the bacterial strain for detecting mutagens by a highly sensitive and convenient back mutation test. The results of this test remain controversial at this time because these are only uncertain means of tracing it back to cancer in animals.

9.5.3   Protection from Chemical Factors.   Toxic environmental agents directly injure the lung or injure it through the immunologic (allergic) response developed by the host. Carcinogenic agents can be inactivated by host metabolism or created by host metabolism (or both). Radiation, alkylating chemicals and oxidants (producing peroxides and free radicals) may alter DNA, causing genetic defects and cancer, and individuals may be born with a genetic constitution that makes them more susceptible to cancer, pulmonary fibrosis, emphysema, or immunologic (allergic) diseases. Pharmaceutical agents can be used to treat infection, allergic disease, and cancer and yet these same agents may cause severe adverse reactions in certain people (e.g., penicillin allergy or immunosuppression).

Industrial plants can be designed with special enclosures and negative pressure exhaust hoods in dangerous areas (Baetjer, 1973b). Certain dusts can be wet down with water (e.g., during drilling in mines and during sandblasting). Goggles and change of clothing can be provided and also mask-type respirators for emergency or short-term use. Medical advice can be made available, e.g., to prevent an inactive tuberculous patient from being placed in a job where silica dust may be inhaled, or to help select the least toxic chemicals for a given industrial process.

Legislation is accomplishing a great deal in protecting the community and occupational environment. The Federal laws governing the contamination of air are embodied in the Clean Air Act of 1970, which includes the Motor Vehicle Air Pollution Act and the Air Quality Act. State and local governments also have laws and surveillance procedures concerning environmental pollutants.

Laws protecting miners are in the Federal Metal and Nonmetallic Mine Safety Act of 1966 and the Coal Mine Health and Safety Act of 1969, administered by the Bureau of Mines of the Department of Interior. Laws protecting other industrial workers are in the Occupational Safety and Health Act of 1970, administered by the Occupational Safety and Health Administration (OSHA) of the U.S. Department of Labor. OSHA (1974) has listed dangerous industrial substances and regulations concerning their use. New regulations are published periodically.

Our knowledge is quite deficient on what levels of hazardous compounds are safe for the public and the industrial worker. In fact, new chemical agents are manufactured even before the existing ones are fully evaluated. Bacterial cultures to detect mutagens are widely used (Ames et al., 1973). Animal experiments help in such determinations, but animals often show different tolerance levels and reactivity than man. Such experiments, usually of short duration, may not detect an effect appearing in man 15 to 30 years after exposure. Individuals differ in their ability to handle toxicants: some are more prone to cancer than others and some become highly allergic to certain chemicals and biological products.

Epidemiological studies are expensive and sometimes inconclusive, for people are mobile throughout the day and over the years, making it difficult to assess their intake of toxicants. Yet with some pollutants, epidemiological studies are our only source of information in the actual conditions of exposure.

The multitude of toxic environmental agents that we all encounter during our lifetime have not been proven to cause a high incidence of overt disease. Human beings, in fact, all animals and plants, often detoxify or adapt to ambient concentrations of such agents. On the other hand, many of these agents may have unrecognized effects, at least in some people. They may lower one's susceptibility to viral and bacterial agents or contribute to the onset of cancers appearing many years after exposure. Living in an industrialized society entails risks and annoyances different from those in a less complex society. We must accept some of these liabilities along with the multitude of assets that our society provides to enhance the comforts and variety in life.

REFERENCES

Alarie, Y. (1973). CRC Crit. Rev. Toxicol. 3, 299.

Altshuler, B., Yarmus, L., Palmes, E. D., and Nelson, N. (1957). AMA Arch. Ind. Health 15, 293.

Altshuler, B., Palmes, E. D., and Nelson, N. (1966). In "Inhaled Particles and Vapors" (C. N. Davies, ed.), p. 323. Pergamon, Oxford.

Ames, B. N., Lee, F. D., and Durston, W. E. (1973). Proc. Natl. Acad. Sci. U.S.A. 70, 782.

Andersen, I. B., Lundqvist, G. R., Jensen, D. L., *et al.* (1974). *Arch. Environ. Health* **28**, 31.

Baetjer, A. M. (1973a). *In* "Preventive Medicine and Public Health" (P. E. Sartwell, ed.), 10th ed., p. 955. Appleton, New York.

Baetjer, A. M. (1973b). *In* "Preventive Medicine and Public Health" (P. E. Sartwell, ed.), 10th ed., p. 891. Appleton, New York.

Bates, D. V., Bell, G. M., Burnham, C. D., Hazucha, M., Mantha, J., Pengelly, L. D., and Silverman, F. (1972). *J. Appl. Physiol.* **32**, 176.

Becklake, M. R. (1976). *Am. Rev. Respir. Dis.* **114**, 187.

Beeckmans, J. M. (1965). *Can. J. Physiol. Pharmacol.* **43**, 157.

Brain, J. D. (1977). *Environ. Health Perspect.* **20**, 113.

Brain, J. D., Sorokin, S. P., and Godleski, J. J. (1977). In "Respiratory Defense Mechanisms" (J. D. Brain, D. F. Proctor, and M. Reid, eds.), p. 42. Dekker, New York.

Brown, J. H., Cook, K. M., Ney, F. G., and Hatch, T. (1950). *Am. J. Public Health* **40**, 150.

Cautrell, E. T., Warr, G. A., Busbee, D. L., and Martin, R. R. (1973). *J. Clin. Invest.* **52**, 1881.

Cole, P. (1974). *In* "Mortality and Morbidity in the U.S." (C. L. Erhardt and J. E. Berlin, eds.), pp. 65–104. Harvard Univ. Press, Cambridge, Massachusetts.

Dalhamn, T. (1956). *Acta Physiol. Scand., Suppl.* No. 123, p. 9.

Dannenberg, A. M., Jr. (1977). *J. of the Reticuloendothel. Soc.* **22**, 273.

Davies, C. N. (1964). *Ann. Occup. Hyg.* **7**, 169.

Davies, C. N. (1973). *In* "Aerosole in Physik, Medizin und Technik," pp. 90–99. Ges. Aerosolforsch., Bad Soden, Fed. Rep. Ger.

Davies, C. N., Heyder, J., and Subba Ramu, M. C. (1972). *J. Appl. Physiol.* **32**, 591.

Dennis, W. L. (1961). *In* "Inhaled Particles and Vapors" (C. N. Davies, ed.), p. 88. Pergamon, Oxford.

Dontenwill, W., Elmenhorst, H., Harke, H. P., Reckzch, G., Weber, K. H., Misfeld, J., and Timm, J. (1970). *Z. Krebsforsch.* **73**, 285.

Falk, H. L. (1977). *In* "Handbook of Physiology, Sect. 9, Reactions to Environmental Agents" (D. H. K. Lee, ed.), Chap. 13. Am. Physiol. Soc., Bethesda, Maryland.

Findeisen, W. (1935). *Pfluegers Arch. Gesamte Physiol. Menschen Tiere* **236**, 367.

Flocchini, R., Cahill, T. A., Shadoan, D. J., Lange, S. J., Eldred, R. A., *et al.* (1976). *Environ. Sci. Technol.* **10**, 76.

Florey, H. W. (1970). "General Pathology," 4th ed. Lloye-Luke, London.

Frank, N. R., Amour, M. O., Worcester, J., and Whittenberger, J. L. (1962). *J. Appl. Physiol.* **17**, 252.

Gee, J. B. L., Smith, G. J., Matthay, R. A., and Zorn, S. K. (1977). *J. Reticuloendothel. Soc.* **22**, 233.

Giacomelli-Maltoni, G., Melandri, C., Prodi, V., and Tarroni, G. (1972). *Am. Ind. Hyg. Assoc. J.* **33**, 603.

Gormely, P. G., and Kennedy, M. (1949). *Proc. R. Ir. Acad., Sect. A* **52A**, 163.

Green, G. M., and Kass, E. H. (1964). *J. Exp. Med.* **119**, 167.

Hammond, E. C., Selikoff, I. J., Lawther, P. L., and Seidman, H. (1976). *Ann. N.Y. Acad. Sci.* **271**, 116.

Hatch, T., and Gross, P. (1964). "Pulmonary Deposition and Retention of Inhaled Aerosols." Academic Press, New York.

Heppleston, A. G. (1963). *Am. J. Pathol.* **42**, 119.

Heyder, J. (1973). *In* "Aerosole in Physik, Medizin, und Technik," p. 122. Ges. Aerosolforsch., Bad Soden, Fed. Rep. Ger.

Heyder, J., Beghart, J., Heigwer, G., Roth, C., and Stahlhofen, W. (1973). *J. Aerosol Sci.* **4**, 191.

Hodgson, T., Jr. (1971). *Environ. Sci. Technol.* **5**, 548.

Hoegg, U. R. (1972). *Environ. Health Perspect.* **2**, 117.

Hoffmann, D., Rathkamp, G., and Wynder, E. L. (1963). *J. Natl. Cancer Inst.* **31**, 627.

Holt, P. G., and Keast, D. (1977). *Bacteriol. Rev.* **41**, 205.

Hounam, R. F., Black, A., and Walsh, M. (1969). *Nature (London)* **221,** 1254.

Iravini, T., and van As, A. (1972). *J. Pathol.* **106,** 81.

Ishikawa, S., Bowden, D. H., Fisher, V., and Wyatt, J. P. (1969). *Arch. Environ. Health* **18** (4), 660.

Kay, K. (1977). *In* "Handbook of Physiology, Sect. 9, Reactions to Environmental Agents" (D. H. K. Lee, ed.), Chap. 11. Am. Physiol. Soc., Bethesda, Maryland.

Keith, C. H., and Derrick, J. C. (1960). *J. Colloid Sci.* **15,** 340.

Landahl, H. D. (1950). *Bull. Math. Biophys.* **12,** 43.

Landahl, H. D. (1963). *Bull. Math. Biophys.* **25,** 29.

Landahl, H. D., Bracewell, T. N., and Lassen, W. H. (1951). *AMA Arch. Ind. Hyg. Occup. Med.* **3,** 359.

Landahl, H. D., Tracewell, T. N., and Lassen, W. H. (1952). *AMA Arch. Ind. Hyg. Occup. Med.* **6,** 508.

Lapp, N. L., Hankinson, J. L., Amerdus, H., and Palmes, E. D. (1975). *Thorax* **30,** 293.

Lave, L. B., and Seskin, E. P. (1970). *Science (Washington, D.C.)* **169,** 723.

Lieberman, J. (1969). *N. Engl. J. Med.* **281,** 279.

Lindsey, A. J. (1962). *In* "Tobacco and Health" (G. Thomas and T. Rosenthal, eds.), Chap. 2. Thomas, Springfield, Illinois.

Lippmann, M. (1970a). *Ann. Otol. Rhinol. Laryngol.* **79,** 519.

Lippmann, M. (1970b). *Am. Ind. Hyg. Assoc. J.* **31,** 138.

Lippmann, M. (1977). *In* "Handbook of Physiology, Sect. 9, Reactions to Environmental Agents" (D. H. K. Lee, ed.), Am. Physiol. Soc., Bethesda, Maryland.

Lippmann, M., and Albert, R. E. (1969). *Am. Ind. Hyg. Assoc. J.* **30,** 257.

Lippmann, M., and Altshuler, B. (1975). *In* "Air Pollution and the Lung" (E. F. Aharonson, ed.), pp. 25–48. Keter Publishing House, Jerusalem, Israel

Lippmann, M., Albert, R. E., and Peterson, H. T. (1970). The regional deposition of inhaled aerosols in man. *Inhaled Part. Vap.* **3** (1), 105.

Lippmann, M., Gurman, J., and Schelsinger, R. (1983). *In* "Aerosols in the Mining and Industrial Work Environments" (V. A. Maple and B. Y. H. Liu, eds.), Vol. 1, p. 119. Ann Arbor Science Press, Ann Arbor, Michigan.

Lippmann, M., Yeates, D., and Albert, R. (1980). *Br. J. Ind. Med.* **37,** 337.

Loo, B. W., French, W., Gatti, R., Goulding, F., Jaklevic, J., Llacer, J., and Thompson, A. (1978). *Atmos. Environ.* **12,** 759.

Macklin, C. C. (1955). *Acta Anat.* **23,** 1.

Märtens, A., and Jacobi, W. (1973). *In* "Aerosole in Physik, Medizin, und Technik, p. 117. Ges. Aerosolforsch., Bad Soden, Fed. Rep. Ger.

Matsuba, K., and Thurlbeck, W. M. (1971). *Am. Rev. Respir. Dis.* **104,** 516.

Miller, F. J., Gardner, D. E., Graham, J., Lee, R., Jr., Wilson, W., and Bachmann, J. (1979). *J. Air Pollut. Control Assoc.* **29,** 610.

Morgan, W. K. C., and Lapp, N. L. (1976). *Am. J. Respir. Dis.* **113,** 531.

Mumpower, R. C., Lewis, J. S., and Toney, G. P. (1962). *Tobacco,* **155,** 30.

National Center for Health Statistics (1964). "Health Survey Procedure: Concepts, Questionnaire Development and Definitions in Health Interview Survey," Publ. No. 1000, Ser. 1, pp. 4, 42–45. U.S. Public Health Serv., Washington, D.C.

National Center for Health Statistics (1970). "Facts of Life and Death," Publ. No. 600, Table 12. U.S. Public Health Serv., Washington, D.C.

Occupational Safety and Health Administration (OSHA), U.S. Dep. of Labor (1974). "Occupational and Health Standards," Federal Register, Vol. 39, pp. 23502–23828. U.S. Gov. Print. Off., Washington, D.C.

Olson, D. E., Sudlow, M. F., Horsfield, K., and Filley, G. F. (1973). *Arch. Int. Med.* **131** (1), 51.

Pattle, R. E. (1961). *In* "Inhaled Particles and Vapors" (C. N. Davies, ed.), pp. 302–309. Pergamon, Oxford.

Pedley, T. J., Schrater, R. C., and Sudlow, M. F. (1971). *J. Fluid Mech.* **46**, 365.

Pepys, J. (1969). "Hypersensitivity Disease of the Lungs due to Fungi and other Organic Dusts," Monographs in Allergy, Vol. 4. Karger, Basel.

Proctor, D. F., Andersen, I., and Lundquist, G. (1973). *Arch. Intern. Med.* **131**, 132.

Reid, D. D. (1964). *Proc. R. Soc. Med.* **57**, 965.

Rose, B., and Phills, A. (1967). *Arch. Environ. Health* **14**, 97.

Rudolf, G., and Heyder, J. (1974). *In* "Aerosole in Natur wissenschaft, Medizin und Technik" (V. Bohlau, ed.), p. 54. Ges. Aerosolforsch, Fed. Rep. Ger.

Saffioti, U., Cefis, F., and Kolb, L. H. (1968). *Cancer Res.* **28**, 104.

Schiller, E. (1961). *In* "Inhaled Particles and Vapours" (C. N. Davies, ed.), p. 342. Pergamon, Oxford.

Schlesinger, R. B., and Lippmann, M. (1972). *Am. Ind. Hyg. Assoc. J.* **33**, 237.

Seaton, A. (1975). *In* "Occupational Lung Diseases" (W. K. C. Morgan and A. Seaton, eds.), p. 112. Saunders, Philadelphia, Pennsylvania.

Seaton, A., and Morgan, W. K. C. (1975). *In* "Occupational Lung Diseases" (W. K. C. Morgan and A. Seaton, eds.), p. 321. Saunders, Philadelphia, Pennsylvania.

Seehofer, F., Hanszen, D., and Schroder, R. (1965). *Beitr. Tabakforsch,* **3**, 135.

Selikoff, I. J., Chung, J., and Hammond, E. C. (1964). *J. Am. Med. Assoc.* **188**, 22.

Selikoff, I. J., Hammond, E. C., and Chung, J. (1968). *J. Am. Med. Assoc.* **204**, 106.

Sorokin, S. P., and Brain, J. D. (1975). *Anat. Rec.* **181**, 581.

Speizer, F. E., and Frank, N. R. (1966). *Arch. Environ. Health* **12**, 725.

Spiegelman, M., and Erhardt, C. L. (1974). *In* "Mortality and Morbidity in the U.S." (C. L. Erhardt and J. E. Berlin, eds.), pp. 1–20. Harvard Univ. Press, Cambridge, Massachusetts.

Stewart, R. D., Peterson, J. E., Baretta, E. D., Bachand, R. T., Hosko, M. J., and Hermann, A. A. (1970). *Arch. Environ. Health* **21**, 154.

Stresemann, E., and von Nieding, G. (1970). *Staub—Reinhalt. Luft* (Engl.) **30**, 33.

Sudlow, M. F., Olson, D. E., and Schroter, R. C. (1971). *In* "Inhaled Particles" (W. H. Walton, ed.), Vol. 3, pp. 19–29. Unwin Brothers (Gresham Press), London.

Task Group on Lung Dynamics (1966). *Health Phys* **12**, 173.

Thakker, D. R., Yagi, Y., Lu, A. Y. H., Levin, W., Conney, A. H., and Jerina, D. M. (1976). *Proc. Natl. Acad. Sci. U.S.A.* **73**, 3381.

Thor, D. E. (1977). *J. Reticuloendothel. Soc.* **22**, 245.

Timbrell, V. (1972a). *In* "Assessment of Airborne Particles" (T. T. Mercer *et al.*, eds.), p. 290. Thomas, Springfield, Illinois.

Timbrell, V. (1972b). *In* "Assessment of Airborne Particles" (T. T. Mercer *et al.*, eds.), p. 429. Thomas, Springfield, Illinois.

Tokuhata, G. K., Dessauer, P., and Pendergrass, E. P. (1970). *Am. J. Public Health* **60**, 441.

U.S. Environmental Protection Agency (1982). "Air Quality Criteria for Particulate Matter and Sulfur Dioxide." Off. Air Qual. Plann. Stand., Research Triangle Park, North Carolina.

U.S. Public Health Service (1969). "Air Quality Criteria for Particulate Matter," p. 109. Natl. Air Pollut. Control Adm., Washington, D.C.

Van Wijk, A. M., and Patterson, H. A. (1950). *J. Ind. Hyg. Toxicol.* **22**, 31.

Veeze, P. (1968). "Rationale and Methods of Early Detection in Lung Cancer." Van Gorcum, Essen, Fed. Rep. Ger.

Williamson, S. J. (1973). "Fundamentals of Air Pollution." Addison-Wesley, Reading, Massachusetts.

Wilson, I. B., and La Mer, V. K. (1948). *J. Ind. Hyg. Toxicol.* **30**, 265.

CHAPTER 10

REGULATION AND CONTROL
OF AEROSOLS

To complete the story of aerosol technology, let us turn to the societal concern for the adverse influences of environmental aerosols, and its consequent actions for control. From the discussion in parts of the text, it is clear that investigations have accumulated considerable evidence for certain environmental effects. The effects may occur either indoors or outdoors in ambient air. These vary in importance, but have direct consequences to human health in association with respiratory disease. They also have elements which affect human welfare ranging from outdoor factors such as interference with manufacturing, material damage from corrosion, vegetation and ecological damage, to aesthetic matters of visibility impairment and degradation of structural surfaces and art objects. All of these involve societal costs which are difficult to assess objectively. However, they have created an emotionally sensitized public, which has continued to press for control of particulate contaminants. In the subsequent discussion, the main themes from Chapters 8 and 9 are followed considering protection of the earth's atmosphere as a "breathable," transparent entity. Attention is focused on the outdoor atmosphere, recognizing that analogous approaches exist to address the indoor and occupational setting. The promise of control in the workplace has been associated with the field of industrial hygiene.

The technological support of regulatory decision making and implementation of aerosol control measures has created a multibillion dollar global industry in the past two decades. Much of the effort has focused on preserving the quality of the earth's atmosphere, particularly in its lowest layers where people live their lives.

The strategy for air pollution control has involved considerable sophistication in blending the regulatory goals with technology. The regulatory structure has reached considerable complexity in the United States as a result of (a) the Federal Clean Air Act (1968) (CAA) and its amendments of 1970 and 1977, and (b) the states adding supplemental legislation to enhance the ideals of the CAA. This chapter begins with elaboration of the philosophy underlying aerosol pollution regulation, based on the experience in the United States. Historically, pollution control has focused on the technology of primary particle emissions. This is discussed next in the chapter, beginning with the environment of the workplace. Finally, recognition is given to the fact that the atmospheric aerosol is highly reactive chemically. As a consequence, control of the secondary contributions to the airborne particle burden becomes of great significance.

10.1   REGULATORY PRINCIPLES

Recognizing the adverse effects of air contamination by aerosols, what can be done to avoid such pollution? There are many reasons why this question has no simple answer. Practically every industrial process adds some contaminants to the air. Domestic activities such as cooking a meal or heating the interior of a house contribute as well. The use of motor vehicles releases not only gaseous contaminants but also particulate matter including metallic particles and asbestos from brake linings and rubber from tire wear. Practically every human activity releases some material into the atmosphere. And for this reason it is generally agreed that air pollution will never be completely eliminated so that a "zero discharge" philosophy is not practical. Thus in a pragmatic sense we should more properly rephrase the question as: What can be done to reduce air pollution to acceptable levels? Here some concepts will be examined which have enjoyed success in at least partially meeting the challenge of pollution control and which show promise of future use. Our purpose is to set the stage for the discussion of the technology for physically controlling the emission of pollutants. We shall be concerned with concepts which can be implemented as legal statutes or economic incentives to encourage the use of the controls.

A well-known industrial physician, T. Hatch (1962), once stated wisely that, "Prevention of community air pollution from industrial operations starts within the factory or mill." Indeed, if efforts are made through process design and internal controls to minimize internal emissions, the external contaminants also will be minimized. Through legislation encouraged by labor organizations, public health services, and industry itself, progress toward protection from exposure to aerosol hazards in the workplace has improved greatly in recent years. The legislation in the United States has culminated in the Occupational Safety and Health Act (1970), which places very broad, stringent requirements on emissions in the work environment (McRae and Whelchel, 1978). Analogous laws have placed regulations and codes governing the domestic, business office environment and motor vehicles relative to heating, ventilation, and air conditioning facilities. Recently increased effort toward restriction and control of tobacco smoking in public places represents a natural evolution for such indoor environments. At the same time, efforts to control ambient air pollution have turned from technology development to action-oriented public abatement programs.

In most countries a *public nuisance* exists when conditions cause discomfort, inconvenience, damage to property, or injury, and the causing of a public nuisance is prohibited by law. If the person or institution responsible for the conditions can be identified, public authorities can seek a court injunction to prevent continuation of the nuisance and impose criminal sanctions. This may be successful if a single, clearly identifiable source is responsible, as when an exhaust stack emits great quantities of large particles which quickly fall out in the immediate neighborhood. But the assessment of responsibility is nearly impossible in the case of community smog, where emissions from more than one source may act synergistically. In addition, prosecution under public nuisance statutes occurs after the fact, and so the procedure is unsatisfactory for establishing preventive measures. When health is concerned, absolute proof of physical injury is difficult to find, as seen from epidemiological analysis of pollution effects. Thus control of pollution sources by use of existing laws governing public nuisances generally has not been found to be sufficient as motivating factors.

Under the common law in most countries, an individual suffering damage from contaminants released by another may seek by *private litigation* to recover the cost of damages he has sustained and cause an injunction to be issued to compel cessation of the harmful conditions. The burden of proof rests upon the complainant; to be successful, he must clearly link the damage he has sustained to the pollutants emitted by the defendant. The complainant may find this expensive, for he must have technical experts available to support his case. Furthermore, legal tradition stipulates that not only must damage be proven but also the actions of the defendant must be shown to be unreasonable. In some instances a court will weigh the costs of improving conditions against the anticipated benefits. For these reasons it is difficult for an individual to seek redress by legal suit. And, as for use of public nuisance statutes, a court suit after the fact is clearly an unsatisfactory procedure for preventing damage and protecting public health.

Thus government at various levels increasingly has taken on the burden of protecting the public interest. This has been approached in a number of ways: through research into the causes and effects of pollution, development of devices for pollution control, establishment of guidelines or standards for air quality, introduction of tax incentives for the installation of control equipment on sources, and, perhaps most important, promulgation and enforcement of ordinances for restricting the emission of contaminants, to name but a few. Most pollution control laws rest on the police powers of the state which derive from the right of people through organized government to protect their health and property. For example, the United States enacted a Clean Air Act in 1968, amended in 1970 and 1977, involving federal authority for abating air pollution which "endangers the health or welfare of any persons." Thus proof that actual injury has been done need not be made before the government can act in the public interest.

With government legislation has come a shift in the burden of proof. An example is the 1970 and 1977 Amendments to the Clean Air Act in the United States, which in certain instances require that, before emissions commence those discharging certain pollutants must show that their actions will not be harmful. Such legislation, when enforced, greatly eases the burden placed on air pollution control authorities. However, this policy has caused considerable delay in modernization and expansion of heavy industry and has greatly magnified the costs of renovation or expansion of industry. This cost in time and money has been borne in the project design and permitting stage, with a probable reduction in new construction.

There is a logical method by which a government has attacked the problem of air pollution. The first step is to define the levels at which a contaminant is considered a pollutant. By this we mean the adoption of a concise statement of what quality for ambient air would ensure that the public health and public welfare would be protected. Such prescriptions are usually called *air quality criteria*. They are determined on the basis of known or suspected adverse effects: toxic effects in humans or animals, damage to vegetation or materials, and degradation of aesthetic aspects of the environment. An air quality criterion indicates the level above which the presence of a pollutant is considered to have an adverse effect. Thus it may indicate a threshold concentration, dosage, and exposure time for each pollutant or combination of pollutants which act synergistically.

An air quality criterion is essentially a technical statement of knowledge. Standing alone, it has no practical significance, for it prescribes no legal standard nor does it indicate how the criterion is to be met. The *desired* levels of air quality may be different from the criterion and are defined by *air quality goals* (Williamson, 1973).

Two factors have traditionally motivated the establishment of air quality goals: (a) public opinion and (b) appreciation by government agencies of demonstrated or suspected adverse effects. The first factor is very much evident today and has been a dominant factor in prompting the adoption of criteria in many nations around the world. The second factor is a recognized and continuing activity of government, for it is the responsibility of government to protect the public health and take cognizance of technical aspects in which the public at large has little expertise. However, it is recognized that in the past many government bodies will not honor this responsibility without the impetus of public opinion. This inertia has stimulated the highly active lobbying of groups representing minority interests in many countries.

In setting goals, there is an inevitable balance of societal values against cost. This occurs for many reasons, but perhaps the prime one is the fact that the *economic cost* of pollution and its control will be borne by the citizen, in taxes and in the prices he pays for goods and services. The more stringent the goal, the more stringent will be the controls on the sources of pollution and the more costly it will be to maintain and police these controls.

In highly polluted regions, there may in fact be a direct economic benefit to the citizen from the first major decrease in pollutant levels, resulting from a reduced toll of material damage and perhaps health expenses. This possibility is illustrated in Fig. 10.1, where a hypothetical case is drawn in which the economic loss from pollution damage decreases as the ambient pollution level decreases. The marginal cost for introducing emission controls generally increases as the controls are made more stringent. For example, the cost of reducing the particulate emissions by a factor of 10 from a power plant stack may be partially or completely offset by the revenue gained from the sale of the collected fly ash as a road paving material or concrete blender. But to reduce the emissions by another factor of 10 would be considerably more expensive, not only because the control equipment would be more sophisticated and costly but also because the plant would obtain only one-tenth again as much sulfur to sell as from the first emission reduction. Thus in Fig. 10.1 the capital and operational costs of control equipment are shown increasing as the pollutant level decreases.

It is clear that the total direct and indirect economic cost to the citizen has

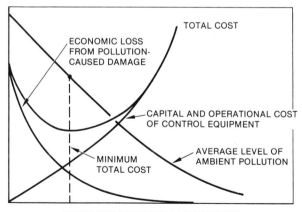

INCREASING STRICTNESS OF CONTROLS

Fig. 10.1. Schematic portrayal of the cost of air pollution and its control. [From Williamson, "Fundamentals of Air Pollution," © 1973. Addison-Wesley, Reading, Massachusetts. Reprinted with permission.]

a minimum value, and to establish an air quality goal on the basis of minimizing the total cost means accepting as that goal the ambient level of pollution corresponding to this minimum cost. Similarly, to achieve this criterion would require an investment in emission controls sufficient to provide the strictness of controls to reach this improvement in pollution burden. Thus a purely economic condition can be used to define an air quality criterion.

The example which was just given in Fig. 10.1 is an oversimplification for practical situations. For example, it has neglected the temporal variations in meteorological conditions which necessitate a more sophisticated means for relating pollution levels to the resulting economic impact. Perhaps it will be found that the yearly average concentration of a pollutant is the important parameter by which pollution levels should be measured, but very likely a more complex measure will prove to be more directly relevant. The simple scheme also has neglected the fact that whereas the economic damage from pollution is generally suffered by those in the immediate vicinity of the source, the cost of controls over the particular industry—if met by an increased price in its products—is borne by the consumers, who may represent quite a different group of people. Thus the money for repairing pollution damage and for emission controls need not come from the same pocket. An accurate economic model for predicting the total cost of pollution will involve many considerations, but the principle illustrated by the simple model is important and is directly relevant to the establishment of air quality goals.

If an air quality goal is to have significance, it must reflect the historical values prevalent in society. Perhaps the economic condition of minimal cost just described is not the determining factor. The population in a locality may believe instead that the preservation of *aesthetic factors* such as minimal odor or good visibility is a more important goal. Thus a community may decide in some circumstances that the air quality goal for suspended particulate matter should be determined on the basis of what prevailing visibility would be considered adverse. But here again there will be a tradeoff between benefit and cost, for maintaining good visibility generally requires much stricter controls over emissions than would be defined by a goal based on minimum cost. Recent pressure from segments of the public has emerged to set up a framework to protect the visual impression of mountains or scenery as seen from great distances, say 100 km or more. Meeting such a goal may require, for example, a tenfold reduction in the aerosol content from the level of emission which permitted a prevailing visibility of only 10 km. The cost of meeting this stringent requirement would be paid in both economic and social terms. Industries would need to filter their exhaust more stringently, farmers could no longer burn refuse in their fields, and perhaps homeowners would not be permitted to burn logs in their home fireplaces.

Another consideration is the fact that aerosol particles from natural sources frequently reduce the visual range and on these occasions there would be little point in keeping the concentration from anthropogenic sources much below the naturally occurring level. And what percentage of a population would ordinarily be at a location where they would have an unobstructed view of more than 10 km on even the clearest days? It would be an unusual community that could justify to itself maintaining a prevailing visibility of 100 km in the face of the economic and social costs. But Americans have decided, through the 1977 Clean Air Act Amendments, to bear the cost and demand improvement in visibility, at least in the National Parks or wilderness setting.

It is generally agreed that an air quality goal ought to be sufficiently strict to safeguard *public health* with some margin of safety. This is a straightforward consideration when pronounced adverse effects and their causes are well-established scientific or medical facts. Unfortunately, neither the threshold levels nor the safety margin for adverse health effects has been documented with any degree of confidence yet. However, when dealing with marginal effects, such as temporary and reversible changes of the chemical equilibria in human pulmonary tissue, there may be no clear indication as to whether the effect is an adverse one. Another ambiguity is associated with the extension of the results of toxicological studies on animals to humans, who may be considerably more—or perhaps less—sensitive to pollutants. There is then an element of *medical opinion* that influences the establishment of air quality goals which are designed to protect health on the side of safety. The degree of conservatism incorporated in a goal is a proper subject for debate, and will be influenced by public preference and expressions of willingness to bear added costs for a more certain measure of protection.

An additional philosophical issue with far-reaching implications is whether a goal ought to be sufficiently stringent to safeguard *all* of the members of a population, including the few very sensitive and susceptible members. This issue has been alternatively phrased in a perhaps more palatable way by asking what percentage of a population would be encouraged to move away to where the air is cleaner. When sufficient medical data become available, this question can be tackled on a more quantitative basis by use of curves for the threshold distributions of a population. At present, there remains very little information of this nature in research results.

An example of the influence of value judgments when agencies set air quality goals is the limited consideration commonly given to some segments of the agricultural community. Certain plants are considerably more sensitive to a given pollutant than are humans, but few goals now in effect give them the required measure of protection. Lettuce, for example, is easily damaged by ozone, and orchids by ethylene; in fact, aside from reduced visibility, plant damage is often the first sign available to the general public that photochemical smog has invaded a region. The foundations for goals have favored the protection of public health but have rarely provided for the protection of all vegetation, on the argument that the additional cost for pollution controls is not justified. Recently more effort has turned to agricultural effects as well as broader terrestrial and aquatic ecological consequences of pollution deposition (Braekke *et al.*, 1976; Hutchinson and Havas, 1980). In particular, the potential ecological threat of acid precipitation caused considerable debate in the late 1970s, and continues today.

In Fig. 10.2 many of the key factors that influence the formulation of air quality goals are summarized. This figure emphasizes the fact that decisions on goals involve an interplay of technical considerations and value judgments. And the value judgments include not only those which derive from society, but properly include medical, scientific, and technical opinions for evaluating the significance and applicability of available data. It should be kept in mind that when we say the object of formulating these goals is usually

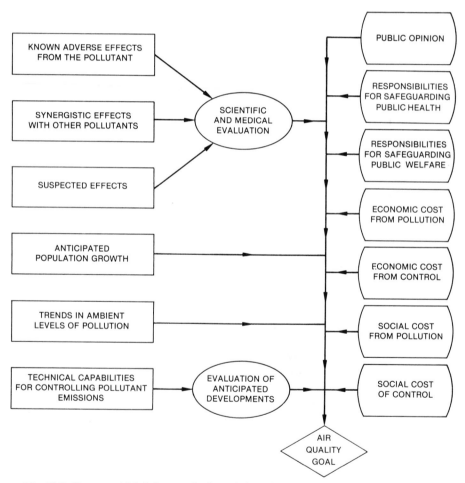

Fig. 10.2. Factors which influence the formulation of an air quality goal or standard for a pollutant. [From Williamson, "Fundamentals of Air Pollution," © 1973. Addison-Wesley, Reading, Massachusetts. Reprinted with permission.]

to seek a balance between benefit and cost, we do not mean just economical cost. The cost in social factors is also important. For example, the imposition of a goal, with anticipation that government agencies will subsequently regulate emissions from sources so as to meet the goal, can more seriously affect some segments of a population than others. As a case in point, suppose that an ambient air quality goal could be met only by making it illegal to operate an automobile which is more than ten years old; the resulting hardship would generally be borne by the poorest members of the population. It is not difficult to imagine other examples which discriminate against certain socioeconomic groups. The element of cost to the citizen has largely been disregarded in most pollution goal setting. It has been assumed that the absence of support for restraint in environmental legislation has meant public support for bearing such costs is present.

10.1.1 Concepts for Criteria and Goals. Three types of criteria have commonly been employed to define air quality: the ambient concentration, the period of exposure to a given average ambient concentration (or dosage), and the total burden of material (a product of concentration and distance, area, or volume). The distance–concentration product involves visual range or turbidity. The area–concentration product involves deposition. The volume–concentration product is less used but emerges in considerations of geochemical cycles.

The distinction between the first two is important. If the concentration is so small that exposure to this maximum level for any anticipated likely duration in time produces a dosage considerably less than the maximum hazardous dosage, then only a restriction on the ambient concentration is significant. On the other hand, if a population will experience an ambient concentration which is generally less than a safe value but is endured for a sustained period such that the dosage exceeds a safe value, then an air quality criterion must contain a restriction on both factors. Usually insufficient data are available to define a dosage, and in many cases it is not clear that such a simple criterion applies to biological systems, since natural processes eliminate much of an ingested pollutant if the concentration is sufficiently low. To account for a dosage factor, ambient criteria for an air quality standard are defined by two parameters: an average concentration and the time over which the average is taken. Because of the focus on ambient concentrations, burden values are normally referenced to concentrations. For example, visual range is calculated in terms of particle mass concentration integrated through a viewer path length, while deposition is calculated in terms of a mass transfer coefficient and concentration in layers of air near a surface (see also Section 7.5.1). With sufficient information, a hierarchy of criteria can be defined for a contaminant with thresholds or limits defined on the basis of different effects.

Instead of an absolute limit on the maximum tolerable concentration, or a limit on the average concentration, criteria can be formulated which permit occasional violations of the adverse level. This has certain practical advantages, since it takes into account the fact that ambient concentrations fluctuate markedly in accordance with changing rates at which contaminants are emitted and with varying meteorological conditions. Infrequent infringements on a safe level may not in fact be harmful or objectionable to a population. A criterion for prevailing visibility, for example, could be established permitting lower visibility on a few days each year with little adverse reaction from the population, even in urban areas having relatively good visibility. The advantage of permitting occasional exceptions is that a simple criterion can be formulated to govern what may most strongly affect people—the frequent exposure to peak pollution levels of lower magnitude than a rare maximum value. Thus criteria have been formulated whereby excursions above a certain level would be considered adverse if they occurred more frequently than, say, 1% of the time, or perhaps three times a year for an hours' duration each. This seeks to control principally the second-highest levels that occur more frequently; it is a hybrid condition, less stringent than an absolute limit, but more than just a condition on the average concentration for the same time interval.

The concept of frequency of occurrence of pollutant levels has been used extensively in current approaches to setting exposure levels and dosages. The application makes use of the fact that meteorological influences on ambient concentration variations are normally much larger than emission changes. Excursion to the limits of high pollutant concentrations follow a more or less standard statistical probability distribution. Definite relationships can be projected between long-term and short-term averages. Conventionally annual average concentrations have been specified for air quality goals, with short-term criteria stipulated for 24-hr, and 3-hr averages, or 1-hr maxima.

10.1.2   Air Quality Standards.   Given the criteria for maintenance of certain levels of pollutant concentrations in time intervals, based on health, welfare, or safety considerations, air quality goals emerge. The goal states the maximum acceptable pollution levels. Neither the criteria nor the goals have significance for legal enforcement of emission controls. However, standard setting formally establishes the foundation for abatement action.

Some governments prescribe legal standards for ambient air quality. In the United States for example they are called *air quality standards,* and there are two types: *primary standards* are based on the protection of health, and *secondary standards* on the protection of public welfare, including protection against known or anticipated effects of air pollution on property, materials, climate, economic values, and personal comfort. The standards indicate ambient levels of pollution that cannot legally be exceeded in a specific geographical region, and they may be imposed by national, state, and local governments. National standards have been adopted by the United States

Environmental Protection Agency (U.S. EPA). How these standards are to be met is left to local governments and the states, with review and approval by the federal agency. It is anticipated that the state and local governments when necessary will impose emission restrictions over at least the most important sources of pollution within their jurisdiction. The formal strategy for achieving the ambient air quality standards is presented in the State Implementation Plan (SIP).

The formulation of air quality standards proceeds in much the same way as that of air quality goals. The considerations indicated in Fig. 10.2 again apply. Of particular importance are the technical factors listed in the lower left-hand portion of the chart, because several questions are crucial in the decision making: Can a standard be met with existing technology? If not, can advance notice of the imposition of a standard be expected to prompt the necessary developments of emission control devices? Standards make sense only if they can be met in principle. Even for a region now meeting the prescribed levels, this is a matter of concern because future growth of population and the introduction of new industries would mean the need for more stringent controls on the new sources, as well as the existing ones, if possible. These issues have been raised in communities faced with achieving air quality standards, and others must be faced as well. For some industries, the alternative to an investment in costly control equipment is termination of operations with a resulting loss of jobs. Hence the imposition of air quality standards and their enforcement means that difficult economic and political decisions must be made, as well as technological ones.

Through the provisions of the U.S. Clean Air Act and its Amendments, there are presently three levels of national air quality standards (NAAQS) for particulate matter. These are (a) the primary standards, (b) the secondary standards, and (c) the incremental limits for change set for prevention of significant deterioration of air quality (PSD). The primary and secondary standards are designated by law along with a standard reference method for measurement. For example, the present *primary* ambient standards for particle concentrations are 75 μg/m^3, 24-hr average, measured with a high-volume air sampler located a fixed distance from the ground, and free of local wind obstructions (Fig. 5.23). The *secondary* standard is an annual geometric mean of 60 μg/m^3 and a 24-hr average of 150 μg/m^3 maximum, determined by high-volume air sampler. As mentioned in Section 9.3.5, the consideration for revision of the NAAQS for suspended particles has resulted in a recommendation for change to the thoracic particle (TP) requirement. The recommended range for TP is in the same range as the current TSP primary standard. Scientists also have suggested to EPA's Administrator that consideration be given to a new secondary standard for finely divided particles based on visual aesthetics and visibility maintenance for aviation safety. Such a secondary standard would be concerned with particles less than 3 μm diameter, and would be a "maintenance" requirement to prevent particles in this range from increasing beyond 8–25 $\mu g/m^3$, on a calendar three-month average over regions extending more than 60 km.

Unlike the primary and secondary standards, the PSD increments of air quality not to be exceeded apply to future conditions. Measurements can establish baseline conditions in an undeveloped area, but compliance with incremental changes requires calculation of effects, usually by applying mathematical models acceptable to the regulator.

The provisions for the prevention of significant deterioration stem from extension of the concepts of clean air maintenance to areas which are presently not affected by pollution. This concept is basically a land use management vehicle, which restricts uncontrolled development of areas such that only increments of increase in pollution levels are acceptable. There are three levels of restriction. Class I designated areas in the United States are pristine in nature and are identified as zones where only very minimal incremental deterioration is projected. These areas include the National Parks and certain wilderness areas. Class II areas are less restricted, and include rural, and undeveloped zones surrounding cities and heavily industrialized zones. Class III refers to those areas which are presently polluted, and have air contamination approaching the national ambient air quality standards. Without certain specific criteria for reference, essentially all non-Class I areas are designated Class II in category. The increments not to be exceeded in suspended particulate concentrations for Class I and II areas according to the U.S. Clean Air Act provisions are (in $\mu g/m^3$)

	Annual average	24-hr maximum
Class I	5	10
Class II	19	37
Class III	37	75

Class III zones would have minimal PSD restrictions; however, no state has adopted any policy for designating Class III areas.

The current U.S. legislation provides that states may set standards that are more stringent than the national levels. Some states have done this based on their judgment of the protection needed for the state's population.

It has been recognized for some time that certain chemical components of the suspended particles are important health factors. Yet the only national ambient standard for a particle constituent is 1.5 $\mu g/m^3$ as elemental lead, maximum arithmetic mean averaged over a calendar quarter. Although the human health basis for a sulfate standard is not supported by recent research, some states have instituted an ambient standard for this particulate component. These include California, Montana, and Pennsylvania. For example, Montana's is 4 $\mu g/m^3$ annual average; California's is 25 $\mu g/m^3$ (24-hr average maximum). In 1980, the basis for California's ambient standard for sulfate was challenged in the courts. Although its scientific basis was found inadequate, and it was overturned, the standard is currently being appealed by the State's air pollution control agency.

10.1.3   Control Measures and Emission Standards.   Given the establishment of ambient air quality standards, the next step is to devise schemes for pollution control to meet the standards. Controls have been considered in two ways: relying on (a) atmospheric processes of dilution, and (b) direct emission restriction from sources.

It has been stated that a way to reduce local ambient levels of pollution is by artificially improving meteorological ventilation. For example, to alleviate the conditions over Los Angeles, giant fans might be installed in tunnels penetrating through the mountains which ring the basin. Unfortunately, a simple calculation will quickly reveal that this would necessitate the expenditure of such great quantities of energy as to be impractical (Heuss *et al.*, 1971). Thus artificially improving ventilation of this type does not hold much promise for the future. This is not to say that one should not take advantage of natural conditions where ventilation is good. If governments can include such factors when locations for industrial zones are established, the effects of the effluent can be minimized. It is also possible to exploit the potential of hot, buoyant exhaust gases to penetrate the atmosphere and rise above the layers of stable air near the surface where mixing may be constrained. If large sources are grouped together, this approach has promise. This may only improve dispersion of pollution locally, but permit pollution to be fumigated to the ground further away from the sources.

It has also been suggested that industrial activity and automobile traffic should be reduced during episodes of high pollution, to reduce the rate of emission of pollutants. Controls then need to be applied only during conditions of atmospheric stagnation. Once ambient levels approach a predetermined value, law enforcement agencies would ensure compliance by industry and the public. Unfortunately, such an idea does not recognize the fact that industrial emissions cannot always be turned on and off as easily as a home stove; the time lag between notification and response will vary from industry to industry. Furthermore, the question of equity in reduction must be answered. Some sources may emit pollutants at a crucial stage in manufacturing and would be more severely hurt economically than others. And who is to say that the shutdown of a power plant may not cause more harm or discomfort when a consequence is that commercial offices can no longer benefit from their air conditioning? Only a few specialized industries might be able to switch to special processes or utilize special control devices that are ordinarily too expensive to use continuously. For these reasons and others, the curtailment of activity to reduce emissions during pollution episodes has been regarded as a measure of last resort.

In the mid-1970s the concept of pollution control at ground level using tall stacks and intermittent operational control based on meteorological forecast became popular. However, this approach recently was abandoned for new sources in the United States because of fears of exacerbating regional pollution conditions from long-range transport processes. Generally, the method is not a practical means to ensure continual compliance with air quality standards. Therefore, emission control measures have been prescribed as the principal means of pollution control in the United States. The problem of pollution control then reduces to the problem of devising emission standards which will continually be in force and can be practically achieved. Such standards or regulations have been implemented for both existing and new sources.

A legal restriction on the amount or conditions of release of a pollutant from any source is an *emission standard*. In principle, a desired level of air quality should be achieved by ensuring that there is a sufficiently low rate of pollutant emissions from all relevant sources.

Setting an emission standard is a relatively simple matter if only one source causes the pollution; a series of measurements of the ambient concentration can be used to deduce by what factor the emissions should be reduced. Then the choice of an emission standard can be decided on the basis of one of several possible strategies, including (a) maintenance of the desired maximum ambient level regardless of consequences, or (b) to require the use of the most efficient existing pollution control devices (best available control technology). If maintenance of the ambient standard cannot be achieved by BACT, two other strategies could be applied in an attempt to improve on the *status quo:* (c) to require slightly stricter controls than are currently feasible in anticipation that they will encourage technical progress toward the lowest achievable emission rate (or, in the very least, that they could be met by the source operating under reduced capacity); or (d) to set up a schedule for future periodic tightening of standards, assuming that research and development will enable them to be met.

Control policy is a considerably more complex problem when multiple sources of different types are involved and when secondary pollutants are produced. This policy is usually worked out through a series of public debates and planning scenarios leading to local and state implementation plans.

The problem of emission controls often has been divided into three aspects. The first involves *stationary sources,* the second concerns *transportation sources* such as automobiles, aircraft, trains, and ships. The third involves *fugitive* emissions such as road dust, dust mining operations, or crushing and grinding operations. In some countries the responsibility over these various sources is divided between local and national agencies. In the United States, local governments such as counties and states have jurisdiction over stationary sources, whereas the federal government has claimed exclusive rights for controlling emissions from automobiles and aircraft. This was done to avoid a multitude of different standards being imposed on sources traveling interstate. It was argued that the federal government could more effectively impose emission standards than could one state or county. For automobiles, the principal emission standard for particles implicitly has been the elimination of leaded fuel in new vehicles. Otherwise no standards have been devised, even though significant quantities of particles are emitted from vehicles.

Fugitive particulate emissions are difficult to control, but are believed to be a major factor in high particulate concentrations in many rural areas especially in the arid western United States, where blowing dust is a prevalent occurrence.

Many types of emission standards have been devised for stationary sources. In most cases they have been applied only to single sources; that is, to a single exhaust stack or to a single furnace with multiple stacks. Some examples of standards for single stacks are listed in Table 10.1. The first column of the table lists the concepts upon which standards can be based. The second column gives one example of possible standards. The specific numbers cited in the examples are meant to be illustrative and should not be interpreted as recommendations. Ordinances prescribing such standards are often tailored to local conditions of meteorological ventilation and the local goals for air quality.

TABLE 10.1

Examples of Particulate Emission Standards for Single Sources

Concept	Example standards
Maximum concentration	Limit maximum concentration of particulates: ≤ 0.7 g/m^3
Maximum concentration, specifying size ranges	More control to avoid effects of light scattering: ≤ 0.3 g/m^3 for radius above 5 μm; ≤ 0.05 g/m^3 for smaller particles
Maximum mass emission rate	Limit the mass per unit time emitted from a facility: Often established in terms of hourly particulate loading (e.g., 50 kg/hr); can be mixed in terms of mass concentration per unit time mass (kg/m^3-hr)
Maximum concentration or mass rate adjusted for effective release height	Permit higher concentrates in plume with a greater rise: ≤ 0.7 g/m^3 for sources greater than 10 MW; ≤ 0.5 g/m^3 for all others
Maximum opacity	Limit the fraction of light absorbed upon traversing a plume: $\leq 20\%$ reduction of the intensity of light (20% opacity)
Coloration	Avoid dense black or gray plumes: plume grayness not to exceed No. 1 on the Ringlemann chart except during starting, but not greater than No. 2 for more than 30 min; prohibition of
Prohibition	refuse disposal by incineration except under prescribed conditions

Several emission concepts illustrated in Table 10.1 warrant detailed examination. For the sake of clarity, we shall denote each by a number.

(1)–(2)   The *maximum concentration* condition is founded on the desirability of diluting contaminants as much as possible before release. The standard can be written in terms of total mass concentration or mass concentration as a function of particle size. This is most important for sources whose effluent is released near the ground because the general public may experience the polluted exhaust only a short time later, before atmospheric turbulence can dilute the contaminants to inconsequential levels. However, the standard is usually applied to large sources such as power plants as well.

(3)   A *maximum mass emission rate* places a limit on the total amount of a pollutant that any source can emit during a prescribed time interval, something that the maximum concentration standard does not accomplish. Limiting the *amount* of released pollutant is important in air basins where pollutants may accumulate as a result of terrain and meteorological factors. Furthermore, diffusion theory indicates that the maximum ground-level concentration of a pollutant downwind from an elevated source, such as a tall exhaust stack, is directly proportional to the mass emission rate of the pollutant, not necessarily to its concentration in the effluent. Thus in some respects the mass emission rate is a more relevant parameter for large sources. Unfortunately, few communities have ordinances which limit the emission rate.

(4)   Both minimum dilution and maximum mass emission rates can be qualified by a parameter which is related to the effective release height of the exhaust plume. This is the *effective stack height,* the sum of the geometric stack height and the additional height attained by the rise of the plume owing to its buoyancy and initial upward momentum. Higher-rising plumes have more time to become diluted by atmospheric turbulence before they touch ground and therefore, in many cases, need not be as severely restricted. An indirect parameter related to plume buoyancy is the rate at which heat is released by combustion in a furnace, often measured in terms of the megawatts of thermal energy released.

Additional concepts can apply to controlling particulate emissions, which provide for separate restrictions and can be placed on particles of different sizes, or plume visibility. Complaints from citizens about the objectionable appearance of dense white plumes or black smoke have resulted in the adoption of criteria specifically aimed to improve the appearance of the air.

(5)   The *maximum opacity* is a measure of how transparent a plume should be. Specifically, it is a limit on the reduction of light intensity that would result when a beam of light were directed transversely through the center of a plume. The measurement can be made by mounting a light source on one wall of an exhaust stack near the exit in such a way that the beam

traverses the interior of the stack and impinges on a photocell on the opposite wall. The decrease in the beam intensity can be deduced from the intensity measured by the photocell and the known original intensity from the light beam. Thus a criterion for 20% maximum opacity requires that the beam intensity be reduced by less than 20% compared with the amount transmitted when the plume is perfectly transparent.

A plume of water droplets can often be distinguished from a plume of particulates by observing the downwind dispersal. Water droplet or "steam" plumes usually end abruptly when conditions are appropriate for the water to evaporate, but particulate plumes become progressively fainter as the particles disperse. The opacity of water droplets is not a measure of the accompanying particulate burden of the effluent.

(6)   The grayness of a plume can be judged by visual observations which compare the plume's appearance to the Ringlemann's standardized gray scale (also Section 8.1). Number 5 on the Ringlemann scale corresponds to dense, black smoke. Number 0 is perfectly white (as distinct from transparent). Both opacity and Ringlemann standards fail to limit the total output of particulate matter because, as noted in Chapter 8, particles of different size affect visible light in different ways.

(7)   For completeness, the concept called *prohibition* is included. This hardly needs explanation, but at the risk of stating the obvious we note that emission control devices may not exist which are sufficient to reduce emissions from certain industries, and these industries might be forced to curtail or terminate their operations. For example, the burning of refuse outdoors, including agricultural burning, releases great numbers of particles that are very effective in scattering light and impairing visibility. Complaints from the public have led in many instances to a complete ban on open burning, affecting not only farmers and construction workers but owners of domestic incinerators as well.

Occasionally an emission standard is written in a manner that permits larger sources to emit more pollutants than smaller ones. This is implicit for example in maximum concentration standards. But some political jurisdictions control emissions of particulate matter by setting the maximum mass emission rate as a fixed percentage of the total hourly weight of all materials introduced into a manufacturing or processing operation, including perhaps liquid and gaseous fuels and air. Without additional safeguards to ensure proper dilution of the effluent, such an emission standard cannot be expected to guarantee attainment of the ambient air quality standards.

Another approach to emission standards is the so-called *bubble concept*. For industrial installations with multiple stacks, a total emission rate for the facility may be specified such that individual stacks may exceed a single-source emission rate, but the entire facility will be in compliance based on summation of emission from all portals.

Emission standards projected by guidelines for the federal government include such terms as new source performance standards (NSPS), best available control technology (BACT), best available retrofit technology (BART), and lowest achievable emission rate (LAER). For certain large sources such as power plants, new installations must comply with certain NSPS requirements for low sulfur fuel usage and emission controls. These are specified by the U.S. EPA (1979).[†] As an example, new coal-fired power plants in the United States are permitted to emit only 1.2 lb of SO_2/million BTU boiler capacity and 0.03 lb of particulates/million BTU capacity. In other circumstances, current emission regulations are technology forcing in that they require continuing design improvement using BACT. This approach includes a cost effectiveness component in decisions for application whereas NSPS is considered a minimum acceptable standard. The BACT concept has been introduced recently in the PSD regulations governing sources which may influence visual impairment in designated Class I areas. If such a source is identified and is less than fifteen years old in 1977, it may be required to install control technology in accordance with BACT. There are circumstances where certain areas in cities and other locations have been found to be in *nonattainment* with respect to one or more ambient air quality standards. In such situations, a new source will be required to use LAER regardless of cost considerations.

Emission standards for fugitive dusts are difficult to specify, but they have been implemented through requirements for watering or paving of road or tilled areas, as well as shrouding of conveyor belts or crushers. Fugitive dust is believed to be responsible for particulate nonattainment in many rural areas. Without effective control possibilities, consideration is being given to restating the suspended-particle standard in terms of finely divided material. A size cutoff of 10 μm diameter for particulate monitors may eliminate spurious large concentration reports and provide a more practical definition of air quality for airborne particles (see also Section 9.3).

The application of emission standards has become increasingly stringent in the past few years with the realization that pollution problems are not only local in nature but geographically regional in effects. This is particularly important in the case of atmospheric aerosol behavior where products of atmospheric reactions are involved. The formation of sulfate, nitrate, or condensed organic material from the gaseous pollutants sulfur dioxide (SO_2), nitrogen oxides (NO, NO_2, or NO_x), and hydrocarbon vapors create special issues for emission standards for particulate matter (Hidy *et al.,* 1980; also Section 10.5). The implications are clear that the gaseous emission standards may have to be reduced below projected levels to achieve air quality standards designated for gases if aerosol particles are to be reduced. This argument has been used recently to rationalize increasingly stringent

---

[†] For additional discussion, see Quarles (1979).

approaches to emission standards for stationary and mobile sources in some states such as California (CARB, 1979), and has emerged as an interstate pollution control issue (U.S. EPA, 1981a).

10.2   IDENTIFICATION OF SOURCES AND CONTROL OF AEROSOLS

10.2.1   Classes of Models.   The process of design for permissible standards within the legislative restrictions is a negotiable one, which results in a series of construction and operating permits. Identification and source apportionment, followed by design and/or retrofit to achieve ambient air quality standards, is done using so-called *air quality models*. Two methods have evolved for such computation; one is receptor based, and the other source based. The former depends on actual measurements of pollution chemistry at locations influenced by different sources. The latter concentrates on sources, either in operation, or contemplated in the future. The approaches are shown conceptually in Fig. 10.3. Source-oriented calculations are normally employed; they estimated ambient concentrations around a source, given prevailing or worst case meteorological conditions. Although the regulatory

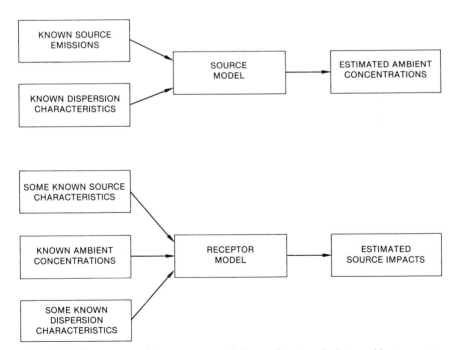

Fig. 10.3. The source model uses source emissions as inputs, calculates ambient concentrations. The receptor model uses ambient concentrations as inputs, calculates source contribution. [From Watson(1980).]

agencies in the United States have avoided certification of "standard" models, they have issued guidelines about acceptability, and offer a set of models which are accepted based on extensive experience in their use (U.S. EPA, 1978). These are updated from time to time as numerical simulation technology improves.

Source-oriented atmospheric dispersion modeling has been the major tool used in attributing ambient concentrations to source emissions. The approach essentially corresponds to a broad group of deterministic techniques in which a synthesis of all known physical and chemical processes influencing an emitted contaminant in the air are provided for to some degree of mathematical approximation. With extensively researched emissions inventories and adequate meteorological information, these models have done a credible job of predicting nonreacting gaseous pollutant concentrations at receptors due to specific sources. Many times, however, predicted gas and particle concentrations do not compare well with ambient measurements (Guldberg and Kern, 1978).

The source-based models have not proved themselves adequate for suspended particulate matter for several reasons. First, particulate matter emissions are widely dispersed and hard to quantify. Whereas most gaseous emissions can be confined to specific points (even auto tailpipes, though numbering in the millions, are identifiable and confined to roadways, garages, and parking lots), particulate matter is produced from point and nonpoint sources; every square meter of surface contacted by the wind is a potential emitter. Emission factors for area sources are hard to estimate and display wide variability. Second, the transport of aerosol is highly dependent on particle size, with larger particles settling out closer to the emission point than smaller ones or impacting out on surfaces such as trees or buildings. The particle size distributions of many aerosol sources are inadequately characterized or highly variable. Even if the distributions were available, few source models have developed size-segregated mechanisms for aerosol particle removal. Finally, the complexities of aerosol chemistry—the condensation and adsorption of organic vapors; the conversion of SO_2 to sulfate species—is not a part of the standard source models. Though these models may potentially provide useful insights, their results must not be relied upon as the definitive statement.

With the development of inexpensive and rapid chemical analysis techniques for dividing ambient and source particulate matter into its components has come another approach, the *receptor model*.

The receptor-oriented model relies on properties of the aerosol which are common to source and receptor and that are unique to specific source types. These properties are chemical composition, size, or other distinctive physical property.

The composition of material from one source type is different from that from another. Auto exhaust particles contain significant amounts of lead and bromine not found in aerosol particles from other sources. Similarly, the shapes and birefringences of certain minerals can identify which aggregate storage pile they came from if the piles contain different minerals.

Receptor models presently in use can be classified into one of three categories: chemical mass balance, multivariate, and microscopic. Each classification will be treated individually, though it will become apparent that they are closely related.

The starting point for the receptor model is the source model. Though the source model may not deliver accurate results under many conditions, its limitations are primarily due to its inability to include every environmentally relevant variable and inadequate measurements for the variables it does include. The mathematical formulations, however, are representative of the way in which particulate matter travels from source to receptor.

In general, the source model states that the contribution of source j to a receptor S_j is the product of an emissions rate E_j and a dispersion factor, D_j:

$$S_j = D_j E_j. \tag{10.1}$$

Various forms for D_j have been proposed (Pasquill, 1974; Seinfeld, 1975; Benarie, 1976), some including provisions for chemical reactions, removal, and specialized topography. None are completely adequate to describe the complicated, random nature of particulate matter travel in the atmosphere. Similarly, E_j is difficult to quantify on an absolute basis for many sources. The advantage of the receptor model is that exact knowledge of D_j and E_j is unnecessary.

In the case of the receptor orientation, a simple mass balance is applied. If a number of sources p exists and there is no interaction between their plumes to cause mass removal, the total mass M of suspended particles measured at the receptor will be a linear sum of the contributions of the individual sources:

$$M = \sum_{j=1}^{p} S_j. \tag{10.2}$$

Similarly, the mass concentration of particle property i, ρ_i, will be

$$\rho_i = \sum_{j=1}^{p} y_{ij} S_j, \tag{10.3}$$

where y_{ij} is the mass fraction of source contribution j possessing property i *at the receptor*. Equations (10.2) and (10.3) are the simplest expressions for the receptor-oriented model.

10.2.2 Source-Oriented Deterministic Models. The discussion of specific air quality models will become clearer by observing that all of the deterministic schemes have certain elements in common. The completeness or detail of any one of these elements varies greatly from model to model, but a diagrammatic representation of the basic structure will clarify the relationships between the various techniques to be presented. Figure 10.4 shows that four streams of input information enter the preprocessing module, namely, source emissions data, background and initial pollution levels, meteorological data, and chemical or deposition rate parameters. The background and initial pollution level may become unimportant if the simulation is run for a sufficiently long period of time and a sufficiently large portion of space. The preprocessing module takes this information and renders it useful for the main computational part of the air quality model. It transmits to the main program the various channels of input data shown on the diagram. Depending on the complexity of the logic, one or more of these channels of data transmission may be unnecessary. Contaminant concentrations constitute the output of any of the models. This again appears in varying degrees of detail, ranging from the single concentration averaged over a long time period for an entire region to hourly maps of concentrations in three dimensions over the region.

Deterministic models make every attempt to use mechanistic principles in determining their structure as outlined in the previous subsection. It must be realized, however, that most deterministic models contain varying degrees of empiricism. Few if any models, for example, utilize turbulent diffusion formulations that are based on first principles, but rather employ measured values of dispersion parameters. The same is true in regard to the atmospheric chemistry formulations. One of the cautions frequently expressed is that if too many disposable parameters exist in deterministic models, they indeed become highly empirical in structure, for all they accomplish is an elegant version of curve fitting.

The reviews by Johnson (1972), Seinfeld (1975), and recently AMS (1981) give helpful guidelines in the classification of models by space and time scale. Perhaps the most fundamental method of classifying models is to categorize them by methodology. Examples of specific methods will be discussed in detail in the next section. Great emphasis on historical data through empirical formulas is found in the methodology of aggregated rollback models (Barth, 1971; NAS, 1974). The rollback models embody the principle that reductions in emissions are reflected by improvements in air quality following a straight line, a curved line, or a complex surface that expresses some proportionality relationship. These models work best for cases where the geographical and temporal distribution of the emitters does not change. The straight-line versions can only apply to the pollutants that do not undergo chemical transformations in the atmosphere.

Dispersion models may take the form of a simple box model (Gifford and Hanna, 1973) or dispersion following a Gaussian formula. Dispersion models using superposition of Gaussian plumes improve the rollback approach for nonreacting species by accounting for geographical distribution of emission sources. Emissions are generally broken down into an array of area elements

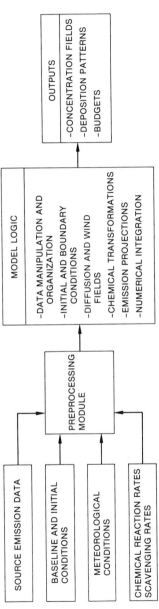

Fig. 10.4. Schematic diagram of input, computational–processing, and output components of air quality models.

or point sources each characterized by an output intensity. The plume from each source contributes in an additive manner to the pollution at any downwind field point. Summation of the contributions over all sources constitutes the superposition aspect of the approach. Computationally this is convenient because it allows sequential consideration of each source element, but this same feature renders the method inapplicable to multireacting systems undergoing chemical transformations. The Gaussian puff approach removes the limitation of steady state assumptions from the calculation and treats discrete emissions as puffs in a quasistationary state, spreading in each time step according to Gaussian formulas.

Species conservation-of-mass models utilize finite-difference techniques combining the effects of diffusion, advection, and sometimes chemistry (Karplus *et al.*, 1958; Reynolds *et al.*, 1973, 1974; Henry *et al.*, 1981). Currently these are the most elaborate simulations applied to air quality analysis. Both time and space are subdivided into cells in these models so that the cumulative effects of emissions, transport, and, when chemistry is included, reactions are simultaneously accounted for. The potential danger in using this class of model is overwhelming the available data base with mathematical detail. This means that the user can be deluded into a high level of confidence by producing large volumes of data output when in reality only very sparse data bases are available for verifying the model. The advantage of finite difference models is potentially greater fidelity where greater detail exists in the input data base and in the validation testing.

Emissions Description.   The quantitative expression of the introduction of primary pollutants into the atmosphere is basic to any source-oriented model. The description of the emissions is generally stated as a geographical, temporal, and chemical distribution that generally requires a rather massive array of numbers. Some simple models need only aggregated numbers found in usual tabulations of emission inventories provided by local authorities. The geographical distribution of pollution to which ambient air quality standards are directed requires disaggregation of emissions. This spatial distribution of emissions is frequently given on a grid or on a traffic flow network system. Similarly, temporal variation of emissions is essential to predict the peak averages that are often mentioned in the standards. Geographical specification requires considerable detail, but can often be achieved using locally available industrial data or using EPA's National Emissions Data System (NEDS).

For purposes of characterization, emissions are generally broken down broadly in terms of stationary sources and mobile or transportation sources. Stationary sources are further broken down by point or area emitters. Typical point sources must include petroleum refineries and electric power plant stacks. Commercial solvent emissions or gasoline marketing emissions may generally be represented by area sources. Recently EPA has adopted a third category of sources called indirect sources. A classification such as this takes into account hybrid sources such as sports arenas or a shopping center where a location is fixed geographically, but the traffic that is generated by the facility or attracted to the facility constitutes the emission source that is combined with the emissions of the facility itself.

Gathering emissions data and putting it in condition for use in air quality models is often one of the most tedious and time consuming parts of their use. For this reason, a preprocessing model is identified as a separate automatic operation in the procedure that is outlined in Fig. 10.4.

Atmospheric Transport Formulation.   The movement of pollution from one place to another and the dilution by atmospheric mixing are both derived from the meteorological conditions of the airshed in question. The air flow patterns are in turn derived from the interaction of the large-scale flow with the topographical detail of the region with regard to altitude variation, roughness of surface, and ground heating characteristics. That part of the air flow–surface interaction mechanisms that influences the pollution level must be taken into account in a model. Rather than computing the local weather as part of the prediction, most models use meteorological measurements to construct atmospheric flow fields that represent on a local scale the driving factor of the transport mechanisms.

With the wind velocity field and the atmospheric dispersion mechanism given, the basic equation is the conservation of mass for an individual species. The equation can be expressed in terms of a mass conservation expression for a given species, as for example, Eq. (3.11).

Strictly speaking, the Gaussian plume computations represent solutions to the species continuity equation for inert particles acting as large gas molecules without growth or coagulation. Even for large "gas" particles, the Gaussian equations are closed-form solutions of the turbulent version of the species mass conservation equation subject to certain simplifying assumptions. The assumptions are reflection of species off the ground (that is, zero flux at the ground), constant value of vertical diffusion coefficient, and large distance from the source compared with the lateral dimensions of the plumes. This solution to the species mass conservation equation is obtained under the assumption that chemical transformation and sink terms are all zero. In some cases an exponential decay factor is applied for reactions which obey first-order kinetics.[†] A typical solution (with the time decay factor) has the form

$$\bar{\rho}_j(x, y, z) = \frac{E_j/\rho_g}{2\pi\sigma_y\sigma_z\bar{u}} \left[\exp\left(-\frac{x}{\bar{u}\tau} - \frac{1}{2}\frac{y}{\sigma_y} - \frac{1}{2}\frac{z - H}{\sigma_z}\right)^2 \right.$$
$$\left. + \exp\left(-\frac{1}{2}\frac{z + H}{\sigma_z}\right)^2 \right], \quad (10.4)$$

[†] An idealized example is the oxidation of SO_2 to form particulate sulfate in the atmosphere. The species j in Eq. (10.4) in this case would be assigned as SO_2.

where $\bar{\rho}_j$ is the time-averaged mass concentration of species j at point x, y, z; E_j the emission rate of species j; ρ_g the air density; H the source height; τ the chemical decay time for first-order reaction; $\sigma_l = (2D_{tl}x/\bar{u})^{1/2}$, D_{tl} is the eddy diffusivity in direction $l = y$ or z; and $\bar{u}$ is the average wind speed taken in direction x. The three coordinate directions x, y, and z are taken to be downwind, crosswind, and vertical. The origin is fixed to the ground at the location of the source.

The origin of the "Gaussian" form is readily seen from the mathematical form of Eq. (10.4). The concentration distribution calculated by this formula has a bell shape of a Gaussian curve, with a maximum concentration at the plume center line.

Chemical Transformation and Deposition. The atmospheric chemical processes undergone by pollutants generally are not readily describable by first-order kinetics. Hence the simple Gaussian plume solution in Eq. (10.4) is inapplicable in most cases in which chemical transformations significantly alter concentrations on a time scale or space scale appropriate to an airshed where complex chemical interactions take place.

The general case must be solved by numerical integration using finite-difference schemes or other approaches to the solution of the mass continuity equation for the species of interest. This requires that a series of partial differential equations for species conversation be solved for each species in the reactive mixture. In reality, however, the number of partial differential equations that must be solved can be reduced by imposing stationary state assumptions. That is, some species are so reactive that their rate of production nearly equals their rate of depletion and these rates may be effectively equated. This being the case, algebraic expressions are employed to relate the stationary state species concentration to all of those reactive compounds and radicals which are responsible for its production and removal.

Generic particle forming mechanisms were summarized in Section 7.4.7. Specific discussions of complex chemical mechanisms in the gas phase are beyond the scope of this investigation. However, examples of working mechanisms for photochemical systems that have been used in models can be found in recent reports (Atkinson *et al.* 1982). Heterogeneous reaction schemes generally have not been used until recently (see, e.g., Henry *et al.*, 1981; Lurmann *et al.*, 1983).

In some cases, calculations require accounting for the surface removal of suspended material by dry deposition processes as well as cloud scavenging. These processes introduce complexities which are incorporated in a highly simplistic, empirical manner (Mueller and Hidy, 1983; Slinn, 1983).

Aggregated Rollback and Box Models. Both linear rollback and modified rollback were used by Barth (1971) to examine motor vehicle emissions goals for standards governing carbon monoxide, hydrocarbons, and oxides of nitrogen in the United States. The *aggregated rollback principle* was suggested and applied to these primary pollutants. The formula employed for linear rollback is

$$FR = [(GF)(PAQ) - DAQ]/[(GF)(PAQ) - B], \qquad (10.5)$$

where FR is the fractional reduction required, GF the growth factor, PAQ the present air quality, DAQ the desired air quality, and B the background concentration.

Linear rollback for emissions of carbon monoxide, hydrocarbons, and oxides of nitrogen emissions involves direct application of Eq. (10.5). The concept also has been applied to changes in precursors involved in secondary pollutants. Best-known of these are the photochemical processes involving hydrocarbons and nitrogen oxides to form ozone and particulate matter.

A form of the aggregated rollback formula has been used in California for estimating reduction in SO_2 emissions for reduction in particulate sulfate. Recognizing the strong relation between photochemical oxidation processes and sulfate formation, Holmes (1975) modified the linear rollback formula [Eq. (10.5)] to account for increased summer SO_2 oxidation resulting from increased photochemical activity. The lack of evidence supporting the proportionality between SO_2 emissions and sulfate levels as discussed in Section 7.3 has led to considerable controversy in the use of the Holmes formula, or other sulfate SO_2 emission rollback models.

The box model is one of the simplest forms of conceptual solutions of the equations for conservation of mass that appears in air quality simulation techniques. Its formulation assumes that the air bounded by the ground and the mixing height is uniform. The model further considers that the source intensity of pollution emanating from the ground is constant and that the wind speed is constant. The consequences of these assumptions result in a simple formula that states that the ambient concentration of some pollutant is directly proportional to the source emission rates and inversely proportional to the product of wind speed and mixing height. The constant of proportionality is determined by mixing parameters or is derived empirically.

The box model approach is embedded in EPA's empirical kinetics model analysis (EKMA) method for oxidant control estimations (U.S. EPA, 1977). The calculations for simulation of peak ozone concentrations expected from a range in hydrocarbon and nitrogen oxide concentration assumes uniform mixing over an air basin, but takes into account a dilution factor resulting from breakup of inversions at midday. Recently, a photochemically based model analogous to the EKMA method has been used to construct a technique for calculating sulfate and nitric acid concentrations as a function of early morning NO_x and HC concentrations. As an example, Lloyd et al. (1981) have reported calculations for reactive hydrocarbon mixtures measured in St. Louis using a scheme of reactions including those in Table 7.24. Using the photochemical mechanism of Atkinson et al. (1982), containing a system of more than 60 coupled reactions, an "EKMA diagram" for maximum ozone concentration as a function of NO_x and nonmethane HC (NMHC) can be derived which is shown in Fig. 10.5. Inspection of results from this calculation indicates that maximum ozone levels theoretically depend on NO_x until NO_x concentrations are below about 0.025 ppm. Likewise, below about 0.025 ppm NO_x maximum ozone levels are largely independent of NMHC concentrations above about 0.4 ppm.

The calculated maximum sulfate concentration due to photochemical reactions, with initial levels of 20 ppb are shown in Fig. 10.6. Here again, the maximum sulfate predicted during the day depends on the initial NO_x and NMHC concentrations. Interestingly, maximum sulfate levels can be realized at low concentrations of *both* NO_x and NMHC if the NMHC/NO_x ratio is about 7. The sulfate production in a photochemical system is dominated by ·OH radical production, which in turn depends on the relative mix of NO_x and NMHC.

Finally, the maximum nitric acid concentrations during a photochemically reactive day can be estimated using the same reaction scheme of Atkinson *et*

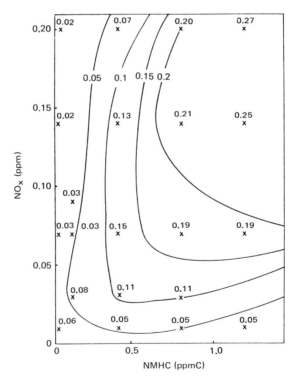

Fig. 10.5.   Calculated daily maximum ozone concentrations as a function of initial (morning) NO_x and NMHC concentrations for the conditions representative of data taken in the St. Louis area. Ozone concentration isopleths are in ppm. [From Lloyd *et al.* (1981).]

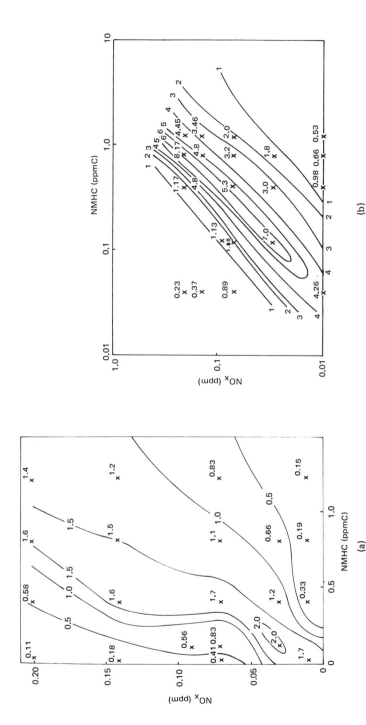

Fig. 10.6. Predicted daily peak SO_4^{2-} values for various initial NMHC and NO_x concentrations for St. Louis conditions and an initial SO_2 concentration of 20 ppb (a). The corresponding daily oxidation rates in %/hr are indicated in (b). (SO_4^{2-} isopleths in ppb.) (From Lloyd *et al.* (1981).

al. (1982). The nitric acid production for a day equivalent to sulfate and ozone production is shown in Fig. 10.7. Like the other reaction products, the maximum nitric acid curves depend on the relative initial NO_x and NMHC concentrations. When maximum levels of sulfate are reached at NMHC/NO_x of 7, one also can expect high concentrations of nitric acid according to this model.

The diagrams in Fig. 10.5–10.7 can be used to estimate the consequences of aggregated reduction in both NO_x and NMHC if initial conditions for SO_2 and other species are fixed at some baseline conditions. The diagrams demonstrate that the rollback of either of the copollutants NO_x and HC may exacerbate a sulfate problem if conditions exist in the regime where NMHC/$NO_x \approx 7$. However, there also are conditions where NMHC reduction causes reduced ozone production, which also should be accompanied by significant reduction in both sulfate and nitric acid without changing ambient SO_2 concentrations.

The EKMA-based calculations have been tested against aerosol data de-

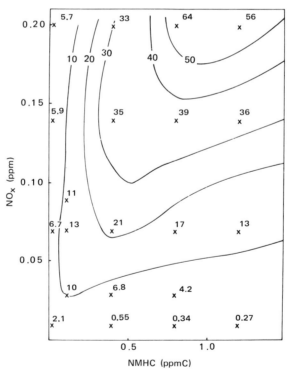

Fig. 10.7. Predicted peak HNO_3 values for various initial NMHC and NO_x concentrations for conditions representative of St. Louis. (HNO_3 isopleths in ppb.) [From Lloyd *et al.* (1981).]

rived from laboratory experiments in smog chambers, but they have not been applied for regulatory purposes in attempting to control chemical components of atmospheric aerosols other than ozone.

Gaussian Solutions. Some of the pioneering work on Gaussian solutions, represented by Eq. (10.4) or its equivalent, was introduced by Sutton (1953), who used the method of sources for solving the species mass conservation equation. The basic "puff" formula was solved by assuming an instantaneous point source. The restrictions of the mathematical solution include no wind shear, no loss of pollutant material at the ground, and steady flow through a field of constant diffusivity in each direction. Cases for instantaneous infinite line and plane sources were also solved. The Gaussian plume formula was obtained for a continuous point source by integrating various puffs. Both the idealized point source and line source formulas are far-field approximations. The point source has the restriction that the downwind distance from the source must be much larger than the crosswind distances. The line source solution has an additional restriction involving diffusion times being small compared with advection times as well as downwind distance being large compared with any crosswind distances characteristic of the plume. These solutions were first obtained by Roberts (1923).

A useful summary of working equations for Gaussian models and values of the dispersion parameters can be found in Turner's (1967) Workbook. These charts and formulas are useful in actual calculations based on Gaussian diffusion equations. The Workbook provides sufficient detail for estimating typical design values. In the text, Turner discussed a range of special topics on plume modeling including fumigation, trapping, dosage, variable averaging times, area sources, topographic effects, and multiple sources.

Martin and Tikvart (1968) used the Gaussian plume equation for a continuous point source as a basis of a multiple-source urban diffusion model. Their predictions were for seasonal and annual averages of air quality. The original calculations using their model were programmed to accommodate 40 different receptor locations. A 16-point wind direction specification, a 6-category wind speed input, and a selection of 105 stability classes were employed in the inputs in forms of frequency distributions. The Air Quality Display Model (AQDM) is based on the Martin–Tikvart approach and is described in a manual (TRW, 1970). The specific model calibration procedure was also discussed. This was one of the first Gaussian models extensively documented for ready access to users. Since then, the application of modeling technology employing Gaussian simulations has become widespread for industrial siting and evaluation of control strategy for nonreactive pollutants. Guidelines for operational models are available from U.S. EPA (1978).

A complicated plume model accounting for the details of the evaluation of the particle by size, and the interaction with photochemical processes has been reported by Eltgroth and Hobbs (1979). As discussed by the authors, this model represents the current state of the art in attempting to account for all elements of atmospheric dynamics involved in plume chemistry.

Finite-Difference Computations.   Simulation of air quality as it is affected by variable diffusivity, time-dependent emission sources, nonlinear chemical reactions, and removal processes necessitates the use of numerical integrations of the species mass conservation equation. Because of limitations of inputs as dispersion data, emission data, or chemical rate data, this approach to the modeling of air pollution may not necessarily assure higher fidelity in estimating concentration distributions. However, the technology definitely holds out the possibility for the incorporation of more of these details as they become known.

An early analog computer study of the solution of the species mass conservation equations was done by Karplus *et al.* (1958). This work consisted of a one-dimensional time-dependent diffusion equation with chemical source terms representing a multicomponent atmospheric kinetic system. An electronic analog computer was used employing one integrator at each node between space cells to handle the combined effects of mass transfer and chemical reaction. Results were obtained for a simple mechanism, but no tests of performance were made with observations.

As another technique, Egan and Mahoney (1972) proposed a two-dimensional time-dependent diffusive and advective model that neglects vertical velocity. Chemical reaction was excluded initially, but was later included for linear oxidation processes to form sulfate (Hidy, 1976). The source term is an effective volume source in the lower grid mesh of the model. Egan and Mahoney (1972) discussed the pseudodiffusion errors that arise with the large grid spacing that is appropriate to urban scale calculations. They further pointed out that this error is orders of magnitude larger than natural diffusion. An example of a puff from a volume source was presented in the paper. The scheme proposed avoids pseudodiffusion by using moments of the concentration distribution in the governing equations. The concentration profile was reconstructed by using the computed first and second moments. Egan and Mahoney applied the model to estimate ground concentrations under different meteorological conditions. Velocity profiles and vertical diffusivity profiles were introduced based on various stability conditions. Two-dimensional time-dependent solutions with variable advective velocity were obtained. Test cases were presented including the effects of wind velocity vector changes with height and the effects of this velocity field distortion on the dispersion of air pollutants. Elevated inversions and time-dependent mixing heights were also investigated. The height variation of the velocity field was shown to be important under stable conditions. Although this model did not treat air chemistry in detail, it was capable of resolving subgrid scale elements because of the moment method employed.

The Egan–Mahoney model later was one of the first used to assess sulfate behavior during regional pollution events over the eastern United States (Hidy *et al.*, 1978). The model was found to be capable of reproducing qualitatively available rural sulfate data.

Reynolds *et al.* (1973) discussed theoretical aspects of urban air pollution modeling in terms of the species mass conservation equations cast into an initial/boundary value problem. Restrictions of the eddy diffusivity formulation for turbulent mixing were reviewed. The vertical coordinate system was mapped between the ground and the inversion base by a linear transformation. The methods were detailed for interpolating discrete data on winds and vertical stability to obtain field values needed for the calculations. The kinetic mechanism for ozone using lumped parameters for hydrocarbon sources was outlined. Eulerian difference equations were integrated numerically and a method of fractional steps was described. Explicit differencing was employed for horizontal coordinates and implicit differencing for the vertical terms of the equations. In a subsequent paper in the series, Reynolds *et al.* (1974) gave a detailed description of an emissions model and the HC–NO_x inventory for ozone formation in the Los Angeles Basin.

Reynolds *et al.* (1974) describe the process of performance evaluation of the airshed photochemical model. Kinetic mechanisms were checked against smog chamber tests yielding branching factors and rate parameters for the simplified lumped-parameter scheme. A microscale model was established to correct for local effects around monitoring station sites. An evaluation procedure involved preparing data, preparing initial and boundary conditions, checking for agreement in dispersion estimation with carbon monoxide behavior, and testing computed versus observed values for reactive pollutants. Results agreed qualitatively with observations; however, no statistical performance measure was used at this stage of the work. This general approach has been extended recently by McRae *et al.* (1982). McRae and Seinfeld (1983) have reported a variety of performance tests for the photochemical oxidant system using the new model. The results indicate improvement over the earlier approach of Reynolds *et al.*

Gradient diffusion scaled by an eddy diffusivity also was assumed in the species mass conservation model of Shir and Shieh (1974). Integration was carried out in the space between the ground and the mixing height with zero fluxes assumed at each boundary. A first-order decay of sulfur dioxide of 1%/h was the only chemical reaction, and it was suggested that this reaction is only important for estimating SO_2 concentrations under low wind conditions. Finite-difference numerical solutions for SO_2 in the St. Louis area were obtained using a second-order central finite-difference scheme for horizontal terms and the Crank–Nicolson technique for the vertical diffusion terms.

Calculation of sulfur dioxide and sulfate concentration distributions over periods from 24 to 72 hr have been done for large metropolitan areas and the greater northeastern United States using a multilayered grid model based on an extension of the Egan and Mahoney (1972) technology. From extensive tests of the Egan–Mahoney formulation, numerical errors in their calculations using the moment method proved too large for this application. An improved technique was adopted which incorporated a fourth-order flux-corrected numerical integration called SHASTA. The revised model and its performance have been reported by Henry *et al.* (1981) and by Mueller and Hidy (1983). This work indicated that the quality of such calculations is highly sensitive to initial conditions, to the estimation of the wind field, and to the numerical integration scheme adopted. Vertical mixing and SO_2 deposition rates also are important in the simulation of observed sulfur oxide concentration distributions on both the urban and regional scales. This grid model accounts for SO_2 emissions injected at different heights; it calculates an air mass conservative wind field and vertical mixing conditions from extrapolation of data. The computations also include diurnal chemistry based on a reaction rate schedule derived from the model of Lloyd *et al.* (1981), and account for the dry deposition of both SO_2 and sulfate. In its present form, the model does not incorporate water cloud and precipitation processes. The calculations were composed with observations of SO_2 and SO_4^{2-} concentration in the greater northeastern United States and in the Los Angeles area. With appropriate initialization, the model was found to simulate such distributions in daily events to a mean difference of $\pm50\%$.

Regional scale budget models calculating SO_2 and SO_4^{2-} concentration as well as dry and wet sulfur deposition in terms of seasonal and annual averages have been reported for western Europe by Eliassen and Saltbones (1975) and Prahm and Christenson (1978). Later analogous models have been applied to conditions in eastern North America, as for example, Shannon (1981) and Johnson *et al.* (1978). These models generally are much less sophisticated than the one for the episode calculation; they employ a gridded computation in a Lagrangian trajectory format or in a pseudospectral computation. Several different techniques have been described recently by Bass (1980). The methods available remain in development. They are useful for research, but generally have not been accepted for regulatory applications as yet.

Evaluation of Performance.   One of the more elusive factors in the selection of air quality models is the determination of their capabilities for quantitative analysis of pollution control options. With the advent of large computers, there exists now the capability to construct highly complex schemes for simulating air pollution behavior. However, verification of the performance of models and determination of their limitations have lagged behind for lack of adequate, comprehensive collections of field measurements. In past evaluations, atmospheric measurements have been taken as an absolute standard without error. This, of course, is certainly not the case. Only recently have methods for quantifying the error bounds become available for atmospheric aerosol data (Mueller and Hidy, 1983). Particularly important to appreciate is the fact that air quality measurements generally are made at a fixed location within fifteen meters of the ground. The representativeness of large volumes of air based on samples taken at one or more stations in an air basin often has been questioned but has not been examined in any systematic way. The representativeness issue is particularly acute when examining regional scale questions. The resolution of regional models often covers an area of 80 × 80 km in a surface layer of 50 m. Thus measurements from a single station sampling in a day a few hundred cubic meters of air are required to represent conditions in as much as 3×10^{11} m^3 of air! With such a difference between the sample and the volume to be described, one must expect large uncertainties from single station measurements representing large air volumes. The use of air quality data for comparison with model estimates should account for the error or uncertainties in both the observational and calculational ways of characterizing chemistry in a large air volume. To date, this has not been attempted in any systematic way.

The current state of development in formalizing evaluation methods for verifying air models has been discussed by Fox (1981). Several different statistical measures have been used, including the mean and variation of the difference between observation and the calculation at a given location (residuals), and the means and variations in the ratio of observed to calculated values.

One of the more comprehensive series of tests of an air quality model was performed on a regional model, as reported by Mueller and Hidy (1983). Measures suggested by Fox (1981) were applied; in addition, the observed geographical sulfur dioxide and sulfate distributions for 24-hr averaged periods were compared over the greater eastern United States. Comparison also was done at stations where hourly SO_2 and three-hourly SO_4^{2-} data were available. Conventional practice often shows a scatter diagram of calculated values versus observed values. An example for the regional model generating 24-hr average conditions at ground level is shown in Fig. 10.8. Test cases for 6 days are shown. The results of these tests indicated that 90% of the pairs of observations and calculations were within a factor of 2 for these simulations. This is considered to be within the current state of the art for numerical models. The performance against shorter-term averages was much poorer for this model. Since such a test on short application has not been attempted in other cases, there is no other "standard" available for comparison.

Another measure of performance was used, which examined the capability of the regional model to reproduce the frequency distribution in occurrence of sulfate concentrations. The results of this comparison are shown in

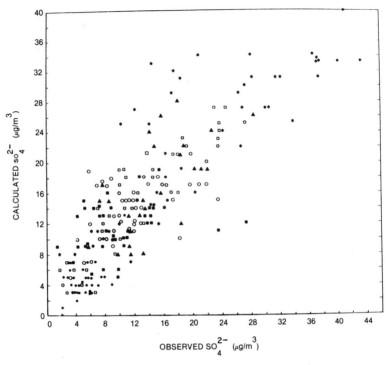

Fig. 10.8. Correlation of observed and calculated 24-hr sulfate concentrations for six days in 1977 and 1978. Observations are from the Sulfate Regional Experiment. Date (symbol): 1/23/78 (▲); 4/14/78 (●); 7/17/78 (■); 7/18/78 (○); 7/21/78 (*); 8/5/77 (□). [From Mueller and Hidy (1983).]

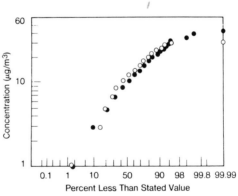

Fig. 10.9. Distribution of observed and calculated 24-hr sulfate concentrations for five days. Observations are from the Sulfate Regional Experiment. Data from 1/23/78, 4/14/78, 7/18/78, 7/21/78, and 8/5/77; 186 data points. ●, observed; ○, calculated. [From Mueller and Hidy (1983).]

Fig. 10.9. In this case, the model simulated the observed occurrence of sulfate values across the region satisfactorily when compared with measurements.

The current status of source-oriented modeling can be summarized by the following rules of thumb for the practitioner.

(a)   For specific design and planning applications, model results can be used with greatest confidence if they are "calibrated" with adjustable parameters using existing emissions, meteorology, terrain features, and air quality data in the geographical area of concern.

(b)   On the average, source-oriented models currently can estimate ambient concentrations to a factor of 2 or 3 or better for nonreactive contaminants, but differences can be much larger depending on terrain, knowledge of meteorology and requirements for inclusion of multiple sources.

(c)   Chemically reactive species such as ozone or particulate sulfate can be calculated to accuracies of 2 to 3 or better over periods of a day or more, given detailed information on NO_x and NMHC precursor emissions and initial ambient conditions.

(d)   Model performance is often viewed in terms of the fidelity in simulating *relative* changes, rather than *absolute* changes.

(e)   The application of source-based simulation techniques for screening or exploratory design requires straightforward application of established methods. However, the detailed emission limitation computations for obtaining regulatory permits of a facility may require considerable expertise and judgment, requiring professional assistance, and detailed analysis of calculations accounting for the facility emissions in the presence of other sources.

10.2.3   Receptor Models and Chemical Mass Balance.

If sufficient information about the chemical composition of particulate matter is available at a given place, the range of potential sources contributing to that chemistry can be deduced, provided knowledge of the source chemistry exists.

Chemical Mass Balance.   If a number of sources p are considered as "source types," that is, grouping sources with similar properties together, a set of equations like Eq. (10.3) can be written for all of the elemental concentrations in particles or gases in the chemical element balance. Perhaps a better name for this class of model might be *chemical mass balance*. Historically, elemental measurements were most readily available. However, recent analytical developments also allow carbon compounds and ions to be among the properties considered (Friedlander, 1973; Heisler *et al.,* 1980a; Gartrell *et al.,* 1980).

If one measures n chemical properties of both source and receptor, n equations of the form of Eq. (10.3) exist. If the number of source types contributing those properties is less than or equal to the number of equations; i.e., $p \leq n$, then the source contributions, the S_j, can be calculated. Four methods of performing this calculation have been proposed, (a) the tracer property, (b) linear programming, (c) ordinary linear least-squares fitting, and (d) effective variance least-squares fitting.

The tracer element method is the simplest. It assumes that each aerosol source type possesses a unique property which is common to no other source type; Eq. (10.3) then reduces to

$$S_j = \rho_t / y_{tj} \qquad (i = t) \qquad (10.6)$$

for each source j with its tracer t. It works well when the tracers meet the following requirements:

(a)   The mass fraction y_{tj} perceived at the receptor is well known and invariant.

(b)   The mass concentration ρ_t can be measured accurately and precisely in the ambient sample.

(c)   The concentration of property t at the receptor comes only from source type j.

These conditions cannot be met completely in practice, and by limiting the model to only one tracer property per source type, valuable information contained in the other aerosol properties is being discarded. Instead of this approach, equations like Eq. (10.3) have been applied to make use of the additional information provided by more than one unique chemical property of a source type, and even that of properties which are not so unique.

The tracer method has been applied in several cases to analyze the origins of particulate matter in cities, as well as rural locations (Miller *et al.*, 1972; Gatz, 1975; Kowalczyk *et al.*, 1978). An example of the results for samples taken in five different communities in California is listed in Table 10.2. The method is generally checked for consistency by comparing the calculated total mass concentration accounted for and the measured value. One can see from Table 10.2 that this consistency test is generally reasonably good, but the Fresno case shows relatively poor agreement for total mass concentration. Recent experience with this method has shown that one can achieve consistency in material balance calculation of ±25% or better. Further, the secondary contribution of sulfate and nitrate is unresolved by source.

The results in Table 10.2 illustrate the range in contributions of the selected natural sources and the anthropogenic sources. In the Los Angeles area (Pasadena, Pomona, and Riverside) the secondary contributors account for a major fraction of the aerosol particles sampled. However, soil dust and sea salt can exceed the concentrations of primary anthropogenic contributions. In the arid California environment, soil dust is always present as a substantial fraction of the ambient particulate matter. Not surprisingly in these samples, water also is an important contributor to the suspended particles along with carbonaceous material.

TABLE 10.2

Calculated Source Contributions[a] for Aerosol Particle Samples Taken at Different Locations in California[b]

Source	Pasadena 9/20/72	Pomona 10/24/72	Riverside 9/10/72	Fresno 9/1/72	San Jose 10/20/72
Sea salt	0.7 ± 0.06	5.7 ± 0.6	1.3 ± 0.1	0[c] ± 0.4	19.4 ± 0.5
Soil dust	19.8 ± 0.1	15.1 ± 0.5	28.5 ± 0.9	51.1 ± 2.8	29.6 ± 1.1
Auto exhaust	5.1 ± 0.15	7.2 ± 0.3	3.9 ± 0.15	2.2 ± 0.1	8.3 ± 0.33
Cement dust	1.4 ± 0.15	3.3 ± 0.6	2.3 ± 0.15	0.5 ± 0.14	4.5 ± 1.3
Fly ash	0.1 ± 0.01	0.2 ± 0.01	0.1 ± 0.01	<0.1 ± 0.01	0.1 ± 0.01
Diesel exhaust[d]	1.4	1.9	0.9	0.6	2.2
Tire dust[e]	0.5	0.7	0.4	0.2	0.8
Industrial and agricultural[d]	4.7	6.6	20.5	37.8	27.6
Aircraft[d]	1.3	1.8	7.4	0.9	4.9
SO_4^{2-}[f]	2.9 ± 0.7	19 ± 5	5.9 ± 1.5	4.2 ± 1.0	16.6 ± 3.3
NO_3^-[f]	4.9 ± 0.4	36.4 ± 2.7	12.9 ± 1.0	7.9 ± 0.6	12.3 ± 0.9
NH_4^+[f]	2.3 ± 0.1	16.3 ± 0.8	5.7 ± 0.3	3.1 ± 0.15	7.2 ± 0.72
Organics[g]	29.6	29.3	24.8	U[h]	U
Water[f]	12 ± 6	18 ± 9	U	U	U
Total mass	86.7	161.5	114.6	108.5	133.5
Measured mass	64 ± 7	180 ± 20	125 ± 14	207 ± 23	189 ± 21

[a] Values in $\mu g/m^3$. Errors associated with sea salt, soil dust, auto exhaust, cement dust, and fly ash are standard errors from the least-squares fit for the chemical element balance; errors associated with SO_4^{2-}, NO_3^-, NH_4^+, and measured total mass concentrations are analytical errors.

[b] From Gartrell *et al.* (1980); courtesy of Wiley.

[c] Values actually found to be slightly negative.

[d] Scaled to auto exhaust based on relative inputs.

[e] Assumed to be 10% of auto exhaust component.

[f] Measured values.

[g] Based on carbon balance.

[h] Unknown.

Linear programming and weighted least-squares solutions provide another method for dealing with the limitation of constraints n greater than the number of unknowns p. Henry (1977) has applied a linear programming algorithm which maximizes the sum of the source contributions subject to certain constraints. This procedure has not developed further since his work was reported.

In the ordinary weighted least-squares method, the most probable values of S_j when $n > p$ are obtained by minimizing the weighted sum of squares of the difference between the measured values of ρ_i and those calculated from Eq. (10.3) weighted by their standard deviation and the analytical uncertainty of the ρ_i measurement. This solution provides the added benefit of being able to propagate the measured uncertainty of the ρ_i through the calculations to estimate a confidence interval around the calculated S_j.

The ordinary weighted least-squares solution is incomplete, however. The ambient particle chemical properties, the ρ_i, are not the only observables on which measurements are made. The chemical properties of the source aerosol, the y_{ij}, are also measured observables, but the errors associated with those measurements are not included in the ordinary weighted least-squares fit.

Watson (1979) has applied the treatment of Britt and Luecke (1973) to the solution of a set of source–receptor relations [Eq. (10.3)]. This solution provided two benefits. First, it propagated a confidence interval around the calculated S_j which reflects the cumulative uncertainty of the input observables. The more precise the measurements of the ambient source property concentrations are, the better the estimate of the source contributions will be. The second benefit provided by this "effective variance" weighting is to give those chemical properties with larger uncertainties, or chemical properties which are not as unique to a source type, less weight in the fitting procedure than those properties having more precise measurements or a truly unique source character.

In a series of simulation studies on artificially generated, randomly perturbed data sets, Watson (1979) showed that these benefits are realized in practice. Here again the contributions by source of secondary particles are poorly characterized.

Another variation of the chemical mass balance which has been used to identify additional source types is the enrichment factors where a b subscript refers to a dominant source-type contribution, usually crustal material, with a tracer element, such as Al or Fe, unique to that source type. Substituting Eq. (10.3) and (10.6) into the expression for enrichment factors where the subscript b refers to the crustal material as the dominant source type:

$$\text{EF}_i' = \text{EF}_i/\text{EF}_0 = 1 + \left(\sum_{j \neq b}^{p} y_{ij} S_j / y_{ib} S_b \right). \tag{10.7}$$

Here, EF_0 is the enrichment factor for a soil dust aerosol cloud.

If no other sources of chemical component i exist, $\text{EF}_i' = 1$. $\text{EF}_i' > 1$ implies sources of component i other than source b. For example, King et al. (1976) noticed a high antimony enrichment at one site in Cleveland. Further investigation revealed the existence of a nearby chemical plant producing antimony compounds.

The source contributions of aerosol formed from gaseous emissions, such as sulfate, nitrate, and certain organic species, cannot be quantified directly by chemical mass balance methods. Watson (1979) proposed scaling to a source tracer to put an upper limit on the contributions of secondary aerosol sources, but this method cannot attribute those contributions to specific emitters. Alternatively, sulfate and nitrate can be scaled linearly to sources of SO_2 and NO_x by an air basin emission inventory as a crude method.

Even given these limitations, users of the chemical mass balance approach feel that the method has been developed to the point where it can be adopted as a basis in state implementation plans for the control of particulate matter pollution. It is conceptually simple, based on actual data, and given the adequate input data for source characterization, it is able to quantitatively apportion contributions of major sources to ambient air quality. The difference between the mass concentration at the receptor and the sum of the source contributions offers a consistency check; if the sum is much greater than the total mass, sources have been left out or some of those included do not contribute. This same check can be applied to chemical components not included in the fitting procedure.

Multivariate Models.   The chemical mass balance model uses the particle property of chemical composition. If source and ambient samples are taken in more than one size range, the size property can be used to separate the contributions of one source from those from another. The mass balance presently has no way to incorporate the variability of ambient concentrations and source emissions. The multivariate methods can do this, at least in principle. Linear regression (Kleinman, 1977), correlation (Moyers *et al.*, 1977), and factor analysis (Hopke *et al.*, 1976; Henry, 1977) are the forms these models have taken.

While the chemical mass balance receptor model was easily derivable from the source model and the explanation of its application has been straightforward thus far, this is not the case for multivariate receptor models.

Watson (1979) has carried through the calculations of the source–receptor model relationship for the correlation and principal component (factor) models. The mathematics has been described but is complex and time consuming to use without a computer for bookkeeping.

Alpert and Hopke (1980) have experimented with an un-normalized correlation matrix with correlations between samples as opposed to the normalized correlation between chemical species. This approach retains more of the information in the data and allows the proportionality among the chemical species to be maintained.

Multivariate models have been successful in identifying source contributions in urban areas. They are not independent of knowledge about source composition since the chemical component associations they reveal must be verified by source emissions analyses. The linear regression model can produce a typical ratio of chemical components in a source, but only under fairly restrictive conditions. The factor or principal component analysis models require source composition estimates to find the self-consistent set of S_j vectors, though it seems that these y_{ij} estimates can be refined as a result of the analysis (Henry, 1977; Alpert and Hopke, 1980).

The most important caveat to be heeded in the use of multivariate models states that although origination in the same source will cause two chemical components to correlate, the converse, that chemical components which correlate must have originated in the same source, is not true.

Microscopic Identification.   Microscopic identification and classification of individual aerosol particles into source categories can be a valuable method for the study of settled and suspended atmospheric particles.

Many different optical and chemical properties of single aerosol particles can be measured, enough to distinguish those originating in one source type from those originating in another. The microscopic analysis receptor model takes the form of the chemical mass balance equation

$$m_i = \sum_{i=1}^{P} y_{ij} S_j, \tag{10.8}$$

where m_i is the percent mass of particle i in the microscopic receptor sample, y_{ij} the fractional mass of particle type i in source type j, and S_j the percentage of total mass measured at the receptor due to source j.

The model corresponding to Eqs. (10.3) and (10.8) might be best termed an aerosol property mass balance to encompass both the chemical mass balance and microscopic receptor models.

The solutions to the set of Eqs. (10.8) are the same as those described by Eqs. (10.3) only through the tracer method (i.e., each receptor particle type is unique to one source type) has been used in the past.

The microscopic receptor model can include many more particle property measurements than those used to date in the chemical mass balance and multivariate models. This method has not yet taken advantage of the mathematical framework developed in the other two models to deal with (a) particle types that are not unique to a given source type, or (b) the ambient and source measurement uncertainties.

The data inputs required for this model are the ambient properties measurements, the m_i, and the source properties measurements, the y_{ij}.

The major limitation of microscopic receptor models is that the analytical method, the classification of particles possessing a defined set of properties, has not been separated from the source apportionment of those particles. Equation (10.8) has not been used in this application. The source identification takes place on recognition of the particle by the microscopist. The particle properties the analyst uses for this identification often are his own judgment (or his laboratory's) and are not necessarily subject to interpretation by another. Source contributions assigned to the same particle sample have varied greatly in intercomparison studies, but without the intermediate particle property classification it is impossible to ascribe the differences to the analytical portion or to the source assignment portion of the process.

The microscopist normally relies on his past knowledge of the properties of aerosol sources without the examination of local source material. For example, a coarse particle sample taken in the vicinity of a coal-fired power plant may show a 20% contribution due to fly ash and an 80% contribution due to minerals. As Brookman and Yocom (1980) point out, the 20% fly ash may not have come from the power plant during the sampling period. An examination of nearby soils could show a 20% fly ash concentration resulting from long-term deposition. Using one of the least-squares or linear programming approaches to the solution of a set of receptor model equations with local source compositions could alleviate this problem.

Though the particle property mass balance suggested by the microscopic models shows promise, it is limited by the lack of a standardized, reproducible analytical method.

Application to Light Extinction Budgets. A logical extension of the chemical mass balance model for source attribution of atmospheric particles concerns the contributions to the extinction coefficient. If only the part of the total extinction coefficient associated with scattering and absorption by particles is used, the chemical mass balance approach can be transferred directly to assignment of extinction sources through an empirical relation like Eq. (8.5b).

This approach has been attempted for simultaneous observations of light extinction and particle chemistry as a function of particle size. Macias *et al.* (1981) have reported results for the rural environment of northwestern Arizona, and Heisler *et al.* (1980b) have discussed the method for a data set taken in Denver, Colorado during the winter of 1978. The two extremes are of interest since they demonstrate this approach for conditions near the Rayleigh scattering limit and under conditions where the air is burdened with pollution.

The Denver results for hazy conditions and for all days sampled are shown in Fig. 10.10. The apportionment of b_{sp} by chemical composition of the fine particles is given in Fig. 10.10a. The major contributions by chemical constituent include ammonium sulfate, ammonium nitrate, organic material, elemental carbon, and water. Contributions of primary inorganic emissions, such as lead halide, and crustal elements are included as "other." Taking the emissions inventory for Denver and associating sources with primary particulate emissions, as well as gaseous emissions of SO_2 and NO_x, the major disaggregation of b_{sp} by source is estimated to be that shown in Fig. 10.10b. Here the city's emissions of carbon including combustion products from diesel engines, autos, residual oil, domestic heating, and wood burning contribute substantially to the light extinction with secondary particles of sulfate and nitrate. On the balance, the urban (Denver) and rural (Arizona) extremes appear to be qualitatively similar in makeup of contributions to light extinction even though the extinction level is very different.

Although the Denver apportionment is considered to be only semiquantitative, the breakdown does show the major importance of certain sources to impairment of visibility, which is not reflected in the conventional emissions inventory of particulate matter as sampled by a total filter sampler. The

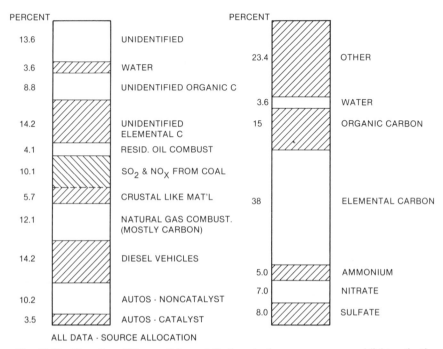

Fig. 10.10. Allocation of fine-particle contributions to the average measured light extinction coefficient b_{ext} of 2.13×10^{-4} m^{-1} for Denver's haze in November–December 1978. Allocation on the left (b) is by source category (based on composition in Table 7.20, for example). Allocation on the right (a) is based on major chemical components of sampled aerosol particles less than 3 µm. [Data from Heisler *et al.* (1980b).]

Denver emissions inventory indicates that blowing dust from road sanding in winter or construction dominates the city's suspended particle concentration. These fugitive dusts are coarse particles and contribute little per unit mass concentration to light extinction. Secondary production of particles from oxidation of SO$_2$ and NO$_x$ is not included in such inventories.

Suspended material collected by a high-volume sampler is heavily weighted by particles larger than 1 µm diameter in standard air monitoring practice. Thus the mass concentration of fine particles and its contributors by source are often largely masked by current air monitoring methods. Available observations of suspended particles and emission inventory are difficult to interpret for visibility effects without application of additional knowledge or supplemental measurements of the fine particles and their sources.

10.3 PARTICLE CONTROL TECHNOLOGY

When the emission limitations for a planned operation are estimated to be in conformance with air quality regulations, a series of process design interactions may take place to ensure compliance. To reduce emissions of existing and new facilities considerable innovation may be required in process and system engineering. In general, gas cleaning and atmospheric emissions can be minimized by reduction to the greatest degree possible of the amount of entering exhaust gas streams. Process, operational, and system control will concentrate contaminants in the smallest possible air volumes. As an engineering principle, this is important since the cost of control equipment is based mainly on the volume of gas that has to be handled, and not the amount of material to be removed. Also most cleaning equipment is more efficient at high concentration all else being equal. Emission control with modern, highly efficient industrial cleaning devices can cost over $2 per cubic meter per second of installed capacity, and entails high, continuous operational and maintenance expense. Thus innovation in the process or system control steps represents major opportunities for minimizing air cleaning control costs. Such practice also can be vital to improving employee safety with regard to exposure inside the plant. Emission reductions or their elimination can be achieved in a variety of ways, including substituting products, changing processes, revising plant layout, or revising internal ventilation systems.

The control of emissions inside a plant is readily achieved in a closed system such as a large boiler or a continuous chemical process. However, many manufacturing operations such as metal plating, metal finishing, or batch process operations are not closed systems. The emissions from open vats or machines can be difficult to control, the usual procedure being through ventilation hoods to collect the relevant material into flow ducts. Occupational hazards also can be minimized in the workplace by concentrating open process emissions employing exhaust hoods locally. Control of fumes and dusts from manufacturing then can be readily achieved by gas cleaning from the hood effluent prior to release outdoors.

Exhaust hood technology for industrial safety applications is described by Alden (1959) and the ACGIH (1974). Typical hood designs are shown in Fig. 10.11. The simplest configuration, a plain open canopy hood is drawn in Fig. 10.11a; a partially walled hooded enclosure is shown in Fig. 10.11b. Fully enclosed hoods also have been built for certain purposes where safety or product control is critical. The key to hood design is the creation of a controlled air velocity that will prevent the escape of contaminants from an exhaust-ventilated enclosure to the workroom air, or to draw inward and capture contaminants at a distance from the hood face. The air flow that will overcome dispersion with a safety factor is called the *control velocity*. In practice, it is adjusted to the least air flow rate that gives satisfactory conditions to maintain gas volumes at a minimum and contaminant burdens at a maximum. Optimum control velocities are determined by several factors which include the nature of the process and the air motion in the neighboring area. For a partially enclosed hood (Fig. 10.11b), the control velocity is the

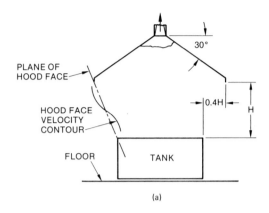

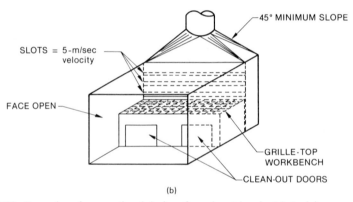

Fig. 10.11. Examples of conventional designs for exhaust hoods: (a) A plain canopy booth, (b) a partially enclosed canopy for a metallizing booth. [From Alden (1959).]

average velocity through the open face. The hood face velocity is normally nonuniform in profile (Fig. 10.11a), but needs to be stable to turbulence generated from movement or activity into and out of enclosures, traffic near the openings, and air currents from other ventilations. The exhaust hood effluent can be treated by filtration or scrubbing as a part of the control measures prior to release to the outdoors.

The control of emissions from a facility can be engineered by examining stepwise several different streams to be treated. The sequence of selection of control equipment is shown in Fig. 10.12. Normally the control requirements are established by emission standards. Given knowledge of process or facility, the collection efficiency is established, which in turn provides the engineer with a measure for selecting alternatives. Available practical aerosol

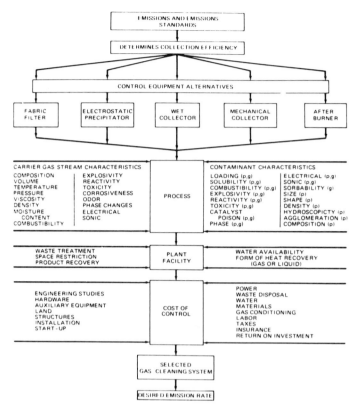

Fig. 10.12. Process for selection of gas cleaning equipment: p, particulate matter; g, gaseous matter. [From Vandergrift *et al.* (1971).]

emission controls are quite limited. They include (a) afterburner combustors, (b) mechanical collectors, (c) wet scrubber or collectors, (d) electrostatic precipitators, and (e) fabric filters. These may be used separately or in combination to achieve the stated control goals. Next the character of the process effluent stream must be taken into account, along with other design constraints of the overall facility. Finally, the cost of technically sound control options must be estimated and compared in the selection process. The effluent stream and contaminant characteristics of concern are listed with the potential facility limitations and cost factors in Fig. 10.12.

The choices for particulate collectors used by different segments of industry are listed in Table 10.3. From inspection of this table, the mechanical collector, including gravity or momentum settlers; cyclones; or electrostatic precipitators generally are preferred for many applications based on an inventory in the mid-1960s. However, filtration and wet scrubbers have become increasingly popular as alternatives in the late 1970s.

TABLE 10.3

Use of Particulate Collectors by Industry[a,b]

Industrial classification	Process	Electro-static pre-cipitator	Mechan-ical collector	Fabric filter	Wet scrubber	Other[c]
Utilities and	Coal	⊕	⊕	—	—	—
industrial	Oil	⊕	⊕	—	—	—
power plants	Natural gas			Not required		
	Lignite	⊕	⊕	—	—	—
	Wood and bark	+	⊕	—	+	—
	Bagasse	—	⊕	—	—	—
	Fluid coke	⊕	+	—	—	+
Pulp and	Kraft	⊕	—	—	⊕	—
paper	Soda	⊕	—	—	⊕	—
	Lime kiln	—	—	—	⊕	—
	Chemical	—	—	—	⊕	—
	Dissolver tank vents	—	⊕	—	—	+
Rock products	Cement	⊕	⊕	⊕	+	—
	Phosphate	⊕	⊕	⊕	⊕	—
	Gypsum	⊕	⊕	⊕	⊕	—
	Alumina	⊕	⊕	⊕	+	—
	Lime	⊕	⊕	+	—	—
	Bauxite	⊕	⊕	—	—	—
	Magnesium oxide	+	+	—	—	—
Steel	Blast furnace	⊕	—	—	⊕	+
	Open hearth	⊕	—	—	+	+
	Basic oxygen furnace	⊕	—	—	⊕	—
	Electric furnace	+	—	⊕	⊕	—
	Sintering	⊕	⊕	—	—	—
	Coke ovens	⊕	—	—	—	+
	Ore roasters	⊕	⊕	—	+	—
	Cupola	+	—	+	⊕	—
	Pyrites roaster	⊕	⊕	—	⊕	—
	Taconite	+	⊕	—	—	—
	Hot scarfing	⊕	—	—	+	—
Mining and	Zinc roaster	⊕	⊕	—	—	—
metallurgical	Zinc smelter	⊕	—	—	—	—
	Copper roaster	⊕	⊕	—	—	—
	Copper reverberatory	⊕	—	—	—	—
	Copper converter	⊕	—	—	—	—
	Lead furnace	—	—	⊕	⊕	—
	Aluminum	⊕	—	—	⊕	+
	Elemental phosphorus	⊕	—	—	—	—
	Ilmenite	⊕	⊕	—	—	—
	Titanium dioxide	+	—	⊕	—	—

TABLE 10.3 (*continued*)

Industrial classification	Process	Electro-static pre-cipitator	Mechan-ical collector	Fabric filter	Wet scrubber	Other[c]
	Molybdenum	+	—	—	—	—
	Sulfuric acid	⊕	—	—	⊕	⊕
	Phosphoric acid	—	—	—	⊕	⊕
	Nitric acid	—	—	—	⊕	⊕
	Ore beneficiation	+	+	+	+	+
Miscellaneous	Refinery catalyst	⊕	⊕	—	—	—
	Coal drying	—	⊕	—	—	—
	Coal mill vents	—	+	⊕	—	—
	Municipal incinerators	+	⊕	—	⊕	+
	Carbon black	+	+	+	—	—
	Apartment incinerators	—	—	—	⊕	—
	Spray drying	—	⊕	⊕	+	—
	Machining operation	—	⊕	⊕	+	+
	Hot coating	—	—	—	⊕	⊕
	Precious metal	⊕	—	⊕	—	—
	Feed and flour milling	—	⊕	⊕	—	—
	Lumber mills	—	⊕	—	—	—
	Wood working	—	⊕	⊕	—	—

[a] From Wilson (1967).
[b] ⊕ = most common, + = also used.
[c] Includes packed towers, mist pads, slag filters, centrifugal exhausters, flame incineration, settling chambers.

10.3.1 Control Efficiency and Related Parameters.

The efficiency of a collector device is an important operating parameter, and is defined as the ratio of the quantity of emissions prevented from entering the ambient air by the control device to the quantity of emissions in the uncontrolled effluent.

Control equipment is designed on the basis of relationships derived from underlying theoretical principles, and from empirical data accumulated from experience in previous operations. The theoretical relationships are useful for guidance, but are often too idealized to provide for accurate performance determinations with given inlet conditions. Therefore, engineering practice normally requires performance testing to ensure objectives are met in design and operation of installed throughput capacity.

The efficiency of gas cleaners is expressed in several ways, including the *control efficiency,* the *penetration,* and *the decontamination factor.* Perhaps the most common means for expressing the efficiency of performance is in terms of the control efficiency. It is defined as the ratio of the quantity of pollutant prevented from entering the atmosphere by the control device to the quantity that would have been emitted (inlet quantity to the gas cleaning device) to the atmosphere had there been no control device. Algebraically, the control efficiency is given by the following equation. Basically it is a direct analogy to theoretically defined collection efficiencies for impaction devices (Section 5.7) or filter media (Section 5.6.3).

$$\mathscr{E} = \text{Collected/Inlet} \equiv C/In = (\text{Inlet} - \text{Outlet})/\text{Inlet} = (In - \theta)/In. \quad (10.9)$$

Penetration is defined as the ratio of the amount of pollutant escaping the gas cleaning device to the amount entering.

$$P_e = \theta/In. \quad (10.10)$$

Hence penetration focuses attention upon the quality of the emission stream, but in reality it is only another way of looking at control efficiency. Since

$$\mathscr{E} = 1 - \theta/In,$$

then

$$\mathscr{E} = 1 - P_e. \quad (10.11)$$

Alternatively, efficiency can be expressed as the decontamination factor DF, which is defined as the ratio of the inlet amount to the outlet amount.

$$DF = In/\theta = 1/(1 - \mathscr{E}) = 1/P_e. \quad (10.12)$$

The logarithm to the base 10 of the decontamination factor is the *decontamination index.*

Where several collectors are arranged in series as in Fig. 10.13, their overall efficiency is given, for k stages, by

$$\mathscr{E}_0 = 1 - (1 - \mathscr{E}_1)(1 - \mathscr{E}_2)(1 - \mathscr{E}_3) \cdots (1 - \mathscr{E}_k) \quad (10.13)$$

where $\mathscr{E}_1$ is the efficiency of the first stage, $\mathscr{E}_2$ of the second, etc. The overall efficiency of the system is then given by the quantity one minus the product of the penetrations for each stage (Spaite and Burckle, 1977).

There are several arrangements of collectors in series used in practice. Where the primary collector acts only as a concentrator to split the gas into two outlet streams, one dirtier and one cleaner than the influent stream, and no dust is removed from the primary collector, removal being effected only from the dirtier gas stream by the secondary collector, $\mathscr{E}_1$ is considered its "efficiency as a concentrator," and the computation becomes

$$\mathscr{E}_0 = \mathscr{E}_1 \mathscr{E}_2.$$

$$\mathscr{E}_0 = \frac{I_1 - \theta_k}{I_1} = 1 - \frac{\theta_k}{I_1}$$

but since $\theta_i = I_i(1 - \mathscr{E}_i)$ and $I_i = \theta_{i-1}$
then $\theta_i = \theta_{i-1}(1 - \mathscr{E}_i)$
and $\mathscr{E}_0 = 1 - (1 - \mathscr{E}_1)(1 - \mathscr{E}_2)(1 - \mathscr{E}_3) \cdots (1 - \mathscr{E}_i) \cdots (1 - \mathscr{E}_k)$ follows

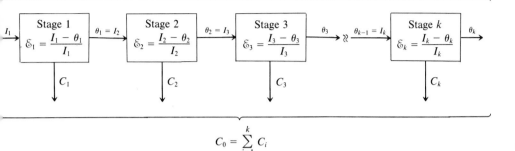

$$C_0 = \sum_{i=1}^{k} C_i$$

Fig. 10.13. Material balance diagram for series collectors. [From Spaite and Burckle (1977).]

In this case $\mathscr{E}_1$ is the ratio of the amount of particulate matter in the concentrated stream to that in the inlet.

For the collector with a purged dust hopper, characterized as a hopper which is placed under suction by a bleed line going to a secondary collector, the equation is

$$\mathscr{E}_0 = \mathscr{E}_1 + Z\mathscr{E}_1(\mathscr{E}_2 - 1),$$

where Z is the fraction of dust collected by the primary collector which leaves the hopper in the purge to the secondary collector.

Another case is that for the skimmed collector, which has a normal dust hopper but in which the cleaned gas stream is split into two streams, one dirtier and one cleaner than the average, for which the equation is

$$\mathscr{E}_0 = \mathscr{E}_1 + X\mathscr{E}_2(1 - \mathscr{E}_1),$$

where X is the percentage of dust leaving the primary collector which goes to the secondary collector.

Since all collection mechanisms depend on particle size, the efficiency of collection is affected in turn by the size of the particles. Spaite and Burckle (1977) indicate that this aspect of particulate control introduces the concept of *fractional efficiency,* the efficiency over a given particle size range with which the particles within that range are collected.

The fractional efficiency is defined as follows: for the jth particle size increment, the fractional efficiency $\mathscr{E}_j$ is the ratio of the amount of particulate matter in the jth increment collected, M_j, to the total amount in the jth increment entering the control device, In_j:

$$\mathscr{E}_j = M_j/In_j. \qquad (10.14)$$

As explained before, it will normally be necessary to calculate the collection efficiency from the inlet and outlet quantities of particulate:

$$\mathcal{E}_j = 1 - \theta_j/In_j,$$

where θ_j is the amount of particulate matter in the jth size interval in the outlet stream of the control device. The overall efficiency for k intervals is

$$\mathcal{E}_0 = (1/In)(\mathcal{E}_1 In_1 + \mathcal{E}_2 In_2 + \cdots + \mathcal{E}_j In_j + \cdots + \mathcal{E} In_k).$$

By appropriate substitution and rearrangements of terms this equation for $\mathcal{E}_0$ can be expressed in terms of θ and M as follows:

$$\mathcal{E}_0 = \left(\sum \frac{\mathcal{E}_j \theta_j}{(1 - \mathcal{E}_j)_k}\right)\bigg/\left(1 + \sum \frac{\mathcal{E}_j \theta_j}{(1 - \mathcal{E}_j)\theta_k}\right), \qquad (10.15)$$

where θ_k is the amount from the last stage.

In another form,

$$\mathcal{E}_0 = \left(\sum \frac{M_j}{\mathcal{E}_j M}\right)^{-1},$$

where M is the total amount collected in all stages.

Just as control efficiency can be expressed for particle size ranges, fractional efficiency can be expressed in terms of penetration. Penetration of a given particle size or size interval through a control device is defined as

$$P_{ej} = 1 - \mathcal{E}_j. \qquad (10.16)$$

The concept of fractional efficiency for the evaluation of particulate collectors is most useful in considering their selection and performance in the fine-particulate range, about 0.01–5 μm in diameter.

10.4   OPERATING PRINCIPLES OF CONTROL DEVICES

The devices that have been manufactured for particulate control represent a wide variety of approaches, with considerable ingenuity in design. However, as in sampling technology (Chapter 5), the actual operating principles are limited to (a) the application of mechanical devices for gravitational or inertial separation, (b) the use of scavenging by liquid sprays, (c) the use of electrical forces for deposition, and (d) the filtration of material through fibrous mats or packed beds. Only in some special circumstances can particles be removed by their combustion in an afterburner. This last type of device makes direct use of the combustion principles discussed in Section 6.4, and will not be treated further.

As seen above, the choice of control device depends on cost as well as efficiency and process factors. The important design parameters include pressure drop through the device, which influences fluid flow costs, and the collection efficiency as a function of space requirements or surface area and power requirements. In the following paragraphs, the main methods are described.

10.4.1 Gravitational Settling. Perhaps the simplest method for removing particles from a gas stream is via gravitational settling. A typical design involves a long horizontal box fitted with dust collection slots leading to hoppers below (e.g., Fig. 10.14). Particles enter the hoppers from the plenum of the settler by the action of gravity. Fallout can be enhanced by adding curvature to the air flow through a series of baffles, at the expanse of additional pressure drop.

Given the nomenclature shown in Fig. 10.14, the control efficiency, given as the weight percentage of particles of settling velocity q_G is, from Eq. (2.2),

$$\mathscr{E} = 100 \, q_G LH/q_p, \tag{10.17}$$

where L is the length of the chamber, H is its height, and q_p is the aerosol

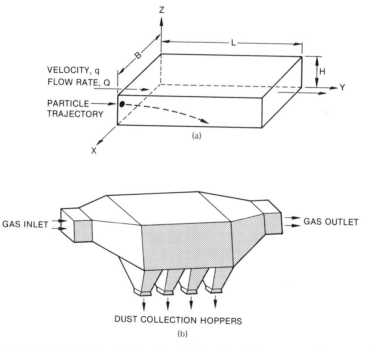

Fig. 10.14. Horizontal settling chamber for collection of particulate matter. Upper panel (a) is a schematic diagram of typical flow of particles. Lower panel (b) illustrates an actual design configuration. [Reprinted from Theodore and Buonicore (1976). Copyright The Chemical Rubber Co., CRC Press, Inc.

flow velocity. If the settling chamber is separated by horizontal settling plates or shelves, as in a horizontal elutriator, the efficiency becomes

$$\mathscr{E} = N q_G W L / q_p, \tag{10.18}$$

where N is the number of shelves and W is the chamber width.

The design of settling chambers becomes complicated when a distribution of different size particles is involved. Iterative methods for sizing are described by Theodore and Buonicore (1976). Typical efficiency curves are shown in Fig. 10.15. As expected, the fractional efficiency decreases dramatically for particles smaller than 100 μm diameter. Such devices are seldom used in practice for dust suspensions of size less than 50 μm diameter. Their most practical use is the removal of very large particles as an aid to the use of gas cleaning equipment placed downstream with greater efficiencies for fine particles.

Careful design of settling chambers is needed to ensure uniform distribution of gas entering and leaving the device so that the theoretical conditions are applicable. The usual designs include gradual transitions, flow splitters, or perforated flow distributors. The flow is normally restricted to 6 m/sec or less to minimize particle reentrainment. Settling chambers with multiple settling plates offer great efficiency per unit volume, but are difficult to clean.

By introduction of flow curvature in settlers, the momentum separation of impaction is used to assist collection. Such devices can be used to split streams of particle-rich and diluted air for further cleaning as in impactor designs for size segregation. Specific design of such devices depends on

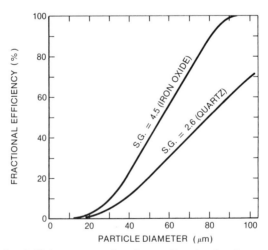

Fig. 10.15. Fractional efficiency curves for a dust settling chamber from a sintering plant as a function of particle size and specific gravity. $H = 3.2$ m; $B = 3.2$ m; $L = 6.5$ m; $Q = 15.7$ m³/sec. [From Jennings (1950).]

empirical testing of curvilinear flow configurations. These devices often involve proprietary geometry; the design is considered an art rather than based on fundamentals.

10.4.2 Centrifugal Separators (Cyclones).

Another inertial separator often chosen for particulate control is the device employing centrifugal forces to spin particles out to a collector surface. Like the gravitational settling chamber, these devices have a minimal pressure drop compared with other collectors. In such a design the aerosol is introduced into a spiraling flow, creating the centrifugal forces acting on particles normal to the principal motion (also Section 5.7.3). The dusty gas flows tangentially into a cylindrical space as shown in Fig. 10.16. The gas spins down into a conical section, then spins upward on the inside of the vortex at the top. The particles experience helical trajectories, gradually moving to the slower flow near the cyclone wall, falling by gravity near the walls to hoppers below.

Different configurations have been employed for industrial cyclone design. Some of these are shown in Fig. 10.17. A conventional form is drawn in Fig. 10.16. Other flow geometries include the tangential inlet, with a peripheral discharge, and arrangements for inducing gas spinning through an axial inlet. Other designs employ a centrifugal blower to induce rotational motion. Unlike the configuration shown in Fig. 5.35a, industrial cyclones are not

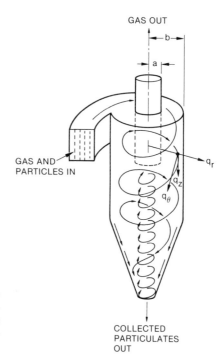

Fig. 10.16. A conventional design centrifugal separator with velocity vectors and geometrical factors shown. [After Theodore and Buonicore (1976). Copyright The Chemical Rubber Co., CRC Press, Inc.]

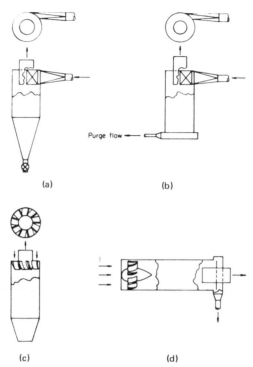

Fig. 10.17. Types of cyclones in common use for industrial separators. (a) Tangential inlet axial discharge; (b) tangential inlet peripheral discharge; (c) axial inlet axial discharge; (d) axial inlet peripheral discharge. [From Caplan (1977); courtesy of Academic Press.]

designed with sharp size cuts in mind. Rather they are designed with flow efficiency and cleanability a primary concern.

The main vortex core is generally smaller than the gas outlet. The radius of the core is between 0.2 and 0.4 times the radius a of the gas outlet. The radius of the maximum tangential velocity is 0.4–0.8 times the radius of the gas outlet. In the annulus between cyclone body, the gas outlet at the top, the tangential gas velocity increases uniformly from the outer wall to the outlet wall. Because the radius of the outlet is greater than the radius of maximum velocity, the gas does not achieve the maximum annular velocity that it attains further down in the main body of the cyclone. There is upward flow in addition along the cyclone wall near the top of the cylinder. This upward drift creates an eddy of secondary flow carrying aerosol upward along the cyclone wall and downward along the gas outlet wall. In this region, particles can be lost into the gas outlet, interfering with the device's efficiency. The longer the gas outlet projection into the cyclone body, the more pronounced the eddy. Elimination of the outlet projection, however, does not eliminate the eddy.

The design of the inlet–outlet configuration thus is a critical feature of the device, and has received considerable attention from engineers. One solution involves the axial inlet cyclone, which eliminates the eddy. If the aerosol particle makes a well-defined set of turns in the annular space before exiting, the number of turns for complete removal of particles of radius R_p is derived from Eq. (5.49). As an alternative calculation, the radius of the smallest particle that can be removed from the aerosol in N_t terms is obtained in Eq. (5.50).

The aerodynamic flow pattern in cyclones is quite complex, and so semiempirical expressions have been developed for performance of these devices. Examples are discussed in Theodore and Buonicore (1976). An example of efficiency curves as a function of inlet velocity is shown in Fig. 10.18. As expected, the higher the inlet velocity the greater the efficiency. However, the efficiency becomes limited at some inlet velocity where higher speeds do not improve performance for a given particle size range. The smaller gas outlet diameter the sharper the efficiency versus inlet velocity curve for a given particle size.

The efficiency of commercial cyclones is also a strong function of particle size as indicated in Fig. 10.19. Generally cyclones are highly efficient collectors for particles larger than 10 μm diameter, but rapidly lose their performance for smaller particles. Comparison between Fig. 10.19 and Fig. 5.35b indicates that this industrial design has a less sharp cutoff in efficiencies than the cyclone adopted for high-volume filter samples.

Typically the results are cast in a form for the diameter of particles of a "cut size" equivalent to 50% efficiency d_e, or

$$d_e \approx [9W/2\pi Nq_i(\rho_p - \rho_g)]^{1/2}, \tag{10.19}$$

where W is the inlet width and q_i is the inlet velocity of the aerosols.

The pressure drop for cyclones has been estimated empirically from the performance of several collectors (Alexander, 1949). Assuming the flow

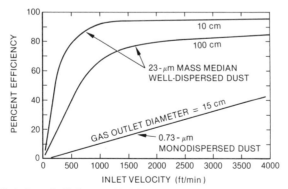

Fig. 10.18. Variation of efficiency with inlet velocity and gas outlet diameter for cyclone separators. [From Caplan (1977); courtesy of Academic Press.]

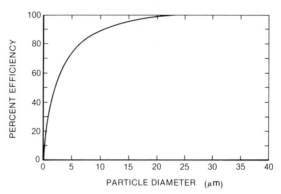

Fig. 10.19. Typical fractional efficiency curve. Air at 21°C; resistance: 76 cm water gauge; load: 162 g/m³; sp. gr.: 2.1. [From Caplan (1977); courtesy of Academic Press.]

resistance is a function of the aerosol inlet area and the outlet area, the cyclone resistance in number of inlet velocity heads,

$$\Delta P = CHW/a^2, \tag{10.20}$$

where C is a factor proportional to the ratio of the outlet gas tube radius to the cyclone radius a/b and H is the inlet height. According to Caplan (1977),

$$C = 4.62 \frac{a}{b} \left\{ \left[\left(\frac{b}{a} \right)^{2n} - 1 \right] \frac{1-n}{n} + f \left(\frac{b}{a} \right)^{2n} \right\};$$

the parameter n varies with body radius b from 0.2 to 0.7 for most configurations, and f is a function of n as shown in the table. The pressure drop varies with the square of the flow rate, which is the reason behind expressing the pressure drop in terms of the number of inlet velocity heads.

			n		
	0	0.2	0.4	0.6	0.8
f	1.90	1.94	2.04	2.21	2.40

The design of cyclones for optimizing efficiency depends not only on geometry and flow factors, but also on the physical properties of the dust to be collected. Aside from particle size, the particle density is an important factor. Other factors that may make the dust powder difficult or easy to handle include its hygroscopicity and electrical resistivity, as well as its capability for erosion and fouling or plugging.

10.4.3   Wet Scrubbers. Cyclones are sometimes operated with liquid sprays to wash particles down the walls and prevent fouling. Spraying also adds the potential for collection of particles on individual droplets through interception, impaction, or diffusional transport. A cyclone operated in this way effectively overlaps the wet scrubber definition.

Wet scrubbers employ a variety of schemes to contact a liquid absorber with gaseous or particulate contaminants in gas. Calvert (1977) has classified scrubbers according to geometrical configuration and mechanism for collection, as listed in Table 10.4. Their form can include centrifugal devices; baffled, packed-plate columns; or moving beds. The mechanisms for capture cover jet or droplet impingement, as well as bubble- or sheet-induced interphase contact. Some schematic forms of different wet contacting devices are shown in Fig. 10.20. They include packed or plate columns, spray towers, Venturi spray collectors, impingement separators, and self-entraining configurations.

The wet scrubber control devices all basically rely on the collection of particles by scavenging mechanisms described by impaction and diffusional transport, the basic theory of which is discussed in Chapter 2. From theoretical considerations, the key to maximizing the collection process in a fixed volume is the maximization of contact surface area and the relative velocity past the collecting surface. Maximization of contact surface area has been achieved using liquid-aerosol flow through packed towers, plate columns, or columns where liquid is sprayed through a volume of dirty gas. High-efficiency devices also have been designed to atomize the contact liquid into an accelerating aerosol stream. These take advantage of improving the droplet collection efficiency by increasing the velocity difference between the spray droplets and the aerosol particles. The flooded disk and Venturi scrubber are devices of this kind. The drops are atomized by injection of the liquid into

TABLE 10.4

Wet Scrubber Design Classification based on Calvert's (1977) Unit Process Scheme

Geometric type	Unit mechanism for particle collection
Plate	Jet impingement, bubbles
Massive packing	Sheets (curved or plane), jet impingement
Fibrous packing	Cylinders
Preformed spray	Drops
Gas atomized spray	Drops, cylinders, sheets
Centrifugal	Sheets
Baffle and secondary flow	Sheets
Impingement and entrainment	Sheets, drops; cylinders, jets
Mechanically aided	Drops, cylinders, sheets
Moving bed	Bubbles, sheets
Combinations	

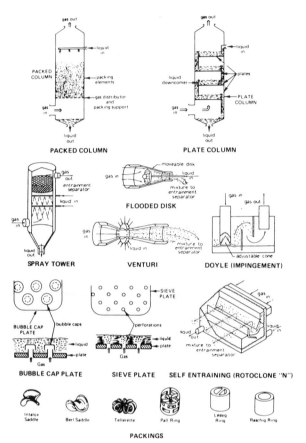

Fig. 10.20. Examples of spray and atomization collectors. [From Calvert (1977); courtesy of Academic Press.]

the accelerating dirty gas either through a spinning disk in the converging section of a pipe or in the throat of a Venturi.

The specific design features for control depend on the aerosol. Packed or plate columns are superior for removal of gaseous components, but tend to be fouled or clog with heavy dust loadings. For particles, spray contactors or Venturi devices are superior collectors, since they avoid plugging from accumulated sludge. Atomizing devices like the flooded disks or Venturis have received increasing attention because of their high efficiency in performance.

Pressure drop is a significant factor in design of wet scrubbers as in other collectors. The pressure drop in contactors is associated with friction from surface contact and fluid acceleration. Friction loss depends strongly on scrubber geometry, but acceleration loss is relatively insensitive to scrubber geometry. Fortunately the latter often dominates the pressure loss.

As a rough approximation, Calvert (1977) reports that the pressure drop in scrubbers due to acceleration in centimeters of water is

$$\Delta P \approx 10^{-3}(q_g - q_l)^2(Q_l/Q_g), \tag{10.21}$$

where q_g and q_l are the linear velocities of the gas and liquid and Q_l and Q_g are the liquid and gas volume flow rates. Friction losses have to be determined experimentally, but can be estimated from correlations such as those given by Eckert (1961).

Scrubber capacity is limited basically to the gas flow at which the amount of liquid entrained by the gas becomes excessive, in other words, when flooding takes place. Flooding occurs in countercurrent packed columns at a gas velocity about 1.5 times that which causes a 1.3-cm water pressure drop.

Entrainment of liquid requires that the drop size be sufficiently small and the gas rate sufficiently high for the drops to be carried out of the device. If drops already exist as in a spray scrubber, entrainment conditions can be estimated from drop dynamics. When shattering of liquid films occurs or drops are initially absent, atomization must be accounted for.

Particle collection performance for scrubbers is often cast in terms of penetration, or

$$P_e \equiv (1 - \mathcal{E}) \equiv \exp(-A_s d_a^\beta), \tag{10.22}$$

where A_s is a constant and d_a is the aerodynamic particle diameter. For packed towers and sieve plate columns, the exponent β has the value of 2. For centrifugal scrubbers $\beta \approx 0.67$. Venturi scrubbers also follow this relation when the Venturi throat impaction parameter, K_{Ve}, is between 1 and 10. In this range, $\beta \approx 2$. The parameter is defined by

$$K_{Ve} = d_a^2 q_g/9\mu_g a,$$

where a is the spray drop diameter. When K_{Ve} is between 1 and 10 μm diameter, the exponent $\beta \approx 2$. The overall penetration for a device scrubber of a size distribution of particles is the sum of all particle penetration values for a given size increment.

The overall penetration for size distribution taken in lognormal form has been calculated for different values of the product of β and the logarithm of the standard deviation σ_g of the distribution. These are shown in Fig. 10.21 as a function of $\beta \ln \sigma_g$ and the ratio of the 50% cutoff particle diameter d_{pe} and the geometric mean particle diameter $\bar{d}_p$. This estimate is best for packed beds and similar devices, but is only an approximation for others.

The penetration for different scrubber configurations has been determined through the combined application of theory for inertial collection and Brownian diffusion, and experimental studies. Specific forms of the penetration formulas based on impaction models include

(a) *Sieved and Impingement Plate Columns.*

$$P_e = \exp(-40FK_p), \tag{10.23}$$

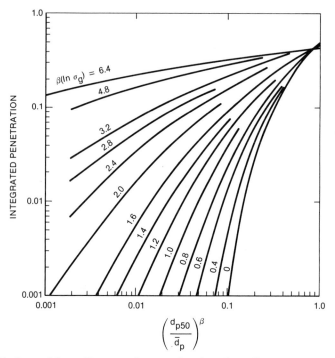

Fig. 10.21. Integral (overall) penetration as a function of cut diameter $d_{pc} \equiv d_{p50}$, particle parameters, and collector characteristics $P = \exp(-A_s d_a)$. [From Calvert (1977); courtesy of Academic Press.]

where F is the foam density, and

$$K_p = q_h d_p^2 / 9 \mu_g \mathcal{L}_h,$$

where q_h is the gas velocity through the sieve hole of diameter $\mathcal{L}_h$. The cut diameter is

$$d_{pc} = 0.4(\mu_g \mathcal{L}_h / q_h F)^{1/2}. \tag{10.24}$$

(b) *Packed Columns.*

$$P_e = \exp(-7.0 H K_p / \varepsilon_v \mathcal{L}_c), \tag{10.25}$$

where H is scrubber height, ε_v is the void fraction, and $\mathcal{L}_c$ is an average packing diameter. The cut diameter is

$$d_{pc} = (\varepsilon_v \mathcal{L}_c^2 \mu_g / \mu_g H)^{1/2}.$$

(c) *Preformed Sprays.* vertical countercurrent flow with inertial impaction yields

$$P_e = \exp[-3 Q_l q_s H \eta / 2 Q_g a (q_s - q_g)], \tag{10.26}$$

where η is the collection efficiency for individual spray droplets (e.g., Fig. 2.3).

Cross-flow inertial impaction gives

$$P_e = \exp(-3Q_1 W\eta/2Q_g a), \tag{10.27}$$

where W is the width of the scrubber.

(d) *Gas-Atomized* (*Cocurrent*) *Sprays* (includes Venturi, disk, orifice, etc.).

$$P_e = \exp[(2Q_1 q_g \rho_1 a/55Q_g \mu_g) F(K_{Ve}, f)], \tag{10.28}$$

where

$$F(K_{Ve}, f) = \frac{1}{K_{Ve}} \left[-0.7 - K_{Ve} f + 1.4\ln\left(\frac{K_{Ve} f + 0.7}{0.7}\right) + \left(\frac{0.49}{0.7 + K_{Ve} f}\right) \right].$$

The factor f is an empirical factor whose value Calvert (1977) recommends as 0.25 for hydrophobic particles. For hydrophilic particles, f may be as high as 0.5. Also, for liquid to gas ratios below 0.2, Calvert (1977) suggests $f = 0.5$ for design calculations.

An example of the performance curves for a Venturi scrubbing device is shown in Fig. 10.22. We see that the penetration increases dramatically in such devices for particles less than 10 μm in diameter. This is expected based on the decrease in droplet collection efficiency with aerosol particle size in the range of 1 μm diameter (e.g., Fig. 2.3). The penetration reaches a limiting low value for large particle collection. This indicates that even

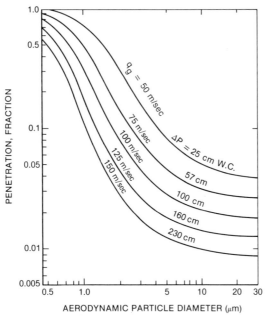

Fig. 10.22. Predicted penetration for Venturi scrubber; $Q_1/Q_g = 1$ lit/m³, $f = 0.25$. [From Calvert (1977); courtesy of Academic Press.]

though the efficiency for single drops collecting large particles approaches unity, the scrubber is limited because there are not enough drops produced to completely sweep the gas stream.

Particle collection by Brownian diffusion to the collecting body is described in terms of the theory for mass transfer (e.g., Section 2.3). A general relationship describing particle deposition in any control device involving turbulent mixing to eliminate concentration gradients normal to the flow outside the diffusion boundary layer has the form for constant deposition velocity k_{BD}

$$P_e = \exp - (k_{BD}S_s/Q_g), \tag{10.29}$$

where S_s is the total collection surface for the scrubber. Here Calvert (1977) gives

$$k_{BD} = 1.13(d_p/t_g)^{1/2},$$

where t_g is the penetration time. For packed columns this is the time for the aerosol to travel one packing diameter. In plate scrubbers involving bubbles rising through liquid, the penetration time for a bubble is roughly that for a bubble to rise one diameter. For spray scrubbers the penetration time is the time required for the aerosol to pass one drop diameter.

Calvert (1977) has shown typical calculated collection efficiencies for plate columns, packed columns, and Venturis. His curves are reproduced as a function of particle size in Fig. 10.23 to illustrate performance. High efficiency of collection is expected for very small particles, and is maximized in a three-plate column. The collection efficiency by diffusion drops rapidly for particles in the regime ≤ 1 μm diameter. Matching the diffusional efficiency with the impaction model reinforces again that the minimum in collection efficiency is in the submicrometer particle range, as discussed before (e.g., Fig. 2.9).

Finally, collection of particles in scrubbers can be improved by employing diffusiophoretic action, forcing particles toward condensing droplets. This flux force effect has been studied in several configurations of cold plate sieve towers or condensing steam nozzles. The results of several investigators for condensing water scrubbers are shown in Fig. 10.24. The fractional penetration dramatically decreases with increasing water vapor content per unit mass of dry gas. The condensable liquid ratio (gram of water vapor per gram of dry gas) is an important operating parameter as well as the number concentration of particles. Particle surface properties, design configuration, and operating conditions also are significant factors in such devices.

10.4.4   Industrial Filtration.   Particle filtration has been found to be one of the most reliable, efficient, and economical methods for gas cleaning. It also is one of the few methods capable of meeting the present particle emission standards in the United States.

The theory of industrial aerosol filtration is well developed and has been

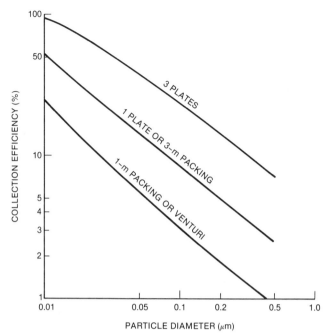

Fig. 10.23. Predicted particle collection by diffusion in plates, packing, and Venturi scrubbers. [From Calvert (1977); courtesy of Academic Press.]

applied to the design of highly efficient particle collection units. Industrial gas filters are broadly classified into two kinds: (a) fabric or cloth filters, or (b) deep packed bed columns. The former is exemplified by fabric bag units such as those in Fig. 10.25, and the latter by a fibrous or matted array, or a gravel packed bed. Fabric filters are used with gas streams with a dust loading of about 1 g/m³; fibrous or packed beds are adopted for much more dilute aerosols, of about 1 mg/m³ loadings.

As for other gas cleaners, performance criteria focus on pressure loss, collection efficiency as well as lifetime, and installation and operating costs. The lifetime of the filter is particularly important from an economic viewpoint because the cost of the filtering medium is a major part of the initial expense and long-term operating expense of a unit.

The fiber materials used for commercial filter media are selected based on several criteria. These include strength per unit mass, high temperature stability; chemical inertness, especially to acid, base, and organic solvent exposure; and cost. Examples of materials used for filter fibers are given in Table 10.5. Materials commonly selected usually have a hairy or downy texture, are capable of withstanding temperatures of 250°C, and have a useful life of a year or so in a particular application. Fabric lifetime depends on operating temperature duration and manner of cleaning or dislodging dust loading, the aerosol characteristics, and the care of installation design.

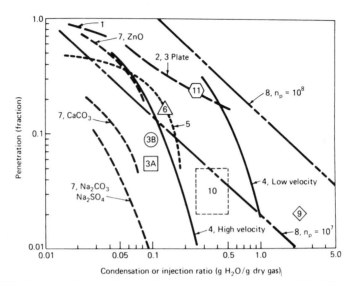

Fig. 10.24. Condensation scrubbing performance; curves are designated for different designs listed below. [From Calvert (1977), with references; courtesy of Academic Press.]

Curve	Scrubber type	d_p (μm)	Particle material	N (#/cm²)
1	Sieve plate (1 cold plate)	0.7	Dibutyl phthalate	5×10^5
2	Sieve plate (3 cold plates)	0.4	Ferric oxide	10^5–10^6
3	(A) Cyclone, or (B) Peabody (1 plate)	<2.0	NaCl	10^3
4	Tubular condenser	?	Nichrome and others	?
5	Steam nozzle + spray + cyclone	1.0	ZnO	10^5–10^6
6	Venturi + 2 sieve plates	1.7	Apatite	$\sim 10^5$
7	Steam nozzle + dry duct		ZnO, CaCO$_3$, Na$_2$CO$_3$, Na$_2$SO$_4$	
8	Sieve plate with alternate hot and cold plates	0.3	Oil	10^5–10^6
9	Steam nozzle + Peabody (5 plates)	0.3	Dioctyl phthalate	2×10^7
10	Tubular condenser	0.1	Iodine	10^3–10^4
11	Vertical netted	0.05	Tin fume	5×10^7

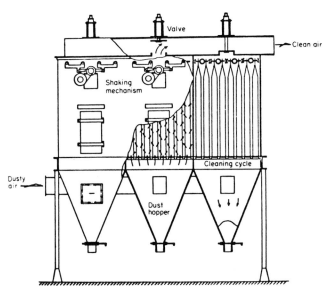

Fig. 10.25. Typical bag filter with mechanical shaking of upper ends. [From Iinoya and Orr (1977); with permission from Academic Press.]

Paper filter media are normally used for laboratory applications, for household vacuum cleaners, engine air cleaners, and the like. They come in a variety of grades and may be mounted as sheets or folds or pleats. Their collection efficiency often exceeds 80% for submicrometer particles. A superior collection efficiency paper, high efficiency paper (HEPA), also has practical application. These filters are used in applications where clean rooms are involved or where bacteria or radioactive particle filtration is needed. At the expense of higher pressure drop than normal paper filters, HEPA filters have >99% collection efficiencies for submicrometer particles.

The electrostatic charging ability of filter media also is important in filtering and cleaning performance. Fibers are classified according to their triboelectric series; dust also can be classified in the same manner. The electrical charge intensity on either the fabric or the dust particles depends on the process conditions and the nature of the materials involved. Charge dissipation is an important factor in design so that fabric bags are sometimes interwoven with stainless-steel wire to combat retained charge and minimize the potential for fire from electrical discharges.

The filter units are designed for maximizing fabric surface area exposure to the aerosol flow. They are also designed for ease in maintenance and cleaning. Filter units are arranged such that mechanical shaking can be used for cleaning. Units are arranged in several compartments (Fig. 10.25). Each compartment will contain several bags; one compartment can be removed from service for bag cleaning without loss of particle collection capability.

TABLE 10.5

Properties of Example Fiber Materials for Industrial Gas Cleaning Applications[a]

Fiber	Physical characteristics				Relative resistance to attack by			Other attribute
	Relative strength	Specific gravity	Normal moisture content (%)	Maximum usable temperature (°F)	Acid	Base	Organic solvent	
Cotton	Strong	1.6	7	180	Poor	Medium	Good	Low cost
Wool	Medium	1.3	15	210	Medium	Poor	Good	—
Paper	Weak	1.5	10	180	Poor	Medium	Good	Low cost
Polyamide (nylon)	Strong	1.1	5	220	Medium	Good	Good[b]	Easy to clean
Polyester (Dacron)	Strong	1.4	0.4	280	Good	Medium	Good[c]	—
Acrylonitrile (Orlon)	Medium	1.2	1	250	Good	Medium	Good[d]	—
Vinylidene chloride	Medium	1.7	10	210	Good	Medium	Good	—
Polyethylene	Strong	1.0	0	250	Medium	Medium	Medium	—
Tetrafluoroethylene	Medium	2.3	0	500	Good	Good	Good	Expensive
Poly(vinyl acetate)	Strong	1.3	5	250	Medium	Good	Poor	—
Glass	Strong	2.5	0	550	Medium	Medium	Good	Poor resistance to abrasion
Graphitized fiber	Weak	2.0	10	500	Medium	Good	Good	Expensive
Asbestos	Weak	3.0	1	500	Medium[e]	Medium	Good	—
"Nomex" nylon	Strong	1.4	5	450	Good	Medium	Good	Poor resistance to moisture

[a] From Iinoya and Orr (1977); courtesy of Academic Press.
[b] Except phenol and formic acid.
[c] Except phenol.
[d] Except heated acetone.
[e] Except SO_2.

Bags are often shaken from their upper ends mechanically, or they can be shaken by rings inside the middle of the bag in combination with a reverse, pulsed air flow.

Flow configurations can be engineered in a variety of different ways. Some of these are shown in Figs. 10.26 and 10.27. Bag houses can be designed either to "push" or "pull" air through the units depending on the nature of the process. The vertical and parallel columnar structure offers optimal surface area for collection combined with use of gravity to collect dust for removal to a hopper.

The pressure loss through a filter medium depends on the particulate matter collected and the filter itself. It is a critical parameter in the design of a filter medium because of its relation to energy requirements for pumping air through the systems. The pressure drop through fabric filters is written in the form

$$\Delta P = \Delta P(\text{filter}) + \Delta P(\text{dust loading}) = (\zeta_0 + \bar{\alpha}\hat{M})q_0\mu_g/g, \quad (10.30)$$

where ζ_0 is the pressure loss coefficient for the clean filter, $\bar{\alpha}$ is the average specific resistance of the collected dust layer, $\hat{M}$ is the mass of collected dust per unit surface, and q_0 is the superficial gas velocity at the face of the filter. The pressure drop for a clean filter is normally small for practical purposes. The values of average specific resistance of collected material depends on the particle size and density, and on the volumetric voids of the fabric. Values typically range between 10^{10} and 10^{11} m/kg. An example cited by Iinoya and Orr (1977) gives for Eq. (10.30) $\zeta_0 = 7 \times 10^7$/m, $\bar{\alpha} = 3 \times 10^{10}$ m/kg, $\hat{M} = 0.12$ kg/m^2, q_0 = m/min, a pressure loss of 115 mm H$_2$O for air at normal pressure and temperature. A correlation of pressure drop of the different media is given in Fig. 10.28 as a function of porosity of the media.

The change in pressure loss with particle accumulation $\hat{M}$ can be deduced from the Kozeny–Carman relationship written

$$\frac{d(\Delta P)}{d\hat{M}} = \frac{180\mu_g q_G(1 - \varepsilon_v)}{g\rho_p d_p^2 \varepsilon_v^3}, \quad (10.31)$$

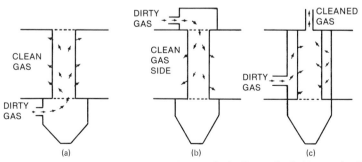

Fig. 10.26. Types of filtering systems: (a) bottom feed, (b) top feed, (c) exterior filtration. From Theodore and Buonicore (1976). Copyright The Chemical Rubber Co., CRC Press, Inc.

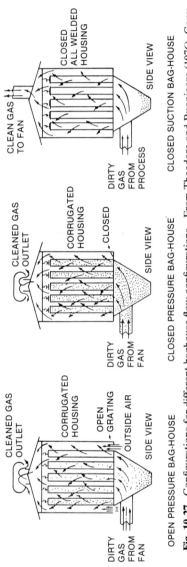

Fig. 10.27. Configurations for different baghouse flow configurations. From Theodore and Buonicore (1976). Copyright the Chemical Rubber Co., CRC Press, Inc.

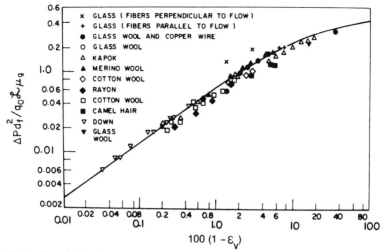

Fig. 10.28. Correlation of mat density $1 - \varepsilon_v$ with pressure drop and flow: O, glass wool; ●, glass wool and copper wire; +, glass (fibers perpendicular to flow); ×, glass (fibers parallel to flow); △, kapok; ▲, Merino wool; ◇, cotton wool; ◆, rayon; □, cotton wool; ■, camel hair; ▽, down; ▼, glass wool. [From Bragg (1981); by courtesy of Wiley.]

where the apparent volumetric void fraction ε_v is obtained from the particle diameter (based on surface area) according to relationship in Fig. 10.29. For example, using the values of q_G and μ_g for air, $\rho_p = 3000$ kg/m³, $d_p = 0.40 \times 10^{-6}$ m, and $\varepsilon_v = 0.96$, Eq. (10.31) gives a value for $d(\Delta P)/d\hat{M}$ of 532 kg/kg or mm H_2O/kg of dust/m² of cloth area. However, the permeability of the dust cake on the cloth may vary with the operating conditions of the bag filter in a way that significantly affects the pressure drop (Iinoya and Orr, 1977).

The following values are common for conventional bag filters under normal operations: $\Delta P = 100$–200 mm H_2O, $\hat{M} = 0.05$–0.03 kg/m², $\bar{\alpha} = 10^{10}$–10^{11} m/kg, $\bar{u} = 0.5$–3 m/min.

The collection efficiency at the beginning of filter use is given in the exponential form

$$\mathcal{E}_0 = 1 - \exp(-4(1 - \varepsilon_v)\mathcal{L}\eta_f/\pi\varepsilon_v d_f), \tag{10.32}$$

where ε_v is the volumetric void fraction of the filter bed, $\mathcal{L}$ is the filter thickness, and d_f is the fiber diameter. The initial collection efficiency of a single fiber is η_f, and is given by Eq. (5.41), for example.

As discussed in Section 5.6.2, filtration depends on the combination of Brownian diffusion to the fibers, impaction, and interception, in the absence of electrical effects. The actual separation of commercial filters often depends on the initial collection of the particle mat. As shown in Fig. 10.30, the filtration efficiency improves during the course of initial particle accumulation because particles are entrapped in the fiber mesh, closing the large, open

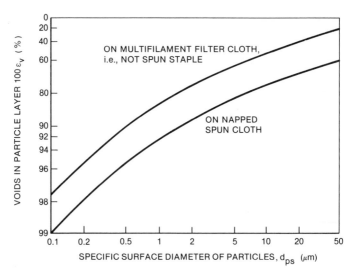

Fig. 10.29. Relation between specific surface diameter of particles to be collected and voids of collected particle layer. [From Kimura and Iinoya (1965).]

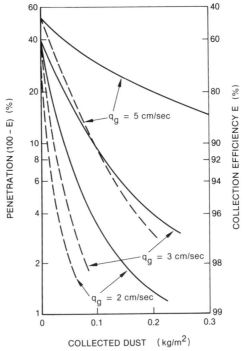

Fig. 10.30. Collection efficiency of fabric filters. The solid lines show cumulative collection efficiencies and the broken lines show instantaneous ones. Polyester cloth and carbon particles, mmd = 1.8 μm (mmd = mass median diameter.) [From Iinoya and Orr (1977); courtesy of Academic Press.]

holes. Once the particle bed is formed across openings in the fabric mat, the collection efficiency will rise to 99% or more. The efficiency will flatten as the dust bed thickens, with an increase in pressure loss when the fabric deposit becomes heavy.

The filtration efficiency also is a function of particle size. Typical curves for removal efficiency in an industrial fabric filter unit is shown in Fig. 10.31. As in laboratory experiments of filter performance (Fig. 5.28), industrial units achieve a minimum collection near 1 μm diameter where neither Brownian diffusion nor impartion–interception is optimum in effectiveness. Even at minimum, however, the filter has an efficiency exceeding 99% (penetration of 1%).

An external electrical field can enhance the particle collection efficiency of filters (Nelson *et al.*, 1978). Electrical forces can draw particles from the gas stream to filter fibers if the two are oppositely charged. Even if only one of the materials is charged, an induced charge will be created on the other producing a polarization force, which will influence collection. Electrostatic charge also induces agglomeration and increases the likelihood of particle entrapment. Branching of particles on filter media shown in Fig. 5.29 is probably an electrical phenomenon.

Application of an electrostatic field to a fibrous filter of a dielectric material by imbedded electrodes enhances the collection efficiency. The dielectric effect can compensate for poor collection efficiency of a filter when

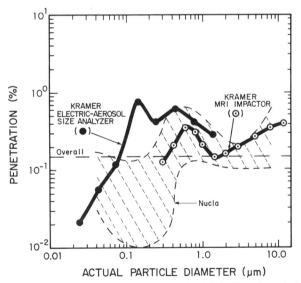

Fig. 10.31. Penetration as a function of particle size for fabric filters installed at two different coal-fired power plants (Kramer, Bellview, Nebraska; Nucla, Nucla, Colorado). [Copyright © 1981 Electric Power Research Institute. Adapted from EPRI report CS-1669, "Kramer Station Fabric Filter Evaluation." Reprinted with permission.]

interception dominates. The theory for this effect can be derived readily for an ideal substrate using Zebel's (1965) model for a cylindrical fiber. The enhancement can be calculated in terms of the ratio P_e/P_{e0} of the filter penetration with and without the presence of an electrical field. The form of this ratio is

$$\ln(P_e/P_{e0}) = -S_f q_E/FAq_0, \tag{10.33}$$

where S_f is the effective fiber surface, q_E is the particle velocity from electrical forces, and FA is the filter face area. Zebel's model yields (Nelson et al., 1978)

$$\ln\left(\frac{P_e}{P_{e0}}\right) = -\frac{4(1-\varepsilon_v)\Omega}{\pi d_f q_0}\left\{\left(\frac{\varepsilon_p-1}{\varepsilon_p+2}\right)\left(\frac{\varepsilon_f-1}{\varepsilon_f+1}\right)\frac{d_p^2 E^2 A}{12\pi\mu_g d_f}\right.$$
$$\left. + \frac{neEA}{3\pi\mu_g d_p}\left[\frac{1+(\varepsilon_f-1)/(\varepsilon_f+1)}{1+neEA/3\pi\mu_g d_p q_g}\right]\right\}. \tag{10.34}$$

This accounts for both the action of electrical forces associated with Coulombic and polarization charge.

An example of the enhancement of collection efficiency is deduced from Fig. 10.32. Here the relative penetration is shown as a function of applied electric field strength. There is a general decrease in penetration (increase in efficiency) over that without the electric field, particularly in the region of the minimum efficiency. For low face velocities the Zebel model is quite accurate, but for larger face velocities, Zebel's theory tends to overestimate the value of this ratio.

10.4.5   Electrostatic Precipitation.   Industrial interest in electrostatic precipitators in the United States can be traced to the investigations of Cottrell (1911). His work was concerned principally with the removal of air pollutants, mainly sulfuric acid mists from copper smelters. As a result of his

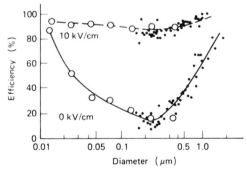

Fig. 10.32. Filter efficiency as a function of particle size with and without an electric field using naturally charged aerosols. ○, electrical mobility analyzer, ●, laser particle counter. [From Bergman et al. (1979).]

pioneering work, the term "Cottrell precipitator" has for many years been used almost interchangeably with electrostatic precipitator. Experimental work by Cottrell in the elimination of acid mists led to the use of precipitators for the collection of metal oxides from the effluents, and this added impetus to the growth of precipitators in the primary smelter industry. This work was followed by the application of precipitators to other nonferrous metal processes, such as lead blast furnaces, ore roasters, and reverberatory furnaces. Success in the nonferrous metals industry was in turn followed by application of precipitators to the collection of dust from cement kilns. From these beginnings, the use of precipitators has expanded to include a wide variety of uses including unique boilers for electric power generation. The principal uses of precipitators today are in gas cleaning applications in which high collection efficiencies of small particles are required for processes that emit large gas volumes. Since the separation force in a precipitator is applied to the particle itself, the energy required for gas cleaning is less than that for equipment in which energy is applied to the entire gas stream. This unique characteristic of precipitators results in lower gas pressure drops and usually lower operating costs than other methods of gas cleaning.

The precipitation process requires (a) a method of providing an electrical charge on a particle, (b) a means of establishing and maintaining an electrical field, and (c) a method for removing the particle from the precipitator.

The process of electrically charging a particle involves the addition of electrons to or removal of electrons from the material or the attachment of ionized gas molecules to the particles. Almost all small particles in nature acquire some charge as a result of naturally occurring radiation, triboelectric effects due to transport through a duct, flame ionization, or other processes. These charges are generally too small to provide effective precipitation, and in all industrial precipitators, charging is accomplished by the attachment of electrical charges produced by an electrical corona. A corona discharge producing negative ions is normally used in precipitators.

The electrical field in an industrial precipitator is provided by the application of a high dc voltage to a dual electrode system (Fig. 10.33). An electrical corona is established in this electrode system, one of which is a small-diameter wire or other configuration which gives the small radius of curvature required to produce a highly nonuniform electric field. The other electrode can be a cylinder concentric with the corona electrode or a plate parallel to the plane of a series of corona wires. The corona generated in the intense electric field region provides the charges necessary for electrical collection. The charged particles entering the region of the electric field are urged toward one of the electrodes where they are collected and held by electrical, mechanical, and molecular forces. Liquid particles, such as acid mists or tars, coalesce on the collection plate and drain away. Solid particles are collected in agglomerates and usually removed by periodic rapping of the collection electrode. This permits the dust to fall into hoppers at the base of

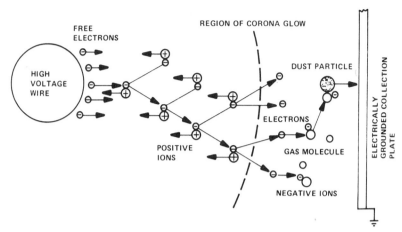

Fig. 10.33. Particle charging process in a corona discharge system. [From Pontius and Smith (1977).]

the precipitator. A large scale industrial installation is shown in Fig. 10.34. Here parallel plates are placed alternately with vertical wire conductors such that dirty gas flows horizontally across them. The dust is collected and falls downward into hoppers below.

In both theory and practice, the operation of precipitators is governed largely by the magnitude of the charge, the electric field, the efficiency of removal from the collectors, and the extent of reentrainment of collected dust. Factors that determine precipitator performance are those which establish limits on each of these parameters.

As discussed earlier in Chapter 5, electric field effects and thermal or diffusion effects are dealt with separately. It was found that the particle charging process may be described adequately for particles less than about 0.2 μm in diameter by a theory based entirely on the thermal motion of the ions, without regard to the presence of an electric field. Such an approach is justified on the grounds that the thermal energy of an ion is much greater than the energy which can be derived from even a very strong electric field acting on an ion in a gas. But the random thermal motion involves frequent collisions between ions and gas molecules, so that the average distance between collisions is only about 0.1 μm. For a charged particle several micrometers in diameter, the effective range of the repulsive force on an ion is many times greater than the mean free path of the ion. On this larger scale, the longer-range directed force due to the applied electric field provides the more effective mechanism for particle charging.

Separate theories for particle charging, each limited in applicability, are useful in delineating the nature of the operative charging mechanisms, but for particle diameters between approximately 0.2 and 2.0 μm neither the

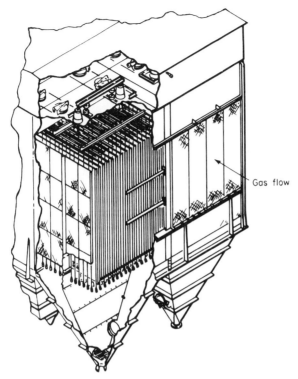

Fig. 10.34. Parallel-plate electrostatic precipitator with pyramid hoppers. [From Oglesby and Nichols (1977a); courtesy of Academic Press.]

field charging theory, the diffusion charging theory, nor a sum of the two provides satisfactory agreement with experimental results (see also Section 5.2).

A semiempirical theory has been derived, taking into account simultaneously the effects of ion diffusion and the applied electric field. Although the electric field makes a small contribution to the kinetic energy of an ion, the most important effect of the field is upon the distribution of ions in the system. In the vicinity of a particle, the electric field consists of that resulting from the applied voltage plus a contribution due to the charge on the particle, polarized by the applied field. The theoretical approach employed is to determine the ion density distribution near a particle in terms of the local electric field, and then calculate statistically the rate at which ions reach the particle as a result of their thermal velocities. The theoretical charging rate cannot, in general, be expressed in a closed, algebraic form, but requires the use of a computer to carry out the calculation.

The charging theory indicates that the total charge accumulated by a particle is strongly dependent upon the electric field strength, the diameter of the particle, the number density of ions, and the residence time of the particle in the charging region. The latter two variables invariably occur as a product in charging theories. Thus the ion density–residence time product is usually treated as a single parameter representing the total exposure of a particle to ions as it passes through the charging region. Other variables which have a significant effect on the particle charging process include the gas temperature, the electrical mobility of the ions in the gas, and the dielectric constant of the particulate material.

The theoretical migration velocity q_E of a particle has a considerable size dependence. Therefore a larger precipitator is required to reach a given efficiency when small particles are collected. Figure 10.35 shows the migration velocities for particles of various sizes for various operating conditions. The range of particle sizes in which field charging and diffusion charging predominate is indicated, together with the size range in which both mechanisms are significant. The migration velocity is adjusted by the Cunningham correction factor for particle diameter below 1 μm. This correction largely accounts for the increase in migration velocity in the very small particle size range.

One consequence of the variation in migration velocity with particle size is that there will be a corresponding change in collection efficiency with particle size. Thus, for a polydisperse dust, the larger particles will tend to be

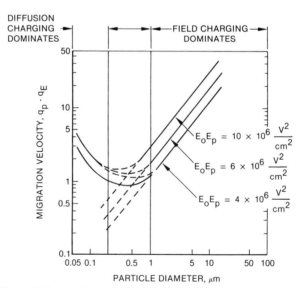

Fig. 10.35. Theoretical migration velocity as a function of particle diameter for field and diffusion charging mechanisms. The parameter E_0 is the average electric field in the interelectrode space, and E_p is the electric field at the collecting electrode. N_0 is the ion concentration in #/cm^3. The intermediate charging regime is shown between the diffusion and field charge regimes. [From Oglesby and Nichols (1977a); courtesy of Academic Press.]

collected in the inlet sections of a precipitator, resulting in a change in the median diameter of the dust as it progresses through the precipitator.

The resistivity of collected particles provides a fundamental limitation to the operation of a precipitator by influencing the electrical field. The configuration of a precipitator is such that the corona current must flow through the collected dust layer to reach the grounded collection electrode. In the case of dry precipitators, this current flow can result in large voltage drops across the dust layer if the electrical resistivity of the dust is high. In many applications, the resistivity of the dust is sufficiently high to impair precipitator performance.

Electrical resistivities of the dusts encountered in industrial gas cleaning applications can differ considerably. Some materials, such as carbon black, have very low resistivity, so that on contact with a grounded metal surface, the particles lose their charge and are easily reentrained into the gas stream. At the opposite extreme, dusts of insulating materials, such as alumina, can have a sufficiently high resistivity that the charge leaks off slowly. In such cases, the electrical force holding the dust to the collection plate can be very high and the voltage drop across the dust layer can be sufficient to cause breakdown of the interstitial gases within the dust layer.

Electrical breakdown of the interstitial gases in the dust layer occurs when the electric field exceeds the breakdown strength of the gases. For most gases encountered in industrial precipitators, breakdown of the precipitated layer occurs when the electric field exceeds about 20 kV/cm. The exact value of the breakdown strength depends on the particle size and extent of packing of the dust and on the gas composition.

If there is no dust layer present, the voltage in a precipitator can be increased with an accompanying increase in current until the gases in the interelectrode region break down. This breakdown takes the form of a sparkover originating at the anode or collecting surface in a negative corona precipitator and propagating to the corona electrode. This condition establishes the maximum current and voltage conditions that can be obtained with the particular electrode geometry, spacing, and gas composition.

When breakdown occurs, the resistance of the localized area around the breakdown is reduced, so that the voltage that was across the dust layer is now applied to the space between the dust surface and the corona wire. This additional voltage can cause breakdown of the gases in the interelectrode region, resulting in a sparkover, provided the voltage across the electrode is sufficient to propagate a spark.

With very high resistivity, electrical breakdown occurs at a voltage that is insufficient to propagate a spark across the interelectrode region. The result of this condition is a continuous breakdown of the dust layer without sparkover between electrodes. This breakdown is analogous to that occurring at the discharge electrode and similarly produces ion–electron pairs. The positive ions flow across the interelectrode region toward the discharge electrode. The net effect is a reduction of the charge on the particles and poor precipitation. This phenomenon is called back corona or reverse ionization and in the dark can be observed as a diffuse glow on the dust surface of the collection electrode.

The mechanism of current conduction in a dust layer has been studied by Bickelhaupt (1974) and others. Two modes of conduction are possible in industrial dusts, depending upon the temperature and composition of the dust and flue gases. At elevated temperatures (above about 200°C) conduction takes place primarily through the bulk of the material. The resistivity of the dust in the temperature region where bulk conduction predominates is referred to as volume resistivity. At lower temperatures, moisture or other substances present in the flue gases are adsorbed and conduction occurs principally along the surface of the dust particles. In the temperature region where conduction takes place along the surface of the dust particle, the resistivity is referred to as surface resistivity.

Figure 10.36 illustrates a typical temperature–resistivity curve showing the regions where volume and surface resistivity predominate and the region

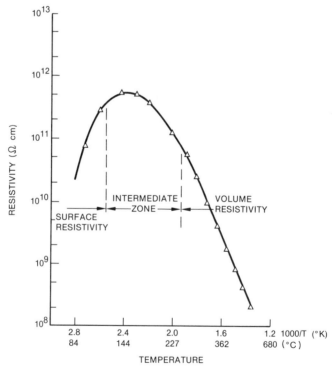

Fig. 10.36. Electrical resistivity of fly ash as a function of temperature. [From Oglesby and Nichols (1977b).]

where volume and surface resistivity predominate and the region where both are significant. The overall resistivity is analogous to that of parallel resistors each of which is temperature dependent. The absolute value of resistivity varies depending upon the method and conditions of measurement, as discussed subsequently. However, the shape of the curve is similar for the various measurement techniques.

In the high temperature region, the resistivity depends on the chemical composition of the material. Bickelhaupt (1974) found that, for fly ashes of generally similar composition, the resistivity decreases with increasing sodium and lithium content, indicating that these ions are the primary charge carriers. The presence of iron causes a further decrease in resistivity, apparently by increasing the percentage of lithium and sodium ions that are capable of participating in the conduction process.

In the low temperature region, resistivity is thought to be an ion transport phenomenon related to the adsorption of water vapor or other conditioning agents present in the flue gas. In the case of fly ash from coal-fired boilers, resistivity is primarily related in an inverse manner to the amount of sulfur trioxide (SO_3) and moisture present in the flue gas. The burning of coal containing sulfur produces sulfur dioxide (SO_2) in quantities dependent on the sulfur content in the coal. Under normal conditions, about 0.5–1% of the SO_2 present is oxidized to SO_3, which serves to reduce the resistivity of the fly ash, if the temperature is low enough for the SO_3 to be adsorbed on the ash. Thus, high-sulfur coals tend to produce ash with lower resistivities than coals with lower sulfur content. In general, lowering the flue-gas temperature increases the rate of SO_3 adsorption, so that the resistivity of fly ash can be controlled to some extent by changes in flue-gas temperature. The presence of large quantities of CaO in the ash apparently has an adverse effect on conductivity. An ash high in lime tends to react initially with the available SO_3 to produce a sulfate which decreases the SO_3 available for conditioning.

For other processes, notably cement kilns and metallurgical furnaces, the principal conditioning agent is adsorbed moisture. Higher moisture content in the flue gases and lower temperature give lower resistivities. Resistivity–temperature relationships can, therefore, be represented as a family of curves with both the volume and surface resistivities changing with particulate and flue-gas composition.

Once collected, particles can be removed from the collection electrode by draining in the case of liquid aerosol, by flushing the plates with a liquid, or by periodically rapping or vibrating the plates in the case of solid particles. Draining of coalesced liquid on the plates constitutes a more or less straightforward process and is common in precipitators used in the removal of acid mist, tar, and similar materials. Removal of solid particles by irrigation of the collection plates has been used extensively in some metallurgical processes, notably blast furnaces, for a number of years. There is at present considerable renewed interest in so-called "wet" precipitators for a number of applications, including final mist elimination from gas scrubbers, collection of fine particles from aluminum production, and collection of high resistivity dust. Irrigation of the collection plates is accomplished by the use of a weir at the top of the plate which provides a flow of fluid down the plate, or by sprays which discharge horizontally or vertically into the interelectrode region.

By far the majority of precipitators in use are of the dry type, in which dust is removed by rapping. Successful removal of dust by rapping involves the formation of a coherent dust which, when dislodged, falls as a sheet or agglomerate. This requires that a dust layer of some appreciable thickness be accumulated between rapping cycles which vary from a few minutes to hours, depending upon dust loadings, dust properties, etc.

There has been surprisingly little information about the mechanism of dust removal in view of its importance to the precipitation process. The factors influencing optimum rapping include the forces that hold the dust layer to the collection surface. These forces are electrical, molecular, and mechanical in nature. Molecular forces include van der Waals' forces at the surface of the particle, modified by molecules adsorbed on the surface. These forces can be influenced by surface conditions owing to the presence of molecular surface layers. Mechanical forces are due primarily to the interlocking of particles and to particle adhesion.

It appears that the phenomenon of collection of agglomerates required to give optimum rapping conditions may be associated with dust resistivity. In the case of high-to-medium resistivity dust, a residual charge remains after the agglomerate is dislodged from the plate. In the case of very low resistivity dust, the residual charge can be low or even of the opposite polarity. In such cases reentrainment losses can be high.

The design of electrostatic precipitators requires considerable experience. However, the principles of design follow the key elements of the operation from particle collection through removal. The effectiveness of collection is measured in terms of an ideal efficiency. The most extensive investigations relating precipitator efficiency to operating parameters have been the work of Anderson (1919) and Deutsch (1922). Anderson (1919) showed experimentally that the control efficiency of a precipitator was described by the empirical equation

$$\mathcal{E} = 1 - \exp(-K_a t). \tag{10.35}$$

The value of the empirical factor K_a depends on the set of operating conditions.

The Deutsch (1922) equation was derived on the basis of theoretical considerations, the principal assumptions being that (a) the particle concentration is uniform through the cross section, (b) the particles are fully charged immediately on entering the precipitator, and (c) there is no loss or reentrainment of the collected particles.

Within the boundary layer near the collection plate, each particle has a migration velocity component q_E in the direction normal to the collection surface. Within the time interval t, particles within a distance $q_E t$ will be precipitated onto the collecting surface.

From these relationships and the concentration of particles in the boundary layer, it can be shown that the efficiency of particle removal is an exponential relationship of the form

$$\mathcal{E} = 1 - \exp[-(S_E/Q_g)q_E], \qquad (10.36)$$

where S_E is the collecting surface area (m²), Q_g the gas volumetric flow rate (m³/sec), and q_E is the particle migration velocity (m/sec). This equation is commonly referred to as the Deutsch equation and is widely used in precipitator design and analysis.

There are several relationships of interest in the use of the Deutsch equation. As derived, the equation predicts only the amount of material reaching the precipitator collection surface and does not consider effects of reentrainment by any of the various mechanisms by which collected dust can be reintroduced into the gas stream. Because it generally describes the exponential relationships between efficiency, collecting area, and gas volume, the Deutsch equation has been used to describe precipitator performance. The expression can be used to calculate the factor q_E from measured efficiencies if the other parameters are known. Migration velocities so calculated have been used by precipitator manufacturers as a basis for precipitator sizing. Used in this sense, the migration velocity is more properly an empirical proportionality constant and should be referred to as the precipitation rate parameter to distinguish it from the more theoretically based migration velocity of the Deutsch equation. The relationship used as an empirical correlation method is often referred to as the Deutsch–Anderson equation since the exponential form was first reported by Anderson (1919).

Typical measured efficiency curves for electrostatic precipitators are shown with theoretical curves in Figs. 10.37 and 10.38. The results correspond to measurements by different techniques taken at two power plants burning different coals. Concentrations of fly ash were measured at the inlet and outlet of each electrostatic precipitator. The average overall mass collection efficiencies were about 99.6% at the plant used for data in Fig. 10.37 and about 99.2% at the second plant (Fig. 10.38). Particle size distributions were also measured at each inlet and outlet and used for calculating fractional collection efficiencies. These figures also show the theoretical fractional collection efficiencies of the precipitators calculated with the use of a computer model of the electrostatic precipitation process (McDonald and Felix, 1977). The model calculates overall mass collection efficiency as a function of specific collecting area (area of the collection electrode per unit volume of gas flow) with the precipitator geometry, electrical conditions, and the measured inlet particle size distributions used as input data to the computer program for each installation. In addition to the theoretical predic-

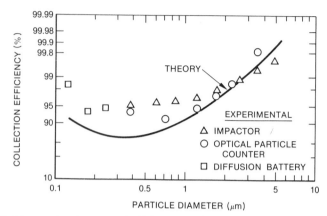

Fig. 10.37. Measured and computed efficiency as a function of particle size for precipitator installation at a plant collecting fly ash from eastern coal burning. [From Oglesby and Nichols (1977b).]

tions, the effect of variation in gas velocity distribution over the cross section of the gas stream was taken into account by including values for the standard deviation σ_g in the gas velocity distribution. The computer program also included procedures for estimating losses in collection efficiency due to gas bypass of the electrified regions of the precipitator and for reentrainment of particles into the effluent gas stream from the deposited fly ash.

The curves in Figs. 10.37 and 10.38 represent fractional collection efficiency values that are predicted by the computer model with σ_g assumed to be 25%, a value which is typical for normal operation of an electrostatic

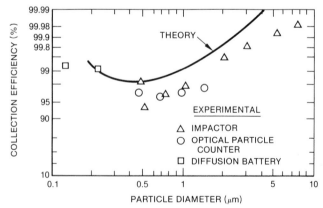

Fig. 10.38. Measured and computer collection efficiency as a function of particle size for collection on the hot gas side of an air heater in a power plant burning ion sulfur western coal. [From Oglesby and Nichols (1977b).]

precipitator, but with no allowance for loss of collection efficiency from gas bypass or fly ash reentrainment.

Figure 10.37 indicates reasonably good agreement between the predicted and observed values characterizing the performance of the precipitator in Plant 1. In contrast, most of the fractional collection efficiency values measured for the precipitator in Plant 2 (Fig. 10.38) were considerably lower than the calculated theoretical values. The difference can be interpreted as the result of a loss of collection due to gas bypass and reentrainment over each of three stages in the precipitator.

Precipitators historically have been sized by analogy with installations of a similar type. The basis for design is the exponential relationship among gas volume, plate area, and efficiency as given by the Deutsch equation [Eq. (10.36)].

The effective migration velocity used is an empirical parameter and includes effects or rapping losses, gas flow distribution, particle size distribution, and dust resistivity, and is properly termed a precipitation rate parameter to distinguish it from the theoretically developed q_E in the Deutsch equation. The exponential relationship has been used in various forms for precipitator sizing. In spite of its various shortcomings, the equation has served as a useful tool in sizing so long as conditions remain about the same for comparable installations.

The method used in sizing of precipitators is to select a value of the effective migration velocity q_E based on experience of the manufacturer or user. For a specified efficiency and gas volume, the plate area can be computed. The values of q_E depend upon many factors and can vary over a considerable range for a given application. Figure 10.39 shows typical values of specific collecting surface area (the ratio of collecting plate area to gas volume) required for 99% collection efficiency for various applications. The general trend is toward a higher specific collecting surface area as the particle size of the dust decreases. The spread in values shows the variations that can be expected for a given application and reflects the range of precipitation rate parameters. Because of this spread, other relationships have been developed which help to narrow the uncertainty in design.

The principal factors influencing the value of migration velocity are particle size, exit dust loading or efficiency, electrical resistivity of the dust, and tendency toward dust reentrainment. Efficiency plays an important role in determining the migration velocity and, hence, the precipitator size.

A second major factor in selecting the precipitation rate parameter is the electrical resistivity of the dust. If the resistivity is high, the greater the allowable charge on the particle, and the longer the particle charging times.

The data from previous installations form the general basis for the selection of the specific value of q_E, and these data must be modified to meet unusual or different circumstances such as abnormally high resistivity or higher efficiency. These procedures have been successful in many types of

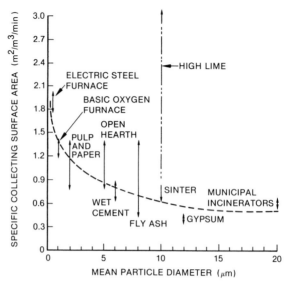

Fig. 10.39. Variation in specific collecting surface area with dust particle size for 99% collection efficiency. [After Oglesby and Nichols (1977a); courtesy of Academic Press.]

applications; however, there have been many installations which have failed to meet design specifications because of inadequate knowledge of those factors which influence performance. With the requirement to meet increasingly severe air pollution regulations, designs have tended to be more conservative, with greater attention being given to process variables.

When dust resistivity is very high, back corona will cause deterioration of precipitator operation at voltages below that for sparking and hence the electrical energization equipment cannot be set for spark rate control.

If a dust with high electrical resistivity is to be collected, the options are (a) design a larger than normal precipitator to accommodate the lower precipitation rate or (b) alter the resistivity so that it is in a more favorable range for precipitation. Generally costs will dictate the latter choice.

The electrical resistivity of a dust can be altered by a change in the operating temperature or by addition of moisture or other conditioning agents.

In the collection of fly ash from electric power boilers, if the usual practice of passing the flue gas through a heat exchanger to heat the combustion air is followed, the temperature of the flue gas is reduced to about 150°C, at which temperature the resistivity of the fly ash may become too high for efficient collection. To overcome this problem, a substantial number of installations have been made in which the precipitator is located ahead of the air heater, where temperatures are generally in the range of 300–350°C. The increased gas volume and the complexity of the duct work required add to the cost of such installations; however, if the volume resistivity of the ash is sufficiently low at these temperatures, the technique is an effective method of combating the high resistivity problem.

At the opposite extreme, flue-gas temperatures can often be reduced with an accompanying reduction in resistivity associated with surface conduction. When low-sulfur coals are being burned, the amount of sulfur trioxide present is low, so that corrosion of the air heater and precipitator may not constitute a problem, unless other corrosive materials, such as chlorine, are present. In such instances, low temperature operation may prove to be an effective and economical method of overcoming high resistivity.

A third method of controlling resistivity is through the addition of chemical conditioning agents to enhance surface conductivity. Historically, additives to flue gas for improving performance have been used in precipitators for the collection of catalyst dust in the petroleum industry. Water spray chambers have been used for the combined purpose of reducing temperature and adding moisture to condition the dust in precipitators used on municipal incinerators, metallurgical furnaces, and cement kilns.

Conditioning of flue gases from electric power boilers has been tried on both pilot and full scale by means of various additives. The most common additive is SO_3, in the stabilized anhydrous form, as vaporized H_2SO_4, or as SO_3 from the catalytic oxidation of SO_2. The quantity of SO_3 required for a given change in resistivity depends upon the composition of the ash. Effectively conditioned ash contains small quantities of free H_2SO_4 on the surface of the particles. As stated previously, a basic ash high in lime (CaO) tends to condition less well than a substantially neutral ash. Although the mechanism has not been definitely established, it has been suggested that the SO_3 deposited on a particle of basic ash probably reacts with water and with calcium ions to form a shell of calcium sulfate, which itself does not materially reduce resistivity. Once this layer of sulfate is formed, adsorption of additional SO_3 forms free H_2SO_4 on the surface, which reduces the dust resistivity.

Ammonia (NH_3) has been used successfully as a conditioning agent for improving fly ash precipitator performance. Reese and Greco (1968) report substantial improvements in performance of a precipitator used on a boiler burning high-sulfur coal (3–4%) with flue-gas temperatures in the vicinity of 125°C. These conditions would normally be expected to produce an ash with very low resistivity. The improvement due to ammonia conditioning appears to be associated with factors other than resistivity changes. Studies of ammonia injection indicate that the ammonia probably reacts with the SO_2 in the flue gases to produce fine particles, probably ammonium sulfate, which alter the space charge electric field, primarily in the inlet section of the precipitator.

Precipitators are generally purchased on the basis of a specified collection efficiency, exit dust loading, stack plume opacity, or all three. Specifications generally include the gas volume to be treated, gas composition, temperature, and information relative to the dust properties or type of fuel burned. To meet these conditions, precipitators must be sized to provide sufficient collecting surface area.

The rapping techniques vary among manufacturers, but they involve either a periodic vibration or impact which dislodges the dust and permits it to fall toward the hopper. The primary requirement for successful rapping is that it accomplishes removal of the dust without excessive reentrainment. This is physically accomplished by (a) adjusting the rapping intensity to prevent powdering or excessive breaking up of the dust layer, (b) adjusting the rapping frequency to give optimum dust layer thickness for the most effective removal, (c) maintaining proper air flow and baffling, and (d) rapping only a small portion of the precipitator at a time.

Sproull (1972) reports that optimum rapping conditions occur if the dust layer slides vertically down the collection plate a distance of perhaps several feet following each rap. Under these conditions, the dust will proceed down the collection plate in discrete steps until it finally falls into the hopper.

Successful rapping should avoid the condition in which the dust layer is allowed to fall freely off the plates; otherwise the dust will reentrain in the gas stream. This requires that the rapping intensity and rapping frequency be adjusted for optimum conditions which are related to the forces holding the dust layer to the collection surface. At the opposite extreme, the optimum rapping conditions must prevent powdering of the dust by too severe a rap or by attempting to remove too thin a dust layer.

It is apparent that a dust layer 0.6–1.2 cm thick falling from a 10-m high plate would reach a rather high terminal velocity in free fall. Thus localized scouring could cause a large percentage of the dust to be reentrained. Further, the fall of so large a quantity of dust into the hopper would generate a large dust cloud, which would be picked up by the gas stream and carried out of the section being rapped.

The effect of reentrainment on overall precipitator performance depends upon the number of precipitator rapping sections, the composite precipitator efficiency, and the rapping loss in each section. The material lost from the inlet section as a result of rapping adds to the inlet dust burden of subsequent sections and is at least partially recollected. Rapping losses from the last section are not collected and appear as direct dust losses. However, the amount of material collected in the last section is small for high efficiency collectors because of the exponential removal rates. Consequently, even though the percentage dust loss due to rapping may be high, its overall effect on performance is not as great as it might appear at first thought.

Oglesby and Nichols (1977a) have analyzed the influence on precipitator performance of various percentage rapping losses for a precipitator with four independently powered sections in series designed for different efficiencies. They find that with rapping losses of 20% per section in a precipitator designed for 99% collection efficiency the efficiency due to rapping loss drops to about 95%. This constitutes a fivefold increase in exit dust loading.

Such a loss in performance would be significant in modern industrial operations.

A standard engineering technique that has been used in precipitator sizing for new applications is to use a small pilot-scale precipitator. There are several approaches to the use of pilot precipitators to derive design data. However, there are also problems with the use of pilot precipitators, since they almost always perform better than full-size units. The primary reasons for this are that in the pilot-scale precipitator (a) the gas flow distribution is almost always better, (b) the degree of sectionalization in flow through collectors is better, and (c) better electrode alignment can be maintained. Because of better sectionalization and alignment, pilot precipitators can generally operate at much higher current densities and voltages than full-scale plants. Thus, if the pilot-scale precipitator is operated spark limited, a scale factor must be applied to arrive at the full-scale precipitator size. This factor can often be quite large and can lead to uncertainties in the sizing of the full-scale plant.

An alternate approach is to set the current density on the pilot-scale precipitator to that expected for the full-scale unit. This technique presupposes that the permissible operating current density is known. Since it is related to dust properties (resistivity and breakdown voltage of the dust layer), these properties must be known in order to have a basis for selecting the operating current density.

10.4.6 Cost–Performance Factors. The design of particulate control systems requires certain technical specifications, but also requires careful performance–cost analysis. In general, costs will be a major consideration in engineering for a given efficiency of particle collection. Performance in terms of energy requirements and collection efficiency for key particle size ranges also will be of concern. Energy requirements often are determined by pumping associated with pressure drop. However, electrical energy loss may be crucial in energy considerations for an electrostatic precipitator. As another factor, wet scrubbers generally operate at lower efficiencies than filters or electrostatic precipitators, particularly for particles in the 0.1–1.0 μm range.

A filter system generally is more efficient than electrostatic precipitation at the expense of greater pressure loss. Because of the minimum in migration velocity for a given electric field in the range between 0.1 and 1.0 μm diameter, the filter system offers significantly better collection performance in this range, other operating factors being equal.

Good practice will involve minimization of control cost while meeting the necessary removal of contamination from the effluent. The cost of the control system is normally segregated into the hardware and installation as compared with operating costs. For any given method, these will vary widely depending upon several factors. Some of them are listed in Table 10.6.

TABLE 10.6

Conditions Affecting Cost of Control Devices Installed[a]

Cost category	Low cost	High cost
Equipment transportation	Minimum distance; simple loading and unloading procedures	Long distance; complex procedure for loading and unloading
Plant age	Hardware designed as an integral part of new plant	Hardware installed into confines of old plant requiring structural or process modification or alteration
Available space	Vacant area for location of control system	Little vacant space requires extensive steel support construction and site preparation
Corrosiveness of aerosol	Noncorrosive gas	Acidic emissions requiring high alloy accessory equipment using special handling and construction techniques
Complexity of startup	Simple startup, no extensive adjustment required	Requires extensive adjustments; testing; considerable down time
Instrumentation	Little required	Complex instrumentation required to assure reliability of control or constant monitoring of gas stream
Guarantee on performance	None needed	Required to assure designed control efficiency
Degree of assembly	Control hardware shipped completely assembled	Control hardware to be assembled and erected in the field
Degree of engineering design	Autonomous "package" control system	Control system requiring extensive integration into process, insulation to correct temperature problem, noise abatement
Utilities	Electricity, water, waste disposal facilities readily available	Electrical and waste treatment facilities must be expanded, water supply must be developed or expanded
Collected waste material handling	No special treatment facilities or handling required	Special treatment facilities and/or handling required
Labor	Low wages in geographical area	Overtime and/or high wages in geographical area

[a] From Vandergrift *et al.* (1971).

Although specific ranges of cost for a given method are difficult to generalize, a feel for the range is shown in Figs. 10.40 and 10.41. In the format, installed capital cost estimates in 1982 dollars are shown. These tend to increase with required efficiency and with air volume to be cleaned. A similar pattern is seen for annualized operating costs shown in Fig. 10.41. Comparison of these results suggests that baghouse filter units are expensive to install and operate, but give very high collection efficiencies, as noted previously. Conventional wet scrubber technology offers the lowest collection

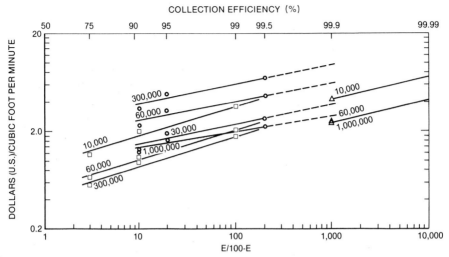

Fig. 10.40. Estimated installed costs in 1982 dollars for control equipment. Parameter on curves: equipment capacity in CFM; △, fabric filter; ○, electrostatic precipitator; □, wet scrubber. [After Shannon *et al.* (1974); scaled linearly to approximate inflation rate since 1973.]

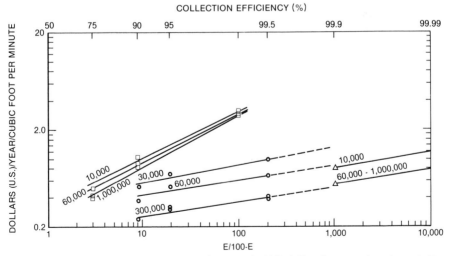

Fig. 10.41. Estimated annualized operating costs in 1982 dollars for control equipment. Parameter on curves: equipment capacity CFM; △, fabric filter; ○, electrostatic precipitator; □, wet scrubber. [After Shannon *et al.* (1974); scaled to approximate inflation rate since 1973.]

efficiencies with somewhat lower installation costs, but comparable or increased operating costs. Electrostatic precipitators have somewhat higher installation costs compared with the other methods shown, but comparable or lower operating costs than baghouse filters. Generally, electrostatic precipitator costs decrease with electrical generating capacity of power plants.

During inflationary periods such as the 1980s operating costs tend to increase, but installation costs will vary year by year where inflationary increases may be counteracted by improved, more cost-effective technology.

10.5 CONTROL OF SECONDARY PARTICLES

The design for control of particulate material derived from atmospheric chemical reactions has been considered recently for meeting air quality standards. This question poses significant problems when the emissions of primary particles are reduced in areas to the range where secondary particles may dominate the ambient conditions. Some cities in the United States are approaching this situation now. The dilemma faced by regulatory agencies involves the determination of reductions required to achieve ambient standards where different gaseous precursors are involved, including sulfur dioxide, nitrogen oxides, and certain organic vapors (U.S. EPA, 1981b). Since conversion of precursors to products occurs with particle-to-gas precursor ratios less than unity, a smaller than one-to-one proportionality between precursor and product is involved. Thus, on an equivalent mole basis, more precursor control may be required to achieve an equivalent reduction in product. The reduction of SO_2 and NO_x by source to emission levels below those required for achievement of the applicable gas standards represents a major increase in costs for gas control.

The difficulty in engineering for the control of products by reduction in precursor gases is illustrated by considering the logical extension of the *rollback concept,* as hypothesized several years ago by Holmes (1975). Accordingly, one would expect that a reduction in aerosol particle precursors such as SO_2 would result in a proportional reduction in particle sulfate. Unfortunately, attempts to verify such logic have not been entirely successful from available emissions and particulate data. As an example, consider the case of the greater Los Angeles area, where power plants represent more than half of the SO_2 emissions reported. Studies of White *et al.* (1978), for example, examined the relationship between actual power plant emissions in the city and sulfate data from monitoring stations downwind on the same day. Because meteorological and chemical variability is so large, their work failed to show any significant relationship between individual source variability and ambient conditions. A similar result was obtained from the Henry and Hidy (1979, 1982) statistical analyses of *urban* sulfate data with sulfur dioxide and other aerometric data. These workers found that factor analysis showed that any significant relation between ambient behavior of SO_2 and particulate sulfate measured the same day at the same location was not present in more than a year of data from Los Angeles, St. Louis, and New York City. This suggests only a limited, direct relationship between local SO_2 precursors and ambient sulfate in these cities. Even in the case of Salt Lake City where SO_2 is dominated by a single large metal smelter source, sulfate variability was accounted for mainly by meteorological factors.

Other data reported by the California Air Resources Board (1976) have shown that trends in annual SO_2 emissions data and ambient sulfate observations do not correlate well in the aggregate. This is shown, for example, in Fig. 10.42. Here, there is no consistent logical link between the two variables which would rationalize the sum of the rollback concept. Later analysis of Chang *et al.* (1979), however, has indicated a statistically rational relation between particulate sulfate concentrations and power plant emissions in the Los Angeles area, given several years of reliable data.

Recently the South Coast Air Quality Management District used an extensive multisource ambient data analysis of Cass (1977) to derive a proposed selective SO_2 emission rollback plan to achieve California's ambient SO_2 and particulate sulfate standards in the Los Angeles area. To date this concept has not been implemented because of disputes about the legal basis for these standards.

Another form of a rollback hypothesis has been suggested recently by Henry (personal communication). Principal component analysis of rural sulfate and other aerometric data taken in 1977–1978 showed that the combinations of physical variables describing sulfate variability apparently were related mainly to seasonal and meteorological factors. However, the association of ambient SO_2 concentrations was significant[†] and was isolated in one or two principal components which were similar in makeup for nine different locations. Since the principal components are independent of one another, by definition, sulfate concentrations on a given day are approximated by the form of Eq. (7.5):

$$[SO_4^{2-}] = \overline{[SO_4^{2-}]} + bP, \tag{10.37}$$

where $\overline{[SO_4^{2-}]}$ is the annual average at the station and b is the (linear) regression coefficient for the component P; P is taken as the principal component containing most of the SO_2 variability associated with sulfate variability. It may not necessarily be a single principal component entirely derived from the data set, but is a component made up of a weighted value of one or more principal components containing the SO_2 variability.

Next, Henry assumes that the value of P can be estimated as a first approximation by the relation

$$aP = ([SO_2] - \overline{[SO_2]})/\sigma_{SO_2}, \tag{10.38}$$

[†] This same argument was not applicable to urban data in four cities because other aerometric factors overshadowed the statistical relation between sulfate and ambient SO_2 concentrations (Henry and Hidy, 1979, 1982).

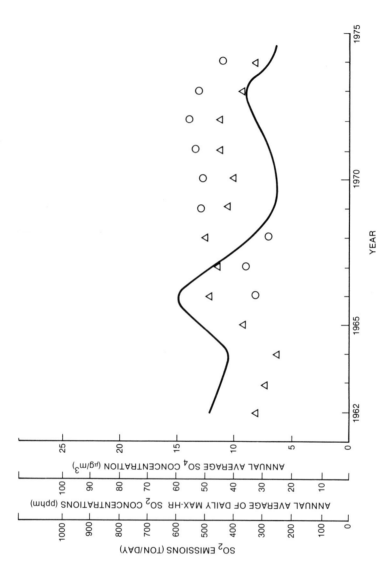

Fig. 10.42. Sulfur dioxide emissions and ambient sulfur dioxide and sulfate concentrations in Los Angeles County 1962–1974. (——), SO$_2$ emissions (Los Angeles County); $\triangle$, SO$_2$ concentrations (composite of data from Los Angeles, Long Beach, and Pasadena—Los Angeles zone); $\bigcirc$, SO$_4$ concentrations (composite of data from Los Angeles, Lennox, and West Los Angeles—Los Angeles zone). [From CARB (1976).]

where $\overline{[SO_2]}$ is the annual mean of SO_2 concentration at the station, a is the loading factor for P, and σ_{SO_2} is the standard deviation of SO_2 variation. This form of an estimator for P can be derived directly from application of matrix algebra to Eq. (7.4) and deriving a least-squares estimation formula for component P (Harmon, 1976). Substituting a form of P from Eq. (10.38) into Eq. (10.37), Henry finds that the portion of sulfate variability associated with a component containing the strongest SO_2 variability is

$$[SO_4^{2-}] = ([SO_4^{2-}] - b\overline{[SO_2]}/a\sigma_{SO_2}) + b[SO_2]/a\sigma_{SO_2}, \quad (10.39)$$

where the proportionality factor $b/a\sigma_{SO_2}$ is in the units $(\mu g\ SO_4^{2-}/m^3)/(ppb\ SO_2)$. Since the loading factor a is in the denominator, it must be maximized for P in the analysis.

Using the principal components derived from the ambient data (Mueller and Hidy, 1983), Henry found, for example, at a characteristic site in Massachusetts,

$$P = 0.76P_3 + 0.62P_4.$$

Here, P_3 and P_4 are the third and fourth principal components derived from analysis of a set of seventeen aerometric variables considered at this site. Numerically, b was determined to be 3.8 $\mu g\ SO_4^{2-}/m^3$, $a = 0.75$, $\sigma_{SO_2} = 4.7$ ppb SO_2, and $\overline{SO_2} = 3.2$ ppb SO_2 for the data set. Calculation of the factor $b/a\sigma_{SO_2}$ gave 1.04 $(\mu g\ SO_4^{2-}/m^3)/(ppb\ SO_2)$. Estimation of this factor for eight other stations located between Indiana and the Atlantic Ocean yielded values between 0.5 and 1.5 $(\mu g\ SO_4^{2-}/m^3)/(ppb\ SO_2)$, with a value of about 1.0 for locations where data was available.

Extension of Henry's hypothesis then implies that reduction in the annual 24-hr arithmetic mean of SO_2 concentration by 1 ppb will reduce 24-hr annual arithmetic mean ambient sulfate concentrations by 1 $\mu g/m^3$, plus or minus about 50% in the eastern United States. If ambient SO_2 concentrations are linearly (1 : 1) proportional to SO_2 emissions, then about four times the emission reduction is required for decrease in ambient sulfate concentration as required for SO_2 reduction in the northeastern United States. Since particulate sulfate is a third or less of the total mass concentration found from high-volume filter samples, major reductions in SO_2 emissions would have little influence on total suspended-particle concentrations as currently regulated if Henry's hypothesis is correct. Presumably this seeming paradox is related to the complex steps to form atmospheric sulfate from a reservoir of ambient SO_2 which mixes, reacts, and is lost by deposition at the ground at varying rates over a large, but limited geographical area.

Reduction in the intensity or occurrence of the regional scale sulfate episodes observed across the eastern United States could offer an alternative approach to regional secondary particulate control. The regional events involving observed sulfate concentrations over horizontal distances exceeding a thousand kilometers occur for several-day periods about one-third of the time especially in summer. Calculations using a numerical air quality model point to major difficulties in reduction of sulfate by reduction in SO_2 emissions at the onset of an episode (Mueller and Hidy, 1983). An illustrative example is shown in Table 10.7. The calculation accounts for air transport mixing, chemical transformation of SO_2, and dry deposition of SO_2 and sulfate (see p. 679). Here a case is taken for conditions corresponding to part of a pollution episode in the greater northeastern United States. The calculated condition for sites selected with increasing distance from the major SO_2 emission centers in the Midwest and Atlantic Coast are indicated. Suppose all the power plants in the region of interest were shut down on the first day of the episode, about two-thirds of the SO_2 emissions would be eliminated according to the regional emission inventory (Mueller and Hidy, 1983). We see that such an action would result in minimal reductions the first day, as seen at sites in the upper Ohio River Valley near large power plant sources. With movement of polluted air away with the prevailing southwest winds and subsequent conversion to sulfate, greater reductions would be observed farther and farther away on subsequent days of the episode. However, the linear proportionality between SO_2 emissions and ambient sulfate generally would not be found by this calculation at any given receptor location.

Detailed calculations for verifying source–receptor relationships between nitrate or secondary organic contributions to particulate matter have not been reported as yet because the transformation chemistry is complex and not well enough understood for parametrization with any confidence.

The results from available air quality models point to the pitfalls in relying on rollback concepts for secondary particle control, as discussed in Section

TABLE 10.7

Comparison of Simulations for 17–19 July 1978 With and Without Electric Utility (UT)
Emissions 24-Hour Average Sulfate Concentrations ($\mu g/m^3$)[a]

	17 July			18 July			19 July		
Region	UT	No UT	Percent decrease	UT	No UT	Percent decrease	UT	No UT	Percent decrease
New York City	15	13	13	19	14	26	19	15	21
Pittsburgh	11	9	18	21	13	28	25	16	56
Adirondacks	6	6	0	5	5	0	21	14	33
Upper Ohio river area	14	8	43	25	12	52	41	15	63
Boston, MA	11	10	9	12	9	25	20	13	35
Washington, D.C.	13	11	15	23	14	39	28	14	50
Central Illinois	19	13	36	26	13	50	34	15	56
Tennessee	10	8	20	20	11	45	27	14	48
South Carolina	10	8	20	16	13	19	18	9	50
Maine	3	3	0	4	4	0	10	7	30

[a] From Mueller and Hidy (1983); courtesy of the Electric Power Research Institute.

10.3. Yet no other methods currently are available for such determinations. Thus the rollback approach will probably be used with all its uncertainties for particulate control strategies for some time to come.

The options for emission control of particulate precursor gases are relatively limited. For sulfur dioxide, these include switching to low-sulfur fuels, fuel desulfurization or washing (particularly coal), or flue-gas desulfurization (FGD) by scrubbing. The latter currently is optimum for sulfur removal from emissions. About 90% SO_2 reduction can be achieved by limestone scrubbers with reasonable reliability, but the cost for their use is a substantial fraction of total costs of power plants and other facilities (OECD, 1981).

Nitrogen oxides can be removed by modification of burner configurations in boilers (Muzio and Arand, 1976) or scrubbing by catalytic reduction with ammonia (Saleem *et al.*, 1979). Burner design offers only limited prospects for control of stationary sources, and the scrubbing technology remains to be used extensively in practice. Preliminary work suggests that such devices can achieve 75–90% removal of NO_x for moderate to large boilers. Solid catalytic converters for NO_x reduction are employed for NO_x reduction on automobiles, and may be employed in the future for small stationary internal combustion devices.

Hydrocarbon vapor reduction generally is accomplished by absorption in charcoal columns, or by catalytic oxidation to carbon dioxide. Essentially no work has been done on selective removal of reactive hydrocarbons potentially involved in aerosol particle production in ambient air. The assumption is made that total hydrocarbon reduction will reduce equally particle precursors and low-molecular-weight vapors.

Given current regulatory practice, the motivation is not strong for significant effort to install special devices to control efficiently reactive gases which are precursors for particle production. Based on cost considerations alone for handling larger gas volumes with low contaminant concentrations, only a limited regulatory effort will be spent in this area until controls for optimizing reduction in gas concentrations is achieved for control to meet ambient air quality standards for gases. The regulatory pressure taking into account widespread visibility impairment and acidic deposition eventually may provide sufficient motivation for human welfare to require incremental reduction programs for SO_2 and NO_x in North America and Europe (U.S. Canadian Working Group, 1981; OECD, 1981).

REFERENCES

Alden, T. L. (1959). "Design of Industrial Exhaust Systems for Dust and Fume Removal," 3rd ed. Industrial Press, New York.
Alexander, R. (1949). *Proc. Australas. Inst. Min. Metall.* No. 152, 202.
Alpert, D. J., and Hopke, P. K. (1980). *Atmos. Environ.* **14**, 1137.
American Conference of Governmental Industrial Hygienists (1974). "Committee on Industrial

Ventilation, Industrial Ventilation, A Manual of Recommended Practice," 13th ed. ACGIH, Lansing, Michigan.

American Meteorological Society (1981). "Air Quality Modeling and the Clean Air Act." Am. Meteorol. Soc., Boston, Massachusetts.

Anderson, E. (1919). *Trans. Am. Inst. Chem. Eng.* **16,** 69.

Atkinson, R., Lloyd, A., and Winges, L. (1982). *Atmos. Environ.* **16,** 1341.

Barth, D. S. (1971). *J. Air Pollut. Control Assoc.* **20,** 519.

Bass, A. (1980). *Proc. J. Conf. Appl. Air Pollut. Meteorol.,* 2nd, Boston, Mass., pp. 193–215.

Benarie, M. (1976). Rep. EPA-600/4-76-055. U.S. Environ. Prot. Agency, Research Triangle Park, North Carolina.

Bergman, W., Hebard, H., Taylor, R., and Lum, B. (1979). *World Filtr. Congr.,* 2nd, London, p. 30.

Bickelhaupt, R. E. (1974). *J. Air Pollut. Control Assoc.* **24,** 251.

Braekke, F. H. (1976). "Impact of Acid Precipitation on Forest and Freshwater Ecosystems," Rep. FR/676. Sur Nedbørs Virkning På Skog og Fisk Proj., Oslo, Norway.

Bragg, G. M. (1981). *In* "Filtration of Particulates, Air Pollution Control" (G. M. Bragg and W. Strauss, eds.), Part 4, p. 276. Wiley, New York.

Britt, H. I., and Luecke, R. H. (1973). *Technometrics* **15,** 233.

Brookman, E. T., and Yocom, J. E. (1980). "Environmental Management: A Case Study in the Use of Aerometric Data for Source Assessment," Rep. for EPA. TRC, Wethersfield, Connecticut (unpublished).

California Air Resources Board (1976). Rep. DTS-76-1. CARB, Sacramento, California.

California Air Resources Board (1979). Staff Rep. 79-2-2. CARB, Sacramento, California.

Calvert, S. (1977). *In* "Air Pollution" (A. C. Stern, ed.), 3rd ed., Vol. 4, p. 257. Academic Press, New York.

Caplan, K. J. (1977). *In* "Air Pollution" (A. C. Stern, ed.), 3rd ed., Vol. 4, p. 98. Academic Press, New York.

Cass, G. (1977). Ph.D. Thesis, California Inst. of Technol., Pasadena.

Chang, N., Zeldin, M. D., and Hone, Y. (1979). Rep. TSC-PD-B592-9, for South. Calif. Edison Co. Technology Service Corp., Santa Monica, California.

Clean Air Act (1968). Public Law 88-206, amended 1970; Public Law 91-604, amended 1977; Public Law 95-604. U.S. Gov. Print. Off., Washington, D.C.

Cottrell, F. G. (1911). *Ind. Eng. Chem.* **3,** 542.

Deutsch, W. (1922). *Ann. Phys. (Leipzig)* **68,** 335.

Dimitriades, B. (1972). *Environ. Sci. Technol.* **3,** 253.

Eckert, J. S. (1961). *Chem. Eng. Prog.* **57,** 54.

Egan, B., and Mahoney, J. R. (1972). *J. Appl. Meteorol.* **11,** 312.

Eliassen, A., and Saltbones, J. (1975). *Atmos. Environ.* **9,** 425.

Eltgroth, M. W., and Hobbs, P. V. (1979). *Atmos. Environ.* **13,** 953.

Ensor, D., Cowen, S., Shendriker, A., Markowski, G., Woffinden, G., Pearson, R., and Scheck, R. (1981). Rep. CS-1669, p. 1–13. Electric Power Research Institute, Palo Alto, California.

Fox, D. (1981). *Bull. Am. Meteorol. Soc.* **62,** 599.

Friedlander, S. K. (1973). *Environ. Sci. Technol.* **7,** 235.

Gartrell, G., Heisler, S. L., and Friedlander, S. K. (1980). *In* "The Character and Origins of Smog Aerosols" (G. M. Hidy and P. K. Mueller, eds.), p. 665. Wiley, New York.

Gatz, D. (1975). *Atmos. Environ.* **9,** 279.

Gifford, F. A., Jr., and Hanna, S. R. (1973). *Atmos. Environ.* **7,** 131.

Guldberg, P. H., and Kern, C. W. (1978). *J. Air Pollut. Control Assoc.* **28,** 907.

Harmon, H. H. (1976). "Modern Factor Analysis," 3rd ed. (revised). Univ. of Chicago Press, Chicago.

Hatch, T. (1962). *In* "Air Pollution" (A. C. Stern, ed.), 1st ed., Vol. 2, p. 211. Academic Press, New York.

Heisler, S. L., Henry, R. C., and Watson, J. G. (1980a). *Air Pollut. Control Assoc. Annu. Meet., 73rd, Montreal* Pap. 80-58.6.

Heisler, S. L., *et al.* (1980b). Rep. P-5417-1 for Motor Vehicle Manuf. Assoc. U.S. Environ. Res. & Technol., Westlake Village, California.

Henry, R. C. (1977). Ph.D. Thesis, Oregon Grad. Cent., Beaverton.

Henry, R. C., and Hidy, G. M. (1979). *Atmos. Environ.* **13**, 1581.

Henry, R. C., and Hidy, G. M. (1982). *Atmos. Environ.* **16**, 929.

Henry, R. C., Godden, D. G., Hidy, G. M., and Lordi, N. J. (1981). Rep. 4339. Sulfate Task Force, Am. Pet. Inst., Washington, D.C.

Heuss, J. M., Nebel, G. J., and Colucci, J. M. (1971). *J. Air Pollut. Control Assoc.* **21**, 535.

Hidy, G. M. (principal investigator) (1976). Rep. EC-125. Electric Power Res. Inst., Palo Alto, California.

Hidy, G. M., Mueller, P. K., and Tong, E. Y. (1978). *Atmos. Environ.* **12**, 735.

Hidy, G. M., and Mueller, P. K., eds. (1980). "The Character and Origins of Smog Aerosols." Wiley, New York.

Holmes, J. R. (1975). Staff Rep. 75-20-3. Calif. Air Resour. Board, Sacramento, California.

Hopke, P. K., Gladney, E. S., Gordon, G., Zoller, W., and Jones, A. (1976) *Atmos. Environ.* **10**, 1015.

Hutchinson, T. C., and Havas, M. (1980). "Effects of Acid Precipitation on Terrestrial Ecosystems." Plenum, New York.

Iinoya, K., and Orr, C., Jr. (1977). *In* "Air Pollution" (A. C. Stern, ed.), 3rd ed., Vol. 4, p. 149. Academic Press, New York.

Jennings, R. G. (1950). *J. Iron Steel Inst., London* **164**, 305.

Johnson, W. B., ed. (1972). *Proc. Interagency Conf. Environ., Livermore, Calif., 1972* p. 114.

Johnson, W. B., Wolf, D. E., and Mancuso, R. L. (1978). *Atmos. Environ.* **12**, 511.

Karplus, W. J., Bekey, G. A., and Pekral, P. J. (1958). *Ind. Eng. Chem.* **50**, 1657.

Kimura, N., and Iinoya, K. (1965). *Kagaku Kogaku* **29**, 166.

King, R. B., Fordyce, J. S., Antoine, A. C., Leibecki, H. F., Neustadter, H. E., and Sidik, S. M. (1976). *J. Air Pollut. Control Assoc.* **26**, 1073.

Kleinman, M. (1977). Ph.D. Thesis, New York Univ., New York.

Kowalczyk, G. S., Choquette, C. E., and Gordon, G. E. (1978). *Atmos. Environ.* **12**, 1143.

Lloyd, A. C., Lurmann, F., and Atkinson, R. (1981). Final Rep. P-A070, API Publ. No. 4348. Sulfate Task Force, Am. Pet. Inst., Washington, D.C.

Lurmann, F., Young, J., and Hidy, G. (1983) *In* "Chemistry of Multiphase Atmospheric Systems." NATO Advanced Study Institute (in press).

McDonald, J., and Felix, L. (1977). EPA Rep. 600/8-77-020b. U.S. Environ. Prot. Agency, Research Triangle Park, North Carolina.

Macias, E., Zwicker, J., Ouimette, J., Hering, S., Friedlander, S., Cahill, T., Kuhlmly, G., and Richards, L. W. (1981). *Atmos. Environ.* **15**, 1971.

McRae, G. J., Goodin, W. R., and Seinfeld, J. H. (1982). *Atmos. Environ.* **16**, 679–696.

McRae, G. J., and Seinfeld, J. H. (1983). *Atmos. Environ.* **17**, 501–522.

McRae, A., and Whelchel, L., eds. (1978). "Toxic Substances Control Source Book," p. 325. Cent. Compliance Inf., Aspen Syst. Corp., Germantown, Maryland.

Martin, D. E., and Tikvart, J. A. (1968). *Air Pollut. Control Assoc. Annu. Meet., St. Paul, MN.* Pap. 68–148.

Miller, M. S., Friedlander, S. K., and Hidy, G. M. (1972). *J. Colloid Interface Sci.* **39**, 165.

Moyers, J., Ranweiler, L. E., Hopf, S. B., and Korte, N. E. (1977). *Environ. Sci. Technol.* **11**, 789.

Mueller, P. K., and Hidy, G. M. (principal investigators) (1983). Rep. EA-1901. Electric Power Res. Inst., Palo Alto, California.

Muzio, L. J., and Arand, J. K. (1976). Rep. FP-253. Electric Power Res. Inst., Palo Alto, California.

National Academy of Sciences (1974). "Air Quality and Automobile Emission Control," Vol. 3, Serial No. 93-24. Prepared for Committee on Public Works, U.S. Senate. U.S. Gov. Print. Off., Washington, D.C.

Nelson, G. O., Bergman, W., Miller, H. H., Taylor, R. D., Richards, C. P., and Biermann, A. H. (1978). *Am. Ind. Hyg. Assoc.* **39** (6), 472.

Occupational Safety and Health (OSHA) Act (1970). 29 U.S. Code 651. U.S. Gov. Print. Off., Washington, D.C.

Oglesby, S., Jr., and Nichols, G. B. (1977a). *In* "Air Pollution" (A. C. Stern, ed.), 3rd ed., Vol. 4, p. 189. Academic Press, New York.

Oglesby, S., Jr., and Nichols, G. (1977b). EPA Rep. 600/8-77-020a. U.S. Environ. Prot. Agency, Research Triangle Park, North Carolina.

Organization for Economic Cooperation and Development (1981). "Cost and Benefits of Sulfur Oxide Control." OECD, Paris.

Pasquill, F. (1974). "Atmospheric Diffusion," 2nd ed. Wiley, New York.

Pontius, D., and Smith, W. (1977). Rep. EPA-600/8-77-020c. U.S. Environ. Prot. Agency, Research Triangle Park, North Carolina.

Prahm, L. P., and Christenson, O. (1977). *J. Appl. Meteorol.* **16,** 896.

Quarles, J. (1979). "Federal Regulations of New Industrial Plants," Rep. J. Quarles, P.O. Box 998, Ben Franklin St., Washington, D.C. (unpublished).

Reese, J. T., and Greco. J. (1968). *J. Air Pollut. Control Assoc.* **18,** 523.

Reynolds, S. D., Roth, P. M., and Seinfeld, J. H. (1973). *Atmos. Environ.* **7,** 1033.

Reynolds, S. D., Liu, M. K., Hecht, T. A., Roth, P. M., and Seinfeld, J. H. (1974). *Atmos. Environ.* **8,** 563.

Roberts, O. F. T. (1923). *Proc. R. Soc. London, Ser. A* **104,** 640.

Saleem, A., Galgano, M., and Inaba, S. (1979). Rep. FP-1109-SR. Electric Power Res. Inst., Palo Alto, California.

Seinfeld, J. H. (1975). "Air Pollution: Physical and Chemical Fundamentals." McGraw-Hill, New York.

Shannon, J. D. (1981). *Atmos. Environ.* **15,** 689.

Shannon, L. J., Gorman, P. G., and Park, W. (1974). Rep. EPA-600/5-74-007. U.S. Environ. Prot. Agency, Research Triangle Park, North Carolina.

Shir, C. C., and Shieh, L. J. (1974). *J. Appl. Meteorol.* **13,** 185.

Slinn, W. G. N. (1983). *In* "Atmospheric Sciences and Power Production" (D. Randerson, ed., U.S. DOE Tech. Inf. Cent., Washington, D.C. (to be published).

Spaite, P., and Burckle, J. O. (1977). *In* "Air Pollution" (A. C. Stern, ed.), 3rd ed., Vol. 4, p. 43. Academic Press, New York.

Sproull, W. T. (1972). *J. Air Pollut. Control Assoc.* **22,** 185.

Sutton, O. G. (1953). "Micrometeorology," pp. 134–140. McGraw-Hill, New York.

Theodore, L., and Buonicore, A. J. (1976). "Industrial Air Pollution Control Equipment for Particulates." CRC Press, Cleveland, Ohio.

TRW Systems Group (1970). "Air Quality Implementation Planning Program, SN-11130, Vol. 1, Operator's Manual," Chap. 4, Thompson Ramo Wooldridge (TRW), El Segundo, California.

Turner, D. B. (1967). Publ. No. 999-AP-26. U.S. Public Health Serv., Washington, D.C.

U.S. Canadian Working Group (3B) (1981). "Emissions, Costs and Engineering Assessment, Performed through U.S.–Canada Memorandum of Intent on Transboundary Air Pollution." U.S. Environ. Prot. Agency, Washington, D.C.

U.S. Environmental Protection Agency (1977). Rep. EPA 450/2-77-021a. U.S. Environ. Prot. Agency, Research Triangle Park, North Carolina.

U.S. Environmental Protection Agency (1978). Rep. EPA-450/2-78-027. U.S. Environ. Prot. Agency, Research Triangle Park, North Carolina.

U.S. Environmental Protection Agency (1979). *Fed. Regist.* **44,** 33580.

U.S. Environmental Protection Agency (1981a). "Petitions of the State of New York, the Commonwealth of Pennsylvania and the State of Maine under Section 126 of the Clean Air Act for Interstate Pollution Abatement," Case Docket No. A-81-09. U.S. Environ. Prot. Agency, Washington, D.C.

U.S. Environmental Protection Agency (1981b). *Fed. Regist.* **46,** 42450.

Vandergrift, A. E., Shannon, L. J., Lawless, E. W., Gorman, P. G., Sailee, E. E., and Reichel, M. (1971). Rep. APTD-0745, p. 548. U.S. Environ. Prot. Agency, Washington, D.C.

Watson, J. (1979). Ph.D. Thesis, Oregon Grad. Cent., Beaverton.

Watson, J. (1980). Rep. EPA Contr. 68-02-2542, Task 8. Environ. Res. & Technol., Concord, Massachusetts.

White, W. W., Heisler, S. L., Henry, R. C., Hidy, G. M., and Straughan, I. (1978). *Atmos. Environ.* **12,** 779.

Williamson, S. J. (1973). "Fundamentals of Air Pollution," Chap. 12. Addison-Wesley, Reading, Massachusetts.

Wilson, E. L. (1967). "Hearings before the Subcommittee on Air and Water Pollution," Committee on Public Works, 90th Congress, U.S. Senate, S.780-Air Pollution-1967. U.S. Gov. Print. Off., Washington, D.C.

Zebel, G. (1965). *J. Colloid Sci.* **20,** 552.

APPENDIX: PARTICLE SIZE DISTRIBUTIONS[†]

Particle size distributions may be fit to a variety of mathematical forms. In practice, one often reduces particle size data into groups of size, from which *histograms* can be drawn. The bar histogram is a convenient means of displaying a particle size analysis. The graph is presented with bar height representing number, volume, or mass of particles of a size grouping. The bar width represents the size range (difference between particle radii). Volume, mass, and number fractions (frequencies), taken as numerical fractions or percentages of the whole sample, are commonly used for the vertical scale of histograms. The height of the bar in a histogram is usually sensitive to the size interval $R_{i+1} - R_i$ chosen: a smaller or a larger size interval may contain a different number of particles or a different weight fraction of the sample. To allow for the particle size variation of several orders of magnitude, the logarithmic size scale is used.

A histogram can be converted into a continuous size spectrum easily. The value n_i represented by the bar height can be divided by the size interval for that bar, $\Delta R_i = R_{i+1} - R_i$. The result has the units of

$$n_i/\Delta R_i \qquad \text{(units of } n)/(\text{units of } R).$$

The mean value of the size interval can be taken as an arithmetic or geometric mean. The geometric mean size $\overline{R}_i$ is

$$\overline{R}_i = (R_i R_{i+1})^{1/2} \qquad (\mu\text{m}). \tag{A.1}$$

[†] Extracted from Lerman (1979).

746

The geometric mean is preferred for aerosols having a broad range of sizes. The results can be plotted as a series of discrete points to which a continuous function can be fitted, giving the equation for dN/dR as a function of R which is identical in form with Eq. (3.2) describing the particle number–radius spectra.

A particle size distribution often encountered for atmospheric particulates, known as the power-law or Pareto distribution, $dN/dR = AR^{-\gamma}$, follows from a special type of volume distribution histogram shown in Fig. A.1.

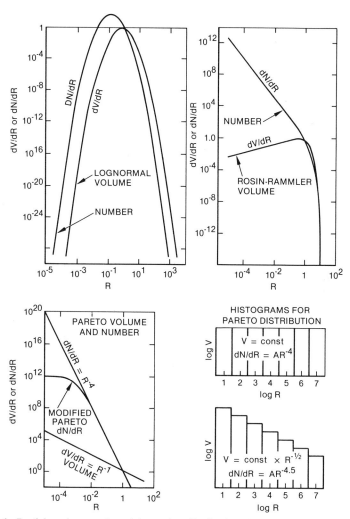

Fig. A.1. Particle volume and particle number distributions based on the lognormal, Rosin–Rammler, and Pareto probability density functions. [From Lerman (1979); reprinted courtesy of Wiley.]

This type of histogram represents the volume distributions in which the volume of particles v_i of each geometric size interval either increases or remains constant or decreases with increasing particle size R_i. Such histograms can be represented by the relationship

$$v_i = \text{const} \times R_i^\alpha \qquad (\mu m^3/cm^3), \qquad (A.2)$$

where α is a constant that can be either positive, zero, or negative. For geometrically increasing size intervals, the ratio of the upper to lower interval R_{i+1}/R_i is constant. If the ratio is denoted $n = R_{i+1}/R_i$, then the geometric mean size of the interval is $\overline{R}_i = R_i n^{1/2}$. The volume v_i is made of n_i particles, all of mean size $\overline{R}_i$. Therefore,

$$v_i = \mathfrak{K}_v \overline{R}_i^3 n_i$$

$$= \mathfrak{K}_v n^{3/2} R^3 n_i \qquad (A.3)$$

where $\mathfrak{K}_v$ is a geometrical shape factor relating the linear size dimension to the volume (for a sphere, $\mathfrak{K}_v = 4\pi/3$).

On the linear scale, the size interval is $\Delta R_i = R_{i+1} - R_i = R_i(n - 1)$. The number of particles per unit of size $N_i/\Delta R_i$ is, from Eqs. (A.2) and (A.3),

$$n_R(R) = \frac{dN}{dR} \simeq \frac{n_i}{\Delta R_i} = \hat{A} R_i^{\alpha-4} \qquad (\text{particles/cm}^3\text{-}\mu m), \qquad (A.4)$$

where $\hat{A}$ is a constant combining the constant coefficients $\mathfrak{K}_v$ and n from the two equations.

A particular case of a flat volume distribution histogram is $v_i = \text{const}$, or $\alpha = 0$. This corresponds to a particle number distribution with a slope of -4 (that is, $dN/dR = \hat{A} R^{-4}$), as shown in Fig. A.1.

Equation (A.4) in logarithmic form becomes

$$\log \frac{dN}{dR} = \log \hat{A} + (\alpha - 4)\log R.$$

A plot of $\log dN/dR$ versus $\log R$ is a straight line with slope $\alpha - 4$, and the constant $\hat{A}$ is equal to the value of dN/dR at $R = 1$. Relatively small departures from the value of $v_i = \text{const}$ are reflected in some scatter of points about a straight line in the plot of $\log dN/dR$ vs. $\log R$. The double-log scale can absorb much of the data scatter. However, it is emphasized that the relationship $dN/dR = \hat{A} R^{-\gamma}$ is best suited to represent the general trends of a particle size distribution, whereas the fine structure of a spectrum is more clearly resolved on a linear scale or in a histogram plot. Deviations from a straight-line log-log plot, or the presence of two or more straight-line segments, each having a different value of the slope γ, can be interpreted as an indicator of such processes as growth or evaporation, differential settling, coagulation fragmentation, or mixing of particles from different sources.

Three frequency distributions have been used extensively to describe the particle number or mass–size spectra of asymmetric shape—the lognormal distribution; the Weibull, or historically more correct, Rosin–Rammler distributions; and the Pareto or power-law distribution. Each of the three distributions and forms derived from the basic distribution types have been applied to particle size spectra of powders, sprays, soils, and airborne dust. The lognormal and the Weibull or Rosin–Rammler distributions are unimodal asymmetric distributions. The Pareto or power-law distribution is a size spectrum decreasing continuously with increasing particle size. Given below are important mathematical and physical characteristics of the three size distributions. A summary of their mathematical relationships and other statistical parameters is given in Table A.1.

In work with atmospheric particulates, one commonly deals with the number percentage or absolute numbers of particles in the size fractions, or particle mass or volume in size fractions. In general, there will be considerable difference in form for the number– and volume–size distributions. But for spheres, at least, the two are interrelated, the latter being the third moment of the former. This relation will not necessarily hold for particle collections of different shapes or for particles of different mass density. Nevertheless, the assumption is often made that the two distributions can be related in a consistent and "universal" way for purposes of particle characterization. The equation for both dV and dN arising from the different frequency distributions are given in the remainder of this section.

Lognormal Distribution. The probability density function

$$f(x)\, dx = (1/x\sigma\sqrt{2\pi}) \exp\left\{- \frac{[(\ln x) - \mu]^2}{2\sigma^2}\right\} dx \tag{A.5}$$

is known as the lognormal distribution, where σ and μ are constants. If a variable χ is introduced instead of ln x (that is, $\chi \equiv \ln x$ and $d\chi \equiv dx/x$), then the lognormal distribution in x transforms into a normal (Gaussian) distribution in χ:

$$f(\chi)\, d\chi = (1/\sigma\sqrt{2\pi}) \exp[-(\chi - \mu)^2/2\sigma^2]\, d\chi, \tag{A.6}$$

where the constants μ and σ have the same significance as in the lognormal distribution, and they are the mean and the standard deviation from the mean of the Gaussian distribution $f(\chi)\, d\chi$. By definition of the probability density function, the integral between the limits of the function's existence is unity, or

$$\int_0^\infty f(x)\, dx = 1, \qquad \int_{-\infty}^{+\infty} f(\chi)\, d\chi = 1.$$

The cumulative normal distribution that gives the probability of occurrence of all the items of size smaller than x, $F_{<x}$, is

$$F_{<x} = \int_{-\infty}^x f(y)\, dy = \frac{1}{\sqrt{\pi}} \int_{y=-\infty}^{y=x} \exp\left(- \frac{(y - \mu)^2}{2\sigma^2}\right) \alpha\, \frac{y - \mu}{2^{1/2}\sigma},$$

$$= \tfrac{1}{2} + \tfrac{1}{2} \operatorname{erf}[(x - \mu)/2^{1/2}\sigma], \tag{A.7}$$

TABLE A.1

The Mean, Median, and Mode of the Frequency Distributions [Probability Density Functions $f(x)dx$] Discussed in the Text[a]

Function	$f(x)dx$ text equation no.	Range of x and constants	Mean x	Median x	Mode x
Gaussian	(A.6)	$-\infty < x < \infty$	μ	μ	μ
Lognormal	(A.5)	$0 < x < \infty$, $\mu > 0,\ \sigma^2 > 0$	$e^{\mu+\sigma^2/2}$	e^{μ}	$e^{\mu-\sigma^2}$
Weibull	(A.13)	$\gamma \leq x < \infty$, $m > 0,\ c > 0$	$\Gamma\!\left(1 + \dfrac{1}{m}\right)$	$\left(\dfrac{\ln 2}{c}\right)^{1/m} + x'$	$\dfrac{1}{c}\left(1 - \dfrac{1}{m}\right)^{1/m} + x'\quad (c > 1)$ $x'\quad (0 < c < 1)$
Rosin–Rammler	(A.16)	$0 \leq x < \infty$, $m > 0,\ c > 0$	$\Gamma\!\left(1 + \dfrac{1}{m}\right)$	$\left(\dfrac{\ln 2}{c}\right)^{1/m}$	$\dfrac{1}{c}\left(1 - \dfrac{1}{m}\right)^{1/m}\quad (c > 1)$ $0\quad (0 < c < 1)$
Pareto or power law	(A.20)	$c \leq x < \infty$, $\gamma > 0,\ c > 0$	$\dfrac{\gamma c}{\gamma - 1}$	$2^{1/c}\,c$	c

[a] From Lerman (1979); courtesy of Wiley.

where y is an integration variable and erf is the error function. Tabulations of both the cumulative probability $F_{<x}$ and the error function are available in many handbooks of mathematical tables. By writing $\mathrm{erf}[(x - \mu)/\sigma\sqrt{2}]$ as a polynomial in $(x - \mu)/\sigma\sqrt{2}$, explicit equations can be written for the cumulative frequency $F_{<x}$ (Lerman, 1979).

Using the same method of integration as in Eq. (A.7), we write the cumulative lognormal probability of all the items smaller than x, $F_{<x}$, as

$$F_{<x} = \int_{y=0}^{y=x} f(y)\ dy = \frac{1}{2} + \frac{1}{2}\ \mathrm{erf}\left(\frac{\ln x - \mu}{2^{1/2}\sigma}\right). \qquad (A.8)$$

The only difference between Eqs. (A.7) and (A.8) is that in the latter, $\ln x$ replaces x. For the lognormal distribution, the limits of x are $0 < x < \infty$, whereas for the normal distribution they are $-\infty < x < +\infty$.

The volume of particles, distributed lognormally as a function of the particle radius R, can be described by the following equation:

$$dV = (\hat{A}/R)\ \exp[-(\ln R - \mu)^2/2\sigma^2]\ dR \qquad (\mu m^3), \qquad (A.9)$$

where $\hat{A}$ is a dimensional constant which includes the product $\sigma\sqrt{2\pi}$ of the lognormal function(if R is taken in μm and dV and $\hat{A}$ are in $\mu m^3/cm^3$). Instead of R, a dimensionless size parameter R/R_0 can be used, where R_0 denotes the lower size of the distribution ($R \geq R_0$). In this case, as R approaches the lower size limit R_0 the quantity dV/dR tends to a finite value of $dV/dR = \hat{A}$ $\exp(-\mu^2/2\sigma^2)$.

In the case of $\mu = 0$ and $\sigma^2 = \frac{1}{2}$, the volume distribution simplifies to

$$dV = (\hat{A}/R)\ \exp[-(\ln R)^2]\ dR. \qquad (A.10)$$

In a double-log form, the preceding equation is

$$\ln \frac{dV}{dR} = \ln \hat{A} - \ln R - (\ln R)^2. \qquad (A.11)$$

The latter produces a nonlinear plot of $\log dV/dR$ against $\log R$.

The volume dV of dN spherical particles of radius between R and $R + dR$ is

$$dV = \tfrac{4}{3}\ \pi R^3\ dN,$$

which in combination with Eq. (A.9) gives the particle number distribution,

$$dN = (\hat{A}'/R^4)\ \exp[-(\ln R - \mu)^2/2\sigma^2]\ dR \qquad (particles), \qquad (A.12)$$

where $\hat{A}'$ is a constant not identical to $\hat{A}$ in (A.11.)

A plot of $\log dN/dR$ versus $\log R$ is nonlinear, as in the case of the particle volume distribution dV/dR. Note that neither dV in Eq. (A.9) nor dN in (A.12) is a probability density function, but they are distribution functions or spectra of the particle volume and particle number, respectively, written as functions of the particle radius.

The mean, median, and mode of the lognormal distribution are listed in Table A.1.

Weibull and Rosin–Rammler Distributions. The probability density function known as the Weibull distribution is

$$f(x) \, dx = cm(x - x')^{m-1} \exp[-c(x - x')^m] \, dx, \qquad (A.13)$$

where c, m, and y are positive constants, and $x > x'$. If $m = 1$, the Weibull distribution becomes an exponential distribution:

$$f(x) \, dx = c \exp[-c(x - x')] \, dx, \qquad (A.14)$$

with the values of $f(x)$ decreasing exponentially with increasing x.

The cumulative distribution of $f(x)$ for all sizes smaller than x is, by integration of Eq. (A.13),

$$F_{<x} = \int_{y=\gamma}^{y=x} f(y) \, dy = 1 - \exp[-c(x - x')^m]. \qquad (A.15)$$

The Swedish physicist W. Weibull derived the probability distribution function with three constant parameters, as given in Eq. (A.13), from his studies of a chain under load: the chances of a chain breaking due to the failure of an individual link increase with an increasing load or, alternatively, the chances of breaking increase with the length of time for which the load has been applied. Weibull recognized the wide applicability of the distribution function to a variety of processes involving fragmentation and comminution of materials. His first contributions were published in the Scandinavian literature in 1939 and subsequently, in 1951, in an English-language periodical. Earlier, however, Eq. (A.13) with only two constant parameters c and m ($y = 0$) was used by Rosin and Rammler (1933), both working in Berlin, to describe the weight distribution of fragments of coal produced by grinding. The statistical properties of the Weibull distribution function were established in the late 1920s, before Rosin and Rammler, and Weibull published their results (Harris, 1971). The Weibull distribution is a special case of a broader class of the so-called generalized gamma distributions, and it has found fairly wide application in studies of fine particles, comminution of natural and man-made materials, and the formation of crystals in solutions.

In Eq. (A.13) for the Weibull distribution, the constant x' is essentially the lower limit of x, because $x \geq x'$ by definition. When x is much larger than x', the latter affects the distribution function $f(x)$ very little. At the lower values of x, the difference $x - x'$ is small, and the shape of the curve $f(x)$ drawn as a function of x can be affected by the choice of x' near the lower end values. With $x' = 0$, the function is known as the Rosin–Rammler distribution,

$$f(x) \, dx = cmx^{m-1} \exp(-cx^m) \, dx. \qquad (A.16)$$

The particle volume distribution dV, obeying the Rosin–Rammler distribution function, is

$$dV = CR^{m-1} \exp(-cR^m)\, dR, \qquad (\mu m^3/cm^3) \qquad (A.17)$$

where R is the particle radius or some characteristic linear dimension and C is a dimensional constant ($\mu m^{3-m}/cm^3$, if R is in μm). If the particle radius is taken as a nondimensional quantity R/R_0, where R_0 is the lower size limit or unity, then A has the dimensions of volume.

The particle number distribution in a sediment, the particle volumes of which obey the Rosin–Rammler function, is

$$dN = C'R^{m-4} \exp(-cR^m)\, dR \qquad (particles/cm^3), \qquad (A.18)$$

where the relationship between dN and dV is $dV = \frac{4}{3}\pi R^3\, dN$, and C' is a dimensional constant not identical to C in Eq. (A.17).

The Weibull distribution has a single mode only if $m > 1$. For $m < 1$, the function $f(x)$ declines steadily from its maximum value at $x = x'$ with increasing x. An example of the Rosin–Rammler or Weibull distribution plot is shown in Fig. A.1. For graphical displays of the cumulative or Rosin–Rammler Weibull distributions, special graph paper is available commercially.

The equations for the mean, median, and mode of the Rosin–Rammler distribution are summarized in Table A.1.

Depending on the values of the parameters c and m in the Weibull distribution, its ascending or descending limb can be similar to the lognormal function with certain values of μ and σ. This means that parts of certain particle size distributions can be satisfactorily fitted by both the Weibull and lognormal functions over some range of R. As a point of interest, the Weibull distribution with the value of $m \simeq 3.3$–3.6 is very similar in shape to the Gaussian distribution curve.

The logarithmic form of Eq. (A.17) for dV,

$$\ln \frac{dV}{dR} = \ln C + (m - 1) \ln R - cR^m \qquad (A.19)$$

shows that a plot of $\log dV/dr$ versus $\log R$ is nonlinear, and this also applies to a plot of $\log dN/dR$ versus $\log R$.

Power-Law or Pareto Distribution. The probability density function

$$f(x)\, dx = \gamma c^\gamma x^{-(\gamma+1)}\, dx, \qquad (A.20)$$

where γ and c are positive constants and $x > c$, is known as the Pareto distribution. The distribution function is attributable to the Italian–Swiss economist V. Pareto, whose original work dealt with the distribution of incomes in a population (Johnson and Kotz, 1970).

The cumulative distribution for all values smaller than x is

$$F_{<x} = \int_c^x f(y)\, dy = 1 - c^\gamma x^{-\gamma}. \qquad (A.21)$$

The Pareto function for the particle number distribution dN can be written as

$$dN = \tilde{A} R^{-\gamma} \, dR \qquad \text{(particles),} \qquad (A.22)$$

where R is a linear dimension of particles and $\tilde{A}$ is a dimensional constant (particle $\mu m^{\gamma-1}/cm^3$ if R is in μm).

The cumulative number of particles $N_{>R}$ of radius greater than R, by integration of Eq. (A.22), is

$$N_{>R} = \int_{R=R}^{R=\infty} dN = \frac{\tilde{A}}{\gamma - 1} R^{-(\gamma-1)} \qquad \text{(particles/cm}^3\text{),} \qquad (A.23)$$

which is valid for $\gamma > 1$.

As the particle radius R approaches zero, the quantity dN/dR increases indefinitely. Thus for all practical purposes, a lower size limit of the distribution must be known or stipulated. With the lower size limit R_0, the cumulative number of particles N of size greater than R_0 and smaller than R ($R \geq R_0$) is, by integration of Eq. (A.22),

$$N_{<R} = \int_{R=R_0}^{R=\infty} dN = \frac{A}{\gamma - 1} \left(\frac{1}{R_0^{\gamma-1}} - \frac{1}{R^{\gamma-1}} \right). \qquad (A.24)$$

The particle volume distribution dV is

$$dV = A R^{3-\gamma} \, dR \qquad (\mu m^3), \qquad (A.25)$$

where constant A is not identical to A in the equation for dN.

If the particle size r varies from a lower limit $R = R_0$ to $R = \infty$, then the total volume and total mass concentrations from the distribution are

$$V = \int_{R_0}^{\infty} dV, \qquad M = \rho_p \int_{R_0}^{\infty} dV. \qquad A.26)$$

These are finite only if $\gamma > 4$. For $\gamma \leq 4$, the integral of Eq. (A.25) increases indefinitely with increasing R.

The logarithmic forms of the particle number and particle volume equations are

$$\log \frac{dN}{dR} = \log \tilde{A} - \gamma \log R, \qquad (A.27)$$

$$\log \frac{dV}{dR} = \log \tilde{A} - (\gamma - 3) \log R. \qquad (A.28)$$

Both equations plot as straight lines in log-log coordinates (Fig. A.1), with a slope of $-\gamma$ for the particle number spectrum and a slope of $-(\gamma - 3)$ for the particle volume spectrum.

The mean and the median of the power-law distribution are listed in Table A.1.

Two modified forms of the Pareto distribution deserve mentioning. One is the form

$$dN = \tilde{A}(R + a)^{-\gamma} \, dR, \qquad (A.29)$$

where a is a positive constant. When R is much smaller than a ($R \ll a$), then dN/dR approaches a constant value $\tilde{A}a^{-\gamma}$ essentially independent of the particle size R.

The other form of the power-law distribution is

$$dN = \tilde{A}R^{-(\gamma-1)} \, d \ln R, \qquad (A.30)$$

which can be plotted in log-log coordinates as $dN/d \log R$ vs. R. Such plots have been introduced by Junge (1963) in studies of particle size distributions of atmospheric aerosols and they are used along with the form dN/dR in the literature dealing with atmospheric particles.

The number of different natural phenomena which can be approximated by the power-law relationship $n(x)$-$\tilde{A}x^{-\gamma}$ is surprisingly large. Atmospheric and stratospheric aerosol particles of the micrometer and submicrometer size obey the power law, at least over some range of the x (particle diameter) values (Junge, 1963); fine-grained sediments and suspended matter in the ocean, in the submillimeter particle size range (Lal and Lerman, 1975); number concentration of asteroids in the asteroid belt, as a function of the asteroid diameter, in the range between 10 and 1000 km (Chapman, 1974); the frequency of oil field occurrences, within a certain geographic area, as a function of the oil field size, a case cited by Johnson and Kotz (1970); the frequency of the crater occurrences on the moon and on Mars, as a function of the crater diameter (Neukum and Wise, 1976); and the number of natural and man-made lakes in Switzerland as a function of the lake surface area (Lerman, 1979).

REFERENCES

Chapman, C. R. (1974). *Geophys. Res. Lett.* **1**, 341.
Harris, C. C. (1971). *Inst. Min. Metall. Trans. C* **80**, 133.
Johnson, N. L., and Kotz, S. (1970). "Continuous Univariate Distributions," Vol. 1. Houghton, Boston, Massachusetts.
Junge, C. E. (1963). "Air Chemistry and Radioactivity." Academic Press, New York.
Lal, D., and Lerman, A. (1975). *J. Geophys. Res.* **80**, 423, 4563.
Lerman, A. (1979). "Geochemical Processes: Water and Sediment Environment," p. 185. Wiley, New York.
Neukum, G., and Wise, D. U. (1976). *Science (Washington, D.C.)* **194**, 1381.
Rosin, P., and Rammler, E. (1933). *J. Inst. Fuel* **7**, 29.

INDEX